METHODS OF NUMERICAL INTEGRATION

SECOND EDITION

This is a volume in
COMPUTER SCIENCE AND APPLIED MATHEMATICS
A Series of Monographs and Textbooks

This series has been renamed
COMPUTER SCIENCE AND SCIENTIFIC COMPUTING

Editor: WERNER RHEINBOLT

A complete list of titles in this series is available from the publisher upon request.

METHODS OF NUMERICAL INTEGRATION

SECOND EDITION

Philip J. Davis

APPLIED MATHEMATICS DIVISION
BROWN UNIVERSITY
PROVIDENCE, RHODE ISLAND

Philip Rabinowitz

DEPARTMENT OF APPLIED MATHEMATICS
THE WEIZMANN INSTITUTE OF SCIENCE
REHOVOT, ISRAEL

ACADEMIC PRESS, INC.
Harcourt Brace Jovanovich, Publishers
Boston San Diego New York
London Sydney Tokyo Toronto

This book is printed on acid-free paper. ∞

ACADEMIC PRESS, INC.
1250 Sixth Avenue, San Diego, CA 92101

United Kingdom Edition published by
ACADEMIC PRESS LIMITED
24–28 Oval Road, London NW1 7DX

Library of Congress Cataloging in Publication Data

Davis, Philip J, Date
Methods of numerical integration.

(Computer science and applied mathematics)
Includes bibliographies and index.
1. Numerical integration. I. Rabinowitz, Philip.
II. Title. III. Series.
QA299.3.D28 1983 515'.624 83-13522
ISBN 0-12-206360-0 (alk. paper)

PRINTED IN THE UNITED STATES OF AMERICA

91 92 93 94 9 8 7 6 5 4 3

To

E. R. and H. F. D.

Contents

Preface to First Edition

The theory and application of integrals is one of the great and central themes of mathematics. For applications, one often requires numerical values. The problem of numerical integration is therefore a basic problem of numerical analysis. It is often an elementary component in a much more complex computation.

Numerical integration is, paradoxically, both simple and exceedingly difficult. It is simple in that it can often be successfully resolved by the simplest of methods. It is difficult in two respects: first, in that it may require an inordinate amount of computing time, verging in some unfavorable situations toward impossibility; second, in that one can be led through it to some of the deepest portions of pure and applied analysis. And not only to analysis but, in fact, to diverse areas of mathematics; for inasmuch as the integral has successfully invaded many areas, it is inevitable that from time to time these areas turn about and contribute their special methods and insights to the problem of how integrals may be computed expeditiously. The problem of numerical integration is also open-ended; no finite collection of techniques is likely to cover all the possibilities that arise and to which an extra bit of ingenuity or of special knowledge may be of great assistance.

This book presents what we think are the major methods of numerical integration. We have tried to produce a balanced work that is both useful to the programmer and stimulating to the theoretician. There are portions of the book where deep results of analysis are derived or are alluded to; yet, it has been our hope that most of the final results have been expressed in a way that is accessible to those with a background only in calculus.

The past generation, under the impact of the electronic computer, has witnessed an enormous productivity in the field of numerical integration. It would not misrepresent the situation to say that a learned journal devoted solely to this topic could have been established. The published work has run the range from subtle computer programs to highly theoretical questions of interest primarily to analysts, both classical and functional. In surveying and compiling material for this book we found ourselves very much in the position of Tristram Shandy who required a year to write up one day of his life.

We wish to thank many friends and colleagues who have provided us with suggestions. Particular thanks go to Dr. Harvey Silverman who helped prepare the material on the Fast Fourier Transform.

Thanks also go to the Weizmann Institute of Science for granting P. J. Davis a Weizmann fellowship during the Spring term of 1970 and to the Office of Naval Research for continued support.

Preface to Second Edition

Even though a journal of numerical integration has not yet appeared on the scene, it is not because of lack of material. Since the publication of the first edition of our book, itself a revision of previous work, there has been considerable activity in the field of numerical integration touching upon almost every section of this book. In addition to the many papers appearing in the ever-increasing number of journals in numerical analysis, several books on numerical integration have appeared, as well as many internal reports and a number of Ph.D. theses, notably those of Genz, Haegemans, Mantel, de Doncker, Malik, and Neumann. In addition, two conferences have been held in Oberwolfach and workshops in Los Alamos and Kyoto. Furthermore, numerical integration subroutines have been included in the IMSL and NAG libraries, and a package of such routines, QUADPACK, has just made its debut.

In preparing this revision, we have tried to remain faithful to the aims of the first edition and have essentially attempted to update the material in view of the fruitful work done in the past eight years. Since we wished to keep the book to a reasonable size, we had to be quite selective and our choice has tended more to the practical rather than the theoretical aspects of the field. Thus, we have considerably expanded Chapter 5, Approximate Integration in Two or More Dimensions, and Chapter 6, Automatic Integration. On the other hand, we have not added much on the subject of optimal quadratures, even though much has been written on this subject. In addition, the work of the Soviet school is not well represented here since much of it is inaccessible to those of us who cannot read Russian mathematics.

In this edition, we have attempted to correct all the errors in the first edition and wish to thank all those who pointed out these errors, in particular, Professors M. Mori and W. Gautschi, for their lists of errata. We also extend our thanks to our many colleagues, especially James Lyness and Uri Ascher, with whom we have had conversations or correspondence for their contributions to this revision. Of course, we alone are responsible for its contents. We hope it will be as well received as our previous efforts.

Chapter 1

Introduction

1.1 Why Numerical Integration?

Numerical integration is the study of how the numerical value of an integral can be found. The beginnings of this subject are to be sought in antiquity. A fine example of ancient numerical integration, but one that is entirely in the spirit of the present volume, is the Greek quadrature of the circle by means of inscribed and circumscribed regular polygons. This process led Archimedes to an upper and lower bound for the value of π. Over the centuries, particularly since the sixteenth century, many methods of numerical integration have been devised.† These include the use of the fundamental theorem of integral calculus, infinite series, functional relationships, differential equations, and integral transforms. Finally, and this is of prime importance in this volume, there is the method of *approximate integration*, wherein an integral is approximated by a linear combination‡ of the

† For a brief history of the older portions of our subject, see Moors [1] and Runge and Willers [1].

‡ Nonlinear combinations occur occasionally, e.g., in extrapolation by rational functions, the epsilon algorithm, and other nonlinear acceleration techniques as well as in inner product integration rules.

values of the integrand

$$\int_a^b f(x)\,dx \approx w_1 f(x_1) + w_2 f(x_2) + \cdots + w_n f(x_n),$$
$$-\infty \le a \le b \le +\infty. \tag{1.1.1}$$

In equation (1.1.1), $x_1, x_2, \ldots, x_n$ are n points or abscissas usually chosen so as to lie in the interval of integration, and the numbers $w_1, w_2, \ldots, w_n$ are n "weights" accompanying these points. Occasionally, values of the derivatives of the integrand appear on the right-hand side of (1.1.1), which is frequently called a *rule of approximate integration*. The terms *mechanical* or *approximate quadrature* are also employed for this type of numerical process.

With an abundance of quite general and sophisticated methods for obtaining values of integrals, one may properly ask why such primitive approximations as those provided by (1.1.1) should be developed and utilized. The answer is very simple: The mathematically sophisticated methods do not always work, and even if they do work, it may not be advantageous to use them. Take, for example, the method embodied in the fundamental theorem of integral calculus. With this method

$$\int_a^b f(x)\,dx = F(b) - F(a), \tag{1.1.2}$$

where $F(x)$ is an indefinite integral (an antiderivative) of $f(x)$. If the indefinite integral is readily available and sufficiently simple, (1.1.2) can provide a most expeditious computation. But, as is well known, the process of integration often leads to new transcendental functions. Thus, the simple integration $\int dx/x$ leads to the logarithm, which is not an algebraic function, whereas the integration $\int e^{-x^2}\,dx$ leads to a function that cannot be expressed in finite terms by combinations of algebraic, logarithmic, or exponential operations. Even if the indefinite integral is an elementary function and can be obtained without undue expenditure of labor, it may be sufficiently complicated for one to pause before applying (1.1.2). Take, for example,

$$\int_0^x \frac{dt}{1+t^4} = \frac{1}{4\sqrt{2}} \log \frac{x^2 + x\sqrt{2} + 1}{x^2 - x\sqrt{2} + 1}$$
$$+ \frac{1}{2\sqrt{2}} \left\{ \arctan \frac{x}{\sqrt{2} - x} + \arctan \frac{x}{\sqrt{2} + x} \right\}. \tag{1.1.3}$$

[In a previous version of this book, formula (1.1.3) was given incorrectly. The present formula is (hopefully) correct. The probability of finding an error—either a mathematical blunder, a typographical error, or an answer that is valid only for certain values of the parameters—in a table of indefinite integrals is higher than one might imagine. Thus, Klerer and Grossman [1] investigated eight very popular tables and this probability seems to be more than 5%. When as a result of this investigation they produced a table of indefinite integrals by means of a computer, the table was itself flawed.]

The number of computations that must be carried out with this "exact" formula is substantial. Notice that to obtain an answer through (1.1.3) we must compute logarithms and inverse tangents—which can be done only with a certain degree of approximation. Methods that appear on the surface to be exact become approximate when reduced to a numerical process.

There is also the possibility of intrinsic difficulties with an elementary formula for the indefinite integral. These include indeterminate forms, branch ambiguities, alternative expressions corresponding to different parameter ranges, etc.

Another difficulty of a more subtle sort may develop in the evaluation of integrals by means of closed expressions. Consider the integral $I_n = \int_0^1 x^n e^{x-1}\,dx$, and suppose that values are desired for $n = 1, 2, \ldots, 20$. Integration by parts yields the recurrence $I_n = 1 - nI_{n-1}$. Since $I_0 = 1 - e^{-1}$, we may compute successively $I_1, I_2, \ldots$. An examination of the integral shows that $0 < I_{n+1} < I_n$ and that $\lim_{n\to\infty} I_n = 0$; yet the computation after behaving decently for the first ten or so values of n develops violent oscillations of increasing amplitude. Some familiarity with the processes of numerical analysis is required to convert good formulas into good numbers.

Another reason for approximate integration occurs when we are solving a functional equation for an unknown function that appears in the integrand of some integral. This will be discussed at greater length in Section 1.16.

A final reason for developing rules of approximate integration is that, in many instances, we are confronted with the problem of integrating *experimental data*, which also includes data generated by a computation. In such cases, theoretical devices may be wholly inapplicable.

References. Babuška, Práger, and Vitásek [1]; Boas and Schoenfeld [1]; Gautschi [1]; Klerer and Grossman [1, 2]; Moors [1, Chap. 8]; Runge and Willers [1]; Squire [4].

1.2 Formal Differentiation and Integration on Computers

The reader should be apprised of computer programs designed to perform indefinite integration. For surveys of recent work in this field, see Moses [2] and Norman and Davenport. While at one time these programs were of more interest from the point of view of artifical intelligence and formula manipulation rather than that of numerical analysis, a symbolic–numeric interface is slowly developing (see Ng). Powerful symbolic integration systems have been incorporated into such symbolic and algebraic computation systems as REDUCE, MACSYMA, and SCRATCHPAD. As for older work, certain problems in quantum mechanics have been solved by extensive symbolic manipulation on a computer. A molecular waveform calculation may require the evaluation of several thousand integrals and may employ hundreds of different formulas. It is claimed that highly accurate results can be obtained in this way. Similarly, multidimensional integrals of coverage have been attacked using the FORMAC formula manipulation system.

Formal differentiation programs are also available on almost all symbolic computation systems. It has provided an important tool in numerical analysis. Thus, it has been combined with Newton's method to solve systems of nonlinear equations and has been used for the numerical solution of ordinary differential equations and the optimization of nonlinear functionals. It has also been used in the analysis of roundoff error as well as the computational verification of the existence of solutions of systems of equations. In many of these applications, it has been combined with interval analysis to bound the error in the results. In this context, it has been applied in numerical integration to determine bounds on the error of a numerical integration rule when this error can be expressed as a constant times the value of some derivative of the integrand at some unknown point in the integration interval. Another application of formal differentiation to approximate integration is that it makes it more feasible to employ formulas involving higher derivatives of the integrand. This has been implemented in the case of the Euler–Maclaurin formula (Section 2.9) by Gray and Rall [3].

References. Davenport [1]; Fletcher and Reeves [1]; Gray and Rall [1–3]; Harrington [1]; Juncosa [1]; Moore [1]; Moses [1, 2]; E. W. Ng [2]; Norman and Davenport [1]; Rall [2–5]; Schiff, Lifson, Pekeris, and Rabinowitz [1]; Slagle [1]; Wactlar and Barnett [1]; Wang [1].

1.3 Numerical Integration and Its Appeal in Mathematics

Despite (or perhaps because of) the simple nature of the problem and the practical value of its methods, numerical integration has been of great interest to the pure mathematician. The most superficial glance at its history will reveal that many masters of mathematics have contributed to this field;

Archimedes, Kepler, Huygens, Newton, Euler, Gauss, Jacobi, Tschebyscheff,† Markoff, Féjer, Pólya, Szegö, Schoenberg, and Sobolev are among them.

Practical problems have a long history of suggesting subtle problems and deep methods of pure theory. In quite recent years, we have applied, for example, equidistributed sequences (which are part of Diophantine approximation theory) to problems of numerical integration in multidimensional regions. The Möbius inversion formula of number theory has been applied to the calculation of Fourier coefficients, and polynomial ideal theory has been used to determine minimal integration rules in several dimensions.

Methods of functional analysis are playing more and more of a role and have yielded much that is of interest, including theories of optimal integration rules. Numerical integration has a deeply theoretical side, which has been widely developed, but we shall have only a taste of this in the present work.

1.4 Limitations of Numerical Integration

The skilled programmer should be acquainted with various analytical techniques for handling integrals. Although it would take us too far afield to discuss analytical methods in this book, they are of great importance. In Appendix 1 we have reproduced an article by M. Abramowitz that will provide an introduction to these methods. The programmer should also be aware of the multitude of integrals whose values have been tabulated and should consider the possibility of reducing his integral to one already known.

As we have seen, numerical integration is usually utilized when analytical techniques fail. Used sensibly and with proper controls, numerical integration can provide satisfactory answers. When used in a blind fashion—and the availability of sophisticated computer programs makes it a temptation to operate blindly—numerical integration may lead to serious errors.

Whenever possible, a problem should be analyzed and put into a proper form before it is run on a computer. An analysis is necessary to establish confidence in the alleged results. But analysis may also be valuable in that it can often establish early in the game ways of carrying out a computation which will save time. *One good thought may be worth a hundred hours on the computer.*‡

† The German version, Tschebyscheff, and the English version, Chebyshev, of the Russian will be used interchangeably throughout this book.

‡ A dramatic example of this is furnished by the rediscovery of the algorithm now known as the Fast Fourier Transform. See Section 3.9.5.

In examples that are given in the various sections, we shall attempt to point out some of the difficulties encountered in numerical integration.

Reference. Chai and Wertz [1].

1.4.5 On the Use of Tabulations and Programs

A tabulation of the values of certain fundamental integrals such as the Gamma function, etc., can be found in the NBS *Handbook of Mathematical Functions.* Careful compilations of definite and indefinite integrals include Gradshteyn and Ryzhik, Gröbner and Hofreiter, and Erdélyi *et al.* Fletcher, Miller, Rosenhead, and Comrie may be consulted with profit. The NBS *Handbook* as well as other tabulations of integrals often contain information on how the integrals may be expeditiously computed for various ranges of the parameter.

A number of books provide polynomial and rational approximations to certain fundamental integrals. These include Hart *et al.* and Luke.

ALGOL‡ and FORTRAN programs which used to be published in the *Communications of the Association for Computing Machinery* (CACM) are now published occasionally in the *Transactions on Mathematical Software of the ACM* (TOMS). These programs are accumulated in the *Collected Algorithms from CACM* which also contains certifications and remarks on each algorithm. It is periodically updated.

The IBM System/360 Scientific Subroutine Packages contain about ten programs in FORTRAN and PL/I for computing special functions which are integrals. The Handbook Series in *Numerische Mathematik* is a source of ALGOL programs. Many programs for numerical integration in several variables are given by Stroud, while a rather complete package of programs for one-dimensional numerical integration appears in QUADPACK. Most of the QUADPACK programs have been incorporated in the NAG library which contains several additional programs. However, the documentation of the NAG library does not include listings of its programs. A similar situation holds with respect to the International Mathematical and Statistical Library (IMSL) which contains a group of programs useful in numerical integration.

‡ ALGOL, or more accurately ALGOL 60, has virtually disappeared from the scene as a language for the dissemination of algorithms. Of the viable descendants of ALGOL, ALGOL 68 has not achieved wide acceptance in the computing community while PASCAL does not have the algorithmic power needed for mathematical software. Consequently, in spite of its many shortcomings, that venerable programming language FORTRAN has become the vehicle for almost all current mathematical software.

References. ACM [1]; Erdélyi *et al.* [1, 2]; Fletcher, Miller, Rosenhead, and Comrie [1]; Gradshteyn and Ryzhik [1]; Gröbner and Hofreiter [1]; Hart *et al.* [1]; IBM [2, 3]; IMSL [1]; Luke [5, 7]; Maričev [1]; NAG [1]; NBS [1]; QUADPACK [F1].

1.5 The Riemann Integral

We shall be dealing entirely with functions that are integrable in the sense of Riemann. In the case of functions of one variable, this concept can be developed as follows. Suppose that $y = f(x)$ is a bounded function on the finite interval $[a, b]$. Partition the interval $[a, b]$ into n subintervals by the points

$$a = x_0 < x_1 < x_2 < \cdots < x_n = b.$$

Let ξ_i be any point in the ith subinterval: $x_{i-1} \le \xi_i \le x_i$, and form the sum

$$S_n = \sum_{i=1}^{n} f(\xi_i)(x_i - x_{i-1}). \tag{1.5.1}$$

Sums of this sort are called *Riemann sums.* Let the maximum length of the subintervals be denoted by Δ: $\Delta = \max_i (x_i - x_{i-1})$, and consider a sequence of sums of type (1.5.1), $S_1, S_2, \ldots,$ whose corresponding maximum subintervals $\Delta_1, \Delta_2, \ldots$ approach 0: $\lim_{m\to\infty} \Delta_m = 0$. If, for *any* sequence of this type and corresponding to any choice of ξ_i, the sequences $\{S_m\}$ have a *common* limit S, then $f(x)$ is said to have the *Riemann integral* S over $[a, b]$:

$$S = \int_a^b f(x)\,dx. \tag{1.5.2}$$

A necessary and sufficient condition that a bounded function $f(x)$ have a Riemann integral is that $f(x)$ be *continuous almost everywhere.* In particular, if $f(x)$ is continuous on $[a, b]$, it has a Riemann integral. Also, if $f(x)$ is bounded on $[a, b]$ and continuous except for a finite number of points of discontinuity, it has a Riemann integral.

The following properties of the Riemann integral are fundamental. It is assumed that f and g are bounded and Riemann-integrable on $[a, b]$.

$$\int_a^a f(x)\,dx = 0, \tag{1.5.3}$$

$$\int_a^b f(x)\,dx = \int_{a+h}^{b+h} f(x-h)\,dx, \tag{1.5.4}$$

$$\int_a^b f(x)\,dx = -\int_b^a f(x)\,dx, \tag{1.5.5}$$

$$\int_a^b f(x)\,dx + \int_b^c f(x)\,dx = \int_a^c f(x)\,dx, \tag{1.5.6}$$

$$\int_a^b cf(x)\,dx = c\int_a^b f(x)\,dx, \tag{1.5.7}$$

$$\int_a^b (f(x) + g(x))\,dx = \int_a^b f(x)\,dx + \int_a^b g(x)\,dx. \tag{1.5.8}$$

If $f(x) \le g(x)$ almost everywhere on $[a, b]$, then

$$\int_a^b f(x)\,dx \le \int_a^b g(x)\,dx. \tag{1.5.9}$$

In particular, if $f(x) \ge 0$ on $[a, b]$, then $\int_a^b f(x)\,dx \ge 0$.

If $f(x)$ and $g(x)$ are both increasing or both decreasing on $[a, b]$, then

$$(b - a)\int_a^b f(x)g(x)\,dx \ge \int_a^b f(x)\,dx \int_a^b g(x)\,dx. \tag{1.5.10}$$

If $f(x)$ and $g(x)$ are of opposite type, then the inequality is reversed.

If f is a bounded Riemann-integrable function on $[a, b]$, then so is $|f|$, and

$$\left|\int_a^b f(x)\,dx\right| \le \int_a^b |f(x)|\,dx. \tag{1.5.11}$$

The Schwarz, Hölder, and Minkowski Inequalities

$$\left|\int_a^b f(x)g(x)\,dx\right| \le \left(\int_a^b |f(x)|^2\,dx\right)^{1/2}\left(\int_a^b |g(x)|^2\,dx\right)^{1/2}. \tag{1.5.12}$$

If $p > 1$ and $p^{-1} + q^{-1} = 1$,

$$\left|\int_a^b f(x)g(x)\,dx\right| \le \left(\int_a^b |f(x)|^p\,dx\right)^{1/p}\left(\int_a^b |g(x)|^q\,dx\right)^{1/q}. \tag{1.5.13}$$

If $p \ge 1$,

$$\left(\int_a^b |f(x) + g(x)|^p\,dx\right)^{1/p} \le \left(\int_a^b |f(x)|^p\,dx\right)^{1/p} + \left(\int_a^b |g(x)|^p\,dx\right)^{1/p}. \tag{1.5.14}$$

First Mean-Value Theorem

Let $f(x)$ be continuous on $a \le x \le b$. Then there exists a value ξ $(a < \xi < b)$ such that

$$\int_a^b f(x)\,dx = (b - a)f(\xi). \tag{1.5.15}$$

If

$$m \leq f(x) \leq M \qquad \text{for} \quad a \leq x \leq b, \tag{1.5.16}$$

then

$$m(b - a) \leq \int_a^b f(x)\, dx \leq M(b - a). \tag{1.5.17}$$

Generalized Mean-Value Theorem

Let $f(x)$ and $g(x)$ be continuous on $a \leq x \leq b$. Let $g(x) \geq 0$ for $a \leq x \leq b$. Then there exists a value ξ $(a < \xi < b)$ such that

$$\int_a^b f(x)g(x)\, dx = f(\xi) \int_a^b g(x)\, dx. \tag{1.5.18}$$

THE FUNDAMENTAL THEOREM OF INTEGRAL CALCULUS. *If $F(x)$ is differentiable on $[a, b]$ and if $F'(x)$ is Riemann-integrable there, then*

$$\int_a^b F'(x)\, dx = F(b) - F(a). \tag{1.5.19}$$

A formulation that is sufficient for our purposes is: If $f(x)$ is continuous on $a \leq x \leq b$ and if $F(x)$ is any indefinite integral of $f(x)$, then

$$\int_a^b f(x)\, dx = F(b) - F(a). \tag{1.5.20}$$

Integration by Parts

$$\int_a^b f(x)g'(x)\, dx = f(b)g(b) - f(a)g(a) - \int_a^b f'(x)g(x)\, dx. \tag{1.5.21}$$

It is sufficient to assume that $f(x)$ and $g(x)$ are continuously differentiable on $a \leq x \leq b$.

A special Riemann sum arises when $[a, b]$ is subdivided into n equal parts and ξ_i is taken at the right-hand end point of its subinterval:

$$S_n = h \sum_{k=1}^{n} f(a + kh), \qquad \text{where} \quad h = (b - a)/n. \tag{1.5.22}$$

If ξ_i is taken at the left-hand end point of its subinterval, we obtain

$$S_n = h \sum_{k=0}^{n-1} f(a + kh). \tag{1.5.23}$$

References. P. Franklin [1, p. 194]; Goldberg [1, Chap. 7]; Hobson [1, Chap. 6].

1.6 Improper Integrals

Integrals whose range or integrand is unbounded are known as *improper integrals*. Such integrals are defined as the limits of certain (proper) integrals.

1 Integrals over $[0, \infty)$

DEFINITION

$$\int_0^\infty f(x)\,dx = \lim_{r\to\infty} \int_0^r f(x)\,dx$$

whenever the latter limit exists. Similar definitions are used for $\int_a^\infty f(x)\,dx$ and for $\int_{-\infty}^a f(x)\,dx$.

2 Integrals over $(-\infty, \infty)$

Here two definitions are employed.

DEFINITION A

$$\int_{-\infty}^\infty f(x)\,dx = \int_{-\infty}^0 f(x)\,dx + \int_0^\infty f(x)\,dx.$$

This is the commonly employed definition.

DEFINITION B

$$\int_{-\infty}^\infty f(x)\,dx = \lim_{r\to\infty} \int_{-r}^r f(x)\,dx.$$

This is known as the *Cauchy Principal Value* of the integral, frequently designated by $P\int_{-\infty}^\infty f(x)\,dx$.

Whenever both limits exist, the limiting values in A and B will be identical, but the limit in B may exist in cases where that in A does not.

3 Unbounded integrands

Assume that $f(x)$ is defined on $(a, b]$ and is unbounded in the neighborhood of $x = a$.

DEFINITION

$$\int_a^b f(x)\,dx = \lim_{r\to a^+} \int_r^b f(x)\,dx$$

whenever the latter limit exists. A similar definition applies to integrands that are unbounded in the neighborhood of the upper limit of integration.

Suppose that $a < c < b$ and $f(x)$ is unbounded in the vicinity of $x = c$. The *Cauchy Principal Value* of the integral, $P\int_a^b f(x)\,dx$, is defined by the limit

$$P\int_a^b f(x)\,dx = \lim_{r\to 0+}\left[\int_a^{c-r} f(x)\,dx + \int_{c+r}^b f(x)\,dx\right].$$

A common principal value integral is the Hilbert transform

$$g(x) = P\int_a^b \frac{f(t)}{t-x}\,dt, \qquad -\infty \le a < b \le \infty, \quad a < x < b. \tag{1.6.1}$$

A sufficient condition for the existence of the Hilbert transform over a finite interval $[a, b]$ is that $f(t)$ satisfy a *Lipschitz or Hölder condition* in $[a, b]$. This means that there are constants $k > 0$ and $0 < \alpha \le 1$ such that for any two points t_1 and t_2 in $[a, b]$ we have

$$|f(t_1) - f(t_2)| \le k|t_1 - t_2|^\alpha. \tag{1.6.2}$$

1.6.1 Hadamard Finite-Part Integrals

If $f(s) \ne 0$, the integral $\int_s^r (x-s)^{\lambda-n} f(x)\,dx$, $-1 < \lambda \le 0$, diverges for all positive integers n. However, in certain physical applications, for example, in the fields of aerodynamics and electron optics, it is important to assign a finite numerical value to expressions involving such integrals. Hadamard developed the notion of the Finite Part (FP) of an integral to deal with such situations. The theory was put on a firmer footing with the introduction of distributions or generalized functions, but we shall not go into this here and refer the interested reader to Gelfand and Shilov and to Schwartz. For our purposes, we shall give a more informal treatment.

Assume first that $\lambda \ne 0$; then the Hadamard FP of an integral, denoted

$$\,⨍_s^r (x-s)^{\lambda-n} f(x)\,dx \tag{1.6.1.1}$$

where $f \in C^n[s, r]$, is defined to be

$$\lim_{t\to s^+}\left[\int_t^r (x-s)^{\lambda-n} f(x)\,dx - g(t)(t-s)^{\lambda-n+1}\right]$$

where $g(t)$ is any function $\in C^n[s, r]$ such that the limit exists. This definition implies that value of the FP is independent of the choice of g. A formula for (1.6.1.1) can be derived by expanding $f(x)$ in a Taylor series to yield

$$\fint_s^r (x-s)^{\lambda-n} f(x)\,dx = \sum_{k=0}^{n-1} f^{(k)}(s)(r-s)^{\lambda-n+k+1}/[(\lambda-n+k+1)k!] + \int_s^r (x-s)^{\lambda-n} R_{n-1}(x,s)\,dx, \qquad s<r \tag{1.6.1.2}$$

where

$$R_{n-1}(x,s) = \frac{1}{(n-1)!}\int_s^x (x-y)^{n-1} f^{(n)}(y)\,dy. \tag{1.6.1.3}$$

For $\lambda = 0$, the situation is more complicated since a logarithmic term also appears and there is a possibility of nonuniqueness of the limit. For example, consider

$$\fint_s^r (x-s)^{-2} f(x)\,dx = \lim_{t\to s^+}\left[\int_t^r (x-s)^{-2} f(x)\,dx - g(t)/(t-s) - h(t)\log(t-s)\right].$$

If we choose $g(t) = f(t)$ and $h(t) = f'(t)$, then the limit exists and is equal to a certain value, say L. However, if we modify $g(t)$ by adding a constant multiple of $t-s$, so that $g(t) = f(t) + A(t-s)$, the limit will still exist but will be equal to $L - A$. Thus, when $\lambda = 0$, we must be more careful in our definition of the FP. A definition leading to a unique value can always be given and leads to the expression

$$\fint_s^r (x-s)^{-n} f(x)\,dx = \sum_{k=0}^{n-2} f^{(k)}(s)(r-s)^{-n+k+1}/[(-n+k+1)k!] + \log(r-s) f^{(n-1)}(s)/(n-1)! + \int_s^r (x-s)^{-n} R_{n-1}(x,s)\,dx, \quad s<r. \tag{1.6.1.4}$$

The last term in (1.6.1.2) and (1.6.1.4) can be integrated by parts to yield further expressions for (1.6.1.1) and these may be found in Kutt [2].

FP integrals have some of the usual properties of regular integrals and some properties that are quite unusual. The following list of rules given by Kutt [1] constitute an informal definition of FP integrals:

1. $\fint$ is a consistent extension of the concept of regular integrals.
2. $\fint$ is additive with respect to the union of integration intervals and is invariant with respect to translation.
3. $\fint$ is a linear and continuous operator on the integrand.
4. $\fint_0^\infty x^{-\mu}\,dx = 0$, $\mu > 1$; $\fint_0^1 x^{-1}\,dx = 0$.

From these rules it follows that for $\mu > 1$,

$$\fint_1^\infty x^{-\mu}\,dx = 1/(\mu - 1) \qquad \text{and} \qquad \fint_0^1 x^{-\mu}\,dx = 1/(1-\mu) < 0.$$

Thus, we see that if $f(x) \geq 0$, $\fint f(x)\,dx$ may be negative. As a result we have that if $f(x) \geq 0$ and $\fint f(x)\,dx = 0$, it does not follow that $f(x) = 0$ almost everywhere. Similarly, if $f(x) \geq g(x)$ in $[s, r]$, the inequality $\fint_s^r f(x)\,dx \geq \fint_s^r g(x)\,dx$ need not hold, nor the inequality $|\fint_s^r f(x)\,dx| \leq \fint_s^r |f(x)|dx$. In general, as far as equalities are concerned, the common rules for ordinary integrals are also valid for FP integrals, but rules concerning inequalities are not applicable.

We now give a few more properties of FP integrals useful in applications. If $r < s$, then

$$\fint_r^s (s-x)^{\lambda-n} f(x)\,dx = \fint_{-s}^{-r} (y+s)^{\lambda-n} f(-y)\,dy, \qquad -s < -r. \tag{1.6.1.5}$$

If we transform the interval $[s, r]$ to $[s', r']$, $s' < r'$, by $y = s' + (x-s)(r'-s')/(r-s)$, then for $\lambda \neq 0$

$$\fint_s^r (x-s)^{\lambda-n} f(x)\,dx = I(s', r')$$

$$\equiv [(r-s)/(r'-s')]^{1+\lambda-n} \fint_{s'}^{r'} (y-s')F(y)\,dy \tag{1.6.1.6}$$

where

$$F(y) = f[s + (y-s')(r-s)/(r'-s')].$$

However, if $\lambda = 0$,

$$\fint_s^r (x-s)^{-n} f(x) dx = I(s', r') + f^{(n-1)}(s) \log[(r-s)/(r'-s')]/(n-1)! \tag{1.6.1.7}$$

Example. $\fint_0^1 dx/x = 0$ but if $y = x/2$, $\fint_0^{1/2} dy/y = -\log 2$.

Finally, we note that other classes of FP integrals exist such as

$$\oint_s^r \frac{f(x)\log^m(x-s)}{(x-s)^{n-\lambda}}\,dx \qquad \text{and} \qquad \oint_s^r \frac{f(x)}{\sin^{n-\lambda}(x-s)}\,dx$$

where m is a positive integer, but we shall not elaborate on this.

References. Gelfand and Shilov [1], Kutt [1, 2], L. Schwartz [1].

1.6.5 The Riemann–Stieltjes Integral

The Riemann–Stieltjes integral occurs with sufficient frequency in numerical work so that we shall give a brief account of it.

Take a partition of the interval $[a, b]$:

$$a = x_0 < x_1 < \cdots < x_n = b. \tag{1.6.5.1}$$

Let a function $f(x)$ be defined on $[a, b]$. If there is a constant M such that

$$\sum_{k=1}^{n} |f(x_k) - f(x_{k-1})| \le M \tag{1.6.5.2}$$

for *all* partitions, then f is said to be of *bounded variation* over $[a, b]$. This class of functions is designated by $BV[a, b]$. The least value M for which (1.6.5.2) is valid for all partitions is called the *variation of* f *over* $[a, b]$ and is designated by $V(f)$ or $V_a^b(f)$.

A function is in class $BV[a, b]$ if and only if it is the difference of two bounded nondecreasing functions. This is occasionally taken as the definition of the class BV. Note that if $f \in BV[a, b]$, then $V_a^t(f)$, the variation of f over $[a, t]$, $a \le t \le b$, is a bounded nondecreasing function of t and hence in class $BV[a, b]$.

Having been given two functions f and g defined on $[a, b]$, a partition, and points ξ_k with $x_{k-1} \le \xi_k \le x_k$, we may form the Riemann–Stieltjes sum

$$S_n = \sum_{k=1}^{n} f(\xi_k)(g(x_k) - g(x_{k-1})). \tag{1.6.5.3}$$

If $g(x) \equiv x$, these sums reduce to ordinary Riemann sums. As in Section 1.5, set $\Delta = \max_k (x_k - x_{k-1})$. This is the *norm* of the partition. Consider a sequence $S_1, S_2, \ldots$ of sums of the form (1.6.5.3) whose corresponding partition norms approach zero in the limit. If, for *any* sequence of this type, and corresponding to any choice of ξ_k, the sequences $\{S_m\}$ have a common limit S, then $f(x)$ is said to have a Riemann–Stieltjes integral with respect to the *integrator function* $g(x)$ and one writes

$$S = \int_a^b f(x)\,dg(x). \tag{1.6.5.4}$$

Sufficient conditions for the Riemann–Stieltjes integral to exist are that $f \in C[a, b]$, $g \in BV[a, b]$ or vice versa.

An extension is to have $f \in C[a, b^*]$, $g \in BV[a, b^*]$ for every b^* with $a < b^* < b$ and $\lim_{b^* \to b} \int_a^{b^*} |f(t)| \, dV_a^t(g) < \infty$. In this case, f is said to be *absolutely integrable* with respect to g.

We list some fundamental properties of the Riemann–Stieltjes integral.

$$\int_a^b (c_1 f_1 + c_2 f_2) \, dg = c_1 \int_a^b f_1 \, dg + c_2 \int_a^b f_2 \, dg; \tag{1.6.5.5}$$

$$\int_a^b f \, d(g + h) = \int_a^b f \, dg + \int_a^b f \, dh; \tag{1.6.5.6}$$

$$\int_a^b f \, dg = \int_a^c f \, dg + \int_c^b f \, dg, \qquad a < c < b. \tag{1.6.5.7}$$

If $f_1 \leq f_2$ on $[a, b]$ and g is nondecreasing there, then

$$\int_a^b f_1 \, dg \leq \int_a^b f_2 \, dg, \tag{1.6.5.8}$$

$$\left| \int_a^b f \, dg \right| \leq (\sup |f|) V_a^b(g). \tag{1.6.5.9}$$

If $f \in C[a, b]$ and $g \in C'[a, b]$, then

$$\int_a^b f \, dg = \int_a^b f g' \, dx. \tag{1.6.5.10}$$

The integral on the right is an ordinary Riemann integral.

If $w(t) \geq 0$ and $\int_a^b w(t) \, dt < \infty$, and if we set $g(x) = \int_a^x w(t) \, dt$, then the Riemann–Stieltjes integral reduces to the weighted Riemann integral:

$$\int_a^b f(x) \, dg(x) = \int_a^b w(x) f(x) \, dx. \tag{1.6.5.11}$$

Integration by Parts

$$\int_a^b f \, dg = f(b)g(b) - f(a)g(a) - \int_a^b g \, df. \tag{1.6.5.12}$$

If $f \in C[a, b]$ and g is nondecreasing, then

$$\int_a^b f \, dg = f(\xi) \int_a^b dg \qquad \text{with } a \leq \xi \leq b \tag{1.6.5.13}$$

(first mean-value theorem).

If f is nondecreasing and $g \in C[a, b]$, then

$$\int_a^b f\,dg = f(a)\int_a^\xi dg + f(b)\int_\xi^b dg, \qquad a \le \xi \le b \tag{1.6.5.14}$$

(second mean-value theorem).

Let $a = t_0 < t_1 < \cdots < t_{n-1} < t_n = b$. Let $g(x)$ have a constant value g_k in (t_{k-1}, t_k). At each point t_k in (a, b), $g(x)$ has a jump $u_k = g_{k+1} - g_k$. At a there may be a jump $u_0 = g_1 - g(a)$, while at b there may be a jump $u_n = g(b) - g_n$. Suppose also that $f(x)$ is continuous at each t_k. Then

$$\int_a^b f\,dg = \sum_{k=0}^{n} u_k f(t_k). \tag{1.6.5.15}$$

Reference. Smirnov [1, Vol. 5, Chap. I].

1.6.8 *The Dominated Integral*

In the definition of the improper integral of an unbounded integrand, there are two limiting processes, one defining the integrals $I_r = \int_r^b f(x)\,dx$ and the other defining $\int_a^b f(x)\,dx = \lim_{r\to a^+} I_r$. However, for a certain class of unbounded integrands, it is possible to define the improper integral by a single limiting process, namely, as the limit of a sequence of suitably restricted Riemann sums. This definition arose by noticing that it is sometimes possible to generate a sequence of numerical integration rules converging to the value of an improper integral, a procedure called "ignoring the singularity" (Section 2.12.7). This concept developed by Osgood and Shisha [1] is called the dominated integral defined as follows.

DEFINITION. The dominated integral of a function defined on (0, 1], if it exists, is the unique number $I(f)$ having the property: For each $\varepsilon > 0$ there exist δ and χ, $0 < \delta < 1$, $0 < \chi < 1$, such that

$$\left| I(f) - \sum_{j=1}^{n} f(\tau_j)(t_j - t_{j-1}) \right| < \varepsilon$$

whenever $0 < t_0 < t_1 < \cdots < t_n = 1$, $t_0 < \chi$, $t_{j-1} \le \tau_j \le t_j$, and $t_{j-1}t_j^{-1} > 1 - \delta$, $j = 1, 2, \ldots, n$.

If $I(f)$ exists, then the improper integral of f exists and equals $I(f)$.

$I(f)$ exists if and only if f is Riemann integrable on each $[a, b] \subset (0, 1]$, and there exists a function h, monotone nonincreasing and improperly Riemann integrable on (0, 1], such that $h(t) \ge |f(t)|$ throughout (0, 1].

Using this concept, Osgood and Shisha [2] were able to show that a function f on (0, 1] is dominantly integrable if and only if every sequence of numerical integration rules of some reasonable, natural form, when applied to f, converges to (the improper Riemann integral) $\int_{0+}^{1} f(t)\,dt$.

In an analogous manner, Haber and Shisha introduced the notion of the simple integral over the infinite interval $[0, \infty)$ so that a subclass of improper integrals over $[0, \infty)$ can be defined by a single limiting process rather than by the double process implied by the definition in Section 1.6.

References. Haber and Shisha [1]; Osgood and Shisha [1, 2].

1.7 The Riemann Integral in Higher Dimensions

We shall give here the definition of the Riemann integral in the case of two dimensions. Extensions are readily made to the higher-dimensional cases. Let R designate the rectangle $a \le x \le b$, $c \le y \le d$. Let the edges be partitioned: $a = x_0 < x_1 < \cdots < x_n = b$, $c = y_0 < y_1 < \cdots < y_m = d$. Designate the subrectangle $x_i \le x \le x_{i+1}$, $y_j \le y \le y_{j+1}$ by R_{ij} and in each R_{ij} choose a point p_{ij}. If G is a grid that partitions R into rectangles R_{ij}, then $d(G)$ will designate the maximum diagonal length of the R_{ij}. The following definition is now made.

The double integral $\iint_R f(x, y)\,dx\,dy$ exists and has the value I if and only if, given an $\varepsilon > 0$, we can find a value $\delta > 0$ such that

$$\left| I - \sum_{i,j} f(p_{ij})(x_{i+1} - x_i)(y_{j+1} - y_j) \right| \le \varepsilon \tag{1.7.1}$$

for any grid G with $d(G) \le \delta$ and for any choice of points p_{ij}.

Sums of the form $\sum f(p_{ij})(x_{i+1} - x_i)(y_{j+1} - y_j)$ are called *two-dimensional Riemann sums.*

If $f(x, y)$ is continuous on R, then $\iint_R f(x, y)\,dx\,dy$ exists. More generally, it exists if $f(x, y)$ is bounded and is continuous in R with the possible exception of a set of points of zero area.

To handle double integrals over bounded regions B that are not rectangular, the following idea is used. Select a rectangle R that contains B. Extend $f(x, y)$ from B to R by defining $f(x, y) = 0$ whenever $(x, y) \notin B$. Finally, define $\iint_B f(x, y)\,dx\,dy = \iint_R f(x, y)\,dx\,dy$.

If B is a bounded set in the plane having an area,† and if $f(x, y)$ is bounded on B and continuous at all interior points of B, then $\iint_B f(x, y)\,dx\,dy$ exists. Its value is independent of the rectangle R selected to contain B.

† Some two-dimensional point sets have no area, but this fine point is of no consequence in numerical work.

The following properties of double integrals are fundamental. Let $f(x, y)$ and $g(x, y)$ be continuous and bounded in a bounded region B that has area. Then

$$\iint_B dx\, dy = \text{area of } B, \tag{1.7.2}$$

$$\iint_B f\, dx\, dy + \iint_B g\, dx\, dy = \iint_B (f + g)\, dx\, dy, \tag{1.7.3}$$

$$\iint_B cf\, dx\, dy = c \iint_B f\, dx\, dy, \tag{1.7.4}$$

$$\left| \iint_B f\, dx\, dy \right| \leq \iint_B |f|\, dx\, dy. \tag{1.7.5}$$

If $f(x, y) \geq 0$ for $(x, y) \in B$, then

$$\iint_B f\, dx\, dy \geq 0. \tag{1.7.6}$$

If $B = B_1 \cup B_2$, and if the common part of B_1 and B_2 has zero area, then

$$\iint_B f\, dx\, dy = \iint_{B_1} f\, dx\, dy + \iint_{B_2} f\, dx\, dy. \tag{1.7.7}$$

Fubini's Theorem

$$\int_a^b \int_c^d f(x, y)\, dx\, dy = \int_a^b \left(\int_c^d f(x, y)\, dy \right) dx$$
$$= \int_c^d \left(\int_a^b f(x, y)\, dx \right) dy. \tag{1.7.8}$$

Reduction to an Iterated Integral

Let $\phi(x)$ and $\psi(x)$ be continuous on $a \leq x \leq b$ and suppose that $\phi(x) \leq \psi(x)$ there. Designate by B the region bounded by $x = a$, $x = b$, $y = \phi(x)$, and $y = \psi(x)$. Then, if $f(x, y)$ is continuous in B,

$$\iint_B f(x, y)\, dx\, dy = \int_a^b dx \int_{\phi(x)}^{\psi(x)} f(x, y)\, dy. \tag{1.7.9}$$

Mean-Value Theorem

Let B be a bounded, open, connected set in the plane. Suppose that $f(x, y)$, $g(x, y)$ are continuous and bounded in B and suppose, further, that $g(x, y) \geq 0$ in B. Then there is a point (ξ, η) in B such that

$$\iint_B f(x, y) g(x, y)\, dx\, dy = f(\xi, \eta) \iint_B g(x, y)\, dx\, dy. \tag{1.7.10}$$

If B is a d-dimensional region, then the integral of $f(x_1, \ldots, x_d)$ over B will be written in one of several alternative ways, namely,

$$\int_B \cdots \int f(x_1, \ldots, x_d)\, dx_1 \cdots dx_d = \int_B \cdots \int f(\mathbf{x})\, d\mathbf{x} = \int_B \cdots \int f\, dV = \int_B f\, dV,$$

where $\mathbf{x} = (x_1, \ldots, x_d)$.

Analogs of the Cauchy Principal Value can be defined in higher dimensions. Suppose that the singularity of the function occurs at the point P_0 of the region B. Designate the disk (or ball) $|P - P_0| \leq r$ by C_r and the region $B - C_r$, consisting of B with C_r deleted, by B_r. Then, one sets

$$P \int_B \int f\, dV = \lim_{r \to 0} \int_{B_r} \int f\, dV. \tag{1.7.11}$$

For further material on this see Mikhlin or Tricomi.

References. Buck [1, Chap. 3]; Mikhlin [1]; Tricomi [1].

In many situations, we are interested in integrating over a set S of dimension $r < d$ in d-dimensional space. Frequently occurring examples of this are integrals of functions defined on the surface of d-dimensional spheres or on one-dimensional space curves. Now the problem of defining such integrals in full generality is very complicated, involving many subtle points in differential geometry. We shall therefore restrict our attention to such sets S that are defined by a parametric representation. Thus, let S be given by the equations

$$\begin{aligned} x_1 &= \phi_1(u_1, \ldots, u_r) \\ &\vdots \\ x_d &= \phi_d(u_1, \ldots, u_r) \end{aligned}$$

where the ϕ_i possess continuous derivatives in a bounded region B of the variables $u_1, \ldots, u_r$. As these range over B, the point $(x_1, \ldots, x_d)$ describes S. From the $r \times d$ rectangular matrix $(\partial x_i / \partial u_j)$, $i = 1, \ldots, r$, $j = 1, \ldots, d$, we form all r-rowed determinants D_k, $k = 1, \ldots, K = \binom{d}{r}$. Then the integral of $f(x_1, \ldots, x_d)$ over S, denoted

$$\int_S \cdots \int f(x_1, \ldots, x_d)\, dS = \int_B \cdots \int g(u_1, \ldots, u_r) \left(\sum_{k=1}^{K} D_k^2 \right)^{1/2} du_1 \cdots du_r$$

where $g(u_1, \ldots, u_r) = f(\phi_1(u_1, \ldots, u_r), \ldots, \phi_d(u_1, \ldots, u_r))$.

Reference. Courant and John [1].

1.8 More General Integrals

The reader who is familiar with more general theories of integration such as that of the Lebesgue integral should observe that the present program is not suitable for the computation of such integrals. We are dealing with approximation of the form

$$\int_a^b f(x)\,dx \approx \sum_{i=1}^{n} w_i f(x_i) \tag{1.8.1}$$

and ultimately (as $n \to \infty$) with the possibility of having

$$\int_a^b f(x)\,dx = \lim_{n\to\infty} \sum_{i=1}^{n} w_{in} f(x_{in}). \tag{1.8.2}$$

We know that such an equation can hold for all Riemann-integrable functions if, for example, we use Riemann sums for the right-hand side. Now, if the functional values of f are changed arbitrarily on the set of points $\{x_{in}\}$ ($i = 1, 2, \ldots, n; n = 1, 2, \ldots$), which is a set of measure zero, then the value of $\int_a^b f(x)\,dx$ as a Lebesgue integral does not change. However, the value of the right-hand side of (1.8.2) will change. This means that computational schemes such as (1.8.2) are not appropriate. For a proper theory of such approximations, one must employ, as elementary data, numbers that are already integrals of the function in question, for example, the moments $\int_a^b x^n f(x)\,dx$ or Fourier coefficients $\int_a^b e^{inx} f(x)\,dx$. However, the ideas of Lebesgue integration have been applied by Burrows to evaluate numerical approximations to one- and multidimensional integrals.

On the other hand, if $w(x)$ is a fixed Lebesgue-integrable weighting function, and if $f(x)$ is restricted to be bounded and Riemann-integrable, then the program $\int_a^b w(x)f(x)\,dx \approx \sum_{i=1}^n w_{in} f(x_{in})$, w_{in} independent of f, makes sense. This approach has been very fruitful in the context of product integration and will be developed in Section 2.5.6

Reference. Burrows [1].

1.9 The Smoothness of Functions and Approximate Integration

It is a general phenomenon pervading all of the theory of interpolation and approximation that the quality of the approximation resulting from common methods depends on the degree of smoothness of the function operated upon. The smoother the function, the closer the approximation, and the more rapid the convergence of a sequence of approximations. The concepts that make "degree of smoothness" more precise are those of continuity, the number of continuous derivatives, their magnitude, analyticity, and extent of analyticity in the complex plane.

Starting at a low level and moving in the direction of increased smoothness, we have

(a) Functions that are bounded and Riemann-integrable over the interval $[a, b]$ (Class $R[a, b]$).
(b) Functions that are of bounded variation over $[a, b]$ (Class $BV[a, b]$).
(c) Functions that are piecewise continuous over the interval in question (Class $PC[a, b]$).
(d) Functions that are continuous over the interval $[a, b]$ (Class $C[a, b]$).
(e) Functions that satisfy a Lipschitz or Hölder condition of order $\alpha \leq 1$ on $[a, b]$ (Class $\text{Lip}_\alpha[a, b]$).
(f) Functions that have a continuous first derivative on the interval $[a, b]$ (Class $C^1[a, b]$ or $C'[a, b]$).
(g) Functions that have a continuous nth derivative on the interval $[a, b]$ (Class $C^n[a, b]$).
(h) Functions that are analytic in a region B containing the interval in its interior (Class A (B)).
(i) Functions that are entire, that is, have a Taylor expansion convergent in $|z| < \infty$.
(j) Functions that are polynomials of degree $\leq n$ (Class $\mathscr{P}_n$).

A class of functions that does not fit into this scheme, but to which we shall refer, is the class $L_p[a, b]$ or $L_p(B)$, $1 \leq p < \infty$, the set of all Lebesgue measurable functions f on $[a, b]$ or B such that $\int_a^b |f(x)|^p \, dx$ or $\int_B |f|^p \, dV$ is finite, as the case may be.

Though approximation theory makes good use of several other classes of functions, the above classes are the principal ones employed in this volume.

1.10 Weight Functions

It is frequently convenient to consider an integral of the form $\int w(x)f(x) \, dx$ instead of the simple integral $\int f(x) \, dx$. Here $w(x)$ is *usually* (but not always) *assumed to be nonnegative* over the interval and *fixed* throughout the discussion. The function $f(x)$, however, may vary. The function $w(x)$ is called a *weight function*, and it is frequently normalized so that

$$\int_a^b w(x) \, dx = 1. \tag{1.10.1}$$

The weight function $w(x)$ will be called *admissible* over the finite or infinite interval $[a, b]$ if $w(x) \geq 0$, $x \in [a, b]$, $\int_a^b w(x) \, dx > 0$ and $\int_a^b w(x)x^k \, dx < \infty$, $k = 0, 1, \ldots$.

The integral $\int_a^b w(x)f(x)\,dx$ can be interpreted as a weighted average of $f(x)$.

It will be assumed throughout this book that the integrand $w(x)f(x)$ is Riemann-integrable, properly or improperly, and that

$$\int_a^b w(x)\,dx < \infty. \tag{1.10.2}$$

The condition (1.10.2) means that we allow only integrable singularities as, for example,

$$\int_0^1 \frac{f(x)}{\sqrt{x}}\,dx,$$

where $f(x)$ is bounded in the neighborhood of zero.

1.11 Some Useful Formulas

We list here some other frequently used formulas.

1 Change of Variable in a Single Integral

Let $g'(u)$ be continuous in the interval $c \le u \le d$, and let $g(c) = a$ and $g(d) = b$. Let $f(x)$ be continuous at all points $x = g(u)$ for $c \le u \le d$. Then

$$\int_a^b f(x)\,dx = \int_c^d f(g(u))g'(u)\,du. \tag{1.11.1}$$

Changes of variable may result in a favorable change of the continuity class of the integrand.

Example. The integrand in $I = \int_0^1 \sqrt{x}\sin x\,dx$ has an infinite second derivative at $x = 0$. Setting $x = u^2$ we obtain $I = 2\int_0^1 u^2 \sin u^2\,du$ where the integrand has no singularity.

One should also be aware of the possibility of the reverse happening.

1.5 Change of Variable in a Cauchy Principal Value

The variable may be changed in a Cauchy principal value integral in the standard way providing certain conditions are met. Let $m(s) \in C^2[\alpha, \beta]$ be monotonic there, and $m(\alpha) = a$, $m(\beta) = b$, $m'(s) \neq 0$ in $[\alpha, \beta]$. Then,

$$P\int_a^b \frac{f(t)}{t - x}\,dt = P\int_\alpha^\beta \frac{f(m(s))m'(s)\,ds}{m(s) - m(\xi)} \tag{1.11.1a}$$

where $x = m(\xi)$.

Reference. Smirnov [1, Vol. 3, pp. 103–106].

2 *Transformation of Multiple Integrals*

Under the transformation

$$x = \phi(u, v), \qquad y = \psi(u, v),$$

we have

$$\iint_B f(x, y)\, dx\, dy = \iint_{B'} f(\phi(u, v), \psi(u, v))\, |J(u, v)|\; du\, dv, \quad (1.11.2)$$

where

$$J(u, v) = \begin{vmatrix} \partial\phi/\partial u & \partial\phi/\partial v \\ \partial\psi/\partial u & \partial\psi/\partial v \end{vmatrix}. \quad (1.11.3)$$

Under the transformation

$$x = \phi(u, v, w),$$
$$y = \psi(u, v, w),$$
$$z = \chi(u, v, w),$$

we have

$$\iiint_V f(x, y, z)\, dx\, dy\, dz$$
$$= \iiint_{V'} f[\phi(u, v, w), \psi(u, v, w), \chi(u, v, w)]\, |J(u, v, w)|\; du\, dv\, dw, \quad (1.11.4)$$

where

$$J(u, v, w) = \begin{vmatrix} \partial\phi/\partial u & \partial\phi/\partial v & \partial\phi/\partial w \\ \partial\psi/\partial u & \partial\psi/\partial v & \partial\psi/\partial w \\ \partial\chi/\partial u & \partial\chi/\partial v & \partial\chi/\partial w \end{vmatrix}.$$

Sufficient conditions for the validity of these transformations are: (1) the regions B and B' (or V and V') correspond to one another in a one-to-one manner; (2) the functions ϕ, ψ, χ have continuous first partial derivatives in B' or V'; and (3) the Jacobian $J \neq 0$ in the region. Similar formulas hold for integrals in higher dimensions.

3 *Jacobians for Various Transformations*

Polar coordinates in the plane

$$x = r\cos\phi,$$
$$y = r\sin\phi, \qquad J(r, \phi) = r.$$

Cylindrical coordinates in 3-space

$$x = r\cos\phi,\qquad y = r\sin\phi,\qquad z = z,\qquad J(r,\phi,z) = r.$$

Spherical coordinates in 3-space

$$x = r\sin\phi\cos\theta,\qquad y = r\sin\phi\sin\theta,\qquad z = r\cos\phi,\qquad J(r,\phi,\theta) = r^2\sin\phi.$$

General spherical coordinates

$$\begin{aligned}
x_1 &= r\sin\theta_1\sin\theta_2\cdots\sin\theta_{n-2}\sin\theta_{n-1},\\
x_2 &= r\sin\theta_1\sin\theta_2\cdots\sin\theta_{n-2}\cos\theta_{n-1},\\
x_3 &= r\sin\theta_1\sin\theta_2\cdots\cos\theta_{n-2},\\
&\ \vdots\\
x_{n-1} &= r\sin\theta_1\cos\theta_2,\\
x_n &= r\cos\theta_1,\\
J &= (-1)^{n(n-1)/2}r^{n-1}\sin^{n-2}\theta_1\cdots\sin\theta_{n-2}.
\end{aligned}$$

The spherical shell of inner radius R_1 and outer radius R_2: $R_1^2 \le \sum_{i=1}^n x_i^2 \le R_2^2$ corresponds to the inequalities

$$R_1 \le r \le R_2,\qquad 0 \le \theta_i \le \pi,\qquad i = 1, 2, \ldots, n-2,\qquad 0 \le \theta_{n-1} \le 2\pi \tag{1.11.5}$$

in this parametrization.

Reference. Buck [1, Chap. 6].

4 *Lagrange Interpolation Formula*

Let $x_0 < x_1 < x_2 \cdots < x_n$ be $n+1$ distinct points, and let there be given $n+1$ arbitrary numbers $v_0, v_1, \ldots, v_n$. Then there exists a unique polynomial $p_n(x)$, of class $\mathscr{P}_n$, such that

$$p_n(x_i) = v_i,\qquad i = 0, 1, \ldots, n. \tag{1.11.6}$$

The polynomial $p_n(x)$ can be given the following explicit representation. Define $n+1$ *fundamental polynomials* $\ell_i(x) \in \mathscr{P}_n$:

$$\ell_i(x) = \frac{(x-x_0)\cdots(x-x_{i-1})(x-x_{i+1})\cdots(x-x_n)}{(x_i-x_0)\cdots(x_i-x_{i-1})(x_i-x_{i+1})\cdots(x_i-x_n)},\qquad i = 0, 1, \ldots, n. \tag{1.11.7}$$

For these polynomials

$$\ell_i(x_j) = \delta_{ij} = \begin{cases} 0 & \text{if } i \neq j, \\ 1 & \text{if } i = j. \end{cases} \tag{1.11.8}$$

Then

$$p_n(x) = \sum_{i=0}^{n} v_i \ell_i(x). \tag{1.11.9}$$

The fundamental polynomials may also be written in the form

$$\ell_i(x) = \frac{\pi(x)}{(x - x_i)\pi'(x_i)}, \tag{1.11.10}$$

where

$$\pi(x) = (x - x_0)(x - x_1) \cdots (x - x_n). \tag{1.11.11}$$

When we wish to interpolate to a function $f(x)$, the polynomial

$$p_n(f; x) = \sum_{i=0}^{n} f(x_i) \frac{\pi(x)}{(x - x_i)\pi'(x_i)} \tag{1.11.12}$$

is the unique member of $\mathscr{P}_n$ for which

$$p_n(f; x_i) = f(x_i), \qquad i = 0, 1, \ldots, n. \tag{1.11.13}$$

The remainder is given by

$$f(x) - p_n(f; x) = \frac{f^{(n+1)}(\xi)}{(n+1)!} \pi(x)$$

where $\min(x, x_0) < \xi < \max(x, x_n)$.

5 Osculatory or Hermite Interpolation

Set

$$L_k(x) = \left[1 - \frac{\pi''(x_k)}{\pi'(x_k)} (x - x_k)\right] \ell_k^2(x),$$

$$M_k(x) = (x - x_k)\ell_k^2(x).$$

These are in $\mathscr{P}_{2n+1}$. Then,

$$p_{2n+1}(x) = \sum_{k=0}^{n} f(x_k)L_k(x) + \sum_{k=0}^{n} f'(x_k)M_k(x) \tag{1.11.14}$$

is the unique member of $\mathscr{P}_{2n+1}$ for which

$$p_{2n+1}(x_k) = f(x_k),$$

$$p'_{2n+1}(x_k) = f'(x_k), \qquad k = 0, 1, \ldots, n. \tag{1.11.15}$$

The remainder is given by

$$f(x) - p_{2n+1}(x) = \frac{f^{(2n+2)}(\xi)}{(2n+2)!}[\pi(x)]^2$$

where $\min(x, x_0) < \xi < \max(x, x_n)$.

The more general interpolation polynomial of degree $n + m + 1$ which interpolates $f(x)$ at $x_0, \ldots, x_n$ and $f'(x)$ at $x_0, \ldots, x_m$, $m < n$, is also of interest. See, for example, Ralston.

6 *Two-Point Taylor Interpolation*

Let a and b be distinct points. The polynomial

$$p_{2n-1}(x) = (x-a)^n \sum_{k=0}^{n-1} \frac{B_k(x-b)^k}{k!} + (x-b)^n \sum_{k=0}^{n-1} \frac{A_k(x-a)^k}{k!} \tag{1.11.16}$$

where

$$A_k = \frac{d^k}{dx^k}\left[\frac{f(x)}{(x-b)^n}\right]_{x=a}, \qquad B_k = \frac{d^k}{dx^k}\left[\frac{f(x)}{(x-a)^n}\right]_{x=b}$$

is the unique member of $\mathscr{P}_{2n-1}$ for which

$$\begin{aligned} &p_{2n-1}(a) = f(a), \quad p'_{2n-1}(a) = f'(a), \quad \ldots, \quad p_{2n-1}^{(n-1)}(a) = f^{(n-1)}(a); \\ &p_{2n-1}(b) = f(b), \quad p'_{2n-1}(b) = f'(b), \quad \ldots, \quad p_{2n-1}^{(n-1)}(b) = f^{(n-1)}(b). \end{aligned} \tag{1.11.17}$$

The remainder is given by

$$f(x) - p_{2n-1}(x) = \frac{f^{(2n)}(\xi)}{(2n)!}(x-a)^n(x-b)^n$$

where

$$\min(x, a) < \xi < \max(x, b).$$

References. Davis [6, pp. 33–37, 67]; Ralston [3, pp. 60–62].

7 *The Taylor Expansion in d Variables*

Several abbreviations will enable us to write this compactly. Let $\mathbf{i}$ designate a d-tuple of nonnegative integers: $\mathbf{i} = (i_1, i_2, \ldots, i_d)$. Define $|\mathbf{i}| = i_1 + i_2 + i_3 + \cdots + i_d$ and $\mathbf{i}! = (i_1!)(i_2!) \cdots (i_d!)$. If $\mathbf{x}$ designates a variable point in d dimensions: $\mathbf{x} = (x_1, x_2, \ldots, x_d)$, set $\mathbf{x}^{\mathbf{i}} = x_1^{i_1} x_2^{i_2} \cdots x_n^{i_n}$. If f is a function of $x_1, x_2, \ldots, x_d$, we define

$$f^{(\mathbf{i})} = \frac{\partial^k f}{(\partial x_1)^{i_1}(\partial x_2)^{i_2} \cdots (\partial x_d)^{i_d}} \tag{1.11.18}$$

where $k = |\mathbf{i}|$. With these definitions, the Taylor expansion of f around $\mathbf{x}_0$ can be written in the form

$$f(\mathbf{x}) = f(\mathbf{x}_0) + \sum_1 \frac{f^{(\mathbf{i})}(\mathbf{x}_0)(\mathbf{x} - \mathbf{x}_0)^{\mathbf{i}}}{\mathbf{i}!} + \sum_2 \frac{f^{(\mathbf{i})}(\mathbf{x}_0)(\mathbf{x} - \mathbf{x}_0)^{\mathbf{i}}}{\mathbf{i}!} + \cdots \tag{1.11.19}$$

where $\sum_1$ designates the sum taken over all $\mathbf{i}$ with $|\mathbf{i}| = 1$, $\sum_2$ is taken over all $\mathbf{i}$ with $|\mathbf{i}| = 2$, etc.

8 Differences

From time to time we shall use the notation for differences. Let there be given a sequence of number $y_0, y_1, \ldots$. The (forward) *differences* of adjacent values are designated by

$$\Delta y_k = y_{k+1} - y_k, \qquad k = 0, 1, \ldots. \tag{1.11.20}$$

Higher differences are defined similarly

$$\begin{aligned} \Delta^2 y_k &= \Delta(\Delta y_k) = \Delta y_{k+1} - \Delta y_k \\ &= (y_{k+2} - y_{k+1}) - (y_{k+1} - y_k). \end{aligned} \tag{1.11.21}$$

In general,

$$\Delta^{n+1} y_k = \Delta(\Delta^n y_k) = \Delta^n y_{k+1} - \Delta^n y_k. \tag{1.11.22}$$

We define $\Delta^0 y_k = y_k$.

We have

$$\Delta^n y_k = \sum_{r=0}^{n} (-1)^{n-r} \binom{n}{r} y_{k+r}, \qquad \binom{n}{r} = \frac{n!}{r!(n-r)!}. \tag{1.11.23}$$

Divided differences of a function f with respect to the abscissas $x_0, x_1, \ldots$ are defined by:

First divided difference:

$$[f(x_0), f(x_1)] = \frac{f(x_0) - f(x_1)}{x_0 - x_1}. \tag{1.11.24}$$

Second divided difference:

$$[f(x_0), f(x_1), f(x_2)] = \frac{[f(x_0), f(x_1)] - [f(x_1), f(x_2)]}{x_0 - x_2}.$$

$$\vdots \tag{1.11.25}$$

kth divided difference:

$$[f(x_0), f(x_1), \ldots, f(x_k)]$$
$$= \frac{[f(x_0), f(x_1), \ldots, f(x_{k-1})] - [f(x_1), f(x_2), \ldots, f(x_k)]}{x_0 - x_k}. \tag{1.11.26}$$

Although it is not apparent from the definition, the divided differences are symmetric functions of their arguments.

9 *Iterated Indefinite Integrals*

$$\int_a^x dt_n \int_a^{t_n} dt_{n-1} \cdots \int_a^{t_3} dt_2 \int_a^{t_2} f(t_1)\, dt_1$$
$$= \frac{1}{(n-1)!} \int_a^x (x - t)^{n-1} f(t)\, dt = \frac{(x-a)^n}{(n-1)!} \int_0^1 t^{n-1} f(x - (x-a)t)\, dt.$$

1.12 Orthogonal Polynomials

Sets of orthogonal polynomials play a considerable role in the theory of numerical integration and we shall therefore mark down the pertinent features and formulas.

Given a real linear space of functions $\mathscr{F}$, an *inner product* (f, g) defined on $\mathscr{F}$ is a bilinear functional of $f, g \in \mathscr{F}$ satisfying the conditions

(a) $(f + g, h) = (f, h) + (g, h)$;
(b) $(\alpha f, g) = \alpha(f, g)$ for α scalar;
(c) $(f, g) = (g, f)$; (1.12.1)
(d) $(f, f) > 0$ if $f \not\equiv 0$.

If $\mathscr{F}$ is a complex linear space of functions, then the requirement (c) is replaced by

(c′) $(f, g) = \overline{(g, f)}$.

The bar designates complex conjugation.

Example 1. Let $\mathscr{F} = C[a, b]$. Let $w(x) \geq 0$ be Riemann-integrable over $[a, b]$ either properly or improperly with $\int_a^b w(x)\,dx > 0$.
Then

$$(f, g) = \int_a^b w(x)f(x)g(x)\,dx \tag{1.12.1a}$$

is an inner product for $\mathscr{F}$.

Example 2. Let B be a bounded simply connected region of the complex plane with a simple rectifiable boundary C. Let $\mathscr{F}$ be the complex linear space of all functions that are analytic in B and continuous on $B + C$. Then

$$(f, g) = \int_C f(z)\overline{g(z)}\,ds, \qquad ds^2 = dx^2 + dy^2,$$

is an inner product for $\mathscr{F}$.

Example 3. Let $\mathscr{F} = \mathscr{P}_n$. Let $x_0, x_1, \ldots, x_n$ be distinct points and $w_0, w_1, \ldots, w_n > 0$. Then

$$(f, g) = \sum_{i=0}^{n} w_i f(x_i)g(x_i)$$

is an inner product for $\mathscr{F}$.

Example 4. Let $\mathscr{F} = \mathscr{P}_{2n-1}$, x_i distinct and $w_i > 0$, $v_i > 0$, then

$$(f, g) = \sum_{i=1}^{n} w_i f(x_i)g(x_i) + v_i f'(x_i)g'(x_i)$$

is an inner product for $\mathscr{F}$.

If $f_0, f_1, \ldots$ is a finite or infinite set of elements of $\mathscr{F}$ such that

$$(f_i, f_j) = 0, \qquad i \neq j, \tag{1.12.2}$$

then the set is called *orthogonal*. If, in addition,

$$(f_i, f_i) = 1, \qquad i = 0, 1, \ldots, \tag{1.12.3}$$

the set is called *orthonormal*.

A set of polynomials $\{f_i\}$ where degree $(f_i) = i$ and which satisfies (1.12.2) is called a set of orthogonal polynomials with respect to the inner product (f, g).

Any linearly independent sequence of functions of $\mathscr{F}$ can be orthogonalized with respect to a given inner product in $\mathscr{F}$. Thus, given an admissible weight function $w(x)$ on a finite or infinite interval $[a, b]$, we may orthonormalize the powers $1, x, x^2, \ldots$ with respect to the inner product (1.12.1a) and arrive at a unique set of polynomials $p_n^*(x)$, degree $(p_n^*) = n$, leading coefficient positive, and

$$\int_a^b w(x)p_n^*(x)p_m^*(x)\,dx = \delta_{mn} = \begin{cases} 0, & m \neq n, \\ 1, & m = n. \end{cases} \tag{1.12.4}$$

The classical orthogonal polynomials correspond to special weights and intervals as follows:

Interval	Weight function	Symbol	Name
$[-1, 1]$	1	$P_n(x)$	Legendre
$[-1, 1]$	$(1-x^2)^{-1/2}$	$T_n(x)$	Tschebyscheff, first kind
$[-1, 1]$	$(1-x^2)^{1/2}$	$U_n(x)$	Tschebyscheff, second kind
$[-1, 1]$	$(1-x^2)^{\mu-1/2}$, $\mu > -1/2$	$C_n^{\mu}(x)$	Ultraspherical or Gegenbauer
$[-1, 1]$	$(1-x)^{\alpha}(1+x)^{\beta}$, $\alpha, \beta > -1$	$P_n^{(\alpha,\beta)}(x)$	Jacobi
$[0, \infty)$	e^{-x}	$L_n(x)$	Laguerre
$[0, \infty)$	$x^{\alpha}e^{-x}$, $\alpha > -1$	$L_n^{(\alpha)}(x)$	Generalized Laguerre
$(-\infty, \infty)$	e^{-x^2}	$H_n(x)$	Hermite

Orthogonal polynomials with respect to the inner product (1.12.1a) satisfy a second type of orthogonality which can be called "discrete" orthogonality. Let $p_0(x), p_1(x), \ldots, p_n(x), p_{n+1}(x)$ be orthonormal polynomials corresponding to the admissible weight function $w(x)$ on $[a, b]$. Suppose that $x_1, x_2, \ldots, x_{n+1}$ are the $n+1$ zeros of $p_{n+1}(x)$ and that $w_1, \ldots, w_{n+1}$ are the corresponding Gaussian weights. (See Section 2.7.) Then, by the fundamental theorem of Gaussian integration,

$$\int_a^b w(x)f(x)\,dx = \sum_{i=1}^{n+1} w_i f(x_i)$$

for all $f \in \mathscr{P}_{2n+1}$. Now for $j, k \leq n$, $p_j(x)p_k(x) \in \mathscr{P}_{2n+1}$ so that

$$\sum_{i=1}^{n+1} w_i p_j(x_i)p_k(x_i) = \int_a^b w(x)p_j(x)p_k(x)\,dx = \delta_{jk}. \tag{1.12.5}$$

Thus, $p_0, p_1, \ldots, p_n$ are orthonormal on the zeros of p_{n+1} with respect to the Gauss weights, that is, with respect to the inner product

$$(f, g) = \sum_{i=1}^{n+1} w_i f(x_i)g(x_i). \tag{1.12.6}$$

We may conclude from this that if one starts with the monomials $1, x, x^2, \ldots, x^n$ and orthonormalizes them with respect to the discrete inner product (1.12.6), the orthonormal polynomials with respect to the continuous inner product (1.12.1a) will result.

THEOREM. *If $p_0(x), p_1(x), \ldots$ are polynomials with*

$$p_n(x) = k_n x^n + \cdots, \qquad k_n > 0,$$

orthogonal with respect to the inner product (f, g), *then we have the recurrence*

$$p_{n+1}(x) = (\gamma_n x - a_{nn})p_n(x) - a_{n,n-1}p_{n-1}(x) - \cdots - a_{n0}\,p_0(x), \qquad n = 0, 1, 2, \ldots \tag{1.12.7}$$

where $p_0(x) \equiv k_0$, $\gamma_n = k_{n+1}/k_n$, and

$$a_{ns} = \gamma_n(xp_n, p_s)/(p_s, p_s), \qquad s = 0, 1, \ldots, n.$$

THEOREM. *If the inner product satisfies the further condition that* $(xf, g) = (f, xg)$, *then the recurrence* (1.12.7) *reduces to the three term recurrence*

$$p_{n+1}(x) = (\gamma_n x - \alpha_n)p_n(x) - \beta_n p_{n-1}(x), \qquad n = 0, 1, \ldots \tag{1.12.8}$$

where we write $p_{-1}(x) = 0$, *and where*

$$\alpha_n = \frac{\gamma_n(xp_n, p_n)}{(p_n, p_n)}, \qquad n = 0, 1, \ldots,$$

$$\beta_n = \frac{\gamma_n(xp_n, p_{n-1})}{(p_{n-1}, p_{n-1})} = \frac{\gamma_n}{\gamma_{n-1}} \frac{(p_n, p_n)}{(p_{n-1}, p_{n-1})}, \qquad n = 1, 2, \ldots.$$

Note that the inner product $\int_a^b w(x)f(x)g(x)\,dx$ satisfies this condition whereas the inner products in Examples 2 and 4 do not in general satisfy it.

The three term recurrence formula for $p_n(x)$ leads to a "backward" recurrence for the efficient and relatively stable computation of an expansion of the form

$$f(x) = \sum_{r=0}^{N} c_r p_r(x).$$

Set

$$B_r = \begin{cases} 0 & \text{for} \quad r > N \\ c_r + (\gamma_{r+1}x - \alpha_{r+1})B_{r+1} - \beta_{r+2}B_{r+2} & \text{for} \quad 0 \le r \le N. \end{cases} \tag{1.12.9}$$

Then

$$f(x) = \gamma_0 B_0 \tag{1.12.10}$$

To compute Tschebyscheff expansions $f(x) = \sum_{r=0}^{N} c_r T_r(x)$, one selects

$$\alpha_r = 0, \qquad \beta_r = 1, \qquad \gamma_0 = \gamma_1 = 1, \qquad \gamma_r = 2 \qquad \text{for} \quad r \ge 2.$$

In addition to recurrence relations, orthogonal polynomials satisfy numerous identities. One of the most useful is the *Christoffel–Darboux identity*

$$\sum_{k=0}^{n} \frac{p_k(x)p_k(y)}{h_k} = \frac{p_{n+1}(x)p_n(y) - p_n(x)p_{n+1}(y)}{\gamma_n h_n(x - y)} \tag{1.12.11}$$

where $h_k = (p_k, p_k)$.

Many important properties of the classical orthogonal polynomials are summarized in the tables in Section 1.13.

For material on sets of orthogonal polynomials in more than one variable, the reader is directed to Stroud.

References. Bergman [1]; Clenshaw [1]; Davis [6]; Freud [1]; Gautschi [3]; Geronimus [1]; Hochstrasser [1]; Nehari [1]; Stroud [18]; Szegö [2]; Tricomi [2].

Functions of the Second Kind

Let $\{p_n(x)\}$ be a set of orthogonal polynomials on $[a, b]$ with respect to the weight function $w(x)$. The functions

$$q_n(z) = \int_a^b w(t)p_n(t)/(z - t)\,dt \tag{1.12.12}$$

are known as the *functions of the second kind* or the *associated functions.* They are regular analytic functions in the complex z plane slit along $[a, b]$.

The ratio $q_n(z)/p_n(z)$ appears as a kernel in a formula for the error in Gaussian integration rules. See (4.6.25). It is of importance to our subject to have asymptotic expressions for this ratio as $n \to \infty$ which are uniform in z.

The following notations are adopted for functions of the second kind in the case of the classical orthogonal polynomials.

Jacobi:

$$\pi_n^{(\alpha,\beta)}(z) = \int_{-1}^{+1} \frac{(1-t)^\alpha(1+t)^\beta P_n^{(\alpha,\beta)}(t)\,dt}{z-t} \tag{1.12.13}$$

Ultraspherical:

$$\pi_n^{(\mu-1/2,\mu-1/2)} \tag{1.12.14}$$

Legendre:

$$\pi_n^{(0,0)}(z) \tag{1.12.15}$$

Tschebyscheff (first kind):

$$\pi_n^{(-1/2,-1/2)}(z) \tag{1.12.16}$$

Tschebyscheff (second kind):

$$\pi_n^{(1/2,1/2)}(z) \tag{1.12.17}$$

Generalized Laguerre:

$$\lambda_n^{(\alpha)}(z) = \int_0^\infty \frac{e^{-t}t^\alpha L_n^{(\alpha)}(t)\,dt}{z-t} \tag{1.12.18}$$

Hermite:

$$\eta_n(z) = \int_{-\infty}^{\infty} \frac{e^{-t^2} H_n(t)\, dt}{z - t} \tag{1.12.19}$$

We have, for $|z|$ sufficiently large, and appropriate constants $c_i = c_{ni}$,

$$\begin{aligned} q_n(z) &= \int_a^b w(t) p_n(t) \sum_{k=0}^{\infty} \frac{t^k}{z^{k+1}}\, dt \\ &= \frac{c_n}{z^{n+1}} + \frac{c_{n+1}}{z^{n+2}} + \cdots. \end{aligned} \tag{1.12.20}$$

The last equality follows from the orthogonality of p_n.

For $a < x < b$, we have the *Sochozki–Plemelj formulas*,

$$\lim_{\varepsilon \to 0^+} (q_n(x - i\varepsilon) - q_n(x + i\varepsilon)) = 2\pi i w(x) p_n(x), \qquad i = \sqrt{-1}, \tag{1.12.21}$$

$$\lim_{\varepsilon \to 0^+} \tfrac{1}{2}(q_n(x - i\varepsilon) + q_n(x + i\varepsilon)) = P \int_a^b w(t) \frac{p_n(t)}{x - t}\, dt \equiv \hat{q}_n(x). \tag{1.12.22}$$

The functions of the second kind, $q_n(z)$, including $\hat{q}_n(x)$ satisfy the same three term recurrence formula (1.12.8) satisfied by the $p_n(x)$ with initial values

$$q_{-1}(z) \equiv 1, \qquad q_0(z) = \int_a^b \frac{w(t)}{z - t}\, dt$$

provided that we define $\beta_0 = \gamma_0(p_0, p_0)k_0$.

References. Elliott [4]; Gakhov [1, Chap. 1, Par. 4.2]; Muskhelishvili [1, Chap. 2, Par. 16]; Szegö [2, p. 78].

1.13 Short Guide to the Orthogonal Polynomials

I

NAME: Legendre SYMBOL: $P_n(x)$ INTERVAL: $[-1, 1]$
WEIGHT: 1 STANDARDIZATION: $P_n(1) = 1$
NORM:

$$\int_{-1}^{+1} (P_n(x))^2\, dx = 2/(2n + 1)$$

EXPLICIT EXPRESSION:

$$P_n(x) = \frac{1}{2^n} \sum_{m=0}^{[n/2]} (-1)^m \binom{n}{m} \binom{2n - 2m}{n} x^{n-2m}$$

RECURRENCE RELATION:

$$(n+1)P_{n+1}(x) = (2n+1)xP_n(x) - nP_{n-1}(x); \qquad P_0 = 1, \quad P_1 = x.$$

DIFFERENTIAL EQUATION:

$$(1-x^2)y'' - 2xy' + n(n+1)y = 0, \qquad y = P_n(x)$$

INDEFINITE INTEGRAL:

$$\int P_n(x)\,dx = \frac{1}{2n+1}[P_{n+1}(x) - P_{n-1}(x)]$$

RODRIGUES' FORMULA:

$$P_n(x) = \frac{(-1)^n}{2^n n!}\frac{d^n}{dx^n}\{(1-x^2)^n\}$$

GENERATING FUNCTION:

$$R^{-1} = \sum_{n=0}^{\infty} P_n(x)z^n, \qquad -1 \le x \le 1, \quad |z| < 1, \qquad R = (1 - 2xz + z^2)^{1/2}$$

INEQUALITY: $|P_n(x)| \le 1, \ -1 \le x \le 1$

ZEROS: $x_{n,m}$ is the mth zero of $P_n(x)$, $x_{n,1} > x_{n,2} > \cdots > x_{n,n}$

$$x_{n,m} = \left(1 - \frac{1}{8n^2} + \frac{1}{8n^3}\right)\cos\frac{4m-1}{4n+2}\pi + O(n^{-4})$$

II

NAME: Tschebyscheff, First Kind SYMBOL: $T_n(x)$
INTERVAL: $[-1, 1]$
WEIGHT: $(1-x^2)^{-1/2}$ STANDARDIZATION: $T_n(1) = 1$
NORM:

$$\int_{-1}^{+1} (1-x^2)^{-1/2}(T_n(x))^2\,dx = \begin{cases} \pi/2, & n \ne 0 \\ \pi, & n = 0 \end{cases}$$

EXPLICIT EXPRESSION:

$$T_n(x) = \frac{n}{2}\sum_{m=0}^{[n/2]} (-1)^m \frac{(n-m-1)!}{m!\,(n-2m)!}(2x)^{n-2m} = \cos(n \arccos x)$$

$$x^{2m} = 2^{1-2m}\sum_{j=0}^{m}\binom{2m}{m-j}T_{2j}(x) - 2^{-2m}\binom{2m}{m}$$

$$x^{2m+1} = 2^{-2m}\sum_{j=0}^{m}\binom{2m+1}{m-j}T_{2j+1}(x)$$

RECURRENCE RELATION:

$$T_{n+1}(x) = 2xT_n(x) - T_{n-1}(x); \qquad T_0 = 1, \quad T_1 = x$$

DIFFERENTIAL EQUATION:

$$(1 - x^2)y'' - xy' + n^2 y = 0, \qquad y = T_n(x)$$

INDEFINITE INTEGRAL:

$$\int T_0 \, dx = T_1, \qquad \int T_1 \, dx = \frac{1}{4} T_2$$

$$\int T_n \, dx = \frac{1}{2}\left[\frac{T_{n+1}(x)}{n+1} - \frac{T_{n-1}(x)}{n-1}\right], \qquad n \geq 2$$

INTEGRAL:

$$\int_{-1}^{1} T_n \, dx = \begin{cases} 2/(1 - n^2), & n \text{ even} \\ 0, & n \text{ odd} \end{cases}$$

RODRIGUES' FORMULA:

$$T_n(x) = \frac{(-1)^n (1 - x^2)^{1/2} \sqrt{\pi}}{2^{n+1}\Gamma(n + \frac{1}{2})} \frac{d^n}{dx^n}\{(1 - x^2)^{n-1/2}\}$$

GENERATING FUNCTION:

$$\frac{1 - xz}{1 - 2xz + z^2} = \sum_{n=0}^{\infty} T_n(x) z^n, \qquad -1 \leq x \leq 1, \quad |z| < 1$$

INEQUALITY: $|T_n(x)| \leq 1, \ -1 \leq x \leq 1.$
ZEROS: $x_{n,m} = \cos((2m - 1)\pi/2n)$
ASYMPTOTIC EXPRESSION: $(n \to \infty)$ Exact.

$$\frac{\pi_n^{(-1/2, -1/2)}(z)}{T_n(z)} = \frac{2\pi}{(z^2 - 1)^{1/2}\{(z + (z^2 - 1)^{1/2})^{2n} + 1\}},$$

$$-\pi < \arg(z \pm 1) < \pi$$

III

NAME: Tschebyscheff, Second Kind SYMBOL: $U_n(x)$
INTERVAL: $[-1, 1]$
WEIGHT: $(1 - x^2)^{1/2}$ STANDARDIZATION: $U_n(1) = n + 1$
NORM:

$$\int_{-1}^{+1} (1 - x^2)^{1/2} [U_n(x)]^2 \, dx = \pi/2$$

EXPLICIT EXPRESSION:

$$U_n(x) = \sum_{m=0}^{[n/2]} (-1)^m \frac{(m-n)!}{m!\,(n-2m)!}(2x)^{n-2m} = \frac{T'_{n+1}(x)}{n+1}$$

$$U_n(\cos\theta) = \frac{\sin(n+1)\theta}{\sin\theta}$$

RECURRENCE RELATION:

$$U_{n+1}(x) = 2xU_n(x) - U_{n-1}(x); \qquad U_0 = 1, \quad U_1 = 2x$$

DIFFERENTIAL EQUATION:

$$(1-x^2)y'' - 3xy' + n(n+2)y = 0, \qquad y = U_n(x)$$

RODRIGUES' FORMULA:

$$U_n(x) = \frac{(-1)^n(n+1)\sqrt{\pi}}{(1-x^2)^{1/2}2^{n+1}\Gamma(n+\frac{3}{2})}\frac{d^n}{dx^n}\{(1-x^2)^{n+1/2}\}$$

INTEGRAL:

$$\int_{-1}^{1} U_k(x)\,dx = \begin{cases} 2/(k+1), & k = 2m \\ 0, & k = 2m+1 \end{cases}$$

GENERATING FUNCTION:

$$\frac{1}{1-2xz+z^2} = \sum_{n=0}^{\infty} U_n(x)z^n, \qquad -1 \le x \le 1, \quad |z| < 1$$

INEQUALITY: $|U_n(x)| \le n+1, \ -1 \le x \le 1$
ZEROS: $x_{n,m} = \cos(m\pi/(n+1))$
ASYMPTOTIC EXPRESSION: $(n \to \infty)$. Exact

$$\frac{\pi_n^{(1/2,\,1/2)}(z)}{U_n(z)} = \frac{2\pi(z^2-1)^{1/2}}{\{(z+(z^2-1)^{1/2})^{2n+2}-1\}}, \qquad -\pi < \arg(z \pm 1) < \pi$$

IV

NAME: Ultraspherical or Gegenbauer* SYMBOL: $C_n^\mu(x)$
INTERVAL: $[-1, 1]$ WEIGHT: $(1-x^2)^{\mu-1/2}, \quad \mu > -\frac{1}{2}, \quad \mu \ne 0$
STANDARDIZATION: $C_n^\mu(1) = \binom{n+2\mu-1}{n}$

* For $\mu = 0$, see under II, the Tschebyscheff polynomials of the first kind.

NORM:

$$\int_{-1}^{1} (1-x^2)^{\mu-1/2}[C_n^\mu(x)]^2\,dx = \frac{\pi 2^{1-2\mu}\Gamma(n+2\mu)}{n!(n+\mu)[\Gamma(\mu)]^2}$$

EXPLICIT EXPRESSION:

$$C_n^\mu(x) = \frac{1}{\Gamma(\mu)}\sum_{m=0}^{[n/2]}(-1)^m\frac{\Gamma(\mu+n-m)}{m!(n-2m)!}(2x)^{n-2m}$$

RECURRENCE RELATION:

$$(n+1)C_{n+1}^\mu(x) = 2(n+\mu)xC_n^\mu(x) - (n+2\mu-1)C_{n-1}^\mu(x);$$

$$C_0^\mu(x) = 1, \qquad C_1^\mu(x) = 2\mu x$$

DIFFERENTIAL EQUATION:

$$(1-x^2)y'' - (2\mu+1)xy' + n(n+2\mu)y = 0, \qquad y = C_n^\mu(x)$$

RODRIGUES' FORMULA:

$$C_n^\mu(x) = \frac{(-1)^n 2^n n!\Gamma(2\mu)\Gamma(\mu+n+\frac{1}{2})}{\Gamma(\mu+\frac{1}{2})\Gamma(n+2\mu)(1-x^2)^{\mu-1/2}}\frac{d^n}{dx^n}\{(1-x^2)^{n+\mu-1/2}\}$$

GENERATING FUNCTION:

$$R^{-2\mu} = \sum_{n=0}^{\infty} C_n^\mu(x)z^n, \qquad -1 \le x \le 1, \quad |z| < 1$$

$$R = (1-2xz+z^2)^{1/2}$$

INEQUALITY:

$$\max_{-1\le x\le 1} |C_n^\mu(x)| = \begin{cases} \binom{n+2\mu-1}{n}, & \mu > 0 \\ |C_n^\mu(x')|, & -\frac{1}{2} < \mu < 0 \end{cases}$$

where $x' = 0$ if $n = 2m$; x' = maximum point nearest zero if $n = 2m+1$.

V

NAME: Jacobi SYMBOL: $P_n^{(\alpha,\beta)}(x)$ INTERVAL: $[-1, 1]$
WEIGHT: $(1-x)^\alpha(1+x)^\beta$; $\alpha, \beta > -1$
STANDARDIZATION:

$$P_n^{(\alpha,\beta)}(1) = \binom{n+\alpha}{n}$$

NORM:

$$\int_{-1}^{+1} (1-x)^{\alpha}(1+x)^{\beta}[P_n^{(\alpha,\beta)}(x)]^2\,dx = \frac{2^{\alpha+\beta+1}\Gamma(n+\alpha+1)\Gamma(n+\beta+1)}{(2n+\alpha+\beta+1)n!\,\Gamma(n+\alpha+\beta+1)}$$

EXPLICIT EXPRESSION:

$$P_n^{(\alpha,\beta)}(x) = \frac{1}{2^n}\sum_{m=0}^{n}\binom{n+\alpha}{m}\binom{n+\beta}{n-m}(x-1)^{n-m}(x+1)^m$$

RECURRENCE RELATION:

$$\begin{aligned}2(n+1)(n+\alpha+\beta+1)(2n+\alpha+\beta)\,P_{n+1}^{(\alpha,\beta)}(x) &= (2n+\alpha+\beta+1)[(\alpha^2-\beta^2)+(2n+\alpha+\beta+2)(2n+\alpha+\beta)x]P_n^{(\alpha,\beta)}(x)\\ &\quad - 2(n+\alpha)(n+\beta)(2n+\alpha+\beta+2)P_{n-1}^{(\alpha,\beta)}(x);\end{aligned}$$

$$P_0^{(\alpha,\beta)} = 1, \qquad P_1^{(\alpha,\beta)} = (1+\tfrac{1}{2}(\alpha+\beta))x + \tfrac{1}{2}(\alpha-\beta)$$

DIFFERENTIAL EQUATION:

$$(1-x^2)y'' + (\beta-\alpha-(\alpha+\beta+2)x)y' + n(n+\alpha+\beta+1)y = 0, \qquad y = P_n^{(\alpha,\beta)}(x)$$

RODRIGUES' FORMULA:

$$P_n^{(\alpha,\beta)}(x) = \frac{(-1)^n}{2^n n!\,(1-x)^{\alpha}(1+x)^{\beta}}\frac{d^n}{dx^n}\{(1-x)^{n+\alpha}(1+x)^{n+\beta}\}$$

GENERATING FUNCTION:

$$R^{-1}(1-z+R)^{-\alpha}(1+z+R)^{-\beta} = \sum_{n=0}^{\infty} 2^{-\alpha-\beta}P_n^{(\alpha,\beta)}(x)z^n$$

$$R = (1-2xz+z^2)^{1/2}, \qquad -1 \le x \le 1, \quad |z| < 1$$

INEQUALITY:

$$\max_{-1\le x\le 1} |P_n^{(\alpha,\beta)}(x)| = \begin{cases}\binom{n+q}{n} \sim n^q & \text{if } q = \max(\alpha,\beta) \ge -\tfrac{1}{2}\\ |P_n^{(\alpha,\beta)}(x')| \sim n^{-1/2} & \text{if } q < -\tfrac{1}{2}\end{cases}$$

where x' is one of the two maximum points nearest $(\beta-\alpha)/(\alpha+\beta+1)$.

ASYMPTOTIC EXPRESSION: $(n \to \infty)$

(i) *z bounded away from the interval* $[-1, 1]$:

$$\frac{\pi_n^{(\alpha,\beta)}(z)}{P_n^{(\alpha,\beta)}(z)} \sim 2^{4n+2(\alpha+\beta+1)} \times \frac{\Gamma(n+\alpha+1)\Gamma(n+\beta+1)\Gamma(n+1)\Gamma(n+\alpha+\beta+1)}{\Gamma(2n+\alpha+\beta+2)\Gamma(2n+\alpha+\beta+1)} \times \frac{(z-1)^\alpha(z+1)^\beta}{[z+(z^2-1)^{1/2}]^{2n+(\alpha+\beta+1)}}$$

where $-\pi < \arg(z-1) < \pi$ and $-\pi < \arg(z+1) < \pi$.

(ii) *all z except for the neighborhood of* $z = -1$:

Defining $z = \cosh 2\zeta$, then for $\alpha \geq 0$,

$$\frac{\pi_n^{(\alpha,\beta)}(z)}{P_n^{(\alpha,\beta)}(z)} \sim 2(z-1)^\alpha(z+1)^\beta \frac{K_\alpha(2k\zeta)}{I_\alpha(2k\zeta)}$$

where $k = n + (\alpha+\beta+1)/2$ and $-\pi < \arg(z \mp 1) < \pi$.

(iii) *all z except for the neighborhood of* $z = 1$:

Defining $z = e^{i\pi}\cosh 2\zeta$, then for $\beta \geq 0$,

$$\frac{\pi_n^{(\alpha,\beta)}(z)}{P_n^{(\alpha,\beta)}(z)} \sim 2e^{-i\pi(\alpha+\beta+1)}(z-1)^\alpha(z+1)^\beta \frac{K_\beta(2k\zeta)}{I_\beta(2k\zeta)}$$

where $k = n + (\alpha+\beta+1)/2$ and $0 < \arg(z \mp 1) < 2\pi$.

VI

NAME: Generalized Laguerre SYMBOL: $L_n^{(\alpha)}(x)$

INTERVAL: $[0, \infty)$

WEIGHT: $x^\alpha e^{-x}$, $\alpha > -1$

STANDARDIZATION: $L_n^{(\alpha)}(x) = ((-1)^n/n!)x^n + \cdots$

NORM:

$$\int_0^\infty x^\alpha e^{-x}(L_n^{(\alpha)}(x))^2\,dx = \frac{\Gamma(n+\alpha+1)}{n!}$$

EXPLICIT EXPRESSION:

$$L_n^{(\alpha)}(x) = \sum_{m=0}^{n} (-1)^m \binom{n+\alpha}{n-m} \frac{1}{m!} x^m$$

RECURRENCE RELATION:

$$(n+1)L_{n+1}^{(\alpha)}(x) = [(2n+\alpha+1)-x]L_n^{(\alpha)}(x) - (n+\alpha)L_{n-1}^{(\alpha)}(x);$$
$$L_0^{(\alpha)} = 1, \quad L_1^{(\alpha)} = 1+\alpha-x$$

DIFFERENTIAL EQUATION:

$$xy'' + (\alpha+1-x)y' + ny = 0, \qquad y = L_n^{(\alpha)}(x)$$

RODRIGUES' FORMULA:

$$L_n^{(\alpha)}(x) = \frac{1}{n!\,x^\alpha e^{-x}} \frac{d^n}{dx^n}\{x^{n+\alpha}e^{-x}\}$$

GENERATING FUNCTION:

$$(1-z)^{-\alpha-1}\exp\left(\frac{xz}{z-1}\right) = \sum_{n=0}^{\infty} L_n^{(\alpha)}(x)z^n, \qquad |z|<1$$

INEQUALITY:

$$|L_n^{(\alpha)}(x)| \le \frac{\Gamma(n+\alpha+1)}{n!\,\Gamma(\alpha+1)}e^{x/2}, \qquad x\ge 0, \quad \alpha \ge 0$$

$$|L_n^{(\alpha)}(x)| \le \left[2 - \frac{\Gamma(\alpha+n+1)}{n!\,\Gamma(\alpha+1)}\right]e^{x/2}, \qquad x \ge 0, \quad -1<\alpha<0$$

ZEROS:

$$x_{n,m} = \frac{j_{\alpha,m}^2}{4k_n}\left(1 + \frac{2(\alpha^2-1)+j_{\alpha,m}^2}{48k_n^2}\right) + O(n^{-5})$$

where $k_n = n + (\alpha+1)/2$ and $j_{\alpha,m}$ is the mth positive zero of the Bessel function $J_\alpha(x)$.

ASYMPTOTIC EXPRESSION: $(n \to \infty)$

(i) *z bounded*, $\alpha \ge 0$:

$$\frac{\lambda_n^{(\alpha)}(z)}{L_n^{(\alpha)}(z)} \sim -2e^{-i\pi\alpha}z^\alpha e^{-z}\frac{K_\alpha(2\sqrt{k}\,e^{-i\pi/2}\sqrt{z})}{I_\alpha(2\sqrt{k}\,e^{-i\pi/2}\sqrt{z})}$$

where $k = n + (\alpha+1)/2$ and $0 < \arg z < 2\pi$.

(ii) *z not necessarily bounded:*

Defining

(a) $k = n + (\alpha+1)/2$,

(b) $t = z/4k$ with $0 < \arg t < 2\pi$ and $0 < \arg(t-1) < 2\pi$,

(c) $\xi = \frac{1}{2}\{t^{1/2}(t-1)^{1/2} - \log(t^{1/2} + (t-1)^{1/2}) + i\pi/2\}$,

then for Re $\xi < 0$ and $z/4k$ not close to 1, we have

$$\frac{\lambda_n^{(\alpha)}(z)}{L_n^{(\alpha)}(z)} \sim -2z^\alpha e^{-z}\frac{K_\alpha(4k\xi e^{-i\pi})}{I_\alpha(4k\xi)}.$$

VII

NAME: Hermite SYMBOL: $H_n(x)$ INTERVAL: $(-\infty, \infty)$
WEIGHT: e^{-x^2} STANDARDIZATION: $H_n(x) = 2^n x^n + \cdots$
NORM:

$$\int_{-\infty}^{+\infty} e^{-x^2}[H_n(x)]^2\, dx = \sqrt{\pi}\, 2^n n!$$

EXPLICIT EXPRESSION:

$$H_n(x) = n! \sum_{m=0}^{[n/2]} (-1)^m \frac{(2x)^{n-2m}}{m!\,(n-2m)!}$$

RECURRENCE RELATION:

$$H_{n+1}(x) = 2xH_n(x) - 2nH_{n-1}(x); \qquad H_0 = 1, \quad H_1 = 2x$$

DIFFERENTIAL EQUATION:

$$y'' - 2xy' + 2ny = 0, \qquad y = H_n(x)$$

RODRIGUES' FORMULA:

$$H_n(x) = (-1)^n e^{x^2} \frac{d^n}{dx^n}(e^{-x^2})$$

GENERATING FUNCTION:

$$e^{2xz - z^2} = \sum_{n=0}^{\infty} \frac{H_n(x) z^n}{n!}$$

INEQUALITY:

$$|H_{2m}(x)| \le e^{x^2/2} 2^{2m} m! \left[2 - \frac{1}{2^{2m}}\binom{2m}{m}\right]$$

$$|H_{2m+1}(x)| \le |x| e^{x^2/2} \frac{(2m+2)!}{(m+1)!}$$

ASYMPTOTIC EXPRESSION: $(n \to \infty)$

In the formulas below, the upper and lower signs correspond to the upper and lower half planes. In the upper half plane we assume $0 < \arg z < \pi$, while in the lower half plane we take $-\pi < \arg z < 0$.

(i) *z bounded*

$$\frac{\eta_n(z)}{H_n(z)} \sim 2\pi e^{\mp i\pi(2n+1)/2}$$

$$\times\, e^{-z^2} \frac{\exp[\pm i(2n+1)^{1/2} z]}{\exp[\mp i(2n+1)^{1/2} z] + (-1)^n \exp[\pm i(2n+1)^{1/2} z]}$$

(ii) *z not necessarily bounded*
Defining
(a) $t = z/(2n+1)^{1/2}$,
(b) $\zeta = \frac{1}{2}t(t^2-1)^{1/2} - \frac{1}{2}\log[t + (t^2-1)^{1/2}] \pm i\pi/4$,
then for $\operatorname{Re}\zeta < 0$ and z not in the neighborhood of either $(2n+1)^{1/2}$ or $-(2n+1)^{1/2}$, we have

$$\frac{\eta_n(z)}{H_n(z)} \sim 2\pi(2n+1)^{(2n+1)/2} \frac{\exp\{-z^2 + z/2[z^2-(2n+1)]^{1/2}\}}{\{z + [z^2-(2n+1)]^{1/2}\}^{2n+1}}.$$

The asymptotic expansions were taken from a summary in Donaldson and Elliott. See also Elliott.

The functions $I_\nu(z)$ and $K_\nu(z)$ appearing in some of these expressions are the modified Bessel Functions:

$$I_\nu(z) = \left(\frac{1}{2}z\right)^\nu \sum_{k=0}^{\infty} \frac{(\frac{1}{4}z^2)^k}{k!\,\Gamma(\nu+k+1)},$$

$$K_\nu(z) = \frac{\pi}{2}\frac{I_{-\nu}(z) - I_\nu(z)}{\sin(\nu\pi)}.$$

See Olver.

References. Elliott [4]; Donaldson and Elliott [1, 2, 3]; Olver [1, p. 374].

1.14 Some Sets of Polynomials Orthogonal over Figures in the Complex Plane

CIRCLE: $|z| = r$. If

$$p_n^*(z) = (2\pi r^{2n+1})^{-1/2} z^n, \tag{1.14.1}$$

then

$$\int_{|z|=r} p_n^*(z)\overline{p_m^*(z)}\, ds = \delta_{mn} \tag{1.14.2}$$

DISK: $|z| \le r$. If

$$p_n^*(z) = \left(\frac{n+1}{\pi}\right)^{1/2} \frac{z^n}{r^{n+1}}, \tag{1.14.3}$$

then

$$\iint_{|z|\le r} p_n^*(z)\overline{p_m^*(z)}\, dx\, dy = \delta_{mn} \tag{1.14.4}$$

ELLIPSE: Let $\mathscr{E}_\rho$ $(\rho > 1)$ designate the ellipse with foci at ± 1 and such that $\rho = a + b =$ sum of semiaxes. The symbol will be used indiscriminately to designate either the boundary or the interior of the ellipse.

If $T_n(z)$ is the Tschebyscheff polynomial of the first kind, we have

$$\int_{\mathscr{E}_\rho} \frac{T_m(z)\overline{T_n(z)}}{|1 - z^2|^{1/2}}\, ds = \mu_m\, \delta_{mn}; \tag{1.14.5}$$

$$\mu_m = \begin{cases} (\pi/2)(\rho^{2m} + \rho^{-2m}), & m > 0, \\ 2\pi, & m = 0. \end{cases} \tag{1.14.6}$$

If

$$p_n^*(z) = 2\left(\frac{n+1}{\pi}\right)^{1/2} (\rho^{2n+2} - \rho^{-2n-2})^{-1/2} U_n(z), \tag{1.14.7}$$

and $U_n(z)$ is the Tschebyscheff polynomial of the second kind, then

$$\iint_{\mathscr{E}_\rho} p_m^*(z)\overline{p_n^*(z)}\, dx\, dy = \delta_{mn}. \tag{1.14.8}$$

1.15 Extrapolation and Speed-Up

In this section we list some standard extrapolation and speed-up methods that are useful in various parts of numerical analysis.

1 Aitken's Δ^2 method

Suppose it is known that a sequence s_n converges to the limit s with geometric rapidity:

$$s = s_n + (c + \varepsilon_n)q^n \tag{1.15.1}$$

where $c \neq 0$, $\varepsilon_n \to 0$, $|q| < 1$. (In the theory of iterations, this type of convergence is sometimes referred to as *linear* convergence.) We suppose, however, that we know *neither* c nor ε_n nor q. We have

$$s - s_{n+1} \approx cq^{n+1} \approx q(s - s_n).$$

Hence,

$$\frac{s - s_{n+1}}{s - s_n} \approx q \approx \frac{s - s_{n+2}}{s - s_{n+1}}.$$

Solving for s we obtain

$$s \approx \frac{s_n s_{n+2} - s_{n+1}^2}{s_{n+2} - 2s_{n+1} + s_n} = s_{n+2} - \frac{(\Delta s_{n+1})^2}{\Delta^2 s_n}.$$

Starting from a given sequence $s_1, s_2, \ldots$, we obtain a second sequence $s'_1, s'_2, \ldots$ by applying the transformation

$$s'_n = \frac{s_n s_{n+2} - s_{n+1}^2}{s_{n+2} - 2s_{n+1} + s_n}. \tag{1.15.2}$$

It can be shown that under the above hypothesis, the transformed sequence converges to s more rapidly than the original sequence. By this is meant

$$\lim_{n \to \infty} \frac{s - s'_n}{s - s_n} = 0. \tag{1.15.3}$$

Shanks has studied certain iterates of the Δ^2 method. He takes as a model sequences of the form

$$s = s_n + \sum_{i=1}^{p} a_i q_i^n \tag{1.15.4}$$

with $|q_i| < 1$.

Implementation is easily based upon the following iteration due to Wynn (referred to by some authors as the *ε-algorithm*). For a given sequence of numbers $s_0, s_1, \ldots$, define the two-dimensional "ε-array"

$$\begin{array}{ccccc}
\varepsilon_{-1}^{(0)} & & & & \\
 & \varepsilon_0^{(0)} & & & \\
\varepsilon_{-1}^{(1)} & & \varepsilon_1^{(0)} & & \\
 & \varepsilon_0^{(1)} & & \varepsilon_2^{(0)} & \cdot \\
\varepsilon_{-1}^{(2)} & & \varepsilon_1^{(1)} & \vdots & \quad \cdot \\
 & \varepsilon_0^{(2)} & \vdots & & \qquad \cdot \\
\varepsilon_{-1}^{(3)} & \vdots & & & \\
\vdots & & & &
\end{array}$$

by means of the relations

$$\varepsilon_{-1}^{(j)} = 0, \qquad \varepsilon_0^{(j)} = s_j,$$
$$\varepsilon_{k+1}^{(j)} = \varepsilon_{k-1}^{(j+1)} + (\varepsilon_k^{(j+1)} - \varepsilon_k^{(j)})^{-1}.$$

This array has the properties:

If $s_j = s + \sum_{i=1}^{p} a_i q_i^j$, $|q_1| > |q_2| > \cdots > |q_p|$, then $\varepsilon_{2p}^{(0)} = s$.
If $|q_i| \ll 1$, then

$$\varepsilon_{2k}^{(j)} \sim s + \frac{a_{k+1}\{(q_{k+1} - q_1) \cdots (q_{k+1} - q_k)\}^2}{\{(1 - q_1) \cdots (1 - q_k)\}^2} q_{k+1}^j.$$

If $s_j = s + \sum_{i=1}^{p} a_i^{(j)} q_i^{(j)}$ where $q_i^{(j)} = q_i^{(j+k_0)}$, k_0 fixed integer, then $\varepsilon_{2k_0 p}^{(0)} = s$.

A FORTRAN program to implement this is TEAS in the IBM Scientific Subroutine Package. Other implementations are the programs SHNK by Kahaner, QEXT in QUADPACK, and C06BAF in the NAG library.

Levin has generalized the ideas behind the ε-algorithm and has derived the nonlinear T-, U-, and V-transformations, which give better results than the ε-algorithm in many cases of interest.

References. Genz [2, 4]; IBM [2]; Kahaner [3, F3]; D. Levin [1]; NAG [1]; QUADPACK [F1]; D. Shanks [1]; Wynn [2].

2 *Richardson Extrapolation*

In the Richardson extrapolation (or the "deferred approach to the limit") for speeding up the convergence of a sequence, we assume that s_n converges to s according to the asymptotic expansion

$$s = s_n + \frac{c}{n^\alpha} + \frac{d}{n^\beta} + \frac{e}{n^\gamma} + \cdots \tag{1.15.5}$$

where

$$0 < \alpha < \beta < \gamma < \cdots.$$

The exponents $\alpha, \beta, \gamma, \ldots$ are regarded as *known* while the coefficients $c, d, e, \ldots$ are unknown. (Actually, we may assume a law of the form

$$s = s_n + c\phi_1(n) + d\phi_2(n) + \cdots \tag{1.15.6}$$

where the $\phi_i(n)$ are regarded as known.)

Write (1.15.5) in the form

$$s = s_n + \frac{c}{n^\alpha} + O\left(\frac{1}{n^\beta}\right) \tag{1.15.7}$$

or

$$s^* = s_n + cn^{-\alpha}. \tag{1.15.7a}$$

This allows us to "extrapolate to infinity" on the basis of two values of n. It is convenient to take the two values as n and σn where σ is a fixed integer ≥ 2. We have from (1.15.7)

$$s = s_{\sigma n} + \frac{c}{\sigma^\alpha n^\alpha} + O\left(\frac{1}{\sigma^\beta n^\beta}\right) \tag{1.15.8}$$

or,

$$s^* = s_{\sigma n} + \frac{c}{\sigma^\alpha n^\alpha}. \tag{1.15.8a}$$

Now multiply (1.15.7a) by $\sigma^{-\alpha}$ and subtract from (1.15.8a). This yields

$$s^* = (\sigma^\alpha s_{\sigma n} - s_n)/(\sigma^\alpha - 1) \tag{1.15.9}$$

The value s^* is often taken as a better value of $s = s_\infty$ than either s_n or $s_{\sigma n}$. The difference $s - s^*$ can be shown to be $O(1/n^\beta)$. A common value taken for σ is 2 in which case

$$s^* = (2^\alpha s_{2n} - s_n)/(2^\alpha - 1). \tag{1.15.10}$$

This defines a sequence to sequence transformation

$$s_n' = (2^\alpha s_{2n} - s_n)/(2^\alpha - 1). \tag{1.15.10'}$$

The first application of this transformation knocks out the term $cn^{-\alpha}$. A second application will knock out the term $dn^{-\beta}$, etc.

In addition to expansions of the form (1.15.5), terms such as $1/n^\alpha(\log n)^\delta$ may appear in error estimates, in which case modifications of the transformation (1.15.10′) must be used.

Example. Refer to the last example in Section 2.1 (p. 56). The error in the trapezoidal rule is approximately c/n^2 so that $\alpha = 2$. Selecting $n = 4$ we have, from the table,

$$s^* = \tfrac{4}{3}s_8 - \tfrac{1}{3}s_4 = .63670541$$

with an error of .00008. We should have to go almost to $n = 128$ in the original sequence to achieve this accuracy.

In general, the computation is organized in the form of a lower triangular table called a T-table with entries T_i^j, $i = 0, \ldots, m$; $j = 0, \ldots, m - i$. In the first column, we set $T_0^j = s_{n_j}$, where we assume that we have evaluated s_n for an increasing sequence of indices $n_0 < n_1 < \cdots < n_m$. Subsequent columns are then computed from the formula

$$T_i^j = \frac{D_{ij} T_{i-1}^{j+1} - T_{i-1}^j}{D_{ij} - 1} = T_{i-1}^{j+1} + \frac{T_{i-1}^{j+1} - T_{i-1}^j}{D_{ij} - 1}, \qquad i = 1, \ldots, m;$$
$$j = 0, \ldots, m - i. \tag{1.15.11}$$

Alternatively, the T-table can be built up row by row with $k = 1, \ldots, m$; $i = 1, \ldots, k; j = k - i$. For the general case (1.15.6), Håvie gives an algorithm for evaluating the D_{ij}. However, in two particular and important cases, explicit expressions for the D_{ij} are known. If $\phi_i(n) = n^{-\alpha i}$ for some fixed α, then $D_{ij} = (n_{i+j}/n_i)^\alpha$, while if $\phi_i(n) = n^{-\gamma_i}$, where $0 < \gamma_1 < \gamma_2 < \cdots < \gamma_m$ and $n_k = n_0 \rho^k$, $\rho > 1$, then $D_{ij} = \rho^{\gamma_i}$.

Sidi considers a generalization of (1.15.6) and studies functions of the form

$$A(y) = A - \sum_{k=0}^{m} \phi_k(y)\beta_k(y) \tag{1.15.12}$$

where the β_k have asymptotic expansions as $y \to \infty$ of the form

$$\beta_k(y) \sim \sum_{i=0}^{\infty} \beta_{ki}\, y^{-ir_k}, \qquad k = 0, 1, \ldots, m. \tag{1.15.13}$$

If we assume that the $\phi_k(y)$ and the r_k are known, then we can approximate A and the β_{ki} by solving the system of linear equations

$$A(y_l) = A_{nj} - \sum_{k=0}^{m} \phi_k(y_l) \sum_{i=0}^{n_k} \bar{\beta}_{ki}\, y_l^{-ir_k}, \qquad j \le l \le j + N \tag{1.15.14}$$

where $n = (n_0, \ldots, n_m)$ is a vector of integers, $N = \sum_{k=0}^{m} (n_k + 1)$, $0 < y_0 < y_1 < \cdots$, and $\lim_{l\to\infty} y_l = \infty$. This is called a generalized Richardson extrapolation process and Sidi gives error bounds and convergence theorems for two processes, one in which n is fixed and $j \to \infty$ and the other in which j is fixed and all $n_k \to \infty$.

For a treatment of Richardson extrapolation in an abstract function space setting, see Pruess.

Finally, we point out that in many discussions of Richardson extrapolation, we deal with $h = 1/n$ and let $h \to 0$ in which case the process is called "extrapolation to zero."

References. Brezinski [3]; Håvie [12]; Pruess [1]; Schneider [1]; Sidi [1, 6].

3 *Other Methods*

A simple and often effective idea advocated by Salzer is to regard the sequence s_n as a polynomial in $1/n$ and then to extrapolate to $n = \infty$ by polynomial extrapolation. This leads to a family of linear formulas. A typical member is

$$s_\infty = -\tfrac{8}{6}s_j + \tfrac{81}{6}s_{j+1} - \tfrac{192}{6}s_{j+2} + \tfrac{125}{6}s_{j+3}. \tag{1.15.15}$$

This uses any four consecutive values of the sequence.

However, beware of the large negative coefficients which can introduce serious roundoff problems.

The Euler transformation is given in Sections 3.8 and 2.10.3, and the B-, G-, and Q-transformations are given in Section 3.2.2.

A method of *rational extrapolation* has been developed by Stoer. There is an extensive literature of extrapolation and speed-up methods. Joyce is a good summary.

Speed-up methods have been used in many diverse areas of numerical analysis. If conditions are right, they may be useful. If hypotheses are not fulfilled, they may ruin a result. In any case, they tend to magnify round off error. In Section 3.8 they are applied to infinite integrals with oscillatory integrands.

More recent work is covered in the two companion volumes of Brezinski, treating both the theoretical and practical aspects of convergence acceleration, including many FORTRAN programs implementing the various methods.

References. Brezinski [1, 2]; Joyce [1]; Salzer [1]; Stoer [1].

1.16 Numerical Integration and the Numerical Solution of Integral Equations

An integral equation is a functional equation in which the unknown function $f(x)$ appears in the integrand of a definite integral. A quite general formulation of an integral equation is

$$\int_a^b h(x, f(x); y, f(y); \lambda)\, dy = 0 \tag{1.16.1}$$

where λ is a scalar parameter. One way in which this equation is solved numerically is by first approximating the integral using a numerical integration rule so that (1.16.1) becomes, for example,

$$\sum_{i=1}^{n} w_i h(x, f(x); y_i, f(y_i); \lambda) + E_n = 0, \tag{1.16.2}$$

where $\sum_{i=1}^{n} w_i = b - a$. We then ignore the error term E_n and replace f by the approximation $\hat{f}$ to get the following equation involving both $\hat{f}(x)$ and the $\hat{f}(y_i)$:

$$\sum_{i=1}^{n} w_i h(x, \hat{f}(x); y_i, \hat{f}(y_i); \lambda) = 0 \tag{1.16.3}$$

which is an implicit equation for $f(x)$. One way to try to solve this equation is to collocate at the points y_i, that is, to require that (1.16.3) holds for $x = y_j, j = 1, \ldots, n$. This yields a system of n equations in the n unknowns $\hat{f}(y_j), j = 1, \ldots, n$:

$$\sum_{i=1}^{n} w_i h(y_j, \hat{f}(y_j); y_i, \hat{f}(y_i); \lambda) = 0, \qquad j = 1, \ldots, n. \tag{1.16.4}$$

These equations are nonlinear in general and if solved, yield approximations $\hat{f}(y_j)$ to $f(y_j)$. If one is then interested in an approximation to the value of $f(\bar{x})$ for some particular value of $\bar{x}$, one can either use an interpolation formula or solve (1.16.3) for $x = \bar{x}$.

We shall illustrate this procedure by considering a particular case, namely, the Fredholm linear integral equation of the second kind. In this case

$$h(x, f(x); y, f(y); \lambda) = f(x)/(b-a) - \lambda K(x, y)f(y) - g(x)/(b-a) \quad (1.16.5)$$

where $K(x, y)$ is called the kernel of the integral equation and $g(x)$ is the forcing term. In more conventional terms the equation is written

$$f(x) - \lambda \int_a^b K(x, y)f(y)\,dy = g(x) \quad (1.16.6)$$

and it is assumed that λ, $K(x, y)$, and $g(x)$ are such that (1.16.6) has a continuous solution, $f(x)$. The procedure outlined above is called Nyström's method and leads to the following equation and system of linear equations:

$$\hat{f}(x) - \lambda \sum_{i=1}^{n} w_i K(x, y_i)\hat{f}(y_i) = g(x), \quad (1.16.7)$$

$$\hat{f}(y_j) - \lambda \sum_{i=1}^{n} w_i K(y_j, y_i)\hat{f}(y_i) = g(y_j), \qquad j = 1, \ldots, n. \quad (1.16.8)$$

Equation (1.16.7) is called the Nyström extension and can be regarded as a natural interpolation formula for $\hat{f}(x)$, once the $\hat{f}(y_i)$ are determined by the solution of (1.16.8).

Now, in general, numerical integration is primarily concerned with the computation of numerical *values* approximating the values of definite integrals $\int_a^b w(x)f(x)\,dx$, where the function $f(x)$ is known either as an expression, a table of values, or a computer subroutine. In many cases this value is computed using an appropriate numerical integration formula

$$\int_a^b w(x)f(x)\,dx = \sum_{i=1}^{n} w_i f(y_i) + E_n \quad (1.16.9)$$

where the weights w_i and abscissas y_i are known in advance. However, it may be the case as in automatic integration, that the weights w_i and abscissas y_i are determined during the course of the computation and depend to a great degree on the behavior of $f(x)$. Such a program is not suitable for our present purpose. Here, our aim is not numerical values but a *numerical integration formula* inasmuch as the integrand contains an unknown component, namely the function $f(x)$. Thus, one of our aims in this book must be that of deriving explicit numerical integration formulas in addition to algorithms for computing approximate values of integrals.

Furthermore, one class of error analyses for the Nyström method measures the error, $f(y_j) - \hat{f}(y_j)$, in terms of the error in approximating the integral $\int_a^b K(y_j, y)f(y)\,dy$ by the sum $\sum_{i=1}^{n} w_i K(y_j, y_i)f(y_i)$. Since the smoothness of $f(x)$ depends on the smoothness of $g(x)$ and $K(x, y)$, we can usually estimate its smoothness and choose an appropriate integration rule.

For further information on the numerical solution of integral equations, the reader is referred to the books by Baker, Delves and Walsh, Atkinson [2], and to the survey by Noble. An algorithm for solving (1.16.6) is presented in Atkinson [3].

References. Atkinson [2, 3]; C. T. H. Baker [2]; Delves and Walsh [1]; Noble [2].

Chapter 2

Approximate Integration over a Finite Interval

2.1 Primitive Rules

By a *primitive rule* of approximate numerical integration we mean either a Riemann sum of the form

$$R_n(f) = R_n = h\sum_{k=1}^{n} f(a+kh), \qquad h = (b-a)/n, \tag{2.1.1}$$

$$\bar{R}_n(f) = \bar{R}_n = h\sum_{k=0}^{n-1} f(a+kh) \tag{2.1.2}$$

(the *rectangular rules*),

$$M_n(f) = M_n = h\sum_{k=0}^{n-1} f(a+(k+\tfrac{1}{2})h) \tag{2.1.3}$$

(the *midpoint rule*), or

$$T_n(f) = T_n = h \sum_{k=0}^{n}{}'' f(a+kh)$$

$$\equiv h\left[\frac{f(a)}{2} + f(a+h) + f(a+2h) + \cdots + f(a+(n-1)h) + \frac{f(b)}{2}\right] \tag{2.1.4}$$

(the *trapezoidal rule*).

(Strictly speaking, the trapezoidal rule is that given by $T_1(f)$, and $T_n(f)$ is actually the compound trapezoidal rule $(n \times T_1)f$ in the notation of Section 2.4. However, we shall use the term trapezoidal rule for $T_n(f)$ as well as for the generalized compound trapezoidal rule

$$T(f; \pi_n) = \sum_{i=1}^{n} \frac{x_i - x_{i-1}}{2} (f(x_{i-1}) + f(x_i))$$

where the partition π_n is given by

$$\pi_n: \quad a = x_0 < x_1 < \cdots < x_n = b.$$

If $x_i = a + (b - a)/n$, then $T(f; \pi_n)$ reduces to $T_n(f)$.

A similar situation holds for the midpoint rule.)

The Riemann sums are simple averages of the function. Note also that the trapezoidal rule is the average of the "right-hand" and the "left-hand" Riemann sums (2.1.1) and (2.1.2), and that for functions periodic over $[a, b]$, $T_n = R_n = \bar{R}_n$. The convergence to the integral of primitive sums is not rapid (exceptions must be made in the case of certain periodic functions) and is governed by the following estimates of error.

DEFINITION. Let $f(x)$ be continuous in $[a, b]$. Then the *modulus of continuity* $w(\delta)$ of $f(x)$ is defined by

$$w(\delta) = \max_{|x_1 - x_2| \leq \delta} |f(x_1) - f(x_2)|, \qquad a \leq x_1, x_2 \leq b. \tag{2.1.5}$$

In other words, the inequality $|x_1 - x_2| \leq \delta$ implies that

$$|f(x_1) - f(x_2)| \leq w(\delta).$$

As the interval δ becomes smaller, the variation of f becomes smaller, so that $\lim_{\delta \to 0} w(\delta) = 0$.

Reference. Davis [6, p. 7].

THEOREM. *Let $f(x)$ be continuous in $[a, b]$. Then*

$$\left| \int_a^b f(x)\, dx - h \sum_{k=1}^{n} f(a + kh) \right| \leq (b - a) w\left(\frac{b - a}{n}\right). \tag{2.1.6}$$

Proof

$$\int_a^b f(x)\, dx - h \sum_{k=1}^{n} f(a + kh) = \sum_{k=1}^{n} \int_{a+(k-1)h}^{a+kh} (f(x) - f(a + kh))\, dx.$$

Now, $|f(x) - f(a + kh)| \le w(h)$ for $a + (k - 1)h \le x \le a + kh$. Hence

$$\left| \int_{a+(k-1)h}^{a+kh} (f(x) - f(a + kh))\, dx \right| \le hw(h).$$

Therefore, by adding n of these inequalities, we obtain

$$\left| \int_a^b f(x)\, dx - h \sum_{k=1}^{n} f(a + kh) \right| \le nhw(h) = (b - a)w\left(\frac{b - a}{n}\right).$$

A similar estimate holds for the Riemann sum (2.1.2). These estimates tell us that for continuous functions the sums (2.1.1) or (2.1.2) approach the integral $\int_a^b f(x)\, dx$ with a rapidity which is *at worst* the rapidity of $w[(b - a)/n]$ approaching 0.

More precise error estimates can be given when restrictions are made on the smoothness of the integrand. We have, for example, the following.

THEOREM. *Let*

$$E_n(f) = E_n = \int_a^b f(x)\, dx - h \sum_{k=1}^{n} f(a + kh). \tag{2.1.7}$$

If $f'(x)$ exists and is bounded in $[a, b]$, then

$$\tfrac{1}{2}h^2 \sum_{k=1}^{n} m_k \le -E_n \le \tfrac{1}{2}h^2 \sum_{k=1}^{n} M_k, \tag{2.1.8}$$

where

$$\begin{aligned} m_k &= \inf f'(x), \\ M_k &= \sup f'(x), \end{aligned} \qquad a + (k - 1)h \le x \le a + kh. \tag{2.1.9}$$

If it is further assumed that $f'(x)$ is integrable over $[a, b]$, then we have

$$\lim_{n \to \infty} nE_n = \frac{b - a}{2} [f(a) - f(b)]. \tag{2.1.10}$$

For such functions, therefore, if we assume that $f(a) \ne f(b)$, the error in the Riemann sum goes to 0 precisely as $1/n$.

Reference. Pólya and Szegö [1, Vol. 1, pp. 37, 195].

The trapezoidal and midpoint rules are exact for linear functions and converge at least as fast as n^{-2}, if we assume that the integrand has a continuous second derivative.

THEOREM. *Let $f(x) \in C^2[a, b]$. Then*

$$\int_a^b f(x)\,dx - h\left[\frac{f(a)}{2} + f(a+h) + \cdots + f(a+(n-1)h) + \frac{f(b)}{2}\right] = -\frac{(b-a)^3}{12n^2} f''(\xi), \qquad a < \xi < b, \tag{2.1.11}$$

$$\int_a^b f(x)\,dx - h\sum_{k=0}^{n-1} f\left(a + \frac{2k+1}{2}h\right) = \frac{(b-a)^3}{24n^2} f''(\xi), \qquad a < \xi < b. \tag{2.1.12}$$

See Section 4.3.

COROLLARY. *If $f(x) \in C^2[a, b]$ and $f''(x) \geq 0$ on $[a, b]$, then*

$$M_n(f) \leq \int_a^b f(x)\,dx \leq T_n(f).$$

This is known as the "bracketing" property.

A similar bracketing property holds when $f \in C^p[a, b]$, $f^{(p)}(x) \geq 0$ on $[a, b]$, and two integration rules exist of the form

$$Q_i f = \sum_{j=1}^{n_i} w_{ij} f(x_{ij}) + c_i f^{(p)}(\xi_i), \qquad a < \xi_i < b, \quad i = 1, 2$$

with $c_1 c_2 < 0$.

The corollary also holds if $f(x)$ is only convex on $[a, b]$. In addition, it holds for $M_n(f)$ and $T_m(f)$ with any pair of integers m, n. From this it follows, since $T_{2n} = \frac{1}{2}(T_n + M_n)$, that $T_n(f) - \int_a^b f(x)\,dx \geq \int_a^b f(x)\,dx - M_n(f)$, i.e., that for convex functions, the midpoint rule is better than the corresponding trapezoidal rule.

For additional pairs of rules which bracket $\int_a^b f(x)\,dx$ when f is convex, see Brass.

References. Brass [1, pp. 60–64]; Hammer [1].

Example. *Riemann Sums.* The high-speed computing machine offers the possibility of carrying out numerical integration by means of primitive formulas, using large values of n. We present three such computations. In each case we have computed

$$\bar{R}_n = \frac{1}{n}\sum_{k=0}^{n-1} f\left(\frac{k}{n}\right) \approx \int_0^1 f(x)\,dx. \tag{2.1.13}$$

n	$f(x) = x$	$f(x) = x^{1/2}$	$f(x) = \sin \pi x$
2	.2500 0000	.3535 5334	.5000 0000
4	.3750 0000	.5182 8297	.6035 5337
8	.4375 0000	.5956 3020	.6284 1740
16	.4687 5000	.6323 3112	.6345 7306
32	.4843 7500	.6499 3387	.6361 0828
64	.4921 8750	.6584 5814	.6364 9176
128	.4960 9364	.6626 1916	.6365 8754
256	.4980 4664	.6646 6308	.6366 1144
512	.4990 2287	.6656 7123	.6366 1644
1024	.4995 0988	.6661 6973	.6366 1790
2048	.4997 5292	.6664 1684	.6366 1782
4096	.4998 7386	.6665 3536	.6366 1043
Exact value	.5000 0000	.6666 6667	.6366 1977

These examples emphasize the very slow rate of convergence of primitive rules. In the first example, the exact error is $1/2n$. For $n = 4096$, the observed error is .000126, whereas the exact error should be .000122. The difference is attributable to roundoff, which, with 4096 addends, has affected the sixth decimal place.

The function $y = x^{1/2}$ does not have a bounded derivative in [0, 1]. For this function, $w(\delta) = \delta^{1/2}$, so that by (2.1.6) the error is at worst $n^{-1/2}$. For $n = 4096$, this amounts to .016. The observed error is .00013. For the function $f(x) = \sin \pi x$, $f(0) = f(1)$ so that the results are identical with those given in the next example to which the reader is referred.

Some further remarks, which apply to the numerical examples in this book, should be made here. In the first place, in order to be in a position to compare the computed value with an exact value, we must take integrals that can be evaluated in an alternative way, for example, through indefinite integrals, transformations, infinite series, tables of special functions, and so on. Although this may give an impression of undesirable simplicity, the examples have in reality been chosen so as to exhibit a variety of behaviors.

In the second place, many of the examples have been "overcomputed." That is, more work has been done than was necessary to obtain the answer to a given number of significant figures. This reflects not only the conscious desire on our part to exhibit the deleterious effects of roundoff but also a frame of mind easily slipped into: Let the machine run a bit longer; it can't hurt.

The examples not credited to other authors were computed on various computers at the Brown University Computation Laboratory and the Department of Applied Mathematics of the Weizmann Institute of Science. In those examples where roundoff pollution is pointed out, the computer used was the IBM 7070. This computer is a decimal computer which works with eight significant figures. However, the results of arithmetic operations are not rounded correctly to eight figures but are truncated or chopped.

Despite the poor convergence properties of the Riemann sums, it must *not* be ruled out as a practical device.† For example, the Riemann sum

$$\sum_{i=1}^{n} f(\zeta_k)(z_i - z_{i-1}) \approx \int_C f(z)\,dz, \qquad \zeta_k \in (z_{i-1}, z_i), \tag{2.1.14}$$

has been used to obtain integrals over arcs in the complex plane. Riemann sums come into their own when the arcs are hard to handle, that is, when they are given by a complicated parametric representation or merely by data points. Primitive situations may require primitive tools.

Riemann sums have also been used in the computation of certain integrals that come up in the solution of partial differential equations over complicated domains by means of the method of least squares. They have been applied to integral equations, too.

References.‡ Davis and Rabinowitz [5, pp. 58–60]; Pfalz [A1]; Rall [1].

Example. Compute $\int_0^1 \sin \pi x \, dx$ by means of the trapezoidal rule.

Number of intervals	$T_n(f)$
2	.5000 0000
4	.6035 5337
8	.6284 1740
16	.6345 7306
32	.6361 0828
64	.6364 9176
128	.6365 8754
256	.6366 1144
512	.6366 1644
1024	.6366 1790
2048	.6366 1782
4096	.6366 1043
8192	.6366 0027
Exact value	.6366 1977 2 = $2\pi^{-1}$

† No mathematical method should ever be ruled out of consideration from computing practice. Changing machine characteristics and more versatile programming languages have restored to good practice numerous mathematical devices previously considered too cumbersome or uneconomical. Thus, whereas current programming tends to avoid the difference calculus approach to numerical analysis, and this is reflected in the selection of material in this book, we may yet see the difference calculus restored.

‡ The symbols A, F, and P preceding reference numbers indicate listings in the Bibliography of ALGOL, FORTRAN, and PL/I Procedures in Appendix 3.

Steady improvement in the answer is observed through $n = 1024$. The theoretical error is $f''(\xi)/12n^2$. Since $|f''(x)| \leq |\pi^2 \sin \pi x| \leq \pi^2$, a theoretical error bound is $\pi^2/12n^2$. At $n = 64$, say, the theoretical error is accordingly $\leq \pi^2/12 \cdot 4096 \approx .00020$. The observed error at $n = 64$ is .000128. Beyond $n = 1024$, the quality of the result deteriorates. (See also Section 4.2.)

2.1.5 *Higher Rules as Riemann Sums*

It follows from the definition (1.5.1) that an integration formula

$$\int_a^b f(x)\,dx \approx \sum_{i=1}^{n} w_i f(\xi_i)$$

will be a Riemann sum if we can find points $a = x_0 < x_1 < \cdots < x_n = b$ such that

$$x_1 - x_0 = w_1, \ldots, x_n - x_{n-1} = w_n \quad \text{and} \quad x_{i-1} \leq \xi_i \leq x_i .$$

Many higher rules qualify as Riemann sums. Thus the trapezoidal rule, Simpson's rule and the (closed) Newton–Cotes rules of order $n = 4, 5, 6, 7$ (see Section 2.5) are all Riemann sums. It has been proved by Baker that the Romberg rules (see Section 6.3) are Riemann sums. The Gauss rules, G_n, of all orders (see Section 2.7) are also Riemann sums. This follows from a theorem of Tschebyscheff.

Finally, the compound rule $(n \times R)f$ given by (2.4.6) and the generalized compound rule $R(f; \pi_n)$ given by (2.4.12) are linear combinations of Riemann sums.

References. C. T. H. Baker [1]; Szegö [2, p. 50].

2.2 *Simpson's Rule*

This rule is very frequently used in obtaining approximate integrals, either in its compound form or as a component in an automatic integration scheme (see Section 6.2). Our late colleague Milton Abramowitz used to say—somewhat in jest—that 95% of all practical work in numerical analysis boiled down to applications of Simpson's rule and linear interpolation.

THEOREM. *Let $f(x) \in C^4[a, b]$; then*

$$\int_a^b f(x)\,dx - \frac{b-a}{6}\left[f(a) + 4f\left(\frac{a+b}{2}\right) + f(b)\right] = -\frac{(b-a)^5}{2880} f^{(4)}(\xi), \qquad a < \xi < b. \qquad (2.2.1)$$

Reference. Davis [6, p. 75].

The Simpson approximation

$$\int_a^b f(x)\,dx \approx \frac{b-a}{6}\left[f(a) + 4f\left(\frac{a+b}{2}\right) + f(b)\right] \tag{2.2.2}$$

is therefore exact for all polynomials of degree three or less.

Simpson's rule is most frequently applied in its extended or *compound* form. (Some authors refer to the compound form of a rule as a *composite* rule.) The interval $[a, b]$ is divided into a number of equal subintervals or *panels* and Simpson's rule is applied to each. In order to describe the final result, it is convenient to employ a somewhat different notation than that of (2.2.2).

Let

$$a = x_0 < x_1 < \cdots < x_{2n-1} < x_{2n} = b$$

be a sequence of equally spaced points in $[a, b]$:

$$x_{i+1} - x_i = h, \qquad i = 0, \ldots, 2n-1. \tag{2.2.3}$$

Set $f_i = f(x_i)$. Then the compound Simpson's rule is

$$\int_{x_0}^{x_{2n}} f(x)\,dx = \frac{h}{3}[f_0 + 4(f_1 + f_3 + \cdots + f_{2n-1})$$

$$+ 2(f_2 + f_4 + \cdots + f_{2n-2}) + f_{2n}] + E_n. \tag{2.2.4}$$

The remainder E_n is given by

$$E_n = -\frac{nh^5}{90} f^{(4)}(\xi), \qquad a < \xi < b. \tag{2.2.5}$$

If N designates the (even) number of subdivisions of $[a, b]$, then $N = 2n$ and $h = (b-a)/N$, so that

$$E_n = -\frac{(b-a)^5}{180N^4} f^{(4)}(\xi), \qquad a < \xi < b. \tag{2.2.6}$$

For functions that have four continuous derivatives, Simpson's rule (or, really, the compound Simpson's rule) converges to the true value of the integral with rapidity N^{-4} at worst. In practice, therefore, one might expect the use of 100 subintervals to return a result with four additional correct decimals to that returned using 10 subintervals.

Example. Integrate the functions

$$f_1(x) = \frac{1}{1+x}, \qquad f_2(x) = \frac{x}{e^x - 1}, \qquad f_3(x) = x^{3/2}, \qquad f_4(x) = x^{1/2}$$

over $[0, 1]$ by Simpson's rule.

n	$f_1(x)$	$f_2(x)$	$f_3(x)$	$f_4(x)$
1	.6944 4444	.7774 9413	.4023 6892	.6380 7119
2	.6932 5395	.7775 0400	.4004 3191	.6565 2627
4	.6931 5450	.7775 0446	.4000 7723	.6630 7925
8	.6931 4759	.7775 0450	.4000 1368	.6653 9813
16	.6931 4708	.7775 0438	.4000 0235	.6662 1804
32	.6931 4683	.7775 0416	.4000 0033	.6665 0782
64	.6931 4670	.7775 0411	.3999 9984	.6666 1024
128	.6931 4664	.7775 0407	.3999 9973	.6666 4641
Exact value	.6931 4718	.7775 0463	.4000 0000	.6666 6667

The theoretical error bound may be easily computed for the first function. We select $n = 8$. Then from (2.2.6) we have

$$|E| \le \frac{1}{180 \cdot 16^4} \max_{0 \le x \le 1} |f^{(4)}(x)|.$$

Now $f^{(4)}(x) = 24(1 + x)^{-5}$, so that $\max_{0 \le x \le 1} |f^{(4)}(x)| = 24$. Therefore

$$|E| \le 24/180 \cdot 16^4 \approx .000002.$$

The observed error at $n = 8$ is .0000004. Note that after $n = 16$, the accuracy of the answer has deteriorated due to roundoff.

The second function is a Debye function. It has only an apparent singularity at $x = 0$, and we have $\lim_{x \to 0} f_2(x) = 1$. The value 1 was inserted at $x = 0$. Since $f_2(x)$ has derivatives of all orders in [0, 1], we can use the error estimate (2.2.6), but we have to compute the fourth derivative of $x/(e^x - 1)$ and estimate its maximum value. This is a troublesome computation; we can avoid it by using the series expansion for $x/(e^x - 1)$. We have

$$f_2(x) = \frac{x}{e^x - 1} = \sum_{n=0}^{\infty} \frac{B_n}{n!} x^n,$$

where B_n is the nth Bernoulli number. Hence

$$\left(\frac{x}{e^x - 1}\right)^{(4)} = B_4 + \frac{B_6}{2!} x^2 + \frac{B_8}{4!} x^4 + \cdots$$

and

$$m = \max_{0 \le x \le 1} |f_2^{(4)}(x)| \le B_4 + \frac{B_6}{2!} + \frac{B_8}{4!} + \cdots \le .05.$$

Selecting $n = 2$, we have $|E| \le .05/180 \cdot 4^4 \approx 10^{-6}$. This high accuracy is borne out by comparison with the exact value.

Reference. NBS Handbook [1, pp. 804, 998].

The functions $f_3(x)$ and $f_4(x)$ are not in class $C^4[0, 1]$; therefore (2.2.6) is not applicable to them. The function f_3 has only one continuous derivative in [0, 1], whereas f_4 lacks even a bounded first derivative. The convergence in the case of f_4 is correspondingly slower. An analysis of f_3 and f_4 may be carried out by use of a method set forth in Section 4.8.

Example. Determine the value of the elliptic integral

$$I = \int_2^3 [(2x^2 + 1)(x^2 - 2)]^{-1/2} \, dx.$$

Using Simpson's rule with an interval of $\frac{1}{10}$ yields $I = .141117$. An alternative computation of this integral can be made from tables of the elliptic integral of the first kind; it yields $I = .141114$.

Reference. NBS Handbook [1, p. 603].

2.3 *Nonequally Spaced Abscissas*

The problem here is to obtain the integral $\int_a^b f(x)\,dx$ when $f(x)$ has been tabulated at nonequally spaced abscissas. This case arises frequently when experimental data are processed but is not likely to arise with functions defined by "formulas" except when $y = f(x)$ is given in parametric form by the pair of formulas $\{x = x(t),\ y = y(t):\ c \le t \le d\}$ and it is inconvenient to evaluate the inverse function $t = t(x)$.

The following rule, based on overlapping parabolas, is frequently employed. It combines both an integrating and a smoothing feature. Let the abscissas be $a = x_0 < x_1 < \cdots < x_n = b$, and let

$$P_2(x_{i-1}, x_i, x_{i+1}) = a_i x^2 + b_i x + c_i \tag{2.3.1}$$

be the quadratic polynomial that interpolates to $f(x)$ at the three consecutive points x_{i-1}, x_i, x_{i+1} $(i = 1, 2, \ldots, n-1)$. Then for $i = 1, 2, \ldots, n-2$ use the approximation

$$\begin{aligned}\int_{x_i}^{x_{i+1}} f(x)\,dx &\approx \tfrac{1}{2}\int_{x_i}^{x_{i+1}} (P_2(x_{i-1}, x_i, x_{i+1}) + P_2(x_i, x_{i+1}, x_{i+2}))\,dx \\ &= \frac{a_i + a_{i+1}}{2}\left(\frac{x_{i+1}^3 - x_i^3}{3}\right) + \frac{b_i + b_{i+1}}{2}\left(\frac{x_{i+1}^2 - x_i^2}{2}\right) \\ &\quad + \frac{c_i + c_{i+1}}{2}(x_{i+1} - x_i).\end{aligned} \tag{2.3.2}$$

Over the first and last intervals no smoothing is done and the approximations

$$\begin{aligned}\int_{x_0}^{x_1} f(x)\,dx &\approx \int_{x_0}^{x_1} P_2(x_0, x_1, x_2)\,dx, \\ \int_{x_{n-1}}^{x_n} f(x)\,dx &\approx \int_{x_{n-1}}^{x_n} P_2(x_{n-2}, x_{n-1}, x_n)\,dx,\end{aligned} \tag{2.3.3}$$

are used.

When a or b do not occur at one of the given abscissas but, say, $x_i < a < x_{i+1}$, then, if $1 \le i \le n-2$, we use the approximation

$$\int_a^{x_{i+1}} f(x)\,dx \approx \tfrac{1}{2}\int_a^{x_{i+1}} (P_2(x_{i-1}, x_i, x_{i+1}) + P_2(x_i, x_{i+1}, x_{i+2}))\,dx. \qquad (2.3.4)$$

If $x_0 < a < x_1$ and also if $a < x_0$, we use the approximation

$$\int_a^{x_1} f(x)\,dx \approx \int_a^{x_1} P_2(x_0, x_1, x_2)\,dx. \qquad (2.3.5)$$

Similar formulas hold for $x_i < b < x_{i+1}$, $1 \le i \le n-2$, $x_{n-1} < b < x_n$, and $x_n < b$. Finally, for $x_{n-1} < a$ or $b < x_1$, we use formulas similar to (2.3.3).

A FORTRAN program, AVINT, which implements this method is given in Appendix 2.

An algorithm using third-degree interpolation has been given by Gill and Miller. This algorithm also computes an error estimate based on an approximation of fourth derivatives by fourth-order divided differences. A FORTRAN version of this algorithm, CUBINT, is given in Appendix 2. In contrast to AVINT, this program is restricted to integrating between two tabulated points and is not applicable when the limits of integration are not among the given abscissas. In the NAG version of this algorithm, the error estimate is added to the result. It is claimed that a dramatic reduction in the error often results from such an addition. The cases in which it does not make an improvement tend to be cases of low accuracy and the modified answer is not significantly inferior to the unmodified one.

The use of interpolation polynomials of degree higher than two or three is *not recommended* when dealing with experimental data on unequal abscissas (or on equal abscissas, for that matter), that is, in situations where the accuracy of the data is significantly less than the accuracy of the available arithmetic.

Questions of error in the integration of experimental data cannot be discussed within the mathematical framework of this book. However, in order to have an idea of what can be accomplished with this method, examples are given where elementary functions have been integrated from data at unequal stations.

References. Boyko, Nassiff, and Pezzullo [1]; Gill and Miller [1]; Hennion [A1]; Kahaner and Wyman [2]; J. M. McNamee [1]; NAG [1]; Secrest [1].

Example. Values of the four functions $x^{1/2}, x^{1/4}, x^{1/8}, x^{1/16}$ were prescribed at $x = 0, .1, .15, .20, .23, .25, .30, .40, .45, .48, .53, .62, .78, .82, .89, .92$, and 1.0 and the functions were integrated over [0, 1] by the methods described above.

	$x^{1/2}$	$x^{1/4}$	$x^{1/8}$	$x^{1/16}$
Exact value	.6666 6667	.8000 0000	.8888 8889	.9411 7647
AVINT value	.6634 2607	.7883 1842	.8676 3089	.9126 5294
CUBINT value	.6641 9429	.7904 9360	.8710 9889	.9169 8532
CUBINT error estimate	.0002 9811	.0009 4167	.0015 7159	.0020 0788

The infinite derivative at $x = 0$ and the large jump in abscissas from $x = 0$ to $x = .1$ joined forces to produce these indifferent results.

The CUBINT values are slightly better than the AVINT values but the error estimates are off by an order of magnitude.

When the point $x = 0$ was omitted, and the integrals computed over [.05, 1] using the remaining 16 points, considerable improvement was obtained.

	$x^{1/2}$	$x^{1/4}$	$x^{1/8}$	$x^{1/16}$
AVINT value	.6593 8498	.7813 4166	.8585 3442	.9022 8430
Exact value	.6592 1311	.7810 8517	.8583 2640	.9021 5293

A similar improvement was shown using CUBINT over the interval [.1, 1]. In addition, the error estimates were of the same order of magnitude as the true error.

	$x^{1/2}$	$x^{1/4}$	$x^{1/8}$	$x^{1/16}$
Exact value	.6455 8482	.7550 1269	.8222 3163	.8596 7395
CUBINT value	.6456 6738	.7552 3301	.8225 7387	.8600 9706
Error	−.0000 8256	−.0002 2032	−.0003 4224	−.0004 2301
CUBINT error estimate	−.0000 6492	−.0001 5573	−.0002 2688	−.0002 7081

2.3.1 Spline Interpolation

As the use of spline functions has entered numerical analysis with considerable éclat in the last two decades, this book would be incomplete without some description of them.

Let the interval $[a, b]$ be divided into n subintervals

$$a = x_0 < x_1 < \cdots < x_{n-1} < x_n = b$$

not necessarily of equal length. By a spline $S(x)$ of degree m we shall mean a function defined on $[a, b]$ which

(a) coincides with a polynomial of class $\mathscr{P}_m$ on each subinterval Δ_i: $[x_{i-1}, x_i]$, $i = 1, 2, \ldots, n$;

(b) is of class $C^{m-1}[a, b]$.

The word *spline* derives from the instrument which is often used by draftsmen in fairing a curve through data points. The notion of a spline can and has been generalized widely. In this book, we shall limit ourselves mainly to splines that are piecewise cubics, i.e., of degree 3.

The abscissas $\{x_i\}$ are called the *nodes* or the *knots* of the spline.

A spline $S(x)$ is said to *interpolate* to the data points $(x_0, y_0), \ldots, (x_n, y_n)$ if

$$S(x_i) = y_i, \qquad i = 0, 1, \ldots, n. \tag{2.3.1.1}$$

A cubic spline is called *periodic* (of period $b - a$) if

$$S(a+) = S(b-), \qquad S'(a+) = S'(b-), \qquad S''(a+) = S''(b-). \tag{2.3.1.2}$$

The advantage in using spline interpolation as opposed to interpolation by polynomials of high degree is that splines tend to have oscillations of smaller amplitude between the interpolating points. A strict formulation of this fact is the following minimum principle due to Holladay.

THEOREM. *Given data points* (x_i, y_i), $i = 0, 1, \ldots, n$. *Of all the functions* $f(x)$ *of class* $C^2[a, b]$ *which interpolate to this data, the interpolating spline* $S(x)$ *that in addition satisfies* $S''(a) = S''(b) = 0$ *minimizes the integral*

$$E = \int_a^b |f''(x)|^2 \, dx \tag{2.3.1.3}$$

and is the unique minimizer in the class.

A suggestive but somewhat inexact way of putting this is that the interpolating spline with end conditions $S''(a) = S''(b) = 0$, which is called the *natural spline*, has the minimum total curvature among all sufficiently smooth interpolating curves.

A similar minimum theorem is valid for splines that satisfy a derivative end condition:

$$S'(a) = y_0', \qquad S'(b) = y_n'.$$

It is possible to derive various representations for interpolating splines. We first follow the arrangement in Ahlberg, Nilson, and Walsh. Set

$$x_j - x_{j-1} = h_j, \qquad j = 1, 2, \ldots, n. \tag{2.3.1.4}$$

Since $S(x)$ is piecewise cubic, $S'(x)$ is piecewise quadratic and $S''(x)$ is piecewise linear and continuous. Hence, we can write

$$S''(x) = M_{j-1}\frac{x_j - x}{h_j} + M_j\frac{x - x_{j-1}}{h_j} \quad \text{on} \quad \Delta_j \tag{2.3.1.5}$$

for certain constants M_j, and where, in fact,

$$S''(x_j) = M_j, \qquad j = 0, 1, \ldots, n. \tag{2.3.1.6}$$

Perform indefinite integration twice on (2.3.1.5) and write the arbitrary linear function in the form indicated. We obtain

$$S(x) = M_{j-1}\frac{(x_j - x)^3}{6h_j} + M_j\frac{(x - x_{j-1})^3}{6h_j} + A_j(x_j - x) + B_j(x - x_{j-1})$$

on Δ_j.

Now use $S(x_{j-1}) = y_{j-1}$ and $S(x_j) = y_j$. This determines A_j and B_j, and yields

$$S(x) = M_{j-1}\frac{(x_j - x)^3}{6h_j} + M_j\frac{(x - x_{j-1})^3}{6h_j} + \left(y_{j-1} - \frac{M_{j-1}h_j^2}{6}\right)\frac{(x_j - x)}{h_j} + \left(y_j - \frac{M_j h_j^2}{6}\right)\frac{(x - x_{j-1})}{h_j}. \tag{2.3.1.7}$$

on Δ_j.

From (2.3.1.7) by differentiating,

$$S'(x) = -M_{j-1}\frac{(x_j - x)^2}{2h_j} + M_j\frac{(x - x_{j-1})^2}{2h_j} + \frac{y_j - y_{j-1}}{h_j} - \frac{M_j - M_{j-1}}{6}h_j \tag{2.3.1.8}$$

on Δ_j. Note the special values

$$S'(x_j^-) = \frac{h_j}{6}M_{j-1} + \frac{h_j}{3}M_j + \frac{y_j - y_{j-1}}{h_j}$$

$$S'(x_j^+) = -\frac{h_{j+1}}{3}M_j - \frac{h_{j+1}}{6}M_{j+1} + \frac{y_{j+1} - y_j}{h_{j+1}} \tag{2.3.1.9}$$

Since $S'(x)$ is required to be continuous, these two values must be identical, and this yields the equations

$$\frac{h_j}{6}M_{j-1} + \frac{h_j + h_{j+1}}{3}M_j + \frac{h_{j+1}}{6}M_{j+1} = \frac{y_{j+1} - y_j}{h_{j+1}} - \frac{y_j - y_{j-1}}{h_j}.$$

$$j = 1, 2, \ldots, n-1 \tag{2.3.1.10}$$

These form a set of $n - 1$ linear equations in $M_0, \ldots, M_n$ so that two more conditions must be added. Once the M's are determined, the interpolation spline is completely determined through (2.3.1.7). We shall abbreviate these equations by setting

$$\lambda_j = \frac{h_{j+1}}{h_j + h_{j+1}}, \qquad \mu_j = 1 - \lambda_j, \qquad j = 1, 2, \ldots, n-1.$$

$$\sigma_j = \frac{y_j - y_{j-1}}{h_j}, \qquad d_j = \frac{6(\sigma_{j+1} - \sigma_j)}{h_j + h_{j+1}}, \tag{2.3.1.11}$$

The equations (2.3.1.10) then become

$$\mu_j M_{j-1} + 2M_j + \lambda_j M_{j+1} = d_j, \qquad j = 1, 2, \ldots, n-1. \tag{2.3.1.12}$$

We shall write two additional conditions in the form

$$\begin{aligned} 2M_0 + \lambda_0 M_1 &= d_0 \\ \mu_n M_{n-1} + 2M_n &= d_n, \end{aligned} \tag{2.3.1.13}$$

and indicate several possible choices of the constants $\lambda_0, d_0, \mu_n, d_n$ and their interpretation. The combined system now becomes

$$\begin{bmatrix} 2 & \lambda_0 & 0 & \cdots & 0 & 0 & 0 \\ \mu_1 & 2 & \lambda_1 & \cdots & 0 & 0 & 0 \\ 0 & \mu_2 & 2 & \cdots & 0 & 0 & 0 \\ \vdots & \vdots & \vdots & & \vdots & \vdots & \vdots \\ 0 & 0 & 0 & \cdots & 2 & \lambda_{n-2} & 0 \\ 0 & 0 & 0 & \cdots & \mu_{n-1} & 2 & \lambda_{n-1} \\ 0 & 0 & 0 & \cdots & 0 & \mu_n & 2 \end{bmatrix} \begin{bmatrix} M_0 \\ M_1 \\ M_2 \\ \vdots \\ M_{n-2} \\ M_{n-1} \\ M_n \end{bmatrix} = \begin{bmatrix} d_0 \\ d_1 \\ d_2 \\ \vdots \\ d_{n-2} \\ d_{n-1} \\ d_n \end{bmatrix} \tag{2.3.1.14}$$

with a tridiagonal matrix.

(a) If one selects $\lambda_0 = d_0 = \mu_n = d_n = 0$, then $M_0 = M_n = 0$. In terms of the spline interpreted as a thin beam, this corresponds to simple support at the ends and yields the natural spline.

(b) We may also specify the slope of the spline at the endpoints. The conditions

$$S'(a) = y'_0 \qquad S'(b) = y'_n \tag{2.3.1.15}$$

are equivalent from (2.3.1.9) to the selection

$$\lambda_0 = 1 = \mu_n,$$

$$d_0 = \frac{6}{h_1}\left(\frac{y_1 - y_0}{h_1} - y'_0\right), \qquad d_n = \frac{6}{h_n}\left(y'_n - \frac{y_n - y_{n-1}}{h_n}\right). \tag{2.3.1.16}$$

If y_0' and y_n' are not available, they may be approximated by derivatives of cubic interpolating polynomials based on the four points closest to the endpoints of the interval.

(c) A third possibility is to choose x_i; $i = 1, \ldots, n-1$, as the spline nodes and to impose the conditions $S(x_0) = y_0$, $S(x_n) = y_n$. In this case, the index j in (2.3.1.12) runs from $j = 2, \ldots, n-2$ and the two additional conditions take the form

$$M_1 + \lambda_1 M_2 = d_1, \qquad \mu_{n-1} M_{n-2} + M_{n-1} = d_{n-1} \tag{2.3.1.17}$$

where

$$\lambda_1 = \frac{h_2 - h_1}{2h_2 + h_1}, \qquad d_1 = \frac{6}{2h_2 + h_1}\left[\frac{h_2 y_0}{h_1(h_1 + h_2)} - \frac{y_1}{h_1} + \frac{y_2}{h_1 + h_2}\right], \tag{2.3.1.18}$$

$$\mu_{n-1} = \frac{h_{n-1} - h_n}{2h_{n-1} + h_n}, \qquad d_{n-1} = \frac{6}{2h_{n-1} + h_n}$$

$$\times \left[\frac{h_{n-1} y_n}{h_n(h_n + h_{n-1})} - \frac{y_{n-1}}{h_n} + \frac{y_{n-2}}{h_n + h_{n-1}}\right].$$

The resulting spline is called a "not-a-knot" spline and appears to be better adapted to the average function that comes up in practice.

Other possibilities are suggested by Forsythe *et al.* and by Behforooz and Papamichael.

We turn next to the computer solution of the equations (2.3.1.14). A *tridiagonal* system of equations in unknowns $x_0, x_1, \ldots, x_n$ is of the form

$$\begin{aligned} b_0 x_0 + c_0 x_1 &= d_0 \\ a_1 x_0 + b_1 x_1 + c_1 x_2 &= d_1 \\ a_2 x_1 + b_2 x_2 + c_2 x_3 &= d_2 \\ &\vdots \\ a_{n-1} x_{n-2} + b_{n-1} x_{n-1} + c_{n-1} x_n &= d_{n-1} \\ a_n x_{n-1} + b_n x_n &= d_n \end{aligned} \tag{2.3.1.19}$$

If we now form recursively for $k = 0, 1, \ldots, n$

$$\begin{aligned} p_k &= a_k q_{k-1} + b_k, & q_{-1} &= 0, \\ q_k &= -c_k / p_k & & \\ u_k &= (d_k - a_k u_{k-1})/p_k, & u_{-1} &= 0, \end{aligned} \tag{2.3.1.20}$$

then successive elimination of $x_0, x_1, \ldots, x_{n-1}$ from the second, third, ..., $(n+1)$st equations leads to the system

$$x_k = q_k x_{k+1} + u_k \qquad (k = 0, 1, \ldots, n-1)$$
$$x_n = u_n . \tag{2.3.1.21}$$

We can now determine $x_n, x_{n-1}, \ldots, x_0$ successively in that order. From the first equation in (2.3.1.11), it appears that $0 < \lambda_j < 1$ and $0 < \mu_j < 1$ for $j = 1, 2, \ldots, n-1$. If, therefore, $|\lambda_0| < 2$, $|\mu_n| < 2$, then the left-hand matrix in (2.3.1.14) will be diagonally dominant and hence nonsingular. In this case, unique solutions to (2.3.1.14) will exist for arbitrary $d_0, \ldots, d_n$. A more penetrating analysis leads to the same conclusion provided only that $\lambda_0 < 4$ and $\mu_n < 4$.

A second representation of an interpolation spline takes the form

$$P_i(x) = a_i + b_i(x - x_i) + c_i(x - x_i)^2 + d_i(x - x_i)^3,$$
$$x_i \le x \le x_{i+1}, \qquad i = 0, 1, \ldots, n-1. \tag{2.3.1.22}$$

This representation is used by the spline routines in Chapter I of the IMSL library.

The determining equations for the natural spline ($M_0 = M_n = 0$) are

$$a_i = y_i$$
$$b_i = \frac{a_{i+1} - a_i}{h_i} - \frac{(2c_i + c_{i+1})}{3} h_i, \qquad i = 0, 1, \ldots, n-1$$
$$d_i = \frac{c_{i+1} - c_i}{3h_i}$$
$$h_{i-1}c_{i-1} + 2(h_{i-1} + h_i)c_i + h_i c_{i+1} = 3\left[\frac{a_{i+1} - a_i}{h_i} - \frac{a_i - a_{i-1}}{h_{i-1}}\right],$$
$$i = 1, 2, \ldots, n-1, \qquad c_0 = c_n = 0. \tag{2.3.1.23}$$

References. Ahlberg, Nilson, and Walsh [1]; Behforooz and Papamichael [1]; Bulirsch and Rutishauser [1]; Forsythe, Malcolm, and Moler [1, pp. 70–79]; Holladay [1]; IMSL [1]; Seidman and Korsan [1].

Define the *plus function* by $x_+ = x$ if $x \ge 0$, $x_+ = 0$ if $x < 0$. Then, a third representation of a cubic spline over $[a, b]$ is given by

$$S(x) = P(x) + \sum_{j=0}^{n-1} c_j(x - x_j)_+^3 \tag{2.3.1.24}$$

where the c_j are constants and $P(x) \in \mathscr{P}_3$.

A numerically more stable representation of splines of arbitrary order n (= degree $n-1$) is given in terms of the (normalized) B-splines defined by

$$N_{nj}(x) = (\lambda_j - \lambda_{j-n})[\lambda_{j-n}, \ldots, \lambda_j]_t(t-x)_+^{n-1}. \qquad (2.3.1.25)$$

Here $\lambda_{j-n} \leq \lambda_{j-n+1} \leq \cdots \leq \lambda_j$ are a subset of the set of knots determining the spline and $[\lambda_{j-n}, \ldots, \lambda_j]_t$ denotes the nth divided difference operator with respect to the variable t. The set of knots need not include any interpolatory points of the function approximated by the spline. The B-splines have the positivity property, $N_{nj}(x) > 0$, $x \in (\lambda_{j-n}, \lambda_j)$ and have compact support, i.e., $N_{nj}(x) = 0$, $x \notin [\lambda_{j-n}, \lambda_j]$. Any spline $s(x)$ defined over the interval $[a, b]$ with knots $a = \lambda_0 < \lambda_1 \leq \lambda_2 \leq \cdots \leq \lambda_N < \lambda_{N+1} = b$ augmented by the exterior knots $\lambda_j = a$, $j < 0$ and $\lambda_j = b$, $j > N+1$, has the B-spline representation

$$s(x) = \sum_{j=1}^{N+n} c_j N_{nj}(x), \qquad x \in [a, b]. \qquad (2.3.1.26)$$

The indefinite integral of a B-spline is given for $x \in [a, b]$ by

$$\int_{-\infty}^{x} N_{nj}(t)\,dt = \begin{cases} 0, & x < \lambda_{j-n}, \\ \dfrac{\lambda_j - \lambda_{j-n}}{n} \displaystyle\sum_{k=j+1}^{j+n} N_{n+1,k}(x), & \lambda_{j-n} \leq x < \lambda_j, \\ \dfrac{\lambda_j - \lambda_{j-n}}{n}, & \lambda_j \leq x. \end{cases}$$

The spline routines in Chapter E02 of the NAG library use the B-spline representation.

References. Cox [1]; de Boor [4]; Gaffney [1]; NAG [1].

2.3.2 Application of Splines to Integration

Fit the data (x_j, y_j) by a spline and integrate the spline formally. From (2.3.1.7) we have

$$\int_{x_{j-1}}^{x_j} S(x)\,dx = \frac{y_{j-1} + y_j}{2} h_j - \frac{M_{j-1} + M_j}{24} h_j^3. \qquad (2.3.2.1)$$

Hence, adding,

$$\int_a^b S(x)\,dx = \sum_{j=1}^{n} \frac{y_{j-1} + y_j}{2} h_j - \sum_{j=1}^{n} \frac{M_{j-1} + M_j}{24} h_j^3. \qquad (2.3.2.2)$$

The formula (2.3.2.2) is the trapezoidal rule with a correction. (Integration rules often reduce to the trapezoidal rule with a variety of corrections.)

By forming $\sum_{j=1}^{p}$, $p \le n$, we obtain the indefinite integral evaluated at the nodes x_p. If the value of the indefinite integral is sought at locations other than x_p, we must add $\int_{x_p}^{x} S(t)\,dt$ to $\sum_{j=1}^{p}$ where $x_p < x < x_{p+1}$.

Example. The example given in Section 2.3 was worked by spline integration using the subroutine SPLINT given in Appendix 2, which is based on (2.3.1.23).

	$x^{1/2}$	$x^{1/4}$	$x^{1/8}$	$x^{1/16}$
Approximate integral	.6629 5088	.7871 7296	.8659 3486	.9106 0317
Exact value	.6666 6667	.8000 0000	.8888 8889	.9411 7647
	Interval: [0, 1]			
Approximate integral	.6591 5694	.7810 0221	.8582 5813	.9021 0945
Exact value	.6592 1311	.7810 8517	.8583 2640	.9021 5293
	Interval: [.05, 1]			

Comparing these figures with those in Section 2.3, we see that for the interval [.05, 1] over which the functions have no singularity, spline integration is better by a factor of around 3. Over the singular interval, the method of overlapping parabolas (AVINT) is slightly better.

It is occasionally required to integrate a function given by a scatter of data. In this case, one fits the data by means of polynomials, splines, or other convenient functions, the fit being made in the sense of least squares or other criteria of minimality. One then integrates formally.

Dierckx [1] has published FORTRAN programs for the smoothing, differentiation, and integration of experimental data using B-splines of any order. The number and position of the knots are chosen automatically by the algorithm. Dierckx [2] has also published a package of FORTRAN programs for the smoothing, differentiation, and integration of functions defined on a rectilinear mesh over a rectangle using bicubic splines defined by

$$s(x, y) = \sum_{i=1}^{g+4} \sum_{j=1}^{h+4} c_{ij} M_i(x) N_j(y) \tag{2.3.2.3}$$

where the M_i and N_j are cubic B-splines.

Least square splines with optimally selected nodes have been programmed by Strauss using techniques of "quadratic programming." See also the two reports by de Boor and Rice which give FORTRAN programs for least squares cubic spline approximation with both fixed and variable nodes. These programs are also available in the IMSL library.

For theoretical results related to approximation by splines, see Schumaker.

References. Ahlberg, Nilson, and Walsh [1]; Bulirsch and Rutishauser [1]; de Boor and Rice [1]; Dierckx [1, 2]; Greville [1, 2]; IMSL [1]; Schumaker [1–3]; Strauss [1].

Werner and Wuytack have proposed the use of so-called regular splines $R(x)$ for numerical integration and recommend their use for integrating functions with a singularity at or near an endpoint of the integration interval. These functions are also in $C^2[a, b]$ but instead of reducing to a cubic polynomial in each interval $[x_i, x_{i+1}]$, reduce to a function of the form

$$u(x) = AZ + B + C(1 + Z/D)^E \qquad \text{or} \qquad u(x) = A + B(1 + Z/C)^E,$$

where $Z = x - x_i$ and E is fixed in the first case. Thus, as with cubic polynomials, $u(x)$ depends on four parameters and these are chosen so that $R(x) \in C^2[a, b]$ and $R(x_i) = f(x_i)$. Algorithms for implementing this program appear in the report by Werner and Zwick.

References. Werner and Wuytack [1]; Werner and Zwick [1].

2.4 *Compound Rules*

A rule of integration may be transferred to an arbitrary interval. Suppose, for example, that we have the rule

$$R(f) = \sum_{k=1}^{n} w_k f(x_k) \approx \int_0^1 f(x)\, dx, \qquad 0 \le x_k \le 1, \tag{2.4.1}$$

which is written for the interval [0, 1]. This rule may be transferred to the interval $[a, b]$ as follows. The transformation $y = a + (b - a)x$ maps [0, 1] onto $[a, b]$. The abscissas x_k become $a + (b - a)x_k$ and the weights w_k become $w_k(b - a)$. Therefore

$$R(f) = (b - a) \sum_{k=1}^{n} w_k f(a + (b - a)x_k) \approx \int_a^b f(x)\, dx. \tag{2.4.2}$$

We shall call two rules by the same name, provided that their abscissas and weights are related by a linear transformation as above. For example, we shall speak of Simpson's rule over the interval [0, 1] and over $[a, b]$.

A *compound* or *composite* rule arises when the interval of integration is subdivided into a number of equal subintervals and a *fixed* rule of integration is applied to each of the subintervals. The trapezoidal rule $T_n(f)$, the midpoint rule $M_n(f)$, and the extended Simpson's rule are examples of compound rules.

DEFINITION. If R designates a fixed rule of approximate integration utilizing m points, then $n \times R$ will designate the rule of mn points which results from dividing the interval of integration into n equal subintervals and applying R to each of them.

Remark. In many cases of importance, when both endpoints of the interval are included among the abscissas in the rule R, the number of points in the $n \times R$ rule is less than mn and is in fact $(m-1)n+1$.

It is often convenient to standardize the rule R so that it applies to the interval $[-1, 1]$ rather than $[0, 1]$:

$$R(f) = \sum_{k=1}^{m} w_k f(t_k) \approx \int_{-1}^{1} f(x)\,dx, \qquad -1 \le t_k \le 1. \tag{2.4.3}$$

Now, let the interval $[a, b]$ be divided into n equal subintervals by $a = x_0 < x_1 < \cdots < x_{n-1} < x_n = b$. Then R applied to the ith subinterval $[x_{i-1}, x_i]$ yields

$$\int_{x_{i-1}}^{x_i} f(x)\,dx \approx \frac{b-a}{2n} \sum_{k=1}^{m} w_k f(y_{ki}), \tag{2.4.4}$$

where

$$y_{ki} = x_{i-1} + \frac{b-a}{2n}(1 + t_k), \qquad i = 1, 2, \ldots, n. \tag{2.4.5}$$

Therefore, by the rule $n \times R$ applied to the interval $[a, b]$ we mean

$$\begin{aligned}(n \times R)(f) &= \sum_{i=1}^{n} \frac{b-a}{2n} \sum_{k=1}^{m} w_k f(y_{ki}) \\ &= \frac{b-a}{2n} \sum_{i=1}^{n} \sum_{k=1}^{m} w_k f(y_{ki}) \approx \int_a^b f(y)\,dy. \end{aligned} \tag{2.4.6}$$

Several theoretical facts should be noted about the effectiveness of a family of compound rules $n \times R$ as $n \to \infty$. In the first place, if the rule R integrates the constant 1 exactly, then $(n \times R)(f)$ will converge to $\int_a^b f(x)\,dx$.

THEOREM. *Let R be a fixed m-point rule such that $R(1) = \int_{-1}^{1} dx = 2$. Let $n \times R$ designate the compound rule on $[a, b]$ as defined above. Then, if $f(x)$ is a bounded, Riemann-integrable function,*

$$\lim_{n \to \infty} (n \times R)(f) = \int_a^b f(x)\,dx. \tag{2.4.7}$$

Proof. From (2.4.6) we have, by rearrangement,

$$(n \times R)(f) = \sum_{k=1}^{m} w_k \frac{b-a}{2n} \sum_{i=1}^{n} f(y_{ki}).$$

For each fixed k,

$$\frac{b-a}{n}\sum_{i=1}^{n} f(y_{ki})$$

is a Riemann sum and, hence, as $n \to \infty$, it has the limit $\int_a^b f(x)\,dx$. Since $R(1) = 2$, $\sum_{k=1}^{m} w_k = 2$ and, therefore, (2.4.7) follows.

In the second place, if R is taken as one of the "usual" sort of rules, then, if we assume sufficient smoothness of f, $(n \times R)(f)$ will approach $\int_a^b f(x)\,dx$ with arithmetic rapidity.

THEOREM. *Let R be a fixed m-point integration rule defined over an interval $[\alpha, \beta]$ and let its error E_R be given by an expression of the form*

$$E_R(f) = c(\beta-\alpha)^{k+1} f^{(k)}(\xi), \qquad \alpha < \xi < \beta, \tag{2.4.8}$$

whenever $f \in C^k[\alpha, \beta]$. (The constant c may depend on R but is independent of α, β, and f.) *Let $n \times R$ be the compound rule formed from R and let $E_{n\times R}$ designate its error. If $f \in C^k[a, b]$, then*

$$\lim_{n\to\infty} n^k E_{n\times R}(f) = c(b-a)^k\,(f^{(k-1)}(b) - f^{(k-1)}(a)). \tag{2.4.9}$$

Proof. Let $E_{i,R}$ designate the error obtained when f is integrated by R over the ith subinterval $[x_{i-1}, x_i]$. Then

$$E_{n\times R}(f) = \sum_{i=1}^{n} E_{i,R}(f) = \sum_{i=1}^{n} c\left(\frac{b-a}{n}\right)^{k+1} f^{(k)}(\xi_i)$$

$$= c\left(\frac{b-a}{n}\right)^{k+1} \sum_{i=1}^{n} f^{(k)}(\xi_i), \qquad x_{i-1} < \xi_i < x_i .$$

Since

$$\lim_{n\to\infty} \frac{b-a}{n} \sum_{i=1}^{n} f^{(k)}(\xi_i) = \int_a^b f^{(k)}(x)\,dx = f^{(k-1)}(b) - f^{(k-1)}(a),$$

equation (2.4.9) follows.

An integration rule that possesses an error expression of the form (2.4.8) is called *simplex*.

References. Davis [4, p. 55]; Duris [1].

Example. Let S designate Simpson's rule. Then, by (2.2.1),

$$E_s(f) = -\tfrac{1}{2880}(b-a)^5 f^{(4)}(\xi)$$

and by (2.4.9),

$$\lim_{n\to\infty} n^4 E_{n\times S}(f) = -\tfrac{1}{2880}(b-a)^4((f^{(3)}(b) - f^{(3)}(a)).$$

For functions of class $C^4[a, b]$, the family of compound Simpson's rules converges with rapidity of n^{-4}.

The more "refined" a rule of approximate integration is, the more certain we must be that it has been applied to a function which is sufficiently smooth. There may be little or no advantage in using a "better" rule for a function that is not smooth.

Example. A striking example of this has been given by Salzer and Levine. The function

$$W(x) = \frac{1}{\pi} \sum_{n=1}^{\infty} 2^{-n} \cos(7^n \pi x) \tag{2.4.10}$$

is a *Weierstrass continuous nondifferentiable function.* The integrals

$$\int_0^{1/10} W(x)\,dx, \ldots, \int_{4/10}^{5/10} W(x)\,dx$$

were computed by the trapezoidal rule and by Simpson's rule using intervals of 0.001. The following values were obtained:

Interval	Exact value	Trapezoidal rule	Error	Simpson's rule	Error
0 to .1	.0189 929	.0189 876	−.0000 053	.0190 144	.0000 215
.1 to .2	−.0414 565	−.0414 380	.0000 185	−.0414 543	.0000 022
.2 to .3	.0308 462	.0308 443	−.0000 019	.0308 514	.0000 052
.3 to .4	.0033 770	.0034 254	.0000 484	.0034 027	.0000 257
.4 to .5	−.0329 802	−.0330 067	−.0000 265	−.0328 827	.0000 975

These results show no advantage whatsoever in using Simpson's rule.

Reference. Salzer and Levine [1].

Compound rules may also be useful for integrating functions with certain types of singularity. See Section 2.12.7.

In some situations it is useful to generalize the notion of a compound rule to allow an arbitrary subdivision of the integration interval $[a, b]$. This enables one to insert points where the integrand varies rapidly. Consider the partition of $[a, b]$

$$\pi_n: \quad a = x_0 < x_1 < \cdots < x_n = b \tag{2.4.11}$$

and let $h_i = x_i - x_{i-1}$, $i = 1, \ldots, n$ and $\Delta_n = \max h_i$. The generalized compound rule $R(f; \pi_n)$ based on $R(f)$ as given in (2.4.3) is defined to be the rule

$$R(f; \pi_n) = \tfrac{1}{2} \sum_{i=1}^{n} h_i \sum_{k=1}^{m} w_i f(y_{ki}) \tag{2.4.12}$$

where

$$y_{ki} = x_{i-1} + h_i(1 + t_k)/2. \tag{2.4.13}$$

For any sequence of partitions $\{\pi_n\}$ such that $\Delta_n \to 0$, we have that $R(f; \pi_n) \to \int_a^b f(x)\,dx$ for all bounded Riemann-integrable functions. Furthermore, if R has an error expression of the form (2.4.8), then if $f \in C^k[a, b]$ the error in using $R(f; \pi_n)$ is bounded by

$$\left| \int_a^b f(x)\,dx - R(f; \pi_n) \right| \le |c|(b-a)M_k \Delta_n^k \tag{2.4.14}$$

where $\max_{a \le x \le b} |f^{(k)}(x)| \le M_k$.

2.5 Integration Formulas of Interpolatory Type

Suppose we want to develop a formula of approximate integration in the form

$$\int_a^b k(x)f(x)\,dx \approx \sum_{j=0}^{n} w_j f(x_j). \tag{2.5.1}$$

The points $x_0, x_1, x_2, \ldots, x_n$ are *fixed and distinct numbers* lying between a and b *prescribed in advance*. The function $k(x)$ is a fixed weight function, not necessarily positive. Two ways suggest themselves for the determination of the constants w_j.

(1) Interpolate to the function $f(x)$ at the $n + 1$ points $x_0, \ldots, x_n$ by a polynomial of class $\mathscr{P}_n$. Then integrate the interpolating polynomial, and express the result in the form (2.5.1).

(2) Select the constants $w_0, w_1, \ldots, w_n$ so that the error

$$E(f) = \int_a^b k(x)f(x)\,dx - \sum_{j=0}^{n} w_j f(x_j) \tag{2.5.2}$$

is zero for $f(x) = 1, x, x^2, \ldots, x^n$.

These two programs lead to the same formula, one which is called an approximate integration formula of *interpolatory type* or, if one wishes to emphasize the presence of the weight function $k(x)$, a *product integration rule*.

THEOREM. *The two ways of determining* $w_0, w_1, \ldots, w_n$, (1) *and* (2), *yield the same numbers.*

Proof. The polynomial of class $\mathscr{P}_n$ that interpolates to $f(x)$ at $x_0, x_1, \ldots, x_n$ is given explicitly by

$$p_n(f; x) = \sum_{j=0}^{n} f(x_j) \frac{w(x)}{w'(x_j)(x - x_j)}, \tag{2.5.3}$$

where

$$w(x) = (x - x_0)(x - x_1) \cdots (x - x_n). \tag{2.5.4}$$

Hence the approximate integration formula is

$$\begin{aligned}\int_a^b k(x)f(x)\,dx &\approx \int_a^b k(x)p_n(f;x)\,dx \\ &= \int_a^b k(x)\sum_{j=0}^{n} f(x_j)\frac{w(x)}{w'(x_j)(x - x_j)}\,dx \\ &= \sum_{j=0}^{n} w_j f(x_j),\end{aligned} \tag{2.5.5}$$

where

$$w_j = \int_a^b k(x)\frac{w(x)\,dx}{w'(x_j)(x - x_j)}, \qquad j = 0, 1, \ldots, n. \tag{2.5.6}$$

We must now prove that

$$E(1) = 0, \quad E(x) = 0, \quad \ldots, \quad E(x^n) = 0 \tag{2.5.7}$$

if and only if the w_j are given by (2.5.6). If w_j are given by (2.5.6), then

$$E(f) = \int_a^b k(x)(f(x) - p_n(f;x))\,dx.$$

But the integrand vanishes identically for $f = 1, x, x^2, \ldots, x^n$, and this leads to (2.5.7). Suppose conversely that (2.5.7) holds. Then

$$\begin{aligned} w_0 + w_1 + \cdots + w_n &= \int_a^b k(x)\,dx = m_0, \\ w_0 x_0 + w_1 x_1 + \cdots + w_n x_n &= \int_a^b k(x)x\,dx = m_1, \\ &\vdots \\ w_0 x_0^n + w_1 x_1^n + \cdots + w_n x_n^n &= \int_a^b k(x)x^n\,dx = m_n. \end{aligned} \tag{2.5.8}$$

The constants m_j are the *moments* of the weighting function $k(x)$. The matrix of this system is the Vandermonde matrix $V(x_0, x_1, \ldots, x_n)$, which is nonsingular if the x_i are distinct. Therefore, there is one and only one solution for the w's. But as we have observed, (2.5.6) is a solution; hence, it must coincide with that given by (2.5.7).

By far the greater part of published algorithms use the class $\mathscr{P}_n$. The popularity of this class is partly explained by the fact that much more is known about it than about other classes. The polynomials form an algebra (if p and q are polynomials then const $\cdot\, p$, $p \pm q$, pq, and $p(q)$ are polynomials). If p is a polynomial, then so are p' and $\int p$. The polynomials are intimately related through power series expansions to analytic functions and through the Weierstrass approximation theorem to continuous functions. Moreover, polynomials often lead to formulas that are easy to compute.

For the rapid computation of the weights of interpolatory integration rules, see Gustafson.

References. Elhay and Kautsky [F1]; Gustafson [2, 3, A1]; Kautsky [1]; Kautsky and Elhay [1].

One must not overlook the possibility of working with other linear spaces of functions possessing a simple basis $\zeta_0(x), \zeta_1(x), \ldots, \zeta_n(x)$. It is desirable that the integrals $\int \zeta_j(x)\,dx$ be computable in closed form. In many cases, the basis functions $\zeta_j(x)$ are selected to have certain prescribed properties, e.g., they may have singularities of a certain type, be periodic, have a common zero, etc.

If the determinant $|\zeta_i(x_j)| \neq 0$ (as it will be in the case of a Tschebyscheff system of functions, see Section 2.7.7), then the two programs outlined above can be carried out and lead to the same result.

A particular case of this has been studied by Engels. He extends Lagrange polynomial interpolation by considering a general set of functions $\lambda_j(x)$, $j = 0, \ldots, n$, satisfying $\lambda_j(x_i) = \delta_{ij}$ so that the function

$$L_n(f; x) = \sum_{i=0}^{n} \lambda_i(x) f(x_i)$$

interpolates $f(x)$ at the points $x_0, x_1, \ldots, x_n$. The weights w_j in (2.5.1) then have the form

$$w_j = \int_a^b k(x)\lambda_j(x)\,dx, \qquad j = 0, 1, \ldots, n.$$

One way to define such functions $\lambda_j(x)$ is via a *nodal function* $\omega_n(x)$ and $n + 1$ *basis functions* $h_i(x)$ that are differentiable and satisfy

$$\left.\begin{array}{ll} \omega_n(x_j) = 0, & 0 < |\omega_n'(x_j)| \\ h_j(x_j) = 0, & 0 < |h_j'(x_j)| \end{array}\right\}, \quad j = 0, \ldots, n,$$
$$h_j(x_k) = 0, \qquad j \neq k.$$

Then

$$\lambda_j(x) = \omega_n(x)h_j'(x_j)/\omega_n'(x_j)h_j(x).$$

One simple choice for $\omega_n(x)$ is given by $\omega_n(x) = \prod_{j=0}^n h_j(x)$. With such a choice of $\omega_n(x)$, where $h_j(x) = x - x_j$, we recover the Lagrange interpolation polynomial $p_n(f; x)$. Other possibilities are $h_j(x) = \sin(\pi(x - x_j)/2)$ and $\omega_n(x)$ as above leading to trigonometric interpolation and $h_j(x) = x - x_j$, $\omega_n(x) = \sum_{j=0}^n (x - x_j)/(1 - xx_j)$ yielding an interpolatory integration rule, which for a particular choice of abscissas becomes the optimal integration rule of Wilf (Section 4.7).

Reference. Engels [13, pp. 60–66].

The *Newton–Cotes integration formulas* are a notable example of formulas of interpolatory type. The interval $[a, b]$ is divided into n subintervals of equal length at the points

$$a,\ a + h,\ a + 2h, \ldots, a + (n - 1)h,\ a + nh = b, \tag{2.5.9}$$

where

$$h = (b - a)/n. \tag{2.5.10}$$

The weight $k(x) \equiv 1$ is used. Then the integration formula of interpolatory type is given by

$$\int_a^b f(x)\,dx \approx \frac{b-a}{n} \sum_{k=0}^n B_{nk} f(a + kh), \tag{2.5.11}$$

where

$$B_{nk} = \frac{n}{b-a} \int_a^b \frac{w(x)\,dx}{(x - a - kh)w'(a + kh)} \tag{2.5.12}$$

with

$$w(x) = (x - a)(x - a - h) \cdots (x - a - nh). \tag{2.5.13}$$

It is convenient to express the integrals in (2.5.12) in a somewhat more standard form. We introduce the variable t by means of

$$x = a + th. \tag{2.5.14}$$

Then

$$\begin{aligned} w(x) &= (th)(th - h) \cdots (th - nh) \\ &= h^{n+1}t(t - 1) \cdots (t - n). \end{aligned} \tag{2.5.15}$$

Now

$$\begin{aligned} w'(a+kh) &= (a+kh-a)(a+kh-(a+h))\cdots(a+kh-(a+(k-1)h)) \\ &\quad \times (a+kh-(a+(k+1)h))\cdots(a+kh-(a+nh)) \\ &= (kh)(k-1)h\cdots(h)(-h)\cdots(k-n)h \\ &= (-1)^{n-k}h^n k!\,(n-k)!. \end{aligned} \tag{2.5.16}$$

Therefore

$$B_{nk} = (-1)^{n-k}\frac{nh}{b-a}\int_0^n \frac{h^{n+1}t(t-1)\cdots(t-n)\,dt}{h(t-k)h^n k!\,(n-k)!} \tag{2.5.17}$$

or

$$\begin{aligned} B_{nk} = \frac{(-1)^{n-k}}{k!\,(n-k)!}\int_0^n & t(t-1)\cdots(t-k+1) \\ & \times (t-k-1)\cdots(t-n)\,dt = B_{n,n-k}. \end{aligned} \tag{2.5.18}$$

The simple trapezoidal rule, which integrates linear functions exactly, and Simpson's rule, which integrates cubics exactly, are the first two rules in the Newton–Cotes family. The next few rules are listed below. The notation is as follows:

$$a = x_0 < x_1 < \cdots < x_{n-1} < x_n = b, \qquad x_{i+1} - x_i = h = (b-a)/n,$$
$$f(x_i) = f_i.$$

Simpson's $\frac{3}{8}$ Rule

$$\int_a^b f(x)\,dx = \frac{3h}{8}(f_0 + 3f_1 + 3f_2 + f_3) + E, \tag{2.5.19}$$

where

$$E = -\tfrac{3}{80}h^5 f^{(4)}(\xi), \qquad a < \xi < b. \tag{2.5.20}$$

Boole's Rule

$$\int_a^b f(x)\,dx = \frac{2h}{45}(7f_0 + 32f_1 + 12f_2 + 32f_3 + 7f_4) + E, \tag{2.5.21}$$

where

$$E = -\frac{8h^7}{945}f^{(6)}(\xi), \qquad a < \xi < b. \tag{2.5.22}$$

Newton–Cotes 6-Point Rule

$$\int_a^b f(x)\,dx = \frac{5h}{288}(19f_0 + 75f_1 + 50f_2 + 50f_3 + 75f_4 + 19f_5) + E, \quad (2.5.23)$$

where

$$E = -\frac{275h^7 f^{(6)}(\xi)}{12096}, \qquad a < \xi < b. \quad (2.5.24)$$

Newton–Cotes 7-Point Rule

$$\int_a^b f(x)\,dx$$

$$= \frac{h}{140}(41f_0 + 216f_1 + 27f_2 + 272f_3 + 27f_4 + 216f_5 + 41f_6) + E \quad (2.5.25)$$

where

$$E = -\frac{9h^9 f^{(8)}(\xi)}{1400}, \qquad a < \xi < b. \quad (2.5.26)$$

Whenever a formula includes an error term of the form $cf^{(n)}(\xi)$, it is tacitly assumed that $f \in C^n$.

It is a feature of the Newton–Cotes rules that if the number of points is odd, say $2k - 1$, the error is of the form

$$E = c_k h^{2k+1} f^{(2k)}(\xi). \quad (2.5.27)$$

If the number of points is even, say $2k$, the error is of the same form:

$$E = d_k h^{2k+1} f^{(2k)}(\xi). \quad (2.5.28)$$

Exact values of the Newton–Cotes coefficients are available for the first twenty or so rules.

Uspensky derived asymptotic expressions for the coefficients B_{nk} in the Newton–Cotes rule. Over the interval [0, 1], they are

$$B_{nk} = \frac{(-1)^{k-1} n!}{k!\,(n-k)!\,n \log^2 n}\left[\frac{1}{k} + \frac{(-1)^n}{n-k}\right]\left[1 + O\left(\frac{1}{\log n}\right)\right], \qquad 1 \le k \le n-1.$$

$$B_{n0} = B_{nn} = \frac{1}{n \log n}\left[1 + O\left(\frac{1}{\log n}\right)\right] \quad (2.5.29)$$

It is clear from this formula that for n sufficiently large the constants B_{nk} are of mixed sign.

Similar formulas for the case of a weighted integral were obtained by Kuzmin. See Krylov.

References. Kopal [1]; Krylov [1]; Krylov and Shulgina [1, p. 21]; J. C. P. Miller [1]; NBS Handbook [1, pp. 886–887]; Uspensky [1, 3].

The following question immediately suggests itself: What happens as $n \to \infty$ when the sequence of n-point Newton–Cotes rules are applied to an integrand? The answer, surprisingly, is that this process may very well be divergent. In fact, it was shown by Pólya that even though the integrand is analytic on the interval of integration, the process may be divergent. On the other hand, it was shown by Davis that the process is convergent if the integrand is an analytic function that is regular in a sufficiently large region of the complex plane containing the interval of integration. Let us draw the ellipse $\mathscr{E}$ whose center is at $(a + b)/2$, whose major axis lies on the x axis, and whose semimajor and semiminor axes have length $\frac{5}{8}(b - a)$ and $\frac{3}{8}(b - a)$, respectively. If $f(z)$ is regular in the closure of $\mathscr{E}$, then the Newton–Cotes process over $[a, b]$ will converge to $\int_a^b f(x)\,dx$ as $n \to \infty$. The convergence is of geometric rapidity.

This result is not the best possible. For a penetrating discussion of the convergence of sequences of interpolatory rules for analytic functions, the reader is referred to Krylov [1, Chap. 12]. It turns out that the "limiting distribution" of the abscissas $\{x_{nk}\}$ holds the key to the minimal region in which analyticity of the integral implies convergence. For the sequence of Newton–Cotes rules, this region is a certain (American) football-like shape whose boundary is the level line of a certain potential function.

Generalizations have been made by Barnhill to interpolatory rules for contours in the complex plane and to Newton–Cotes rules with derivative data. See Section 2.8.

When n is large, the Newton–Cotes n-point coefficients are large and are of mixed sign. Since this may lead to large losses of significance by cancellation, a high-order Newton–Cotes rule must be used with caution, and the use of a sequence of Newton–Cotes rules of increasing order has little or nothing to recommend it.

References. Barnhill [1]; Davis [3]; Krylov [1, pp. 248–249]; Pólya [1].

Example. The value of the integral $4 \int_0^1 dx/(1 + x^2) = \pi$ was obtained by a sequence of Newton–Cotes rules with 2–21 points. The printouts were as follows.

2	3.0000 000	12	3.1415 925
3	3.1333 333	13	3.1415 926
4	3.1384 615	14	3.1415 920
5	3.1421 176	15	3.1415 932
6	3.1418 781	16	3.1415 925
7	3.1415 708	17	3.1415 962
8	3.1415 789	18	3.1415 935
9	3.1415 925	19	3.1415 896
10	3.1415 926	20	3.1415 920
11	3.1415 925	21	3.1415 775
		Exact value	3.1415 926

The function $f(z) = 1/(1 + z^2)$ has singularities only at $z = \pm i$. It is therefore regular in the ellipse $\mathscr{E}$ and, theoretically, convergence to the proper value should take place. The improvement is rapid up to $n = 10$. No progress is made beyond this point.

Reference. Ohringer [1].

2.5.4 *Inner Product Integration Rules*

In some situations it may be desirable to decompose an integrand $f(x)$ into two factors, $f(x) = g(x)h(x)$, each exhibiting a different type of behavior. For instance, $g(x)$ may be well behaved over the entire integration interval $[a, b]$, but $h(x)$ may exhibit some sort of singular behavior or be highly oscillatory in $[a, b]$. Alternatively, $g(x)$ may be irregular in one part of the interval with $h(x)$ having a different kind of irregularity in another part. It may then pay to treat $g(x)$ and $h(x)$ separately. This leads to the concept of an inner product integration rule (IPIR) defined by

$$\int_a^b w(x)g(x)h(x)\,dx \simeq \sum_{i=0}^{m} \sum_{j=0}^{n} a_{ij}\, g(x_i)h(y_j) \equiv \mathbf{g}^{\mathrm{T}} A \mathbf{h} \qquad (2.5.4.1)$$

where $w(x)$ is an arbitrary weight function, $\mathbf{g}$ is the vector $(g(x_0), \ldots, g(x_m))^{\mathrm{T}}$, $\mathbf{h} = (h(y_0), \ldots, h(y_n))^{\mathrm{T}}$ and A is the $(m+1) \times (n+1)$ matrix (a_{ij}). Without loss of generality we may assume that $m \geq n$. We remark that an IPIR may also be useful in integrating the product of two functions defined by tabular values at two different sets of points.

An IPIR is said to be symmetric if $n = m$, $x_i = y_i$, $i = 0, \ldots, m$, and $A = A^{\mathrm{T}}$. If A is a diagonal matrix, $A = \mathrm{diag}(w_0, \ldots, w_m)$, then the symmetric IPIR reduces to an ordinary integration rule, so that the latter is a particular case of an IPIR.

The most accessible type of IPIR is the interpolatory IPIR. In this case

$$a_{ij} = \int_a^b w(x) l_i(x) L_j(x)\,dx, \qquad i = 0, \ldots, m; \quad j = 0, \ldots, n \qquad (2.5.4.2)$$

where the l_i and L_j are the fundamental Lagrange polynomials

$$l_i(x) = \frac{\pi(x)}{(x - x_i)\pi'(x_i)}, \qquad L_j(x) = \frac{\rho(x)}{(x - y_j)\rho'(y_j)} \tag{2.5.4.3}$$

with $\pi(x) = \prod_{i=0}^{m} (x - x_i)$ and $\rho(x) = \prod_{j=0}^{n} (x - y_j)$. An interpolatory IPIR is exact for all functions f of joint degree (m, n), i.e., for all $f = gh$ such that $g \in \mathscr{P}_m$ and $h \in \mathscr{P}_n$. Conversely, any IPIR of the form (2.5.4.1) which is exact for all functions of joint degree (m, n) is interpolatory. Furthermore, if $w(x) \geq 0$, $\int_a^b w(x) > 0$, as we shall henceforth assume, then if $m = n$, the interpolatory IPIR (2.5.4.1) cannot be exact for all functions of joint degree $(m + 1 + l, m + 1 - l)$, $l = 0, 1, \ldots, m + 1$. To see this, we simply take $f(x) = \pi^2(x)$, and set $g(x) = \pi(x) \prod_{i=0}^{l-1} (x - x_i)$, $h(x) = \prod_{i=l}^{m} (x - x_i)$ so that $\int_a^b w(x)f(x)\,dx > 0$ while $\mathbf{g}^{\mathrm{T}} A\mathbf{h} = 0$. Similarly, if $m > n$, such a rule cannot be exact for all f of joint degree $(m - l, n + 1 + l)$, $l = 0, \ldots, m$. For $l = 0, \ldots, n$ take $g(x) = (x - a)^{m-n-1} \prod_{j=l}^{n} (x - y_j)$ and $h(x) = \rho(x) \prod_{j=0}^{l-1} (x - y_j)$ while for $l = n + 1, \ldots, m$ take $g(x) = (x - a)^{m-l}$, $h(x) = (x - a)^{l-n-1}\rho^2(x)$. For $w(x) = 1$ and equally spaced points x_i and y_j, the so-called Newton–Cotes case, Duris [4, F1] has given a simpler method to generate the a_{ij} and a FORTRAN implementation.

If $\pi(x)$ and $\rho(x)$ belong to the sequence of orthogonal polynomials over $[a, b]$ with respect to $w(x)$, then we have a Gauss-type IPIR (cf. Section 2.7). If $m = n$, the IPIR reduces to the usual Gauss rule (2.7.7). If $m > n$, the rule is exact for all functions of joint degree (k, l) where $0 \leq k \leq m$, $0 \leq l \leq n$; $m + 1 \leq k \leq 2m + 1$, $0 \leq l \leq \min(n, 2m + 1 - k)$; and $n + 1 \leq l \leq 2n + 1$, $0 \leq k \leq 2n + 1 - l$. McGrath has given a simpler algorithm to compute the a_{ij} for these Gauss-type IPIRs.

Restricting ourselves now to the case $w(x) = 1$, we see that we can shift an IPIR to any interval $[c, d]$ by the formula

$$\int_c^d g(x)h(x)\,dx \simeq [(d - c)/(b - a)]\hat{\mathbf{g}}A\hat{\mathbf{h}} \tag{2.5.4.4}$$

where $\hat{\mathbf{g}} = (g(t_0), \ldots, g(t_m))^{\mathrm{T}}$, $\hat{\mathbf{h}} = (h(s_0), \ldots, h(s_n))$,

$$t_i = \left(\frac{d - c}{b - a}\right)x_i + \frac{bc - ad}{b - a}, \qquad \text{and} \qquad s_j = \left(\frac{d - c}{b - a}\right)y_j + \frac{bc - ad}{b - a}.$$

We can also form the composite IPIR with p panels

$$\int_a^b f(x)\,dx \simeq \frac{1}{p} \sum_{k=1}^{p} \mathbf{g}_k^{\mathrm{T}} A\mathbf{h}_k$$

where the $\mathbf{g}_k$ and $\mathbf{h}_k$ have the obvious definitions. As we let p tend to infinity, the resulting sequence of approximations converges to the integral for all Riemann integrable functions g and h provided only that the basic IPIR $\mathbf{g}^T A \mathbf{h}$ is exact for all functions of joint degree (0,0). The more complicated problem of forming a (p, r)-composite rule involving p copies of the x_i and r copies of the y_j is treated by Duris and Lyness. For the Newton–Cotes case, a program is given by Duris [F2].

The idea of the IPIR can be extended to the decomposition of $f(x)$ into three or more factors. Also, one can require that an IFIP be accurate if $g(x) = \sum_{i=0}^{m} a_i \phi_i(x)$ and $h(x) = \sum_{j=0}^{n} b_j \psi_j(x)$ where $\{\phi_i\}$ and $\{\psi_j\}$ are two sets of linearly independent functions. Finally, IPIRs in several dimensions have been suggested.

References. Boland [1–3, F1–F4]; Boland and Duris [1]; Boland and Haymond [1]; Bunse [1]; Duris [3, 4, F1–F3]; Duris and Lyness [1]; Gautschi [9]; Gribble [1–3]; Hunkins [1]; McGrath [1, 2].

Example (Duris). Approximate

$$I = \int_0^{2\pi} x \sin 20x \cos 50x \, dx = 2\pi/105 = .05983986.$$

The IPIR, $Q(f, g)$, corresponding to Simpson's rule over [0, 1] has the form

$$Q(f, g) = (f(0), f(\tfrac{1}{2}), f(1))A(g(0), g(\tfrac{1}{2}), g(1))^T$$

where

$$A = \frac{1}{15}\begin{bmatrix} 4 & 2 & -1 \\ 2 & 16 & 2 \\ -1 & 2 & 4 \end{bmatrix}.$$

A composite rule $Q^{(k,m)}(f, g)$ based on $Q(f, g)$ was applied with k subintervals for $f(x) = x \sin 20x$ and $m = 2.5k$ subintervals for $g(x) = \cos 50x$. The results are as follows.

k	$Q^{(k,m)}(f, g)$
60	.06409230
80	.05238701
100	.05796191
150	.05953581

The number of function evaluations required using $Q^{(k,m)}(f, g)$ is $2k + 1$ evaluations of f and $5k + 1$ evaluations of g.

I was also evaluated by Filon's method (Section 2.10.2) with the rules F_i, where the step length $h = 2\pi/20i$. The following results were obtained.

i	F_i
7	.09287746
9	.03912549
11	.05346164

The number of function evaluations using F_i is $20i + 1$ evaluations of both f and g.

Reference. Duris [4].

2.5.5 Interpolatory Integration Formulas on Special Sets of Abscissas. Fejér's Rules

It is occasionally useful, both theoretically and practically, to have interpolatory formulas on sets of abscissas other than the equidistant set. A common choice is the set of zeros of an orthogonal polynomial. Let the inner product $(f, g) = \int_a^b w(x)f(x)g(x)\,dx$, $w(x) \geq 0$, generate a set of orthogonal polynomials $p_0(x), p_1(x), \ldots$. Let $x_1, \ldots, x_n$ be the n zeros of $p_n(x)$. With these abscissas, we can work out interpolatory integration formulas corresponding to the integral $\int_a^b k(x)f(x)\,dx$; that is, rules such that

$$\int_a^b k(x)f(x)\,dx = \sum_{k=1}^{n} w_k f(x_k), \qquad f \in \mathscr{P}_{n-1}. \tag{2.5.5.1}$$

Note that when $k(x) \equiv w(x)$, we are dealing with the Gauss case (Section 2.7), and then the above formula is valid for $f \in \mathscr{P}_{2n-1}$.

The two cases most often studied and used were introduced by Fejér:

(1) Interval: $[-1, 1]$, $k(x) \equiv 1$, $w(x) = (1 - x^2)^{-1/2}$, $p_n(x) = T_n(x) =$ Tchebyscheff polynomial of first kind.

(2) Interval: $[-1, 1]$, $k(x) \equiv 1$, $w(x) = (1 - x^2)^{1/2}$, $p_n(x) = U_n(x) =$ Tchebyscheff polynomial of second kind.

The zeros of the respective polynomials have simple trigonometric expressions and explicit formulas can be worked out for the weights w_k.

(1) Since $T_n(x) = \cos(n \arccos x)$, its zeros are

$$x_{nk} = x_k = \cos\theta_k, \qquad \theta_k = (2k-1)\pi/2n, \qquad k = 1, 2, \ldots, n.$$

Since the rules are to be interpolatory, it follows from (2.5.6) that

$$w_{nk} \equiv w_k = \frac{1}{T_n'(x_k)} \int_{-1}^{1} \frac{T_n(x)\,dx}{x - x_k}. \tag{2.5.5.2}$$

If we set $y = x_k$ in the Christoffel–Darboux formula (1.12.11) for the T_n, we obtain

$$1 + 2\sum_{m=1}^{n-1} T_m(x)T_m(x_k) = -\frac{T_n(x)T_{n+1}(x_k)}{x - x_k},$$

so that (2.5.5.2) can be written as

$$w_k = -\frac{2}{T_n'(x_k)T_{n+1}(x_k)}\left\{1 + \sum_{m=1}^{n-1} T_m(x_k)\int_{-1}^{1} T_m(x)\,dx\right\}. \quad (2.5.5.3)$$

Since $T_n'(\cos\theta) = n(\sin n\theta)/\sin\theta$, we find

$$T_n'(x_k) = T_n'(\cos\theta_k) = (-1)^{k-1} n/\sin\theta_k,$$

$$T_{n+1}(x_k) = \cos(n+1)\theta_k = (-1)^k \sin\theta_k.$$

$$\int_{-1}^{1} T_m(x)\,dx = \int_0^{\pi} \cos m\theta \sin\theta\,d\theta = \begin{cases} 2/(1-m^2), & m \text{ even},\\ 0, & m \text{ odd}.\end{cases}$$

Insert these values in (2.5.5.3) and obtain

$$w_k = \frac{2}{n}\left\{1 - 2\sum_{m=1}^{[n/2]} \frac{\cos(2m\theta_k)}{4m^2 - 1}\right\}. \quad (2.5.5.4)$$

Fejér's first integration formula is therefore

$$\int_{-1}^{1} f(x)\,dx = \sum_{k=1}^{n} w_k f\left(\cos\frac{2k-1}{2n}\pi\right), \qquad f \in \mathscr{P}_{n-1} \quad (2.5.5.5)$$

with w_k given by (2.5.5.4).

(2) A similar analysis with $U_n(\cos\theta) = \sin(n+1)\theta/\sin\theta$ yields the rule

$$\int_{-1}^{1} f(x)\,dx = \sum_{k=1}^{n} w_k f\left(\cos\frac{k}{n+1}\pi\right), \qquad f \in \mathscr{P}_{n-1} \quad (2.5.5.6)$$

where

$$w_k = \frac{2}{n+1}\left\{1 - 2\sum_{m=1}^{[(n-1)/2]} \frac{\cos(2m\theta_k)}{4m^2 - 1} - \frac{1}{p}\cos(p+1)\theta_k\right\} \quad (2.5.5.7)$$

where $\theta_k = k\pi/(n+1)$ and where $p = 2[(n+1)/2] - 1$ is the largest odd number $\leq n$. This can be written in the alternative form

$$w_k = \frac{4\sin\theta_k}{n+1}\sum_{m=1}^{[(n+1)/2]} \frac{\sin(2m-1)\theta_k}{2m-1} \quad (2.5.5.8)$$

In both (2.5.5.5) and (2.5.5.6), the weights w_k can be proved to be positive, and the rules therefore fall under the remarks of Section 2.7.8. As $n \to \infty$, the approximate integrations converge to $\int_{-1}^{1} f(x)\,dx$ for all $f \in R[-1, 1]$ and, in fact, for certain singular functions.

Gautschi has shown that if $f(x)$ has monotonic singularities at $x = \pm 1$, then convergence takes place. The rate of convergence depends upon the precise nature of the singularity. It is rapid for logarithmic singularities but may be slow for square root singularities.

The Tschebyscheff polynomials of the first and second kinds are special cases of the *ultraspherical polynomials* $C_n^\alpha(x)$. For $-\frac{1}{2} < \alpha \leq 2$, the weights arising from interpolatory integration on the zeros of $C_n^\alpha(x)$ are known to be positive. See Szegö and Askey and Fitch.

For the more general case of Jacobi polynomials $P_n^{(\alpha,\beta)}(x)$, Askey has shown that the weights arising from interpolatory integration on the zeros of $P_n^{(\alpha,\beta)}(x)$ are positive if $\alpha, \beta \geq 0$ and $\alpha + \beta \leq 1$, if $\alpha = \beta + 1$ and $-1 < \beta \leq \frac{1}{2}$ and if $\beta = \alpha + 1$ and $-1 < \alpha \leq \frac{1}{2}$ in addition to the above-mentioned case $\alpha = \beta$, $-1 < \alpha \leq \frac{3}{2}$. On the other hand, Lether *et al.* have shown the existence of negative weights for some values of n and certain pairs (α, β) satisfying $-1 < \alpha, \beta \leq \frac{3}{2}$.

(3) *The Clenshaw–Curtis and Basu rules*

These interpolatory rules arise by adjoining the endpoints -1 and 1 to the abscissas of the Fejér $(n-2)$-point rules. Explicitly, the abscissas of the Clenshaw–Curtis rule are given by

$$x_k = \cos\left(\frac{k-1}{n-1}\pi\right), \qquad k = 1, \ldots, n$$

and those of the Basu rule by

$$x_1 = 1, \qquad x_n = -1, \qquad x_k = \cos\left(\frac{2k-3}{2n-2}\pi\right), \qquad k = 2, \ldots, n-1.$$

The Clenshaw–Curtis rules are by far the most prevalent of the four symmetric interpolatory integration rules based on trigonometric points. The abscissas are the extreme points of $T_{n-1}(x)$ on $[-1, 1]$ and the weights w_k, which are all positive, are given by

$$w_1 = w_n = \begin{cases} 1/(n-1)^2, & n \text{ even}, \\ 1/n(n-2), & n \text{ odd}, \end{cases}$$

$$w_k = \left\{1 - \sum_{j=1}^{[(n-1)/2]}{}^{*} \frac{2}{4j^2-1} \cos\frac{2j(k-1)}{n-1}\pi\right\}, \qquad k = 2, \ldots, n-1.$$

The notation $\sum^*$ means that the last term in the sum should be halved if n is odd.

The weights of the Basu rule are not all positive; however, the sequence of sums of the magnitudes of the weights is uniformly bounded so that the Basu rules converge for all $f \in C[-1, 1]$.

Let m be the smallest even integer $\geq n$ and assume that $f \in C^m[-1, 1]$. Then we can bound the error in any of the above four integration rules by

$$\frac{c_m}{m!} \max_{-1 \leq x \leq 1} |f^{(m)}(x)|.$$

The best possible relation for the c_m as $m \to \infty$ is as follows:

Fejér's second rule:	$O(2^{-m}m^{-1})$,
Fejér's first rule:	$O(2^{-m}m^{-2})$,
Clenshaw–Curtis rule:	$O(2^{-m}m^{-3})$,
Basu rule:	$O(2^{-m}m^{-4})$.

In discussions of the Clenshaw–Curtis method, the abscissas $\cos[(k-1)\pi/(n-1)]$ are sometimes called the "practical" abscissas while $\cos[(2k-1)\pi/2n]$ are the "classical" abscissas. The abscissas $\cos[k\pi/(n+1)]$ are the "Filippi" abscissas.

References. Askey [1, 2]; Askey and Fitch [1]; Basu [2]; Brass [1]; Brass and Schmeisser [2]; Clenshaw and Curtis [1]; Elliott [3]; Fejér [1]; Feldheim [1]; Filippi [1, 4]; Gautschi [2]; Lether, Wilhelmsen, and Frazier [1]; Pólya [1]; Riess and Johnson [2]; Sottas [1]; Szegö [2]; Wagner [1].

2.5.6 Product Integration Rules

The sets of "classical" abscissas

$$T_1 = \{\cos(2k+1)\pi/2(n+1),\ k = 0, \ldots, n\}$$

and "practical" abscissas

$$T_2 = \{\cos k\pi/n,\ k = 0, \ldots, n\}$$

are optimal choices in the context of product integration from the point of view of convergence. A third such set is

$$T_3 = \{\cos 2k\pi/(2n+1),\ k = 0, \ldots, n\}.$$

Consider the integral

$$I(kf) = \int_{-1}^{1} k(x)f(x)\,dx \tag{2.5.6.1}$$

and the product integration rules

$$Q_n f = \sum_{i=0}^{n} w_i(k)f(x_i) \tag{2.5.6.2}$$

where the $w_i(k)$ are chosen so that $Q_n f = I(kf)$, $f \in \mathscr{P}_n$. Sloan and Smith have shown that if the set of abscissas $\{x_i\}$ in (2.5.6.2) is T_1, T_2, or T_3, then $\{Q_n f\} \to I(kf)$ for all $f \in R[-1, 1]$ provided only that $k \in L_p[-1, 1]$ for some $p > 1$. The case $\{x_i\} = T_1$ is a special case of the more general result that if the x_i are the zeros of the Jacobi polynominal $P_{n+1}^{(\alpha,\beta)}(x)$, then $\{Q_n f\} \to I(kf)$ for all $f \in R[-1, 1]$ if

$$\int_{-1}^{1} |k(x)(1 - x)^{-\max[(2\alpha+1)/4,\, 0]}(1 + x)^{-\max[(2\beta+1)/4,\, 0]}|^p \, dx < \infty$$

for some $p > 1$. If the x_i are the zeros of orthogonal polynomials with respect to the admissible weight function $w(x)$, then we have convergence for all $f \in R[-1, 1]$ if

$$\int_{-1}^{1} \frac{k^2(x)}{w(x)} \, dx < \infty .$$

The weights $w_i(k)$ can be evaluated using the modified moments

$$M_j(k) = \int_{-1}^{1} k(x) p_j(x) \, dx, \qquad j = 0, \ldots, n \tag{2.5.6.3}$$

where $\{p_j(x),\, j = 0, \ldots, n\}$ is a set of orthogonal polynomials with respect to the discrete inner product

$$(g, h)_n = \sum_{i=0}^{n} \mu_i \, g(x_i) h(x_i), \qquad \mu_i > 0 . \tag{2.5.6.4}$$

In this case, the Lagrange interpolation polynomial $p_n(f; x)$ can be written as

$$p_n(f; x) = \sum_{j=0}^{n} d_j p_j(x) \tag{2.5.6.5}$$

where $d_j = (f, p_j)_n / (p_j, p_j)_n$ so that

$$Q_n f = \sum_{j=0}^{n} d_j M_j(k). \tag{2.5.6.6}$$

From this it follows that

$$w_i(k) = \mu_i \sum_{j=0}^{n} M_j(k) p_j(x_i) / (p_j, p_j)_n . \tag{2.5.6.7}$$

Either (2.5.6.2) or (2.5.6.6) can be used to evaluate $Q_n f$, whichever is more convenient.

If $\{p_m(x)\}$ is a sequence of polynomials orthogonal with respect to the inner product

$$(g, h) = \int_{-1}^{1} w(x)g(x)h(x)\,dx$$

and if the x_i are the zeros of $p_{n+1}(x)$, then the set $\{p_j(x), j = 0, \ldots, n\}$ is also orthogonal with respect to the discrete inner product (2.5.6.4) where the μ_i are the corresponding Gauss weights (see Section 2.7). If the set $\{x_i\} = T_2$, then the Tschebyscheff polynomials $T_j(x)$ are orthogonal with respect to (2.5.6.4) if $\mu_0 = \mu_n = \frac{1}{2}$, $\mu_i = 1$, $i = 1, \ldots, n-1$, and similarly for $\{x_i\} = T_3$ with $\mu_0 = \frac{1}{2}$, $\mu_i = 1$, $i = 1, \ldots, n$. In cases T_1–T_3, closed forms for the weights are given by Sloan and Smith [2].

Piessens and Branders have given recurrence relations for evaluating the modified moments $M_j(k)$ with respect to the Tschebyscheff polynomials $T_j(x)$ for various functions $k(x)$.

Example. Let $k(x) = |x - \lambda|^\alpha$, $\alpha > -1$, $\lambda \in (-1, 1)$. Then the modified moments

$$M_j = M_j(k) = \int_{-1}^{1} |x - \lambda|^\alpha T_j(x)\,dx$$

satisfy the recurrence relation

$$\left[1 + \frac{1+\alpha}{j+1}\right] M_{j+1} - 2\lambda M_j + \left[1 - \frac{1+\alpha}{j-1}\right] M_{j-1}$$
$$= -\left(\frac{2}{j^2 - 1}\right)\left[(1-\lambda)^{1+\alpha} - (-1)^j (1+\lambda)^{1+\alpha}\right], \qquad j \geq 2,$$

with

$$M_0 = [(1-\lambda)^{1+\alpha} + (1+\lambda)^{1+\alpha}]/(1+\alpha),$$
$$M_1 = (M_0 - [(1-\lambda)^{1+\alpha} + (1+\lambda)^{1+\alpha}])/(2+\alpha),$$
$$M_2 = (4\lambda M_1 + 2[(1-\lambda)^{1+\alpha} + (1+\lambda)^{1+\alpha}])/(3+\alpha) - M_0.$$

Paget has studied generalized product integration in which an admissible weight function $w(x)$ is included in addition to the arbitrary weight $k(x)$. In this case, the results above apply with the modified moments given by

$$M_j(k) = \int_{-1}^{1} w(x)k(x)p_j(x)\,dx$$

where the $\{p_j(x)\}$ are orthogonal with respect to $w(x)$. For the Jacobi weight $w(x) = (1-x)^\alpha(1+x)^\beta$ and for various functions $k(x)$, Paget determines three term recurrence relations for the $M_j(k)$ of the form

$g_{k+1} = u_k g_k + v_k g_{k-1}$. Using these, one can apply the backward recurrence formulas

$$b_{n+1} = b_n = 0, \qquad b_k = d_k + u_k b_{k+1} + v_{k+1} b_{k+2}, \qquad Q_n f = b_0$$

to evaluate $Q_n f$ as given by (2.5.6.6).

Examples. $w(x) = 1$, $\{x_i, \mu_i\}$-Gauss–Legendre points and weights

a) $k(x) = \exp(i\lambda x)$, $\quad \lambda$ real,

$u_k = i(2k+1)/\lambda, \qquad v_k = 1,$

$Q_n f = 2(b_0 \sin \lambda - i\, b_1 \cos \lambda)/\lambda;$

(b) $k(x) = \log|x - \lambda|, \qquad -1 < \lambda < 1,$

$u_k = (2k+1)\lambda/(k+2), \qquad v_k = -(k-1)/(k+2),$

$Q_n f = (b_0 - \frac{1}{2}b_1)(1+\lambda)\log(1+\lambda) + (b_0 + \frac{1}{2}b_1)(1-\lambda)\log(1-\lambda) + 2b_2/3 - 2b_0;$

(c) $k(x) = |x - \lambda|^\alpha, \qquad \alpha > -1, \quad -1 < \lambda < 1,$

$u_k = (2k+1)\lambda/(k+\alpha+2), \qquad v_k = -(k-\alpha-1)/(k+\alpha+2),$

$$Q_n f = \left(\frac{b_0}{\alpha+1} + \frac{b_1}{\alpha+2}\right)(1-\lambda)^{\alpha+1} + \left(\frac{b_0}{\alpha+1} - \frac{b_1}{\alpha+2}\right)(1+\lambda)^{\alpha+1}.$$

References. Beard [1]; Elliott and Paget [2, 3]; Gatteschi [2]; Kussmaul [1]; Lehman, Parke, and Maximon [1]; Paget [1]; Piessens and Branders [2, 6]; Schneider [2]; Sloan [2, 3]; Sloan and Smith [1–3]; Smith and Sloan [1]; A. Young [1, 2].

2.5.8 The Method of Approximation

A companion piece to the method of interpolation is the method of approximation. Here the idea is to approximate the integrand $f(x)$ by a sequence of functions, each of which can be integrated in closed form. The integral of the approximation then constitutes an approximation to the integral. It is probably most convenient to represent the approximation as a series

$$f(x) = a_1\phi_1(x) + \cdots + a_n\phi_n(x) + \varepsilon_n(x) \tag{2.5.8.1}$$

where $\varepsilon_n(x)$ is small over the range of integration.

Example 1. *Bernstein Polynomials.* If $f(x) \in C[0, 1]$, then the Bernstein polynomial of order n for $f(x)$ is defined by

$$B_n(f; x) = \sum_{k=0}^{n} \binom{n}{k} f\left(\frac{k}{n}\right) x^k (1-x)^{n-k}. \tag{2.5.8.2}$$

Now $B_n(f; x) \to f(x)$ uniformly in [0, 1] as $n \to \infty$. Hence,

$$\int_0^1 f(x)\,dx \approx \int_0^1 B_n(f; x)\,dx$$
$$= \sum_{k=0}^{n} \binom{n}{k} f\left(\frac{k}{n}\right) \int_0^1 x^k(1-x)^{n-k}\,dx$$
$$= \frac{1}{n+1} \sum_{k=0}^{n} f\left(\frac{k}{n}\right).$$

Thus, integrating the Bernstein polynomial approximation is equivalent to averaging the function at the equidistant stations $0, 1/n, \ldots, n/n$.

Example 2. *Taylor Polynomials.* If $f(z)$ is analytic in $|z| \le R$, it can be approximated in the form

$$f(z) = f(0) + f'(0)z + \frac{f''(0)}{2!} z^2 + \cdots + \frac{f^{(n)}(0)}{n!} z^n + \varepsilon_n(z). \tag{2.5.8.3}$$

Hence, assuming that $R \ge 1$, we have

$$\frac{1}{2}\int_{-1}^{+1} f(x)\,dx \approx f(0) + \frac{f''(0)}{3!} + \frac{f^{(4)}(0)}{5!} + \cdots + \frac{f^{(2m)}(0)}{(2m+1)!}, \qquad m = \left[\frac{n}{2}\right]. \tag{2.5.8.4}$$

This idea has been developed by Lyness who uses contour integration in the complex plane to evaluate $f^{(k)}(0)/k!$.

A good choice for the $\phi_i(x)$ are orthogonal polynomials over the interval of integration. This idea will be expounded subsequently in separate sections.

If it were advantageous or otherwise convenient, one might integrate "best" approximations to functions. Say we approximate $f(x)$ over $[a, b]$ by a polynomial $a_0 + a_1 x + \cdots + a_n x^n$ so that the deviation

$$\delta = \max_{a \le x \le b} |f(x) - (a_0 + a_1 x + \cdots + a_n x^n)|$$

has been minimized from among all polynomials in $\mathscr{P}_n$. There are many programs available for doing this. One or two additional statements may be added to such a program which will provide either the definite or indefinite integral. On the interval $[a, b]$, the answer will be good to within $(b - a)\,\delta$. It is naturally far more convenient to approximate from linear than from nonlinear families. This would rule out the use of rational approximations.

One may also use tabulated approximations to the elementary transcendental functions as building blocks on which to base approximations of definite or indefinite integrals. If the manipulations involve lengthy or tedious polynomial algebra (as may happen in several variables), formula manipulation programs of the FORMAC type may be brought into play.

Example 3. Compute

$$I = \int_0^1 x^{1/2}\Gamma(x + 2)\,dx.$$

Here $\Gamma(x)$ is the Gamma function. Using an approximation obtained from Hart *et al.*,

$$\Gamma(x + 2) = .99910836 + .4497361x + .2855737x^2 + .2646888x^3 + \varepsilon(x)$$

where

$$|\varepsilon(x)| \le .0009 \quad \text{on} \quad [0, 1].$$

Hence,

$$I = \tfrac{2}{3}(.99910836) + \tfrac{2}{5}(.4497361) + \tfrac{2}{7}(.2855737) + \tfrac{2}{9}(.2646888) + \eta$$

where

$$|\eta| \le \tfrac{2}{3}(.0009) = .0006;$$

$$I \approx .986.$$

References. Hart *et al.* [1]; Lyness [6]; Lyness and Sande [F1].

2.6 Integration Formulas of Open Type

An integration formula

$$\int_a^b f(x)\,dx \approx \sum_{k=1}^{n} w_k f(x_k), \qquad x_1 < x_2 < \cdots < x_n \tag{2.6.1}$$

is said to be of *closed type* if the function $f(x)$ is evaluated at the end points of the interval, that is, if $x_1 = a$ and $x_n = b$. An integration formula is said to be of *open type* if both of the end points are omitted from the evaluation. A formula of open type, so to speak, performs an extrapolation of the function to the whole interval and then integrates forward and backward beyond the last known values.

Formulas of open type are of use in the integration of ordinary differential equations. We list several such formulas, sometimes called Steffensen's formulas.

Two-point Formula

$$\int_a^b f(x)\,dx = \frac{3h}{2}[f(a + h) + f(a + 2h)] + \frac{3h^3}{4} f''(\xi),$$

$$h = \frac{b - a}{3}, \qquad a < \xi < b. \tag{2.6.2}$$

Three-point Formula

$$\int_a^b f(x)\,dx = \frac{4h}{3}[2f(a+h) - f(a+2h) + 2f(a+3h)] + \frac{14h^5}{45}f^{(4)}(\xi),$$

$$h = \frac{b-a}{4}, \qquad a < \xi < b. \tag{2.6.3}$$

Four-point Formula

$$\int_a^b f(x)\,dx = \frac{5h}{24}[11f(a+h) + f(a+2h) + f(a+3h) + 11f(a+4h)]$$

$$+ \frac{95h^5}{144}f^{(4)}(\xi), \qquad h = \frac{b-a}{5}, \qquad a < \xi < b, \tag{2.6.4}$$

In addition to these, there is the

Midpoint Formula

$$\int_a^{a+h} f(x)\,dx = hf\left(a + \frac{h}{2}\right) + \frac{1}{24}h^3 f''(\xi), \qquad a < \xi < a + h, \tag{2.6.5}$$

mentioned before. It is the first of a series of formulas of *Maclaurin type*:

$$\int_a^b f(x)\,dx \approx h \sum_{k=1}^{n} w_{kn} f\left(a + \frac{2k-1}{2}h\right), \qquad h = \frac{b-a}{n}. \tag{2.6.6}$$

References. J. C. P. Miller [1]; Rubbert [1].

A note of caution must be sounded for the use of formulas of open type. The error estimates given in (2.6.2), (2.6.3), and (2.6.4) are valid provided that the integrands $f(x)$ are, respectively, in classes $C^2[a, b]$, $C^4[a, b]$, and $C^4[a, b]$. When the value of the function at an end point of the interval cannot be computed because of a singularity, the use of such formulas can lead to serious error.

On the other hand, for functions whose derivatives have singularities at the end points, open formulas are more effective than corresponding closed formulas. This is an instance of the principle of "avoiding the singularity."

Open formulas are not necessarily restricted to equidistant points. For example, the Gauss formulas are of open type. The Lobatto formulas are closed, while the Radau formulas are half-open (see Sections 2.7 and 2.7.1).

References. Hildebrand [1, p. 88]; Kunz [1, p. 148]; Rabinowitz [9].

Formulas of open and half-open type are used as "predictors" in the solution of differential equations. A long list of such formulas may be found in Rosser.

Reference. Rosser [2].

2.6.1 *Generalized Midpoint Rule*

In this section we indicate an interesting generalization of the midpoint rule (2.1.3) or (2.6.5) to weighted integrals. The resulting rule has equal weights, is exact for functions of class $\mathscr{P}_1$, but will not in general have equally spaced abscissas.

Let a weight function $w(x)$ satisfy $w(x) > 0$, and be normalized so that $\int_a^b w(x)\,dx = 1$. The function

$$y = H(x) = \int_a^x w(t)\,dt \tag{2.6.1.1}$$

will therefore be monotonically increasing from $H(a) = 0$ to $H(b) = 1$. We can therefore define an inverse function $L(y)$ on $[0, 1]$ which increases monotonically from $L(0) = a$ to $L(1) = b$. For an integer $N > 1$, set

$$x_i = N \int_{i/N}^{(i+1)/N} L(y)\,dy, \qquad i = 0, 1, \ldots, N-1. \tag{2.6.1.2}$$

The values x_i will be monotonically increasing and located in (a, b). The generalized midpoint rule is defined by

$$\int_a^b w(x) f(x)\,dx = N^{-1} \sum_{i=0}^{N-1} f(x_i) + E. \tag{2.6.1.3}$$

Clearly, the rule integrates constants exactly, and since $\int_a^b x w(x)\,dx = \int_0^1 L(y)\,dy = N^{-1}[x_0 + \cdots + x_{N-1}]$, it integrates the function x and hence all linear functions. It can be shown that

$$E = \tfrac{1}{2} C_N f''(\xi), \qquad a < \xi < b. \tag{2.6.1.4}$$

where

$$C_N = \int_a^b x^2 w(x)\,dx - N^{-1} \sum_{i=0}^{N-1} x_i^2.$$

References. Jagerman [1]; F. Stetter [3].

2.6.2 *Integration Rules with Abscissas outside Integration Interval*

While this book stresses rules with abscissas *inside* the interval of integration, there are situations in numerical analysis where it is required to have abscissas *outside* this interval.

A discussion of such rules can be found in Berezin and Zhidkov, while Rosser lists many of the rules of this type that are of interest in the numerical integration of differential equations.

One context in which the use of exterior points is appropriate is in the tabulation of an indefinite integral. If we wish to tabulate

$$F(x) = \int_a^x f(t)\,dt \tag{2.6.2.1}$$

at the equally spaced points $a = x_0, x_1, \ldots, x_n = b$, we can do this by writing

$$F(x_m) = F(x_{m-1}) + \int_{x_{m-1}}^{x_m} f(t)\,dt, \qquad m = 1, \ldots, n \tag{2.6.2.2}$$

and evaluating the integrals numerically. When the interval $I_m = [x_{m-1}, x_m]$ is interior to $[a, b]$, we can use points outside I_m in the evaluation of $\int_{x_{m-1}}^{x_m} f(t)\,dt$ because $f(t)$ is defined over the entire interval $[a, b]$. The advantage is that we can use the same abscissa many times, once as an interior point and several times as exterior points, both to the left and to the right of I_m.

Example. (Krylov). Assume that $f \in C^{10}[a, b]$ and that the interval $[x_{m-2}, x_{m+3}] \subset [a, b]$. Define $\alpha_k = x_k + ch$, $\beta_k = x_{k+1} - ch$, $c = .23963\ 00931$, $h = (b - a)/n$. Then

$$\begin{aligned} F(x_{m+1}) = F(x_m) &+ .48259\ 37250h[f(\alpha_m) + f(\beta_m)] + .01797\ 22221h[f(\alpha_{m+1}) + f(\beta_{m-1})] \\ &- .00057\ 82647h[f(\alpha_{m-1}) + f(\beta_{m+1})] + .00001\ 23177h[f(\alpha_{m+2}) + f(\beta_{m-2})] + R_m \end{aligned} \tag{2.6.2.3}$$

where

$$|R_m| \le 3 \times 10^{-9} h^{11} \max_{\beta_{m-2} \le x \le \alpha_{m+2}} |f^{(10)}(x)|.$$

References. Berezin and Zhidkov [1]; Engels [13, pp. 328–331]; Krylov [1, Chap. 15]; Rosser [2].

2.7 *Integration Rules of Gauss Type*

Let $w(x)$ be a fixed admissible weight function defined on the interval $[a, b]$, which may be finite or infinite. The integral

$$\int_a^b w(x)f(x)g(x)\,dx = (f, g) \tag{2.7.1}$$

is known as the *inner product* of the functions $f(x)$ and $g(x)$ over the interval $[a, b]$ with respect to the weight $w(x)$.

Two functions f and g are called *orthogonal* over $[a, b]$ with respect to the weight $w(x)$ if

$$(f, g) = \int_a^b w(x) f(x) g(x)\,dx = 0. \tag{2.7.2}$$

For a given weight $w(x)$, it is possible to define a sequence of polynomials $p_0(x), p_1(x), \ldots$, which are orthogonal and in which $p_n(x)$ is of exact degree n:

$$(p_m, p_n) = \int_a^b w(x) p_m(x) p_n(x)\,dx = 0, \qquad m \neq n. \tag{2.7.3}$$

By multiplying each $p_n(x)$ by an appropriate constant we can produce a set of polynomials p_n^*, which are *orthonormal*:

$$(p_m^*, p_n^*) = \int_a^b w(x) p_m^*(x) p_n^*(x)\,dx = \delta_{mn} = \begin{cases} 0 & \text{if } m \neq n, \\ 1 & \text{if } m = n. \end{cases} \tag{2.7.4}$$

The leading coefficient of p_n^* can be taken as positive:

$$p_n^*(x) = k_n x^n + \cdots, \qquad k_n > 0. \tag{2.7.5}$$

A description of the most commonly used orthogonal polynomials is given in Sections 1.12 and 1.13.

THEOREM. *The zeros of (real) orthogonal polynomials are real, simple, and located in the interior of* $[a, b]$.

THEOREM. *The orthonormal polynomials* $p_n^*(x)$ *satisfy a three-term recurrence relationship.*

$$p_n^*(x) = (a_n x + b_n) p_{n-1}^*(x) - c_n p_{n-2}^*(x), \qquad n = 1, 2, 3, \ldots, \tag{2.7.6}$$

$$a_n, c_n \neq 0 \quad \textit{and} \quad p_{-1}^*(x) = 0, \qquad p_0^*(x) = \left(\int_a^b w(x)\,dx\right)^{1/2}.$$

The following recurrence is particularly convenient for systematic computation.

$$\begin{aligned} p_{-1}(x) &= 0, \\ p_0(x) &= 1, \\ &\vdots \\ p_{n+1}(x) &= x p_n^*(x) - (x p_n^*, p_n^*) p_n^*(x) - (p_n, p_n)^{1/2} p_{n-1}^*(x), \\ p_n^*(x) &= p_n(x)/(p_n, p_n)^{1/2}, \qquad n = 0, 1, 2, \ldots. \end{aligned}$$

References. Davis [6, pp. 167–168, 234–255]; NBS Handbook [1, Chap. 22].

If n distinct points $x_1, \ldots, x_n$ of the interval $[a, b]$ are specified in advance, then we know that we can find coefficients $w_1, w_2, \ldots, w_n$ such that the rule

$$\int_a^b w(x)f(x)\,dx \approx \sum_{k=1}^{n} w_k f(x_k)$$

will be exact for all polynomials of class $\mathscr{P}_{n-1}$. If we treat both the x's and the w's as $2n$ unknowns, and determine them carefully, perhaps we can arrange matters so that the rule will be exact for polynomials of class $\mathscr{P}_{2n-1}$, that is, for all linear combinations of the $2n$ powers $1, x, x^2, \ldots, x^{2n-1}$. This is possible, and the solution is intimately related to the orthogonal polynomials generated by the weight function $w(x)$.

THEOREM. *Let $w(x)$ be an admissible weight function defined on $[a, b]$ with corresponding orthonormal polynomials $p_n^*(x)$. Let the zeros of $p_n^*(x)$ be*

$$a < x_1 < x_2 < \cdots < x_n < b.$$

Then we can find positive constants $w_1, w_2, \ldots, w_n$ such that

$$\int_a^b w(x)p(x)\,dx = \sum_{k=1}^{n} w_k p(x_k) \tag{2.7.7}$$

whenever $p(x)$ is a polynomial of class $\mathscr{P}_{2n-1}$.

The weights w_k have the explicit representation

$$w_k = -\frac{k_{n+1}}{k_n}\frac{1}{p_{n+1}^*(x_k)p_n^{*\prime}(x_k)}. \tag{2.7.8}$$

An alternative expression for the weights is given by (2.7.5.9). Lether has used this formula, which is more stable but requires more computation.

When abscissas and weights have been determined as in this theorem, we say that the resulting integration rule is of *Gauss type*. We shall frequently refer to the n-point Gauss rule with the weight $w(x) \equiv 1$ by the symbol G_n.

It should be observed that if $w(x)$ is a symmetric function, i.e.,

$$w(\tfrac{1}{2}(a + b) + x) = w(\tfrac{1}{2}(a + b) - x),$$

then the abscissas x_k are located symmetrically in the interval $[a, b]$ and the weights corresponding to symmetric points are equal. Thus, in this case, half the work of the computation of these fundamental constants can be saved (as well as the punching of input data).

The error incurred in Gaussian integration is governed by the following estimate.

THEOREM. *Let $w(x)$, $x_1, \ldots, x_n$, $w_1, \ldots, w_n$ be as in the previous theorem. Then, if $f(x) \in C^{2n}[a, b]$,*

$$E_n(f) = \int_a^b w(x) f(x)\, dx - \sum_{k=1}^{n} w_k f(x_k)$$

$$= \frac{f^{(2n)}(\xi)}{(2n)!\, k_n^2}, \qquad a < \xi < b. \tag{2.7.9}$$

COROLLARY. *In the case of the Jacobi weight*

$$w(x) = (1 - x)^\alpha (1 + x)^\beta, \qquad \alpha > -1, \qquad \beta > -1,$$

over the interval $[-1, 1]$, the error is given by

$$E_n(f) = \frac{2^{2n+\alpha+\beta+1}\Gamma(n + \alpha + 1)\Gamma(n + \beta + 1)\Gamma(n + \alpha + \beta + 1)n!}{\Gamma(2n + \alpha + \beta + 1)\Gamma(2n + \alpha + \beta + 2)(2n)!} f^{(2n)}(\xi),$$
$$-1 < \xi < 1. \tag{2.7.10}$$

COROLLARY. *In the case of the weight $w(x) \equiv 1$ over $[-1, 1]$* (that is, $\alpha = 0$, $\beta = 0$) *the error is given by*

$$E_n(f) = \frac{2^{2n+1}(n!)^4}{(2n + 1)[(2n)!]^3} f^{(2n)}(\xi), \qquad -1 < \xi < 1. \tag{2.7.11}$$

Over the interval $[a, b]$ and for the weight $w(x) \equiv 1$, the error is given by

$$E_n(f) = \frac{(b - a)^{2n+1}(n!)^4}{(2n + 1)[(2n)!]^3} f^{(2n)}(\xi), \qquad a < \xi < b. \tag{2.7.12}$$

COROLLARY. *In the case of the weight $w(x) = (1 - x^2)^{-1/2}$ over $[-1, 1]$, we have the Gauss–Tschebyscheff integration rule of the first kind*

$$\int_{-1}^{1} (1 - x^2)^{-1/2} f(x)\, dx = \frac{\pi}{n} \sum_{i=1}^{n} f(x_i) + \frac{\pi}{2^{2n-1}(2n)!} f^{(2n)}(\xi), \qquad -1 < \xi < 1 \tag{2.7.13}$$

where $x_i = \cos((2i - 1)\pi/2n)$, $i = 1, \ldots, n$.

For $w(x) = (1 - x^2)^{1/2}$ over $[-1, 1]$, we have the Gauss–Tschebyscheff integration rule of the second kind

$$\int_{-1}^{1} (1 - x^2)^{1/2} f(x)\, dx = \sum_{i=1}^{n} w_i f(x_i) + \frac{\pi}{2^{2n+1}(2n)!} f^{(2n)}(\xi), \qquad 1 < \xi < 1 \tag{2.7.14}$$

where

$$w_i = \frac{\pi}{n+1} \sin^2 \frac{i\pi}{n+1} \qquad \text{and} \qquad x_i = \cos \frac{i\pi}{n+1}, \qquad i = 1, \ldots, n.$$

The error term (2.7.12) is a special case of the following general result.

THEOREM. *Given an integration rule over a finite interval* $[u, v]$ *of the form*

$$\int_u^v f(x)\, dx = \sum_{i=1}^n w_i f(x_i) + c_n f^{(k_n)}(\xi), \qquad u < \xi < v, \tag{2.7.15}$$

then, over the finite interval $[a, b]$, *we have the rule*

$$\int_a^b f(y)\, dy = A \sum_{i=1}^n w_i f(y_i) + c_n A^{k_n+1} f^{(k_n)}(\eta), \qquad a < \eta < b \tag{2.7.16}$$

where $A = (b - a)/(v - u)$ and $y_i = a + A(x_i - u)$, $i = 1, \ldots, n$.

Since no integration rule of the type $\sum_{k=1}^n w_k f(x_k)$ can integrate exactly the function $\prod_{k=1}^n (x - x_k)^2 \in \mathscr{P}_{2n}$, we see that Gauss rules are best† in the sense that they integrate exactly polynomials of as high degree as possible with a formula of the type $\sum_{k=1}^n w_k f(x_k)$. This optimality often carries over to near-optimality when this term has been defined in the language of normed spaces (see Section 4.7). The positive weights are also useful in keeping down roundoff error.

There are several disadvantages to Gauss rules. The weights and abscissas of the Gauss rules are generally irrational numbers. If computing is done by hand, it is an error-liable nuisance to deal with many digits, and so in years gone by, the Gauss rules were not popular. Digital computers, on the other hand, do not distinguish between "simple" numbers such as .5000000 and more "complicated" numbers such as .577350269. The Gauss rules which integrate exactly polynomials of maximal degree are excellent for large classes of functions arising in practice and are now very popular. But the old difficulty of rational versus irrational still comes into play in that the preparation of a program for Gauss integration requires the typing up and checking of many "irrational" numbers. On the other hand, the abscissas and weights for G_n may be generated as needed by the program GRULE given in Appendix 2, and programs for computing

† The use of the term *best rule* is common in the literature of numerical analysis. But it should be taken with a large dose of salt. After all, *any* rule of approximate integration $\sum_{k=1}^n w_k f(x_k)$, no matter how w_k and x_k are chosen, will integrate *exactly* an infinite dimensional family of functions. What is best for me may not be best for thee, and the careful expositor will make clear the sense in which *best* is used.

Gauss–Jacobi, Laguerre, and Hermite rules (see Section 3.6) are given by Stroud and Secrest and in the NAG library.

Secondly, the weights and abscissas of rules of any order are distinct from those of any other order (except that zero appears as an abscissa in all rules of odd order). Thus, in proceeding from a computation of $G_n(f)$ to $G_m(f)$ with $m > n$, almost all the information obtained in computing the former value is discarded. As a partial answer to this objection, there is a device, due to Kronrod and developed by Patterson, which enables us to add new abscissas to a given set of abscissas to produce a new rule of higher—but not optimally higher—accuracy. (See Section 2.7.1.1.)

References. Davis [6, pp. 343–344]; Kronrod [1]; Lether [9]; NAG [1]; Patterson [1, F1]; Stroud and Secrest [2].

Consult Appendix 4.3 for a partial list of tabulations of Gauss abscissas and weights.

Example

Rule	$\int_0^1 f(x)\,dx$	$\int_0^1 x^{1/2}\,dx$	$\int_0^1 x^{3/2}\,dx$	$\int_0^1 \frac{2\,dx}{2+\sin 10\pi x}$
G_2	.2261 0879	.6738 8734	.3987 7398	1.0312 8099
G_3	.3092 5767	.6691 7963	.3998 1241	1.0217 8819
G_4	.3461 1571	.6678 2765	.3999 5038	1.1978 1330
G_{10}	.3302 4750	.6667 5604	.3999 9934	1.1614 6323
G_{16}	.2949 6118	.6666 8963	.3999 9993	1.1980 6091
G_{32}	.3120 5239	.6666 6967	.4000 0000	1.1522 6418
G_{48}	.3071 4034	.6666 6757	.4000 0000	1.1548 2215
Exact value	.3068 5282	.6666 6667	.4000 0000	1.1547 0054

G_2 = Gauss 2-point formula; G_3, 3-point; etc.
($f(x) = (x+2)^{-1}$, $0 \le x \le e-2$; $f(x) = 0$, $e-2 \le x \le 1$.)

Rule	$\int_0^1 \frac{dx}{1+x^4}$	$\int_0^1 \frac{dx}{1+x}$	$\int_0^1 \frac{dx}{1+e^x}$	$\int_0^1 \frac{x\,dx}{e^x-1}$
G_2	.8595 2249	.6923 0769	.3799 0887	.7775 1164
G_3	.8675 1847	.6931 2169	.3798 8531	.7775 0462
G_4	.8669 5566	.6931 4642	.3798 8549	.7775 0463
G_{10}	.8669 7299	.6931 4718	.3798 8549	.7775 0463
G_{16}	.8669 7299	.6931 4718	.3798 8549	.7775 0463
Exact value	.8669 7299	.6931 4718	.3798 8549	.7775 0463

G_2 = Gauss 2-point formula; G_3, 3-point; etc.

These examples, reading from left to right, are in order of increasing smoothness of integrand in the sense explained in Section 1.9. The function in the first column is piecewise continuous. It is in $R[0, 1]$ but not in $C[0, 1]$. The function $x^{1/2}$, which has a singularity at $x = 0$, is in $C[0, 1]$ but is not in $C^1[0, 1]$. The function $x^{3/2}$, which also has a singularity at $x = 0$, is in $C^1[0, 1]$ but is not in $C^2[0, 1]$.

The function $2/(2 + \sin 10\pi x)$ is analytic on $[0, 1]$ but has a singularity at $x = -(.05 + .0418i)$. The function $(1 + x^4)^{-1}$ has four singularities on $|z| = 1$ while $(1 + x)^{-1}$ has one singularity on $|z| = 1$. The function $(1 + e^x)^{-1}$ has a singularity at $x = \pi i$. The function $x(e^x - 1)^{-1}$ has an apparent singularity at $x = 0$ but, in reality, is analytic in $|z| < 2\pi$.

Notice that the rapidity of convergence increases as we move from the left-hand columns to the right. The results to the far left are quite poor. Of course, the piecewise continuous function should be integrated in two parts. Our point here is merely to exhibit the quality of the convergence. The difficulties in integrating a rapidly oscillating function such as $(2 + \sin 10\pi x)^{-1}$ are apparent from the figures.

2.7.1 *Integration Formulas of Gauss Type with Preassigned Abscissas*

Next we consider integration formulas of Gauss type with a certain number of preassigned abscissas. By this we mean a formula of the type

$$\int_a^b w(x)f(x)\,dx \approx \sum_{k=1}^{m} a_k f(y_k) + \sum_{k=1}^{n} w_k f(x_k), \qquad (2.7.1.1)$$

where the abscissas y_k are fixed and prescribed in advance and where the $m + 2n$ constants a_k, w_k, and x_k are to be determined so that the rule is exact for polynomials of the highest possible degree (that is, $m + 2n - 1$). We introduce the polynomials

$$r(x) = (x - y_1)(x - y_2)\cdots(x - y_m),$$
$$s(x) = (x - x_1)(x - x_2)\cdots(x - x_n). \qquad (2.7.1.2)$$

THEOREM. *The rule (2.7.1.1) is exact for all polynomials that are of degree $\le m + 2n - 1$ if and only if*

(a) *it is exact for all polynomials of degree $\le m + n - 1$,*
(b) $\int_a^b w(x)r(x)s(x)p(x)\,dx = 0$ *for every polynomial $p(x)$ of degree $\le n - 1$.*

Proof. The necessity of (a) is trivial. Let $p(x)$ have degree $\le n - 1$. Then $t(x) = r(x)s(x)p(x)$ is a polynomial of degree $\le m + 2n - 1$. Hence, if (2.7.1.1) is exact for such polynomials, then

$$\int_a^b w(x)t(x)\,dx = \sum_{k=1}^{m} a_k t(y_k) + \sum_{k=1}^{n} w_k t(x_k).$$

But $t(y_k) = 0$ $(k = 1, 2, \ldots, m)$ and $t(x_k) = 0$ $(k = 1, 2, \ldots, n)$. Thus the necessity of (b) follows.

Suppose, conversely, that (a) and (b) hold. Let $t(x)$ be a polynomial of degree $\leq m + 2n - 1$. Then it can be written in the form $t(x) = r(x)s(x)q(x) + v(x)$, where $q(x)$ is a polynomial of degree $\leq n - 1$ and $v(x)$ is a polynomial of degree $\leq m + n - 1$. Note that $t(y_k) = v(y_k)$ $(k = 1, 2, \ldots, m)$ and $t(x_k) = v(x_k)$ $(k = 1, 2, \ldots, n)$. Then

$$\int_a^b w(x)t(x)\,dx = \int_a^b w(x)[r(x)s(x)q(x) + v(x)]\,dx$$

$$= \int_a^b w(x)v(x)\,dx,$$

inasmuch as the first part vanishes, by (b). But, by (a),

$$\int_a^b w(x)v(x)\,dx = \sum_{k=1}^{m} a_k v(y_k) + \sum_{k=1}^{n} w_k v(x_k)$$

$$= \sum_{k=1}^{m} a_k t(y_k) + \sum_{k=1}^{n} w_k t(x_k).$$

This proves the sufficiency.

In order to proceed numerically, it is clear that we must determine $s(x) = (x - x_1) \cdots (x - x_n)$ as one of a family of polynomials that are orthogonal on $[a, b]$ with respect to the weight $w(x)r(x)$. Toward this end, a theorem of Christoffel is of importance.

THEOREM. *Let $p_n(x)$, $n = 0, 1, \ldots$ be orthonormal polynomials on $[a, b]$ with respect to the admissible weight $w(x)$. Let $r(x) = (x - y_1)(x - y_2) \cdots (x - y_m) \geq 0$ on $[a, b]$ and suppose that the y_i are distinct. Suppose that $q_n(x)$, $n = 0, 1, \ldots$ are orthogonal polynomials over $[a, b]$ with respect to $w(x)r(x)$. Then*

$$r(x)q_n(x) = \begin{vmatrix} p_n(x) & p_{n+1}(x) & \cdots & p_{n+m}(x) \\ p_n(y_1) & p_{n+1}(y_1) & \cdots & p_{n+m}(y_1) \\ \vdots & & & \vdots \\ p_n(y_m) & p_{n+1}(y_m) & \cdots & p_{n+m}(y_m) \end{vmatrix}. \tag{2.7.1.3}$$

Proof. It is clear that the right-hand side of (2.7.1.3) is in $\mathscr{P}_{n+m}$. Moreover, it vanishes for $x = y_1$, $x = y_2$, $\ldots$, $x = y_m$ (equal rows in the determinant). Hence, it can be written in the form $r(x)q_n(x)$, where $q_n(x) \in \mathscr{P}_n$. We show next that $q_n(x)$ is orthogonal to all members of class $\mathscr{P}_{n-1}$ with respect to the weight $w(x)r(x)$. Let $q(x) \in \mathscr{P}_{n-1}$. Then, since $r(x)q_n(x)$ is obviously a certain linear combination of $p_n(x), \ldots, p_{n+m}(x)$,

$$r(x)q_n(x) = c_1 p_n(x) + \cdots + c_{m+1} p_{n+m}(x).$$

Therefore we have

$$\int_a^b w(x)r(x)q_n(x)q(x)\,dx = \int_a^b w(x)[c_1 p_n(x) + \cdots + c_{m+1}p_{n+m}(x)]q(x)\,dx = 0, \tag{2.7.1.4}$$

in view of the fact that each orthonormal polynomial is orthogonal to polynomials of lower degree.

Finally, we show that the degree of $q_n(x)$ is precisely n. This will follow if we show that the coefficient of $p_{n+m}(x)$ does not vanish. Now, this coefficient is

$$c_{m+1} = \begin{vmatrix} p_n(y_1) & p_{n+1}(y_1) & \cdots & p_{n+m-1}(y_1) \\ \vdots & & & \vdots \\ p_n(y_m) & p_{n+1}(y_m) & \cdots & p_{n+m-1}(y_m) \end{vmatrix}. \tag{2.7.1.5}$$

If $c_{m+1} = 0$, we could find constants $d_1, d_2, \ldots, d_m$, not all zero, such that the polynomial

$$s(x) = d_1 p_n(x) + d_2 p_{n+1}(x) + \cdots + d_m p_{n+m-1}(x) \tag{2.7.1.6}$$

would vanish at $x = y_1$, $x = y_2, \ldots, x = y_m$. The polynomial $s(x)$ would therefore be of the form $s(x) = r(x)t(x)$, where $t(x) \in \mathscr{P}_{n-1}$. Furthermore $s(x)$ is obviously orthogonal to all elements of $\mathscr{P}_{n-1}$. Hence

$$0 = \int_a^b w(x)s(x)t(x)\,dx = \int_a^b w(x)r(x)t^2(x)\,dx.$$

Therefore $t(x) \equiv 0$, and this contradicts the fact that the d_i do not all vanish.

The two most common rules of Gauss type with preassigned abscissas have the preassigned abscissas at the end points of the interval. The weight $w(x) \equiv 1$ is used. These rules are called Radau and Lobatto integration. The specific formulas are as follows.

THEOREM (*Radau integration*). *Let* $f(x) \in C^{2n-1}[-1, 1]$. *Then*

$$\int_{-1}^{+1} f(x)\,dx = \frac{2}{n^2}f(-1) + \sum_{j=1}^{n-1} w_j f(x_j) + E, \tag{2.7.1.7}$$

where x_j *is the* jth *zero of*

$$\frac{P_{n-1}(x) + P_n(x)}{x - 1}, \qquad P(x) = \textit{Legendre polynomial}, \tag{2.7.1.8}$$

where

$$w_j = \frac{1}{n^2}\frac{1 - x_j}{[P_{n-1}(x_j)]^2} = \frac{1}{1 - x_j}\frac{1}{[P'_{n-1}(x_j)]^2}, \tag{2.7.1.9}$$

and where

$$E = E(f) = \frac{2^{2n-1}n}{[(2n-1)!]^3}[(n-1)!]^4 f^{(2n-1)}(\xi), \qquad -1 < \xi < 1. \tag{2.7.1.10}$$

THEOREM (*Lobatto integration*). *Let $f(x) \in C^{2n-2}[-1, 1]$. Then*

$$\int_{-1}^{+1} f(x)\,dx = \frac{2}{n(n-1)}[f(1) + f(-1)] + \sum_{j=2}^{n-1} w_j f(x_j) + E, \tag{2.7.1.11}$$

*where x_j is the $(j-1)$*th *zero of $P'_{n-1}(x)$, $P(x)$ = Legendre polynomial. Here*

$$w_j = \frac{2}{n(n-1)[P_{n-1}(x_j)]^2}, \qquad x_j \neq \pm 1, \tag{2.7.1.12}$$

and

$$E = E(f) = \frac{-n(n-1)^3 2^{2n-1}[(n-2)!]^4}{(2n-1)[(2n-2)!]^3} f^{(2n-2)}(\xi), \qquad -1 < \xi < 1. \tag{2.7.1.13}$$

Another rule of interest, corresponding to $w(x) = (1 - x^2)^{-1/2}$, is the Lobatto–Tschebyscheff integration rule of the first kind

$$\int_{-1}^{1} (1 - x^2)^{-1/2} f(x)\,dx = \frac{\pi}{n-1} \sum_{i=1}^{n}{}'' f(x_i) - \frac{\pi}{2^{2n-3}(2n-2)!} f^{(2n-2)}(\xi),$$
$$-1 < \xi < 1 \tag{2.7.1.14}$$

where $x_i = \cos[(i-1)\pi/(n-1)]$, $i = 1, \ldots, n$, and the double prime indicates that the first and last term of the sum are to be halved.

The approximate integration formulas of Radau and Lobatto are of use in the following situations.

(1) $f(\pm 1) = 0$ or any other known value.

(2) $f(x)$ displays peculiar behavior at $x = \pm 1$ such as an apparent singularity.

(3) If an $(n \times L)$-rule is set up (L for Lobatto), the functional values at the end of the $(k-1)$th panel coincide with those at the beginning of the kth panel; hence, the number of functional evaluations is reduced.

(4) The Radau formula is useful for solving the ordinary differential equation

$$y' = f(x, y) \tag{2.7.1.15}$$

as follows:

$$y(1) = y(-1) + \int_{-1}^{1} f(x, y)\,dx$$

$$\approx y(-1) + \frac{2}{n^2} f(-1, y(-1)) + \sum_{i=1}^{n-1} w_i f(x_i, y(x_i)). \qquad (2.7.1.16)$$

Here the x_i are the Radau abscissas and the w_i are the corresponding weights. If the values $f(x_i, y(x_i))$ are computed by some standard method, then the above approximation will "improve" the value of $y(1)$. See also Gates for further applications to differential equations.

(5) A Lobatto rule has been applied to the numerical solution of linear differential equations and Volterra's integral equation.

(6) In the numerical solution of ordinary differential equations, collocation with piecewise polynomials of degree k at k Gauss, Radau, or Lobatto points per subinterval leads to implicit Runge–Kutta formulas. The Radau points have particularly useful stability properties for stiff initial value problems (L-stability or stiff stability). Gauss and Lobatto points lead only to A-stable methods but have an advantage in solving boundary value problems because of their symmetry.

Example

Rule	$\int_0^1 x^{1/2}\,dx$	$\int_0^1 x^{3/2}\,dx$	$\int_0^1 dx/(1+x)$	$\int_0^1 dx/(1+x^4)$
L_4	0.6568 2580	0.4003 5217	0.6931 8182	0.8662 6092
L_{10}	0.6661 9841	0.4000 0199	0.6931 4718	0.8669 7299
R_6	0.6648 0585	0.4000 2032	0.6931 4718	0.8669 7523
$\bar{R}_6$	0.6671 5566	0.3999 8857	0.6931 4718	0.8669 7059
Exact value	0.6666 6667	0.4000 0000	0.6931 4718	0.8669 72987

Rule	$\int_0^1 dx/(1+e^x)$	$\int_0^1 x\,dx/(e^x-1)$	$\int_0^1 \frac{2\,dx}{2+\sin 10\pi x}$
L_4	0.3798 8574	0.7775 0465	1.1072 9967
L_{10}	0.3798 8549	0.7775 0463	1.1911 9517
R_6	0.3798 8549	0.7775 0463	1.3258 4956
$\bar{R}_6$	0.3798 8549	0.7775 0463	0.8793 0050
Exact value	0.3798 8549	0.7775 0463	1.1547 0054

Note: L_4 = Lobatto 4-point formula, R_6 = Radau 6-point left-hand formula,
L_{10} = Lobatto 10-point formula, $\bar{R}_6$ = Radau 6-point right-hand formula.

References. Ascher and Weiss [1]; Bakker [1]; Butcher [1]; Christoffel [1]; Gates [1]; Gautschi [9]; Henrici [1, pp. 101–102]; Hilderbrand [1, Chap. 8]; Ioakimidis [2]; Jain and Sharma [1]; Kopal [1, Chap. 7]; Michels [1]; Prothero and Robinson [1]; Rabinowitz [1]; Scherer [1]; Shohat [2]; Stroud and Secrest [2]; Szegö [2, pp. 29–30].

Gauss-type integration formulas involving derivatives,

$$\int_a^b f(x)\,dx \approx \sum_{k=1}^{m} \sum_{j=1}^{n_k} A_{kj} f^{(j)}(a_k), \qquad -\infty < a < b < \infty, \quad (2.7.1.17)$$

have also been developed. See Section 4.3.5.

References. Gautschi [9]; Ghizzetti and Ossicini [1]; Popoviciu [1]; Rebolia [1]; Stancu and Stroud [1]; Stroud and Stancu [1]; L. Tchakaloff [1].

A glance at the error terms for the n-point Gauss rule G_n and the $(n+1)$-point Lobatto rule L_{n+1} shows that the coefficients of $f^{(2n)}(\xi)$ in these two rules are almost equal and of opposite sign. Thus, if $f^{(2n)}$ does not change sign in the integration interval, the value of the integral $I(f)$ is "bracketed" between the two values. The weighted mean

$$\frac{n+1}{2n+1} G_n + \frac{n}{2n+1} L_{n+1}$$

will usually be a better approximation to $I(f)$ than either G_n or L_{n+1}. For large values of n and for well-behaved functions, this will generally be true.

2.7.1.1 *The Kronrod Scheme*

One of the objections to the use of the family of Gauss rules is that when one proceeds from G_m to G_n with $n > m$, all functional values except possibly 0 are discarded inasmuch as the abscissas of G_n do not include those of G_m. This objection can be overcome to some extent by making use of the first theorem of Section 2.7.1 combined with a certain set of polynomials introduced by Stieltjes. If we start with m points $y_1, \ldots, y_m$ and add *any* n points $x_1, \ldots, x_n$ distinct from the y's, we can, of course, obtain an interpolatory rule for the combined set of abscissas $(y_1, \ldots, y_m; x_1, \ldots, x_n)$. Such a rule will be exact for $\mathscr{P}_{m+n-1}$ at least. However, the above-mentioned theorem suggests the possibility of selecting $x_1, \ldots, x_n$ so that polynomials of maximal degree will be integrated exactly. Kronrod starts with an n-point Gauss rule G_n, adds $n+1$ abscissas and arrives at a rule exact for $\mathscr{P}_{3n+1}$ if n is even and $\mathscr{P}_{3n+2}$ if n is odd. One price that must be paid, however, is that the weights associated with G_n are not conserved in the process of extension. Here are the theoretical details.

Let $p_n(t)$ designate the Legendre polynomials. From (1.12.20), and for appropriate constants b, c, d, e,

$$q_n(z) = \int_{-1}^{+1} \frac{p_n(t)}{z-t}\,dt = \frac{1}{z^{n+1}}\left(c_n + \frac{c_{n+1}}{z} + \cdots\right),$$

so that

$$\begin{aligned}\frac{1}{q_n(z)} &= z^{n+1}\left(d_n + \frac{d_{n+1}}{z} + \cdots\right)\\ &= E_n(z) + \frac{e_1}{z} + \frac{e_2}{z^2} + \cdots.\end{aligned} \tag{2.7.1.1.1}$$

where $E_n(z) \in \mathscr{P}_{n+1}$.

It follows that

$$q_n(z)E_n(z) = 1 + \frac{b_1}{z^{n+2}} + \frac{b_2}{z^{n+3}} + \cdots. \tag{2.7.1.1.2}$$

Now the functions of the second kind $q_n(z)$ satisfy a three term recurrence of the form

$$zq_n(z) = \alpha_n q_{n-1}(z) + \beta_n q_{n+1}(z); \qquad zq_0(z) = 1 + q_1(z). \tag{2.7.1.1.3}$$

The function $q_n(z)E_n(z)$ is a linear combination of $q_n(z)$, $zq_n(z), \ldots, z^{n+1}q_n(z)$ and hence by (2.7.1.1.3) a linear combination of $1, q_0(z), q_1(z), \ldots, q_{2n+1}(z)$. But in view of (2.7.1.1.2) and (1.12.20), $q_0, \ldots, q_n$ must drop out leaving us with

$$q_n(z)E_n(z) = 1 + h_0 q_{n+1}(z) + h_1 q_{n+2}(z) + \cdots + h_n q_{2n+1}(z). \tag{2.7.1.1.4}$$

Applying (1.12.21) to this equation, we obtain

$$p_n(x)E_n(x) = h'_0 p_{n+1}(x) + h'_1 p_{n+2}(x) + \cdots + h'_n p_{2n+1}(x). \tag{2.7.1.1.5}$$

If we multiply this by x^k, $0 \le k \le n$ and integrate, then in view of the orthogonality of $p_{n+j}(x)$ we obtain

$$\int_{-1}^{+1} p_n(x)E_n(x)x^k\,dx = 0, \qquad 0 \le k \le n. \tag{2.7.1.1.6}$$

In addition to the orthogonality of the Stieltjes polynomial $E_n(x)$ with respect to the oscillatory weight factor $p_n(x)$, we need to know this crucial fact established by Szegö:

THEOREM. *The zeros of $E_n(x)$ are real, simple, and located in the interior of $[-1, 1]$. They are separated by the zeros of $p_n(x)$.*

Designate the zeros of $p_n(x)$ by $y_1, \ldots, y_n$ and those of $E_n(x)$ by $x_1, \ldots, x_{n+1}$. From the first theorem of Section 2.7.1 (with $m = n, n = n + 1$, $a = -1, b = 1, w(x) \equiv 1, r(x) = p_n(x), s(x) = E_n(x)$), it follows that

$$\int_{-1}^{1} f(x)\,dx \approx \sum_{k=1}^{n} a_k f(y_k) + \sum_{k=1}^{n+1} b_k f(x_k) \qquad (2.7.1.1.7)$$

will be exact for $f \in \mathscr{P}_{3n+1}$ if and only if it is exact for $f \in \mathscr{P}_{2n}$. Since the $2n + 1$ abscissas are distinct, we may determine weights a_k and b_k by interpolation or otherwise which make (2.7.1.1.7) exact for $\mathscr{P}_{2n}$ and hence for $\mathscr{P}_{3n+1}$. Monegato [2] has shown that these weights are always positive.

Tables of abscissas and weights and details of computation can be found in Kronrod. More stable algorithms have been given by Patterson and by Piessens and Branders. The latter have published a FORTRAN program implementing their algorithm.

Kronrod extensions exist not only for the usual Gauss rules ($w(x) = 1$) but also for Gauss rules with respect to the weight function $(1 - x^2)^\mu$, $-\frac{1}{2} \le \mu \le \frac{3}{2}$, as well as for Lobatto integration with respect to the same weight function with $-1 < \mu \le \frac{1}{2}$. In the Lobatto case, the Kronrod extension to an $(n + 1)$-point rule has $2n + 1$ points and is exact for $f \in \mathscr{P}_{3n}$, n odd, and $f \in \mathscr{P}_{3n+1}$, n even. For the Radau case, Kronrod extensions have been computed for values of n up to 40 but no theoretical existence results are known. Kronrod extensions to Gauss–Laguerre and Gauss–Hermite rules probably do not exist except for isolated small values of n.

For $\mu = \frac{1}{2}$, the Gauss rule is the Gauss–Chebyshev rule of the second kind. Its Kronrod extension is also such a rule with $2n + 1$ points and hence is exact for $f \in \mathscr{P}_{4n+1}$. Similarly, for $\mu = -\frac{1}{2}$, the Gauss rule is the Gauss–Chebyshev rule of the first kind, and its Kronrod extension is a Lobatto–Chebyshev rule with $2n + 1$ points which is exact for $f \in \mathscr{P}_{4n-1}$. This rule coincides with the Kronrod extension of the $(n + 1)$-point Lobatto–Chebyshev rule. Apart from these special cases, the Kronrod extensions of Gauss and Lobatto rules with weight function $(1 - x^2)^\mu$ are of exact degree $3n + 1$, $3n + 2$ (Gauss), and $3n$, $3n + 1$ (Lobatto) as the case may be.

Suppose that one has computed $G_n(f)$ and augmented the abscissas by $n + 1$ points to arrive at a Kronrod scheme $K_{2n+1}(f)$. Suppose, further, that these two values do not agree sufficiently. How should one proceed? Patterson suggests that one iterate the procedure (via the first theorem of Section 2.7.1) by adding $2n + 2$ abscissas and arriving at a rule P_{4n+3} which will be exact for $\mathscr{P}_{6n+5}$. The existence theory here is incomplete, but the sequence $G_3, K_7, P_{15}, P_{31}, \ldots, P_{255}$ has been computed. Each P_n is exact for $\mathscr{P}_{(3n+1)/2}$. Abscissas and weights can be found in Patterson.

Similarly, starting with G_{10} and K_{21}, P_{43} and P_{87} have been computed for use in one of the programs in QUADPACK (Piessens *et al.*).

References. Atkinson [4, pp. 243–248]; Baratella [1]; Barrucand [2]; Cranley and Patterson [3]; Kronrod [1]; Monegato [1–4, 6]; Patterson [1, F1]; Piessens and Branders [3, F1]; Piessens, de Doncker, Überhuber, and Kahaner [1]; Rabinowitz [12]; Szegö [1].

Another idea advocated by Patterson is to build up successively toward a final high-accuracy rule of $2^r + 1$ points. Suppose, as an example, we set as our goal a Gauss or Lobatto rule of 65 points. Take the Gauss or Lobatto abscissas and successively strike out every second one. In this way we obtain interlaced subsets of 3, 5, 9, 17, 33 points. Using these sets of points as abscissas, determine the weights corresponding to interpolatory rules which will integrate exactly functions in class $\mathscr{P}_3, \mathscr{P}_5, \mathscr{P}_9, \mathscr{P}_{17}, \mathscr{P}_{33}$. Experience has shown these weights to be nonnegative.

The weights corresponding to these (and all other) interpolatory rules may be computed numerically as follows. Designate the abscissas by $x_1, \ldots, x_n$. From (2.5.6), (1.11.10), (1.11.7), the weights are given by

$$w_i = \int \ell_i(x)\, dx$$

where

$$\ell_i(x) = \frac{(x - x_1)\cdots(x - x_{i-1})(x - x_{i+1})\cdots(x - x_n)}{(x_i - x_1)\cdots(x_i - x_{i-1})(x_i - x_{i+1})\cdots(x_i - x_n)}.$$

Now apply a Gauss rule of order at least $[(n + 1)/2]$.

Reference. Patterson [2]; Rabinowitz, Kautsky, and Elhay [1].

A FORMAC program for evaluating the weights and error terms for interpolatory rules corresponding to prescribed weighting functions has been given by Engels and Meuer. If the moments of the weighting function can be evaluated as rational numbers, the program yields exact numbers.

Reference. Engels and Meuer [1].

2.7.2 *The Algebraic Approach to the Gauss Integration Formulas*

Let $w(x)$ be admissible. It is desired to find the $2n$ values $w_1, w_2, \ldots, w_n$, $x_1, x_2, \ldots, x_n$ such that

$$\int_a^b w(x)f(x)\, dx = \sum_{i=1}^{n} w_i f(x_i) \tag{2.7.2.1}$$

for $f(x) = 1, x, x^2, \ldots, x^{2n-1}$. Writing out each of these conditions, and using the abbreviation $\int_a^b w(x)x^j\,dx = m_j$, we obtain the system below.

$$\begin{aligned}
m_0 &= w_1 + w_2 + \cdots + w_n, \\
m_1 &= w_1 x_1 + w_2 x_2 + \cdots + w_n x_n, \\
m_2 &= w_1 x_1^2 + w_2 x_2^2 + \cdots + w_n x_n^2, \\
&\vdots \\
m_{2n-1} &= w_1 x_1^{2n-1} + w_2 x_2^{2n-1} + \cdots + w_n x_n^{2n-1}.
\end{aligned} \tag{2.7.2.2}$$

We define

$$p(x) = (x - x_1)(x - x_2)\cdots(x - x_n) = \sum_{k=0}^{n} c_k x^k, \qquad c_n = 1. \tag{2.7.2.3}$$

Now we have the following:

$$\begin{aligned}
m_0 c_0 &= w_1 c_0 + w_2 c_0 + \cdots + w_n c_0, \\
m_1 c_1 &= w_1 c_1 x_1 + w_2 c_1 x_2 + \cdots + w_n c_1 x_n, \\
m_2 c_2 &= w_1 c_2 x_1^2 + w_2 c_2 x_2^2 + \cdots + w_n c_2 x_n^2, \\
&\vdots \\
m_n c_n &= w_1 c_n x_1^n + w_2 c_n x_2^n + \cdots + w_n c_n x_n^n.
\end{aligned} \tag{2.7.2.4}$$

Hence

$$\begin{aligned}
m_0 c_0 + m_1 c_1 + m_2 c_2 + \cdots + m_n c_n = {} & w_1(c_0 + c_1 x_1 + \cdots + c_n x_1^n) \\
& + w_2(c_0 + c_1 x_2 + \cdots + c_n x_2^n) \\
& + \cdots + w_n(c_0 + c_1 x_n + \cdots + c_n x_n^n).
\end{aligned}$$

Since a typical term in parentheses is $\sum_{k=0}^{n} c_k x_i^k = p(x_i) = 0$, we have

$$m_0 c_0 + m_1 c_1 + \cdots + m_n c_n = 0.$$

Treating the second, third, ..., $(n + 1)$th equations of (2.7.2.2) in the same way, we obtain $m_1 c_0 + m_2 c_1 + \cdots + m_{n+1} c_n = 0$. Proceeding similarly, we obtain the system (since $c_n = 1$) of n equations in the quantities $c_0, \ldots, c_{n-1}$:

$$\begin{aligned}
m_0 c_0 + m_1 c_1 + \cdots + m_{n-1} c_{n-1} &= -m_n, \\
m_1 c_0 + m_2 c_1 + \cdots + m_n c_{n-1} &= -m_{n+1}, \\
m_2 c_0 + m_3 c_1 + \cdots + m_{n+1} c_{n-1} &= -m_{n+2}, \\
&\vdots \\
m_{n-1} c_0 + m_n c_1 + \cdots + m_{2n-2} c_{n-1} &= -m_{2n-1}.
\end{aligned} \tag{2.7.2.5}$$

The determinant of this system is

$$d = |m_{i+j}| = \left| \int_a^b w(x)x^i \cdot x^j \, dx \right|,$$

which is the Gram determinant of the functions $1, x, x^2, \ldots, x^{n-1}$. In view of the fact that these functions are linearly independent, it follows that $d \neq 0$ (see, for example, Davis [6, p. 178]). Hence, the system (2.7.2.5) may be solved uniquely for the constants $c_0, \ldots, c_{n-1}$. From (2.7.2.3) we may therefore solve for the abscissas $x_1, \ldots, x_n$ [which, as we know, will be real, simple, and located in (a, b)]. Having determined the abscissas, we may find the weights from (2.7.2.2).

A modification of this device may be used to obtain formulas of Gauss type with one or more fixed points. For example, let us find the $2n - 1$ values $w_1, w_2, \ldots, w_n, x_2, x_3, \ldots, x_n$ such that

$$\int_a^b w(x)f(x)\,dx = w_1 f(x_1) + \sum_{k=2}^{n} w_k f(x_k) \tag{2.7.2.6}$$

is valid for $f(x) = 1, x, x^2, \ldots, x^{2n-2}$. It is assumed that x_1 is a *fixed* number satisfying $x_1 \geq b$ or $x_1 \leq a$.

The defining equations are now

$$m_k = w_1 x_1^k + \sum_{i=2}^{n} w_i x_i^k, \qquad k = 0, 1, \ldots, 2n - 2. \tag{2.7.2.7}$$

By subtracting appropriate multiples of equations (2.7.2.7) from one another, we obtain

$$\mu_k = m_{k+1} - x_1 m_k = \sum_{i=2}^{n} w_i(x_i - x_1)x_i^k, \qquad k = 0, 1, \ldots, 2n - 3. \tag{2.7.2.8}$$

If we now set

$$p(x) = (x - x_2)(x - x_3)\cdots(x - x_n) = \sum_{k=0}^{n-1} c_k x^k \tag{2.7.2.9}$$

and proceed as before, we have

$$\sum_{k=0}^{n-1} c_k \mu_{k+j} = \sum_{i=2}^{n} w_i(x_i - x_1)p(x_i) = 0, \qquad j = 0, 1, \ldots, n - 2. \tag{2.7.2.10}$$

The determinant of this system $d = |\mu_{k+j}|$ is

$$\det\left(\int_a^b w(x)(x^{k+j+1} - x_1 x^{k+j})\,dx\right) = \det\left(\int_a^b w(x)(x - x_1)x^k \cdot x^j \, dx\right).$$

This is the Gram determinant of the powers 1, x, x^2, ... with respect to the semidefinite weighting function $w(x)(x - x_1)$. Hence, the determinant does not vanish. The system (2.7.2.10) may therefore be solved, and we can proceed as before. This method of determining Gauss integration formulas is not advocated for numerical work inasmuch as the relevant matrices are often ill-conditioned.

Furthermore, when n is large, the evaluation of a polynomial in the form $\sum c_k x^k$ may be an ill-conditioned process so that its zeros cannot be determined accurately. This ill-conditioning is very noticeable in the case of the classical orthogonal polynomials discussed in Section 1.13, where the coefficients c_k alternate in sign and grow rapidly in magnitude as n increases. A remedy, suggested by Sack and Donovan, is to start with modified moments

$$\int_a^b w(x)q_j(x)\,dx = m_j^*$$

and then work with

$$\sum c_j^* q_j(x).$$

Here $\{q_j(x)\}$ is some convenient set of polynomials, usually orthogonal polynomials.

References. Gautschi [3, 5]; Hildebrand [1, pp. 351–353]; Rice [1]; Sack and Donovan [1]; Wilkinson [1].

2.7.3 *Determination of Gaussian Integration Formulas through Orthogonal Polynomials*

A good way of determining numerically the Gauss-type integration formulas is through the three-term recurrence formula for orthogonal polynomials. The zeros of the polynomials are real and simple, and can be calculated expeditiously via Newton–Raphson iteration or any polynomial root-finding algorithm designed to work well when all the zeros are real and simple. An algorithm that does not require good initial values and computes all the zeros simultaneously is desirable. However, note that while in theory the zeros are simple, in many cases they tend to cluster around the endpoints of the interval, so that in practice it may appear that there are multiple zeros.

In numerous instances, for example in the case of the Jacobi weight function $(1 - x)^\alpha(1 + x)^\beta$, the three-term recurrence is known explicitly. To see how this method works, let us consider the classical Gauss case

($\alpha = \beta = 0$). The Legendre polynomials, $P_n(x)$, are orthogonal over $[-1, 1]$ and satisfy the recurrence

$$
\begin{aligned}
P_0(x) &= 1, \\
P_1(x) &= x, \\
nP_n(x) &= (2n-1)xP_{n-1}(x) - (n-1)P_{n-2}(x), \qquad n \geq 2.
\end{aligned} \tag{2.7.3.1}
$$

Recurrence in the form

$$
\begin{aligned}
P_{n+1}(x) &= [xP_n - P_{n-1}] - \frac{1}{n+1}[xP_n - P_{n-1}] + xP_n \\
&\equiv \frac{n}{n+1}[xP_n(x) - P_{n-1}(x)] + xP_n(x)
\end{aligned} \tag{2.7.3.2}
$$

—which saves a multiplication—provides a stable mode of computing the Legendre polynomials. (See Lebedev and Baburin.) The zeros of $P_n(x)$, x_{kn}, are given approximately by

$$x_{kn} \approx x_{kn}^{(0)} = \cos\theta_{kn}, \qquad k = 1, 2, \ldots, n,$$

where

$$\frac{j_k}{[(n+\frac{1}{2})^2 + \frac{1}{4}c]^{1/2}} < \theta_{kn} < \frac{j_k}{n+\frac{1}{2}}, \qquad k = 1, 2, \ldots, n. \tag{2.7.3.3}$$

Here j_k are the successive zeros of the Bessel function $J_0(x)$, and $c = 1 - (2/\pi)^2$. (For further information on the asymptotic distribution of abscissas and weights of Gaussian type, see Davis and Rabinowitz [6], Ossicini, Whitney. See also Section 1.13.)

A sequence of approximations to each zero, $x_{kn}^{(i)}$, is defined by the Newton iteration scheme

$$x_{kn}^{(i+1)} = x_{kn}^{(i)} - \frac{P_n(x_{kn}^{(i)})}{P_n'(x_{kn}^{(i)})}, \qquad i = 0, 1, 2, \ldots. \tag{2.7.3.4}$$

The derivative $P_n'(x)$ may be computed through the orthogonal polynomials themselves:

$$(1 - x^2)P_n'(x) = n[P_{n-1}(x) - xP_n(x)]. \tag{2.7.3.5}$$

If $P_n^*(x)$ are the *normalized* Legendre polynomials, that is if $\int_{-1}^{1} (P_n^*(x))^2\,dx = 1$, and if

$$P_n^*(x) = k_n x^n + \cdots, \tag{2.7.3.6}$$

then the Gaussian weights w_{kn} are given by

$$w_{kn} = -\frac{k_{n+1}}{k_n} \frac{1}{P^*_{n+1}(x_{kn}) P^{*\prime}_n(x_{kn})}. \tag{2.7.3.7}$$

The alternative formula

$$w_{kn} = \frac{2(1 - x^2_{kn})}{[nP_{n-1}(x_{kn})]^2} \tag{2.7.3.8}$$

may be used.

A refinement of the Newton scheme may also be used conveniently. We can use, instead of (2.7.3.4), the iteration

$$x^{(i+1)}_{kn} = x^{(i)}_{kn} - \frac{P_n(x^{(i)}_{kn})}{P'_n(x^{(i)}_{kn})}\left(1 + \frac{P_n(x^{(i)}_{kn})}{P'_n(x^{(i)}_{kn})} \frac{P''_n(x^{(i)}_{kn})}{2P'_n(x^{(i)}_{kn})}\right), \tag{2.7.3.9}$$

which is more rapidly (in fact, cubically) convergent. In view of the fact that orthogonal polynomials satisfy a second-order linear differential equation, the values of P''_n are available as linear combinations of P_n and P'_n. In a similar fashion, formulas with even higher-order convergence can be used. These use higher derivatives of P_n which can be easily computed as linear combinations of lower-order derivatives.

A particularly efficient fifth-order iteration scheme was developed and used by Lether. This scheme has been generalized by Gatteschi for ultraspherical polynomials.

An alternative approximation to (2.7.3.3) that is more suitable for a general subroutine for computing Gauss rules is given by

$$\theta_{kn} = \frac{4k - 1}{4n + 2}\pi + \frac{n - 1}{8n^3} \cot \frac{4k - 1}{4n + 2}\pi + O(n^{-4}). \tag{2.7.3.3a}$$

or equivalently

$$x_{kn} = (1 - \tfrac{1}{8}n^{-2} + \tfrac{1}{8}n^{-3}) \cos \frac{4k - 1}{4n + 2}\pi + O(n^{-4}). \tag{2.7.3.3b}$$

More accurate starting values are given by Lether for Legendre polynomials and by Gatteschi for ultraspherical polynomials. In particular, these are applicable to the computation of the Lobatto integration rules (2.7.1.11).

Using the starting values given by (2.7.3.3b) and the iteration scheme (2.7.3.9), it requires one iteration to compute Gauss abscissas correct to single precision (7 figures) and two iterations for double precision accuracy. For the computation of the weights, (2.7.3.8) is used with $P_{n-1}(x_{kn})$ approximated by a Taylor series using the values of P_n and P_{n-1} computed in the last iteration and a formula similar to (2.7.3.5) to compute P'_{n-1}. The subroutine GRULE implementing these ideas is given in Appendix 2.†

References. Cadete [1]; Davis [6, pp. 246–253]; Davis and Rabinowitz [3, 4, 6]; Engels [13]; Gatteschi [1]; Gautschi [8]; Hofsommer [1]; Lebedev and Baburin [1]; Lether [9]; Ossicini [1]; Shao, Chen, and Frank [2]; Szegö [2]; Tricomi [2, p. 185]; Whitney [1]; Wynn [1].

For weight functions $w(x)$ whose related theory of orthogonal polynomials is not well developed, there are several difficulties that must be met. In the first place, while there is a three-term recurrence formula for all sets of orthogonal polynomials, its form may not be known in advance and might have to be calculated step by step. A convenient way of doing this is as follows, in which we shall use the inner product notation $(f, g) = \int_a^b w(x)f(x)g(x)\,dx$. We write the orthogonal polynomials as

$$p_n(x) = A_n x^n + \cdots, \qquad n = 0, 1, \ldots, \tag{2.7.3.10}$$

where the constants $A_0, A_1, \ldots$ may be selected to suit our convenience. The selection $A_0 = 1$, $A_n = 2^{n-1}$, $n \geq 1$, for example, corresponds to the Tschebyscheff polynomials $T_n(x)$ under the normalization $T_n(1) = 1$ and is probably a good choice for computation. Then we have

$$p_0(x) = A_0, \qquad p_1(x) = A_1\left(x - \frac{(xp_0, p_0)}{(p_0, p_0)}\right),$$

$$\begin{aligned} p_n(x) = {} & \frac{A_n}{A_{n-1}}\left(x - \frac{(xp_{n-1}, p_{n-1})}{(p_{n-1}, p_{n-1})}\right)p_{n-1}(x) \\ & - \frac{A_n A_{n-2}}{A_{n-1}^2}\frac{(p_{n-1}, p_{n-1})}{(p_{n-2}, p_{n-2})}p_{n-2}(x), \qquad n = 2, 3, \ldots. \end{aligned} \tag{2.7.3.11}$$

The inner products that occur in this recusion may all be evaluated as linear combinations of the basic quantities

$$(x^i, x^j) = (x^{i+j}, 1) = \int_a^b w(x)x^{i+j}\,dx = m_{i+j}.$$

† A word is in order on the use of programs to generate rules in numerical analysis. On the plus side, it may be much simpler and more amenable to checking to insert a program that generates (let us say) Gauss weights and abscissas than laboriously to punch up precomputed values. Such a rule generator clearly can have great flexibility. On the minus side, the practice is accompanied by a loss of control. Programs are produced which are more and more general. The casual user may therefore have to browse through a six-page imperfectly written set of instructions to find out what values to assign the key parameters in *his* special case. Such programs may not have enough "fools' exits" built in or may be of such generality that fringe areas of applicability and possible pitfalls defy description.

These are the moments of the weight function $w(x)$. It may be that the moments m_i are evaluable in closed form. This is the case with

$$\int_0^1 \left(\log\frac{1}{x}\right)x^n\,dx = \frac{1}{(n+1)^2}.$$

But if $w(x)$ is sufficiently complicated, for example if

$$m_n = \int_0^1 \frac{x^n}{\sqrt{x} + e^x}\,dx,$$

then special analysis or a numerical "bootstrap" operation may be required to determine these basic numbers with sufficient accuracy.

Another complication is that unusual weights $w(x)$ may induce an unknown or an unusual asymptotic distribution of zeros of the orthogonal polynomials in $[a, b]$. It may require an initial scanning of the interval to locate the zeros crudely and, hence, in such a case the bisection routine for the solution of equations may be appropriate. One can also take advantage of the "separation" property of the zeros of successive orthogonal polynomials.

THEOREM. *Let $a < x_1 < x_2 < \cdots < x_n < b$ be the zeros of the orthogonal polynomial $p_n(x)$. Then in each interval (a, x_1), (x_1, x_2), $\ldots$, (x_{n-1}, x_n), (x_n, b) there is precisely one zero of the orthogonal polynomial $p_{n+1}(x)$.*

Proof. If $p_n^*(x) = k_n x^n + \cdots$, $(k_n > 0)$ are the orthonormalized polynomials, then by the Christoffel–Darboux formula (1.12.11) we obtain

$$\sum_{k=0}^{n} (p_k^*(x))^2 = \frac{k_n}{k_{n+1}} (p_{n+1}^{*\prime}(x)p_n^*(x) - p_n^{*\prime}(x)p_{n+1}^*(x)). \qquad (2.7.3.12)$$

This implies that

$$p_{n+1}'(x)p_n(x) - p_n'(x)p_{n+1}(x) > 0. \qquad (2.7.3.13)$$

It follows from (2.7.3.13) that if x_i and x_{i+1} are two adjacent zeros of $p_n(x)$, then

$$p_n'(x_i)p_{n+1}(x_i) < 0,$$
$$p_n'(x_{i+1})p_{n+1}(x_{i+1}) < 0. \qquad (2.7.3.14)$$

Since the zeros of p_n are simple, $p_n'(x_i)$ and $p_n'(x_{i+1})$ must have opposite signs and from (2.7.3.14), therefore, $p_{n+1}(x_i)$ and $p_{n+1}(x_{i+1})$ must have opposite signs. Thus $p_{n+1}(x)$ must have at least one zero between x_i and x_{i+1}. This accounts for at least $n - 1$ zeros of $p_{n+1}(x)$.

Now, $p_n'(x_n) > 0$ and hence from (2.7.3.14), $p_{n+1}(x_n) < 0$. Since $p_{n+1}(b) > 0$, there must be a zero of $p_{n+1}(x)$ between x_n and b. A similar argument shows that there must be a zero of $p_{n+1}(x)$ between a and x_1. Thus, each of the $n + 1$ intervals $(a, x_1), (x_1, x_2), \ldots, (x_{n-1}, x_n), (x_n, b)$ has at least one zero and, hence, precisely one zero of $p_{n+1}(x)$.

Gautschi [3, A1] suggests that the inner product $(f, g) = \int_a^b w(x)f(x)g(x)\,dx$ be evaluated by a rule

$$(f, g) \approx R_N(wfg) = \sum_{k=1}^{N} w_{Nk} w(x_{Nk}) f(x_{Nk}) g(x_{Nk}). \qquad (2.7.3.15)$$

On the basis of these approximate values of the inner products, the orthogonal polynomials corresponding to $w(x)$ may be computed approximately by recurrence and hence, approximate values for the abscissas and weights of the Gauss integration formula for $w(x)$ can be obtained. As $N \to \infty$, these approximations will improve; or, early approximate values may be used as starting values in the solution by Newton's method of the system

$$\sum_{j=1}^{n} w_j q_k(x_j) = m_k^*, \qquad k = 0, 1, \ldots, 2n - 1,$$

where the m_k^* are the moments relative to a convenient set of polynomials $\{q_k(x)\}$.

Gautschi [3, A1] observes that the rule R_N ought to be such that its convergence is reasonably fast even when $w(x)$ has singularities, that it be easy to generate for large N, and that the interval $[x_{N1}, x_{NN}]$ spanned by its abscissas should include, insofar as possible, the abscissas $\{x_{nk}\}$ under determination. One should take, of course, $N \gg n$. As a compromise choice, it is recommended that rules of interpolatory type based upon the zeros of the Tchebyscheff polynomial $T_N(x)$ be used. Written for the interval $[-1, 1]$, this formula is

$$\int_{-1}^{1} \phi(x)\,dx \approx R_N(\phi) = \sum_{k=1}^{N} w_{Nk} \phi(x_{Nk})$$

where

$$x_{Nk} = \cos\theta_{Nk}, \qquad \theta_{Nk} = \frac{2k-1}{2N}\pi, \qquad k = 1, 2, \ldots, N$$

$$w_{Nk} = \frac{2}{N}\left\{1 - 2\sum_{m=1}^{[N/2]} \frac{\cos(2m\theta_{Nk})}{4m^2 - 1}\right\} \qquad (2.7.3.16)$$

See Section 2.5.5.

References. Davis [6, pp. 238–239]; Gautschi [3, 5, 9, A1]; T. E. Price [2]; Szegö [2, p. 46].

2.7.4 Determination of Gaussian Integration Formulas through the Use of Continued Fractions

It is also possible to obtain Gaussian integration formulas beginning with a knowledge of the coefficients in the continued fraction expansion of the transform

$$\int_a^b \frac{w(x)}{z-x}\,dx = \frac{1|}{|z} - \frac{q_1|}{|1} - \frac{e_1|}{|z} - \frac{q_2|}{|1} - \frac{e_2|}{|z} - \cdots. \qquad (2.7.4.1)$$

Let $R_n(z)/S_n(z)$ be the successive convergents of this continued fraction. Then $S_n(z)$ are the polynomials orthogonal on $[a, b]$ with respect to $w(x)$. The poles of these convergents are therefore the abscissas x_{kn} of the Gauss integration rule of order n. Moreover

$$\frac{R_n(z)}{S_n(z)} = \text{const} \sum_{k=1}^{n} \frac{w_{kn}}{z - x_{kn}}. \qquad (2.7.4.2)$$

When a modified Q-D algorithm is applied to these convergents, it yields the abscissas and weights. We cannot enter into the details here. Thacher reports numerical instability for $\int_0^\infty x^{-1/2}e^{-x}f(x)\,dx$, using this method.

References. Rutishauser [1, A1]; Szegö [2, pp. 54–57].

2.7.5 Computation of Gauss Rules by Eigenvalue Methods

Eigenvalue theory and programs may be brought to bear on this problem. If $p_0^*, p_1^*, \ldots, p_N^*$ are a set of real orthonormal polynomials, then they satisfy the three-term recurrence

$$p_n^*(x) = (a_n x + b_n)p_{n-1}^*(x) - c_n p_{n-2}^*(x), \qquad n = 2, 3, \ldots, N. \qquad (2.7.5.1)$$

If the first two coefficients of p_n^* are designated by k_n and s_n:

$$p_n^*(x) = k_n x^n + s_n x^{n-1} + \cdots \qquad (2.7.5.2)$$

it is known that

$$a_n = \frac{k_n}{k_{n-1}}, \qquad b_n = a_n\left(\frac{s_n}{k_n} - \frac{s_{n-1}}{k_{n-1}}\right), \qquad c_n = \frac{k_n k_{n-2}}{k_{n-1}^2}. \qquad (2.7.5.3)$$

We may write (2.7.5.1) in the form

$$x p_{n-1}^*(x) = \frac{1}{a_n} p_n^*(x) - \frac{b_n}{a_n} p_{n-1}^*(x) + \frac{c_n}{a_n} p_{n-2}^*(x),$$
$$n = 2, 3, \ldots, N. \qquad (2.7.5.4)$$

Since $1/a_n = k_{n-1}/k_n$ and $c_n/a_n = k_{n-2}/k_{n-1}$, we may write (2.7.5.4) in the form

$$xp_{n-1}^*(x) = \alpha_n p_n^*(x) + \beta_n p_{n-1}^*(x) + \alpha_{n-1} p_{n-2}^*(x) \tag{2.7.5.5}$$

where $\alpha_n = k_{n-1}/k_n$ and $\beta_n = -b_n/a_n$.

Introduce the matrix notation

$$P(x) = [p_0^*(x), p_1^*(x), \ldots, p_{N-1}^*(x)]^{\mathrm{T}}$$

$$E = [0, 0, \ldots, 1]^{\mathrm{T}} \tag{2.7.5.6}$$

$$J = \begin{bmatrix} \beta_1 & \alpha_1 & 0 & 0 & \cdots & \\ \alpha_1 & \beta_2 & \alpha_2 & 0 & \cdots & \\ 0 & \alpha_2 & \beta_3 & \alpha_3 & \cdots & \\ \vdots & & & & & \\ 0 & 0 & & & \cdots & \beta_N \end{bmatrix}.$$

Superscript T designates the transpose. Then, (2.7.5.5) takes the form

$$xP(x) = JP(x) + \alpha_N p_N^*(x)E. \tag{2.7.5.7}$$

From this equation it is clear that x_i are the zeros of $p_N^*(x)$ if and only if

$$x_i P(x_i) = JP(x_i). \tag{2.7.5.8}$$

In other words, x_i are the eigenvalues of the symmetric tridiagonal matrix J (and hence are real) and $P(x_i)$ are the corresponding eigenvectors.

We now wish to express the Gauss weights in terms of these eigenvectors. Begin with the known representation (see, e.g., Szegö)

$$1/w_j = \sum_{n=0}^{N} (p_n^*(x_j))^2, \qquad j = 1, 2, \ldots, N. \tag{2.7.5.9}$$

Since $p_N^*(x_i) = 0$, we obtain

$$1 = w_j \sum_{n=0}^{N-1} (p_n^*(x_j))^2 = w_j[P(x_j)]^{\mathrm{T}}[P(x_j)]. \tag{2.7.5.10}$$

It follows from this that

$$Q(x_j) = w_j^{1/2} P(x_j) \tag{2.7.5.11}$$

are the *normalized* eigenvectors of J corresponding to the eigenvalue x_j. If we write

$$Q(x_j) = (q_{0j}, \ldots, q_{N-1,j})^{\mathrm{T}}$$

and select out the first components of both sides of (2.7.5.11), we obtain

$$q_{0j} = w_j^{1/2} p_0^*(x_j).$$

But $p_0^*(x)$ is merely the constant $1/m_0^{1/2}$, $m_0 = \int_a^b w(x)\,dx$. Hence,

$$w_j = m_0 q_{0j}^2 \tag{2.7.5.12}$$

expresses the weights in terms of the first components of the normalized eigenvectors of J.

Golub and Welsch recommend using the Q-R algorithm of Francis to obtain the eigenvalues and eigenvectors of J.

Gauss Rules from Moments

It may be that we do not know the three-term recurrence relationship for our orthonormal polynomials explicitly, but only have available the moments

$$m_k = \int_a^b w(x)x^k\,dx, \qquad k = 0, 1, \ldots, 2N. \tag{2.7.5.13}$$

Write

$$M = \begin{bmatrix} m_0 & m_1 & \cdots & m_N \\ m_1 & m_2 & \cdots & m_{N+1} \\ \vdots & & & \\ m_N & m_{N+1} & \cdots & m_{2N} \end{bmatrix}, \tag{2.7.5.14}$$

so that the symmetric matrix M is known. Let D_j designate the principal minors of M, i.e.,

$$D_j = \begin{vmatrix} m_0 & m_1 & \cdots & m_j \\ m_1 & m_2 & \cdots & m_{j+1} \\ \vdots & & & \\ m_j & m_{j+1} & \cdots & m_{2j} \end{vmatrix}, \qquad j = 0, 1, \ldots, N;$$

$D_{-1} = 1$, while

$$F_j = \begin{vmatrix} m_0 & m_1 & \cdots & m_{j-1} & m_{j+1} \\ m_1 & m_2 & \cdots & m_j & m_{j+2} \\ \vdots & & & & \\ m_j & m_{j+1} & & m_{2j-1} & m_{2j+1} \end{vmatrix}, \qquad j = 1, 2, \ldots, N-1;$$

$F_{-1} = 0$, $F_0 = m_1$.

Then, it can be shown that the three-term recurrence for the orthogonal polynomials corresponding to the weight $w(x)$ is given by

$$xp_j(x) = \beta_j p_{j+1}(x) + \alpha_j p_j(x) + \beta_{j-1} p_{j-1}(x), \qquad j = 1, 2, \ldots, N,$$

where

$$\alpha_j = \frac{F_{j-1}}{D_{j-1}} - \frac{F_{j-2}}{D_{j-2}}, \qquad j = 1, 2, \ldots, N$$

$$\beta_j = \frac{(D_{j-2}D_j)^{1/2}}{D_{j-1}}, \qquad j = 1, 2, \ldots, N.$$

Other expressions for α_j, β_j can be obtained in terms of the Cholesky decomposition of M: $M = R^T R$. Golub and Welsch advocate its use. Golub has generalized this method to the case where several abscissas are preassigned.

References. Capovani, Ghelardoni, and Lombardi [1, 2]; Golub [1]; Golub and Welsch [1]; Kautsky and Golub [1]; Szegö [2, p. 48]; Wilf [3, Chap. 2].

Since the process of determining the coefficients in the three-term recurrence from the moments m_k is ill-conditioned in many cases, Gautschi [5] suggested that one start from modified moments of the form

$$v_k = \int_a^b w(x)\pi_k(x)\, dx$$

where the $\pi_k(x)$ satisfy the three-term recurrence

$$\pi_{-1}(x) = 0, \qquad \pi_0(x) = 1,$$

$$\pi_{k+1}(x) = (x - \alpha_k)\pi_k(x) - \beta_k \pi_{k-1}(x), \qquad k \geq 0.$$

Note that the $\pi_k(x)$ need not be orthogonal and that α_k and β_k may vanish, in which case v_k reduces to the ordinary moment m_k. However, for a well-conditioned process, the $\pi_k(x)$ will usually be orthogonal polynomials with respect to some admissible weight function $\omega(x)$. For the interval [0, 1], the shifted Legendre and Chebyshev polynomials are good choices. In these and more general cases, Gatteschi, Piessens and Branders, and Lewanowicz among others have given formulas and algorithms for evaluating the modified moments of a wide variety of weight functions $w(x)$.

Once we have v_k, we can proceed to evaluate the coefficients a_k and b_k in the three-term recurrence for the monic polynomials $p_k(x)$ orthogonal with respect to $w(x)$

$$p_{-1}(x) = 0, \qquad p_0(x) = 1,$$

$$p_{k+1}(x) = (x - a_k)p_k(x) - b_k p_{k-1}(x), \quad k = 0, 1, 2, \ldots, n-1,$$

using the following algorithm of Wheeler:

$$\sigma_{-1,l} = 0, \qquad l = 1, 2, \ldots, 2n-2,$$

$$\sigma_{0,l} = v_l, \qquad l = 0, 1, \ldots, 2n-1,$$

$$a_0 = \alpha_0 + (v_1/v_0), \qquad b_0 = 0;$$

$$\sigma_{k,l} = \sigma_{k-1,l+1} - (a_{k-1} - \alpha_l)\sigma_{k-1,l} - b_{k-1}\sigma_{k-2,l} + b_l\sigma_{k-1,l-1},$$
$$l = k, k+1, \ldots, 2n-k-1,$$

$$a_k = \alpha_k - \frac{\sigma_{k-1,k}}{\sigma_{k-1,k-1}} + \frac{\sigma_{k,k+1}}{\sigma_{k,k}}, \qquad b_k = \frac{\sigma_{k,k}}{\sigma_{k-1,k-1}}, \qquad k = 1, 2, \ldots, n-1.$$

The $\sigma_{k,l}$ are the so-called mixed moments

$$\sigma_{k,l} = \int_a^b w(x)p_k(x)\pi_l(x)\,dx$$

so that

$$h_k = \int_a^b w(x)p_k^2(x)\,dx = \sigma_{k,k}.$$

Once we have the three-term recurrence, we can proceed using the Golub–Welsch algorithm or any available polynomial root-finding program. In a comparison by Gautschi [8], an eigenvalue approach was more efficient. An algorithm for computing Gauss integration rules using the modified moments $\int_{-1}^{1} w(x)T_k(x)\,dx$ and an efficient root finding program is given by Laurie and Rolfes.

References. Boujot [1]; Boujot and Maroni [1]; Gatteschi [2]; Gautschi [5, 8]; Laurie and Rolfes [F1]; Lewanowicz [1]; Piessens and Branders [2, 6]; Sack and Donovan [1]; Wheeler [1].

2.7.6 Rules of Gauss Type for Weighting Functions of Mixed Sign.

If $w(x) \geq 0$ and $w > 0$ on, say, a subinterval, then as we have seen, abscissas x_k and weights w_k may be found such that $\sum_{k=1}^{n} w_k f(x_k) = \int_a^b w(x)f(x)\,dx$, for all $f \in \mathscr{P}_{2n-1}$. The condition $w(x) \geq 0$ is not necessary; in certain circumstances it is possible to find abscissas in $[a, b]$ and weights of mixed sign such that the rule is exact for polynomials of maximal degree. One such case occurs in connection with the Kronrod scheme (Section 2.7.1.1) where for $w(x) = P_{n-1}(x)$, the Legendre polynomial of degree $n-1$, we have

$$\int_{-1}^{1} P_{n-1}(x)f(x)\,dx = \sum_{i=1}^{n} w_i f(x_i), \qquad f \in \mathscr{P}_{2n-1}, \tag{2.7.6.1}$$

if the x_i are the zeros of the Stieltjes polynomial $E_{n-1}(x)$ and the weights are interpolatory.

A second case is where the weighting function $w(x)$ is *odd* over $[-a, a]$: $w(x) = -w(-x)$ and $w(x) > 0$ in $(0, a]$. For, assuming oddness, we have

$$\int_{-a}^{a} xw(x)g(x^2)\,dx = \int_0^{a^2} w(u^{1/2})g(u)\,du. \tag{2.7.6.2}$$

Now let u_k and w_k be the Gauss abscissas and weights with M points corresponding to the (nonnegative) weighting function $w(u^{1/2})$ over $[0, a^2]$. Thus,

$$\int_{-a}^{a} w(x)x^{2r+1}\,dx = \sum_{j=1}^{M} w_j u_j^r = \frac{1}{2}\left(\sum_{j=1}^{M} \frac{w_j}{x_j} x_j^{2r+1} + \sum_{j=1}^{M} \frac{w_j}{(-x_j)}(-x_j)^{2r+1}\right),$$
$$0 \le r \le 2M - 1. \tag{2.7.6.3}$$

where we have written $x_j = u_j^{1/2}$. Now, we get the even powers "free" by symmetry since

$$0 = \int_{-a}^{a} w(x)x^{2r}\,dx = \frac{1}{2}\left(\sum_{j=1}^{M} \frac{w_j}{x_j} x_j^{2r} + \sum_{j=1}^{M} \frac{w_j}{(-x_j)}(-x_j)^{2r}\right),$$
$$0 \le r \le 2M. \tag{2.7.6.4}$$

Therefore,

$$\int_{-a}^{a} w(x)f(x)\,dx = \sum_{j=1}^{2M} w_j^* f(x_j^*), \qquad f \in \mathscr{P}_{4M}, \tag{2.7.6.5}$$

where

$$\begin{aligned} x_j^* &= x_j = u_j^{1/2} & w_j^* &= \tfrac{1}{2}w_j/x_j \\ x_{j+M}^* &= -x_j & w_{j+M}^* &= -w_j^* \qquad j = 1, 2, \ldots, M. \end{aligned} \tag{2.7.6.6}$$

References. Monegato [5]; Piessens [4, 5]; Stroud and Secrest [2, pp. 2–12]; Struble [1].

2.7.7 *Tschebyscheff Systems and Numerical Integration*

A theory has been developed which not only extends the possibilities of Gaussian integration to certain systems of functions that are not polynomials, but also places it within an interesting geometric framework. For detailed exposition of this theory, the reader is referred to the book of Karlin and Studden.

DEFINITION 1. A set $\{u_1(t), u_2(t), \ldots, u_n(t)\}$ of n real functions continuous on the closed interval $[a, b]$ is called a *Tschebyscheff system* over $[a, b]$ if for any set of n distinct points $a \le t_1 < t_2 < \cdots < t_n \le b$, the following inequality holds:

$$\varepsilon_n \begin{vmatrix} u_1(t_1) & \cdots & u_1(t_n) \\ \vdots & & \\ u_n(t_1) & \cdots & u_n(t_n) \end{vmatrix} > 0 \qquad (2.7.7.1)$$

where $\varepsilon_n = \pm 1$ and is independent of the selection of points.

There are many known Tschebyscheff systems of functions. Karlin and Studden contains an extensive list.

THEOREM 1. *Every function in the real space spanned by* $\{u_1(t), \ldots, u_n(t)\}$ *has at most* $n - 1$ *distinct roots in* $[a, b]$.

A "confluent" form of the inequality (2.7.7.1) is contained in the next result.

THEOREM 2. *Let* $\{u_1, \ldots, u_n\}$ *be a Tschebyscheff system over* $[a, b]$ *such that* $u_i(t) \in C^{p-1}[a, b]$, $i = 1, 2, \ldots, n$. *Then, for any set of k distinct points* $a \le t_1 < t_2 < \cdots < t_k \le b$ *with multiplicities* $\beta_1 + 1, \beta_2 + 1, \ldots, \beta_k + 1$ *where* $0 \le \beta_i \le p - 1$, $i = 1, 2, \ldots, k$, *and* $\sum_{i=1}^{k} (\beta_i + 1) = n$, *the following inequality holds:*

$$\varepsilon_n \begin{vmatrix} u_1(t_1) & u_1'(t_1) & \cdots & u_1^{(\beta_1)}(t_1) & u_1(t_2) & \cdots & u_1^{(\beta_k)}(t_k) \\ u_2(t_1) & \cdots & & & & & \\ \vdots & & & & & & \\ u_n(t_1) & u_n'(t_1) & \cdots & & & & u_n^{(\beta_k)}(t_k) \end{vmatrix} \ge 0. \qquad (2.7.7.2)$$

DEFINITION 2. The *moment space* M_n of a Tschebyscheff system $\{u_1, u_2, \ldots, u_n\}$ over $[a, b]$, is the set of all points $c = (c_1, c_2, \ldots, c_n) \in E^n$ of the form

$$c_i = \int_a^b u_i(t)\, d\sigma(t), \qquad i = 1, 2, \ldots, n, \qquad (2.7.7.3)$$

where $\sigma(t)$ traverses the set of all nondecreasing right-continuous functions of bounded variation.

DEFINITION 3. By $\mathbb{C}_n$ we shall mean the curve in E^n given in parametric form by the equation

$$\mathbb{C}_n\colon \{u_1(t), u_2(t), \ldots, u_n(t)\}; \qquad a \le t \le b.$$

THEOREM 3. M_n *is a closed convex cone. It is the convex conical hull of the curve* $\mathbb{C}_n$. *This means that every point in* M_n *can be represented as a finite linear combination with positive coefficients of points on* $\mathbb{C}_n$.

Let us expand this further. A point in M_n has the form $(\int_a^b u_1(t)\, d\sigma(t), \ldots, \int_a^b u_n(t)\, d\sigma(t))$. A point on $\mathbb{C}_n$ has the form $(u_1(t_k), \ldots, u_n(t_k))$. The theorem therefore asserts that

$$\left(\int_a^b u_1(t)\, d\sigma(t), \ldots, \int_a^b u_n(t)\, d\sigma(t)\right) = \sum_{k=1}^{r} A_k(u_1(t_k), \ldots, u_n(t_k)) \tag{2.7.7.4}$$

with $r \leq n$ and $A_k > 0$, $k = 1, 2, \ldots, r$. Thus,

$$R(f) = \sum_{k=1}^{r} A_k f(t_k) \tag{2.7.7.5}$$

is an integration rule with positive coefficients which is precise whenever f is a linear combination of $u_1(t), \ldots, u_n(t)$.

DEFINITION 4. A *representation* of a point $c \in M_n$ is a finite linear combination with positive coefficients of points of $\mathbb{C}_n$ for which the identity (2.7.7.4) holds.

The values $t_1, t_2, \ldots, t_r$ which yield the points of $\mathbb{C}_n$ are called the *roots* of the representation.

DEFINITION 5. The *index of a representation* is the number of roots of the representation where the roots $t = a$ and $t = b$ are each counted as half a root.

DEFINITION 6. The *index* $I(c)$ *of a point* $c \in M_n$ is the least index among all the indices of representations of c.

THEOREM 4. *A point $c^\circ \in M_n$ is a boundary point of M_n $[c^\circ \in \mathrm{Bd}(M_n)]$ if and only if $I(c^\circ) < n/2$.*

THEOREM 5. *A point $c \in M_n$ is an interior point of M_n $[c \in \mathrm{Int}(M_n)]$ if for every $g(t) = \sum_{i=1}^n a_i u_i(t) \geq 0$ with $\sum_{i=1}^n a_i^2 > 0$, the inequality $\sum_{i=1}^n a_i c_i > 0$ holds.*

If we now specialize to $d\sigma(t) = w(t)\, dt$, the following result can be derived from Theorems 1 and 5.

COROLLARY 1. *Let $w(t) \geq 0$ on $[a, b]$ and $\neq 0$ on a set of positive measure. Let the point $c = (c_1, \ldots, c_n)$ be given by*

$$c_i = \int_a^b w(t) u_i(t)\, dt, \qquad i = 1, 2, \ldots, n.$$

Then c is an interior point of M_n.

DEFINITION 7. A representation of a point $c \in \mathrm{Int}(M_n)$ is called a *principal representation* if its index is $n/2$.

THEOREM 6. *To every point $c \in \mathrm{Int}(M_n)$, there exist two principal representations. When $n = \text{even} = 2m$, one principal representation has m roots in the interior (a, b) while the other has $m + 1$ roots of which two are endpoints. When $n = \text{odd} = 2m + 1$, each principal representation has $m + 1$ roots of which one is an endpoint.*

If we now consider the set of powers $\{1, t, t^2, \ldots, t^{n-1}\}$, then, for any n, these form a Tschebyscheff system over any finite interval $[a, b]$. Hence, any $w(t)$ satisfying the condition of Corollary 1 defines an interior point in the moment space of n dimensions. Therefore for $n = 2m$, there exists a principal representation with index m such that

$$\int_a^b w(t)t^j \, dt = \sum_{i=1}^{m} A_i t_i^j, \qquad j = 0, 1, \ldots, 2m - 1. \tag{2.7.7.6}$$

The points t_i interior to the interval are the Gauss abscissas, while the positive coefficients A_i are the Gauss weights corresponding to the weighting function $w(t)$. The second principal representation which includes the end points a and b among its roots corresponds to a Lobatto rule with $m + 1$ points. For $n = 2m + 1$, the two principal representations correspond to the two Radau-type rules with $m + 1$ points.

Similar rules of integration hold for any Tschebyscheff system on $[a, b]$.

For the numerical computation of principal representations of arbitrary Tschebyscheff systems, see Gustafson.

References. Gustafson [5]; Harris and Evans [1]; Karlin and Studden [1].

2.7.8 *Convergence of Gaussian Rules*

The Gaussian rules G_n have the remarkable property that as $n \to \infty$, $G_n(f)$ converges to $\int f \, dx$ for a very extensive class of functions.

THEOREM. *Let $f(x) \in C[-1, 1]$, and suppose that*

$$E_{G_n}(f) = \int_{-1}^{1} f(x) \, dx - G_n(f).$$

Then

$$\lim_{n \to \infty} E_{G_n}(f) = 0. \tag{2.7.8.1}$$

Proof. Let an $\varepsilon > 0$ be given. Then by the Weierstrass approximation theorem (see, for example, Davis [6, p. 107]) we can find a polynomial $p_m(x)$ such that $|f(x) - p_m(x)| \leq \varepsilon$ for $-1 \leq x \leq 1$. Now we know that

$$E_{G_n}(f) = E_{G_n}(p_m) + E_{G_n}(f - p_m).$$

Since p_m is a polynomial, then for all n such that $m \leq 2n - 1$, we have $E_{G_n}(p_m) = 0$. Hence, for all n sufficiently large, we obtain

$$\begin{aligned} |E_{G_n}(f)| &= |E_{G_n}(f - p_m)| \\ &= \left| \int_{-1}^{+1} (f - p_m)\,dx - \sum_{k=1}^{n} w_{nk}(f(x_{nk}) - p_m(x_{nk})) \right| \\ &\leq \int_{-1}^{+1} |f(x) - p_m(x)|\,dx + \sum_{k=1}^{n} w_{nk}\,|f(x_{nk}) - p_m(x_{nk})| \\ &\leq \varepsilon \int_{-1}^{+1} dx + \varepsilon \sum_{k=1}^{n} w_{nk} = 4\varepsilon. \end{aligned}$$

However, it is possible to prove still more: if f is a bounded Riemann-integrable function on $[-1, 1]$, that is, if $f \in R[-1, 1]$, then $\lim_{n\to\infty} E_{G_n}(f) = 0$ is also valid. To prove this we need several lemmas.

LEMMA. *Let $f(x) \in R[a, b]$. Then, given $\varepsilon > 0$, we can find two piecewise continuous functions $s(x)$ and $S(x)$ such that*

$$s(x) \leq f(x) \leq S(x), \tag{2.7.8.2}$$

$$\int_a^b (S(x) - s(x))\,dx \leq \varepsilon. \tag{2.7.8.3}$$

Proof. Introduce a partition $a = x_0 < x_1 < \cdots < x_n = b$. Now set $M_k = \sup_{x_{k-1} \leq x \leq x_k} f(x)$, $m_k = \inf_{x_{k-1} \leq x \leq x_k} f(x)$, and form

$$U = \sum_{k=1}^{n} M_k(x_k - x_{k-1}), \qquad L = \sum_{k=1}^{n} m_k(x_k - x_{k-1}).$$

By the definition of the Riemann integral we can take a partition so fine that $U - L \leq \varepsilon$. Then define

$$\begin{aligned} S(x) &= M_k \quad \text{for} \quad x_{k-1} \leq x < x_k, \qquad k = 1, 2, \ldots, n; \\ S(b) &= M_n. \end{aligned} \tag{2.7.8.4}$$

Define $s(x)$ similarly with m_k. Now, $\int_a^b S(x)\,dx = U$ and $\int_a^b s(x)\,dx = L$. Hence, $\int_a^b (S(x) - s(x))\,dx \leq \varepsilon$.

We shall next prove that $G_n(f)$ converges to the "proper" answer when f is a simple piecewise constant function.

LEMMA. *Let* $-1 < a < b < 1$ *and set*

$$\begin{aligned} f(x) &= 0, & -1 \le x < a, \\ f(x) &= 1, & a \le x < b, \\ f(x) &= 0, & b \le x \le 1. \end{aligned} \tag{2.7.8.5}$$

Then $\lim_{n\to\infty} E_{G_n}(f) = 0$.

Proof. Take an ε so small that $-1 < a - \varepsilon$, $b + \varepsilon < 1$, and $a + \varepsilon < b - \varepsilon$. Construct two continuous functions ϕ and ψ (depending on ε) as follows.

$$\begin{aligned} \phi(x) &= 0, & -1 \le x \le a - \varepsilon, \\ \phi(x) &= \text{linear}, & a - \varepsilon \le x \le a, \\ \phi(x) &= 1, & a \le x \le b, \\ \phi(x) &= \text{linear}, & b \le x \le b + \varepsilon, \\ \phi(x) &= 0, & b + \varepsilon \le x \le 1. \end{aligned} \tag{2.7.8.6}$$

$$\begin{aligned} \psi(x) &= 0, & -1 \le x \le a, \\ \psi(x) &= \text{linear}, & a \le x \le a + \varepsilon, \\ \psi(x) &= 1, & a + \varepsilon \le x \le b - \varepsilon, \\ \psi(x) &= \text{linear}, & b - \varepsilon \le x \le b, \\ \psi(x) &= 0, & b \le x \le 1. \end{aligned} \tag{2.7.8.7}$$

Then we have $\phi(x) \ge f(x) \ge \psi(x)$. Since the Gauss rule G_n has positive weights, then

$$G_n(\phi) \ge G_n(f) \ge G_n(\psi).$$

In this inequality, allow $n \to \infty$. Since ϕ and ψ are continuous, we have

$$\int_{-1}^{1} \phi(x)\,dx \ge \limsup_{n\to\infty} G_n(f) \ge \liminf_{n\to\infty} G_n(f) \ge \int_{-1}^{1} \psi(x)\,dx.$$

But

$$\int_{-1}^{1} \phi(x)\,dx = b - a + \varepsilon \quad \text{and} \quad \int_{-1}^{1} \psi(x)\,dx = b - a - \varepsilon.$$

Therefore, since ε may be chosen arbitrarily small, it follows that

$$\lim_{n\to\infty} G_n(f) = b - a = \int_{-1}^{1} f(x)\,dx.$$

COROLLARY. *If $f(x)$ is piecewise constant on $[-1, 1]$, then*

$$\lim_{n\to\infty} E_{G_n}(f) = 0 \quad or \quad \lim_{n\to\infty} G_n(f) = \int_{-1}^{1} f(x)\,dx. \tag{2.7.8.8}$$

Proof. Every piecewise constant function is a finite linear combination of the simple sort of function just investigated.

THEOREM. *Let $f(x) \in R[-1, 1]$. Then*

$$\lim_{n\to\infty} E_{G_n}(f) = 0. \tag{2.7.8.9}$$

Proof. Given ε. Construct piecewise constant functions $s(x)$ and $S(x)$ as in the first lemma. Since $s(x) \le f(x) \le S(x)$, we have

$$\int_{-1}^{1} s(x)\,dx \le \int_{-1}^{1} f(x)\,dx \le \int_{-1}^{1} S(x)\,dx$$

and $G_n(s) \le G_n(f) \le G_n(S)$. Allow $n \to \infty$; then by the corollary, we know that

$$G_n(s) \to \int_{-1}^{1} s(x)\,dx \quad \text{and} \quad G_n(S) \to \int_{-1}^{1} S(x)\,dx.$$

Therefore

$$\int_{-1}^{1} s(x)\,dx \le \liminf_{n\to\infty} G_n(f) \le \limsup_{n\to\infty} G_n(f) \le \int_{-1}^{1} S(x)\,dx.$$

Hence

$$\int_{-1}^{1} f(x)\,dx - \varepsilon \le \liminf_{n\to\infty} G_n(f) \le \limsup_{n\to\infty} G_n(f) \le \int_{-1}^{1} f(x)\,dx + \varepsilon.$$

Since ε is arbitrary, it follows that

$$\lim_{n\to\infty} G_n(f) = \int_{-1}^{1} f(x)\,dx.$$

An inspection of the proof of this theorem shows that the only properties of G_n used were that the sequence $\{E_{G_n}(f)\}$ tends to zero for all $f \in C[-1, 1]$ and that the weights in every rule were all positive. Hence, we have the result that any sequence of integration rules with positive weights which converges for all continuous functions converges for all Riemann-integrable functions.

A second proof of this theorem can be based upon the fact (see Section 2.1.5) that Gauss rules are Riemann sums. It should be observed that a similar theorem holds for integrals of the form $\int_a^b w(x)f(x)\,dx$, where $-\infty < a < b < \infty$.

A theorem of great generality which deals with the convergence of a family of rules is due to Pólya.

THEOREM. *Let*

$$L_n(f) = \sum_{k=1}^{n} w_{nk} f(x_{nk}), \qquad a \le x_{nk} \le b.$$

Then

$$\lim_{n\to\infty} L_n(f) = \int_a^b f(x)\,dx \quad \textit{for all} \quad f \in C[a, b] \tag{2.7.8.10}$$

if and only if

$$\lim_{n\to\infty} L_n(x^k) = \int_a^b x^k\,dx, \qquad k = 0, 1, \ldots \tag{2.7.8.11}$$

and

$$\sum_{k=1}^{n} |w_{nk}| \le M, \qquad n = 1, 2, \ldots \tag{2.7.8.12}$$

for some constant M.

If the weights w_{nk} are positive (as they are in the Gauss case), then (2.7.8.11) implies (2.7.8.12). For, in this case,

$$\sum_{k=1}^{n} |w_{nk}| = \sum_{k=1}^{n} w_{nk} = L_n(1)$$

and $L_n(1)$ is bounded since $L_n(1) \to b - a$. If the integration formula is of interpolatory type and the weights are positive, then (2.7.8.11) and, hence, (2.7.8.10) will be fulfilled. In fact, by the remarks after the previous theorem, (2.7.8.10) will hold for all $f \in R[a, b]$.

Corresponding theorems hold for weighted integrals and for multidimensional integrals. An abstract formulation is possible through the "principle of uniform boundedness."

We should note that a family of approximate integration formulas,

$$L_n(f) = \sum_{k=1}^{n} w_{nk} f(x_{nk}), \qquad a \le x_{nk} \le b, \tag{2.7.8.13}$$

which converges for all functions of class $C[a, b]$, will not automatically converge for all functions of class $R[a, b]$. This fact is brought out by the selection of weights and abscissas

$$\begin{aligned} x_{nj} &= (j-1)/n, \qquad j = 1, 2, \ldots, n, \\ w_{n1} &= 1, \\ w_{n2} &= -1, \\ w_{nj} &= 1/n, \qquad j = 3, 4, \ldots, n. \end{aligned} \tag{2.7.8.14}$$

This family of rules integrates all continuous functions in $[0, 1]$ properly in the limit but fails to integrate the function

$$\begin{aligned} f(0) &= 1, \\ f(x) &= 0, \qquad 0 < x \le 1. \end{aligned}$$

Necessary and sufficient conditions for convergence for all functions of class $R[a, b]$ were given by Pólya. Let I designate the sum of a finite number of intervals (disjoint or not) located in (a, b). Let $m(I)$ be the sum of the lengths of the individual intervals of I. The notation $\sum_I |w_{nk}|$ will designate the sum taken over those w_{nk} for which $x_{nk} \in I$. Set

$$\Delta(I) = \limsup_{n\to\infty} \sum_I |w_{nk}|. \tag{2.7.8.15}$$

The set function $\Delta(I)$ can be shown to be nonnegative: $\Delta(I) \ge 0$, monotone: $\Delta(I_1) \le \Delta(I_1 + I_2)$, and subadditive: $\Delta(I_1 + I_2) \le \Delta(I_1) + \Delta(I_2)$. It is called *semicontinuous* if for any sequence $I_1 \supset I_2 \supset I_3 \supset \cdots$ with $m(I_n) \to 0$, we have

$$\lim_{n\to\infty} \Delta(I_n) = 0. \tag{2.7.8.16}$$

THEOREM. *If*

$$\lim_{n\to\infty} \sum_{k=1}^{n} w_{nk} f(x_{nk}) = \int_a^b f(x)\, dx$$

holds for all $f \in C[a, b]$, *it holds for all* $f \in R[a, b]$ *if and only if* $\Delta(I)$ *is semicontinuous.*

Necessary and sufficient conditions for narrower classes such as $C^r[a, b]$ can be found in Krylov.

For further material on convergence theory, particularly for infinite intervals, see Freud.

Before leaving this subject, it is interesting to note that when dealing with a family of rules that integrate continuous functions properly in the limit, *nothing positive* can be said about the degree of convergence. In fact, Lipow and Stenger have demonstrated the following theorem.

Suppose that $L_n(f)$ converges to $\int_a^b f(x)\,dx$ for all $f \in C[a, b]$ and suppose that $\varepsilon_n \geq 0$ is any sequence with $\lim_{n\to\infty} \varepsilon_n = 0$. Then, one can find an $f \in C[a, b]$ such that $|E_n(f)| = |L_n(f) - \int_a^b f(x)\,dx| \geq \varepsilon_n$.

References. Davis [6, pp. 353–355]; Féjer [1]; Freud [1, Chap. 3]; Krylov [1, pp. 266–273]; Lipow and Stenger [1]; Pólya [1].

2.8 *Integration Rules Using Derivative Data*

In this book, the emphasis has been on approximate rules of integration that involve functional values. However, it is possible to derive approximate rules that make use of other sorts of functional information. For example, from the two-point Taylor interpolation formula (see Section 1.11.6), we have

$$\int_a^b f(x)\,dx = \frac{h}{2}[f(a) + f(b)] + \frac{h^2}{12}[f'(a) - f'(b)] + \frac{h^5}{720} f^{(4)}(\xi),$$
$$a < \xi < b, \qquad h = b - a. \tag{2.8.1}$$

This rule makes use of the derivative values at a and b in addition to the values of the function itself. In cases where derivative information is readily available, for example with functions that satisfy a differential equation, such rules may be used to advantage.

One should note also that in compounding such rules the weights of the first derivative at the interior points cancel and hence we need evaluate the derivatives only at the endpoints of the interval to achieve a substantial increase in accuracy. Thus, the trapezoidal rule with "end correction" is

$$\int_a^b f(x)\,dx$$
$$= h\left[\frac{f(a)}{2} + f(a + h) + f(a + 2h) + \cdots + f(a + (n - 1)h) + \frac{f(b)}{2}\right]$$
$$+ \frac{h^2}{12}[f'(a) - f'(b)] + E, \qquad h = \frac{b - a}{n}, \tag{2.8.2}$$

where

$$E \leq \frac{1}{720} h^4(b - a) \max_{a \leq x \leq b} |f^{(4)}(x)|. \tag{2.8.3}$$

The following formula makes use of f, f', f'' at two points:

$$\int_a^b f(x)\,dx = \frac{h}{2}[f(a)+f(b)] + \frac{h^2}{10}[f'(a)-f'(b)] + \frac{h^3}{120}[f''(a)+f''(b)] - \frac{h^7}{100{,}800}f^{(6)}(\xi),$$
$$a < \xi < b, \qquad h = b - a. \tag{2.8.4}$$

There are similar rules of Simpson's and of Gauss type.

References. Ghizzetti and Ossicini [1]; IBM [2, pp. 293–295]; Kowalewski [1, p. 130]; Turan [1].

Krylov and Arljuk have extended the theorems on convergence of integration rules (see Section 2.7.8) to rules involving derivatives:

$$R_n(f) = \sum_{k=1}^{n}\sum_{j=0}^{m} A_{njk} f^{(j)}(x_{nk}) \approx \int_a^b w(x)f(x)\,dx. \tag{2.8.5}$$

The generalized Newton–Cotes rules with derivative data converge properly as the degree of the rule $\to\infty$ provided that the integrand $f(z)$ is regular in the region

$$\operatorname{Re}[(z-b)\log(z-b) - (z-a)\log(z-a) + (b-a)\log(b-a)] < 0.$$

Cf. Section 2.5. See Donaldson and Elliott.

Derivative rules are disadvantageous if derivatives are not easily computed, and they are less "automatic" in that more pencil work is involved in setting them up. For these reasons, we shall merely provide a few references.

References. Cicenia [1, 2]; Donaldson and Elliott [2]; Flinn [1]; Golub and Kautsky [1]; Gray and Rall [1–3]; Hammer and Wicke [1]; Håvie [5]; Kress [6, 7]; Krylov and Arljuk [1]; Lambert and Mitchell [1]; Lanczos [2]; Lotkin [1]; Patterson [3]; Sack [1]; Salzer [2]; Schoenberg [1]; Squire [2]; Stancu and Stroud [2]; Stroud and Stancu [1].

2.8.1 Integration Formulas of Hermite or Osculatory Type

If we integrate the Hermite interpolation formula (1.11.14), we obtain an integration rule of the form

$$\int_a^b w(x)f(x)\,dx \approx \sum_{k=0}^{n} a_k f(x_k) + \sum_{k=0}^{n} b_k f'(x_k) \tag{2.8.1.1}$$

where

$$a_k = \int_a^b w(x)L_k(x)\,dx, \qquad b_k = \int_a^b w(x)M_k(x)\,dx. \tag{2.8.1.2}$$

The rule is exact for $f \in \mathscr{P}_{2n+1}$. Similar formulas may be developed with higher derivatives.

Reference. Bulirsch and Rutishauser [1].

2.9 *Integration of Periodic Functions*

Under certain conditions, the trapezoidal rule gives surprisingly good results when it is applied to *periodic* functions, much better in fact than what might have been predicted from the error estimate (2.1.11). We begin our discussion of this possibility by first developing the Euler–Maclaurin formula. This formula may be regarded as an extension of the trapezoidal rule.

The following identity is easily verified by integration by parts:

$$\tfrac{1}{2}[f(k) + f(k+1)] = \int_k^{k+1} f(x)\,dx + \int_k^{k+1} (x - [x] - \tfrac{1}{2})f'(x)\,dx. \tag{2.9.1}$$

Here

$$[x] = \text{largest integer} \le x.$$

We set

$$P_1(x) = x - [x] - \tfrac{1}{2}. \tag{2.9.2}$$

The function $P_1(x)$ is a piecewise linear, periodic function with period 1. Writing equation (2.9.1) with $k = 0, 1, \ldots, n-1$, and adding, we obtain

$$\tfrac{1}{2}f(0) + f(1) + \cdots + f(n-1) + \tfrac{1}{2}f(n) = \int_0^n f(x)\,dx + \int_0^n P_1(x)f'(x)\,dx. \tag{2.9.3}$$

Notice that the left-hand side of (2.9.3) is a trapezoidal sum.

The function $P_1(x)$ has the Fourier expansion

$$P_1(x) = -\sum_{n=1}^{\infty} \frac{2\sin 2\pi nx}{2\pi n}. \tag{2.9.4}$$

We integrate this series formally and set, by way of definition,

$$P_2(x) = \sum_{n=1}^{\infty} \frac{2 \cos 2\pi nx}{(2\pi n)^2}. \tag{2.9.5}$$

Repeating this, we have

$$P_3(x) = \sum_{n=1}^{\infty} \frac{2 \sin 2\pi nx}{(2\pi n)^3}. \tag{2.9.6}$$

In general, for $j = 1, 2, \ldots,$

$$P_{2j}(x) = (-1)^{j-1} \sum_{n=1}^{\infty} \frac{2 \cos 2\pi nx}{(2\pi n)^{2j}},$$

$$P_{2j+1}(x) = (-1)^{j-1} \sum_{n=1}^{\infty} \frac{2 \sin 2\pi nx}{(2\pi n)^{2j+1}}. \tag{2.9.7}$$

The function $P_n(x)$ is related to $\bar{B}_n(x)$, the periodic extension of the Bernoulli polynomial $B_n(x)$, by the equation $P_n(x) = \bar{B}_n(x)/n!$. Bernoulli polynomials, defined by the generating function

$$\frac{te^{xt}}{e^t - 1} = \sum_{n=0}^{\infty} B_n(x) \frac{t^n}{n!}, \tag{2.9.8}$$

appear very frequently in the context of numerical integration. For properties of $B_n(x)$, see the NBS Handbook and Krylov.

The following properties of $P_n(x)$ should be noted: $P_n(x)$ is a piecewise polynomial of degree n, and it is a periodic function of period 1. Moreover

$$P'_{n+1}(x) = P_n(x), \tag{2.9.9}$$

$$P_{2j+1}(0) = P_{2j+1}(1) = 0, \qquad j = 1, 2, \ldots, \tag{2.9.10}$$

$$P_{2j}(0) = P_{2j}(1) = (-1)^{j-1} \sum_{n=1}^{\infty} \frac{2}{(2n\pi)^{2j}} = \frac{B_{2j}}{(2j)!}, \qquad j \geq 1. \tag{2.9.11}$$

The constants B_{2j} are the *Bernoulli numbers* and have the values

$$B_2 = \tfrac{1}{6}, \qquad B_4 = -\tfrac{1}{30}, \qquad B_6 = \tfrac{1}{42}, \qquad B_8 = -\tfrac{1}{30}, \ldots. \tag{2.9.11a}$$

Further integration by parts yields

$$\int_0^n P_1(x) f'(x)\, dx = P_2(x) f'(x) \Big|_0^n - \int_0^n P_2(x) f''(x)\, dx$$

$$= \frac{B_2}{2!} [f'(n) - f'(0)] - \int_0^n P_2(x) f''(x)\, dx \tag{2.9.12}$$

and

$$\int_0^n P_2(x)f''(x)\,dx = P_3(x)f''(x)\Big|_0^n - \int_0^n P_3(x)f'''(x)\,dx$$

$$= -\int_0^n P_3(x)f'''(x)\,dx. \tag{2.9.13}$$

This process of integration by parts can be carried out again and again and, if equations (2.9.3), (2.9.12), (2.9.13), and similarly developed equations are combined, we obtain

$$\begin{aligned}
&\tfrac{1}{2}f(0) + f(1) + \cdots + f(n-1) + \tfrac{1}{2}f(n) \\
&\quad = \int_0^n f(x)\,dx + \frac{B_2}{2!}[f'(n) - f'(0)] + \frac{B_4}{4!}[f'''(n) - f'''(0)] \\
&\quad + \cdots + \frac{B_{2k}}{(2k)!}[f^{(2k-1)}(n) - f^{(2k-1)}(0)] \\
&\quad + \int_0^n P_{2k+1}(x)f^{(2k+1)}(x)\,dx.
\end{aligned} \tag{2.9.14}$$

This establishes the Euler–Maclaurin summation formula, which follows.

THEOREM. *Let $f(x) \in C^{2k+1}[0, n]$; then formula (2.9.14) is valid.*

COROLLARY. *Let $g(x) \in C^{2k+1}[a, b]$. Set $h = (b - a)/n$. Then*

$$\begin{aligned}
&h[\tfrac{1}{2}g(a) + g(a+h) + g(a+2h) + \cdots + g(a+(n-1)h) + \tfrac{1}{2}g(b)] \\
&\quad = \int_a^b g(x)\,dx + \frac{B_2}{2!}h^2[g'(b) - g'(a)] + \frac{B_4}{4!}h^4[g'''(b) - g'''(a)] \\
&\quad + \cdots + \frac{B_{2k}}{(2k)!}h^{2k}[g^{(2k-1)}(b) - g^{(2k-1)}(a)] \\
&\quad + h^{2k+1}\int_a^b P_{2k+1}\left(n\frac{x-a}{b-a}\right)g^{(2k+1)}(x)\,dx.
\end{aligned} \tag{2.9.15}$$

Proof. Apply the Euler–Maclaurin summation to $g(a + hx)$.

Formula (2.9.15) is convenient because it refers to a fixed interval.

THEOREM. *Let* $g(x) \in C^{(2k+1)}[a, b]$, $g'(a) = g'(b)$, $g'''(a) = g'''(b)$, $\ldots$, $g^{(2k-1)}(a) = g^{(2k-1)}(b)$, *and let* $|g^{(2k+1)}(x)| \leq M$ *for* $a \leq x \leq b$. *If* T_n *designates the trapezoidal sum, that is, if*

$$T_n(g) = h[\tfrac{1}{2}g(a) + g(a + h) + \cdots + g(a + (n - 1)h) + \tfrac{1}{2}g(b)],$$

$$h = (b - a)/n,$$

then

$$\left| \int_a^b g(x)\, dx - T_n(g) \right| \leq \frac{C}{n^{2k+1}}. \tag{2.9.16}$$

The constant C is independent of n and may be taken as

$$C = M(b - a)^{2k+2} 2^{-2k} \pi^{-2k-1} \zeta(2k + 1),$$

where $\zeta(k) = \sum_{j=1}^{\infty} j^{-k}$ *is the Riemann zeta function.*

Proof. Under the above hypothesis we have, from (2.9.15),

$$\int_a^b g(x)\, dx - T_n(g) = -h^{2k+1} \int_a^b P_{2k+1}\left(n \frac{x - a}{b - a}\right) g^{(2k+1)}(x)\, dx.$$

Hence

$$\left| \int_a^b g(x)\, dx - T_n(g) \right| \leq h^{2k+1} \int_a^b \left| P_{2k+1}\left(n \frac{x - a}{b - a}\right) \right| |g^{(2k+1)}(x)|\, dx.$$

Now

$$P_{2k+1}(t) = (-1)^{k-1} \sum_{j=1}^{\infty} 2 \frac{\sin 2\pi jt}{(2\pi j)^{2k+1}}.$$

Hence

$$|P_{2k+1}(t)| \leq \sum_{j=1}^{\infty} \frac{2}{(2\pi j)^{2k+1}} = 2^{-2k} \pi^{-2k-1} \zeta(2k + 1). \tag{2.9.17}$$

Combining these inequalities, we obtain the stated result.

The conditions of this theorem are fulfilled, for example, when $g(x)$ is a periodic function of a high degree of smoothness in $(-\infty, \infty)$.

COROLLARY. *Let* $g(x)$ *have period* 2π *and be of class* $C^{2k+1}(-\infty, \infty)$ *with* $|g^{(2k+1)}(x)| \leq M$. *Then,*

$$\left| \int_0^{2\pi} g(x)\, dx - T_n(g) \right| \leq 4\pi M \zeta(2k + 1)/n^{2k+1}. \tag{2.9.18}$$

Proof. In this case, $g'(0) = g'(2\pi), \ldots, g^{(2k-1)}(0) = g^{(2k-1)}(2\pi)$.

If $g(x)$ is a periodic function with period $b - a$, $g(a) = g(b)$ and, hence, the trapezoidal rule becomes

$$T_n(g) = \frac{b-a}{n}[g(a) + g(a+h) + \cdots + g(a + (n-1)h)],$$

which is a *simple average* of the functional values at equally spaced points.

If we set $b - a = p$ and

$$E_{T_n}(f) = \frac{p}{n}\sum_{k=0}^{n-1} f\left(\frac{k}{n}p\right) - \int_0^p f(x)\,dx, \tag{2.9.19}$$

it is easily verified that

$$E_{T_n}\left(\exp\frac{2\pi ijx}{p}\right) = \begin{cases} p, & j \neq 0, \quad n \mid j, \\ 0, & \text{otherwise}, \end{cases} \qquad i = \sqrt{-1}. \tag{2.9.20}$$

This means that the trapezoidal rule T_n is exact for the $2n$ periodic functions 1, $\sin x$, $\cos x, \ldots, \sin(n-1)x$, $\cos(n-1)x$, $\sin nx$. A further consequence of (2.9.20) is that if a periodic function $f(x)$ has the Fourier expansion

$$f(x) = \frac{1}{2}\alpha_0 + \sum_{n=1}^{\infty} \alpha_n \cos\frac{2\pi nx}{p} + \beta_n \sin\frac{2\pi nx}{p}, \tag{2.9.21}$$

then, assuming the series converges uniformly, it follows that

$$E_{T_n}(f(x)) = p\sum_{k=1}^{\infty} \alpha_{kn} = p(\alpha_n + \alpha_{2n} + \alpha_{3n} + \cdots). \tag{2.9.22}$$

This is *Poisson's summation formula*, frequently written in the form

$$\int_0^p f(x)\,dx = h\left\{\frac{1}{2}[f(0) + f(p)] + \sum_{k=1}^{n-1} f(kh)\right\} - p\sum_{k=1}^{\infty} g\left(\frac{2k\pi}{h}\right),$$

$$h = \frac{p}{n}, \qquad g(t) = \int_0^p f(x)\cos tx\,dx. \tag{2.9.23}$$

If in (2.9.15) one allows $k \to \infty$ and ignores the remainder on the right-hand side, an asymptotic expansion is obtained:

$$T_n(g) - \int_a^b g(x)\,dx \sim \frac{B_2}{2!}h^2[g'(b) - g'(a)] + \frac{B_4}{4!}h^4[g'''(b) - g'''(a)] + \cdots. \tag{2.9.24}$$

A similar asymptotic formula may be obtained for the midpoint rule:

$$M_n(g) - \int_a^b g(x)\,dx \sim \frac{C_2}{2!}h^2[g'(b) - g'(a)] + \frac{C_4}{4!}h^4[g'''(b) - g'''(a)] + \cdots \tag{2.9.25}$$

where $C_{2n} = -(1 - 2^{1-2n})B_{2n}$.

For generalizations of the Euler–Maclaurin formula that are valid for functions with certain types of singularity, see the work of Navot. See also Section 2.9.5.

For Euler–Maclaurin formulas corresponding to the weighting functions e^{kx}, $\sin kx$, $\cos kx$, $(\sin kx)/x$, $(x - x_0)^a e^{kx}$, see Chakravarti.

For the analysis of the trapezoidal error for periodic analytic functions, see Section 4.6.5.

References. Chakravarti [1]; Davis [4]; Fettis [2]; Hämmerlin [1]; Kowalewski [1, Chap. 3]; Krylov [1, Chap. 1]; Lohmann [1]; Lyness and Ninham [1]; NBS Handbook [1, Chap. 23]; Navot [1–4]; Ninham and Lyness [1]; Walsh and Sewell [1].

Example. The function $(1 + \sigma \sin 2j\pi x)^{-1}$ has period 1 and, if $|\sigma| < 1$, has derivatives of all orders on $-\infty < x < \infty$. Hence, the convergence of the trapezoidal rule should be better than n^{-k} for *all* integers k.

Number of points	$(1 + \frac{1}{2}\sin 2\pi x)^{-1}$	$(1 + \frac{1}{2}\sin 10\pi x)^{-1}$
2	0.9999 9995	0.9999 9980
4	1.1666 6660	1.1666 6650
8	1.1547 6180	1.1547 6180
16	1.1547 0050	1.1547 0050
32	1.1547 0040	1.1547 0040
64	1.1547 0030	1.1547 0020
128	1.1547 0010	1.1547 0010
256	1.1546 9960	1.1546 9950
512	1.1546 9880	1.1546 9840
1024	1.1546 9790	1.1546 9760
2048	1.1546 9440	1.1546 9400
4096	1.1546 8600	1.1546 8240
Exact value	1.1547 0054	1.1547 0054

As predicted, the convergence is very rapid; seven decimals are achieved when $n = 16$. Thereafter, the quality of the results deteriorates due to roundoff. This emphasizes the fact that when a high-precision rule is employed, care must be taken not to use too fine a mesh or the noise level in the output will rise. Note that the oscillation of $\sin 10\pi x$ is not sufficiently rapid to induce any significant difference in the numerical integration.

It is interesting to compare the results of $(1 + \frac{1}{2}\sin 10\pi x)^{-1}$ integrated by the trapezoidal rule and integrated by the Gauss rule. With 16 points, the trapezoidal rule is greatly superior. See page 100.

Example. For rapid convergence with the trapezoidal rule it is not necessary to have a periodic function; it is sufficient to have one for which $f'(a) = f'(b)$, $f'''(a) = f'''(b)$, ..., $f^{(2k-1)}(a) = f^{(2k-1)}(b)$. The function $f(x) = \exp[x^2(1-x)^2]$ has this behavior with $a = 0$, $b = 1$, $k = 1$ since $f'(0) = f'(1) = 0$ and $f'''(0) = -12 = -f'''(1)$. Here are the results of determining $\int_0^1 \exp[x^2(1-x)^2]\,dx$ by the trapezoidal rule.

n	Trapezoidal rule
2	1.0322 472
4	1.0340 143
8	1.0341 329
16	1.0341 405
32	1.0341 409
64	1.0341 407
128	1.0341 407
256	1.0341 401
512	1.0341 389
1024	1.0341 379
Exact value	1.0341 4105

Eight figures are achieved at $n = 32$. With $n = 32$, the term involving h^4 in (2.9.15) predicts an error of 3×10^{-8}. Note the subsequent contamination by roundoff.

Example. (Imhof). Another example of this kind is provided by the integral

$$J_k(t) = \frac{1}{\pi}\int_0^{\pi} \cos(t \sin x - kx)\,dx.$$

$J_k(t)$ is the *Bessel function.* The values below are for $k = 1$, $t = 8$.

n	Trapezoidal rule	Simpson's rule
6	.2123 144	.1673 824
8	.2343 632	.2991 192
16	.2346 363	.2347 274
Exact value	.2346 3634 7	

Reference. Imhof [1].

Stroud and Kohli subsequently investigated this integral numerically by the trapezoidal rule, Simpson's rule, the Gauss rule, and the Romberg rule (see Section 6.3) and concluded that the trapezoidal rule is the most accurate of these for the same number of functional evaluations. The trapezoidal rule appears to be a reasonable alternative method for computing $J_n(x)$. Typical number of points required: for accuracies of 10^{-8} over the range $0 \le x \le 20$; $J_0(x)$, 22 points; $J_{10}(x)$, 26 points; $J_{20}(x)$, 30 points.

Reference. Stroud and Kohli [1].

Approximate integration can lead to useful functional approximations. Here is an example that also involves the Bessel functions. It is known that

$$J_k(z) = ((i)^k/\pi)\int_0^\pi \exp[iz\cos t]\cos kt\,dt. \tag{2.9.26}$$

In the case $k = 0$, we apply the trapezoidal rule with $h = \pi/6$. This leads to

$$\begin{aligned}
J_0(z) &\approx \frac{1}{6}\left\{\frac{1}{2}\exp[iz] + \exp\left[iz\cos\left(\frac{\pi}{6}\right)\right] + \exp\left[iz\cos\left(\frac{\pi}{3}\right)\right] + \exp\left[iz\cos\left(\frac{\pi}{2}\right)\right]\right.\\
&\quad \left. + \exp\left[iz\cos\left(\frac{2\pi}{3}\right)\right] + \exp\left[iz\cos\left(\frac{5\pi}{6}\right)\right] + \frac{1}{2}\exp[-iz]\right\}\\
&= \frac{1}{6}\left[\cos z + 2\cos\left(\frac{\sqrt{3}}{2}z\right) + 2\cos\left(\frac{z}{2}\right) + 1\right].
\end{aligned} \tag{2.9.27}$$

Over the range $0 \le z \le 2$, this approximation is found to be accurate to within 10^{-8}.

Another example involving Bessel functions $J_\nu(x)$ of arbitrary order $\nu > -\frac{1}{2}$ has been given by Baratella *et al.* This uses the representation

$$J_\nu(x) = \frac{(x/2)^\nu}{\sqrt{\pi}\Gamma(\nu + \frac{1}{2})}\int_{-1}^{1}(1 - t^2)^{\nu - 1/2}\cos xt\,dt, \qquad \nu > -\tfrac{1}{2}. \tag{2.9.28}$$

If we apply to the integral in (2.9.28) an n-point Gauss rule based on the zeros of the ultraspherical polynomial $C_n^\nu(x)$, we find that the error term has the form

$$(-1)^n a_{n\nu} x^{2n}\cos x\xi, \qquad -1 < \xi < 1 \tag{2.9.29}$$

where

$$a_{n\nu} = \frac{\pi\Gamma(n + 2\nu)n!}{2^{2n+2\nu-1}\Gamma(n + \nu)\Gamma(n + \nu + 1)(2n)!}. \tag{2.9.30}$$

Hence, a bound on the error can be readily computed; in particular, if $|x| \le 1$, the magnitude of the error is bounded by $|a_{n\nu}|$.

Luke gives a detailed discussion of the error when the trapezoidal rule is applied to the integral (2.9.26), as well as to the generalized elliptic integral

$$\int_0^{\pi/2}\frac{dt}{(1 - \nu^2\sin^2 t)(1 - k^2\sin^2 t)^\omega}, \qquad 0 < k < 1,$$

$$\nu^2 < 1, \quad \omega < 1, \ne 0,$$

the modified Bessel function

$$\int_0^\infty e^{-z\cosh t}\cosh \nu t\, dt, \qquad \nu = 0, 1,$$

and the integral

$$\int_0^\infty e^{-t^2}t^{2n}(2z + t^2)^{n-1/2}\, dt, \qquad n = 0, 1, \ldots, \ |\arg z| < \pi.$$

Further references will be found in his work.

References. Baratella, Garetto, and Vinardi [1]; Fettis [1]; Hunter [1, 3]; Jagerman [1]; Kitahara and Yano [1]; Luke [4, 6, 7, pp. 214–226]; NBS Handbook [1, Chap. 9]; Temme [1].

2.9.2 The IMT Rule

The IMT rule, proposed by Iri, Moriguti, and Takasawa, is based upon the idea of transforming the independent variable in such a way that all the derivatives of the new integrand vanish at both end points of the integration interval. A trapezoidal rule is then applied to the new integrand and under proper conditions, the high degree of accuracy implied by (2.9.16) will be operative.

Let

$$\begin{aligned} \phi(t) &= \exp(-at^{-p} - b(1-t)^{-q}), \qquad p, q \geq 1, \quad a, b > 0, \\ \psi(x) &= K^{-1}\int_0^x \phi(t)\, dt, \qquad K = \int_0^1 \phi(t)\, dt. \end{aligned} \tag{2.9.2.1}$$

The function $\psi(x)$ is monotonic increasing, performing a one–one transformation of [0, 1] onto itself. Consequently,

$$I = \int_0^1 f(x)\, dx = K^{-1}\int_0^1 \phi(x) f(\psi(x))\, dx = K^{-1}\int_0^1 g(x)\, dx.$$

Since all the derivatives of $\phi(x)$ vanish at $x = 0, 1$, it follows that if $f(x) \in C^n[0, 1]$, $g^{(j)}(0) = g^{(j)}(1) = 0$ for $j = 0, 1, \ldots, n$.

Applying the trapezoidal and the midpoint rules to g and noting that $\phi(0) = \phi(1) = 0$, we obtain

$$\begin{aligned} I &= \frac{1}{Kn}\sum_{j=1}^{n-1} \phi\left(\frac{j}{n}\right) f\left(\psi\left(\frac{j}{n}\right)\right) + E_1 \\ &= \frac{1}{Kn}\sum_{j=1}^{n} \phi\left(\frac{2j-1}{2n}\right) f\left(\psi\left(\frac{2j-1}{2n}\right)\right) + E_2 . \end{aligned} \tag{2.9.2.2}$$

Here E_1 and E_2 are the errors in the trapezoidal and midpoint rules respectively. Equation (2.9.2.2) yields the new class of rules

$$\int_0^1 f(x)\,dx \approx \sum_{j=1}^{n-1} w_j f(x_j) \equiv \hat{T}_n(f) \tag{2.9.2.3}$$

$$\int_0^1 f(x)\,dx \approx \sum_{j=1}^{n} \tilde{w}_j f(\tilde{x}_j) \equiv \hat{M}_n(f) \tag{2.9.2.4}$$

where

$$x_j = \psi\left(\frac{j}{n}\right), \qquad \tilde{x}_j = \psi\left(\frac{2j-1}{2n}\right),$$

$$w_j = \frac{1}{Kn}\phi\left(\frac{j}{n}\right), \qquad \tilde{w}_j = \frac{1}{Kn}\phi\left(\frac{2j-1}{2n}\right). \tag{2.9.2.5}$$

As in the usual trapezoidal rule,

$$\hat{T}_{2n}(f) = \tfrac{1}{2}(\hat{T}_n(f) + \hat{M}_n(f)).$$

The values of p and q are usually taken to be both equal either to 1 or to 2. If $p = q$ and $a = b$, the rule is symmetric about $x = \frac{1}{2}$. Good results have been achieved with $a = b \in [4, 16]$.

The abscissas x_j, $\tilde{x}_j$ and weights w_j, $\tilde{w}_j$ can be precomputed. To compute, e.g., x_j, it suffices to work with $K^{-1}\int_{j/n}^{(j+1)/n} \phi(x)\,dx$, accumulating these values. It has been found that the use of G_{16} is quite adequate for this purpose. Alternatively, the x_j and w_j may be evaluated as needed using a Tschebyscheff expansion for the x_j. This has been done in the implementation of the IMT rule in one and two dimensions by de Doncker and her co-workers.

The IMT rule appears to be useful in evaluating integrals with endpoint singularities (the transformation often regularizes them) or with real poles near the interval of integration. It is not suitable for integrands of low continuity class inside the interval or for highly oscillatory integrands.

Very often, $\hat{T}_n(f)$ and $\hat{M}_n(f)$ will bracket I for moderate values of n so that if $|\hat{T}_n - \hat{M}_n| < \varepsilon$, it is highly probable that $|\hat{T}_{n+1} - I| < \varepsilon$. Thus, these rules can form the basis of an automatic integration scheme.

An interesting feature of this rule is that it fails to integrate constants exactly. We have $\sum_{j=1}^{n} w_j =$ the evaluation of $K^{-1}\int_0^1 \phi(x)\,dx$ by the trapezoidal rule.

For $n = 128$, this sum is 1 to 20 figures.

Another transformation suggested by Mori which gives faster decay of the integrand at the endpoints is

$$\phi(t) = \tanh(a \sinh[b/(t^2 - 1)]), \qquad a, b > 0, \tag{2.9.2.6}$$

mapping $(-1, 1)$ onto itself. Hence

$$\int_{-1}^{1} f(x)\,dx = \int_{-1}^{1} f(\phi(t))\phi'(t)\,dt \simeq \frac{2}{N}\sum_{n=1}^{N-1} f(\phi(-1+nh))\phi'(-1+nh),$$

$$h = \frac{2}{N}. \qquad (2.9.2.7)$$

The recommended values of the parameters are $a = \pi/2$, $b = \pi/4$.

References. de Doncker and Piessens [1]; Dixon [1]; Iri, Moriguti, and Takasawa [1]; Mori [1]; Murota and Iri [1]; Robinson and de Doncker [1]; Takahasi and Mori [1, 3]; Toda and Ono [1].

2.9.5 Euler–Maclaurin Formulas for Integrands with Endpoint Singularities

If $f(x)$ has period 1 and satisfies the sufficient conditions

(a) $\lim_{\varepsilon \to 0} \frac{1}{2}[f(x+\varepsilon) + f(x-\varepsilon)] = f(x)$,
(b) $\int_0^1 |f(x)|\,dx < \infty$,
(c) $f'(x^+)$ and $f'(x^-)$ exist,

then the Fourier expansion of f converges pointwise to f and we can write

$$f(x) = \sum_{k=-\infty}^{\infty} \exp(-2\pi ikx) \int_0^1 f(t)\exp(2\pi ikt)\,dt. \qquad (2.9.5.1)$$

Designate by $M_n(f)$, $T_n(f)$, and $I(f)$ the midpoint rule, the trapezoidal rule, and $\int_0^1 f(t)\,dt$ respectively. Then, on the basis of the identity (2.9.20), and a similar identity for M_n, it is easily shown that

$$E_{M_n}(f) = M_n(f) - I(f) = \sum_{k=-\infty}^{\infty}{}' (-1)^k \int_0^1 f(t)\exp(2\pi iknt)\,dt,$$

$$E_{T_n}(f) = T_n(f) - I(f) = \sum_{k=-\infty}^{\infty}{}' \int_0^1 f(t)\exp(2\pi iknt)\,dt.$$

The notation $\sum'$ means that the term $k = 0$ is omitted from the sum.

We are interested in the case

$$f(x) = x^{\beta}(1-x)^{\omega}h(x), \qquad \beta > -1, \qquad \omega > -1 \qquad (2.9.5.2)$$

where $h(x) \in C^{N-1}[0, 1]$. In the case of the rule T_n, we must assume $\beta, \omega > 0$. On the basis of asymptotic expansions for the Fourier transform

$\int_0^1 f(t)\exp(2\pi iknt)\,dt$, Lyness and Ninham show that $E_{M_n}(f)$ and $E_{T_n}(f)$ both have asymptotic expansions of the form

$$\left.\begin{matrix} E_{M_n}(f) \\ E_{T_n}(f) \end{matrix}\right\} = \sum_{s=0}^{N-1} \frac{a_s}{n^{\beta+s+1}} + \sum_{s=0}^{N-1} \frac{b_s}{n^{\omega+s+1}} + O(n^{-N}). \tag{2.9.5.3}$$

The constants a_s and b_s are independent of n. In some investigations it is important to identify these constants more closely. Set $\psi_0(x) = (1-x)^\omega h(x)$ and $\psi_1(x) = x^\beta h(x)$. It is clear that $\psi_0(x)$ is $N-1$ times differentiable at $x=0$ while $\psi_1(x)$ is $N-1$ times differentiable at $x=1$. Let $\zeta(s)$ be the Riemann zeta function $\zeta(s) = \sum_{n=1}^{\infty} 1/n^s$ (Re $s > 1$) continued analytically by means of the functional equation

$$\zeta(1-s) = \frac{2(s-1)!}{(2\pi)^s}\cos\left(\frac{\pi s}{2}\right)\zeta(s), \qquad s \neq 1. \tag{2.9.5.4}$$

These particular values are to be noted:

$$\zeta(2p) = \frac{(2\pi)^{2p}}{2(2p)!}|B_{2p}|, \qquad \zeta(-p) = -\frac{1}{p+1}B_{p+1} \tag{2.9.5.5}$$

where p is a positive integer and the B's are the Bernoulli numbers (2.9.11a) augmented by the definition $B_1 = -\frac{1}{2}$, $B_{2p+1} = 0$, $p = 1, 2, \ldots$.

Then, for $\beta > 0$,

$$E_{T_n}(f) = \sum_{s=s_0}^{N-1} \frac{\psi_0^{(s)}(0)\zeta(-\beta-s)}{s!\,n^{\beta+s+1}} + \sum_{s=s_1}^{N-1} \frac{(-1)^s\psi_1^{(s)}(1)\zeta(-\omega-s)}{s!\,n^{\omega+s+1}} + O(n^{-N}). \tag{2.9.5.6}$$

In this formula, s_0 and $s_1 = 0$, unless $\beta = 0$ or $\omega = 0$ in which case s_0 and s_1 respectively equal 1.

Moreover,

$$E_{M_n}(f) = \sum_{s=0}^{N-1} \frac{\psi_0^{(s)}(0)}{s!}\,\frac{(2^{-\beta-s}-1)\zeta(-\beta-s)}{n^{\beta+s+1}} + \sum_{s=0}^{N-1} \frac{(-1)^s\psi_1^{(s)}(1)}{s!}\,\frac{(2^{-\omega-s}-1)\zeta(-\omega-s)}{n^{\omega+s+1}} + O(n^{-N}). \tag{2.9.5.7}$$

It should be observed that $\zeta(-2p) = 0$ if p is a positive integer, and that the expansions above coincide with the usual Euler–Maclaurin expansions in the case of no singularities.

Occasionally, one wants the leading term in the case $\int_0^1 x^\beta h(x)\,dx$, $0 < \beta < 1$. Here $\omega = 0$ and $\psi_0(x) = h(x)$. We have

$$E_{M_n}(f) = \frac{h(0)(2^{-\beta} - 1)\zeta(-\beta)}{n^{\beta+1}} + \cdots \tag{2.9.5.8}$$

while

$$E_{T_n}(f) = \frac{h(0)\zeta(-\beta)}{n^{\beta+1}} + \cdots. \tag{2.9.5.9}$$

Thus,

$$\frac{E_{M_n}(f)}{E_{T_n}(f)} = (2^{-\beta} - 1) + \cdots. \tag{2.9.5.10}$$

As $\beta \to 0$ this ratio goes to 0 so that the midpoint rule is much more effective than the trapezoidal rule for strong algebraic singularities. It is an experimental fact experienced throughout the whole of numerical integration that it is more effective to use rules which avoid the singularities of the integrand. (The word singularity is used in the sense of branchpoint singularity etc. and not merely an infinity of the integrand.)

Lyness and Ninham also present further details for integrands $f(x)$ having algebraic–logarithmic singularities of the type

$$f(x) = x^\beta(1 - x)^\omega |x - x_1|^\gamma \operatorname{sgn}(x - x_2)(\log|x - x_3|)^n \tag{2.9.5.11}$$

etc. In these cases, the identification of the constants becomes more difficult. These authors make use of Fourier transform methods and generalized function theory.

Making use of Laplace transform methods and contour integration, Waterman, Yos, and Abodeely derive asymptotic expressions for algebraic singularities as well as for singularities of the type $\int_0^1 x^{-1/2} e^{-1/x} f(x)\,dx$.

References. Fox [2]; Fox and Hayes [1]; Lyness and Ninham [1]; Navot [1]; Ninham and Lyness [1]; Rabinowitz [11]; Waterman, Yos, and Abodeely [1].

2.10 Integration of Rapidly Oscillatory Functions

By a *rapidly oscillatory integrand* we mean one with numerous (more than 10) local maxima and minima over the range of integration.

The principal examples of rapidly oscillatory integrands occur in various transforms. There is the *Fourier transform*:

$$\int_a^b f(x) \cos nx\,dx, \qquad \int_a^b f(x) \sin nx\,dx$$

or in complex form,

$$\int_a^b f(x) e^{isx}\,dx.$$

There is the *Fourier–Bessel transform*

$$\int_0^1 f(x)xJ_n(\gamma_m x)\,dx,$$

where $0 < \gamma_1 < \gamma_2 < \cdots$ are the zeros of the Bessel function $J_n(x)$.

What is usually desired is not the value of an isolated integral, but a whole family of such integrals where the oscillations increase. We may take the general form to be

$$I(t) = \int_a^b f(x)K(x, t)\,dx, \qquad -\infty \leq a < b \leq \infty, \tag{2.10.1}$$

where $K(x, t)$ is an oscillatory kernel and $f(x)$ is the "nonoscillatory" part.

Numerical integration of oscillatory integrands is beset with difficulties peculiar to it. For example, consider the determination of

$$\int_0^{2\pi} f(x) \cos tx\,dx.$$

As $t \to \infty$, the graph of $f(x) \cos tx$ will consist of plus areas and minus areas of nearly equal size and the resulting cancellation of area is attended by a loss of significance. Furthermore, as $t \to \infty$, the function $f(x) \cos tx$ looks less and less like a polynomial of low degree, and this suggests that special methods should be developed.

THEOREM. *If $f(x) \in C[0, 2\pi]$ and*

$$a_n = \int_0^{2\pi} f(x) \cos nx\,dx, \qquad b_n = \int_0^{2\pi} f(x) \sin nx\,dx,$$

then

$$|a_n|, |b_n| \leq w(\pi/n), \qquad n = 1, 2, \ldots$$

where $w(\delta)$ is the modulus of continuity of $f(x)$ on $[0, 2\pi]$ (see Section 2.1).

Fourier integrals may occasionally be treated by means of repeated integration by parts.

THEOREM. *Let $f(x) \in C^n[a, b]$. Then*

$$\begin{aligned} I(s) &= \int_a^b e^{isx} f(x)\,dx \\ &= e^{isb} \sum_{k=0}^{n-1} i^{k-1} f^{(k)}(b) s^{-k-1} - e^{isa} \sum_{k=0}^{n-1} i^{k-1} f^{(k)}(a) s^{-k-1} \\ &\quad + (-is)^{-n} \int_a^b e^{isx} f^{(n)}(x)\,dx. \end{aligned} \tag{2.10.2}$$

Proof. Integrate by parts n times.

COROLLARY. *This theorem is true when* $a = -\infty$, $b = \infty$ *(or both), provided that* $f^{(k)}(x) \to 0$ *as* $x \to -\infty$ *(or* $x \to \infty$*) for each* $k = 0, 1, \ldots, n-1$ *and provided that*

$$\int_a^b |f^{(n)}(x)|\, dx < \infty. \tag{2.10.3}$$

For example,

$$\int_a^\infty e^{isx} f(x)\, dx = -e^{isa} \sum_{k=0}^{n-1} i^{k-1} f^{(k)}(a) s^{-k-1} + (-is)^{-n} \int_a^\infty e^{isx} f^{(n)}(x)\, dx. \tag{2.10.4}$$

This theorem provides an asymptotic expansion of $I(s)$ in negative powers of s. It may be useful numerically when the parameter is large.

Example. $I_n = \int_0^\pi e^{x^2} \cos nx\, dx$. Integration by parts three times yields

$$I_n = \frac{(-1)^n}{n^2} 2\pi e^{\pi^2} + \frac{2}{n^3} \int_0^\pi (xe^{x^2})'' \sin nx\, dx.$$

Since the integral on the right approaches 0 as $n \to \infty$, we have

$$I_n = \frac{(-1)^n}{n^2} 2\pi e^{\pi^2} + \frac{2\varepsilon_n}{n^3}$$

with $\lim_{n\to\infty} \varepsilon_n = 0$.

Generalizations can be obtained in which $f(x)$ is allowed to possess singularities at $x = a$ or $x = b$.

THEOREM. *Let* $f(x) \in C^n[a, b]$. *Let* $0 < \lambda \le 1$, $0 < \mu \le 1$. *Then,*

$$I(s) = \int_a^b e^{ixs}(x-a)^{\lambda-1}(b-x)^{\mu-1} f(x)\, dx = B_n(s) - A_n(s) + O(s^{-n}) \tag{2.10.5}$$

where

$$A_n(s) = \sum_{k=0}^{n-1} \frac{\Gamma(k+\lambda)}{k!} e^{\pi i(k+\lambda-2)/2} s^{-k-\lambda} e^{isa} \times \frac{d^k}{da^k}[(b-a)^{\mu-1} f(a)]$$

$$B_n(s) = \sum_{k=0}^{n-1} \frac{\Gamma(k+\mu)}{k!} e^{\pi i(k-\mu)/2} s^{-k-\mu} e^{isb} \times \frac{d^k}{db^k}[(b-a)^{\lambda-1} f(b)]. \tag{2.10.6}$$

In the case when $\lambda = \mu = 1$, the error term becomes $o(s^{-n})$ as shown by (2.10.2).

The expansion (2.10.2) is made the basis of a method of numerical approximation of Fourier transforms by Stetter. Stetter's program is to obtain a polynomial approximation to $I(s)$ in powers of $1/s$:

$$I(s) \approx a_0 + \frac{a_1}{s} + \frac{a_2}{s^2} + \cdots + \frac{a_q}{s^q}. \tag{2.10.7}$$

The constants $a_0, \ldots, a_q$ are determined from a knowledge of alternating rectangular sums and alternating trapezoidal sums, i.e., from

$$R_n^A(f) = h \sum_{k=0}^{n-1} (-1)^k f(a + kh)$$

and

$$T_n^A(f) = h\left[\frac{f(a)}{2} - f(a+h) + f(a+2h) - \cdots\right].$$

References. Erdélyi [1]; Lighthill [1]; H. J. Stetter [1].

2.10.1 *Integration between the Zeros*

In this very simple method, the zeros of the oscillatory part of the integrand are located: $a \le x_1 < x_2 < \cdots < x_p \le b$ and each subintegral $\int_{x_i}^{x_{i+1}}$ is evaluated by a rule. It is advantageous to use a rule that employs the values of the integrand at the endpoints of the integration interval. Since the integrand is zero at these points, more accuracy is obtained without additional computation. A Lobatto rule would seem particularly good for this purpose.

Let us write

$$\int_0^{2\pi} f(x) \sin nx \, dx = \sum_{k=0}^{2n-1} \int_{k\pi/n}^{(k+1)\pi/n} f(x) \sin nx \, dx. \tag{2.10.1.1}$$

In each of the integrals in the right-hand sum, the integrand vanishes at the endpoints. Hence

$$\int_{k\pi/n}^{(k+1)\pi/n} f(x) \sin nx \, dx$$

may be expeditiously computed by use of a Lobatto rule. For example, a 5-point Lobatto rule (2 endpoints and 3 interior points) can be carried out with three functional evaluations per interval. A similar reduction can be made for the integral $\int_0^{2\pi} f(x) \cos nx \, dx$.

An alternative development has been worked out by Price for finite Fourier integrals. Let us consider, for example,

$$I(k) = \int_0^{\pi} g(t) \sin kt \, dt = \frac{1}{k} \int_0^{k\pi} g\left(\frac{u}{k}\right) \sin u \, du. \tag{2.10.1.2}$$

If k is an integer, we have

$$I(k) = \frac{1}{k} \sum_{j=1}^{k} \int_{(j-1)\pi}^{j\pi} g\left(\frac{u}{k}\right) \sin u \, du. \tag{2.10.1.3}$$

This throws the burden onto integrals of the form $\int_{(j-1)\pi}^{j\pi} f(y) \sin y \, dy$ for which the following formula of interpolatory type can be obtained:

$$\begin{aligned}
(-1)^{j-1} & \int_{(j-1)\pi}^{j\pi} f(y) \sin y \, dy \\
&= H_1[f((j-1)\pi) + f(j\pi)] \\
&\quad + H_2[f((j-\tfrac{3}{4})\pi) + f((j-\tfrac{1}{4})\pi)] + H_3 f((j-\tfrac{1}{2})\pi) \\
&\quad + .01002 f^{(4)}(\xi), \qquad (j-1)\pi < \xi < j\pi,
\end{aligned} \tag{2.10.1.4}$$

$$H_1 = 1 + \frac{2}{\pi} - \frac{16}{\pi^2} \approx .0154\,8083,$$

$$H_2 = -\frac{8}{\pi} + \frac{32}{\pi^2} \approx .6957\,9879,$$

$$H_3 = \frac{12}{\pi} - \frac{32}{\pi^2} \approx .5774\,4076.$$

We note that when (2.10.1.4) is substituted into (2.10.1.3), the terms involving H_1 "telescope" into the sum $H_1[g(0) - (-1)^k g(\pi)]$. Note also that this formula requires the evaluation only of f at the given abscissas and does not require any computation of $\sin x$. However, this scheme requires separate treatment for each value of n; Filon's method (Section 2.10.2) uses the same values of the function for all values of n.

A similar method has been proposed by van de Vooren and van Linde. They compute a series of approximations and use extrapolation to improve their results.

References. J. F. Price [1]; van de Vooren and van Linde [1].

2.10.2 *Use of Approximation: Filon's Method for Finite Fourier Integrals*

Suppose that we can write

$$f(x) = a_1\phi_1(x) + a_2\phi_2(x) + \cdots + a_n\phi_n(x) + \varepsilon(x), \qquad a \le x \le b, \tag{2.10.2.1}$$

where $\varepsilon(x)$ is small over $a \le x \le b$ and where the transforms

$$\psi_k(t) = \int_a^b \phi_k(x)K(t, x)\,dx, \qquad k = 1, 2, \ldots, n \tag{2.10.2.2}$$

can be computed explicitly in elementary terms. (This will be the case, for example, where $\phi_k(x) = x^k$ and $K(x, t) = e^{itx}$.) Then

$$\begin{aligned} I(t) = \int_a^b f(x)K(x, t)\,dx &= \sum_{k=1}^{n} a_k\psi_k(t) + \int_a^b \varepsilon(x)K(x, t)\,dx \\ &\approx \sum_{k=1}^{n} a_k\psi_k(t). \end{aligned} \tag{2.10.2.3}$$

This program was worked out by Filon, who approximates $f(x)$ by parabolic arcs.

Consider the integral

$$I(k) = \int_a^b f(t)\cos kt\,dt. \tag{2.10.2.4}$$

In Filon's method, the interval $[a, b]$ is divided into $2N$ subintervals of equal length h:

$$h = (b - a)/2N. \tag{2.10.2.5}$$

Over each double subinterval, $f(t)$ is approximated by a parabola obtained by interpolation to $f(t)$ at the mesh points. For parabolic $f(t)$, the Fourier integrals can be computed explicitly by integration by parts. This program leads to the following rules of approximate integration. Let

$$\begin{aligned} C_{2n} = {}& \tfrac{1}{2}f(a)\cos ka + f(a + 2h)\cos k(a + 2h) \\ & + f(a + 4h)\cos k(a + 4h) + \cdots + \tfrac{1}{2}f(b)\cos kb, \end{aligned} \tag{2.10.2.6}$$

$$\begin{aligned} C_{2n-1} = {}& f(a + h)\cos k(a + h) + f(a + 3h)\cos k(a + 3h) \\ & + \cdots + f(b - h)\cos k(b - h). \end{aligned} \tag{2.10.2.7}$$

Similarly, define S_{2n} and S_{2n-1} as the corresponding sums formed from $f(t) \sin kt$. Further, let

$$\theta = kh = k(b - a)/2N, \tag{2.10.2.8}$$

and

$$\begin{aligned} \alpha &= \alpha(\theta) = (\theta^2 + \theta \sin\theta \cos\theta - 2\sin^2\theta)/\theta^3, \\ \beta &= \beta(\theta) = 2[\theta(1 + \cos^2\theta) - 2\sin\theta\cos\theta]/\theta^3, \\ \gamma &= \gamma(\theta) = 4(\sin\theta - \theta\cos\theta)/\theta^3. \end{aligned} \tag{2.10.2.9}$$

Then

$$\begin{aligned} \int_a^b f(t)\cos kt\,dt &\approx h\{\alpha[f(b)\sin kb - f(a)\sin ka] + \beta C_{2n} + \gamma C_{2n-1}\}, \\ \int_a^b f(t)\sin kt\,dt &\approx h\{-\alpha[f(b)\cos kb - f(a)\cos ka] + \beta S_{2n} + \gamma S_{2n-1}\}. \end{aligned} \tag{2.10.2.10}$$

It should be noted that for small θ, the functions α, β and γ have the Taylor expansions

$$\begin{aligned} \alpha(\theta) &= \tfrac{2}{45}\theta^3 - \tfrac{2}{315}\theta^5 + \tfrac{2}{4725}\theta^7 + \cdots, \\ \beta(\theta) &= \tfrac{2}{3} + \tfrac{2}{15}\theta^2 - \tfrac{4}{105}\theta^4 + \tfrac{2}{567}\theta^6 + \cdots, \\ \gamma(\theta) &= \tfrac{4}{3} - \tfrac{2}{15}\theta^2 + \tfrac{1}{210}\theta^4 - \tfrac{1}{11340}\theta^6 + \cdots. \end{aligned} \tag{2.10.2.11}$$

Thus for $\theta = 0$, Filon's rule reduces to an $N \times S$ rule.

The expressions (2.10.2.9) are indeterminate for $\theta = 0$ so that when θ is small, their use should be avoided in favor of alternative expressions such as (2.10.2.11). The value $\theta = \frac{1}{6}$ has been recommended as the switchover point for a machine having a 44-bit mantissa.

It is generally advisable in numerical work to keep the parameter θ smaller than 1. This means that for large values of k, we are compelled to take small values of h. However, if $f(x)$ can be approximated well by a piecewise quadratic function $p_2(x)$ using a coarse mesh, then θ need not be small at all. Indeed, if $\max_{a \le x \le b} |f(x) - p_2(x)| \le \varepsilon$, then a uniform bound on the error in Filon's method is given by $(b - a)\varepsilon$.

Let E_S and E_C designate respectively the error in Filon's sine and cosine formulas. Set

$$H(\theta) = \left|\frac{\sin\theta}{3\theta^2} + \frac{\cos\theta}{\theta^3} - \frac{\sin\theta}{\theta^4}\right| \tag{2.10.2.12a}$$

$$M = \max_{a \le x \le b} |f^{(3)}(x)|. \tag{2.10.2.12b}$$

Then, assuming that $\theta < 1$, it has been shown that

$$\left.\begin{matrix} |E_S| \\ |E_C| \end{matrix}\right\} \leq (b-a)MH(\theta)h^3 + O(h^4). \tag{2.10.2.12c}$$

For an exact expression for the error, see Håvie, who derives Filon's method as a special case of computing an expansion for the integral

$$\int_a^b f(x)e^{igx}\,dx, \qquad g \text{ arbitrary}$$

using the expansion for $f(x)$ in terms of Bernoulli polynomials.

An extension of Filon's method has been made by Flinn, who uses fifth-degree polynomials, found by interpolating to the values of $f(x)$ and of its first derivative at the above points.

Luke extends Filon by approximating $f(x)$ by piecewise polynomials of higher degree.

References. Buyst and Schotsmans [1]; Chase and Fosdick [1, F1]; Filon [1]; Flinn [1]; Fosdick [1]; Hamming [1]; Håvie [7]; Kruglikova [1]; Luke [2]; Teijelo [A1]; Zhileikin [1].

Examples. In the following examples, G_{32} designates a Gauss 32-point rule; $2n \times L_4$ and $2n \times L_5$ designate, respectively, the result of a 4- and 5-point Lobatto rule inserted into each arch of the sine curve; F_i designates the Filon rule with $h = 2\pi/in$. Thus, in F_{11},

$$nh = \theta = 2\pi/11 \approx .57.$$

"Price" designates the rule (2.10.1.4). For a given n, it requires $6n + 2$ evaluations of $f(x)$ but no evaluations of $\sin nx$.

It is interesting to note the poor quality of G_{32} as n becomes large. However, for small n, a high-order Gauss rule seems to be competitive, as is seen in the following tables.

$$\int_0^{2\pi} x \cos x \sin nx\,dx = \begin{cases} -\pi/2, & n = 1, \\ -2n\pi/(n^2-1) & n \neq 1. \end{cases}$$

n	Exact	G_{32}	$2n \times L_4$	$2n \times L_5$
1	−1.5707 9633	−1.5704 811	−2.4368 252	
2	−4.1887 902	−4.1842 807	−4.7246 829	
4	−1.6755 161	−1.6756 476	−1.8613 529	
10	−.6346 6518	−.6340 2069	−.7020 6954	−.5587 594
20	−.3149 4663	−1.2092 524	−.3481 8404	−.2778 962
30	−.2096 7243	−1.5822 272	−.2317 7723	−.1850 8448

	F_7	F_9	F_{11}	Price
10	−.6346 6469	−.6346 6497	−.6346 6508	−.6346 6486
20	−.3149 4462	−.3149 4463	−.3149 4463	−.3149 4662
30	−.2096 7248	−.2096 7248	−.2096 7248	−.2096 7248

Here $f(x) = x \cos x$ is very smooth and Filon yields good results.

$$\int_0^{2\pi} x \cos 50x \sin nx \, dx = \frac{2n\pi}{2500 - n^2} \qquad (n \neq 50).$$

n	Exact	G_{32}	$2n \times L_4$	$2n \times L_5$
1	.0025 1428	2.1561 858	.5631 2612	
2	.0050 3460	.8756 9672	4.7048 779	
4	.0101 1785	−.0974 2242	.6975 8325	
10	.0261 7994	.1058 5082	9.3325 313	.3305 4736
20	.0598 3986	−.8128 6374	.1110 445	−.3301 2624
30	.1178 0972	−.6453 4403	.1639 1873	−.0304 516

	F_7	F_9	F_{11}	Price
10	.1104 9126	.2979 8208	.2905 7620	.1448 4073
20	.0928 7746	.0391 2549	.0534 6164	.0051 1070
30	.1020 5656	.1134 3113	.1160 0732	.1101 3379

Here $f(x) = x \cos 50x$ is highly oscillatory so that the polynomial rules do poorly even for low n. In F_{11}, $nh = .57$ so that with $n = 10$, $h = .057$. One arch of the curve $f(x)$ has width $\pi/50 \approx .063$ and approximation by parabolic arches of width $2 \times .057$ would be poor.

$$\int_0^{2\pi} \frac{x \sin nx}{(1 - x^2/4\pi^2)^{1/2}} dx = 2\pi^3 J_1(2\pi n) \approx 62.012553 J_1(2\pi n).$$

Note the singularity at $x = 2\pi$.

n	Exact	G_{32}	$2n \times L_4$	$2n \times L_5$
1	−13.1704	−13.1715	−13.9910	
2	−9.5829	−9.5829	−10.2183	
4	−6.8761	−6.8778	−7.3418	
10	−4.3876	−4.3942	−4.6872	−4.4390
20	−3.1118	−1.2104	−3.3245	−3.1460
30	−2.5433	−7.5194	−2.7171	−2.5706

	F_7	F_9	F_{11}	Price
10	−4.0049	−4.1616	−4.2347	−4.0998
20	−2.8411	−2.9519	−3.0037	−2.9082
30	−2.3223	−2.4128	−2.4550	−2.3771

$$\int_0^{2\pi} \log x \sin nx \, dx = -\frac{1}{n}[\gamma + \log 2n\pi - \mathrm{Ci}(2n\pi)], \qquad \gamma \approx .5772\ 1566\ 5,$$

$$\mathrm{Ci} = \text{cosine integral}.$$

Note the singularity at $x = 0$.

n	Exact	G_{32}	$2n \times L_4$	$2n \times L_5$
1	−2.4377	−2.4378	−2.4380	
2	−1.5572	−1.5570	−1.5880	
4	−.9507	−.9507	−.9830	
10	−.4718	−.4721	−.4941	−.4410
20	−.2705	.1237	−.2853	−.2511
30	−.1939	−1.1143	−.2051	−.1793

	F_7	F_9	F_{11}	Price
10	−.4569	−.4646	−.4676	−.4617
20	−.2622	−.2666	−.2683	−.2650
30	−.1880	−.1911	−.1923	−.1899

If instead of parabolic arches we use polygonal arches, we obtain a "Filon–trapezoidal" rule. This is cruder than standard Filon, but has also been found useful. Write the integral in complex form

$$I(k) = \int_{-T}^{T} f(t)e^{ikt} \, dt$$

and set $h = T/N$. Then,

$$I(k) \approx h \sum_{n=-N}^{N} \omega_n e^{iknh} f(nh) \tag{2.10.2.13}$$

where

$$\omega_{-N} = (1 + ikh - e^{ikh})/k^2h^2$$

$$\omega_n = \left(\left(\sin\frac{kh}{2}\right) \Big/ \frac{kh}{2}\right)^2, \qquad n \neq \pm N$$

$$\omega_N = (1 - ikh - e^{-ikh})/k^2h^2. \tag{2.10.2.14}$$

Reference. Tuck [1].

If a cubic spline $S(x)$ is passed through selected functional values and the resulting integrations are carried out formally, we obtain a rule which will be called the Filon–spline rule. Refer to the notation of Section 2.3.1. It is advisable, if possible, to use the boundary conditions

$$S'(a) = f'(a), \qquad S'(b) = f'(b).$$

If these derivatives are not available, one can approximate them by differentiating the appropriate cubic interpolation polynomial.

Integration by parts yields

$$\begin{aligned}\int_a^b & S(x) \cos kx \, dx \\ &= \left[\frac{1}{k} S(x) \sin kx + \frac{1}{k^2} S'(x) \cos kx - \frac{1}{k^3} S''(x) \sin kx\right]_a^b \\ &\quad + \frac{1}{k^3} \int_a^b S'''(x) \sin kx \, dx. \end{aligned} \tag{2.10.2.15}$$

These expressions may be evaluated in terms of the quantities M_i using (2.3.1.7), (2.3.1.8), and (2.3.1.5). In particular, we have

$$\frac{1}{k^3} \int_a^b S'''(x) \sin kx \, dx = -\frac{1}{k^4} \sum_{j=1}^{n} \frac{M_j - M_{j-1}}{h_j} (\cos kx_j - \cos kx_{j-1}). \tag{2.10.2.16}$$

Similar formulas hold for $\int_a^b S(x) \sin kx \, dx$.

If we assume that the interval is divided into n equal parts, explicit formulas, analogous to those of Filon, can be obtained. These formulas are lengthy and the reader is referred to the articles by Einarsson.

On the basis of error estimates for spline approximation obtained by Birkhoff and deBoor, we can obtain the following error estimate for the error in the Filon–Spline rule. (We assume that the boundary conditions $S'(a) = f'(a)$, $S'(b) = f'(b)$ are used.)

$$\left| \int_a^b f(x) \cos kx \, dx - \int_a^b S(x) \cos kx \, dx \right| \le h^4 \frac{990}{7} (b - a) \max_{a \le x \le b} |f^{(4)}(x)|. \tag{2.10.2.17}$$

If the spline boundary conditions are $M_0 = M_n = 0$, Filon is superior.

Marsden and Taylor have extended this work and derived integration rules of the form

$$\int_a^b f(x)e^{iwx}\,dx = hA\left[B\sum_{j=0}^{n}{}'' f(x_j)e^{iwx_j} - \sum_{r=0}^{k} i^{r+1}h^r C_r(f^{(r)}(b)e^{iwb} - f^{(r)}(a)e^{iwa})\right] + Rf \quad (2.10.2.18)$$

where $h = (b-a)/n$, $x_j = a + jh$, and A, B, C_r are real functions of wh and k chosen so that $Rf = 0$ whenever f is a spline of degree k with knots at the points x_j if k is odd and at the points $x_j + h/2$ if k is even. The expressions for A, B, and C_r are listed for $k = 0, \ldots, 7$ and formulas are given for their determination for any k.

Dierckx and Piessens discuss the computation of Fourier coefficients of cubic B-splines; for B-splines of arbitrary degree, see Marti and Neuman [1].

Example. Einarsson reports the following experience with the integral $I = \int_2^{12} e^{-x}\cos kx\,dx$. $h = 10/N$. S designates the spline method while F is Filon. The entries are the *errors* $\times 10^8$.

k	I	$N =$	16	32	64	128	256
0.0005	0.1353 2896	S	−2841	−180	−14	−7	−14
		F	10959	708	42	−1	−7
0.005	0.1353 1223	S	−2841	−180	−14	−7	−14
		F	10961	707	42	−1	−7
0.05	0.1336 4461	S	−2807	−177	−14	−7	−14
		F	11155	720	43	0	−7
0.5	0.0129 3988	S	−343	−18	−1	−1	−2
		F	20351	1325	83	4	0
5.0	0.0097 9102	S	−589	−13	−1	−1	−1
		F	50331	−2963	−166	−10	−1
50.0	0.0014 1670	S	2	0	0	0	0
		F	−165	14	−35	−14	1

References. Birkhoff and de Boor [1]; Dierckx and Piessens [1]; Einarsson [1, 2, F1]; Lax and Agrawal [1]; Marsden and Taylor [1]; Marti [1]; E. Neuman [1,2].

2.10.3 *Application of Speed-Up Methods*

It is also possible to break up an oscillatory integral into a series of integrands of alternating sign and to apply a transformation designed to "speed up" the convergence of such a series.

The following formula, which is a variation of the much employed Euler transformation,† was developed by Longman.

Let

$$S = v_0 - v_1 + v_2 - \cdots + (-1)^n v_n. \tag{2.10.3.1}$$

Write

$$S(x) = v_0 - v_1 x + v_2 x^2 - \cdots + (-1)^n v_n x^n. \tag{2.10.3.2}$$

Then

$$\begin{aligned}(1 + x)S(x) &= v_0 - (v_1 - v_0)x + (v_2 - v_1)x^2 - \cdots + (-1)^n(v_n - v_{n-1})x^n \\ &\quad + (-1)^n v_n x^{n+1} \\ &= v_0 - (\Delta v_0)x + (\Delta v_1)x^2 - \cdots + (-1)^n(\Delta v_{n-1})x^n \\ &\quad + (-1)^n v_n x^{n+1}.\end{aligned} \tag{2.10.3.3}$$

Here we use the notation

$$\begin{aligned}\Delta v_k &= v_{k+1} - v_k, \\ \Delta^{r+1} v_k &= \Delta^r v_{k+1} - \Delta^r v_k.\end{aligned} \tag{2.10.3.4}$$

From (2.10.3.3) we obtain

$$\begin{aligned}S(x) = &\frac{v_0 + (-1)^n v_n x^{n+1}}{1 + x} \\ &- y[\Delta v_0 - (\Delta v_1)x + (\Delta v_2)x^2 - \cdots + (-1)^{n-1}(\Delta v_{n-1})x^{n-1}],\end{aligned} \tag{2.10.3.5}$$

where

$$y = x/(1 + x). \tag{2.10.3.6}$$

Applying this transformation again to the bracketed series in (2.10.3.5), we obtain

$$\begin{aligned}S(x) = &\frac{v_0 + (-1)^n v_n x^{n+1}}{1 + x} - \frac{\Delta v_0 + (-1)^{n-1}(\Delta v_{n-1})x^n}{1 + x} y \\ &+ y^2[\Delta^2 v_0 - (\Delta^2 v_1)x + (\Delta^2 v_2)x^2 - \cdots + (-1)^{n-2}(\Delta^2 v_{n-2})x^{n-2}].\end{aligned} \tag{2.10.3.7}$$

† For the Euler transformation, see Section 3.8.

If $p \le n$, then p applications yield

$$S(x) = \frac{v_0 - y\Delta v_0 + y^2\Delta^2 v_0 - \cdots + (-1)^{p-1}y^{p-1}\Delta^{p-1}v_0}{1+x}$$
$$+ \frac{(-1)^n[v_n x^{n+1} + (\Delta v_{n-1})x^n y + (\Delta^2 v_{n-2})x^{n-1}y^2 + \cdots + (\Delta^{p-1}v_{n-p+1})x^{n-p+2}y^{p-1}]}{1+x}$$
$$+ (-1)^p y^p[\Delta^p v_0 - (\Delta^p v_1)x + (\Delta^p v_2)x^2 - \cdots + (-1)^{n-p}(\Delta^p v_{n-p})x^{n-p}]. \quad (2.10.3.8)$$

Set $x = 1$, and we obtain the identity

$$S = \tfrac{1}{2}v_0 - \tfrac{1}{4}\Delta v_0 + \tfrac{1}{8}\Delta^2 v_0 - \cdots + (-1)^{p-1}2^{-p}\Delta^{p-1}v_0$$
$$+ (-1)^n[\tfrac{1}{2}v_n + \tfrac{1}{4}\Delta v_{n-1} + \tfrac{1}{8}\Delta^2 v_{n-2} + \cdots + 2^{-p}\Delta^{p-1}v_{n-p+1}]$$
$$+ 2^{-p}(-1)^p[\Delta^p v_0 - \Delta^p v_1 + \Delta^p v_2 - \cdots + (-1)^{n-p}\Delta^p v_{n-p}], \qquad p \le n. \quad (2.10.3.9)$$

Assuming now that n and p are large and that the high-order differences are small, we neglect the last bracket and obtain

$$S \approx \tfrac{1}{2}v_0 - \tfrac{1}{4}\Delta v_0 + \tfrac{1}{8}\Delta^2 v_0 - \cdots$$
$$+ (-1)^n[\tfrac{1}{2}v_n + \tfrac{1}{4}\Delta v_{n-1} + \tfrac{1}{8}\Delta^2 v_{n-2} + \cdots]. \quad (2.10.3.10)$$

Example. (Longman). Determine $I = \int_0^{100\pi} (100^2\pi^2 - x^2)^{1/2} \sin x \, dx$. Write

$$I = \sum_{r=0}^{99} (-1)^r \int_0^\pi [100^2\pi^2 - (r\pi + x)^2]^{1/2} \sin x \, dx$$
$$= \sum_{r=0}^{99} (-1)^r v_r,$$

where v_r are the integrals in the above sum. For values of r near 0 and near 99, the quantities v_r were computed by use of a 16-point Gauss rule.

r	v_r	r	v_r
1	628.30915	89	280.25486
2	628.24630	90	267.27464
3	628.12061	91	253.47487
4	627.93204	92	238.71325
5	627.68049	93	222.79836
6	627.36594	94	205.46181
⋮	⋮	95	186.30583
85	325.85292	96	164.30583
86	315.26077	97	139.47917
87	304.17027	98	108.12528
88	292.52472	99	60.96022

Since v_r changes rapidly near $r = 99$, it is more expeditious to write

$$I = \sum_{r=0}^{85} (-1)^r v_r + \sum_{r=86}^{99} (-1)^r v_r$$

and to apply the summation formula (2.10.3.10) only to the first sum above. This yields (we omit the computation of the differences)

$$I = 298.43558.$$

The exact value is $I = 50\pi^2 H_1(100\pi) = 298.435716$, where H_1 is the Struve function, so that disagreement occurs only in the last two places of the approximation.

References. Kukarin and Novikova [1]; Longman [4]; NBS Handbook [1, p. 496]; Rosser [1]; Shanks [1].

2.10.4 *Use of Gauss Rules for Fourier Coefficients*

Two ways have been suggested.

(1) Define positive weighting functions

$$\left.\begin{aligned} c_m(x) &= \tfrac{1}{2}(1 + \cos m\pi x) \\ s_m(x) &= \tfrac{1}{2}(1 + \sin m\pi x) \end{aligned}\right\} \quad -1 \le x \le 1, \qquad m = 0, 1, \ldots .$$

Then

$$\begin{aligned} (1/2\pi) \int_{-\pi}^{\pi} f(x) \cos mx \, dx &= \int_{-1}^{1} f(\pi x) c_m(x) \, dx - \tfrac{1}{2} \int_{-1}^{1} f(\pi x) \, dx; \\ (1/2\pi) \int_{-\pi}^{\pi} f(x) \sin mx \, dx &= \int_{-1}^{1} f(\pi x) s_m(x) \, dx - \tfrac{1}{2} \int_{-1}^{1} f(\pi x) \, dx. \end{aligned} \tag{2.10.4.1}$$

Gauss-type rules can now be obtained for the first integrals on the right-hand side.

(2) Write for the sine transform

$$\int_0^1 \sin 2\pi nx \, f(x) \, dx = \sum_{j=1}^{n} \int_{-\pi}^{\pi} \sin u \, f\left[\frac{(u + \pi)j}{2n\pi}\right] du \tag{2.10.4.2}$$

and use the method described in Section 2.7.6 for the odd weighting function $\sin x$ to obtain Gauss-type rules of the form

$$\int_{-\pi}^{\pi} \sin x \, g(x) \, dx \simeq \sum_{k=1}^{N} w_k[g(x_k) - g(-x_k)]. \tag{2.10.4.3}$$

For the cosine transform

$$\int_0^1 \cos 2\pi nx\, f(x)\, dx = \sum_{j=1}^{n} \int_{-\pi}^{\pi} \cos u\, f\left[\frac{(u+\pi)j}{2n\pi}\right] du \quad (2.10.4.4)$$

this approach does not succeed. However, Gauss-type rules of the form

$$\int_{-\pi}^{\pi} \cos x\, g(x)\, dx \simeq \sum_{k=1}^{N} w_k[g(x_k) + g(-x_k)] + w_{N+1}g(0) \quad (2.10.4.5)$$

have been tabulated by Piessens and Branders for selected values of N. For N of the form $3M - 1$, one of the abscissas is larger than π. Furthermore, the weights w_k are not all positive for any $N > 2$.

References. Gautschi [5]; Hopp [1]; Mikloško [1–4]; Piessens [4, 9]; Piessens and Branders [5]; Piessens and Haegemans [1]; Zamfirescu [1].

2.10.5 Use of Tschebyscheff and Legendre Expansions

Piessens and Poleunis have developed the following method for the integration of Fourier coefficients which they claim gives high accuracy at low cost. The idea is to use the relations

$$\int_{-1}^{1} \frac{\sin mxT_n(x)\, dx}{(1-x^2)^{1/2}} = \begin{cases} 0, & n = 2k, \\ (-1)^k \pi J_n(m), & n = 2k+1; \end{cases} \quad (2.10.5.1)$$

$$\int_{-1}^{1} \frac{\cos mxT_n(x)\, dx}{(1-x^2)^{1/2}} = \begin{cases} (-1)^k \pi J_n(m), & n = 2k, \\ 0, & n = 2k+1. \end{cases} \quad (2.10.5.2)$$

If we then have the expansion

$$(1-x^2)^{1/2} f(x) = \sum_{n=0}^{\infty}{}' C_n T_n(x), \quad (2.10.5.3)$$

we have that

$$S(m) = \int_{-1}^{1} f(x) \sin mx\, dx = \pi \sum_{k=0}^{\infty} C_{2k+1}(-1)^k J_{2k+1}(m) \quad (2.10.5.4)$$

$$C(m) = \int_{-1}^{1} f(x) \cos mx\, dx = \pi \sum_{k=0}^{\infty}{}' C_{2k}(-1)^k J_{2k}(m). \quad (2.10.5.5)$$

The C_n are approximated using the α_n given by (2.13.1.10) as follows:

$$C_{2k} \approx -\frac{2}{\pi}\sum_{i=0}^{[N/2]}{}' \alpha_{2i}\left[\frac{1}{(2i+2k)^2-1}+\frac{1}{(2i-2k)^2-1}\right], \tag{2.10.5.6}$$

$$C_{2k+1} \approx -\frac{2}{\pi}\sum_{i=0}^{[(N-1)/2]} \alpha_{2i+1}\left[\frac{1}{(2i+2k+2)^2-1}+\frac{1}{(2i-2k)^2-1}\right], \tag{2.10.5.7}$$

where N is chosen sufficiently large. Note that m is not necessarily an integer. Note also that once the C_n are computed, we can compute $S(m)$ and $C(m)$ for as many values of m as are needed with very little additional computation.

The same approach can also be used to yield the following approximations:

$$\int_{-1}^{1} e^{mx} f(x)\,dx \simeq \pi \sum_{k=0}^{M}{}' C_k I_k(m), \tag{2.10.5.8}$$

$$\int_{-1}^{1} e^{mx^2} f(x)\,dx \simeq 2\pi e^{m/2} \sum_{k=0}^{M}{}' C_{2k} I_k(m/2). \tag{2.10.5.9}$$

All the sums involving Bessel functions (J_n) or modified Bessel functions (I_n) of the same argument can be evaluated by using a modification of the backward recurrence formulas (1.12.9).

If the function $f(x)$ is smooth, then the convergence of (2.10.5.3) will be slow inasmuch as we have introduced a nonsmooth component. The Tschebyscheff expansion of $f(x)$ itself given by (2.13.1.7) will converge more rapidly. Algorithms for computing $S(m)$ and $C(m)$ based on (2.13.1.7) or more precisely on the finite approximation (2.13.1.13) are given by Patterson and by Alaylioglu *et al.*

An alternative approach is to use the expansion of $f(x)$ in Legendre polynomials

$$f(x) = \sum_{k=0}^{\infty} c_k P_k(x), \qquad c_k = \frac{2k+1}{2}\int_{-1}^{1} f(x) P_k(x)\,dx. \tag{2.10.5.10}$$

The advantage of this approach is that we have the closed form

$$\int_{-1}^{1} e^{imx} P_k(x)\,dx = i^k (2\pi/m)^{1/2} J_{k+1/2}(m) \tag{2.10.5.11}$$

which yields the expansions

$$S(m) = (2\pi/m)^{1/2} \sum_{k=0}^{\infty} c_{2k+1}(-1)^k J_{2k+3/2}(m), \tag{2.10.5.12}$$

$$C(m) = (2\pi/m)^{1/2} \sum_{k=0}^{\infty} c_{2k}(-1)^k J_{2k+1/2}(m). \tag{2.10.5.13}$$

In practice, the coefficients c_k are approximated in one of several ways, three possibilities appearing in Bakhvalov and Vasileva, Littlewood and Zakian, and Piessens [F3], and the series of $S(m)$ and $C(m)$ are truncated at a suitably chosen point. The subsequent evaluation of the series in the spherical Bessel functions $J_{k+1/2}(x)$ using the three term recurrence relation for $J_{k+1/2}(x)$ is discussed by Littlewood and Zakian. Error bounds are given by Smith and by Patterson, who discusses various additional possibilities.

For the use of these ideas in the case where $f(x)$ is given by a table of values on a set of equally or unequally spaced points, see Piessens [11, 12].

References. Alaylioglu, Evans, and Hyslop [3]; Bakhvalov and Vasileva [1]; Hunter and Parsons [1]; Littlewood and Zakian [1]; Patterson [4]; Piessens [11, 12, F3]; Piessens and Branders [7]; Piessens and Poulenis [1]; H. V. Smith [1].

2.10.6 MIPS Methods

This abbreviation stands for Möbius inversion of Poisson summation formula and is a method for the computation of Fourier coefficients proposed by Goldberg and Varga and greatly elaborated by Lyness. We shall mark down the relevant formulas for the cosine coefficient.

Let $n \geq 1$ be an arbitrary positive integer. Let $K_1, K_2, \ldots, K_n$ be arbitrary numbers. Let $\mu_1, \mu_2, \ldots$ be the Möbius numbers which occur in the theory of numbers. They are defined by

$$\mu_1 = 1$$

$$\mu_n = \begin{cases} 0 & \text{if } n \text{ is divisible by a square} > 1, \\ (-1)^k & \text{if } n \text{ is the product of } k \text{ distinct primes.} \end{cases} \tag{2.10.6.1}$$

The first ten Möbius numbers are 1, -1, -1, 0, -1, 1, -1, 0, 0, 1. Let

$$I(f) = \int_0^1 f(x)\,dx, \qquad T_s(f) = \frac{1}{s}\left[\frac{1}{2}f(0) + \sum_{j=1}^{s-1} f\left(\frac{j}{s}\right) + \frac{1}{2}f(1)\right],$$

$$\zeta(k) = \sum_{j=1}^{\infty} j^{-k}.$$

Define quantities $E_s(f)$ by

$$E_s(f) = T_s(f) - I(f) - \frac{\zeta(2)K_2}{s^2} - \cdots - \frac{\zeta(2n)K_{2n}}{s^{2n}}. \tag{2.10.6.2}$$

Then, for any $f \in C[0, 1]$ we have the identity

$$2\int_0^1 f(x)\cos 2\pi mx\,dx = \frac{K_2}{m^2} + \frac{K_4}{m^4} + \cdots + \frac{K_{2n}}{m^{2n}} + \sum_{s=1}^{\infty} \mu_s E_s(f). \tag{2.10.6.3}$$

If, now, f is sufficiently differentiable and we make the selection

$$K_{2q} = \frac{2(-1)^{q-1}(f^{(2q-1)}(1) - f^{(2q-1)}(0))}{(2\pi)^{2q}}, \tag{2.10.6.4}$$

then it can be shown that the sth term in the sum in (2.10.6.3) is of the order of s^{-2n-2}.

This method has the advantage of providing uniform accuracy with respect to m (if we shut the sum off in such a way that $\sum_{s=N}^{\infty} |E_s(f)| < \varepsilon$). It has the considerable disadvantage of requiring the derivatives of f at the end points.

For the numerical implementation of this method, the reader is referred to a series of articles by Lyness. Lyness also considers modifications necessary to handle piecewise continuous functions as well as functions that exhibit large peaks due to poles in the complex plane.

References. Goldberg and Varga [1]; Lyness [4, 9, 10].

2.10.7 *Conversion to a Boundary-Value Problem*

Levin considers integrals of the form

$$I(f; q) = \int_a^b f(x)e^{iq(x)}\,dx. \tag{2.10.7.1}$$

If $|q'(x)| \gg b - a$, this integral is highly oscillatory. He observes that if f were of the form

$$f(x) = iq'(x)p(x) + p'(x), \qquad a \le x \le b, \tag{2.10.7.2}$$

then the integral could be evaluated directly as

$$I(f; q) = p(b)e^{iq(b)} - p(a)e^{iq(a)}. \tag{2.10.7.3}$$

Equation (2.10.7.2) can be considered as a differential equation for $p(x)$, of which the general solution is given by

$$p(x) = e^{-iq(x)}\left[\int_a^x f(t)e^{iq(t)}\,dt + c\right]. \tag{2.10.7.4}$$

This does not help much in computing $I(f; q)$ since the evaluation of (2.10.7.4) involves integrals of the form (2.10.7.1). Furthermore, $p(x)$ is as oscillatory as the integrand in (2.10.7.1). However, Levin shows that if f and q' are slowly oscillatory, then there exists a slowly oscillatory solution p_0 of (2.10.7.2). If we write the general solution of (2.10.7.2) as

$$p(x) = p_0(x) + de^{-iq(x)} \tag{2.10.7.5}$$

we see that

$$I(f;q) = p_0(b)e^{iq(b)} - p_0(a)e^{iq(a)} \tag{2.10.7.6}$$

so that all that is needed are the values $p_0(a)$ and $p_0(b)$. To find these, we solve (2.10.7.2) by collocation using a set $\{u_k(x)\colon k = 1, \ldots, n\}$ of slowly oscillatory basis functions. Then, in the approximate solution of (2.10.7.2)

$$p_n(x) = \sum_{k=1}^{n} \alpha_k u_k(x), \tag{2.10.7.7}$$

the coefficients α_k are chosen so that

$$\sum_{k=1}^{n} \alpha_k(iq'(x_j)u_k(x_j) + u'(x_j)) = f(x_j), \qquad j = 1, \ldots, n \tag{2.10.7.8}$$

at a suitably chosen set of distinct points $\{x_j\colon = j = 1, \ldots, n\}$ that include the endpoints a and b. Usually, equidistant points are chosen. Once the α_k are available, an approximation I_n to $I(f;q)$ is given by

$$I_n = \sum_{k=1}^{n} \alpha_k(u_k(b)e^{iq(b)} - u_k(a)e^{iq(a)}). \tag{2.10.7.9}$$

Example (Levin). Evaluate

$$I = \int_0^1 \sin x \, e^{i500(x+x^2)}\, dx.$$

Using $u_k(x) = x^{k-1}$, $k = 1, \ldots, n$, and $x_j = (j-1)/(n-1)$, $j = 1, \ldots, n$,

$$\operatorname{Re} I_5 = 4.6610(-4) \qquad \operatorname{Re} I_{10} = 4.5986(-4).$$

With $n = 10$, results correct to 5 significant figures were achieved at a much smaller cost in computing time than with the automatic routine CADRE (Appendix 2).

Reference. D. Levin [2].

2.10.8 *Use of the Lanczos Representation*

We shall assume in this section that $f(x)$ is analytic in a region containing the interval [0, 1]. Then, for any positive integer $p > 1$, we can decompose $f(x)$ into the sum of a polynomial $h_{p-1}(x)$ of degree $p-1$ and a remainder $g_p(x)$ in such a way that the Fourier coefficients $C_r g_p$ and $S_r g_p$ converge at the rate r^{-p}. Here, for any function g,

$$C_r g + iS_r g = \tfrac{1}{2}\int_0^1 g(x)\exp(2\pi i r x)\, dx, \qquad r \geq 1, \tag{2.10.8.1}$$

$$C_0 g = Ig = \int_0^1 g(x)\, dx, \qquad S_0 g \equiv 0. \tag{2.10.8.2}$$

The polynomial h_{p-1} is given by

$$h_{p-1}(x) = \sum_{q=1}^{p-1} \varphi^{(q-1)}(0)B_q(x)/q!, \tag{2.10.8.3}$$

where $\varphi(x) = f(x+1) - f(x)$ and $B_q(x)$ is the Bernoulli polynomial defined by (2.9.8). The remainder g_p has the expansion

$$g_p(x) \equiv f(x) - h_{p-1}(x) = \sum_{r=0}^{\infty} C_r g_p \cos 2\pi rx + S_r g_p \sin 2\pi rx. \tag{2.10.8.4}$$

Note that since $Ih_{p-1} = 0$, $Ig_p = If$.

An approximation to $f(x)$ is given by

$$\hat{f}(x) = \hat{h}_{p-1}(x) + \hat{g}_p(x) = \sum_{q=1}^{p-1} \lambda_q B_q(x)/q!$$
$$+ \sum_{r=0}^{m} \hat{C}_r \tilde{g}_p \cos 2\pi rx + \hat{S}_r \tilde{g}_p \sin 2\pi rx \tag{2.10.8.5}$$

where λ_q is an approximation to $\varphi^{(q-1)}(0)$ and $\hat{C}_r \tilde{g}_p$ and $\hat{S}_r \tilde{g}_p$ are approximations to the Fourier coefficients of $\tilde{g}_p \equiv f - \hat{h}_{p-1}$. If the derivatives of $f(x)$ are available, then we can set $\lambda_q = \varphi^{(q-1)}(0)$ so that $\hat{h}_{p-1} = h_{p-1}$. However, in general the λ_p are determined using approximations based on polynomial interpolation. Surprisingly, Lyness [13] has shown that such approximations do not degrade the accuracy of the process to a great extent provided that p is not too large. Lyness [19] recommends $p = 7$.

If $f(x)$ can be evaluated for any argument in [0, 1], then the Fourier coefficients $C_r \tilde{g}_p$ and $S_r \tilde{g}_p$ are best approximated using the trapezoidal rule because $\tilde{g}_p$ is almost periodic. For a chosen value of m, a trapezoidal rule with $2m$ panels should be used. If m is a power of 2, an FFT program (Section 3.9.5) may be used to implement the computation. Lyness [19] recommends $m = 8$ for moderate accuracy and $m = 16$ or 32 for high accuracy or difficult integrands. If $f(x)$ is specified by a table of values $\{x_i, y_i: i = 0, 1, \ldots, N\}$, one determines the trigonometric polynomial $\hat{T}_n(x)$ of order $[N/2]$ which interpolates the values $y_i - \hat{h}_{p-1}(x_i)$ at the data points,

$$\hat{T}_n(x) = a_0 + \sum_{r=1}^{[N/2]} a_r \cos 2\pi rx + b_r \sin 2\pi rx \tag{2.10.8.6}$$

and uses the a_r and b_r as approximations to the Fourier coefficients. Note that a_0 is a very good albeit expensive approximation to the integral of a function known at a set of arbitrarily spaced points (cf. Section 2.3).

Once we have the coefficients in the expansion (2.10.8.5), we can approximate any integral of the form $\int_0^1 \exp(-\gamma x)f(x)\,dx$ by $\int_0^1 \exp(-\gamma x)\hat{f}(x)\,dx$

for any complex value γ and evaluate the latter integral analytically. For that matter, we can approximate any integral of the form $\int_A^B \exp(-\gamma x) f(x)\, dx$ with $0 \le A < B \le 1$. We shall give here the formulas for the interval [0, 1]. Note that in this case, if $\gamma = -2\pi ir$, we get the Fourier coefficients of $\hat{f}$ which take the form

$$C_r\hat{f} = C_r\hat{h} + \hat{C}_r\tilde{g}, \qquad S_r\hat{f} = S_r\hat{h} + \hat{S}_r\tilde{g}.$$

The formulas for the integral $J_q(\gamma) = \int_0^1 \exp(-\gamma x)(B_q(x)/q!)\, dx$ are

$$J_0(0) = 1, \qquad J_q(0) = 0, \qquad q \ge 1,$$

$$J_q(\gamma) = (1 - e^{-\gamma}) \sum_{\substack{l=0 \\ l \text{ even}}}^{q} \frac{B_l}{l!} \gamma^{l-q-1} - \frac{\gamma^{-q}}{2}(e^{-\gamma} + 1), \qquad |\gamma| > 0, \quad (2.10.8.7)$$

where B_l is the lth Bernoulli number, $B_l = B_l(1)$. If γ is small, the alternative expansion may be used:

$$J_q(\gamma) = -(1 - e^{-\gamma}) \sum_{l=q+1}^{\infty} \frac{B_l}{l!} \gamma^{l-q-1}, \qquad |\gamma| < 2\pi. \quad (2.10.8.8)$$

The formulas for the integrals involving trigonometric functions are simplified by using the exponential form. In this case

$$w_r(\gamma) \equiv \int_0^1 e^{-\gamma x} e^{2\pi irx}\, dx = (e^{-\gamma} - 1)/(2\pi ir - \gamma), \qquad \gamma \ne 2\pi ir. \quad (2.10.8.9)$$

If γ is close to $2\pi ir$, $\gamma = 2\pi ir + \varepsilon$, use the expansion

$$w_r(\gamma) = 1 - \varepsilon/2! + \varepsilon^2/3! - \cdots.$$

Note that for $p = 2$, no derivatives of $f(x)$ are involved and the method reduces to that proposed by Abramovici for Fourier coefficients.

References. Abramovici [1]; Einarsson [3]; Lanczos [3]; Lyness [13, 19]; K.-C. Ng [1].

2.10.9 *Special Methods*

The schemes just presented should not camouflage the fact that difficult integrals requiring special study and treatment may occur from time to time. For example, the integral

$$C(x, y, \beta) = \frac{1}{2\beta} \int_{-\beta}^{\beta} \cos(x \cos\theta - y \sin\theta) \cos\theta\, d\theta,$$

where x and y are large, is in this category and has been analyzed by Hartree.

Reference. Hartree [1].

Burnett and Soroka present a numerical technique that yields an efficient computer algorithm for evaluating large quantities of integrals of the form

$$I(r; p) = \int_{\alpha(p)}^{\beta(p)} f(r; p; t)\, dt$$

where p represents one or more parameters in addition to the "oscillatory" parameter r.

Reference. Burnett and Soroka [1].

2.11 Contour Integrals

By a contour integral, we mean an integral of the form $\int_C f(x, y)\, dx$, $\int_C f(x, y)\, dy$, or $\int_C f(x, y)\, ds$, extended over a contour C in the xy plane. If the contour C can be conveniently parameterized as

$$C: \begin{cases} x = x(t), \\ y = y(t), \end{cases} \qquad t_0 \le t \le t_1, \tag{2.11.1}$$

these integrals can be transformed to ordinary integrals of a single variable

$$\int_{t_0}^{t_1} f(x(t), y(t)) \frac{dx}{dt}\, dt, \qquad \int_{t_0}^{t_1} f(x(t), y(t)) \frac{dy}{dt}\, dt,$$

$$\int_{t_0}^{t_1} f(x(t), y(t)) \frac{ds}{dt}\, dt, \tag{2.11.2}$$

and can then be treated by the rules discussed.

An interesting special case occurs when C is a closed contour. It may be possible to write a parametric representation in terms of a central angle θ. The integrand is then a periodic function of θ and, if sufficiently smooth, can be very accurately integrated by use of a mean value as developed in Section 2.9. Thus, if $\mathscr{E}$ designates the ellipse

$$x = a \cos \theta, \quad y = b \sin \theta, \qquad 0 \le \theta \le 2\pi,$$

then

$$\begin{aligned} &\int_{\mathscr{E}} f(x, y)\, ds \\ &\quad = \int_0^{2\pi} f(a \cos \theta, b \sin \theta)(a^2 \sin^2 \theta + b^2 \cos^2 \theta)^{1/2}\, d\theta \\ &\quad \approx \frac{2\pi}{N} \sum_{k=0}^{N-1} f\left(a \cos \frac{2\pi k}{N}, b \sin \frac{2\pi k}{N}\right)\left(a^2 \sin^2 \frac{2\pi k}{N} + b^2 \cos^2 \frac{2\pi k}{N}\right)^{1/2} \end{aligned} \tag{2.11.3}$$

Example. Compute the length of the perimeters of ellipses of various eccentricities, using 60 points.

a	b	Approximate value of $\frac{1}{4}$ perimeter	Exact value
1.0	.80	1.4180 830	1.4180 834
1.0	.40	1.1506 554	1.1506 556
1.0	.20	1.0505 019	1.0505 022
1.0	.10	1.0159 888	1.0159 935

Ichida and Kiyono have carried this idea further and applied it to the approximation of the d-dimensional integral

$$I = \int \cdots \int_D g(x_1, \ldots, x_d)\, dx_1 \cdots dx_d \tag{2.11.4}$$

where D is a d-dimensional closed region with boundary

$$r = u(\theta_1, \ldots, \theta_{d-1}). \tag{2.11.5}$$

For example, in the two-dimensional case,

$$I = \int\int_D g(x, y)\, dx\, dy \tag{2.11.6}$$

where D is a simply-connected region with boundary $r = u(\theta)$ and $u(\theta)$ is a single-valued function of θ, $0 \le \theta < 2\pi$. Using polar coordinates, this integral transforms to

$$I = \int_0^{2\pi} \int_0^1 g(Ru(\theta)\cos\theta, Ru(\theta)\sin\theta) R u^2(\theta)\, dR\, d\theta. \tag{2.11.7}$$

The integration in the θ variable is done using a trapezoidal rule.

2.11.1 Contour Integrals in the Complex Plane

Suppose that $z = x + iy$ and $f(z)$ is a function of a complex variable. Write $f(z) = R(x, y) + iI(x, y)$. Then

$$\int_C f(z)\, dz = \int_C (R + iI)(dx + i\, dy)$$

$$= \int_C (R\, dx - I\, dy) + i \int_C (I\, dx + R\, dy). \tag{2.11.1.1}$$

In this way, a complex contour integral may be expressed as the sum of real contour integrals and can be treated as before.

If $f(z)$ is a *regular analytic* function in a simply connected region B that contains an open contour C in its interior, then, by Cauchy's theorem in the theory of analytic functions, the contour C may be replaced by any other contour that lies in B and starts and ends at the same points as C. In particular, under proper conditions of regularity, C may be replaced by the straight line joining its end points, or by two straight lines—one parallel to the x axis and one to the y axis.

If $f(z)$ is analytic, it is also possible to give rules of approximate integration which involve values of the function that are off the contour C. The following formula of interpolatory type has been derived by Birkhoff and Young:

$$\int_{z_0-h}^{z_0+h} f(z)\,dz = \frac{h}{15}\{24f(z_0) + 4[f(z_0+h) + f(z_0-h)] - [f(z_0+ih) + f(z_0-ih)]\} + R. \qquad (2.11.1.2)$$

The remainder R satisfies

$$|R| \le \tfrac{1}{1890}|h|^7 \max_{z \in S} |f^{(6)}(z)|, \qquad (2.11.1.3)$$

where S denotes the square whose vertices are $z_0 + i^k h$ $(k = 0, 1, 2, 3;$ $i = \sqrt{-1})$.

However, as pointed out by Lether, a more accurate method for evaluating an integral over the complex line segment $[z_0 - h, z_0 + h]$ is simply by transforming an integration rule on $[-1, 1]$, for example, a Gauss rule, to this segment. Thus, suppose we have an integration rule of the form

$$\int_{-1}^{1} F(u)\,du = \sum_{k=1}^{n} A_k F(a_k) + E_n(F) \qquad (2.11.1.4)$$

where the A_k and a_k may be complex. Then

$$\int_{z_0-h}^{z_0+h} f(z)\,dz = \sum_{k=1}^{n} w_k f(z_k) + R_n(f) \qquad (2.11.1.5)$$

where $w_k = hA_k$, $z_k = ha_k + z_0$, and $R_n(f) = hE_n[f(hu + z_0)]$. The z_k are on the contour of integration if and only if the $a_k \in [-1, 1]$.

When C is the circle $|z| = r$, the contour integral has the form

$$I = \frac{1}{2\pi i}\int_C f(z)\,dz = \frac{r}{2\pi}\int_0^{2\pi} e^{i\theta} f(re^{i\theta})\,d\theta$$

$$= r\int_0^1 e^{2\pi i t} f(re^{2\pi i t})\,dt. \qquad (2.11.1.6)$$

Inasmuch as the integrand has period 1, it may be conveniently and accurately integrated by means of the trapezoidal rule T_N:

$$I \approx \frac{r}{N} \sum_{k=1}^{N} e^{2\pi ik/N} f(re^{2\pi ik/N}). \tag{2.11.1.7}$$

In view of identities (2.9.20), it follows that this rule is exact whenever $f \in \mathscr{P}_{N-1}$.

If $f(z)$ is analytic in $|z| \le r$, then, by Cauchy's theorem,

$$f(z) = \sum_{j=0}^{\infty} a_j z^j = \sum_{j=0}^{\infty} \frac{f^{(j)}(0) z^j}{j!},$$

where

$$\frac{f^{(j)}(0)}{j!} = a_j = \frac{1}{2\pi i} \int_C \frac{f(z)\,dz}{z^{j+1}} = \frac{1}{r^j} \int_0^1 \frac{f(re^{2\pi it})\,dt}{e^{2\pi ijt}}$$

$$\approx \frac{1}{r^j N} \sum_{k=1}^{N} f(re^{2\pi ik/N}) e^{-2\pi ijk/N} \equiv \tilde{a}_j. \tag{2.11.1.8}$$

Again, this rule is exact for $f \in \mathscr{P}_{N-1}$ for $j = 0, 1, \ldots, N-1$. Note that the same values of f are used to approximate all the a_j.

This gives us rules for derivatives, and if we integrate the power series we shall obtain integration rules in terms of complex functional values. We have

$$\int_{-r}^{r} f(z)\,dz = 2r \sum_{j=0,\,\text{even}}^{\infty} \frac{r^j a_j}{j+1}. \tag{2.11.1.9}$$

Truncating the infinite series and replacing a_j by $\tilde{a}_j$ we obtain the rule

$$\int_{-r}^{r} f(r)\,dz \approx Q_N(f) = 2r \sum_{j=0,\,\text{even}}^{N-1} \frac{r^j \tilde{a}_j}{j+1}. \tag{2.11.1.10}$$

which is exact for $f \in \mathscr{P}_{N-1}$. The modified rule

$$\bar{Q}_N(f) = Q_N(f) + \frac{2rN}{N+1}(f(0) - \tilde{a}_0) \tag{2.11.1.11}$$

requires one more functional evaluation and can be shown to be exact for $f \in \mathscr{P}_{N+1}$. The rule (2.11.1.2) is essentially $\bar{Q}_4$.

Indefinite integration can be carried out similarly.

Contour integration has been used to locate the zeros of an analytic function via the argument principle. It has also been used to find the zeros and poles of meromorphic functions.

FORTRAN programs to compute $\tilde{a}_j$ have been given by Lyness and Sande.

References. Abd-elall *et al.* [1]; Birkhoff and Young [1]; Elliott and Donaldson [1]; Fornaro [1]; Ichida and Kiyono [1]; Lether [7]; Lyness [6]; Lyness and Delves [1]; Lyness and Sande [F1]; Maskell and Sack [1]; Pfalz [A1]; D. Young [1].

2.12 *Improper Integrals (Finite Interval)*

By an *improper* integral is generally meant one in which the integrand is *unbounded* over the interval of integration. However, it is occasionally useful to include in this category integrands that possess an apparent or removable singularity in the interval of integration.

Improper integrals occur with great frequency in computational work and must be handled by special devices.

Let us assume that the integral to be evaluated is in the form $\int_0^1 f(x)\,dx$, where $f(x)$ is continuous in $0 < x \leq 1$ but not in $0 \leq x \leq 1$. For example, $f(x)$ may be unbounded in the vicinity of $x = 0$.

2.12.05 *Apparent (or Removable) Singularities*

The functional form of the integrand may possess apparent singularities at one or more values of the integration variable. Typical examples: $(\sin x)/x$ at $x = 0$, $\cot x - x^{-1}$ at $x = 0$. We assume that the programmer is aware of the apparent singularity and its location, say x_0, and that he knows the value $\lim_{x\to x_0} f(x) = y_0$. In writing the function evaluation part of his integration program, he should therefore take care to use the value y_0 for $f(x_0)$, or for any abscissa within ε of x_0.

2.12.1 *Proceeding to the Limit*

The basic definition

$$\int_0^1 f(x)\,dx = \lim_{r\to 0+} \int_r^1 f(x)\,dx \tag{2.12.1.1}$$

suggests a primitive mode of procedure. Let $1 > r_1 > r_2 > \cdots$ be a sequence of points that converge to 0, for example $r_n = 2^{-n}$. Write

$$\int_0^1 f(x)\,dx = \int_{r_1}^1 f(x)\,dx + \int_{r_2}^{r_1} f(x)\,dx + \int_{r_3}^{r_2} f(x)\,dx + \cdots. \tag{2.12.1.2}$$

Each of the integrals on the right-hand side is proper, and the evaluations are terminated when $|\int_{r_{n+1}}^{r_n} f(x)\,dx| \leq \varepsilon$. This is only a practical criterion and is not correct theoretically.

Example

$$I = \int_0^1 \frac{dx}{x^{1/2} + x^{1/3}}, \qquad I_n = \int_{r_n}^1 \frac{dx}{x^{1/2} + x^{1/3}}, \qquad r_n = 2^{-n}.$$

n	I_n	Number of functional evaluations
1	.2849 2598	9
2	.4744 8022	18
4	.6832 3927	44
8	.8128 0497	80
16	.8402 9678	176
32	.8411 1612	344
40	.8411 1663	432
Exact value	.8411 1692	

Each integral $\int_{r_n}^{r_{n-1}}$ was computed (presumably correct to five figures) by an adaptive modification of the Romberg integration scheme available on the Brown University computer. (See Section 6.3, p. 442.)

Squire has suggested using extrapolation to accelerate the convergence of the sequence $\{I_n\}$.

Herrero has shown how to implement the process of proceeding to the limit on a programmable hand calculator.

References. Evans, Hyslop, and Morgan [1]; Herrero [1]; Squire [6].

2.12.2 Truncation of the Interval

In some instances it might be possible to obtain an estimate of $\int_0^r f(x)\,dx$ without much difficulty. If $\left|\int_0^r f(x)\,dx\right| \le \varepsilon$, then we can simply evaluate the proper integral $\int_r^1 f(x)\,dx$.

Example. Estimate

$$\int_0^r \frac{g(x)}{x^{1/2} + x^{1/3}}\,dx,$$

where $g(x)$ is in $C[0, 1]$ and satisfies $|g(x)| \le 1$. Since $x^{1/2} \le x^{1/3}$ in $[0, 1]$, we have

$$\left|\frac{g(x)}{x^{1/2} + x^{1/3}}\right| \le \frac{1}{2x^{1/2}}.$$

Hence

$$\left|\int_0^r \frac{g(x)}{x^{1/2} + x^{1/3}}\,dx\right| \le \frac{1}{2}\int_0^r \frac{dx}{x^{1/2}} = r^{1/2}.$$

This suggests that we take $r \le 10^{-6}$ for an accuracy of 10^{-3}.

2.12.3 *Change of Variable*

A change of variable which will eliminate the singularity can sometimes be found. For example, if $f(x) \in C[0, 1]$, the change of variable $t^n = x$ transforms the integral

$$\int_0^1 x^{-1/n} f(x)\, dx, \qquad n \geq 2,$$

into $n \int_0^1 f(t^n) t^{n-2}\, dt$, which is a proper integral.

If the improper integral $\int_0^1 f(x) \log x\, dx, f(x) \in C[0, 1], f(0) \neq 0$ is treated by the obvious substitution $t = -\log x$, we obtain $-\int_0^\infty te^{-t} f(e^{-t})\, dt$, an integral with an infinite range of integration. This sort of transformation may result only in exchanging one kind of difficulty for another.

If the integrand is bounded, but has a low order of continuity, it may also be desirable to change the variable. For example, $I = \int_0^a x^{p/q} f(x)\, dx$ where p and q are positive integers in lowest terms. Setting $x = t^q$ yields

$$I = q \int_0^{a^{1/q}} t^{p+q-1} f(t^q)\, dt.$$

Other useful transformations are

$$\int_{-1}^{1} \frac{f(x)\, dx}{(1 - x^2)^{1/2}} = \int_0^\pi f(\cos t)\, dt,$$

$$\int_0^1 \frac{f(x)\, dx}{(x(1 - x))^{1/2}} = 2 \int_0^{\pi/2} f(\sin^2 t)\, dt.$$

References. Evans, Forbes, and Hyslop [1]; Hunter [2].

2.12.4 *Elimination of the Singularity*

It may be possible to "subtract out" the singularity. For example, evaluate

$$\int_0^1 \frac{\cos x}{x^{1/2}}\, dx.$$

Write

$$\int_0^1 \frac{\cos x}{x^{1/2}}\, dx = \int_0^1 \frac{dx}{x^{1/2}} + \int_0^1 \frac{\cos x - 1}{x^{1/2}}\, dx = 2 + \int_0^1 \frac{\cos x - 1}{x^{1/2}}\, dx.$$

Now, since $\cos x - 1 \simeq x^2/2$ near $x = 0$, this last integrand is now in $C[0, 1]$. For accurate numerical work, it would be better to subtract off more and write

$$\int_0^1 \frac{\cos x}{x^{1/2}}\, dx = \int_0^1 \frac{1 - \frac{1}{2}x^2}{x^{1/2}}\, dx + \int_0^1 \frac{\cos x - 1 + \frac{1}{2}x^2}{x^{1/2}}\, dx.$$

With this simple integrand, the whole expansion of $\cos x$ can, of course, be used.

The method of eliminating the singularity may be particularly valuable in the case of an indefinite integral. Let

$$I(x) = \int_0^x \frac{e^{-t}\,dt}{1-t}. \tag{2.12.4.1}$$

In the neighborhood of $t = 1$, the integrand behaves like $e^{-1}/(1-t)$. Hence we write

$$\begin{aligned} I(x) &= e^{-1}\int_0^x \frac{dt}{1-t} + \int_0^x \left(\frac{e^{-t}-e^{-1}}{1-t}\right) dt \\ &= -e^{-1}\log(1-x) + \int_0^x \left(\frac{e^{-t}-e^{-1}}{1-t}\right) dt. \end{aligned} \tag{2.12.4.2}$$

The second integrand is now in $C[0, 1]$.

The general method is to subtract from the singular integrand $f(x)$ a function $g(x)$ whose integral is known in closed form and is such that $f(x) - g(x)$ is no longer singular. The function $g(x)$ must therefore mimic the behavior of $f(x)$ closely at its singular point.

Integration by parts may be used from time to time. For example, integrating by parts, we have

$$\int_0^1 \frac{\cos x}{x^{1/2}}\,dx = 2x^{1/2}\cos x \Big|_0^1 + 2\int_0^1 x^{1/2}\sin x\,dx.$$

The burden is now on the integrand $x^{1/2}\sin x$, which is no longer unbounded (but has an unbounded derivative).

References. Abramowitz [1]; Fröberg [1, pp. 172–175]; Kantorovitch [1]; Krylov [1, Section 11.2]; Mineur [1, Chap. 13].

2.12.5 Integration Formulas of Interpolatory Type

Let $w(x)$ be a fixed function with a singularity in the neighborhood of $x = 0$, but such that $\int_0^1 w(x)x^k\,dx$ exists for $k = 0, 1, \ldots, n$. For a given sequence of abscissas $0 < x_0 < x_1 < \cdots < x_n \le 1$, we can determine weights w_i such that

$$\int_0^1 w(x)f(x)\,dx = \sum_{i=0}^{n} w_i f(x_i) \tag{2.12.5.1}$$

whenever $f(x) \in \mathscr{P}_n$. This leads to the approximate integration formula

$$\int_0^1 w(x)f(x)\,dx \approx \sum_{i=0}^{n} w_i f(x_i). \tag{2.12.5.2}$$

For example, let $w(x) = x^{-1/2}$, $x_0 = \frac{1}{3}$, $x_1 = \frac{2}{3}$, $x_2 = 1$. Then the w_i are determined by the linear system

$$w_1 + w_2 + w_3 = \int_0^1 x^{-1/2}\,dx = 2,$$

$$\tfrac{1}{3}w_1 + \tfrac{2}{3}w_2 + w_3 = \int_0^1 x^{-1/2}x\,dx = \tfrac{2}{3},$$

$$\tfrac{1}{9}w_1 + \tfrac{4}{9}w_2 + w_3 = \int_0^1 x^{-1/2}x^2\,dx = \tfrac{2}{5}. \tag{2.12.5.3}$$

This leads to the rule

$$\int_0^1 x^{-1/2}f(x)\,dx \approx \tfrac{14}{5}f(\tfrac{1}{3}) - \tfrac{8}{5}f(\tfrac{2}{3}) + \tfrac{4}{5}f(1). \tag{2.12.5.4}$$

It may be more convenient to use it in the form

$$\int_0^r x^{-1/2}f(x)\,dx \approx r^{1/2}(\tfrac{14}{5}f(\tfrac{1}{3}r) - \tfrac{8}{5}f(\tfrac{2}{3}r) + \tfrac{4}{5}f(r)). \tag{2.12.5.5}$$

As a second example, we quote the formula of A. Young,

$$\int_{-1}^1 \frac{f(x)}{(1-x^2)^{1/2}}\,dx \approx \frac{\pi}{6}\left[f(-1) + 2f\left(-\frac{1}{2}\right) + 2f\left(\frac{1}{2}\right) + f(1)\right]. \tag{2.12.5.6}$$

The reader is referred to Section 2.5.6 on product integration, which is relevant in this context.

References. Abramowitz [1]; Kaplan [1]; Luke [1, 3]; A. Young [1].

2.12.5.1 *Singularities off but near the Interval of Integration*

It has been observed that the presence of singularities off but near the interval of integration may affect the accuracy of an integration rule adversely. If the nature and the location of the singularity can be determined, it may be useful to work out special rules tailored to the situation. For example, if $y(x)$ has a pole of order r at the point x^*, we may precompute and apply rules that are exact for functions of the form $y(x) = p_n(x)/(x - x^*)^r$, $p_n(x) \in \mathscr{P}_n$.

Monegato and others have suggested the use of integration rules designed for Cauchy principal value integrals (Section 2.12.8) for integrands with real simple poles near the interval of integration.

References. Chawla and Jayarajan [1]; Eisner [1]; Monegato [7].

For singularities on or near the imaginary axis, the following procedure, based on the conformal mapping of the connected ovals of Cassini onto the circle, has been recommended by Rabinowitz and Richter.

Let x_i, w_i, $i = 1, \ldots, n$, be the Gauss abscissas and weights. Then the rule

$$\int_{-1}^{1} f(x)\,dx \approx \sum_{i=1}^{n} w_i^* f(x_i^*) \equiv G_n^*(f)$$

where

$$x_i^* = \left(\frac{a^2}{1 + a^2 - x_i^2}\right)^{1/2} x_i, \qquad w_i^* = \left(\frac{a^2}{(1 + a^2 - x_i^2)^3}\right)^{1/2} (1 + a^2) w_i, \qquad i = 1, \ldots, n,$$

is exact for the functions

$$g_k(x) = \frac{x^k}{(x^2 + a^2)^{(k+3)/2}}, \qquad k = 0, 1, \ldots, 2n - 1$$

and hence for all functions of the form

$$g(x) = \frac{p(x)}{(x^2 + a^2)^{n+1/2}}, \qquad p(x) \in \mathscr{P}_{2n-1}.$$

Therefore, a sequence of such rules will converge much more rapidly than a sequence of the corresponding Gauss rules for any function $f(x)$ which has a rapidly convergent expansion in terms of the $g_k(x)$ for some value of a. This will be true, for example, if $f(x)$ has a pole in the neighborhood of $\pm ia$.

Example

$$f(x) = 1/(a^2 + x^2)^2$$

	$a = .01$		$a = .1$	
n	E_{G_n}	$E_{G_n^*}$	E_{G_n}	$E_{G_n^*}$
8	9.9(−1)	−8.6(−4)	7.5(−1)	−5.3(−4)
16	9.9(−1)	−1.1(−4)	3.0(−1)	−2.9(−5)
24	9.9(−1)	−3.4(−5)	8.7(−2)	−2.9(−6)
32	9.9(−1)	−1.4(−5)	2.3(−2)	−3.6(−7)

E denotes the relative error.

Reference. Rabinowitz and Richter [5].

Another possibility proposed by Lether is to subtract out the complex singularities. For example, if $f(z)$ has poles of order τ_j at a_j, $j = 1, \ldots, m$, and if b_{rj} are the Laurent coefficients in the principal part of f around a_j, then, if Q_n is any suitable integration rule with error term R_n,

$$\int_{-1}^{1} w(x)f(x)\,dx = -\sum_{j=1}^{m}\sum_{r=1}^{\tau_j} b_{rj}T^{(r-1)}(a_j)/(r-1)! + Q_n(\varphi) + R_n(\varphi).$$

Here

$$T(z) = \int_{-1}^{1} \frac{w(x)}{z-x}\,dx, \qquad z \notin [-1, 1],$$

and

$$\varphi(x) = f(x) - \sum_{j=1}^{m}\sum_{r=1}^{\tau_j} b_{rj}(x-a_j)^{-r}.$$

Example. Let $f(x) = F(x)/\pi(x)$ where $\pi(x) = \prod_{j=1}^{m}(x - a_j)$, and the a_j are m distinct real or complex numbers $\notin [-1, 1]$. Then

$$\int_{-1}^{1} \frac{w(x)F(x)}{\pi(x)}\,dx \simeq \sum_{k=1}^{n} \frac{w_k F(x_k)}{\pi(x_k)} - \sum_{j=1}^{m} \frac{k_n(a_j)}{\pi'(a_j)} F(a_j)$$

where

$$k_n(z) = T(z) - \sum_{k=1}^{n} \frac{w_k}{z - x_k}$$

and the x_k and w_k are the abscissas and weights of an integration rule with respect to the weight function $w(x)$.

Example. Let x_k and w_k be the Gauss abscissas and weights. Then

$$\int_{-1}^{1} \frac{F(x)}{x^2+\alpha^2}\,dx \simeq \sum_{k=1}^{n} \frac{w_k}{x_k^2+\alpha^2}(F(x_k) - \mathrm{Re}[F(i\alpha)]) + \frac{2}{\alpha}\arctan\frac{1}{\alpha}(\mathrm{Re}[F(i\alpha)]).$$

References. Lether [8, 11].

2.12.6 *Formulas of Gauss Type*

Singularities may also be accommodated by means of Gauss-type formulas. The integral is written in the form

$$I = \int_a^b w(x)f(x)\,dx, \tag{2.12.6.1}$$

where $w(x)$ is a fixed nonnegative weight function. The moments $\int_a^b w(x)x^n\,dx$ are assumed to exist for $n = 0, 1, 2, \ldots$, but $w(x)$ may have one or more singularities in the interval $[a, b]$.

The most thoroughly investigated case is that of the Jacobi weight

$$w(x) = (1 - x)^{\alpha}(1 + x)^{\beta}, \qquad \alpha > -1, \qquad \beta > -1. \qquad (2.12.6.2)$$

If either $\alpha < 0$ or $\beta < 0$, we have an unbounded singularity. The Gauss–Jacobi formulas were discussed in Section 2.7 and explicit formulas for the weight functions $w(x) = (1 - x^2)^{\pm 1/2}$ are given by (2.7.13), (2.7.14), and (2.7.1.14). We note some additional special cases.

$$\int_0^1 (1 - x)^{1/2} f(x)\, dx = \sum_{k=1}^{n} w_k f(x_k) + \frac{2^{4n+3}[(2n + 1)!]^4}{(2n)!\,(4n + 3)[(4n + 2)!]^2} f^{(2n)}(\xi), \qquad 0 < \xi < 1. \qquad (2.12.6.3)$$

In this formula, $x_k = 1 - \xi_k^2$, where ξ_k is the kth positive zero of the Legendre polynomial $P_{2n+1}(x)$, and $w_k = 2\xi_k^2 w_k^{(2n+1)}$, where $w_k^{(2n+1)}$ is the weight corresponding to ξ_k in the rule G_{2n+1}.

$$\int_0^1 \frac{f(x)}{(1 - x)^{1/2}}\, dx = \sum_{k=1}^{n} w_k f(x_k) + \frac{2^{4n+1}}{4n + 1} \frac{[(2n)!]^3}{[(4n)!]^2} f^{(2n)}(\xi), \qquad 0 < \xi < 1. \qquad (2.12.6.4)$$

Here $x_k = 1 - \xi_k^2$, where ξ_k is the kth positive zero of $P_{2n}(x)$, and $w_k = 2w_k^{(2n)}$ where $w_k^{(2n)}$ is the weight corresponding to ξ_k in the rule G_{2n}.

$$\int_0^1 \left(\frac{x}{1 - x}\right)^{1/2} f(x)\, dx = \sum_{k=1}^{n} w_k f(x_k) + \frac{\pi}{(2n)!\, 2^{4n+1}} f^{(2n)}(\xi), \qquad 0 < \xi < 1; \qquad (2.12.6.5)$$

$$x_k = \cos^2 \frac{2k - 1}{2n + 1} \frac{\pi}{2}, \qquad w_k = \frac{2\pi}{2n + 1} x_k .$$

Weights and abscissas for Gaussian formulas for the integral $\int_0^1 \log(1/x) f(x)\, dx$ can be found in Anderson. Similar tables for $\int_0^h \log x\, f(x)\, dx$ can be found in Price.

Additional tables of Gaussian formulas involving singular weight functions appear in Piessens and Branders [5].

For another way of dealing with the Jacobi weight (2.12.6.2), see the work of Piessens and his collaborators.

References. Anderson [1]; Branders and Piessens [1]; Krylov, Lugin, and Ianovitch [1]; Krylov and Pal'tsev [1]; Kutt [4]; NBS Handbook [1, p. 920]; Piessens and Branders [2, 5]; Piessens, Mertens, and Branders [1]; J. F. Price [1]; Sidi [2]; Stroud and Secrest [2].

2.12.7 *Ignoring the Singularity*

It is also possible to ignore the fact that the integrand has a singularity and merely use standard rules for approximate integration. Suppose we would like to compute $\int_0^1 f(x)\,dx$, where $f(x)$ is unbounded in the neighborhood of $x = 0$. We can arbitrarily set $f(0) = 0$ (or any other value) and use a sequence of trapezoidal rules or any sequence of rules. Or we might use a sequence of rules that do not involve the value of $f(x)$ at $x = 0$ (a better idea).

Example

$$\int_0^1 \frac{dx}{x^{1/2}} = 2.0$$

$32 \times S$	1.8427	G_2	1.65068
$64 \times S$	1.8887	G_3	1.75086
$128 \times S$	1.9213	G_4	1.80634
$256 \times S$	1.9444	G_{10}	1.91706
$512 \times S$	1.9606	G_{16}	1.94722
$1024 \times S$	1.9721	G_{32}	1.97321
	S = Simpson	G = Gauss	

The convergence of $n \times S$ is very slow, and the comparison with G_n is striking.

The method of "ignoring the singularity" *may not work if the integrand is oscillatory.*

Example

$$\int_0^1 \frac{1}{x}\sin\frac{1}{x}\,dx = \int_1^\infty \frac{\sin x}{x}\,dx = .624713$$

$32 \times S$	2.3123
$64 \times S$	1.6946
$128 \times S$	−.6083
$256 \times S$	1.2181
$512 \times S$	.7215
$1024 \times S$	.3178

No pattern of convergence is discernible from these computations. However, if the integrand is monotonic in a neighborhood of its singularity, or even if it can be majorized by a monotonic integrable function, it can be shown that the method is convergent to the proper answer.

For simplicity, suppose that we integrate over [0, 1] and the singularity occurs at $x = 1$. Let M designate the class of functions $g(x) \geq 0$, which are continuous and nondecreasing in [0, 1) and such that $\lim_{t \to 1^-} \int_0^t g(x)\,dx < \infty$. Let BM designate the class of functions $f(x)$ that are continuous in [0, 1) and such that for each f we can find a $g \in M$ with $|f(x)| \leq g(x)$ in [0, 1). (For example, $f(x) = (1 - x)^{-1/2} \sin(1 - x)^{-1/2}$ is in BM but not in M.)

Let R designate a fixed m-point rule of approximate integration in [0, 1]:

$$R(f) = \sum_{k=1}^{m} w_k f(x_k),$$

with $0 \leq x_1 < x_2 < \cdots < x_m \leq 1$, $\sum_{k=1}^{m} w_k = 1$.

Let R_n designate the compound rule that arises by applying R to each of the n subintervals $[0, 1/n], [1/n, 2/n], \ldots, [(n - 1)/n, 1]$. (As before, we take an arbitrary value, say 0, for $f(1)$ if it occurs.)

THEOREM. *If $f(x) \in BM$, then*

$$\lim_{n \to \infty} R_n(f) = \int_0^1 f(x)\,dx. \tag{2.12.7.1}$$

More general sequences of rules have been treated by Rabinowitz. Let

$$Q_n(f) = \sum_{k=1}^{n} w_{nk} f(x_{nk}) \tag{2.12.7.2}$$

be a sequence of rules with

$$0 \leq x_{nn} < x_{n,n-1} < \cdots < x_{n1} < x_{n0} = 1 \tag{2.12.7.3}$$

and such that

$$\lim_{n \to \infty} Q_n(f) = \int_0^1 f(x)\,dx \tag{2.12.7.4}$$

for all $f \in C[0, 1]$. Suppose, further, that the weights and abscissas are such that for some $c > 0$,

$$|w_{nk}| \leq c(x_{n,k-1} - x_{nk}) \tag{2.12.7.5}$$

for all sufficiently large n and for all k such that

$$|x_{nk} - 1| < \delta \tag{2.12.7.6}$$

for some fixed $\delta > 0$.

THEOREM. *Under the above hypotheses on* Q_n,

$$\lim_{n\to\infty} Q_n(f) = \int_0^1 f(x)\,dx \tag{2.12.7.7}$$

for any $f \in BM$.

A similar theorem holds for closed rules of the form

$$\hat{Q}_n(f) = \sum_{k=0}^{n} \hat{w}_{nk} f(x_{nk}). \tag{2.12.7.8}$$

The inequality (2.12.7.5) can be shown to hold in the cases of the composite rules above, Gauss rules, and Romberg rules. It also holds for the two rules of Féjer described in Section 2.5.5. If $w(x) = (1-x)^\alpha(1+x)^\beta$, $-\frac{1}{2} \le \alpha$, $\beta \le \frac{1}{2}$, if GJ_n designates the Gauss rule with weight $w(x)$, and if $w(x)f(x)$ is in class BM over $[-1, 1]$, then

$$\lim_{n\to\infty} GJ_n(f) = \int_{-1}^{+1} w(x)f(x)\,dx. \tag{2.12.7.9}$$

Miller made the extension from integrands of class M to class BM and has given error estimates in some instances.

Rabinowitz and Sloan have generalized these results to the case of product integration.

Despite these positive results, the process of ignoring the singularity is a tricky business and should be avoided wherever possible.

References. Anselone and Opfer [1]; Davis and Rabinowitz [7]; el-Tom [2]; Feldstein and Miller [1]; Gautschi [2, 3]; Mikloško [5]; R. K. Miller [1]; Pólya and Szegö [1]; Rabinowitz [3, 9, 14]; Rabinowitz and Sloan [1].

2.12.8 *Numerical Evaluation of the Cauchy Principal Value*

Suppose that $a < c < b$ and that $f(x)$ is unbounded in the neighborhood of $x = c$. Suppose, however, that the Cauchy principal value of $\int_a^b f(x)\,dx$,

$$\lim_{r\to 0+} \left[\int_a^{c-r} f(x)\,dx + \int_{c+r}^{b} f(x)\,dx\right],$$

exists. The following analytical device is occasionally useful to obtain this value. It is no restriction to take $c = 0$ and to take the integral in the form $\int_{-a}^{a} f(x)\,dx$. Set

$$g(x) = \tfrac{1}{2}[f(x) - f(-x)], \qquad h(x) = \tfrac{1}{2}[f(x) + f(-x)]. \tag{2.12.8.1}$$

Then

$$f(x) = g(x) + h(x), \tag{2.12.8.2}$$

where $g(x)$ is an odd function,

$$g(x) = -g(-x), \tag{2.12.8.3}$$

and $h(x)$ is an even function,

$$h(x) = h(-x). \tag{2.12.8.4}$$

Hence

$$\begin{aligned}\int_{-a}^{-r} f(x)\,dx + \int_{r}^{a} f(x)\,dx &= \int_{-a}^{-r} g(x)\,dx + \int_{r}^{a} g(x)\,dx + \int_{-a}^{-r} h(x)\,dx + \int_{r}^{a} h(x)\,dx \\ &= 2\int_{r}^{a} h(x)\,dx. \end{aligned}\tag{2.12.8.5}$$

Therefore,

$$\begin{aligned}P\int_{-a}^{a} f(x)\,dx &= 2\lim_{r\to 0+}\int_{r}^{a} h(x)\,dx = 2\int_{0}^{a} h(x)\,dx \\ &= \int_{0}^{a} (f(x) + f(-x))\,dx. \end{aligned}\tag{2.12.8.6}$$

It is possible that $h(x)$ has no singularity at $x = 0$. In most cases, however, this device reduces the determination of a Cauchy principal value to that of an ordinary integral with a singularity at $x = 0$.

The same device can be used for Cauchy principal values at ∞. We have

$$P\int_{-\infty}^{\infty} f(x)\,dx = 2\int_{0}^{\infty} h(x)\,dx. \tag{2.12.8.7}$$

Examples

(1)
$$P\int_{-1}^{1} \frac{dx}{x}$$

Here

$$h(x) = \frac{1}{2}\left(\frac{1}{x} - \frac{1}{x}\right) = 0,$$

so that

$$P\int_{-1}^{1} \frac{dx}{x} = 0.$$

(2) $$P\int_{-1}^{1}\frac{e^x}{x}\,dx$$

Here

$$h(x)=\frac{1}{2}\left(\frac{e^x}{x}+\frac{e^{-x}}{-x}\right)=\frac{1}{x}\sinh x.$$

Hence

$$P\int_{-1}^{1}\frac{e^x}{x}\,dx=2\int_{0}^{1}\frac{\sinh x}{x}\,dx.$$

The function $\sinh x/x$ has only an apparent singularity at $x = 0$.

The method of subtracting the singularity may also be used. Suppose we consider the *Hilbert transform* of $f(t)$

$$I(x)=I=P\int_a^b\frac{f(t)}{t-x}\,dt,\qquad a<x<b. \tag{2.12.8.8}$$

We have

$$\begin{aligned}I=P\int_a^b\frac{f(t)}{t-x}\,dt&=\int_a^b\frac{f(t)-f(x)}{t-x}\,dt+f(x)P\int_a^b\frac{dt}{t-x}\\&=\int_a^b\frac{f(t)-f(x)}{t-x}\,dt+f(x)\log\frac{b-x}{x-a}.\end{aligned}\tag{2.12.8.9}$$

If we assume that the function

$$\phi(t,x)=\frac{f(t)-f(x)}{t-x}\tag{2.12.8.10}$$

is of class C^1 for fixed x and variable t, then $\phi(x, x) = f'(x)$ and the integral $\int_a^b \phi(t, x)\,dt$ in (2.12.8.9) has no difficulties associated with it.

It may be useful to consider

$$\int_{x-h}^{x+h}\phi(t,x)\,dt=\int_{-h}^{h}\frac{f(t+x)-f(x)}{t}\,dt.\tag{2.12.8.11}$$

If $f(t)$ can be expanded in a Taylor series at $t = x$, then we have

$$\begin{aligned}\int_{x-h}^{x+h}\phi(t,x)\,dt&=\int_{-h}^{h}\left(f'(x)+\frac{tf''(x)}{2!}+\frac{t^2f'''(x)}{3!}+\cdots\right)dt\\&=2hf'(x)+\frac{h^3}{9}f'''(x)+\cdots.\end{aligned}\tag{2.12.8.12}$$

References. Bareiss and Neuman [1]; Ioakimidis [1]; Longman [3]; Stewart [1].

Interpolatory-type and Gauss-type formulas can be developed for Cauchy principal value integrals. Thus, Price has worked out the following rules. The first rule is a nine-point formula (the coefficient of $f(0)$ is 0), which is exact for $f(x) \in \mathscr{P}_8$.

$$P\int_{-1}^{1} \frac{f(x)\,dx}{x} \approx A[f(1) - f(-1)] + B\left[f\left(\frac{3}{4}\right) - f\left(-\frac{3}{4}\right)\right] + C\left[f\left(\frac{1}{2}\right) - f\left(-\frac{1}{2}\right)\right] + D\left[f\left(\frac{1}{4}\right) - f\left(-\frac{1}{4}\right)\right]. \tag{2.12.8.13}$$

Here

$$\begin{aligned} A &= \tfrac{2459}{33075} \approx .07434\,618292 \\ B &= \tfrac{1856}{3675} \approx .50503\,40136 \\ C &= \tfrac{592}{4725} \approx .12529\,10052 \\ D &= \tfrac{9152}{4725} \approx 1.93693\,1217, \end{aligned} \tag{2.12.8.14}$$

The second rule is a four-point formula exact for $f(x) \in \mathscr{P}_8$ and, hence, is of Gauss type.

$$P\int_{-1}^{1} \frac{f(x)}{x}\,dx \approx w_1[f(x_1) - f(-x_1)] + w_2[f(x_2) - f(-x_2)]. \tag{2.12.8.15}$$

Here

$$\begin{aligned} w_1 &= .4039\,4864, & x_1 &= .8611\,3631, \\ w_2 &= 1.9181\,8095, & x_2 &= .3399\,8104. \end{aligned} \tag{2.12.8.16}$$

The weight function $1/x$ is not of one sign over $[-1, 1]$, and this is reflected in the fact that the weights in the rule (2.12.8.15) are not of one sign.

The abscissas here are the positive abscissas x_i in G_4 while the weights are w_i/x_i where w_i are the weights of G_4 corresponding to the x_i. This is a general situation. Here are the details. From (2.12.8.6),

$$P\int_{-1}^{1} \frac{f(x)}{x}\,dx = \int_{0}^{1} \frac{f(x) - f(-x)}{x}\,dx. \tag{2.12.8.17}$$

If $f(x)$ is differentiable on $[-1, 1]$ sufficiently many times, then the function $\psi(x) = [f(x) - f(-x)]/x$ will be even and differentiable one less time.

Suppose now that $x_i, w_i, i = 1, \ldots, 2n, x_i = -x_{2n-i}, w_i = w_{2n-i}$, are the abscissas and weights of a Gauss rule on $[-1, 1]$. Then

$$\int_{0}^{1} \psi(x)\,dx \approx \sum_{i=1}^{n} w_i \psi(x_i) = \sum_{i=1}^{n} w_i \left[\frac{f(x_i) - f(-x_i)}{x_i}\right]. \tag{2.12.8.18}$$

Thus,

$$P\int_{-1}^{1}\frac{f(x)}{x}\,dx \approx \sum_{i=1}^{2n} w_i\frac{f(x_i)}{x_i} = \sum_{i=1}^{2n}\frac{w_i}{x_i}f(x_i). \qquad (2.12.8.19)$$

This formula can also be derived as a special instance of Section 2.7.6.

Formula (2.12.8.19) is also a particular case of a more general Gauss-type method involving functions of the second kind (Section 1.12). Consider an admissible weight function $w(x)$ on $[a, b]$ and the corresponding orthogonal polynomials $P_n(x) = k_n x^n + \cdots$, $k_n > 0$, $n = 0, 1, \ldots$. The functions of the second kind $Q_n(\lambda)$ are given by

$$Q_n(\lambda) = P\int_a^b w(x)\frac{P_n(x)}{x-\lambda}\,dx, \qquad a < \lambda < b. \qquad (2.12.8.20)$$

We are interested in the Cauchy principal value integral

$$I(f; \lambda) = P\int_a^b w(x)\frac{f(x)}{x-\lambda}\,dx, \qquad a < \lambda < b. \qquad (2.12.8.21)$$

Let $\{x_i, w_i; i = 1, \ldots, n\}$ be the Gauss abscissas and weights appropriate to $w(x)$ and let $L_n(x; f)$ be the Lagrange polynomial of degree n interpolating to $f(x)$ at the points $x_1, \ldots, x_n$ and λ, assumed distinct from the x_i. Then

$$L_n(x; f) = \sum_{i=1}^{n} l_i(x)\frac{(x-\lambda)}{(x_i-\lambda)}f(x_i) + \frac{P_n(x)}{P_n(\lambda)}f(\lambda) \qquad (2.12.8.22)$$

where $l_i(x) = P_n(x)/P_n'(x_i)(x - x_i)$. Hence

$$I(f; \lambda) \simeq I(L_n; \lambda) = \sum_{i=1}^{n} w_i\frac{f(x_i)}{x_i-\lambda} + \left(P\int_a^b \frac{w(x)P_n(x)}{x-\lambda}\,dx\right)\frac{f(\lambda)}{P_n(\lambda)}$$

$$= \sum_{i=1}^{n} w_i\frac{f(x_i)}{x_i-\lambda} + \frac{Q_n(\lambda)}{P_n(\lambda)}f(\lambda). \qquad (2.12.8.23)$$

If $f \in C^{2n+1}[a, b]$, then the error in the rule is given by

$$f^{(2n+1)}(\xi)/(2n)!k_n^2, \qquad a < \xi < b.$$

For the case $w(x) = 1$ and $[a, b] = [-1, 1]$, $Q_n(0) = 0$ if n is even so that (2.12.8.23) reduces to (2.12.8.19). If λ coincides with one of the x_i, (2.12.8.23) must be modified to include the value of the derivative $f'(\lambda)$. See Elliott and Paget for the resulting formula. Formulas for the Kronrod extension of (2.12.8.23) are given by Rabinowitz [13].

The evaluation of $I(f; \lambda)$ can also be approached from the point of view of product integration (Section 2.5.6). In this case the modified moments $P\int_a^b w(x)P_n(x)/(x-\lambda)\,dx$ are just the functions of the second kind.

For the case $w(x) = 1, [a,b] = [-1,1]$, and trigonometric points, the modified moments

$$M_j(\lambda) = P\int_{-1}^{1} \frac{T_j(x)}{x-\lambda}\,dx, \qquad -1 < \lambda < 1 \tag{2.12.8.24}$$

with respect to the Tschebyscheff polynomials $T_j(x)$ are given by Branders and Piessens and used in programs by them.

For additional approaches to this problem, see the survey paper by Rabinowitz [10].

References. Acharya and Das [1]; Branders and Piessens [1]; Chawla and Jayarajan [1]; Chawla and Kumar [1]; Chawla and Ramakrishnan [1]; Delves [1]; Elliot and Paget [1, 4]; Gautschi [9]; Hunter [5, 6]; Kumar [1]; Kutt [3]; Lebedev and Baburin [1]; Maroni [1]; Morawitz [1]; Noble and Beighton [1]; Paget [1]; Paget and Elliot [1]; Piessens [5]; Piessens, van Roy-Branders and Mertens [1]; J. F. Price [1]; QUADPACK [F1]; Rabinowitz [10, 13]; Sloan [1]; van der Sluis and Zweerus [1]; Wesseling [1].

A special treatment of the Cauchy principal value integral,

$$T(f) = \frac{1}{2\pi}\int_{-\pi}^{\pi} \cot\frac{1}{2}(\theta - \phi) f(\phi)\,d\phi, \tag{2.12.8.25}$$

which occurs in aerodynamics can be found in Serbin. A deeper study of the numerical treatment of Hilbert transforms is given in Bareiss and Neuman.

Atkinson discusses the numerical evaluation of the Cauchy transform

$$T\phi(z) = \frac{1}{\pi i}\int_{\Gamma} \frac{\phi(\zeta)\,d\zeta}{\zeta - z}, \qquad z \in \Gamma \tag{2.12.8.26}$$

in which Γ is a simple closed curve in the complex plane and the integral is defined as a principal value.

References. Atkinson [1]; Bareiss and Neuman [1]; Serbin [1].

2.12.9 *Integrals of Absolute Values*

Integrands of the form $|f(x)|$ occur with sufficient frequency so that a word is in order about their evaluation. Of course, if $f(x)$ does not change sign over the interval of integration, then $\int |f(x)|\,dx = \pm\int f(x)\,dx$ and we are in the usual case.

If $f(x)$ changes sign at x_0, then the continuity class of $f(x)$ may be lowered when the absolute value is taken. Thus, $(x - x_0)^{2n+1}$ is of class C^∞ at x_0 while $|(x - x_0)^{2n+1}|$ lacks a $(2n+1)$th derivative there.

If possible, it is advised to locate the zeros of $f(x)$ and to integrate between them. If this is not convenient, and a sequence of ordinary rules are used, one must be prepared for slow convergence.

Similar remarks apply to the absolute value of a complex function

$$|f(x)| = ((\mathrm{Re}\, f)^2 + (\mathrm{Im}\, f)^2)^{1/2}$$

if f has a zero on the path of integration.

Example. Compute $\int_{-1}^{1} |(x^2 - \frac{1}{4})(x - \frac{1}{4})|\, dx = 503/1536 = .3274739583$ using Gauss rules.

$$G_{36} = .32798767 \qquad G_{64} = .32753771$$

$$G_{48} = .32776175 \qquad G_{96} = .32753845$$

Convergence is bad, especially from 64 to 96.

Analysis of a very deep sort is required to derive the asymptotics of the error in G_n applied to such an integrand. Some indications of this will be found in Section 4.6.1.

For a discussion of the difficulties in the computation of $\int_{-1}^{1} |K_r(t)|\, dt$, $1 \le r < 2n$, where $K_r(t)$ is a Peano kernel (see Section 4.3) associated with an n-point Gauss rule, see Stroud and Secrest.

Reference. Stroud and Secrest [2, pp. 65–66].

2.12.10 *Divergent Integrals*

Consider the integral $I(\alpha) = \int_0^1 x^\alpha\, dx$ as a prototype. For $\alpha \ge 0$, the integral is proper. For $-1 < \alpha < 0$, the integral is improper but is convergent and the value $I(\alpha) = 1/(\alpha + 1)$ is obtained for all $\alpha > -1$. For $\alpha \le -1$, the integral diverges. On the other hand, the function $1/(\alpha + 1)$ may be continued analytically to all values of $\alpha \ne -1$. If the midpoint rule is applied to the integrand for $\alpha < -1$, all terms are positive, while the first term is $(1/n)(1/2n)^\alpha = (1/2^\alpha)(1/n^{1+\alpha})$. Since $1 + \alpha < 0$, it is clear that the sequence of midpoint rules $M_n(x^\alpha) \to +\infty$. By a deeper analysis of midpoint rule errors, it is possible to "sum" this sequence to the value given by the analytic continuation. Speed-up methods may also produce a summation to this value.

Reference. Ninham [1].

2.12.11 *Hadamard Finite-Part Integrals*

In order to evaluate the finite-part integral

$$\oint_s^r (x - s)^{-a} f(x)\, dx, \qquad a \ge 1, \quad f(s) \ne 0 \tag{2.12.11.1}$$

it suffices by (1.6.1.6) and (1.6.1.7) to have a numerical integration rule for the integral

$$I(f;a) = \int\!\!\!\!\!\!=_0^1 f(y)y^{-a}\,dy, \qquad a \geq 1, \quad f(0) \neq 0. \tag{2.12.11.2}$$

To this end, we approximate $I(f;a)$ by an interpolatory rule as in Section 2.5 based on the moments $I(y^{k-1};a)$, $k = 1, \ldots, n$, which may be evaluated by the formula

$$I(y^{k-1};a) = \int\!\!\!\!\!\!=_0^1 y^{k-1-a}\,dy = \begin{cases} 0, & k = a+1, \\ (k-a)^{-1}, & k \neq a+1. \end{cases} \tag{2.12.11.3}$$

We thus end up with the approximation

$$I(f;a) \simeq \sum_{i=1}^{n} w_i(a)f(x_i) \tag{2.12.11.4}$$

where the weights $w_i(a)$ are the solution of the system

$$\sum_{i=1}^{n} w_i(a)x_i^{k-1} = I(y^{k-1};a), \qquad k = 1, \ldots, n \tag{2.12.11.5}$$

and the x_i are any preassigned set of n points.

Kutt has tabulated values of $w_i(a)$ when the x_i are taken to be the equally spaced points $(i-1)/n$, $i = 1, \ldots, n$, for $n = 3, \ldots, 20$ and $a = 1$, $\frac{4}{3}, \frac{3}{2}, \frac{5}{3}$, 2, 3, 4, 5. It turns out that the magnitudes of the weights increase rapidly with increasing n and a so that at the extreme case $n = 20$, $a = 5$, some of the weights have magnitude of the order of 10^{10}. This implies that formula (2.12.11.4) must be evaluated with high precision. Kutt has also computed Gauss-type formulas of the form (2.12.11.4) where both the weights $w_i(a)$ and abscissas x_i are computed from (2.12.11.5) and where k runs from 1 to $2n$. However, the resulting formulas involve complex abscissas and weights.

Paget [2] has chosen the x_i to be the zeros of the shifted Legendre polynomial $P_n^*(x) = P_n(2x-1)$ so that if y_i are the abscissas in the n-point Gauss rule

$$\int_{-1}^{1} g(x)\,dx = \sum_{i=1}^{n} \mu_i g(y_i) + E_n(g), \tag{2.12.11.6}$$

then $x_i = (y_i + 1)/2$, $i = 1, \ldots, n$. In this case, the weights $w_i(a)$ are given by

$$w_i(a) = \mu_i \sum_{k=0}^{n-1} (k + \tfrac{1}{2})P_k(y_i)b_k(a), \qquad i = 1, \ldots, n \tag{2.12.11.7}$$

where

$$b_k(a) = \fint_0^1 y^{-a} P_k^*(y)\, dy, \qquad k = 0, \ldots, n-1. \tag{2.12.11.8}$$

Paget gives closed expressions for the $b_k(a)$ as well as a simple algorithm for evaluating the $w_i(a)$.

Both Kutt and Paget give bounds for the error in terms of higher derivatives of $f(x)$.

Example. (Paget [2]). Evaluate

$$I(f; 2) = \fint_0^1 \frac{y^{-2}\, dy}{[(y-2)^2 + 1]^{1/2}} = \frac{2\sqrt{5}}{25} (\log[10\sqrt{10} - 30] - 1) - \frac{\sqrt{2}}{5}.$$

For $n = 3$, the error using either Kutt's or Paget's points was about 10^{-2}, while for $n = 10$, it was less than 10^{-6}.

Paget [3] has also studied the particular finite-part integral $\fint_a^b w(x) f(x) \times (x - s)^{-2}\, dx$, $s \in (a, b)$, which can also be expressed as the derivative with respect to s of the Cauchy principal value integral $P \int_a^b w(x) f(x)(x - s)^{-1}\, dx$. He differentiates the numerical approximation (2.12.8.23) to the latter integral to get an approximation to the former. This approximation appears to have the disadvantage that it involves the derivative of f at s, $f'(s)$. However, a closer look at (1.6.1.7) shows that whenever a is an integer in (2.12.11.1), values of derivatives of f at s will be required.

References. Kutt [1]; Paget [2, 3].

2.13 *Indefinite Integration*

The problem here is the computation of

$$F(x) = \int_a^x f(t)\, dt, \qquad a \le x \le b. \tag{2.13.1}$$

We shall also consider briefly the more complicated indefinite integral in which the integrand also depends on x:

$$F(x) = \int_a^x f(x, t)\, dt, \qquad a \le x \le b. \tag{2.13.2}$$

The integral (2.13.1) may be considered from several points of view. These include (1) regarding (2.13.1) as a definite integral over a variable range and (2) regarding $F(x)$ as the solution to the differential equation

$$\frac{dF}{dx} = f(x), \qquad F(a) = 0. \tag{2.13.3}$$

A simple (and often very satisfactory) approach consists in dividing the interval of integration $a \leq x \leq b$ into a set of subintervals and applying a rule of approximate integration to each subinterval. Simpson's rule is widely used for this purpose despite the fact that it advances the integration two steps at a time. To get started, we may use, for example, the following rule (of "overhanging" type):

$$\int_a^{a+h} f(x)\,dx \approx \frac{h}{12}[5f(a) + 8f(a+h) - f(a+2h)]. \qquad (2.13.4)$$

For the use of other integration rules with exterior points in indefinite integration, see Section 2.6.2.

Indefinite Integration via Differential Equations

Some of the standard techniques reduce to familiar rules. Consider, for example, the classical Runge–Kutta method for the solution of

$$\frac{dy}{dx} = g(x, y), \qquad y(x_0) = y_0\,. \qquad (2.13.5)$$

The relevant formulas are

$$y_{m+1} = y_m + \frac{h}{6}(k_1 + 2k_2 + 2k_3 + k_4),$$

$$k_1 = g(x_m, y_m), \qquad k_3 = g\left(x_m + \frac{h}{2}, y_m + \frac{hk_2}{2}\right),$$

$$k_2 = g\left(x_m + \frac{h}{2}, y_m + \frac{hk_1}{2}\right), \qquad k_4 = g(x_m + h, y_m + hk_3). \qquad (2.13.6)$$

Here y_m is the computed value of the solution at x_m and $x_{m+1} - x_m = h$. When $g(x, y) = f(x)$, these formulas reduce to

$$y_{m+1} = y_m + \frac{h}{6}\left[f(x_m) + 4f\left(x_m + \frac{h}{2}\right) + f(x_m + h)\right]. \qquad (2.13.7)$$

This is equivalent to

$$\int_{x_m}^{x_{m+1}} f(x)\,dx \approx \frac{h}{6}\left[f(x_m) + 4f\left(x_m + \frac{h}{2}\right) + f(x_m + h)\right], \qquad (2.13.8)$$

which is merely Simpson's rule applied to $[x_m, x_{m+1}]$.

Other Runge–Kutta rules may reduce to higher order integration formulas when g is independent of y. For example, King has developed Runge–Kutta rules which reduce to 5th-order Radau and 6th-order Lobatto integration rules.

A general *multistep method* for indefinite integration would consist in computing the value of the integral at the next step, y_{n+1}, in terms of values of the integral previously computed, $y_n, y_{n-1}, \ldots$, and of the integrand $f(x_{n+1}), f(x_n), f(x_{n-1}), \ldots$. For example, a formula of this type

$$y_{n+1} = \tfrac{1}{8}[9y_n - y_{n-2} + 3h(f(x_{n+1}) + 2f(x_n) - f(x_{n-1}))] - \tfrac{1}{40}h^4(f^{(5)}(\xi)), \tag{2.13.9}$$

is sometimes used. Multistep methods are of interest principally in the solution of differential equations. They lead immediately to questions of numerical stability, and the subject will not be pursued here.

The corrector formula in the predictor–corrector pair

$$P: \bar{y}_{m+1} = y_{m-1} + 2hf(x_m, y_m),$$

$$C: y_{m+1} = y_m + \frac{h}{2}[f(x_m, y_m) + f(x_{m+1}, \bar{y}_{m+1})], \tag{2.13.10}$$

reduces to the trapezoidal rule.

When the indefinite integral is desired, it may be convenient to exhibit the computation as a simple recurrence. For example, the trapezoidal rule may be written in the form

$$y_{n+1} = y_n + \frac{h}{2}[f(x_n) + f(x_{n+1})]. \tag{2.13.11}$$

Example. Compute

$$y(x) = \int_0^x e^t\,dt = e^x - 1 \quad \text{for} \quad x = 0(.1)1,$$

using the trapezoidal rule.

x	y_n	Error $= y_n - y$
.0	.0000	.0000
.1	.1053	.0001
.2	.2216	.0002
.3	.3501	.0003
.4	.4922	.0004
.5	.6493	.0005
.6	.8228	.0007
.7	1.0146	.0008
.8	1.2266	.0010
.9	1.4608	.0012
1.0	1.7197	.0014

More sophisticated packages for the numerical solution of ordinary differential equations are now available and may be used to tabulate an indefinite integral. See, for example, the well-documented program by Shampine and Gordon.

References. Hamming [1, Chap. 13]; King [1]; Krylov [1, Chaps. 13–16]; Shampine and Gordon [1].

2.13.1 Application of Approximation Theory; Tschebyscheff Series

Suppose that it is desired to compute

$$F(x) = \int_a^x f(t)\,dt, \qquad a \le x \le b, \qquad -\infty < a < b < \infty, \tag{2.13.1.1}$$

for a given integrand $f(x)$. Suppose, further, that we are in possession of an approximation to $f(x)$:

$$f(x) = a_0\phi_0(x) + a_1\phi_1(x) + \cdots + a_n\phi_n(x) + \varepsilon(x), \qquad a \le x \le b, \tag{2.13.1.2}$$

where

$$|\varepsilon(x)| \le \varepsilon, \qquad a \le x \le b \tag{2.13.1.3}$$

and where each of the approximating functions $\phi_i(x)$ has an indefinite integral

$$\psi_i(x) = \int_a^x \phi_i(t)\,dt, \tag{2.13.1.4}$$

which is simple to handle. Then, integrating (2.13.1.2), we obtain

$$F(x) = \int_a^x f(t)\,dt = a_0\psi_0(x) + a_1\psi_1(x) + \cdots + a_n\psi_n(x) + \eta(x), \qquad a \le x \le b, \tag{2.13.1.5}$$

where

$$|\eta(x)| = \left|\int_a^x \varepsilon(t)\,dt\right| \le (b - a)\varepsilon. \tag{2.13.1.6}$$

The closed form (2.13.1.5) of the indefinite integral may be very convenient for further manipulation.

We may, for example, arrive at an approximation of the form (2.13.1.2) by truncating the Taylor expansion $f(x) = \sum_{n=0}^{\infty} a_n x^n$, $a_n = f^{(n)}(0)/n!$. If it is too difficult to obtain the coefficients a_n by formal manipulation or through the use of derivatives, they may be approximated numerically by the use of contour integration. This has been explained in Section 2.11.1. In principle, we may employ any sort of approximation of form (2.13.1.2), but in practice it turns out to be particularly convenient to use expansions in terms of orthogonal polynomials, especially the Tschebyscheff polynomials $T_n(x)$. Important facts about these polynomials have been tabulated in Sections 1.12, 1.13,II.

It can be shown that under mild conditions on the function $f(x)$, for example if $f(x)$ is continuous and of bounded variation in $[-1, 1]$, then $f(x)$ may be expanded in a uniformly convergent series of T's:

$$f(x) = \tfrac{1}{2}a_0 + a_1 T_1(x) + a_2 T_2(x) \cdots. \tag{2.13.1.7}$$

The constants a_r are the "Fourier–Tschebyscheff" coefficients of $f(x)$ and are given by the formula

$$a_r = \frac{2}{\pi}\int_{-1}^{1} \frac{f(x)T_r(x)}{(1-x^2)^{1/2}}\,dx = \frac{2}{\pi}\int_0^{\pi} f(\cos\theta)\cos r\theta\, d\theta. \tag{2.13.1.8}$$

For many functions the sequence $a_0, a_1, \ldots$ decreases to zero rapidly. Furthermore, the partial sum $\tfrac{1}{2}a_0 + a_1 T_1(x) + \cdots + a_N T_N(x)$ is a polynomial of degree $\le N$, which is very close to the best approximation to $f(x)$ by polynomials $p_N(x)$ of this degree, approximation being measured in the sense of $\max_{-1\le x\le 1} |f(x) - p_N(x)|$. (See, for example, Davis.)

In any case, we have the estimate

$$|f(x) - (\tfrac{1}{2}a_0 + \cdots + a_N T_N(x))| = \left|\sum_{j=N+1}^{\infty} a_j T_j(x)\right| \le \sum_{j=N+1}^{\infty} |a_j|, \qquad -1 \le x \le 1. \tag{2.13.1.9}$$

For a general sort of function, the coefficients a_k cannot be evaluated in terms of closed expressions, and we must seek an approximation to them. What we shall do essentially is to apply a trapezoidal rule of a certain order n to the second integral in (2.13.1.8).

We define constants $\alpha_0, \alpha_1, \ldots, \alpha_n$ by means of

$$\begin{aligned}\alpha_r &= \frac{2}{n}\sum_{j=0}^{n}{}'' f(x_j)T_r(x_j) = \frac{2}{n}\sum_{j=0}^{n}{}'' f\left(\cos\frac{\pi j}{n}\right)\cos\frac{\pi r j}{n} \\ &= \frac{2}{n}\sum_{j=0}^{n}{}'' f(x_j)T_j(x_r).\end{aligned} \tag{2.13.1.10}$$

To see how α_r differs from a_r, we have, from (2.13.1.10),

$$\begin{aligned}\alpha_r &= \frac{2}{n}\sum_{s=0}^{\infty}{}' a_s \sum_{j=0}^{n}{}'' T_s(x_j)T_r(x_j) \\ &= a_r + a_{2n-r} + a_{2n+r} + a_{4n-r} + \cdots.\end{aligned} \tag{2.13.1.11}$$

The single prime on the sigma means that the term with subscript $s = 0$ is halved. The double prime on the sigma indicates that the terms with subscripts $j = 0$ and $j = n$ are to be halved.

If n is sufficiently large and if the coefficients $a_0, a_1, \ldots$ decrease rapidly, we have very closely

$$f(x) \simeq \tfrac{1}{2}a_0 + a_1 T_1(x) + \cdots + a_n T_n(x). \tag{2.13.1.12}$$

In view of (2.13.1.11), this in turn is approximated by

$$f(x) \simeq \tfrac{1}{2}\alpha_0 + \alpha_1 T_1(x) + \cdots + \alpha_{n-1} T_{n-1}(x) + \tfrac{1}{2}\alpha_n T_n(x). \tag{2.13.1.13}$$

The coefficients α_r are available computationally through (2.13.1.10) using backward recurrence on the T_j (Section 1.12).

For n sufficiently large, the FFT (Section 3.9.5) may be profitably employed in their evaluation.

If we integrate the uniformly convergent series (2.13.1.7) term by term, we obtain

$$\int_{-1}^{x} f(t)\,dt = \frac{a_0}{2} T_1(x) + \frac{a_1}{4} T_2(x) + \sum_{r=2}^{\infty} \frac{a_r}{2}\left(\frac{T_{r+1}(x)}{r+1} - \frac{T_{r-1}(x)}{r-1}\right)$$

$$+ \text{const} = \sum_{r=0}^{\infty} A_r T_r(x) \tag{2.13.1.14}$$

where

$$A_r = \frac{a_{r-1} - a_{r+1}}{2r}, \qquad r > 0 \tag{2.13.1.15}$$

and

$$A_0 = \sum_{r=1}^{\infty} (-1)^{r+1} A_r .$$

The corresponding approximation is given by

$$\int_{-1}^{x} f(t)\,dt \simeq \frac{1}{2}\left\{\sum_{r=1}^{n-1} \frac{\alpha_{r-1} - \alpha_{r+1}}{r} (T_r(x) - (-1)^r) + \frac{\alpha_{n-1}}{n} T_n(x)\right.$$

$$\left. + \frac{\alpha_n}{2}\left(\frac{T_{n+1}(x)}{n+1} + \frac{T_{n-1}(x)}{n-1}\right) - (-1)^n\left(\frac{\alpha_{n-1}}{n} + \frac{n\alpha_n}{n^2-1}\right)\right\}. \tag{2.13.1.16}$$

For $x = 1$, this reduces to the Clenshaw–Curtis rule (Section 2.5.5).

As we have just seen, the evaluations of (2.13.1.8) by the trapezoidal rule are accomplished through the evaluations of f on the "practical" abscissas

$\cos \pi j/n$. If the integrals were evaluated by the midpoint rule, we would obtain a formula on the "classical" abscissas $\cos((2k-1)\pi/2n)$. This would not be as convenient as the "practical" case because if the interval is halved, the previously computed values cannot be used. Furthermore, the coefficients cannot be computed by backward recurrence.

The Filippi abscissas $\cos(k\pi/(n+1))$ result when $f(x)$ is expanded in the Tschebycheff polynomials of the 2nd kind $U_n(x)$. See Section 6.4. Here again the computations are as convenient as in the practical case, and appear to give more accurate results.

References. Clenshaw [1]; Clenshaw and Curtis [1]; Davis [6, pp. 60–64, 174–175, Chap. 10]; Filippi [1, 4]; Kearfott [1]; Lanczos [1]; Wright [1].

2.13.2 *Indefinite Integration and Approximation*

Indefinite integrals or integrals with a parameter may be used to obtain approximations to functions.

Example

$$\arctan x = \int_0^x \frac{dt}{1+t^2}.$$

Change the interval of integration to $[-1, 1]$ and obtain

$$\arctan x = 2x \int_{-1}^{1} \frac{du}{4 + x^2(u+1)^2}.$$

Now use a 5-point Gauss rule on this definite integral:

$$\arctan x \approx 2x\left[\frac{w_1}{4+(1+x_1)^2x^2} + \frac{w_2}{4+(1+x_2)^2x^2} + \frac{2-2w_1-2w_2}{4+x^2} + \frac{w_2}{4+(1-x_2)^2x^2} + \frac{w_1}{4+(1-x_1)^2x^2}\right], \tag{2.13.2.1}$$

where

$$x_1 = .9061\,7985, \qquad w_1 = .2369\,2689,$$
$$x_2 = .5384\,6931, \qquad w_2 = .4786\,2867.$$

This yields an approximation to $\arctan x$, which is accurate to at least seven figures over $0 \le x \le 1$.

Example. The *exponential integral* is defined by

$$E_m(z) = \int_z^\infty e^{-t}t^{-m}\,dt. \tag{2.13.2.2}$$

We can write

$$\phi(z) = e^z E_m(z) = \int_0^\infty \frac{e^{-u}}{(u+z)^m}\,du. \tag{2.13.2.3}$$

If we apply the Gauss–Laguerre formula (see Section 3.6) to this integral, we obtain

$$\phi(z) \approx \sum_{k=1}^{n} \frac{w_k}{(x_k + z)^m} \tag{2.13.2.4}$$

where x_k and w_k are the abscissas and weights in the n-point Gauss–Laguerre formula.

If $z = x + iy$ and R designates the error in this approximation, it has been shown by Todd that

$$|\mathrm{Re}\, R|, \quad |\mathrm{Im}\, R| \le \frac{(n!)^2}{(x^2+y^2)^{n+1/2}} \qquad \text{for} \quad x \ge 0,$$

and

$$\le \frac{(n!)^2}{y^{2n+1}} \qquad \text{for} \quad x \le 0.$$

Reference. Todd [3].

For further examples, see Section 2.9.

Spline approximations may be obtained through the use of (4.3.8) and the remainder formula which follows.

Example

$$e^x = 1 + x + \frac{x^2}{2!} + \frac{x^3}{3!} + \frac{1}{3!}\int_0^1 e^t (x-t)_+^3\,dt.$$

$$\equiv P_3(x) + R_3(x).$$

Evaluating the integral $R_3(x)$ approximately by means of a five-point compound Simpson's rule, we obtain an approximation to $R_3(x)$ as a cubic spline:

$$\begin{aligned} R_3(x) \approx \tfrac{1}{6}[\tfrac{1}{12}(e^0(x-0)_+^3) &+ \tfrac{1}{3}(e^{.25}(x-.25)_+^3) \\ &+ \tfrac{1}{6}(e^{.5}(x-.5)_+^3) + \tfrac{1}{3}(e^{.75}(x-.75)_+^3) \\ &+ \tfrac{1}{12}(e^1(x-1)_+^3)] \equiv S_3(x). \end{aligned}$$

The cubic spline $P_3(x) + S_3(x)$ approximates e^x to within 8×10^{-5} over $[0, 1]$.

Expansions other than Taylor may be used here. For example, the error in polynomial interpolation is expressible as the integral of a spline (see Davis), so that the same device is applicable.

Reference. Davis [6, p. 71].

Barrett has obtained rational approximations to the transforms

$$f(z) = \int_{-1}^{1} \frac{w(x)\,dx}{z - x}, \qquad z \notin [-1, 1] \tag{2.13.2.5}$$

by applying Gauss rules with abscissas x_k and weights w_k to the integral. A rational approximation

$$f(z) \approx \sum_{k=1}^{n} \frac{w_k\, w(x_k)}{z - x_k} \equiv S_n(z) \tag{2.13.2.6}$$

results. The degree of approximation of $f(z)$ by $S_n(z)$ is studied.

Reference. Barrett [3].

2.13.3 *Indefinite Integration of Nonequally Spaced Data*

We may use the trapezoidal rule or, if something "better" is desired, the rule based on overlapping parabolas explained in Section 2.3.

The use of splines has been explained in Section 2.3.1. A FORTRAN program for indefinite integration by splines has been given in Appendix 2 as SPLINT. For repeated integrals of a function defined on nonequally spaced abscissas, see Thacher.

References. Schweikert [1]; Thacher [5].

2.13.4 *Computation of Integrals of the Form* $\int_a^x f(x, t)\, dt$

In general, the integral must be treated as a sequence of definite integrals for $x_0, x_1, \ldots$. Some computational economies can be obtained when $f(x, t)$ is of a certain form. For example, suppose we have to compute the *convolution integral* $\int_0^x K(x-t)f(t)\, dt$, over the range $0 \le x \le 1$. We make a table of $f(t)$ over the range $0 \le t \le 1$ and of $K(x)$ over the range $0 \le x \le 1$. Referring to the values in this table, we avoid duplication of computing functional values. Convolutions may also be computed by techniques involving the Fast Fourier Transform. See Section 3.9.5.

Some tips on integration when the integrand is of the form $f(x) \cdot g(x, \alpha)$ where α is a parameter can be found in Rabinowitz and in Tompa. In this case, the values of $f(x)$ are computed once and stored. They are then available for all values of α.

References. Rabinowitz [2]; Tompa [1].

Chapter 3

Approximate Integration over Infinite Intervals

3.1 Change of Variable

The substitution $x = e^{-y}$ changes the interval $0 \le y \le \infty$ into the interval $0 \le x \le 1$. Hence, we have the formula

$$\int_0^\infty f(y)\,dy = \int_0^1 \frac{f(-\log x)}{x}\,dx = \int_0^1 \frac{g(x)}{x}\,dx. \tag{3.1.1}$$

This reduces an integral over an infinite range to one over a finite range. If $g(x)/x$ is bounded in the neighborhood of $x = 0$, then the second integral will be proper. If not, the integral will be improper and we have only exchanged one sort of difficulty for another. An alternative form of this transformation is

$$\int_0^\infty e^{-y} f(y)\,dy = \int_0^1 f\left(\log \frac{1}{x}\right) dx. \tag{3.1.2}$$

There is some numerical evidence that the transformation (3.1.1) is valuable if $f(y)$ is of exponential type, i.e., if $|f(y)| \le e^{-ky}, 0 \le y < \infty$. When a sequence of functions of this type was transformed and the resulting functions integrated over [0, 1] by the automatic integration routine CADRE (see Section 6.3) satisfactory results were obtained. This was not the case with integrands of power type.

The transformation (3.1.1) is, of course, a special case of a general procedure. Let $t(x)$ be any function that is in $C^1[0, \infty)$ and is monotonic there. Furthermore, let $t(x)$ satisfy either $t(0) = 1$, $t(\infty) = 0$ or $t(0) = 0$, $t(\infty) = 1$. Then we have

$$\int_0^\infty f(x)\,dx = \int_0^1 f(x(t)) \left|\frac{dx}{dt}\right| dt. \tag{3.1.3}$$

Formula (3.1.1) results from the use of $t(x) = e^{-x}$. Other possibilities are $t(x) = x/(1 + x)$ and $t(x) = \tanh x$. As in (3.1.1), the resulting integrals are usually improper.

Setting $x = a + (1 + t)/(1 - t)$ de Doncker and Piessens have that

$$\int_a^\infty f(x)\,dx = \int_{-1}^1 f\left(a + \frac{1+t}{1-t}\right) \frac{2}{(1-t)^2}\,dt \tag{3.1.4}$$

and then apply an IMT rule (Section 2.9.2) which alleviates the singularity at the endpoint, $t = 1$. Piessens *et al.* make a similar transformation to yield

$$\int_a^\infty f(x)\,dx = \int_0^1 f\left(a + \frac{1-t}{t}\right) \frac{1}{t^2}\,dt \tag{3.1.5}$$

and then evaluate the finite integral by an adaptive scheme based on the Gauss–Kronrod pair (G_7, K_{15}) together with the ε-algorithm.

For transformations of the interval $-\infty \le x \le \infty$ into $-1 \le t \le 1$, the substitution $x = t/(1 - t^{2p})$ for integral $p > 0$ is sometimes useful (see Section 5.1.5). When $f(x)$ is regular at $\pm\infty$, the substitution $x = \tan t$ yields

$$\int_{-\infty}^\infty f(x)\,dx = \int_{-\pi/2}^{\pi/2} \frac{f(\tan t)}{\cos^2 t}\,dt \tag{3.1.6}$$

which can be evaluated quite efficiently by the trapezoidal rule inasmuch as the integrand is a periodic function in t (cf. Section 2.9).

Additional transformations that are useful include

$$\int_0^{\pi/2} f(x)\,dx = \int_0^\infty f(x) \frac{\sin x}{x}\,dx, \tag{3.1.7}$$

provided that

$$f(x + \pi) = f(x) \quad \text{and} \quad f(x) = f(-x),$$

and

$$\int_0^{\pi/2} f(x)\cos x\,dx = \int_0^{\infty} f(x)\frac{\sin x}{x}\,dx, \tag{3.1.8}$$

provided that

$$f(x+\pi) = -f(x) \quad \text{and} \quad f(x) = f(-x).$$

A one-parameter family of transformations has been suggested by Squire. We split the range $[0, \infty)$ into two intervals $[0, S]$ and $[S, \infty)$ and set $t = x/S$ in the first interval, $t = S/x$ in the second. This gives

$$\int_0^{\infty} f(x)\,dx = S\int_0^1 [f(St) + t^{-2}f(S/t)]\,dt. \tag{3.1.9}$$

Any open n-point rule $\{w_i, x_i : i = 1, \ldots, n\}$ on $[0, 1]$ is converted into a $2n$-point rule $\{w_i^*, x_i^* : i = 1, \ldots, 2n\}$ on $[0, \infty)$ as follows:

$$w_i^* = Sw_i, \qquad x_i^* = Sx_i, \qquad i = 1, \ldots, n;$$

$$w_i^* = Sw_{i-n}/x_{i-n}^2, \qquad x_i^* = S/x_{i-n}, \qquad i = n+1, \ldots, 2n. \tag{3.1.10}$$

The S should be chosen as a point where $f(x)$ starts to behave like a series in $1/x$. Of course, it is not necessary to treat both integrands in (3.1.9) by the same rule. It is just a convenience.

Examples. The following examples were computed by evaluating the two integrals $\int_0^1 f(x)\,dx$ and $\int_0^1 x^{-2}f(1/x)\,dx$ $(S = 1)$ using the automatic integration routine CADRE (Section 6.3). In the second integral the value of the integrand at $x = 0$ was set equal to 0. The target accuracy was nine figures.

$f(x)$	Number of functional evaluations	Was accuracy achieved?
$\dfrac{e^{-x}}{1+x^4}$	145	Yes
$\dfrac{1}{(x+2)\log^2(x+2)}$	937	No
$\dfrac{\sin(x+2)}{x+2}$	553	No
$\dfrac{1}{x^{1/2}(1+x)}$	1281	Yes
$e^{-x}\sin x$	353	Yes

References. de Doncker and Piessens [1]; Piessens, de Doncker, Überhuber, and Kahaner [1]; Squire [3, 4]; Takahasi and Mori [3].

3.2 *Proceeding to the Limit*

The basic definition

$$\int_0^\infty f(x)\,dx = \lim_{r\to\infty} \int_0^r f(x)\,dx \tag{3.2.1}$$

suggests a primitive mode of procedure. Let $0 < r_0 < r_1 < \cdots$ be a sequence of numbers that converge to ∞. Write

$$\int_0^\infty f(x)\,dx = \int_0^{r_0} f(x)\,dx + \int_{r_0}^{r_1} f(x)\,dx + \cdots. \tag{3.2.2}$$

Each of the integrals on the right-hand side is proper, and the evaluations are terminated when $|\int_{r_n}^{r_{n+1}} f(x)\,dx| \le \varepsilon$. This is only a practical termination criterion and is not correct theoretically. For example, when the divergent integral $\int_1^\infty dx/x$ is evaluated by such a procedure, a finite answer will be printed out. The interval is frequently doubled at each step; that is, $r_n = 2^n$. The idea behind this selection is that if an arithmetic sequence is used ($r_n = cn$) the contribution of each additional step may be too insignificant to be worth a special computation. Furthermore, it may be less than ε, thus stopping the process.

Example

$$I_n = \int_0^{r_n} \frac{e^{-x}}{1+x^4}\,dx, \qquad r_n = 2^n.$$

n	I_n	Number of functional evaluations
0	.5720 2582	35
1	.6274 5952	52
2	.6304 3990	100
3	.6304 7761	178
4	.6304 7766	322
Exact	.6304 7783	

3.2.1 *Speed-Up of Convergence*

The method just described may be speeded up if one can obtain a reasonably good asymptotic expansion for the tail $\int_r^\infty f(x)\,dx$. We would expect for example, using the first term of the shifted Laguerre rule (3.6.4a) that

$\int_r^\infty e^{-x}/(1 + x^4)\,dx \sim ce^{-r}/(1 + r^4)$. We shall now use Richardson's extrapolation to infinity in the form

$$I'_n = \frac{I_n\phi(r_{n+1}) - I_{n+1}\phi(r_n)}{\phi(r_{n+1}) - \phi(r_n)}, \qquad \phi(r) = \frac{e^{-r}}{1 + r^4}, \qquad r_n = 2^n.$$

For the example above, this yields

n	I'_n
0	.6299 6722
1	.6304 6682
2	.6304 7765
3	.6304 7766

Note that I'_1 is much better than I_2 and I'_2 is almost identical to I_4.

Speed-up methods using the epsilon algorithm have been applied to integrals over an infinite range with limited success.

Reference. Chisholm, Genz, and Rowlands [1].

3.2.2 *Nonlinear Transformations*

Speed-up methods similar to those used for accelerating the convergence of slowly convergent series may be applied to infinite integrals. Let $f(x) \in C[a, \infty)$ and assume that

$$F(t) = \int_a^t f(x)\,dx$$

converges to S as $t \to \infty$. A function $F_1(t)$ is said to converge more rapidly to S than $F(t)$ if

$$\lim_{t\to\infty} \frac{S - F_1(t)}{S - F(t)} = 0.$$

Let us now define $R(t; k) = f(t + k)/f(t)$, $k > 0$ for $f(t) \neq 0$ and denote by $R(k)$, the limit, $\lim_{t\to\infty} R(t; k)$, if it exists. We are now in a position to define the G-transform of F and give some of its properties.

DEFINITION. The G-transform of F is given by

$$G[F; t, k] = \frac{F(t + k) - R(t, k)F(t)}{1 - R(t, k)}, \qquad R(t, k) \neq 1. \tag{3.2.2.1}$$

For any k such that $R(k) \neq 1$, $G[F; t, k]$ converges to S. If $R(k) \neq 0, 1$, the convergence is more rapid than that of $F(t + k)$.

A limiting case of the G-transform, which also converges more rapidly than $F(t + k)$, may be defined in which $R(t, k)$ is replaced by $R(k)$ in (3.2.2.1).

Example

$$F(t) = \int_0^t \frac{\sin x}{x}\, dx \to \frac{\pi}{2} = 1.5707\,963\ldots,$$

$$R(t, k) = \frac{\sin(t + k)}{\sin t} \frac{t}{t + k}.$$

With $k = \pi$, $R(t; \pi) = -t/(t + \pi) \to -1 = R(\pi)$.

$$G[F; t, \pi] = \frac{t + \pi}{2t + \pi} \int_0^{t+\pi} \frac{\sin x}{x}\, dx + \frac{t}{2t + \pi} \int_0^t \frac{\sin x}{x}\, dx.$$

For $t = 9\pi$, $G \sim 1.5707\,886$. Using the Euler transformation (3.8.2) on the same range yields 1.5707 911.

In case $R(k) = 1$, we may be able to achieve better convergence with the Q-transform which, when defined, converges more rapidly than $F(t + 1)$ and $G[F; t, 1]$. Let $q = \lim_{t\to\infty} t[1 - R(t, 1)]$. If $q \neq 1$, we define the Q-transform of $F(t)$ by

$$Q[F; t] = (qG[F; t, 1] - F(t))/(q - 1).$$

Example

$$F(t) = \int_0^t dx/(1 + x)^2 = 1 - 1/(t + 1).$$

In this case we can work everything out analytically. We have that $R(t, 1) = ((t + 1)/(t + 2))^2$, $q = 2$, $S = 1$, $G[F; t, 1] = 1 - 1/(2t + 3)$, and $Q[F; t, 1] = 1 + 1/(t + 1)(2t + 3)$.

Similar to the G-transform is the B-transform. Let $\rho(t; k) = kf(kt)/f(t)$, $k > 1, f(t) \neq 0$, and let $\rho(k) = \lim_{t\to\infty} \rho(t; k)$ exist. The B-transform is given by

$$B[F; t, k] = \frac{F(kt) - \rho(t, k)F(t)}{1 - \rho(t, k)}, \qquad \rho(t, k) \neq 1. \tag{3.2.2.2}$$

If $\rho(k) \neq 1$, $B[F; t, k]$ converges to S more rapidly than $F(t)$ while if $\rho(k) \neq 0, 1$, convergence is more rapid than that of $F(kt)$. As previously, a limiting case may be defined in which $\rho(t, k)$ is replaced by $\rho(k) \neq 0, 1$ in (3.2.2.2). This transformation also converges more rapidly than $F(kt)$.

Example

$$F(t) = \int_0^t \frac{dx}{1 + x^2} \to \frac{\pi}{2}.$$

For $k = 1.1$, $t = 18$, $B(F; t, k) = 1.5709\,9504$ with an error less than .0002.

Further generalizations, theoretical results, and examples are given in the references.

References. Atchison [1]; Atchison and Gray [1]; Gray and Atchison [1–3]; Gray, Atchison, and McWilliams [1]; Gray and Schucany [1]; Levin and Sidi [1]; McWilliams and Thompson [1].

3.3 Truncation of the Infinite Interval

We may also reduce the infinite interval to a finite interval by ignoring the "tail" of the integrand. Rigorous application of this method requires that the analyst be able to estimate this tail by some simple analytical device.

Example. Determine numerically $\int_0^\infty e^{-x^2}\,dx$. We estimate $\int_k^\infty e^{-x^2}\,dx$. For $x \ge k$, we have $x^2 \ge kx$. Hence

$$\int_k^\infty e^{-x^2}\,dx \le \int_k^\infty e^{-kx}\,dx = e^{-k^2}/k.$$

For $k = 4$, we have $e^{-k^2}/k \approx 10^{-8}$. For a seven-figure computation, it suffices therefore to evaluate $\int_0^4 e^{-x^2}\,dx$ by some standard method.

Example. Determine numerically

$$\int_0^\infty \frac{\sin x}{1 + x^2}\,dx.$$

We have

$$\left|\int_{2k\pi}^\infty \frac{\sin x}{1 + x^2}\,dx\right| = |r_1 + r_2 + \cdots|, \quad \text{where} \quad r_j = \int_{(2k+j-1)\pi}^{(2k+j)\pi} \frac{\sin x}{1 + x^2}\,dx.$$

Since $r_{2n} < 0$, $r_{2n+1} > 0$, and $|r_1| > |r_2| > \cdots$, we have

$$|r_1 + r_2 + \cdots| < r_1 = \int_{2k\pi}^{(2k+1)\pi} \frac{\sin x\,dx}{1 + x^2} < \int_{2k\pi}^{(2k+1)\pi} \frac{dx}{x^2} < \frac{1}{4\pi k^2}.$$

For a truncation error of 10^{-4}, this analysis suggests that $k \approx 28$.

3.3.1 Reducing the "Intensity" of the Singularity

Suppose it is possible to express the integrand $f(x)$ in the form $f(x) = g(x) + r(x)$ where $\int_0^\infty g(x)\,dx$ is available in closed form or from other sources and where the remainder $r(x) \to 0$ more rapidly than $f(x)$ as $x \to \infty$.

The burden of evaluation is now thrown to $\int_0^\infty r(x)\,dx$ which presumably will be attended by less numerical difficulty. Repeated application of this principle can lead to convergent or asymptotic expansions.

Rice has suggested subtracting the following simple integrals when appropriate.

$$\int_0^\infty u^{-1}\sin au\,du = \begin{cases} \pi/2, & a>0, \\ -\pi/2, & a<0, \end{cases} \tag{3.3.1.1}$$

$$\int_0^\infty u^{-1}[\cos au - e^{-au}]\,du = 0, \tag{3.3.1.2}$$

$$\int_0^\infty \frac{\cos au}{1+u^2}\,du = \frac{\pi}{2}e^{-|a|}. \tag{3.3.1.3}$$

Example

$$\int_0^\infty \frac{x}{1+x^2}\sin x\,dx = \int_0^\infty \frac{\sin x}{x}\,dx - \int_0^\infty \frac{1}{1+x^2}\frac{\sin x}{x}\,dx$$
$$= \frac{\pi}{2} - \int_0^\infty \frac{1}{1+x^2}\frac{\sin x}{x}\,dx.$$

When the integrand is asymptotic to $u^{-r}\exp(iau)$ where $r>0$ and a is real, Rice recommends the more general family of integrals

$$\int_0^\infty u^{-n-\varepsilon}(e^{-\alpha u} - e^{-\beta u})^n\,du = \frac{\Gamma(1-\varepsilon)}{(\varepsilon)_n}\sum_{k=0}^{n}(-1)^{n-k}\binom{n}{k} \times [(n-k)\alpha + k\beta]^{n+\varepsilon-1}, \tag{3.3.1.4}$$

$$\int_0^\infty \left(\frac{e^{-\alpha u} - e^{-\beta u}}{u}\right)^n du = \frac{1}{(n-1)!}\sum_{k=0}^{n}(-1)^{n-k}\binom{n}{k} \times [(n-k)\alpha + k\beta]^{n-1}\log[(n-k)\alpha + k\beta] \tag{3.3.1.5}$$

where n is a positive integer and n, α, β, and ε are such that the integrals converge. In (3.3.1.4), ε is not an integer, $(\varepsilon)_0 = 1$, and $(\varepsilon)_n = \varepsilon(\varepsilon+1)\cdots(\varepsilon+n-1)$. In these formulas, we set $\alpha = -ia/n$, $\beta = |a|/n$, and $n+\varepsilon = r$ with $\varepsilon < -1$.

Examples (Rice).

1. $$I = \int_0^\infty (1+u^2)^{-1/2}e^{iu}\,du.$$

Taking first $n = 1$, $\alpha = -i$, $\beta = 1$ and then $n = 3$, $\alpha = -i/3$, $\beta = \frac{1}{3}$, we get using (3.3.1.5)

$$I = \int_0^\infty du\left[(1+u^2)^{-1/2}e^{iu} - \left(\frac{e^{iu}-e^{-u}}{u}\right) + \frac{1}{2}\left(\frac{e^{iu/3}-e^{-u/3}}{u}\right)^3\right] + \frac{i\pi}{2}$$
$$-\frac{1}{2}\frac{1}{2!}\sum_{k=0}^{3}\binom{3}{k}(-1)^{3-k}[(3-k)(-i/3)+k/3]^2\log[(3-k)(-i/3)+k/3].$$

The integral was evaluated by the trapezoidal rule after the change of variable, $u = a\log[1+e^{x/a}]$ with $a = 6$, using 52 points and $h = 4$ and gave I correct to 5 decimals.

2. $$I = \int_0^\infty (1+u^2)^{-5/6}e^{iu}\,du.$$

Using (3.3.1.4) with $\varepsilon = -\frac{4}{3}$, $n = 3$, $\alpha = -i/3$, $\beta = \frac{1}{3}$ we get

$$I = \int_0^\infty du[(1+u^2)^{-5/6}e^{iu} - u^{-5/3}(e^{iu/3}-e^{-u/3})^3]$$
$$+\frac{\Gamma(\frac{7}{3})}{(-\frac{4}{3})(-\frac{1}{3})(\frac{2}{3})}\sum_{k=0}^{3}(-1)^{3-k}\binom{3}{k}[(3-k)(-i/3)+(k/3)]^{2/3}.$$

The integrand now converges as $u^{-11/3}$.

Reference. S. O. Rice [2].

3.4 *Primitive Rules for the Infinite Interval*

The simple right-hand Riemann sum approximation takes on the form

$$\int_0^\infty f(x)\,dx \approx R_R(f;h) = h[f(h)+f(2h)+\cdots] = h\sum_{k=1}^{\infty} f(kh) \quad (3.4.1)$$

in the case of a singly infinite integral. The corresponding left-hand Riemann sum is

$$R_L(f;h) = hf(0) + R_R(f;h),$$

while the average of the two is the trapezoidal rule

$$T(f;h) = \tfrac{1}{2}hf(0) + R_R(f;h) = h\sum_{k=0}^{\infty}{}' f(kh).$$

In the case of the doubly infinite interval one has the rectangular approximation

$$\int_{-\infty}^{\infty} f(x)\,dx \approx R(f;h) = h\sum_{k=-\infty}^{\infty} f(kh), \quad (3.4.1a)$$

and, since

$$\int_{-\infty}^{\infty} f(x + x_0)\,dx = \int_{-\infty}^{\infty} f(x)\,dx,$$

the shifted rectangular approximation

$$\int_{-\infty}^{\infty} f(x)\,dx \approx h \sum_{k=-\infty}^{\infty} f(kh + x_0), \qquad -\infty < x_0 < \infty.$$

THEOREM. *Let $f(x)$ be monotonic for $x \geq 0$ and suppose that $\int_0^\infty f(x)\,dx$ exists. Then*

$$\lim_{h\to 0} R_{\mathrm{L}}(f; h) = \lim_{h\to 0} R_{\mathrm{R}}(f; h) = \int_0^\infty f(x)\,dx. \tag{3.4.2}$$

Proof. Since $\int_0^\infty f(x)\,dx$ exists, and $f(x)$ is monotonic, it follows that $\lim_{x\to\infty} f(x) = 0$. Furthermore, $f(x)$ must have the same sign throughout $x \geq 0$. Without loss of generality, we can assume $f(x)$ to be positive and decreasing. Since $f(x)$ is decreasing,

$$0 \leq \int_h^{(n+1)h} f(x)\,dx \leq h[f(h) + f(2h) + \cdots + f(nh)] \leq \int_0^{nh} f(x)\,dx. \tag{3.4.3}$$

Allow $n \to \infty$, and we have

$$\int_h^\infty f(x)\,dx \leq h \sum_{k=1}^{\infty} f(kh) \leq \int_0^\infty f(x)\,dx. \tag{3.4.4}$$

Allowing $h \to 0$, we obtain (3.4.2).

COROLLARY. *The theorem holds if $f(x)$ is ultimately monotonic, that is, if there is an $x_0 \geq 0$ such that $f(x)$ is monotonic for $x \geq x_0$.*

From the inequality (3.4.4) we obtain the error estimate

$$0 \leq \int_0^\infty f(x)\,dx - h \sum_{k=1}^{\infty} f(kh) \leq \int_0^h f(x)\,dx \leq hf(0). \tag{3.4.5}$$

This is an indication of very slow convergence in general.

Under certain circumstances, the trapezoidal rule can give very good approximations to infinite integrals. For the development of this result, we make use of the Euler–Maclaurin formula.

THEOREM. *Let a and k be fixed and let $f(x) \in C^{2k+1}[a, b]$ for all $b \geq a$. Suppose, further, that $\int_a^\infty f(x)\,dx$ exists, that*

$$M = \int_a^\infty |f^{(2k+1)}(x)|\,dx < \infty,$$

and that

$$f'(a) = f'''(a) = \cdots = f^{(2k-1)}(a) = 0,$$
$$f'(\infty) = f'''(\infty) = \cdots = f^{(2k-1)}(\infty) = 0. \tag{3.4.6}$$

Then, for fixed $h > 0$,

$$\left| \int_a^\infty f(x)\,dx - h\left[\frac{1}{2}f(a) + f(a+h) + f(a+2h) + \cdots\right] \right|$$
$$\leq \frac{h^{2k+1} M \zeta(2k+1)}{2^{2k}\pi^{2k+1}}. \tag{3.4.7}$$

Here $\zeta(k) = \sum_{j=1}^{\infty} j^{-k}$ *is the Riemann zeta function.*

Proof. From (2.9.15) we have

$$\Big| h[\tfrac{1}{2}f(a) + f(a+h) + \cdots + f(a+(n-1)h)$$
$$+ \tfrac{1}{2}f(a+nh)] - \int_a^{a+nh} f(x)\,dx \Big|$$
$$\leq \left|\frac{B_2}{2!} h^2[f'(a+nh) - f'(a)]\right|$$
$$+ \cdots + \left|\frac{B_{2k}}{(2k)!} h^{2k}[f^{(2k-1)}(a+nh) - f^{(2k-1)}(a)]\right|$$
$$+ h^{2k+1} \int_a^{a+nh} \left| P_{2k+1}\left(\frac{x-a}{h}\right) \right| |f^{(2k+1)}(x)|\,dx.$$

Now allow $n \to \infty$, and use conditions (3.4.6), the definition of M, and the inequality (2.9.17) for the periodic function $P_{2k+1}(x)$.

This theorem tells us that if the integrand and all of its odd-order derivatives up to order $2k-1$ vanish at both ends of an infinite interval, then, as $h \to 0$, the trapezoidal rule will converge to the proper answer with the rapidity of h^{2k+1}. If all odd-order derivatives vanish, then the rapidity exceeds h^{2k+1} for *all* k. A similar theorem can be formulated for integrals of the type $\int_{-\infty}^{\infty} f(x)\,dx$.

The Poisson formula related to (3.4.7) is

$$\int_0^\infty f(x)\,dx = h\left\{\frac{1}{2}f(0) + \sum_{k=1}^{\infty} f(kh)\right\} - 2\sum_{k=1}^{\infty} g\left(\frac{2k\pi}{h}\right), \tag{3.4.8}$$

where

$$g(x) = \int_0^\infty f(t)\cos xt\,dt. \tag{3.4.9}$$

Example (Hartree)

$$I = \int_0^\infty e^{-x^2}\,dx = \tfrac{1}{2}\sqrt{\pi}.$$

h	Trapezoidal rule
1.1	.88674
1.0	.88632 0
.9	.88623 598
.8	.88622 72808
.7	.88622 69285
.6	.88622 69254 8
.5	.88622 69254 5
Exact value	.88622 69254 5

Of course, if the values of the derivatives of odd order do not vanish at $x = a$ but are known there, then the error bound (3.4.7) will hold for the trapezoidal rule modified by the appropriate terms in the Euler–Maclaurin expansion. Another modification of the trapezoidal rule has been proposed by Silliman. He considers the class $S_{2m-1}(R^+)$ of semicardinal splines $S(x)$ of degree $2m - 1$ which are splines satisfying

$$S(x) \in C^{2m-2}(R^+),$$

$$S(x) \in \mathscr{P}_{2m-1} \quad \text{in each interval} \quad (v, v+1), \quad v = 0, 1, \ldots$$

where $R^+ = [0, \infty)$, and derives a set of integration rules of the form

$$\int_0^\infty f(x)\,dx = T + \sum_{v=0}^{\infty} h_v^{(m)} f(v) + \sum_{j=1}^{m-1} A_j^{(m)} f^{(j)}(0) + Rf \qquad (3.4.10)$$

such that $Rf = 0$ if $f \in S_{2m-1}(R^+) \cap L_1(R^+)$. Here T is the trapezoidal sum $\sum_{v=0}^{\prime\infty} f(v)$ and $L_1(R^+)$ is the class of functions such that $\int_0^\infty |f(x)|\,dx < \infty$. For $m = 1$, (3.4.10) reduces to $T + Rf$ while for $m = 2$, (3.4.10) becomes the Euler–Maclaurin rule $T + f'(0)/12 + Rf$. For $m = 3, 4, 5$ values of $A_j^{(m)}$ and $h_v^{(m)}$ are tabulated by Silliman.

A similar modification of the trapezoidal rule involving only function values was proposed by Schoenberg and Silliman who considered the class S_{2m-1}^+ of natural semicardinal splines $S(x)$ of degree $2m - 1$ which are splines satisfying

$$S(x) \in C^{2m-2}(-\infty, \infty),$$

$$S(x) \in \mathscr{P}_{2m-1} \quad \text{in each interval} \quad (v, v+1), \quad v = 0, 1, \ldots,$$

$$S(x) \in \mathscr{P}_{m-1} \quad \text{in} \quad (-\infty, 0).$$

They derive a set of integration rules of the form

$$\int_0^\infty f(x)\,dx = T + \sum_{\nu=0}^{\infty} \hat{h}_\nu^{(m)} f(\nu) + R(f) \tag{3.4.11}$$

such that $Rf = 0$ for $f \in S_{2m-1}^{+} \cap L_1(R^+)$, and tabulate the values of $\hat{h}_\nu^{(m)}$ for $m = 2(1)\,7$.

Formulas (3.4.10) and (3.4.11) are called for when $f(x)$ can be well approximated by the appropriate spline.

Using contour integration, Martensen derived a useful error estimate for the rectangular rule (3.4.1a). Suppose that $f(z)$ is analytic in the infinite strip $0 \le \operatorname{Im} z \le s$, is real on the real axis, and converges to 0 uniformly in the strip as $z \to \infty$. Then, $\int_{-\infty}^{\infty} f(x)\,dx$ exists, and $h \sum_{k=-\infty}^{\infty} f(kh)$ converges for all $h > 0$. If we write

$$\int_{-\infty}^{\infty} f(x)\,dx = h \sum_{k=-\infty}^{\infty} f(kh) + E_h(f), \tag{3.4.12}$$

we have

$$E_h(f) = \operatorname{Re}\left\{ \int_{-\infty+is}^{\infty+is} f(z)\left(1 + \frac{1}{i}\cot\frac{\pi z}{h}\right) dx \right\}, \tag{3.4.13}$$

and

$$|E_h(f)| \le \frac{2}{\exp(2\pi s/h) - 1} \int_{-\infty+is}^{\infty+is} |f(z)|\,dx. \tag{3.4.14}$$

The convergence is therefore extremely rapid for this type of function, and the rapidity increases with the width of the strip of analyticity.

Using (3.4.14), Engels has shown that for every $f \in C(-\infty, \infty)$ such that $If = \int_{-\infty}^{\infty} f(x)\,dx$ exists, the rectangle rule $R(f; h)$ converges to If as $h \to 0$.

A rectangular rule may be used to approximate the Hilbert Transform

$$g(x_0) = P \int_{-\infty}^{\infty} \frac{f(x)}{x - x_0}\,dx.$$

This yields

$$g(x_0) = 2 \sum_{\substack{k=-\infty \\ k,\,\text{odd}}}^{\infty} \frac{f(x_0 + kh)}{k} + E_h(f). \tag{3.4.15}$$

Under the above analyticity and growth conditions, Kress and Martensen have shown that

$$|E_h(f)| \le s^{-1}(\coth(\pi s/2h) - 1) \int_{-\infty+is}^{\infty+is} |f(z)|\,dx, \tag{3.4.16}$$

an error estimate that is independent of x_0.

Example (Milne)

$$I = \int_{-\infty}^{\infty} \operatorname{sech}^2 x \, dx = 2.0.$$

h	Trapezoidal rule
1.0	2.00408 43212
.8	2.00043 29224
.6	2.00000 94492
.4	2.00000 00036
.2	2.00000 00000
Exact value	2.00000 00000

The computation of an infinite integral by the trapezoidal rule usually requires that several values of h be used. For each value of h, the trapezoidal rule is an infinite series, $T(h) = h \sum_{j=0}^{\prime\infty} f(jh)$, which must be truncated at some value N to yield $T(h, N)$. The strategy to be used in selecting sequences h_i and N_i is not a simple matter since we are confronted here with the problem of evaluating an iterated limit numerically. One possibility is to choose N_i such that $h_i N_i \to \infty$ as $h_i \to 0$. (For example, $N_i h_i^2 \to$ constant.) Another possibility is to choose N_i so that $T(h_i, N_i) = T(h_i)$ to within ε_i where $\varepsilon_i \to 0$. If h_i is chosen to be $h_{i-1}/2$, then the latter strategy will not involve any superfluous computations of $f(x)$ since all the values used in $T(h_{i-1})$ are needed for $T(h_i)$.

References. Engels [13, p. 343]; Goodwin [1]; Hartree [2, p. 116]; Kress [1, 3, 8]; Kress and Martensen [1]; Martensen [2]; Milne [2]; Schoenberg and Silliman [1]; Silliman [3].

3.4.5 *Conversions to the Whole Axis*

In view of the often very rapid convergence of the rectangular rule over $(-\infty, \infty)$, Schwartz suggests that integrals over finite or semi-infinite intervals may be treated by converting the interval to $(-\infty, \infty)$ by the transforms

$$x = 1/(1 + e^{-y}), \qquad 0 \le x \le 1; \qquad -\infty \le y \le \infty$$

$$x = (e^y - 1)/(e^y + 1), \qquad -1 \le x \le 1; \qquad -\infty \le y \le \infty$$

$$x = e^y, \qquad 0 \le x < \infty; \qquad -\infty \le y \le \infty.$$

In order to provide a more rapid decay of the integrand at $\pm\infty$, the further transformation

$$\int_{-\infty}^{\infty} f(x)\,dx = \int_{-\infty}^{\infty} g(y)\,dy \tag{3.4.5.1}$$

with $g(y) = (e^y + e^{-y})f(e^y - e^{-y})$ may be employed one or more times.

Explicitly, these transformations lead to the rules

$$\int_{-1}^{1} f(x)\,dx = h\sum_{m=-\infty}^{\infty} \frac{2e^{hm}}{(1+e^{hm})^2} f\left(\frac{e^{hm}-1}{e^{hm}+1}\right) + E_h(f), \tag{3.4.5.2}$$

$$\int_{0}^{\infty} f(x)\,dx = h\sum_{m=-\infty}^{\infty} e^{hm} f(e^{hm}) + E_h(f). \tag{3.4.5.3}$$

Stenger has obtained error estimates analogous to (3.4.12) for (3.4.1a), (3.4.5.2), (3.4.5.3) but under somewhat different assumptions as to the growth of $f(x)$. A strategy for the truncation of these infinite series amounting to $N_i h_i^2 \to$ constant is indicated. Convergence then takes place with rapidity $\exp(-\text{const}\cdot N^{1/2})$.

Squire discusses the problems in implementing (3.4.5.2) and (3.4.5.3) on a computer and gives two computer programs and examples illustrating the range of applicability of these two rules. He has also described the implementation of (3.4.5.2) on a programmable hand calculator.

The integration rule (3.4.5.2), also known as the tanh rule, has been analyzed by Haber in the context of the Hardy space H_2. This space consists of the functions analytic in the unit disk with Taylor coefficients that satisfy $\sum_{n=0}^{\infty} |a_n|^2 < \infty$. With the inner product

$$(f, g) = \frac{1}{2\pi}\int_{|z|=1} f(z)\overline{g(z)}\,|dz|, \tag{3.4.5.4}$$

H_2 is a Hilbert space and the norm of a function $f \in H_2$ is given by $\|f\| = (f, f)^{1/2}$. If we now define the error functional $E_{N,h}$ for the truncated approximation

$$E_{N,h}f = \int_{-1}^{1} f(x)\,dx - h\sum_{m=-N}^{N} \frac{2e^{hm}}{(1+e^{hm})^2} f\left(\frac{e^{hm}-1}{e^{hm}+1}\right), \tag{3.4.5.5}$$

then $E_{N,h}$ is a bounded linear functional on H_2 with norm $\|E_{N,h}\|$ (see Section 4.7) and $|E_{N,h}f| \le \|E_{N,h}\|\cdot\|f\|$. Haber shows that if $h \to 0$ and $N \to \infty$ simultaneously in such a manner that $Nh \to \infty$, then

$$\|E_{N,h}\|^2 = 4\pi^2 e^{-2\pi^2/h} + 8\log 2e^{-Nh} + O(e^{-4\pi^2/h}) + O(he^{-Nh}), \tag{3.4.5.6}$$

and that a near-optimal choice of h is $\hat{h}(N) = \pi\sqrt{2/N} - 1/N$ for which $\| E_{N,\hat{h}} \| \sim 5.440e^{-(\pi/\sqrt{2})\sqrt{N}}$.

Additional rules which have a more rapid rate of convergence than $O(e^{-c\sqrt{N}})$ are given by Takahasi and Mori. These include two rules based on transformations using the error function

$$\operatorname{erf} u = \frac{2}{\sqrt{\pi}} \int_0^u \exp(-t^2)\, dt, \tag{3.4.5.7}$$

$$\int_{-1}^{1} f(x)\, dx = \frac{2}{\sqrt{\pi}} \int_{-\infty}^{\infty} f(\operatorname{erf} u) e^{-u^2}\, du = \frac{2h}{\sqrt{\pi}} \sum_{m=-\infty}^{\infty} f(\operatorname{erf} mh) e^{-m^2h^2} + E_h f, \tag{3.4.5.8}$$

$$\int_0^{\infty} f(x)\, dx = \frac{2h}{\sqrt{\pi}} \sum_{m=-\infty}^{\infty} f\left(\log \frac{2}{1 - \operatorname{erf} mh}\right) \frac{e^{-m^2h^2}}{1 - \operatorname{erf} mh} + E_h f, \tag{3.4.5.9}$$

which converge as $\exp(-cN^{2/3})$ and the so called double exponential (DE) rules which converge as $\exp(-cN/\log N)$. These latter rules are based on transformations which cause the transformed integral to decay as $\exp(-(\pi/2)\exp|u|)$ as $|u| \to \infty$, hence the name. They are essentially the same as the rules resulting from following one of the transformations listed at the beginning of this section by (3.4.5.1). We give here the transformations of the integrals, from which the rules follow as above. In the next three integrals, $z = (\pi/2) \sinh u$:

$$\int_{-1}^{1} f(x)\, dx = \frac{\pi}{2} \int_{-\infty}^{\infty} f(\tanh z) \frac{\cosh u}{\cosh^2 z}\, du, \tag{3.4.5.10}$$

$$\int_0^{\infty} f(x)\, dx = \frac{\pi}{2} \int_{-\infty}^{\infty} f(\exp z) \cosh u \exp z\, du, \tag{3.4.5.11}$$

$$\int_{-\infty}^{\infty} f(x)\, dx = \frac{\pi}{2} \int_{-\infty}^{\infty} f(\sinh z) \cosh u \cosh z\, du. \tag{3.4.5.12}$$

A further DE rule applicable if $f(x) = f_1(x)e^{-x}$ arises from the transformation

$$\int_0^{\infty} f(x)\, dx = \int_{-\infty}^{\infty} f(w)(1 + e^{-u}) w\, du \tag{3.4.5.13}$$

where $w = \exp(u - \exp(-u))$. Murota and Iri compare various refinements of the IMT rule (Section 2.9.2) and the DE rule (3.4.5.10) while Toda and Ono discuss the problems in implementing DE rules on a computer and give many numerical examples.

References. Beighton and Noble [1]; Haber [12]; Kress [1, 3]; Murota and Iri [1]; S. O. Rice [1]; C. Schwartz [2]; Squire [7, 8, 10]; Stenger [5, 6]; Takahasi and Mori [3, 4]; Toda and Ono [1].

Examples of the Rule (3.4.5.3)

h	N	hN	$e^{-x}/(1+x^4)$	$1/(x+2)\log^2(x+2)$	$1/(x+1)^{1.25}$
.5	4	2	.5313 3265	.9350 4602	1.6798 3447
.25	16	4	.6144 8495	1.1856 6696	2.5637 0950
.125	64	8	.6301 6295	1.3183 4855	3.4668 0033
.0625	256	16	.6304 7773	1.3803 1684	3.9273 0821
.03125	1024	32	.6304 7783	1.4114 6029	3.9986 6338
.015625	4096	64	.6304 7783	1.4270 7195	3.9999 9955
	Exact		.6304 7783	1.4426 9495	4.0000 0000

After a singly infinite integral is transformed to a doubly infinite one by the transformation $x = e^y$, it may be further transformed by (3.4.5.1). A rather complicated rule is produced when the rectangular rule is applied to this new integral. Numerical experience with sequences of such rules has shown convergence to be slower than with (3.4.5.3) except in cases where the original integrand tends to zero sufficiently slowly ($x^{-1.25}$ or slower). Furthermore, overflow often rears its ugly head. For conditionally convergent integrals, (3.4.5.3) is ineffective and the new rule does not improve matters.

The rules (3.4.1), (3.4.5.3) use abscissas with a constant ratio of interval lengths. Clendenin quite generally uses the abscissas

$$\begin{aligned} x_{n+1} &= x_n + A^n h, \qquad A > 1, \quad h > 0 \\ x_0 &= 0 \end{aligned} \tag{3.4.5.14}$$

to compute $\int_0^\infty f(x)\,dx$. Quadratic interpolation is used over x_{n-1}, x_n, x_{n+1} and then integrated to provide an estimate for $\int_{x_n}^{x_{n+1}} f(x)\,dx$. Having made one computation with a given A and h, if one then selects

$$A' = A^{1/2}, \qquad h' = h/(1 + A^{1/2}),$$

the resulting set of abscissas will constitute a refinement containing the first set. Clendenin provides explicit formulas, an error estimate in terms of A, h, and f''', and some numerical comparisons with Laguerre methods.

Reference. Clendenin [2].

3.4.6 *Use of Whittaker Cardinal Functions*

Stenger has made an extensive study of the Whittaker cardinal function

$$C(x; f, h) = \sum_{k=-\infty}^{\infty} f(kh)S(x; k, h) \tag{3.4.6.1}$$

where the $S(x; k, h)$ are the so-called sinc functions,

$$S(x; k, h) = \frac{\sin[(\pi/h)(x - kh)]}{(\pi/h)(x - kh)}. \tag{3.4.6.2}$$

The interest of this function, also called the sinc function expansion of $f(x)$, is that

$$\int_{-\infty}^{\infty} C(x; f, h)\, dx = h \sum_{k=-\infty}^{\infty} f(kh). \tag{3.4.6.3}$$

Thus, the rectangular rule can be looked upon as the integral of either a piecewise linear function or a smooth function with many interesting properties. Furthermore, the error in the rectangular approximation (3.4.1a) depends on how well $f(x)$ can be approximated by $C(x; f, h)$. The best results are achieved if

$$|f(x)| \le Ae^{-\alpha|x|}, \qquad -\infty < x < \infty \tag{3.4.6.4}$$

where A and α are positive constants.

Before stating some integration rules based on the sinc function expansion, we make the following definition.

DEFINITION. Let $d > 0$ and $\mathscr{D}_d$ denote the domain

$$\mathscr{D}_d = \{z = x + iy: |y| < d\}. \tag{3.4.6.5}$$

Then $B(\mathscr{D}_d)$ denotes the family of all functions analytic in $\mathscr{D}_d$ such that

$$\int_{-d}^{d} |f(x + iy)|\, dy \to 0 \qquad \text{as} \quad x \to \pm\infty$$

and such that $N(f, \mathscr{D}_d) < \infty$, where

$$N(f, \mathscr{D}_d) = \lim_{y \to d} \int_{-\infty}^{\infty} (|f(x + iy)| + |f(x - iy|)\, dx. \tag{3.4.6.6}$$

THEOREM. *Let $f \in B(\mathscr{D}_d)$ for some $d > 0$ and let*

$$\eta_N(f, h) = \int_{-\infty}^{\infty} f(x)\, dx - h \sum_{k=-N}^{N} f(kh) \tag{3.4.6.7}$$

and

$$\eta(f, h) = \lim_{N \to \infty} \eta_N(f, h).$$

Then

$$|\eta(f, h)| \le \frac{1}{2} \frac{e^{-\pi d/h}}{\sinh(\pi d/h)} N(f, \mathscr{D}_d). \tag{3.4.6.8}$$

If in addition, f satisfies (3.4.6.4), *then by taking*

$$h = [2\pi d/(\alpha N)]^{1/2} \tag{3.4.6.9}$$

we have

$$|\eta_N(f, h)| \le A_1 g(\alpha, d, N) \tag{3.4.6.10}$$

where

$$g(\alpha, d, N) = \exp[-(2\pi d\alpha N)^{1/2}] \tag{3.4.6.11}$$

and A_1 denotes a constant that depends only on f, d, and α.

THEOREM. *Let $f \in B(\mathscr{D}_d)$ for some $d > 0$ and let*

$$\delta_N(w; f, h) = \begin{cases} \displaystyle\int_{-\infty}^{\infty} f(t)e^{iwt} - h \sum_{j=-N}^{N} f(jh)e^{ijhw}, & |w| \le \pi/h, \\ \displaystyle\int_{-\infty}^{\infty} f(t)e^{iwt}, & |w| > \pi/h, \end{cases} \tag{3.4.6.12}$$

and

$$\delta(w; f, h) = \lim_{N\to\infty} \delta_N(w; f, h). \tag{3.4.6.13}$$

Then

$$|\delta(w; f, h)| \le \frac{1}{2} \frac{N(f, \mathscr{D}_d) \exp(-d(\pi/h - |w|))}{\sinh(\pi d/h)}. \tag{3.4.6.14}$$

If f satisfies (3.4.6.4) *and h is chosen as in* (3.4.6.9), *then*

$$|\delta_N(w; f, h)| \le A_1 g(\alpha, d/2, N); \qquad |w| \le \pi/h. \tag{3.4.6.15}$$

THEOREM. *Let $f \in B(\mathscr{D}_d)$ for some $d > 0$ and satisfy* (3.4.6.4) *and let $h = [\pi d/(\alpha N)]^{1/2}$. Then there exists a constant A_1 depending only on f, d, and α such that, for all x,*

$$\left| \frac{1}{\pi i} P \int_{-\infty}^{\infty} \frac{f(t)}{t - x} dt - i \sum_{k=-N}^{N} f(kh) \frac{1 - \cos[(\pi/h)(x - kh)]}{(\pi/h)(x - kh)} \right| \le A_1 N^{1/2} g(\alpha, d/2, N). \tag{3.4.6.16}$$

In order to derive formulas for the semi-infinite interval $[0, \infty)$ and the finite interval $[-1, 1]$, we must transform these intervals into $(-\infty, \infty)$ such that certain conditions hold. Before studying these transformations we make the following definition.

DEFINITION. Let $\mathscr{D}$ be a simply connected domain with boundary $\partial\mathscr{D}$ and let a and b be distinct points of $\partial\mathscr{D}$. Let ϕ be a conformal map of $\mathscr{D}$ onto $\mathscr{D}_d$ such that $\phi(a) = -\infty$ and $\phi(b) = \infty$ and let $\psi = \phi^{-1}$ denote the inverse map. Then $B(\mathscr{D})$ denotes the family of all functions F analytic in $\mathscr{D}$ such that

$$\int_{\psi(u+L)} |F(z)\,dz| \to 0 \qquad \text{as} \quad |u| \to \infty$$

where $L = \{iy\colon y \text{ real}, |y| \le d\}$ and such that

$$N(F, \mathscr{D}) = \lim_{C_1 \to \partial\mathscr{D}} \inf_{C_1 \subset \mathscr{D}} \int_{C_1} |F(z)\,dz| < \infty.$$

If we now take

$$\mathscr{D} = \left\{z\colon \left|\arg\left(\frac{1+z}{1-z}\right)\right| < d\right\},$$

assume that $F \in B(\mathscr{D})$ satisfies $|F(x)| \le C(1 - x^2)^{\alpha-1}$ for $x \in [-1, 1]$, $\alpha > 0$, and $C > 0$, and take h as in (3.4.6.9), then

$$\left|\int_{-1}^{1} F(x)\,dx - h \sum_{k=-N}^{N} \frac{2e^{kh}}{(1+e^{kh})^2} F\left(\frac{e^{kh}-1}{e^{kh}+1}\right)\right| \le C_1 g(\alpha, d, N) \tag{3.4.6.17}$$

where C_1 denotes a constant depending only on F, α, and d.

If we take $\mathscr{D}$ to be the sector $\{z\colon |\arg z| < d\}$ and assume that $F \in B(\mathscr{D})$ satisfies

$$|F(x)| \le \begin{cases} Cx^{\alpha-1} & \text{if } 0 \le x \le 1, \\ Cx^{-\alpha-1} & \text{if } 1 \le x \le \infty, \end{cases} \tag{3.4.6.18}$$

then by taking h as in (3.4.6.9), we get the integration rule

$$\left|\int_{0}^{\infty} F(x)\,dx - h \sum_{k=-N}^{N} e^{kh} F(e^{kh})\right| \le C_1 g(\alpha, d, N), \tag{3.4.6.19}$$

while if we take $\mathscr{D}$ to be the region defined by

$$\mathscr{D} = \{z\colon |\arg \sinh(z)| < d\}, \qquad 0 < d < \pi/2,$$

and assume that $F \in B(\mathscr{D})$ satisfies

$$|F(x)| \le \left.\begin{cases} Cx^{\alpha-1} & \text{if } 0 \le x \le 1 \\ Ce^{-\alpha x} & \text{if } 1 \le x \le \infty \end{cases}\right\}, \qquad \alpha > 0, \tag{3.4.6.20}$$

then with h as in (3.4.6.9), we get the integration rule

$$\left|\int_{0}^{\infty} F(x)\,dx - h \sum_{k=-N}^{N} \frac{1}{\sqrt{1+e^{-2kh}}} F\{\log[e^{kh} + \sqrt{1+e^{2kh}}]\}\right| \le C_1 g(\alpha, d, N). \tag{3.4.6.21}$$

Integration rule (3.4.6.17) yields accurate results for the integration of functions F such as $F(x) = (1-x)^{-1/3}(1+x)^{-3/5}\log(1-x)$ or $F(x) = (1-x)^{-4}\exp\{-2/(1-x)\}$, while (3.4.6.19) does an accurate job of integrating functions F such as $F(x) = x^{\alpha-1}/(1+x)^{2\alpha}$, or $F(x) = x^{-3/2}\sin(x/2)e^{-x}$. The rule (3.4.6.21) is suitable for integrating F such as

$$F(x) = x^{-1/2}\log[1 - \sin x/x]e^{-x/2}, \quad F(x) = x^{-2/7}e^{-x^2},$$

or

$$F(x) = x^5 \exp(-[(x-5)^2 + 2]^{1/2} - 1/x^2)\sin(3x).$$

Stenger also recommends (3.4.6.21) for approximating the Fourier and Hankel transforms, $\int_0^\infty F(t)e^{iwt}\,dt$ and $\int_0^\infty F(t)J_\nu(\lambda t)$, respectively, and (3.4.6.19) for the Laplace and Mellin transforms, $\int_0^\infty F(t)e^{-st}\,dt$ and $\int_0^\infty F(t)t^{\lambda-1}\,dt$. For additional applications of sinc function expansions to indefinite integrals, singular integrals and Cauchy principal value integrals as well as to other areas of numerical analysis, the reader is referred to Stenger, where he will also find a discussion of computer algorithms and pitfalls. For a computer program for integrating over any finite or infinite interval based on the results of this section see Sikorski *et al.*

References. Kearfott [1]; Lund [1]; Sikorski and Stenger [1]; Sikorski, Stenger, and Schwing [F1]; Stenger [9].

3.5 Formulas of Interpolatory Type

Formulas of the interpolatory type may be developed for infinite intervals as well as for finite intervals. One example will suffice to show how this can be done. Suppose it is desired to obtain a formula for $\int_0^\infty e^{-x}f(x)\,dx$ in terms of the values $f(0), f(h), \ldots, f(nh)$. Write the formal *Newton Series*

$$f(x) \sim \sum_{k=0}^{\infty} \frac{\Delta^k f(0)}{k!\,h^k}(x)(x-h)\cdots(x-(k-1)h). \tag{3.5.1}$$

Finite segments of this series interpolate to $f(x)$ at $x = 0, h, \ldots, nh$. Then, formally at least,

$$\begin{aligned}\int_0^\infty e^{-x}f(x)\,dx &\sim \int_0^\infty e^{-x}\sum_{k=0}^{\infty}\frac{\Delta^k f(0)}{k!\,h^k}x(x-h)\cdots(x-(k-1)h)\,dx\\ &\sim \sum_{k=0}^{\infty}\frac{\Delta^k f(0)}{k!\,h^k}\int_0^\infty e^{-x}x(x-h)\cdots(x-(k-1)h)\,dx\\ &= \sum_{k=0}^{\infty}\frac{\Delta^k f(0)m_k}{k!\,h^k},\end{aligned} \tag{3.5.2}$$

where

$$m_k = \int_0^\infty x(x-h)\cdots(x-(k-1)h)e^{-x}\,dx$$
$$= h^{k+1}\int_0^\infty t(t-1)\cdots(t-(k-1))e^{-ht}\,dt. \tag{3.5.3}$$

The constants m_k are now precomputed (recurrence is convenient), and we have the approximate integration formula

$$\int_0^\infty e^{-x}f(x)\,dx \approx \sum_{k=0}^{n} \frac{\Delta^k f(0)m_k}{k!\,h^k}. \tag{3.5.4}$$

It is clear that similar formulas can be developed with other interpolation formulas of the difference calculus such as Stirling's formula, and for weights other than e^{-x}. If an unequal spacing of abscissas is desired, Lagrange interpolation would be indicated.

Example. Compute numerically

$$\int_0^\infty \frac{e^{-x}}{2x+100}\,dx,$$

using $h = \frac{1}{2}$.

n	
0	.0100 0000
1	.0098 0198
2	.0098 0780
3	.0098 0757
Exact	.0098 0757

In view of the fact that a Newton series converges only for a very restricted class of functions $f(x)$, we cannot expect that, as $n \to \infty$, the above integration rule will converge to the proper value with any frequency. Numerical experiments with a variety of integrals yield very few cases of success.

References. Burgoyne [1]; Davis [6, pp. 51–52].

3.5.1 *Product Integration Rules for the Infinite Interval*

Interpolatory rules for the evaluation of

$$I(f) = \int_a^\infty k(x)f(x)\,dx, \tag{3.5.1.1}$$

where $a = -\infty$ or $a = 0$ as the case may be, which are based on a suitable choice of basis functions $\{\phi_j\}$ and interpolation points $\{t_{in}\}$ can be derived which will converge for an extensive class of functions f and k. In some cases this can be done by transforming the infinite interval to the standard interval $[-1, 1]$ and using the theory developed in Section 2.5.6 for $[-1, 1]$.

Example. (Sloan) Let $a = 0$ and consider $I(f)$ where $f(t) = O(t^{-1})$ as $t \to \infty$ and $I((t+\alpha)^{-1}) < \infty$ for some suitable scaling parameter $\alpha > 0$. If we choose as basis functions

$$\phi_j(t) = \frac{1}{t+\alpha}\left(\frac{t}{t+\alpha}\right)^{t-1}, \qquad j \geq 1 \tag{3.5.1.2}$$

or preferably for computational reasons, the equivalent set

$$\phi_j(t) = \frac{1}{t+\alpha}\, T_{j-1}\left(\frac{t-\alpha}{t+\alpha}\right) \tag{3.5.1.3}$$

where the $T_k(x)$ are the Tschebyscheff polynomials, and as interpolation points

$$t_{in} = \alpha \tan^2\left(\frac{2i-1}{4n}\pi\right), \qquad i = 1, \ldots, n, \tag{3.5.1.4}$$

then the ensuing product integration rule becomes

$$Q_n(f) = \sum_{i=1}^{n} w_{in} f(t_{in}) \tag{3.5.1.5}$$

where the $w_{in} = w_{in}(k)$ satisfy the linear system

$$\sum_{i=1}^{n} w_{in}\,\phi_j(t_{in}) = I(\phi_j), \qquad j = 1, \ldots, n. \tag{3.5.1.6}$$

If k is such that for some $p > 1$

$$\int_0^\infty \left|\frac{k(x)}{t+\alpha}\right|^p (t+\alpha)^{2p-2}\,dt < \infty, \tag{3.5.1.7}$$

then $Q_n(t)$ converges to $I(f)$ provided that $f(t)(t+\alpha) \in C[0, \infty)$ and $\lim_{t\to\infty} f(t)(t+\alpha) < \infty$. A sufficient condition that (3.5.1.7) holds is that $\int_0^T |k(t)|^p\,dt < \infty$ for some T and $p > 1$ and that $|k(t)| \leq Bt^{-\varepsilon}$ for some B, $\varepsilon > 0$, and all $t \geq T$.

If we choose the interpolatory points to be the zeros t_{in} of the generalized Laguerre polynomials $L_n^{(\alpha)}(x)$ when $a = 0$ and of the Hermite polynomials $H_n(x)$ when $a = -\infty$ and choose the basis functions to be the powers of x, then the weights w_{in} in (3.5.1.5) have the form

$$w_{in} = \int_a^\infty k(x) l_{in}(x)\,dx, \qquad i = 1, \ldots, n \tag{3.5.1.8}$$

where

$$l_{in}(x) = \frac{P(x)}{P_n'(t_{in})(x - t_{in})}, \qquad i = 1, \ldots, n \tag{3.5.1.9}$$

and $P_n(x) = L_n^{(\alpha)}(x)$ or $H_n(x)$ as the case may be. Our underlying assumption here is that all the moments $\int_a^\infty k(x)x^m\,dx$ are finite. An alternative expression for $Q_n(f)$ in terms of the modified moments $I(P_j) = \int_a^\infty k(x)P_j(x)\,dx$ is given by

$$Q_n(f) = \sum_{j=0}^{n-1} c_j I(P_j) \tag{3.5.1.10}$$

where

$$c_j = \sum_{i=1}^{n} \mu_{in} f(t_{in}) P_j(t_{in}) \Big/ \sum_{i=1}^{n} \mu_{in} P_j^2(t_{in}) \tag{3.5.1.11}$$

and the μ_{in} are the Gauss–Laguerre or Gauss–Hermite weights corresponding to the zeros t_{in} (see Section 3.6). For these rules Lubinsky and Sidi have shown the following convergence results.

THEOREM. *Let k satisfy* $\int_a^\infty k^2(x)/w(x)\,dx < \infty$ *where* $w(x)$ *is the Laguerre weight* $x^\alpha e^{-x}$ *or the Hermite weight* e^{-x^2} *as the case may be. Let* $f \in C(a, \infty)$ *satisfy the following conditions*:

(1) *If* $a = 0$, $\lim_{x\to 0} f(x)x^{(1+\alpha-\delta)/2} = 0$ *for some* $\delta > 0$.

(2) $\lim_{x\to\infty} f(x)e^{-\eta x} = 0$ or $\lim_{|x|\to\infty} f(x)e^{-\eta x^2} = 0$ *as the case may be for some* η, $0 < \eta < \frac{1}{2}$.

Then $Q_n(f) \to I(f)$. *Furthermore*

$$\sum_{i=1}^{n} |w_{in}| f(t_{in}) \to \int_a^\infty |k(x)| f(x)\,dx.$$

References. Lubinsky and Sidi [1]; Sloan [3].

3.6 Gaussian Formulas for the Infinite Interval

The ultimate in the interpolatory formula is, of course, the formula of Gauss type

$$\int_0^\infty w(x)f(x)\,dx \approx \sum_{k=1}^{n} w_k f(x_k) \tag{3.6.1}$$

or

$$\int_{-\infty}^\infty w(x)f(x)\,dx \approx \sum_{k=1}^{n} w_k f(x_k), \tag{3.6.2}$$

where the x_k and the w_k have been determined so that the formula is exact for functions $f(x)$ of class $\mathscr{P}_{2n-1}$. The general theory of the Gauss formula has already been discussed in Section 2.7. A set of tables can be developed for any weight function with $\int w(x)\,dx < \infty$, but the most widely employed Gauss-type formulas are the Laguerre formulas and the Hermite formulas.

Laguerre Formula

$$\int_0^\infty e^{-x} f(x)\,dx = \sum_{k=1}^{n} w_k f(x_k) + \frac{(n!)^2}{(2n)!} f^{(2n)}(\xi), \qquad 0 < \xi < \infty. \quad (3.6.3)$$

Here the abscissas x_k are the zeros of the Laguerre polynomials $L_n(x) \equiv L_n^{(0)}(x)$ and

$$w_k = x_k/[L_{n+1}(x_k)]^2. \quad (3.6.4)$$

Shifted Laguerre

If we change the variable to $x + a$, a "shifted" Laguerre formula is obtained:

$$\int_a^\infty e^{-x} f(x)\,dx = e^{-a} \sum_{k=1}^{n} w_k f(x_k + a) + \frac{(n!)^2}{(2n)!} f^{(2n)}(\xi),$$
$$-\infty < a < \xi < \infty. \quad (3.6.4a)$$

Generalized Laguerre Formula

A more general weight function is $w(x) = x^\alpha e^{-x}$, $\alpha > -1$. We have

$$\int_0^\infty x^\alpha e^{-x} f(x)\,dx = \sum_{k=1}^{n} w_k f(x_k) + \frac{n!\,\Gamma(n + \alpha + 1)}{(2n)!} f^{(2n)}(\xi), \qquad 0 < \xi < \infty. \quad (3.6.5)$$

The abscissas x_k are the zeros of the generalized or associated Laguerre polynomial $L_n^{(\alpha)}(x)$ and

$$w_k = \Gamma(n + \alpha + 1) x_k / n! [L_{n+1}^{(\alpha)}(x_k)]^2. \quad (3.6.6)$$

Radau–Laguerre Formula

Corresponding to the Radau rule (2.7.1.7) with a preassigned abscissa at one endpoint of the integration interval, we have

$$\int_0^\infty x^\alpha e^{-x} f(x)\,dx = \frac{(n-1)!\,\Gamma(\alpha+1)\Gamma(\alpha+2)}{\Gamma(n+\alpha+1)} f(0) + \sum_{k=1}^{n-1} w_k f(x_k)$$
$$+ \frac{(n-1)!\,\Gamma(n+\alpha+1)}{(2n-1)!} f^{(2n-1)}(\xi), \qquad 0 < \xi < \infty. \quad (3.6.7)$$

The abscissas x_k are the zeros of $L_{n-1}^{(\alpha+1)}(x)$ and

$$w_k = \Gamma(n+\alpha)/(n-1)!(n+\alpha)[L_{n-1}^{(\alpha)}(x_k)]^2. \tag{3.6.8}$$

For $\alpha = 0$, the formulas simplify to

$$\int_0^\infty e^{-x} f(x)\,dx = \frac{1}{n} f(0) + \sum_{k=1}^{n-1} w_k f(x_k) + \frac{(n-1)!n!}{(2n-1)!} f^{(2n-1)}(\xi), \quad 0 < \xi < \infty \tag{3.6.9}$$

where the abscissas x_k are the zeros of $L_{n-1}^{(1)}(x) = -L_n'(x)$ and

$$w_k = 1/n[L_{n-1}(x_k)]^2. \tag{3.6.10}$$

Hermite Formula

$$\int_{-\infty}^\infty e^{-x^2} f(x)\,dx = \sum_{k=1}^{n} w_k f(x_k) + \frac{n!\sqrt{\pi}}{2^n(2n)!} f^{(2n)}(\xi), \quad -\infty < \xi < \infty. \tag{3.6.11}$$

The abscissas x_k are the zeros of the Hermite polynomial $H_n(x)$, where

$$H_n(x) = 2^n x^n + \cdots$$

and

$$w_k = 2^{n+1} n! \sqrt{\pi}/[H_{n+1}(x_k)]^2 \tag{3.6.12}$$

The Hermite abscissas and weights are related to the generalized Laguerre abscissas and weights for $\alpha = \pm\frac{1}{2}$. In fact, for the generalized Hermite rules

$$\int_{-\infty}^\infty |x|^{2\lambda} e^{-x^2} f(x)\,dx \approx \sum_{k=1}^{n} w_k^{(\lambda)} f(x_k^{(\lambda)}), \tag{3.6.13}$$

the $w_k^{(\lambda)}$ and $x_k^{(\lambda)}$ are given in terms of the generalized Laguerre abscissas and weights as follows:

For $n = 2m$,

$$x_k^{(\lambda)} = -x_{n-k}^{(\lambda)} = (\tilde{x}_{k,m}^{(\lambda-.5)})^{1/2}, \qquad k = 1, \ldots, m.$$

$$w_k^{(\lambda)} = w_{n-k}^{(\lambda)} = \tfrac{1}{2}\tilde{w}_{k,m}^{(\lambda-.5)}, \tag{3.6.14}$$

For $n = 2m + 1$,

$$x_k^{(\lambda)} = -x_{n-k}^{(\lambda)} = (\tilde{x}_{k,m}^{(\lambda+.5)})^{1/2}, \qquad k = 1, \ldots, m.$$

$$w_k^{(\lambda)} = w_{n-k}^{(\lambda)} = \tfrac{1}{2}\tilde{w}_{k,m}^{(\lambda+.5)},$$

$$x_{m+1}^{(\lambda)} = 0, \qquad w_{m+1}^{(\lambda)} = \Gamma(\lambda + \tfrac{1}{2}) - 2\sum_{k=1}^{m} w_k^{(\lambda)}, \tag{3.6.15}$$

Here $\tilde{x}_{k,m}^{(\alpha)}$ and $\tilde{w}_{k,m}^{(\alpha)}$ denote the abscissas and weights in (3.6.5).

For the computation of abscissas and weights of Laguerre and Hermite rules, any of the methods discussed in Section 2.7 are applicable. In particular, the method of Golub and Welsch (Section 2.7.5) is easy to apply since the three-term recurrence relations for $L_n^{(\alpha)}(x)$ and $H_n(x)$ are available. Specific programs for Laguerre and Hermite rules are given by Stroud and Secrest and by Capovani *et al.*

Information on Gaussian formulas for $\int_{-\infty}^{\infty} (1 + x^2)^{-k-1} f(x)\, dx$ can be found in Harper and in Haber, while Gauss rules for $\int_0^\infty e^{-x^2} f(x)\, dx$ can be found in Steen, Byrne, and Gelbard.

For convergent integrals of the form $\int_0^\infty f(x)\, dx$ and $\int_{-\infty}^{\infty} f(x)\, dx$, we may also use the Laguerre and Hermite integration formulas, for example in the Laguerre case, by writing

$$\int_0^\infty f(x)\, dx = \int_0^\infty e^{-x} e^x f(x)\, dx \approx \sum_{k=1}^{n} w_k e^{x_k} f(x_k)$$

$$= \sum_{k=1}^{n} V_k f(x_k), \quad \text{where} \quad V_k = w_k e^{x_k}. \tag{3.6.16}$$

In many tabulations of zeros and weights for Laguerre and Hermite integration, V_k is tabulated along with x_k and w_k. We must be careful to use this method only when the function $e^x f(x)$ has a polynomial-like behavior.

For functions $f(x)$ that can be approximated by a sum of negative exponentials

$$f(x) \approx \sum_{k=1}^{m} a_k e^{-kx},$$

Stroud and Secrest recommend the rule

$$\int_0^\infty f(x)\, dx \approx \sum_{i=1}^{n} w_i^* f(x_i^*) \tag{3.6.17}$$

where

$$w_i^* = w_i/(1 + x_i), \qquad x_i^* = -\log((1 + x_i)/2) \tag{3.6.18}$$

and the x_i, w_i are the Gauss abscissas and weights.

If the function $f(x)$ is approximable by sums of the form

$$f(x) \approx \sum_{k=1}^{m} a_k (1 + x)^{-k},$$

the rule

$$\int_0^\infty f(x)\, dx \approx \sum_{i=1}^{n} u_i f(v_i) \tag{3.6.19}$$

where

$$u_i = 2w_i/(1 + x_i)^2, \qquad v_i = (1 - x_i)/(1 + x_i) \tag{3.6.20}$$

is applicable. Here also the x_i, w_i are the Gauss abscissas and weights.

References. Aizenshtat, Krylov, and Metleskii [1]; Baburin and Lebedev [1]; Baratella and Vinardi [1]; Berger [2, 3]; Capovani, Ghelardoni, and Lombardi [1, 2]; Concus, Cassatt, Jaehnig, and Melby [1]; Haber [1]; Harper [1]; NBS Handbook [1, Chap. 25]; Rabinowitz and Weiss [1]; Shao, Chen, and Frank [1, 2]; Shizgal [1]; Steen, Byrne, and Gelbard [1]; Stroud and Secrest [1, 2, F2, F3].

3.6.1 A Mixed Method

If the integrand is smooth for large values of the argument, but not for early values, the following mixed method may be advantageous. Write $\int_0^\infty = \int_0^{r_1} + \int_{r_1}^\infty$ and compute the first integral on the right by a rule for a finite interval and the second integral by a shifted Laguerre rule (3.6.4a). Now compare the value obtained against $\int_0^{r_1} + \int_{r_1}^{r_2} + \int_{r_2}^\infty$, etc.

References. Chisholm, Genz and Rowlands [1]; Wolberg [1, p. 100].

Basu and Kundu have developed analogs of the Clenshaw–Curtis and Filippi methods (Section 6.4) for $\int_0^\infty e^{-x} f(x)\, dx$ that are a little less accurate than the Gauss–Laguerre rule using the same number of points. However, they have the same advantages as these methods in that the abscissas and weights are easy to compute, an error estimate is available, and one can generate a sequence of approximations, each of which uses all previously computed function values.

References. Basu and Kundu [1, 2].

3.6.2 Cauchy Principal Value Integrals

The Cauchy principal value integrals

$$I(f; a) = P\int_0^\infty e^{-x}(f(x)/(x-a))\, dx, \qquad 0 < a < \infty, \tag{3.6.2.1}$$

and

$$I(f; b) = P\int_{-\infty}^\infty e^{-x^2}(f(x)/(x-b))\, dx, \qquad -\infty < b < \infty, \tag{3.6.2.2}$$

can be converted to regular integrals and evaluated by the Laguerre and Hermite formulas, respectively, using the identities

$$I(f; a) = \int_0^\infty e^{-x}(f(x) - f(a))/(x - a)\,dx + f(a)I(1; a) \qquad (3.6.2.3)$$

and

$$I(f; b) = \int_{-\infty}^\infty e^{-x^2}(f(x) - f(b))/(x - b)\,dx + f(b)I(1; b) \qquad (3.6.2.4)$$

where

$$I(1; a) = P\int_0^\infty e^{-x}(x - a)^{-1}\,dx = -e^{-a}\left[\gamma + \log a + \sum_{n=1}^{\infty} (a^n/(n!n))\right] \qquad (3.6.2.5)$$

(γ, Euler's constant, $= .5772156659\ldots$) and

$$I(1; b) = P\int_{-\infty}^\infty e^{-x^2}(x - b)^{-1}\,dx = -e^{-b^2}\sum_{n=0}^{\infty} ((2b)^{2n+1}\Gamma(n + \tfrac{1}{2})/(2n + 1)!)$$

$$= -\pi^{1/2}be^{-b^2}\int_{-1}^{1} e^{b^2x^2}\,dx. \qquad (3.6.2.6)$$

Reference. Kumar [1].

3.7 *Convergence of Formulas of Gauss Type for Singly and Doubly Infinite Intervals*

Consider the family of Laguerre formulas

$$L_n(f) = \sum_{k=1}^{n} w_{nk} f(x_{nk}) \approx \int_0^\infty e^{-x}x^\alpha f(x)\,dx, \qquad \alpha > -1, \qquad (3.7.1)$$

which are exact for $f(x) \in \mathscr{P}_{2n-1}$.

THEOREM. *If for all sufficiently large values of* x *the function* $f(x)$ *satisfies the inequality*

$$|f(x)| \le \frac{e^x}{x^{\alpha+1+\rho}}, \qquad \textit{for some } \rho > 0, \qquad (3.7.2)$$

then

$$\lim_{n\to\infty} L_n(f) = \int_0^\infty e^{-x}x^\alpha f(x)\,dx. \qquad (3.7.3)$$

A similar convergence theorem holds for the Hermite formulas

$$H_n(f) = \sum_{k=1}^{n} w_{nk} f(x_{nk}) \approx \int_{-\infty}^{\infty} e^{-x^2} f(x)\, dx. \tag{3.7.4}$$

THEOREM. *If for all sufficiently large values of* $|x|$, $f(x)$ *satisfies the inequality*

$$|f(x)| \le \frac{e^{x^2}}{|x|^{1+\rho}}, \qquad \text{for some } \rho > 0, \tag{3.7.5}$$

then

$$\lim_{n\to\infty} H_n(f) = \int_{-\infty}^{\infty} e^{-x^2} f(x)\, dx. \tag{3.7.6}$$

The proofs of these theorems, together with some generalizations, can be found in the work of Uspensky.

For entire functions $f(z) = \sum_{n=0}^{\infty} b_n z^n$ satisfying certain conditions on the coefficients b_n, Lubinsky has proved geometric convergence of $L_n(f)$ and $H_n(f)$.

THEOREM. *Let*

$$A = \limsup_{n\to\infty}(|b_n|^{1/n} n/2) \tag{3.7.7}$$

and

$$B = \limsup_{n\to\infty}(|b_n|^{1/n}\sqrt{n/2}), \tag{3.7.8}$$

then for sufficiently large n, if $A < 1$

$$\left| \int_0^{\infty} e^{-x} x^{\alpha} f(x)\, dx - L_n(f) \right| \le A^{2n} \tag{3.7.9}$$

and if $B < 1$

$$\left| \int_{-\infty}^{\infty} e^{-x^2} f(x)\, dx - H_n(f) \right| \le B^{2n}. \tag{3.7.10}$$

References. Lubinsky [1]; Uspensky [2].

Examples

1 $\quad e^{-2}\displaystyle\int_2^{\infty} \frac{dx}{x(\log x)^2}$

2 $\quad e^{-2}\displaystyle\int_2^{\infty} \frac{dx}{x(\log x)^{3/2}}$

3 $\quad e^{-2}\int_2^\infty \frac{dx}{x^{1.01}}$

4 $\quad e^{-2}\int_2^\infty \left(\frac{\sin x}{x}\right) dx$ (Sine integral)

5 $\quad e^{-2}\int_2^\infty \cos\left(\frac{\pi}{2}x^2\right) dx$ (Fresnel integral)

6 $\quad e^{-2}\int_2^\infty e^{-x^2}\, dx$ (Complementary error function)

7 $\quad e^{-2}\int_2^\infty \frac{\sin(x-1)\, dx}{\sqrt{x(x-2)}}$ (Bessel function)

8 $\quad e^{-2}\int_2^\infty \frac{x\, dx}{(e^x-1)}$ (Debye function)

Integral	L_4	L_8	L_{16}
1	.1451 0750	.1554 3187	.1662 3627
2	.1610 1337	.1783 2886	.1914 2399
3	.2701 1936	.3587 1939	.4499 6932
4	−.0587 1937 6	−.0407 9735 8	−.0392 5869 6
5	−1.3992 326	−2.0529 382	−.0678 5954 5
6	.0005 1218 446	.0005 6386 851	.0005 6100 775
7	.0367 1888 3	.0395 0364 6	.0970 8306 4
8	.0583 3517 7	.0583 3484 7	.0583 3484 1

Integral	L_{32}	Exact
1	.1670 8562	.1952 4753
2	.2016 3572	.3251 0855
3	.5414 5344	13.628
4	−.0002 3993 672	−.0046 984
5	1.1197 176	.0015 8973
6	.0005 6103 72	.0005 6103 71
7	.1007 0835	.1626 6891
8		.0583 349

L_4 = Laguerre 4-point rule; L_8, 8-point; etc.

The only integrands on this list that are covered by the convergence theorem are 3, 6, and 8. The last two indeed exhibit strong convergence toward the proper answer. Integrand 3 is something of a numerical joke inasmuch as it is scarcely distinguishable from the divergent integrand x^{-1}. The values for 1 and 2 are increasing monotonically and might conceivably be convergent, though slowly, to the correct value. The values for 4 and 5 are very bad.

3.8 Oscillatory Integrands

In computing the value of an integral whose integrand oscillates over $[0, \infty)$ it may be useful to compute the positive and negative contributions individually and to sum the resulting infinite series. This series, however, may be slowly convergent. Note that a Lobatto rule (Section 2.7.1) is most effective for integrating between the zeros of the integrand as only n function evaluations are needed to integrate exactly polynomials of degree $2n + 1$ whereas $n + 1$ evaluations are needed using a Gauss rule.

Numerous devices that can transform a slowly convergent series into one more rapidly convergent are available. We make a slight digression here to describe the *Euler transformation*, which is the one most commonly employed.

The formal transformation is most expeditiously derived by means of the calculus of finite differences. Let $\Delta u_0 = u_1 - u_0$, $\Delta^2 u_0 = \Delta(\Delta u_0) = \Delta(u_1 - u_0) = u_2 - 2u_1 + u_0$, and so on. Let $Eu_0 = u_1$, $E^2 u_0 = u_2$, and so on. We have $E = \Delta + I$, where I is the identity operator. Then, with these operators, we may write formally

$$\begin{aligned} u_0 - u_1 + u_2 - \cdots &= u_0 - Eu_0 + E^2 u_0 - E^3 u_0 + \cdots \\ &= (I - E + E^2 - \cdots)u_0 = (I + E)^{-1} u_0 \\ &= (2I + E - I)^{-1} u_0 = (2I + \Delta)^{-1} u_0 \\ &= \tfrac{1}{2}(I + \tfrac{1}{2}\Delta)^{-1} u_0 \\ &= \tfrac{1}{2}u_0 - \tfrac{1}{4}\Delta u_0 + \tfrac{1}{8}\Delta^2 u_0 - \tfrac{1}{16}\Delta^3 u_0 + \cdots. \end{aligned} \tag{3.8.1}$$

The Euler transformation is

$$u_0 - u_1 + u_2 - \cdots = \tfrac{1}{2}u_0 - \tfrac{1}{4}\Delta u_0 + \tfrac{1}{8}\Delta^2 u_0 - \cdots. \tag{3.8.2}$$

It can be proved that if the left-hand series is convergent, the right-hand series is also convergent and to the same value. In numerous cases of practical interest (but not always), the right-hand series will converge more rapidly than the left-hand series.

Given the series $u_0 - u_1 + \cdots$, it is not always desirable to start the transformation with u_0 but with some later term, say u_m, so that

$$\sum_{i=0}^{\infty} (-1)^i u_i = \sum_{i=0}^{m-1} (-1)^i u_i + (-1)^m[\tfrac{1}{2}u_m - \tfrac{1}{4}\Delta u_m + \tfrac{1}{8}\Delta^2 u_m - \cdots].$$

A FORTRAN program that implements the Euler transformation while choosing the most suitable value of m is given in the IBM Scientific Subroutine Package.

Example. The slowly convergent series

$$S = \frac{1}{\log 2} - \frac{1}{\log 3} + \frac{1}{\log 4} - \cdots$$

may be speeded up by the use of the Euler transformation. Writing

$$S = \left(\frac{1}{\log 2} - \frac{1}{\log 3}\right) + \frac{1}{\log 4} - \cdots,$$

we apply the transformation to the infinite series

$$S' = \frac{1}{\log 4} - \frac{1}{\log 5} + \cdots.$$

We have $u_0 = .721348$, $\Delta u_0 = -.100013$, $\Delta^2 u_0 = .036788$, $\Delta^3 u_0 = -.01778$, $\Delta^4 u_0 = .00998$. Therefore

$$S' = \tfrac{1}{2}(.721348) + \tfrac{1}{4}(.100013) + \tfrac{1}{8}(.036788) + \tfrac{1}{16}(.01778) + \tfrac{1}{32}(.00998) + \cdots = .3917.$$

Hence, $S = .9242$.

In the next example, the Euler transformation is applied to an integral with an oscillatory integrand.

Example (Longman)

$$I = \int_0^\infty J_0(2x)\frac{f(x)}{g(x)}\,dx,$$

where

$$f(x) = x(x^2 + \tfrac{1}{3})^{1/2}[2x^2 \exp\{-\tfrac{1}{5}(x^2 + 1)^{1/2}\} - (2x^2 + 1)\exp\{-\tfrac{1}{5}(x^2 + \tfrac{1}{3})^{1/2}\}],$$

$$g(x) = (2x^2 + 1)^2 - 4x^2(x^2 + \tfrac{1}{3})^{1/2}(x^2 + 1)^{1/2}.$$

The zeros $x_1, x_2, \ldots$ of the Bessel function $J_0(2x)$ are readily available. See, for example, the NBS Handbook.

Set

$$x_0 = 0 \qquad \text{and} \qquad u_i = (-1)^{i+1}\int_{x_i}^{x_{i+1}}.$$

Performing approximate integration between zeros, we find $u_0 = .145234$, $u_1 = .206401$, $u_2 = .150723$, $u_3 = .108661$, $u_4 = .079288$, $u_5 = .058351$, $u_6 = .043165$, $u_7 = .032028$. In order to deal with small differences, we begin the Euler transformation with the third term. Therefore

$$\begin{aligned}\int_0^\infty &\approx -.145234 + .206401 - [\tfrac{1}{2}(.150723) + \tfrac{1}{4}(.042062)\\ &\quad + \tfrac{1}{8}(.012689) + \tfrac{1}{16}(.004253) + \tfrac{1}{32}(.001568) + \tfrac{1}{64}(.000585)]\\ &= -.02662.\end{aligned}$$

Gustafson turns this procedure around and applies numerical integration to speed up the convergence of the series $s = \sum_{k=1}^{\infty} a_k$. He assumes that a_k can be written in the form

$$a_k = \phi(1/k)m_k \tag{3.8.3}$$

where $\sum_{k=1}^{\infty} m_k$ is convergent and where $\phi(z)$ is analytic in a region that contains $[0, 1]$ in its interior. Define the function

$$\alpha(x) = \sum_{k=[x^{-1}]+1}^{\infty} m_k . \tag{3.8.4}$$

This function has jumps at $x = 1/k$, $k = 1, 2, \ldots$ and is of bounded variation. Moreover

$$s = a_1 \phi(1) + \int_0^1 \phi(x)\, d\alpha(x).$$

Numerical methods are now applied to this integral.

References. Cornille [1]; Gustafson [1, 4]; Hurwitz, Pfeiffer, and Zweifel [1]; Hurwitz and Zweifel [1]; IBM [2]; Knopp [1, pp. 244–246]; Longman [1, 2]; Lubkin [1]; NBS Handbook [1, p. 409]; D. Shanks [1].

The use of other transformations such as the ε-transformation, iterations of Aitken's Δ^2 method, and Levin's V-transformation has been reported to work effectively in accelerating the convergence of sequences of partial sums of series resulting from the integration of oscillatory integrals. The V-transformation is defined by

$$V_{kn} = \sum_{j=0}^{k} c(j, k, n) A_{n+j} \Big/ \sum_{j=0}^{k} c(j, k, n) \tag{3.8.5}$$

where

$$c(j, k, n) = (-1)^j \binom{k}{j} \left(\frac{n+j}{n+k} \right)^{k-1} (a_{n+j}^{-1} - a_{n+j+1}^{-1}) \tag{3.8.6}$$

and $A_m = \sum_{i=1}^{m} a_i$. Usually the diagonal transformation V_{k1} is used.

Example. (Levin). Consider the slowly convergent series

$$\sum_{j=1}^{\infty} (-1)^{j+1} / \sqrt{j} = .604898643.$$

Applying the V-transformation to the sequence of partial sums A_k, we obtain the following results:

k	A_k	V_{k1}
3	.870243488	.604933952
5	.817457083	.604898762
7	.787173266	.604898643

References. Alaylioglu, Evans, and Hyslop [1]; Blakemore, Evans, and Hyslop [1]; Hillion and Nurdin [1]; D. Levin [1].

Sidi has developed an extrapolation method for evaluating a very general class of oscillatory integrals provided the asymptotic behavior of the integrand is known. To this end we make the following definition.

DEFINITION. A function $\alpha(x)$ defined for $x > a \geq 0$ belongs to the set $A^{(\gamma)}$ if it is infinitely differentiable for all $x > a$ and if as $x \to \infty$, it has an asymptotic expansion of the form

$$\alpha(x) \sim x^{\gamma} \sum_{i=0}^{\infty} \alpha_i / x^i \tag{3.8.7}$$

and so do all its derivatives, these being obtained by differentiating the right-hand side of (3.8.7) term by term.

Sidi's method works for integrals

$$If = \int_a^{\infty} f(t)\, dt, \qquad a \geq 0 \tag{3.8.8}$$

where

$$f(t) = u(\theta(t)) \exp(\phi(t)) h(t) \tag{3.8.9}$$

and $u(x)$ is either $\sin(x)$ or $\cos(x)$, $\theta \in A^{(m)}$, m an integer >0, $\phi \in A^{(k)}$, k an integer ≥ 0 such that if $k \geq 1$, $\lim_{x\to\infty} \phi(x) = -\infty$, and $h \in A^{(\gamma)}$ for some γ such that $If < \infty$. (If $k \geq 1$, γ may be arbitrary while if $k = 0$, $\gamma < m - 1$.) If we define $\theta_m \in \mathscr{P}_m$ and $\phi_k \in \mathscr{P}_k$ to be the polynomial parts of θ and ϕ, respectively, and $\sigma = \min\{-m + 1, -k + 1\}$, then it can be shown that

$$\int_x^{\infty} f(t)\, dt = x^{\sigma+\gamma} \exp(\phi_k(x))[\cos(\theta_m(x))b(x) + \sin(\theta_m(x))c(x)] \tag{3.8.10}$$

where $b(x)$, $c(x) \in A^{(0)}$. More generally, if

$$f_j(t) = u(\theta_j(t)) \exp(\phi_j(t)) h_j(t), \qquad j = 1, \ldots, r \tag{3.8.11}$$

where the polynomial parts of all the θ_j are identical and those of all the ϕ_j are identical and $\gamma_i - \gamma_k$ is an integer ($h_j \in A^{(\gamma_j)}$), then with $\gamma = \max(\gamma_1, \ldots, \gamma_r)$, (3.8.10) holds for $f(x) = \sum_{j=1}^{r} f_j(x)$. This generalization is useful when dealing with Bessel functions.

We now let x_0 be the smallest zero of $\sin(\theta_m(x))$ greater than a so that x_0 is a root of $\theta_m(x) - q\pi = 0$ for some integer q, and then determine $x_0 < x_1 < x_2 < \cdots$ where x_i is a root of $\theta_m(x) - (q + i)\pi = 0$. Then

$$If = F(x_i) + \psi(x_i) b(x_i), \qquad i = 0, 1, 2, \ldots \tag{3.8.12}$$

where

$$F(x) = \int_a^x f(t)\, dt \tag{3.8.13a}$$

and

$$\psi(x) = \cos(\theta_m(x))x^{\sigma+\gamma}\exp(\phi_k(x)). \tag{3.8.13b}$$

In a similar fashion we can reverse the roles of sin and cos, starting with x_0 as the root of $\theta_m(x) - (q + \frac{1}{2})\pi = 0$.

Since $b \in A^{(0)}$ so that $b(x) \sim \sum_{j=0}^{\infty} b_j/x^j$, we can determine a sequence of approximations $\{W_n\}$ to If by solving the system of equations

$$W_n = F(x_i) + \psi(x_i)\sum_{j=0}^{n} \bar{b}_j/x_i^j, \qquad i = 0, \ldots, n+1, \quad n = 1, 2, \ldots \tag{3.8.14}$$

where the $\bar{b}_j$ are approximations to the b_j. Since we are usually not interested in the $\bar{b}_j$, we can determine the W_n by Sidi's W-algorithm as follows:

Let

$$N_{-1}^{(s)} = 1/\psi(x_s), \quad M_{-1}^{(s)} = F(x_s)N_{-1}^{(s)}, \qquad s = 0, 1, 2, \ldots.$$

Compute

$$M_k^{(s)} = (M_{k-1}^{(s)} - M_{k-1}^{(s+1)})/(x_s^{-1} - x_{s+k+1}^{-1}), \qquad k = 0, 1, 2, \ldots,$$

$$N_k^{(s)} = (N_{k-1}^{(s)} - N_{k-1}^{(s+1)})/(x_s^{-1} - x_{s+k+1}^{-1}), \qquad s = 0, 1, 2, \ldots.$$

Then $W_n = M_n^{(0)}/N_n^{(0)}$. Sidi shows that $|If - W_n| = O(n^{-\mu})$ for all $\mu > 0$.

Example. (Sidi [5]). Use the above method to evaluate

$$I = \int_0^{\infty} \sin(\pi/t^2)\cos(\pi t^2/4)\,dt/t^2 = \frac{1}{4\sqrt{2}}(e^{-\pi} - 1) = -.1691374816.$$

Since the integrand has an infinite number of oscillations as $t \to 0$, we divide the range of integration into $(0, 1)$ and $(1, \infty)$ and transform $(0, 1)$ into $(1, \infty)$ to yield the two integrals

$$I = I_1 + I_2 = \int_1^{\infty} f_1(t)\,dt + \int_1^{\infty} f_2(t)\,dt$$

where

$$f_1(t) = \sin(\pi/t^2)\cos(\pi t^2/4)t^{-2},$$

$$f_2(t) = \sin(\pi t^2)\cos(\pi/4t^2).$$

$f_1(t)$ is of the form (3.8.9) with $u(x) = \cos x$, $\theta(t) = \pi t^2/4 = \theta_2(t)$, $\phi(t) = 0$, and $h \in A^{(-4)}$ so that $\gamma = -4$. Since $m = 2$ and $k = 0$, $\sigma = -1$ so that $\psi(x) = \cos(\pi x^2/4)x^{-5}$. With $x_i = 2\sqrt{i+1}$, $i = 0, 1, \ldots$, $\psi(x_i) = (-1)^{i+1}x_i^{-5}$. For $f_2(t)$, $u(x) = \sin x$, $\theta(t) = \pi t^2 = \theta_2(t)$, $\phi(t) = 0$, and

$h \in A^{(0)}$. Again $\sigma = -1$ and now $\psi(x) = \cos(\pi x^2)/x$. With $x_i = \sqrt{i+2}$, $\psi(x_i) = (-1)^i/x_i$, $i = 0, 1, \ldots$. The sums S_n of the approximation to $I_1 + I_2$ listed below show rapid convergence.

n	S_n
1	−.16899
3	−.1691378
5	−.1691374808
7	−.1691374816

References. Sidi [3, 5].

A novel extrapolation procedure has been proposed by Lugannani and Rice for the evaluation of the slowly convergent integral

$$F(y) = \int_0^\infty e^{-iuy} f(u)\,du. \tag{3.8.15}$$

In this procedure, we evaluate

$$F(y, \sigma) = \int_0^\infty \exp[-iuy - \tfrac{1}{2}\sigma^2 u^2] f(u)\,du \tag{3.8.16}$$

for a decreasing sequence of values σ_k, $k = 0, 1, \ldots$ tending to zero. These integrals can be evaluated more easily than $F(y)$ since the integrands converge to zero much more rapidly than the integrand in (3.8.15). Since it can be shown that

$$F(y, \sigma) = F(y) + \sum_{n=1}^{N-1} C_{2n}(y)\sigma^{2n} + O(\sigma^{2N}), \tag{3.8.17}$$

we can apply Richardson extrapolation (Section 1.15) to the sequence $F(y, \sigma_k)$. This idea is applicable to the general case when the integral $I = \int_0^\infty f(u)\,du$ converges slowly. However, since we may not have an expansion in σ^2 as in (3.8.17) but of a more general form

$$I(\sigma) \equiv \int_0^\infty e^{-\sigma^2 u^2/2} f(u)\,du = I + \sum_{n=1}^{N-1} C_n \sigma^{\gamma_n} + O(\sigma^{\gamma_N}) \tag{3.8.18}$$

it may be necessary to use the method of Lynch to determine the exponents in (3.8.18) (cf. Section 6.3).

Alternatively, we could use the epsilon algorithm (Section 1.15).

References. Lugannani and Rice [1]; Lynch [1].

3.9 The Fourier Transform

The numerical computation of the Fourier transform

$$\hat{f}(w) = \int_{-\infty}^{\infty} f(x)e^{iwx}\,dx, \qquad -\infty < w < \infty, \tag{3.9.1}$$

may be attended by considerable difficulties due to the infinite range and the oscillatory integrand.

If we set

$$\psi(x) = f(x) + f(-x), \qquad \phi(x) = f(x) - f(-x), \tag{3.9.2}$$

we may write (3.9.1) as $\hat{f}(w) = C(w) + iS(w)$ where

$$C(w) = \int_0^{\infty} \psi(x) \cos wx\,dx,$$

$$S(w) = \int_0^{\infty} \phi(x) \sin wx\,dx. \tag{3.9.3}$$

The functions $C(w)$ and $S(w)$ are the Fourier cosine and sine transforms respectively.

It is often convenient to divide (3.9.1) into three integrals:

$$\hat{f} = \int_{-\infty}^{-k} + \int_{-k}^{k} + \int_{k}^{\infty}. \tag{3.9.4}$$

The integral over the finite range may be treated by the methods of Chapter 2. If w is very small, e^{iwx} is slowly oscillatory and no special methods may be called for; if w is larger, one can use the methods of Section 2.10.

The tails, that is, $\int_{-\infty}^{-k}$, $\int_{k}^{\infty}$, may be treated quite crudely if $f(x)$ decreases very rapidly as $x \to \pm\infty$ and if we can estimate

$$\int_{-\infty}^{-k} |f(x)|\,dx, \qquad \int_{k}^{\infty} |f(x)|\,dx.$$

The asymptotic series

$$-e^{-iwk} \int_k^{\infty} e^{iwx} f(x)\,dx \sim \frac{f(k)}{iw} - \frac{f'(k)}{(iw)^2} + \frac{f''(k)}{(iw)^3} - \cdots \tag{3.9.5}$$

might prove convenient.

Gustafson and Dahlquist have experimented with an approximation method for dealing with the tails in a more effective and systematic manner.

For fixed k, write a tail operator

$$Tf(w) = e^{-iwk} \int_k^\infty e^{iwx} f(x)\, dx. \tag{3.9.6}$$

Now approximate $f(x)$ over the range $[k, \infty)$ by exponential sums

$$f(x) \approx \sum_{j=1}^{q} m_j e^{-x_j x} \equiv f_A(x) \tag{3.9.7}$$

where the constants m_j and x_j are to be determined by some method. Insertion of (3.9.7) into (3.9.6) yields

$$Tf(w) \approx \sum_{j=1}^{q} m_j \frac{e^{-x_j k}}{(x_j - iw)}, \tag{3.9.8}$$

so that the tail is approximated by a fairly simple rational function of w. After examining a number of strategies for the determination of m_j and x_j, Gustafson and Dahlquist recommend the following.

The exponential sums (3.9.7) are matched to $f(x)$ by interpolation at n equidistant stations

$$y_r = y_0 + rh, \qquad f_A(y_r) = f(y_r), \qquad r = 0, 1, \ldots, n-1. \tag{3.9.9}$$

The mesh size h is dependent upon w, decreasing as w increases. In many cases, three values of h have been found sufficient to cover the range $1 \le w \le 10^4$. If one selects $q = n$ and the damping constants x_j are chosen in advance, then the interpolation (3.9.9) leads to a linear system. Recommended values for x_j are

$$x_j = \frac{1}{h} \log \frac{2}{1 + u_j} \tag{3.9.10}$$

where the u_j are the zeros of the nth Jacobi polynomial corresponding to the weighting function $(u + 1)^{16}$ over $-1 \le u \le 1$. The rationale for this selection of x_j as well as experimental results on the slowly convergent integral $\int_0^\infty \{e^{iwx}/[1 + \log(1 + x)]\}\, dx$ are given in their paper.

The Fourier transform may, of course, be treated without splitting off the tails. For w large, the asymptotic expression (3.9.5) with $k = 0$ may prove useful. Simple approximation to $C(w)$ and $S(w)$ of the form (3.4.1)

$$C(w) \approx C(w; h) = h \sum_{k=0}^{\infty} \psi((k + \tfrac{1}{2})h) \cos(w(k + \tfrac{1}{2})h) \tag{3.9.11}$$

$$S(w) \approx S(w; h) = h \sum_{k=1}^{\infty} \phi(kh) \sin(wkh) \tag{3.9.12}$$

have been recommended, and the claim has been made that these rules are at least as good (in a certain sense) as any other rule that uses equally spaced data. Note that in order to apply (3.9.11) and (3.9.12) we must select both h and a value N for the number of terms taken in the series. In the interplay between the sizes of w, N, and h there is considerable room for numerical ambiguity. A good strategy for fixed w seems to be: starting from low values of N and h, increase N and decrease h simultaneously in such a way that Nh increases.

De Balbine and Franklin work out error expressions but as they involve an infinite series of values of the transform, they may not always be convenient to interpret. These authors also establish that under certain conditions on the smoothness of ϕ and ψ, the Euler transform of (3.9.11) and (3.9.12) converges for all w and can serve to speed up the summation of these series.

References. de Balbine and Franklin [1]; Gustafson and Dahlquist [1]; Pantis [1].

Examples of Rule (3.9.12)

$$\int_0^\infty e^{-x} \sin wx \, dx = \frac{w}{1 + w^2}.$$

h	N	hN	$w = 1$	$w = 10$	$w = 100$
1.0	1	1	.309560	−.200134	−.186282
.5	4	2	.472989	−.256549	−.311559
.25	16	4	.505997	.042792	−.405104
.125	64	8	.498577	.085634	−.413781
.0625	256	16	.494675	.095734	−.414021
.03125	1024	32	.499919	.098195	.000130
	Exact		.500000	.099010	.009999

$$\int_0^\infty \frac{x}{1 + x^2} \sin wx \, dx = \frac{\pi}{2} e^{-w}.$$

h	N	hN	$w = 1$	$w = 10$	$w = 100$
1.0	100	100	.560590	−.077682	−.939329
.5	400	200	.574397	−.120044	−.915394
.25	1600	400	.578909	−.000067	−.921895
.125	6400	800	.578492	.000142	−.925506
	Exact		.577864	.000071	.000000

$$\int_0^\infty \frac{\sin^2(\frac{1}{2}x)}{x} \sin wx\, dx = \begin{cases} \pi/4, & w < 1, \\ \pi/8, & w = 1, \\ 0, & w > 1. \end{cases}$$

h	N	hN	$w = 1$	$w = 10$	$w = 100$
1.0	100	100	.388990	−.000415	−.798415
.5	400	200	.390936	−.000238	−.778956
.25	1600	400	.393015	−.000092	−.783191
.125	6400	800	.392936	.000058	−.786609
	Exact		.392699	.000000	.000000

By linearly interpolating the function $f(x)$ at the points $0, h, 2h, \ldots$ and integrating the resulting expression over each of the intervals $[kh, (k+1)h]$, $k = 0, 1, 2, \ldots$, Krylov and Skoblya obtain the approximations

$$I_c \equiv \int_0^\infty f(x) \cos wx\, dx \approx \frac{2 - 2\cos wh}{w^2 h} \sum_{k=0}^{\infty}{}' \cos kwh\, f(kh) \tag{3.9.13}$$

$$I_s \equiv \int_0^\infty f(x) \sin wx\, dx \approx \frac{wh - \sin wh}{w^2 h} f(0) + \frac{2 - 2\cos wh}{w^2 h} \sum_{k=1}^{\infty} \sin kwh\, f(kh). \tag{3.9.14}$$

By interpolating over two, three, and four subintervals using polynomials of degree 2, 3, and 4 respectively, they obtain successively more complicated formulas for I_c and I_s which involve the values of f at the points kh. They also give two sets of formulas for I_c and I_s which involve values of both $f(x)$ and $f'(x)$ at the points kh.

It appears that (3.9.14) is better than (3.9.12) and numerical experience bears this out for the case of slowly decreasing functions. However the formula which is correct for second degree polynomials did not yield any significant improvement over (3.9.14) and in some cases even gave worse results.

Additional methods and formulas are given by Krylov and his co-workers.

Examples of Rule (3.9.14)

$$\int_0^\infty e^{-x} \sin wx \, dx = \frac{w}{1 + w^2}.$$

h	N	hN	$w = 1$	$w = 10$	$w = 100$
1.0	1	1	.443137	.098079	.010046
.5	4	2	.504365	.104476	.010044
.25	16	4	.513751	.100725	.010042
.125	64	8	.500531	.099130	.010041
.0625	256	16	.500163	.099042	.010041
.03125	1024	32	.500041	.099018	.010000
	Exact		.500000	.099010	.009999

$$\int_0^\infty \frac{x}{1 + x^2} \sin wx \, dx = \frac{\pi}{2} e^{-w}.$$

h	N	hN	$w = 1$	$w = 10$	$w = 100$
1.0	100	100	.515404	−.002857	−.000026
.5	400	200	.562529	−.006879	−.000026
.25	1600	400	.575901	−.000039	−.000026
.125	6400	800	.577739	.000125	−.000026
	Exact		.577864	.000071	.000000

$$\int_0^\infty \frac{\sin^2(\frac{1}{2}x)}{x} \sin wx \, dx = \begin{cases} \pi/4, & w < 1, \\ \pi/8, & w = 1, \\ 0, & w > 1. \end{cases}$$

h	N	hN	$w = 1$	$w = 10$	$w = 100$
1.0	100	100	.357636	−.000015	−.000022
.5	400	200	.382859	−.000014	−.000022
.25	1600	400	.390973	−.000053	−.000022
.125	6400	800	.392424	−.000051	−.000022
	Exact		.392699	.000000	.000000

Silliman has generalized (3.9.13) and (3.9.14) by approximating $f(x)$ by a spline of degree $2m - 1$ instead of the linear spline used to derive those rules. To this end, we define the three functions

$$\psi_n(t) = \left(\frac{2 \sin t/2}{t}\right)^n, \tag{3.9.15}$$

$$\phi_n(t) = \sum_{j=-\infty}^{\infty} \psi_n(t + 2\pi j) = \left(2 \sin \frac{t}{2}\right)^n \sum_{\nu=-\infty}^{\infty} \frac{(-1)^{\nu n}}{(t + 2\pi\nu)^n}, \tag{3.9.16}$$

and

$$\rho_n(t) = \left(2 \sin \frac{t}{2}\right)^n \sum_{\nu=-\infty}^{\infty} \frac{1}{(t + 2\pi\nu)^n}. \tag{3.9.17}$$

We now consider the set of functions $f(x)$ such that $f \in C^{2m}(A) \cap L_1(A)$, $f^{(2m)} \in L_1(A)$, and $f(x)$, $f^{(2m)}(x) \to 0$ as $|x| \to \infty$ where A denotes the real axis $(-\infty, \infty)$ or the semi-infinite line $[0, \infty)$ as the case may be and $L_1(A)$ is the set of functions g such that

$$\|g\|_1 = \int_A |g(x)|\, dx < \infty. \tag{3.9.18}$$

The following results now hold. For all h and t such that $-\pi/h \le t \le \pi/h$,

$$\int_{-\infty}^{\infty} f(x)e^{ixt}\, dx = \frac{\psi_{2m}(th)}{\phi_{2m}(th)} h \sum_{\nu=-\infty}^{\infty} f(\nu h)e^{i\nu th} + Rf, \tag{3.9.19}$$

$$\int_0^{\infty} f(x) \cos xt\, dx = \frac{\psi_{2m}(th)}{\phi_{2m}(th)} h \left\{\frac{1}{2} f(0) + \sum_{\nu=1}^{\infty} f(\nu h) \cos \nu th\right\}$$
$$+ \sum_{j=1}^{m-1} \frac{(-1)^j}{t^{2j}} \left[1 - \frac{\phi_{2j}(th)\, \psi_{2m-2j}(th)}{\phi_{2m}(th)}\right] f^{(2j-1)}(0) + Rf, \tag{3.9.20}$$

$$\int_0^{\infty} f(x) \sin xt\, dx = \frac{\psi_{2m}(th)}{\phi_{2m}(th)} h \sum_{\nu=1}^{\infty} f(\nu h) \sin \nu th$$
$$+ \sum_{j=0}^{m-1} \frac{(-1)^j}{t^{2j+1}} \left[1 - \frac{\psi_{2m-1-2j}(th)\, \rho_{2j+1}(th)}{\phi_{2m}(th)}\right] f^{(2j)}(0) + Rf \tag{3.9.21}$$

where a bound on Rf is given by

$$|Rf| \le A_m (h/\pi)^{2m} \|f^{(2m)}\|_1 \tag{3.9.22}$$

with

$$A_m = 2\left(1 + 2 \sum_{\nu=1}^{\infty} \frac{1}{(2\nu + 1)^{2m}}\right) < 3 \qquad \text{for} \quad m = 1, 2, \ldots. \tag{3.9.23}$$

Clendenin has given a method for computing Fourier integrals based on the following approximation for the cosine integral:

$$\int_{t_0}^{t_0+Nh} f(t) \cos \omega t\, dt \approx \{4/(\omega^2 h)\}(-1)^j \left[\frac{1}{2} f(t_0) + \sum_{n=1}^{N-1} (-1)^n f(t_0 + nh)\right.$$
$$\left. + \frac{1}{2}(-1)^N f(t_0 + Nh)\right], \tag{3.9.24}$$

and a slightly more complicated formula for the sine integral. In the above formula, $h = |2k + 1|\pi/\omega$, k an arbitrary integer, while $j = \omega t_0/\pi$ must be an integer. Since k may be quite large, this has the advantage of not requiring several function evaluations in each period. In addition, it is not necessary to compute sines and cosines. The necessary details for the implementation of this method are given in the reference.

References. Clendenin [1]; Krylov and Kruglikova [1]; Krylov and Skoblya [1]; Silliman [1, 2].

Many authors write the Fourier integrals as infinite series of the form

$$C(w) = \frac{1}{w}\int_0^{\pi/2} \cos x\, f\left(\frac{x}{w}\right) dx + \frac{1}{w}\sum_{k=1}^{\infty} (-1)^k \int_{-\pi/2}^{\pi/2} \cos x\, f\left(\frac{x + k\pi}{w}\right) dx, \tag{3.9.25}$$

$$S(w) = \frac{1}{w}\sum_{k=0}^{\infty} (-1)^k \int_{-\pi/2}^{\pi/2} \cos x\, f\left[\frac{x + (k + \frac{1}{2})\pi}{w}\right] dx. \tag{3.9.26}$$

They then proceed in one of two ways. In one case, they approximate the integrals for a sequence of values of k and then apply one of the speedup methods mentioned in Section 3.8. The individual integrals can be evaluated efficiently using a Gauss rule based on the weight functions $\cos x$, thus avoiding the necessity to evaluate $\cos x$ each time $f(x)$ is computed. A short table of abscissas and weights for this Gauss rule is given in Blakemore *et al.* Additional values may be extracted from the ALGOL program of Haegemans and Piessens. If these abscissas and weights are not available, the next best choice is a Lobatto rule inasmuch as the integrands vanish at the endpoints of the integration interval. The first integral in (3.9.25) must be treated separately by any standard method.

In the second approach, they choose an appropriate integration rule $Q_n f = \sum_{i=1}^n a_i f(x_i)$, where either $\cos x$ is incorporated into the a_i or a_i is of the form $b_i \cos x_i$, and rewrite say $S(w)$ as

$$\begin{aligned} S(w) &= \frac{1}{w}\int_{-\pi/2}^{\pi/2} \cos x \sum_{k=0}^{\infty} (-1)^k f\left[\frac{x + (k + \frac{1}{2})\pi}{w}\right] dx \\ &\simeq \frac{1}{w}\sum_{i=1}^{n} a_i \sum_{k=0}^{\infty} (-1)^k f\left[\frac{x_i + (k + \frac{1}{2})\pi}{w}\right]. \end{aligned} \tag{3.9.27}$$

They then apply a convergence acceleration algorithm to each of the series

$$\sum_{k=0}^{\infty} (-1)^k f\left[\frac{x_i + (k + \frac{1}{2})\pi}{w}\right].$$

Blakemore *et al.* have compared these two approaches and concluded that the former approach is superior.

One situation in which the latter method may be advantageous is when $f(x)$ is even in the cosine case or odd in the sine case. One can then symmetrize the integrand to yield a periodic function so that the integrals can be evaluated efficiently by a trapezoidal rule.

References. Blakemore, Evans, and Hyslop [1]; Haegemans and Piessens [A1]; Piessens and Haegemans [1]; Squire [5, 6].

3.9.1 *Use of Laguerre and Hermite Expansions*

Since the Fourier transforms of the Laguerre and Hermite polynomials multiplied by their associated weight functions are known in closed form as follows:

$$V_k(w) = \frac{k!}{\Gamma(k+1+\alpha)} \int_0^\infty x^\alpha e^{-x} L_k^{(\alpha)}(x) e^{iwx}\, dx = i^{1+\alpha} w^k \left(\frac{w-i}{1+w^2} \right)^{k+1+\alpha}, \tag{3.9.1.1}$$

$$U_k(w) = \int_{-\infty}^{\infty} e^{-x^2} H_k(x) e^{iwx}\, dx = i^k \pi^{1/2} w^k e^{-w^2/4}, \tag{3.9.1.2}$$

Patterson suggested approximating the Fourier integrals

$$L(w) = \int_0^\infty x^\alpha e^{-x} f(x) e^{iwx}\, dx, \tag{3.9.1.3}$$

$$H(w) = \int_{-\infty}^{\infty} e^{-x^2} f(x) e^{iwx}\, dx \tag{3.9.1.4}$$

by expanding $f(x)$ in a series of Laguerre or Hermite polynomials as the case may be. Thus, if

$$f(x) \simeq \sum_{k=0}^{n} b_k L_k^{(\alpha)}(x) k!/\Gamma(k+1+\alpha) \tag{3.9.1.5}$$

where

$$b_k = \int_0^\infty x^\alpha e^{-x} L_k^{(\alpha)}(x) f(x)\, dx, \tag{3.9.1.6}$$

then

$$L(w) \simeq \sum_{k=0}^{n} b_k V_k(w), \tag{3.9.1.7}$$

while if

$$f(x) \simeq \pi^{-1/2} \sum_{k=0}^{n} q_k H_k(x) \tag{3.9.1.8}$$

where

$$q_k = \frac{1}{2^k k!} \int_{-\infty}^{\infty} e^{-x^2} H_k(x) f(x)\, dx, \tag{3.9.1.9}$$

then

$$H(w) \simeq e^{-w^2/4} \sum_{k=0}^{n} i^k q^k w^k. \tag{3.9.1.10}$$

The b_k and q_k are usually approximated by an $(n+1)$-point Gauss–Laguerre or Gauss–Hermite rule, respectively.

For $\alpha = 0$, the real and imaginary parts of $L(w)$ give the approximations

$$I_c = \int_0^{\infty} e^{-x} f(x) \cos w(x)\, dx \simeq \sum_{k=0}^{n} b_k C_k, \tag{3.9.1.11}$$

$$I_s = \int_0^{\infty} e^{-x} f(x) \sin w(x)\, dx \simeq \sum_{k=0}^{n} b_k S_k \tag{3.9.1.12}$$

where

$$C_0 = 1/(1+w^2), \qquad S_0 = wC_0,$$
$$C_k = S_0(wC_{k-1} + S_{k-1}), \qquad S_k = S_0(wS_{k-1} - C_{k-1}), \qquad k > 0.$$

In practice, one usually gets better accuracy by breaking the integral into the ranges $[0, a]$ and $[a, \infty)$ so that, for example,

$$\begin{aligned} I &= \int_0^{\infty} e^{-x} f(x) e^{iwx}\, dx \\ &= \frac{a}{2} e^{iwa/2} \int_{-1}^{1} e^{iwx/2} e^{-(x+1)a/2} f((x+1)a/2)\, dx \\ &\quad + e^{iwa} e^{-a} \int_0^{\infty} e^{-x} f(x+a) e^{iwx}\, dx = I_1 + I_2. \end{aligned} \tag{3.9.1.13}$$

Example (Patterson). Evaluate

$$I_0 = \int_0^{\infty} 10e^{-x} \sin(16\pi x)/(1+x^2)\, dx = .1990228.$$

We used (3.9.1.12) to approximate I_0 and also I_2 with $a = 3$; I_1 was approximated using a Legendre expansion (Section 2.10.5). The n_j give the number of integrand evaluations used to approximate I_j.

I_0	n_0	I_1	n_1	I_2	n_2	$I_1 + I_2$
.1989	36	.19875	5	.981(−3)	7	.199731
.19907	48	.1980405	11	.98738(−3)	9	.1990279
.1990418	56	.1980335	17	.989377(−3)	17	.1990229
.1990238	69	.1980334	36	.9894175(−3)	36	.1990228

A similar approach using the Hermite functions

$$h_n(x) = \pi^{1/4}(2^n n!)^{-1/2} e^{-x^2/2} H_n(x) \tag{3.9.1.14}$$

was proposed by Eberlein and by Walter and Schultz. In this case

$$\int_{-\infty}^{\infty} h_n(x) e^{iwx}\, dx = \sqrt{2\pi} i^n h_n(w) \tag{3.9.1.15}$$

so that if

$$f(x) = \sum_{k=0}^{\infty} a_k h_k(x) \tag{3.9.1.16}$$

where

$$a_n = \int_{-\infty}^{\infty} f(x) h_n(x)\, dx, \tag{3.9.1.17}$$

then

$$\hat{f}(w) = \sqrt{2\pi} \sum_{k=0}^{\infty} i^k a_k h_k(w). \tag{3.9.1.18}$$

Since

$$a_n = \pi^{-1/4}(2^n n!)^{-1/2} \int_{-\infty}^{\infty} e^{-x^2} H_n(x) g(x)\, dx \tag{3.9.1.19}$$

where $g(x) = e^{x^2/2} f(x)$, we can approximate a_n, as before, by a Gauss–Hermite integration rule.

Eberlein shows that if f is absolutely continuous, i.e., $f(x) = f(0) + \int_0^x f'(t) dt$ $(-\infty < x < \infty)$ and $f' \in L_2$ and if $xf \in L_2$, then the process with exact a_n converges. However, the best way to approximate the a_n is still an open question. If $f(x)$ is only available at a discrete set of points, $t_0, t_1, \ldots, t_N$, we can approximate the a_n by interpolation so that the a_n are the solution of the system of linear equations

$$\sum_{n=0}^{N} a_n h_n(t_m) = f(t_m), \qquad m = 0, \ldots, N. \tag{3.9.1.20}$$

Weber turns this process around and expands the Fourier transform $\hat{f}(w)$ in Laguerre functions

$$\phi_n(w) = e^{-w/2} L_n(w). \tag{3.9.1.21}$$

Since the Laguerre functions are defined in $[0, \infty)$ and we are interested in the range $(-\infty, \infty)$, we define a new set of functions $\psi_n(w)$ as follows

$$\psi_n(w) = \begin{cases} \phi_n(w), & w \geq 0; \quad 0, \quad w < 0; \quad n \geq 0, \\ -\phi_{-n-1}(w), & w < 0; \quad 0, \quad w \geq 0; \quad n < 0. \end{cases} \tag{3.9.1.22}$$

If we now let

$$\hat{f}(w) = 2\pi \sum_{n=-\infty}^{\infty} C_n \psi_n(w), \tag{3.9.1.23}$$

then the C_n can be approximated by

$$C_n^M = \frac{1}{2M+1} \sum_{j=0}^{2M} g(y_j) e^{-iny_j}, \qquad y_j = \frac{2\pi j}{2M+1}, \qquad j = 0, \ldots, 2M \tag{3.9.1.24}$$

where

$$g(y) = \frac{1}{1 - e^{iy}} f\left(\frac{1}{2i} \frac{1 + e^{iy}}{1 - e^{iy}}\right). \tag{3.9.1.25}$$

The advantage of this formulation is that the C_n^M can be evaluated using the FFT (Section 3.9.5). If f and $\hat{f}$ are sufficiently smooth and $\hat{f}$ falls off quickly enough as $|t| \to \infty$, then this method is fast and accurate.

References. Eberlein [1]; Patterson [4]; Walter and Schultz [1]; Weber [1].

3.9.5 *The Discrete Fourier Transform and Fast Fourier Transform Methods*

3.9.5.1 *Introduction*

The Fourier transform of g is given by the integral

$$G(f) = \mathscr{F}(g) = \int_{-\infty}^{\infty} g(t) e^{2\pi i f t}\, dt, \qquad -\infty < f < \infty. \tag{1.1}†$$

The function $g(t)$ is often a signal and t is referred to as the "time domain" while f is the "frequency domain." The integral is often approximated by simple averages evaluated at equally spaced abscissas. This leads to the

† Throughout Section 3.9.5, we shall suppress the first three digits in labeling equations. Thus, (3.9.5.1.1) is written as (1.1).

notion of the *Discrete Fourier Transform* (DFT). The DFT is a linear transformation of a finite sequence of length N. Normally, such a transformation will require N^2 multiplications. However, in the case of the DFT, if N is selected as a power of 2 (or as a highly composite integer), one may take advantage of the symmetries inherent in the matrix of the transformation to remove redundant operations and to reduce the computation to the order of $N \log_2 N$ multiplications. Such an algorithm is known as a *Fast Fourier Transform* (FFT).

The algorithm is particularly useful since it offers the saving of a significant number of numerical operations over conventional methods. As a result, the FFT has gained wide acceptance in many areas of data handling and interpretation, linear system analysis, stochastic analysis, and digital signal and image processing.

The FFT algorithm can be traced back to Runge. A history of the algorithm may be found in Cooley, Lewis, and Welch [2]. However, the recent explosion of the use of the method is generally credited to the appearance of a paper by Cooley and Tukey in 1965. There is now a burgeoning literature devoted solely to this subject and its many applications.

References. Bingham, Godfrey, and Tukey [1]; Bloomfield [1]; Bracewell [1]; Brigham [1]; Brigham and Morrow [1]; Cochran *et al.* [1]; Cooley, Lewis, and Welch [1–3]; Cooley and Tukey [1]; Davis [13]; Elliott and Rao [1]; Gold and Rader [1]; Kaporin [1]; Nussbaumer [1]; Runge [1]; Welch [1]; Whelchel and Guinn [1].

The study of the most efficient method for evaluating the DFT has become a subject in the recently developed theory of arithmetic complexity of computations. In this context, Winograd has proposed a method which is more efficient than the FFT for evaluating the DFT and is especially advantageous in the multidimensional case.

References. Silverman [2]; Winograd [1–3]; Zohar [1, 2].

3.9.5.2 The Relationship of the DFT to the Fourier Transform

Select an interval of length T. Then we may rewrite (1.1) as

$$G(f) = \sum_{k=-\infty}^{\infty} \int_0^T g(t + kT)e^{2\pi i f(t+kT)}\, dt. \tag{2.1}$$

Interchanging summation and integration,

$$G(f) = \int_0^T \left[\sum_{k=-\infty}^{\infty} g(t + kT)e^{2\pi i f kT} \right] e^{2\pi i f t}\, dt. \tag{2.2}$$

We shall evaluate the transform only for the discrete values of f:

$$f = f_r = \frac{r}{T}, \qquad r = 0, \pm 1, \pm 2, \ldots. \tag{2.3}$$

Since $e^{2\pi i f_r kT} = 1$, the transform becomes

$$G(f_r) = \int_0^T \left[\sum_{k=-\infty}^{\infty} g(t + kT) \right] e^{2\pi i f_r t} \, dt. \tag{2.4}$$

If we set

$$g_p(t) = \sum_{k=-\infty}^{\infty} g(t + kT), \qquad -\infty < t < \infty, \qquad T > 0, \tag{2.5}$$

the transform becomes

$$G(f_r) = \int_0^T g_p(t) e^{2\pi i f_r t} \, dt. \tag{2.6}$$

For any $g(t)$, the function $g_p(t)$ is periodic of period T. The function $g_p(t)$ coincides with $g(t)$ over the interval $[0, T)$ if and only if $g(t)$ vanishes outside this interval. It is then the periodic extension of that portion of $g(t)$ in $[0, T)$. If $g(t)$ is nonzero over a larger interval than $[0, T)$, then an overlap or "aliasing" occurs.

We shall next assume that

$$g(t) = 0 \quad \text{for} \quad t \notin [0, T). \tag{2.7}$$

We then have

$$G(f_r) = \int_0^T g(t) e^{2\pi i f_r t} \, dt. \tag{2.8}$$

We abbreviate $G(f_r)$ by $G(r)$. This integral is now approximated by using the trapezoidal rule with mesh T/N. Then (2.8) becomes

$$\begin{aligned} G(r) &\approx \frac{T}{N} \sum_{k=0}^{N-1} g\left(\frac{kT}{N}\right) e^{2\pi i (r/T)(kT/N)} \\ &= \frac{T}{N} \sum_{k=0}^{N-1} g(k) e^{2\pi i rk/N} \end{aligned} \tag{2.9}$$

where we use the abbreviation $g(k) = g(kT/N)$. If we limit the values of r to $0, 1, 2, \ldots, N-1$, then (2.9) transforms the sequence $g(0), g(1), \ldots, g(N-1)$ into $G(0), G(1), \ldots, G(N-1)$ and is called the *Discrete Fourier Transform* (DFT).

3.9.5.3 The DFT and its Properties

The DFT may be considered as an entity in its own right unrelated to the Continuous Fourier Transform. It is useful to suppress the factor T/N in (2.9) and write simply

$$G(r) = \sum_{k=0}^{N-1} g(k)w^{rk}, \qquad r = 0, 1, \ldots, N-1, \qquad w = e^{2\pi i/N}. \tag{3.1}$$

We shall consider this as our basic linear transformation mapping the sequence $g(0), g(1), \ldots, g(N-1)$ onto the sequence $G(0), G(1), \ldots, G(N-1)$.

In view of the identity

$$\sum_{r=0}^{N-1} w^{r(j-k)} = \frac{1 - w^{N(j-k)}}{1 - w^{j-k}} = \begin{cases} N & \text{if } j = k, \\ 0 & \text{if } j \neq k, \end{cases} \tag{3.2}$$

we have

$$\begin{aligned} \sum_{r=0}^{N-1} G(r)w^{-kr} &= \sum_{r=0}^{N-1} w^{-kr} \sum_{j=0}^{N-1} g(j)w^{jr} \\ &= \sum_{j=0}^{N-1} g(j) \sum_{r=0}^{N-1} w^{r(j-k)} = Ng(k). \end{aligned} \tag{3.3}$$

Hence,

$$g(k) = N^{-1} \sum_{r=0}^{N-1} G(r)w^{-kr}, \qquad k = 0, 1, \ldots, N-1 \tag{3.4}$$

gives us the inverse transformation. The pair of equations (3.1), (3.4) defines the DFT ordinarily used for most FFT implementations. It is often convenient to express this in matrix notation. Introduce the symmetric matrix

$$D = \begin{bmatrix} 1 & 1 & 1 & \cdots & 1 \\ 1 & w & w^2 & \cdots & w^{N-1} \\ 1 & w^2 & w^4 & \cdots & w^{2(N-1)} \\ \vdots & \vdots & \vdots & \vdots & \vdots \\ 1 & w^{N-1} & w^{2(N-1)} & \cdots & w^{(N-1)^2} \end{bmatrix}. \tag{3.5}$$

In view of the periodicity of w^k, there are only N distinct elements in D and it can be written alternatively as

$$D = \begin{bmatrix} 1 & 1 & 1 & \cdots & 1 \\ 1 & w & w^2 & \cdots & w^{N-1} \\ 1 & w^2 & w^4 & \cdots & w^{N-2} \\ \vdots & \vdots & \vdots & \vdots & \vdots \\ 1 & w^{N-1} & w^{N-2} & \cdots & w^1 \end{bmatrix}. \tag{3.6}$$

Introducing the column vectors $g = (g(0), g(1), \ldots, g(N-1))^{\mathrm{T}}$, $G = (G(0), G(1), \ldots, G(N-1))^{\mathrm{T}}$, we can write (3.1) as

$$G = Dg. \tag{3.7}$$

From (3.4), if we set

$$B = \frac{1}{N}\begin{bmatrix} 1 & 1 & \cdots & 1 \\ 1 & w^{-1} & \cdots & w^{-(N-1)} \\ \vdots & \vdots & \vdots & \vdots \\ 1 & w^{-(N-1)} & \cdots & w^{-1} \end{bmatrix}, \tag{3.8}$$

we have

$$g = BG. \tag{3.9}$$

Therefore $B = D^{-1}$. Since $w\bar{w} = 1$, $B = (1/N)D^*$, where * represents the complex conjugate transpose. We have therefore

$$D^{-1} = (1/N)D^* \tag{3.10}$$

This relation implies that $N^{-1/2}D$ is a unitary matrix.

The Discrete Fourier Transform has many of the properties exhibited by the Continuous Fourier Transform. Several of these will now be discussed.

1 *Parseval's Theorem*

Let $G = Dg$ and $H = Dh$. Then $G^* = g^*D^* = Ng^*D^{-1}$. Hence,

$$G^*H = Ng^*h. \tag{3.11}$$

In particular, if $g = h$, we have

$$\|G\|^2 = G^*G = Ng^*g = N\|g^*\|^2, \tag{3.12}$$

the notation $\| \;\|$ designating the usual vector norm. Equation (3.12) shows that the "energy" of the vector G in the frequency domain differs only by the constant factor N from the energy of the corresponding vector g in the time domain.

2 *The Shift Theorem*

Let the vector $g(0), g(1), \ldots, g(N-1)$ be "shifted" by k units:

$$g_k(j) = g(j+k) \tag{3.13}$$

where we have taken $j + k$ mod N (or extended g periodically). Then

$$G_k(r) = \sum_{j=0}^{N-1} g_k(j)w^{jr} = \sum_{l=k}^{N-1+k} g(l)w^{(l-k)r}$$

$$= w^{-kr} \sum_{l=k}^{N-1+k} g(l)w^{lr} = w^{-kr} \sum_{l=0}^{N-1} g(l)w^{lr}.$$

Thus,

$$G_k(r) = w^{-kr}G(r). \tag{3.14}$$

3 *Hermitian Symmetry*

A vector V is said to be *Hermitian symmetric* if $V(t) = \overline{V(N-t)}$. It is said to be *Hermitian antisymmetric* if $V(t) = -\overline{V(N-t)}$. If g is real, then G is Hermitian symmetric. If g is Hermitian symmetric, then G is real. If g is imaginary, then G is Hermitian antisymmetric and vice versa.

For example, let $g(0), g(1), \ldots, g(N-1)$ be real, N even. Then

$$G\left(\frac{N}{2} + r\right) = \sum_{j=0}^{N-1} (-1)^j g(j)w^{jr}$$

$$G\left(\frac{N}{2} - r\right) = \sum_{j=0}^{N-1} (-1)^j g(j)w^{-jr}.$$

Hence

$$G\left(\frac{N}{2} + r\right) = \overline{G\left(\frac{N}{2} - r\right)}, \qquad r = 0, 1, \ldots, \frac{N}{2} - 1. \tag{3.15}$$

The symmetry property is very important in computing the DFT for two-dimensional real arrays or for very long sequences. For a real N-vector, N storage locations are required. However, the complex transform would seem to require $2N$ storage locations. In efficient programs for computation, only the elements $r = 0, 1, \ldots, N/2$ for $G(r)$ are computed for g real. Thus, only $2(N/2 + 1)$ storage locations are required. However, as $G(0)$ and $G(N/2)$ are always real, only N storage locations are necessary. In the inverse transform, knowing the result is to be real, an efficient program which has only stored the coefficients $G(0), \ldots, G(N/2)$ can compute $G(N/2 + 1), \ldots, G(N-1)$ by (3.15).

4 *Differences*

As an application of (3.14), consider the first difference

$$\Delta g(j) = g(j + 1) - g(j). \tag{3.16}$$

By (3.14), the DFT of $\Delta g(j)$ is $(w^{-r} - 1)G(r)$ so that

$$\Delta g(j) = \frac{1}{N} \sum_{r=0}^{N-1} (w^{-r} - 1)G(r)w^{-jr}$$

$$= -\frac{1}{N} \sum_{r=0}^{N-1} \left[2iG(r)w^{-r/2} \sin \frac{\pi r}{N}\right] w^{-jr}. \tag{3.17}$$

This identity and its extension to higher-order differences can be made the basis for the solution of certain linear difference equations by DFT techniques.

5 *Convolution Theorem*

The *convolution* of two functions $g(t)$, $h(t)$ is defined by

$$g(t)*h(t) = f(t) = \int_{-\infty}^{\infty} g(x)h(t - x)\,dx. \tag{3.18}$$

The operation $*$ is both commutative and associative. The *cross correlation* of g and h is defined by

$$R_{gh}(t) = \int_{-\infty}^{\infty} g(x)h(x - t)\,dx \tag{3.19}$$

and one has $R_{gh}(t) = g(t)*h(-t)$. The *autocorrelation* of g is R_{gg}.

The convolution theorem for the Continuous Fourier Transform says that the transform of the convolution of two functions equals the product of the transforms:

$$\mathscr{F}(g*h) = \mathscr{F}(g) \cdot \mathscr{F}(h). \tag{3.20}$$

Toward obtaining a convolution theorem for the DFT, we first derive an identity for circulant matrices.

Suppose that C_0 is the column vector

$$\begin{pmatrix} \gamma_0 \\ \gamma_1 \\ \vdots \\ \gamma_{N-1} \end{pmatrix}.$$

Let $C_1, C_2, \ldots, C_{N-1}$ be column vectors obtained from C_0 by cyclic forward permutations or shifts of the elements, i.e.,

$$C_1 = \begin{pmatrix} \gamma_{N-1} \\ \gamma_0 \\ \gamma_1 \\ \vdots \\ \gamma_{N-2} \end{pmatrix}, \qquad C_2 = \begin{pmatrix} \gamma_{N-2} \\ \gamma_{N-1} \\ \gamma_0 \\ \vdots \\ \gamma_{N-3} \end{pmatrix}, \ldots.$$

The $N \times N$ matrix

$$C = (C_0, C_1, \ldots, C_{N-1}) = \begin{pmatrix} \gamma_0 & \gamma_{N-1} & \cdots & \gamma_1 \\ \gamma_1 & \gamma_0 & \cdots & \gamma_2 \\ \vdots & \vdots & & \vdots \\ & & & \gamma_{N-1} \\ \gamma_{N-1} & \gamma_{N-2} & \cdots & \gamma_0 \end{pmatrix}$$

is known as the *circulant matrix* generated by C_0. Let the DFT of C_0 be $\hat{C}_0$ with components $\hat{\gamma}_0, \hat{\gamma}_1, \ldots, \hat{\gamma}_{N-1}$. Then, *the eigenvalues of C are the components of $\hat{C}_0$ and C is diagonalized by the* DFT *matrix D:*

$$C = D^{-1}\Lambda(\hat{C}_0)D = \frac{1}{N}D^*\Lambda(\hat{C}_0)D. \tag{3.21}$$

Here we have written $\Lambda(\hat{C}_0)$ for the diagonal matrix whose diagonal elements are the components of $\hat{C}_0$.

Proof

$$DC = D(C_0, C_1, \ldots, C_{N-1}) = (DC_0, DC_1, \ldots, DC_{N-1}).$$

By the shift theorem, this equals

$$\begin{bmatrix} \hat{\gamma}_0 & \hat{\gamma}_0 & & \hat{\gamma}_0 \\ \hat{\gamma}_1 & w\hat{\gamma}_1 & & w^{N-1}\hat{\gamma}_1 \\ \vdots & w^2\hat{\gamma}_2 & \cdots & w^{2(N-1)}\hat{\gamma}_2 \\ & \vdots & & \vdots \\ \hat{\gamma}_{N-1} & w^{N-1}\hat{\gamma}_{N-1} & & w^{(N-1)^2}\hat{\gamma}_{N-1} \end{bmatrix}$$

$$= \begin{pmatrix} \hat{\gamma}_0 & & & 0 \\ & \hat{\gamma}_1 & & \\ & & \ddots & \\ 0 & & & \\ & & & \hat{\gamma}_{N-1} \end{pmatrix} \begin{pmatrix} 1 & 1 & 1 & & 1 \\ 1 & w & w^2 & \cdots & w^{N-1} \\ 1 & w^2 & w^4 & & \\ \vdots & \vdots & \vdots & & \vdots \end{pmatrix}$$

$$= \Lambda(\hat{C}_0)D = \Lambda D.$$

Hence, $C = D^{-1}\Lambda D = (1/N)D^*\Lambda D$.

Suppose that we have two infinite sequences $g(j)$, $h(j)$, $-\infty < j < \infty$. Their *convolution* is defined as

$$g(k) * h(k) = f(k) = \sum_{j=-\infty}^{\infty} g(j)h(k-j), \qquad -\infty < k < \infty. \tag{3.22}$$

In applications, g is often an *impulse response* while h is the *input*. If now, $g(j)$ vanishes outside the finite interval $[0, N-1]$, then we have

$$f(k) = \sum_{j=0}^{N-1} g(j)h(k-j). \tag{3.23}$$

Suppose we want to obtain $f(0), f(1), \ldots, f(p-1)$. Write

$$f_p = \begin{bmatrix} f(0) \\ f(1) \\ \vdots \\ f(p-1) \end{bmatrix},$$

then

$$f_p = \begin{bmatrix} g(N-1) & g(N-2) & g(N-3) & \cdots & g(0) & 0 & 0 & \cdots & 0 & 0 \\ 0 & g(N-1) & g(N-2) & \cdots & g(1) & g(0) & 0 & \cdots & 0 & 0 \\ 0 & 0 & g(N-1) & \cdots & g(2) & g(1) & g(0) & \cdots & 0 & 0 \\ \vdots & \vdots & \vdots & \vdots & & & & \vdots & \vdots & \vdots \\ 0 & 0 & 0 & & & \cdots & & & g(1) & g(0) \end{bmatrix} \begin{bmatrix} h(-(N-1)) \\ h(-(N-2)) \\ \vdots \\ h(-1) \\ h(0) \\ h(1) \\ \vdots \\ h(p-1) \end{bmatrix} \tag{3.24}$$

where we take the matrix to be of order $p \times (N+p-1)$ and the column vector to be of length $N+p-1$. The matrix has nearly circulant form and can be made into a circulant by augmentation. Define the $(N+p) \times (N+p)$ circulant matrix Γ by means of

$$\Gamma = \begin{bmatrix} 0 & g(N-1) & g(N-2) & \cdots & g(1) & g(0) & 0 & 0 & \cdots & 0 & 0 & 0 \\ 0 & 0 & g(N-1) & \cdots & g(2) & g(1) & g(0) & 0 & \cdots & 0 & 0 & 0 \\ 0 & 0 & 0 & \cdots & g(3) & g(2) & g(1) & g(0) & \cdots & 0 & 0 & 0 \\ \vdots & \vdots & \vdots & & & & & & & \vdots & \vdots & \vdots \\ 0 & 0 & 0 & \cdots & & & & \cdots & \cdots & g(2) & g(1) & g(0) \\ g(0) & 0 & 0 & & \cdots & 0 & \cdots & & \cdots & g(3) & g(2) & g(1) \\ g(1) & g(0) & 0 & \cdots & & 0 & \cdots & & & g(4) & g(3) & g(2) \\ \vdots & \vdots & & & & & & & & \vdots & \vdots & \vdots \\ g(N-1) & g(N-2) & g(N-3) & \cdots & & 0 & \cdots & & & 0 & 0 & 0 \end{bmatrix}$$

This is the circulant generated by the vector

$$g = \begin{bmatrix} 0 \\ 0 \\ \vdots \\ 0 \\ g(0) \\ \vdots \\ g(N-1) \end{bmatrix}$$

of length $N + p$. If $\tilde{h}$ designates the column vector of length $N + p$:

$$\tilde{h} = \begin{bmatrix} h(-N) \\ h(-(N-1)) \\ \vdots \\ h(-1) \\ h(0) \\ h(1) \\ \vdots \\ h(p-1) \end{bmatrix}$$

and if $\tilde{f}$ designates the column vector of length $N + p$ given in block form by

$$\tilde{f} = \begin{bmatrix} f_p \\ --- \\ f_{N-1} \end{bmatrix}$$

where f_{N-1} is a column vector of length $N - 1$, then

$$\tilde{f} = \Gamma \tilde{h}. \tag{3.25}$$

Now apply the $N + p$ point DFT with matrix D:

$$\tilde{F} = D\tilde{f} = D\Gamma\tilde{h}.$$

Let G designate the DFT of the column vector g. Then by (3.21), we have

$$\tilde{F} = \Lambda(G)D\tilde{h} = \Lambda(G)\tilde{H} = G\tilde{H}. \tag{3.26}$$

This relationship is the *DFT Convolution Theorem.*

3.9.5.4 *The Fast Fourier Transform* (FFT)

Write the DFT as

$$F(k) = \sum_{r=0}^{N-1} f(r)w^{kr}, \qquad w = e^{2\pi i/N}, \; 0 \le k \le N - 1. \tag{4.1}$$

Assume that the integer N can be factored as $N = n_1 n_2$. Then upon division by n_1, the integer k can be exhibited as

$$k = k_1 n_1 + k_0 \quad \text{where} \quad 0 \le k_0 \le n_1 - 1 \quad \text{and} \quad 0 \le k_1 \le n_2 - 1.$$

Similarly for r:

$$r = r_1 n_2 + r_0 \quad \text{where} \quad 0 \le r_0 \le n_2 - 1 \quad \text{and} \quad 0 \le r_1 \le n_1 - 1.$$

With this decomposition, we can regard $f(r)$ as a function of r_0 and r_1 which we write as $f(r_1, r_0)$ and $F(k)$ can be regarded as a function of k_1, k_0 which we write as $F(k_1, k_0)$. The DFT now takes the form

$$F(k) = F(k_1, k_0) = \sum_{r_0=0}^{n_2-1} \sum_{r_1=0}^{n_1-1} f(r_1, r_0) w^{k(r_1 n_2 + r_0)}.$$

Now,

$$w^{kr_1 n_2} = w^{(k_1 n_1 + k_0) r_1 n_2} = w^{k_0 r_1 n_2}$$

since w raised to any integral multiple of N is equal to 1. Therefore

$$F(k) = \sum_{r_0=0}^{n_2-1} \sum_{r_1=0}^{n_1-1} f(r_1, r_0) w^{k_0 r_1 n_2} w^{kr_0}.$$

Write the inner sum as

$$f_1(k_0, r_0) = \sum_{r_1=0}^{n_1-1} f(r_1, r_0) w^{k_0 r_1 n_2}. \tag{4.2}$$

Then we have

$$F(k) = F(k_1, k_0) = \sum_{r_0=0}^{n_2-1} f_1(k_0, r_0) w^{(k_1 n_1 + k_0) r_0}. \tag{4.3}$$

For each value of k, (4.3) requires n_2 operations where by an operation we mean a complex multiplication and addition. There are N values of k, hence a total of Nn_2 operations. However, there are N values of f_1 to be computed, and by (4.2), each requires n_1 operations. Thus Nn_1 operations are required for f_1 making a grand total of $N(n_2 + n_1)$ operations.

This procedure may be iterated, and if $N = n_1 n_2 \cdots n_m$, the transform may be computed in $N(n_1 + n_2 + \cdots + n_m)$ operations. Suppose, in particular, that $N = 2^p$. In this case, the number of operations equals $2N \log_2 N$. The exponents r and k may be expressed in binary notation:

$$r = r_{p-1} 2^{p-1} + r_{p-2} 2^{p-2} + \cdots + r_0 = (r_{p-1}, \ldots, r_0),$$

and

$$k = k_{p-1} 2^{p-1} + k_{p-2} 2^{p-2} + \cdots + k_0 = (k_{p-1}, \ldots, k_0),$$

where each r_n and k_n has the value 0 or 1. The transform can now be written as

$$F(k) = F(k_{p-1}, k_{p-2}, \ldots, k_0) = \sum_{r_0=0}^{1} \sum_{r_1=0}^{1} \cdots \sum_{r_{p-1}=0}^{1} f(r_{p-1}, r_{p-2}, \ldots, r_0) w^{rk}.$$

Expand the exponent of w and ignore integral powers of $e^{2\pi i}$. The transform becomes

$$F(k) = F(k_{p-1}, \ldots, k_0)$$
$$= \sum_{r_0=0}^{1} \sum_{r_1=0}^{1} \cdots \sum_{r_{p-2}=0}^{1} \left[\sum_{r_{p-1}=0}^{1} f(r_{p-1}, \ldots, r_0) w^{k_0 r_{p-1} 2^{p-1}} \right] w^{k(r_{p-2} 2^{p-2} + \cdots + r_0)}.$$

The expression in brackets is a function of $r_{p-2}, r_{p-3}, \ldots, r_0$ and k_0 and can be written $f_1(k_0, r_{p-2}, \ldots, r_0)$. In general, at the lth step, the function

$$f_l(k_0, \ldots, k_{l-1}, r_{p-l-1}, \ldots, r_0)$$
$$= \sum_{r_{p-l}=0}^{1} f_{l-1}(k_0, \ldots, k_{l-2}, r_{p-l}, \ldots, r_0) w^{(k_{l-1} 2^{l-1} + \cdots + k_0) r_{p-l} 2^{p-l}} \tag{4.4}$$

is obtained. The final step yields

$$F(k_{p-1}, \ldots, k_0) = f_p(k_0, k_1, \ldots, k_{p-1}). \tag{4.5}$$

Notice that the index of the final result has its digits in reverse order and the data must be unscrambled, by performing a "bit reversal" on the indices, to arrange the results in the correct order.

Most FFT programs have been written for $N = 2^p$, but savings can be achieved for any integer N that is highly factorable. It has been shown that it is best theoretically to select $N = 3^p$. However, there are practical reasons for the selection $N = 2^p$.

FFT programs all have three portions. In the first portion the coefficients w^i are calculated. In simple implementations these are calculated as needed. However, in efficient implementations, such as those of the IBM Scientific Subroutine Package, a table of coefficients is stored in $(N/4) - 1$ locations. Each is a value of

$$\sin(2\pi\sigma/N) \equiv \alpha(\sigma), \qquad \sigma = 0, 1, \ldots, N/4. \tag{4.6}$$

Then, the following relations are used where needed in the calculation:

$$w^k = \alpha(N/4 - k) + i\alpha(k), \qquad k = 0, \ldots, N/4, \tag{4.7}$$
$$w^k = -\alpha(k - N/4) + i\alpha(N/2 - k), \qquad k = N/4, \ldots, N/2, \tag{4.8}$$
$$w^k = -w^{k-N/2}, \qquad k = N/2, \ldots, N. \tag{4.9}$$

A considerable amount of computation is saved by this procedure; for in most cases, many FFTs of the same order are used in a program. A provision is made that only in the first time through need the sine table be calculated. It should be observed, in passing, that there is no need to multiply by w^0, or by unity, so that many complex multiplications may be saved there. The second and third segments of the program may be calculated in any sequence.

If the time domain vector is to be operated upon in sequential order, the second part of the program will compute the transform. If the algorithm works in a manner similar to this section, the time domain vector must be rearranged before programming the transform so that the frequency vector will come out in sequential order. Rearrangement must be done either to the time domain vector or the frequency domain vector for all in-place algorithms,† but may be avoided if extra memory is used.

The rearrangement procedure is known as "bit-reversing." Take, for example, the case $N = 2^3$. For the frequency domain vector to come out in order, the time domain sequence must have indices (0, 4, 2, 6, 1, 5, 3, 7). If each index is represented by a three bit binary number, then the rearrangement is accomplished by scanning the sequential order (0, 1, 2, 3, 4, 5, 6, 7) and reading each binary number from left to right, i.e.,

$$
\begin{array}{c|ccccccc|c}
0 & 0 & 0 & 0 \to 0 & 0 & 0 & 0 \\
1 & 0 & 0 & 1 \to 1 & 0 & 0 & 4 \\
2 & 0 & 1 & 0 \to 0 & 1 & 0 & 2 \\
3 & 0 & 1 & 1 \to 1 & 1 & 0 & 6
\end{array}
\qquad
\begin{array}{c|ccccccc|c}
4 & 1 & 0 & 0 \to 0 & 0 & 1 & 1 \\
5 & 1 & 0 & 1 \to 1 & 0 & 1 & 5 \\
6 & 1 & 1 & 0 \to 0 & 1 & 1 & 3 \\
7 & 1 & 1 & 1 \to 1 & 1 & 1 & 7
\end{array}
$$

In some languages the bit-reversing portion of the program is most easily stored as a small table. Generally about $2N$ or $3N$ simple replacement operations are necessary.

To compute the inverse DFT, the same program may be used. Only the signs on the coefficients need be changed.

Among the important considerations in the selection of an FFT program are (a) speed required, (b) maximum size of transform to be performed (core or core-disk), (c) length of program, (d) limitations on N (e.g., $N = 2^p$).

A compact FORTRAN program for evaluating a complex FFT for the case $N = 2^m$ is given by Cooley *et al.* Programs for evaluating FFTs based on algorithms by Singleton are available in Chapter C06 of the NAG library and Chapter F of the IMSL library. Algorithmic formulations of two methods for evaluating FFTs appear in Stoer and Bulirsch.

References. Conte and de Boor [F1]; Cooley, Lewis, and Welch [3, p. 419]; Digital Signal Processing Committee [1]; IBM [2, pp. 276–283]; IMSL [1]; Monro [F1, F2]; Monro and Branch [F1]; NAG [1]; Singleton [1, A1–A3]; Stoer and Bulirsch [1, pp. 78–84].

† By an "in-place algorithm" is meant one in which the number of storage locations used for processing the data is small and is independent of the number of pieces of data processed.

3.9.5.5 *Applications of the One-Dimensional DFT*

Spectral Analysis

The DFT may be used for general spectral analysis. It has proven to be of great value for statistical analysis through its inclusion in general purpose packages such as SASE (Statistical Analysis for a Sequence of Events) (see Lewis). A direct application to a data sequence is calculated, and what is known as the "Periodogram" (see Welch) results. Note that the periodogram is *not* the actual spectrum, but, rather, an approximation under certain conditions. It is evident, however, that if the parameters N, for a sequence of discrete events, or N and Δt, for continuous data, are properly selected, then desired frequency domain effects will be seen. It is known that the frequencies k/N for $k = 0, 1, 2, \ldots, N/2$ are of interest. From these $N/2 + 1$ points, the cumulative log periodogram and other frequency-dependent curves for statistical analysis may be calculated.

Correlation/Convolution

The calculation of the correlation or convolution of two sequences has already been discussed. Given sequences $g(i)$, $h(i)$, $i = 0, 1, \ldots, N - 1$, one may calculate the convolution $f(k)$ of length p, defined in (3.23), by means of the DFT as follows:

(1) Set up $\tilde{h}$ of length $N + p$.
(2) Transform $\tilde{h}$ yielding $\tilde{H}$.
(3) Calculate the product in (3.26). One should note that the product vector may be calculated by multiplying the $N + p$ DFT coefficients of $\tilde{H}$, term-by-term by the $N + p$ coefficients of the diagonal of $\Lambda(G)$. This may be obtained by taking the transform of g, $g \to G$.
(4) Perform the inverse transform of the resultant vector yielding $\tilde{f}$. The desired sequence of length p will be the p values following the first, i.e.,

$$f_p(i) = \tilde{f}(i + 1), \qquad i = 0, 1, \ldots, p - 1. \tag{5.1}$$

It is required, in general, to perform DFTs of order $N + p$ to determine a p length sequence f when g is nonzero for a length N. Three transforms are necessary. Therefore about $3(N + p)\log_2(N + p)$ complex multiplications and additions are necessary. Direct calculation [Equation (3.23)] requires Np multiplications and additions. For

$$p \equiv kN \tag{5.2}$$

the number of points N for which the calculation speeds of the two methods are equivalent, under the condition that a complex multiplication–addition takes about four times as long as a real multiplication–addition, is given by the solution to

$$kN/12(k+1) = \log_2(k+1)N. \tag{5.3}$$

A graphical solution of (5.3) shows that the break-even point lies between $N = 128$ and $N = 256$, for $k > \frac{1}{2}$. Therefore, using the FFT method for one-dimensional correlations or convolutions is advised only when $g(i)$ is a fairly long sequence.

Deconvolution

The problem of deconvolution is to determine the sequence $g(i)$ from the sequences $f(i)$ and $h(i)$ assuming the relation (3.23). N is given. A direct application of the DFT will usually lead to an incorrect result.

Consider the matrix form of (3.25). Clearly, one is looking for the generating vector of a circulant matrix, so that the DFT may be applied. One expects, therefore, the relation (3.26) to hold, i.e.,

$$G(r) = \tilde{F}(r)/\tilde{H}(r), \qquad r = 0, 1, \ldots, N + p. \tag{5.4}$$

The transform $\tilde{F}(r)$ must come from a given sequence $f(i)$. However, the $N + p$ dimensional vector $\tilde{f}(i)$ is *not specified* completely from $f(i)$ except under special conditions.

One can show that G may be found directly from (5.4) if the sequence $\tilde{h}$ has the following properties:

(a) $\tilde{h}$ has all nonzero DFT coefficients (i.e., $\tilde{h}$ is not a finite sum of oscillations).

(b) The values $h(-N), h(-N+1), \ldots, h(0)$ are all identically zero. This implies that $f(i) \equiv \tilde{f}(i)$, $i = N, N-1, \ldots, p-1$.

Solutions to the deconvolution problem when h has no interval of length N identically zero may be found by using iteration. The problem is discussed fully in Silverman and Pearson, and Silverman.

Measurement data are, in many cases, changed by the characteristics of the measurement device. This change is often described by a convolution. In this case, one is interested in determining a segment of $h(i)$ given g and $f(i)$.

Once again, a segment of length p may be determined from a form of (5.4), i.e.,

$$\tilde{H}(r) = \tilde{F}(r)/G(r), \qquad r = 0, 1, \ldots, N + p \tag{5.5}$$

only under the conditions

$\tilde{f}_p(-N), \ldots, \tilde{f}_p(0)$ are all identically zero and
g has all nonzero DFT coefficients.

While these conditions limit the use of (5.5), numerous important cases are included.

Applications to Numerical Integration

An application to the method of Clenshaw and Curtis will be found in Section 6.4. For the use of the FFT in the evaluation of Fourier transforms using Laguerre functions, see Section 3.9.1.

In computing the Taylor coefficients of an analytic function by means of a contour integral in the complex plane, Lyness and Sande replace the integral by a trapezoidal approximation and compute the resulting sums by means of the FFT.

Osgood defines a new class of transforms, the lattice point transforms, and shows how an n-dimensional lattice point transform allows one to approximate an n-dimensional convolution integral to greater accuracy than that achieved by using an n-dimensional DFT. Moreover, there is no loss in speed since the FFT can also be applied to the evaluation of these transforms.

References. Lewis [1]; Lyness and Sande [F1]; Osgood [2]; Silverman [1]; Silverman and Pearson [1]; Welch [1].

Applications to Numerical Analysis

Henrici has pointed out the following areas (mainly) in computational complex analysis in which the FFT can be fruitfully applied: calculation of Fourier coefficients using attenuation factors; solution of Symm's integral equation in numerical conformal mapping; trigonometric interpolation; determination of conjugate periodic functions and their application to Theodorsen's integral equation for the conformal mapping of simply and of doubly connected regions; determination of Laurent coefficients with applications to numerical differentiation, generating functions, and the numerical inversion of Laplace transforms; determination of the "density" of the zeros of high-degree polynomials. He also discusses convolution and its application to time series analysis, to the multiplication of polynomials and of large integers, and to fast Poisson solvers.

Delves and his co-workers have based their programs for solving Fredholm linear integral equations of the second kind on the FFT. An important component of these programs involves the multiplication of the Tschebyscheff series representing two functions. Let

$$f(x) \simeq \hat{f}(x) = \sum_{i=0}^{n}{}' a_i T_i(x), \tag{5.6}$$

$$g(x) \simeq \hat{g}(x) = \sum_{i=0}^{n}{}' b_i T_i(x), \tag{5.7}$$

and

$$h(x) = f(x)g(x) \simeq \hat{h}(x) = \sum_{i=0}^{n}{}' c_i T_i(x). \tag{5.8}$$

Then the coefficients c_i are evaluated in a round-about fashion using three applications of the FFT. First $\tilde{f}(x_j)$ and $\hat{g}(x_j)$, $x_j = \cos \pi j/n$, $j = 0, \ldots, n$, are evaluated using the FFT. Then $\hat{h}(x_j) = \hat{f}(x_j)\hat{g}(x_j)$, $j = 0, \ldots, n$. Finally, the FFT is used to evaluate

$$c_i = \frac{2}{n} \sum_{j=0}^{n}{}'' \hat{h}(x_j)T_i(x_j) = \frac{2}{n} \sum_{j=0}^{n}{}'' \hat{h}(x_j)T_j(x_i), \qquad i = 0, \ldots, n. \tag{5.9}$$

They have also applied the FFT to achieve a fast implementation of the global element method for solving elliptic partial differential equations.

References. Delves [3]; Delves and Abd-Elal [1]; Delves and Philips [1]; Henrici [3]; Monro [1].

3.9.5.6 *The Two-Dimensional DFT*

The two-dimensional FFT has found wide use as a tool for image processing by digital computer and for pattern recognition. See Andrews and Pratt, Anuta, Rosenfeld. The two-dimensional DFT is defined by

$$F(r, s) = \sum_{i=0}^{N-1} \sum_{j=0}^{M-1} f(i, j)w^{ir+js}, \qquad 0 \le r \le N-1, \qquad 0 \le s \le M-1. \tag{6.1}$$

Equation (6.1) may be written

$$F(r, s) = \sum_{i=0}^{N-1} \left[\sum_{j=0}^{M-1} f(i, j)w^{js} \right] w^{ir} = \sum_{i=0}^{N-1} f'(i, s)w^{ir}. \tag{6.2}$$

The two-dimensional DFT may be computed, therefore, by fixing i at the N distinct values and taking the M-point DFT N times to obtain the function $f'(i, s)$. This involves about $NM \log_2 M$ complex multiplication–additions using the one-dimensional FFT.

Next, s may be fixed at each of M values and, taking the N point DFT M times, one obtains $F(r, s)$. This involves about another $NM \log_2 N$ operations. Therefore, the two-dimensional FFT requires about

$$MN(\log_2 N + \log_2 M) = NM \log_2 NM \tag{6.3}$$

operations.

One very important application is the case of two-dimensional convolution (here let $M = N$),

$$f(i, j) = \sum_{k=0}^{N-1} \sum_{l=0}^{N-1} g(k, l)h(i - k, j - l), \qquad 0 \le i, j \le p - N \tag{6.4}$$

where $h(k, l)$ is given over $0 \le k, l \le p - 1$, $p > N$.

Here the convolution may be computed if $g(k, l)$ is embedded in zeros (analogous to the one-dimensional case), and a transform of the $p \times p$ array for g is computed; therefore $3p^2 \log_2 p^2$ complex operations are required. Direct computation of (6.4) requires $(p - N)^2N^2$ operations. Thus, including the factor of four as in the part on spectral analysis, one must compare $12p^2 \log_2 p^2$ against $(p - N)^2N^2$. For the important case where $N = p/2$, the advantage in using the FFT method is when p is 64 or more. A discussion on the advantages and disadvantages of two-dimensional DFT convolution may be found in Silverman and Barnea and Silverman.

References. Andrews and Pratt [1]; Anuta [1]; Barnea and Silverman [1]; Nussbaumer [2]; Rosenfeld [1]; Silverman [1].

3.9.5.7 Error in DFT Convolution

Errors in performing a DFT convolution or correlation come from two sources. First, there is the error incurred in sampling continuous functions. Error also results from roundoff in the FFTs and in multiplications in the frequency domain.

The error in approximation of a Continuous Fourier Transform by the DFT comes from three sources. First, one assumes that the function is nonzero only over a finite interval. Should this not be the case, aliasing error occurs. The magnitude of this error can easily be bounded by comparing some norm for the function on the intervals, $(-\infty, 0]$, (T, ∞) with its norm over $(0, T]$. For example, if we use the norm defined by

$$\|X(t)\|_{ab} \equiv \left[\int_a^b X^2(t)\, dt\right]^{1/2}, \tag{7.1}$$

then

$$\text{aliasing error} \le \frac{(\|X(t)\|_{-\infty, 0} + \|X(t)\|_{T, \infty})}{\|X(t)\|_{0, T}}. \tag{7.2}$$

A second aliasing error occurs as a result of working with only a finite number of frequencies.

A third source of error comes from the trapezoidal rule. This is proportional to the second derivative of the function being approximated:

$$E \sim (Nh^3/12)f''(nh). \tag{7.3}$$

The DFT is a reasonable approximation to the Continuous Fourier Transform, therefore, only if all three of these sources of error are small. These three error criteria severely influence the selection of the two parameters Δt and T. The integration interval T must be selected to optimize two things. First, it must cause the error due to aliasing in the time domain to be small by including the major portion of the region of interest. It must also be selected large enough so that the sampling in the frequency domain will be fine enough. The parameter Δt must also be selected to optimize two things. It must be selected to minimize the approximation error of the trapezoidal integration, while at the same time ensuring that the range in the frequency domain is large. Of course, there is a computation cost associated with making T large and Δt small. Therefore, one must take some care to consider error and computational efficiency in the selection of the DFT parameters.

The errors due to roundoff in calculation of the convolution by DFT are somewhat more difficult to analyze. It is evident that if one performs discrete correlation directly, rounding in the multiplicative process can yield an accumulative error proportional to the size M of the correlation sum and the degree of precision of the computer.

An appropriately written DFT program does not multiply when the coefficient is w^0. The fact that each element of the transform is made up from the sum of products of the data with a set of equispaced phase terms implies that the rounding error of the forward FFTs will turn out to have zero mean. This fact will also hold for the case of the inverse transform. This averaging property of the rounding error will usually more than make up for any loss in precision due to the need to use rounded phase terms and perform complex multiplications.

Experience has shown that rounding error does not generally accumulate in the performance of DFT convolutions when floating-point arithmetic is used.

3.10 The Laplace Transform and Its Numerical Inversion

The theory of the Laplace transform is a world of its own; we shall be able to describe only a small portion of the vast literature that has developed.

An integral of the form

$$F(s) = \int_0^\infty e^{-st} f(t)\, dt = L(f) \tag{3.10.1}$$

is called the *Laplace transform* of f. Other forms include the Laplace–Stieltjes transform: $\int_0^\infty e^{-st}\, dg(t)$, where $g(t)$ is in $BV[0, R]$ for all $R > 0$, and the *bilateral Laplace transform* $\int_{-\infty}^{\infty} e^{-st}f(t)\,dt$. In what follows, we limit ourselves to the form (3.10.1). We consider that t is a real variable while s is a complex variable. $F(s)$ is often called the *generating function* while $f(t)$ is the *determining function.*

Generally speaking, a Laplace integral is convergent for all values of s lying in a half plane $\mathrm{Re}(s) > \sigma$, and defines a single-valued analytic function $F(s)$ there. Analytic continuation to a larger region may be possible. If $f(t)$ grows at most exponentially, i.e., if $f(t)$ satisfies an inequality of the form

$$|f(t)| \le Me^{\sigma t} \qquad (t \to \infty) \tag{3.10.2}$$

and if for all $T > 0$ one has

$$\int_0^T |f(t)|\,dt < \infty, \tag{3.10.3}$$

then the Laplace integral (3.10.1) converges for all $\mathrm{Re}(s) > \sigma$. Among the elementary properties of the Laplace transform we cite

Linearity:

$$L(af + bg) = aL(f) + bL(g). \tag{3.10.4}$$

Shift Theorem:

$$L(e^{-at}f(t)) = F(s + a). \tag{3.10.5}$$

Derivative Theorem:

$$L(f^{(n)}(t)) = s^nL(f) - s^{n-1}f(0) - \cdots - f^{(n-1)}(0) \tag{3.10.6}$$

Convolution Theorem:

$$L(f * g) = L(f)L(g) \tag{3.10.7}$$

By the *convolution* $f*g$ is meant

$$f * g = h(t) = \int_0^t f(u)g(t-u)\,du. \tag{3.10.8}$$

Complex Inversion Theorem:

$$\frac{1}{2}(f(t_0^+) + f(t_0^-)) = \lim_{T\to\infty} \frac{1}{2\pi i}\int_{c-iT}^{c+iT} F(s)e^{st_0}\,ds. \tag{3.10.9}$$

Sufficient conditions under which (3.10.9) is valid are that the integral (3.10.1) converge absolutely along the line $\mathrm{Re}(s) = c$ and that $f(t)$ is of bounded variation in some neighborhood of t_0. The integration in (3.10.9) is carried out along the line L: $\mathrm{Re}(s) = c$ in the complex plane. One often sees the inversion formula written in the imprecise form

$$f(t) = (2\pi i)^{-1}\int_{c-i\infty}^{c+i\infty} F(s)e^{st}\,ds. \tag{3.10.10}$$

If one sets $s = c + i\sigma$, $-\infty < \sigma < \infty$, then (3.10.10) becomes the Fourier integral

$$f(t) = \frac{e^{ct}}{2\pi} \int_{-\infty}^{\infty} F(c + i\sigma) e^{i\sigma t} \, dt. \tag{3.10.11}$$

For theoretical material, see Widder. For tables of Laplace transforms, see Erdélyi.

References. Erdélyi *et al.* [2]; Widder [3].

3.10.1 Numerical Computation of the Direct Laplace Transform

To compute the direct Laplace transform, write

$$\int_0^{\infty} e^{-st} f(t) \, dt = \frac{1}{s} \int_0^{\infty} e^{-u} f\left(\frac{u}{s}\right) du \tag{3.10.1.1}$$

and use the Laguerre formula (3.6.3). This yields

$$\int_0^{\infty} e^{-st} f(t) \, dt \approx \frac{1}{s} \sum_{k=1}^{n} w_k f\left(\frac{x_k}{s}\right). \tag{3.10.1.2}$$

This approximation assumes that $f(t)$ is "polynomial in character" over $[0, \infty)$.

If $f(t)$ is bounded and nonconstant, then f is not really of polynomial character. The following device has been used by Berger with good effect. Take a Laguerre formula of very high order (say 500). The Laguerre weights approach zero with tremendous rapidity. Therefore one needs only use the first few abscissas (say the first 20 or 30) in the formula (3.10.1.2).

The G and B transforms (see Section 3.2.2) have been applied to the numerical computation of Laplace transforms.

References. Berger [1]; Gray and Atchinson [3].

3.10.2 Numerical Inversion of the Laplace Transform

In many applications, it is desired to find the determining function $f(t)$ given the generating function $F(s)$, usually along the real axis. This is not always possible theoretically since only a limited class of functions are generating functions. For example, a generating function must be an analytic function. But if there is an $f(t)$ giving rise to an $F(s)$, then apart from minor normalizations, the inverse $f(t)$ will be unique.

Numerical inversion is a notoriously ill-conditioned process. To see why this is so, consider the identity $L(\sin wt) = w/(s^2 + w^2)$. The generating function is always bounded by $1/w$, yet the determining function oscillates between ± 1 no matter how small $1/w$ is. Therefore, a small noise level in $F(s)$ might be amplified considerably when we pass to $f(t)$.

Laplace inversion is the solution of an integral equation of the first type and general methods are applicable. We shall elaborate special methods.

(1) *The method of approximation.* This proceeds by assuming that $f(t)$ can be approximated over $[0, \infty)$ by a finite series of the form

$$f(t) \approx \sum_{i=1}^{n} a_i \zeta_i(t). \tag{3.10.2.1}$$

Assume next that the transforms $L(\zeta_i(t)) = \psi_i(s)$ are known explicitly and are fairly simple. For example, $\zeta_i(t) = e^{\lambda_i t}$. Then

$$F(s) \approx \sum_{i=1}^{n} a_i \psi_i(s). \tag{3.10.2.2}$$

The problem then is to determine constants a_i so that the right-hand side of (3.10.2.2) approximates $F(s)$ over a certain range of values. This approximation may be carried out by interpolation, by least squares approximation, or by approximation in some other norm such as minimax approximation. The range of values or the data points over which we perform the approximation is of some importance here, and one should give heed to the fact that the behavior of $F(s)$ at $s = \infty$ determines that of $f(t)$ at $t = 0$. We can therefore "play" with four things, the special functions ψ_i, n, the norm, and the range in the s variable.

(2) *The method of Bellman, Kalaba, and Lockett.* Make the change of variable $x = e^{-t}$ in (3.10.1) yielding

$$F(s) = \int_0^1 x^{s-1} g(x)\, dx \tag{3.10.2.3}$$

where

$$g(x) = f(-\log x). \tag{3.10.2.4}$$

Now integrate the right-hand side of (3.10.2.3) approximately by a Gauss rule of order n on $(0, 1)$. This yields

$$F(s) \approx \sum_{i=1}^{n} w_i x_i^{s-1} g(x_i), \tag{3.10.2.5}$$

and if we allow s to take on the values $s = 1, \ldots, n$, we obtain the linear system

$$\sum_{i=1}^{n} w_i x_i^k g(x_i) = F(k+1), \qquad k = 0, 1, \ldots, n-1 \tag{3.10.2.6}$$

or, introducing

$$y_i = w_i g(x_i), \qquad i = 1, 2, \ldots, n,$$

$$\sum_{i=1}^{n} x_i^k y_i = F(k+1), \qquad k = 0, 1, \ldots, n-1. \tag{3.10.2.7}$$

If the elements in the inverse of the Vandermonde matrix (x_i^k) are designated by q_{kj}, this system may be solved in the form

$$y_j = \sum_{k=0}^{n-1} q_{kj} F(k+1).$$

By a device similar to that used in Section 2.7.2, one can show that the elements q_{kj} are the coefficients of the fundamental interpolation polynomials $Q_n(x)/(x - x_j)Q_n'(x_j)$ where the $Q_n(x)$ are the Legendre polynomials shifted to [0, 1]. Bellman, Kalaba, and Lockett tabulate the elements of the inverse matrix for $n = 2(1)15$.

An equivalent approach which yields a continuous approximation to $g(x)$, equal to $g(x_i)$ at the Gauss points, is given by Schoenberg. If we let $\mu_i = F(i+1)$, $i = 0, 1, \ldots$, then

$$g(x) \simeq \sum_{\nu=0}^{n-1} c_\nu P_\nu(2x - 1) \tag{3.10.2.8}$$

where

$$c_\nu = (2\nu + 1) \sum_{i=0}^{\nu} (-1)^{\nu+1} \binom{\nu + i}{\nu} \binom{\nu}{i} \mu_i, \qquad \nu = 0, \ldots, n-1 \tag{3.10.2.9}$$

and the P_ν are the (shifted) Legendre polynomials. However, the coefficients of the μ_i in (3.10.2.9) grow rapidly in magnitude and alternate in sign so that there is a progressive loss of accuracy in the computation of the c_ν.

(3) *Gaussing the Bromwich Integral.* According to the Bromwich inversion formula (3.10.9), we have

$$f(t) = (2\pi i)^{-1} \int_L e^{st} F(s)\, ds \tag{3.10.2.10}$$

where one takes L to be any vertical line $\mathrm{Re}(s) = c =$ constant lying to the right of the abscissa of convergence of the original Laplace integral $F(s) = \int_0^\infty e^{-st} f(t)\, dt$. For fixed t, make the change of variable $st = u$ and set

$$G(u) = u^s F(u/t) \tag{3.10.2.11}$$

so that (3.10.2.10) becomes

$$tf(t) = (2\pi i)^{-1} \int_{L*} e^u u^{-s} G(u)\, du \tag{3.10.2.12}$$

where L^* is now $\text{Re}(u) = ct$. Salzer showed it is possible to "Gauss" this complex integral, that is, we can find complex abscissas u_k and complex weights w_k so that

$$(2\pi i)^{-1} \int_{L^*} e^u u^{-s} G(u)\, du = \sum_{k=1}^{n} w_k G(u_k) \tag{3.10.2.13}$$

whenever $G(u)$ is a polynomial of degree $\leq 2n - 1$ in the variable $1/u$. The relevant orthogonal polynomials are

$$p_n(1/u) = (-1)^n e^{-u} u^{n+s-1} \frac{d^n}{du^n}\left(\frac{e^u}{u^{n+s+1}}\right) \tag{3.10.2.14}$$

for which

$$(2\pi i)^{-1} \int_{L^*} e^u u^{-s} p_n(1/u) u^{-k}\, du = 0 \tag{3.10.2.15}$$

for $k = 0, 1, \ldots, n - 1$. Note that the polynomials depend upon s. Tabulations of the abscissas and weights can be found in Stroud and Secrest, Krylov and Skoblya for $s = 1$; Skoblya, Piessens for several additional values of s.

For values of u that are large, the weights are large, reflecting the ill-conditioning of the process.

Piessens has observed further that it is possible to "Kronrod the Gauss" (see Section 2.7.1.1), so that starting with G_n, an n-point rule of precision $2n - 1$, one adds $n + 1$ abscissas and arrives at G_n^*, a $(2n + 1)$-point rule of precision $3n + 1$. Limited tables are in Piessens.

Example (Piessens). Given $F(s) = (1 + s^2)^{-1/2}$. Compute $L^{-1}F$. The determining function is $f(t) = J_0(t)$ (the Bessel function of order zero).

t	Exact Value	G_8	G_8^*
1.0	0.7651 976	0.76525	0.76519
2.0	0.2238 907	0.22403	0.22405
4.0	−0.3971 498	−0.39732	−0.39722
6.0	0.1506 452	0.15104	0.15077
8.0	0.1716 508	0.17162	0.17160
10.0	−0.2459 357	−0.24436	−0.24579
12.0	0.0476 893	0.02722	0.04762
14.0	0.1710 734	0.18925	0.17098
16.0	−0.1748 990		−0.17419
18.0	−0.0133 558		0.00393
20.0	0.1670 246		0.05395

For a recent survey and comparison of some 14 different methods for numerical inversion of Laplace transforms, see Davies and Martin who remark "It can be argued that Laplace transform inversion is still more an art than a science."

References. Bellman, Kalaba, and Lockett [1]; Davies and Martin [1]; Kiefer and Weiss [1]; Krylov and Skoblya [1]; Miller and Guy [1]; Papoulis [1]; Piessens [1–3, 6, 8, F1, A6]; Piessens and Branders [1]; Salzer [4, 6, 7]; Schoenberg [6]; Shirtliffe and Stephenson [1]; Skoblya [1]; Spinelli [1]; Stehfest [A1]; Talbot [1]; Weillon [A1].

Chapter 4

Error Analysis

4.1 Types of Error

In the preceding chapters, we have dealt with errors from time to time, treating them informally. In the present chapter we deal with them on a more systematic basis.

In approximate integration, we usually replace the value of an integral $\int_a^b f(x)\,dx$ by a finite sum $\sum_{i=1}^{n} w_i f(x_i)$, and in so doing we incur two sorts of error. In the first place, there is the *truncation error* E that arises from the fact that the sum is only approximately equal to the integral:

$$\int_a^b f(x)\,dx = \sum_{i=1}^{n} w_i f(x_i) + E. \qquad (4.1.1)$$

In the second place, there is the *roundoff error* R which arises from the fact that we have computed $\sum_{i=1}^{n} w_i f(x_i)$ only approximately, due to the limitation of accuracy of the computer. We have produced and accepted a value Σ^* such that

$$\Sigma^* = \sum_{i=1}^{n} w_i f(x_i) + R. \qquad (4.1.2)$$

The estimate of total error is therefore

$$\left| \int_a^b f(x)\, dx - \Sigma^* \right| \le |E| + |R|. \tag{4.1.3}$$

In the analysis of the present chapter, we assume that f is a function defined mathematically (that is, f does not consist of experimental data) and that we are able to compute f to within the accuracy of the computer word length. We do not present error analyses based either on statistical assumptions of roundoff or on the assumption that the function f is extracted from an ensemble of functions with certain statistical properties.

In practice, roundoff error is usually negligible; but it should be pointed out that with the high speeds of present computers, there is a temptation to use large values of n in the sum (4.1.1) or too small a tolerance ε in the case of an automatic integration and this may take the roundoff error out of the negligible category. For this reason, the principal concern of this chapter is with the truncation error. It is precisely in the area of truncation-error analysis that some of the most brilliant contributions to the theory of approximate integration lie.

In laboratory practice exact error analysis is usually omitted. There are many reasons for this: (1) it is difficult or impossible to carry out; (2) many of the mathematical theorems are irrelevant to machine computation; (3) the estimates are too pessimistic; (4) it is replaced by an automatic or approximate sort of analysis. Error analysis is the tithe that intelligence demands of action, but it is rarely paid.

4.2 Roundoff Error for a Fixed Integration Rule

In this section, we shall analyze the effect of roundoff in the computation of the rule

$$\int_a^b f(x)\, dx \approx \sum_{k=1}^{n} w_k f(x_k). \tag{4.2.1}$$

We are particularly interested in what happens when n becomes large. The analysis below follows, in large measure, the analysis given by Wilkinson of a general roundoff error.

We assume that the right-hand sum in (4.2.1) is computed on a computer that works with t binary (or decimal) digits and that the computation is carried out in *floating-point* arithmetic. We assume, further, that our computer has a single-precision accumulator. If $fl(x_1 + x_2)$ designates the result of adding x_2 to x_1 in the floating-point mode, then it can be shown that

$$fl(x_1 + x_2) = x_1(1 + \varepsilon') + x_2(1 + \varepsilon''), \tag{4.2.2}$$

where

$$|\varepsilon'|, \ |\varepsilon''| \leq (\tfrac{3}{2})2^{-t} \tag{4.2.3a}$$

if one uses a binary machine, or

$$|\varepsilon'|, \ |\varepsilon''| \leq (5.5)10^{-t} \tag{4.2.3b}$$

if one uses a decimal machine.

If $fl(x_1 x_2)$ correspondingly is the result of multiplying x_2 by x_1 in floating, then it can be shown that

$$fl(x_1 x_2) = x_1 x_2(1 + \varepsilon'''), \tag{4.2.4}$$

where

$$|\varepsilon'''| \leq 2^{-t} \tag{4.2.5a}$$

for a binary machine, or

$$|\varepsilon'''| \leq \tfrac{1}{2}(10^{1-t}) \tag{4.2.5b}$$

for a decimal machine.

With these basic inequalities at hand, we are in a position to analyze the error made in computing a sum of products (i.e., an inner product) such as that which appears in (4.2.1).

We let

$$s_n = fl(a_1 b_1 + a_2 b_2 + \cdots + a_n b_n). \tag{4.2.6}$$

By this is meant that we compute the products $a_i b_i$ in floating and then add them in floating in the order indicated. The computation (4.2.6) is really an abbreviation for the following computation indicated recursively:

$$\begin{aligned} s_1 &= t_1 = fl(a_1 b_1), \\ t_r &= fl(a_r b_r), \\ s_r &= fl(s_{r-1} + t_r), \qquad r > 1. \end{aligned} \tag{4.2.7}$$

Using (4.2.4) and (4.2.2), we obtain

$$\begin{aligned} t_r &= a_r b_r(1 + \xi_r), \\ s_r &= s_{r-1}(1 + \eta_r') + t_r(1 + \eta_r''), \end{aligned} \tag{4.2.8}$$

where the quantities ξ_r, η_r', and η_r'' satisfy

$$|\eta_r'|, \ |\eta_r''| \leq \Omega_1, \qquad |\xi_r| \leq \Omega_2, \tag{4.2.9}$$

with

$$\Omega_1 = (\tfrac{3}{2})2^{-t}, \qquad \Omega_2 = 2^{-t}$$

for a binary machine,† or

$$\Omega_1 = (5.5)10^{-t}, \qquad \Omega_2 = \tfrac{1}{2}(10)^{1-t}$$

for a decimal machine.

By recurrence we find

$$s_n = a_1 b_1(1 + \varepsilon_1) + a_2 b_2(1 + \varepsilon_2) + \cdots + a_n b_n(1 + \varepsilon_n), \tag{4.2.10}$$

where

$$\begin{aligned}
(1 + \varepsilon_1) &= (1 + \xi_1)(1 + \eta'_2)\cdots(1 + \eta'_n),\\
(1 + \varepsilon_r) &= (1 + \xi_r)(1 + \eta''_r)(1 + \eta'_{r+1})\cdots(1 + \eta'_n), \qquad r = 2, \ldots, n - 1,\\
(1 + \varepsilon_n) &= (1 + \xi_n)(1 + \eta''_n).
\end{aligned} \tag{4.2.11}$$

Using (4.2.9) and (4.2.11), we have

$$\begin{aligned}
(1 - \Omega_2)(1 - \Omega_1)^{n-1} &\le 1 + \varepsilon_1 \le (1 + \Omega_2)(1 + \Omega_1)^{n-1},\\
(1 - \Omega_2)(1 - \Omega_1)^{n-r+1} &\le 1 + \varepsilon_r \le (1 + \Omega_2)(1 + \Omega_1)^{n-r+1},\\
&\qquad r = 2, 3, \ldots, n.
\end{aligned} \tag{4.2.12}$$

This implies the uniform estimate for $r = 1, 2, \ldots, n$:

$$(1 - \Omega_2)(1 - \Omega_1)^{n-r+1} \le 1 + \varepsilon_r \le (1 + \Omega_2)(1 + \Omega_1)^{n-r+1}. \tag{4.2.13}$$

Now

$$\begin{aligned}
(1 + \Omega)^k &= 1 + k\Omega + \frac{k(k-1)}{2!}\Omega^2 + \cdots\\
&= 1 + k\Omega\left(1 + \frac{k-1}{2!}\Omega + \frac{(k-1)(k-2)}{3!}\Omega^2 + \cdots\right)\\
&\le 1 + k\Omega\left(1 + \frac{k}{2!}\Omega + \frac{k^2}{3!}\Omega^2 + \cdots\right)\\
&= 1 + k\Omega\left(\frac{e^{k\Omega} - 1}{k\Omega}\right).
\end{aligned} \tag{4.2.14}$$

Similarly,

$$(1 - \Omega)^k \ge 1 - k\Omega\left(\frac{e^{k\Omega} - 1}{k\Omega}\right). \tag{4.2.15}$$

If we assume that

$$n\Omega_1 \le \tfrac{1}{10} \tag{4.2.16}$$

† On some binary computers, one must take $\Omega_1 = 2^{2-t}$, $\Omega_2 = 2^{1-t}$.

(which will certainly be the case with the n's occurring in approximate integration), then since $(e^{.1} - 1)/.1 \leq 1.06$, we have, from (4.2.13), (4.2.14), (4.2.15),

$$(1 - \Omega_2)[1 - (n - r + 1)\Omega_1(1.06)] \leq 1 + \varepsilon_r$$
$$\leq (1 + \Omega_2)[1 + (n - r + 1)\Omega_1(1.06)]. \tag{4.2.17}$$

Hence

$$|\varepsilon_r| \leq \Omega_2 + (n - r + 1)\Omega_1(1 + \Omega_2)(1.06), \qquad r = 1, 2, \ldots, n. \tag{4.2.18}$$

Now, from (4.2.6) and (4.2.10), we have

$$\sum_{i=1}^{n} a_i b_i - \mathscr{f}\left(\sum_{i=1}^{n} a_i b_i\right) = \sum_{i=1}^{n} a_i b_i \varepsilon_i . \tag{4.2.19}$$

This identity will be applied to $\sum_{i=1}^{n} w_i f(x_i)$. We designate by $\bar{f}_i$ the result of computing $f(x_i)$ in the floating mode. We shall assume that

$$\bar{f}_i = f(x_i)(1 + \theta_i), \qquad i = 1, 2, \ldots, n, \tag{4.2.20}$$

where

$$|\theta_i| \leq \theta, \qquad i = 1, 2, \ldots, n, \tag{4.2.21}$$

and θ is a small quantity of the order of magnitude of Ω_1. (This assumption may not be fulfilled in practice. In the first place, there may be roundoff error in the computation of the abscissas and, as we run through the abscissas, the error may propagate, affecting the computation of $f(x)$ adversely. In the second place, f itself may be given by a complicated formula and substantial roundoff error may result. In computing higher dimensional integrals these errors are compounded even more. A larger number of points are used and care must be taken in interpreting the results.) We have

$$R = \sum_{i=1}^{n} w_i f(x_i) - \mathscr{f}\left(\sum_{i=1}^{n} w_i \bar{f}_i\right). \tag{4.2.22}$$

By (4.2.19), we obtain

$$R = \sum_{i=1}^{n} w_i f(x_i) - \sum_{i=1}^{n} w_i \bar{f}_i + \sum_{i=1}^{n} w_i \bar{f}_i \varepsilon_i . \tag{4.2.23}$$

Therefore, by (4.2.20) and (4.2.21), it follows that

$$|R| \leq \theta \sum_{i=1}^{n} |w_i| \, |f(x_i)| + (1 + \theta) \sum_{i=1}^{n} |w_i| \, |f(x_i)| \, |\varepsilon_i| . \tag{4.2.24}$$

If we set

$$M = \max_{a \le x \le b} |f(x)|, \tag{4.2.25}$$

then (4.2.24) reduces to

$$|R| \le \theta M \sum_{i=1}^{n} |w_i| + (1+\theta) M \sum_{i=1}^{n} |w_i|\,|\varepsilon_i|, \tag{4.2.26}$$

and, from (4.2.18),

$$|R| \le \theta M \sum_{i=1}^{n} |w_i| + (1+\theta) M \Omega_2 \sum_{i=1}^{n} |w_i| + (1.06)(1+\theta) M \Omega_1 (1+\Omega_2) \sum_{i=1}^{n} |w_i| (n-i+1). \tag{4.2.27}$$

This is as far as we can carry the analysis without further assumptions as to the nature of the rule (4.2.1). At this point we therefore assume that

$$w_i \ge 0, \qquad \sum_{i=1}^{n} w_i = b - a, \quad \text{and} \quad w_i \le \frac{A}{n}, \qquad i = 1, 2, \ldots, n, \tag{4.2.28}$$

where A is a constant independent of n. These assumptions are fulfilled for the trapezoidal, Simpson's, and all compound rules formed from simple rules with positive coefficients. We have from (4.2.27) our final estimate:

$$|R| \le \theta(b-a)M + (1+\theta)M\Omega_2(b-a) + 1.06(1+\theta)M\Omega_1(1+\Omega_2)A\frac{(n+1)}{2}. \tag{4.2.29}$$

This analysis was based on exact weights. But if the relative error in each weight is bounded by θ', then the term $M(1+\theta)n\theta'$ should be added to the right-hand sides of (4.2.27) and (4.2.29).

The conclusion to be reached from this analysis is that roundoff error in approximate integration can be expected to grow, at worst, as the first power of n. In ordinary circumstances, $n \approx 10^1$ to 10^3, so that no great damage is done. Some machines have the facility of accumulating inner products in double precision. In such a case, the roundoff will be even less. On the other hand, if vast numbers of abscissas are used (as sometimes occurs in multidimensional integrals), some thought must be given to the roundoff error.

The appearance of $|w_i|$ and $\sum_{i=1}^{n} |w_i|$ in formula (4.2.27) shows that rules with both plus and minus weights are less favorable in regard to roundoff. The quantity $\sum_{i=1}^{n} |w_i|$ in such instances may become quite large as $n \to \infty$. However, if we are dealing with a family of integration rules that converges to the value of the integral for all continuous functions, then the theorem of Pólya (Section 2.7.8) shows that the sum $\sum_{i=1}^{n} |w_i|$ is bounded as $n \to \infty$. The error estimate just carried out is conservative in that it assumes the worst possible build-up of error. A statistical approach might yield an estimate $\sim n^{1/2}$.

Lapshin has carried out a statistical investigation of roundoff error in evaluating the rectangular rule (2.1.2). He assumes he is computing on a fixed-point machine with smallest number ε. Certain statistical assumptions about independence are made and that $\max |f(x)| < 1$, $A_1 \le f''(x) \le A_2$. Then, if the number of points N in the rule is such that $\varepsilon = N^{-\beta}$ for some $1 < \beta < 2$, the expected error is $O(N^{2-\beta})$. This, of course, is asymptotically smaller than (4.2.29).

The total error committed in the approximation (4.2.1) consists of the roundoff error plus the truncation error. For the usual rules employed, the truncation error is $\sim n^{-1}$ or better. As n increases, roundoff goes up and truncation goes down, the most favorable selection of n being at some intermediate value.

"Interval analysis" may also be applied to the discussion of roundoff error. See Nickel, Moore.

References. Henrici [2, pp. 41–58]; Lapshin [1]; Lyness [7, 8, F1]; McCracken and Dorn [1, pp. 166–171], Moore [1]; Nickel [1]; Wilkinson [1, pp. 7–19].

Example. Integrate $\int_0^1 \sin \pi x \, dx$, using the trapezoidal rule (see Section 2.1).

Number of intervals	Total error
2	.1366 1977
4	.0330 6640
8	.0082 0237
16	.0020 4671
32	.0005 1149
64	.0001 2801
128	.0000 3223
256	.0000 0833
512	.0000 0333
1024	.0000 0187
2048	.0000 0195
4096	.0000 0934
8192	.0000 1950

In this example, $a = 0$, $b = 1$, $A = 1$, $\theta = \Omega_1 = 5.5 \times 10^{-8}$, $\Omega_2 = \frac{1}{2}(10^{-7})$, $M = 1$. Hence an upper bound for $|R|$ is approximately $[.53(n + 1) + 2]5.5 \times 10^{-8}$. The truncation error E is bounded by $|E| \le \pi^2/12n^2$. For $n = 8192$, $|E| \approx 1.2 \times 10^{-8}$, so that the observed error at this stage is largely roundoff. The predicted roundoff error bound at $n = 8192$ is $|R| \le .00024$, and this overestimates the observed error of .00002 by an order of magnitude. Note that the total error bound is of the form $c_1 + c_2 n + c_3 n^{-2}$ and that the predicted behavior of steady improvement followed by deterioration due to roundoff is borne out.

4.2.1 *Alleviation of Roundoff Error*

(a) *Roundoff error in computing abscissas.* Equally spaced abscissas $a + jh$, $h = (b - a)/n, j = 0, 1, \ldots, n$ are very often employed in the computation of integrals and in the solution of differential equations. Commonplace programming programs this recursively as

$$a + jh = (a + (j - 1)h) + h. \tag{4.2.1.1}$$

But it is better to set

$$a + jh = a + [j(b - a)]/n. \tag{4.2.1.2}$$

at the expense of one multiplication and one division.

Here, roughly, is the reason for the improvement. In the first method, the machine does not compute $h = (b - a)/n$ exactly, but $\bar{h} = h + \varepsilon$ where ε designates a roundoff error. Hence, $a + jh$ is computed as $a + jh + j\varepsilon$ with a possible error buildup of $n\varepsilon$. In the second method $j(b - a)$ is computed as $j(b - a) + \varepsilon$ so that $a + [j(b - a)]/n$ is computed as

$$a + [j(b - a) + \varepsilon]/n + \varepsilon = a + jh + (\varepsilon/n) + \varepsilon.$$

The error is now only $(1 + 1/n)\varepsilon$.

A more careful analysis of the error of (4.2.1.2) using (4.2.2), (4.2.4), and a similar equation for division, namely

$$fl(x_1/x_2) = x_1/x_2(1 + \varepsilon''')$$

where ε''' is bounded as in (4.2.5), and assuming that

$$fl(b - a) = (b - a)(1 + \varepsilon''''),$$

leads to the result

$$a + jh - fl\left[a + \frac{j(b - a)}{n}\right] = a\varepsilon_4 + \frac{j(b - a)}{n}(\varepsilon_1 + \varepsilon_2 + \varepsilon_3 + \varepsilon_5)$$

where $|\varepsilon_i| \le 2^{-t}, i = 1, 2, 3,$ $|\varepsilon_i| \le (\frac{3}{2})2^{-t}, i = 4, 5.$

On the other hand, if we were to compute $a + jh$ by

$$a + jh = [(n - j)a + jb]/n \tag{4.2.1.3}$$

we would find that the error has the form

$$\frac{(n - j)a}{n}(\varepsilon_1 + \varepsilon_3 + \varepsilon_5) + \frac{jb}{n}(\varepsilon_2 + \varepsilon_3 + \varepsilon_4).$$

Whereas (4.2.1.2) emphasizes accuracy in the vicinity of the point a at the expense of accuracy near point b, (4.2.1.3) spreads the error more evenly throughout the interval.

(b) *Errors in the computation of function values.* Assuming that the computed abscissas are acceptable, every effort should be made to provide good function values $f(x_k)$. It is beyond the scope of this book to discuss algorithms and programs for functions, but several remarks will be made. Only certified or time-tested functional subroutines should be used. (Hair-raising stories have been told of subroutines that are "inaccurate at many an argument.") Roundoff errors in function values can cause serious difficulties in certain automatic integration routines (see Sections 4.2.3, 4.2.4, and 6.2).

Furthermore, the reader should be cognizant of the fact that functional expressions that are equivalent mathematically may not be equivalent computationally even when "certified" subroutines are used.

In the following eye-opening example, the same sum employed in approximating an integral was inadvertently computed in three different ways and three widely differing answers were obtained.

Example

$$f_1(x) = \frac{\sin x}{x}, \qquad f_2(x) = \left(\frac{\sin x}{x}\right)^{1.0}, \qquad f_3(x) = \frac{\sin x}{x^{1.0}}.$$

$$s_i = h \sum_{m=-4096}^{4096} (e^{mh} + e^{-mh}) z_m f_i(z_m), \qquad i = 1, 2, 3,$$

where

$$z_m = \exp(e^{mh} - e^{-mh}), \qquad h = 1.5625 \times 10^{-3}.$$

s_1	s_2	s_3
1.5216	29.9713	2.4559

Finally, it is our opinion that certain high-level computer languages such as APL offer the user seductively easy programming possibilities, and by concealing many numerical processes lead inevitably away from hi-fi numerical practice.

(c) *Roundoff error in forming sums.* A number of methods can be employed to reduce roundoff error when the number of summands is very large. The methods are based upon the fact that when a small number is added to a large number a part of the accuracy inhering in the former will be lost. Successive summands in the integration rules are of the same order of magnitude and hence should be combined one with the other. Set $S = \sum_{j=1}^{p} a_j$.

Method A. This is the commonplace method. Write $t_1 = a_1$,

$$t_{k+1} = t_k + a_{k+1}, \qquad k = 1, 2, \ldots, p-1, \qquad S = t_p. \tag{4.2.1.4}$$

Method B. This method should be clear from the accompanying diagram.

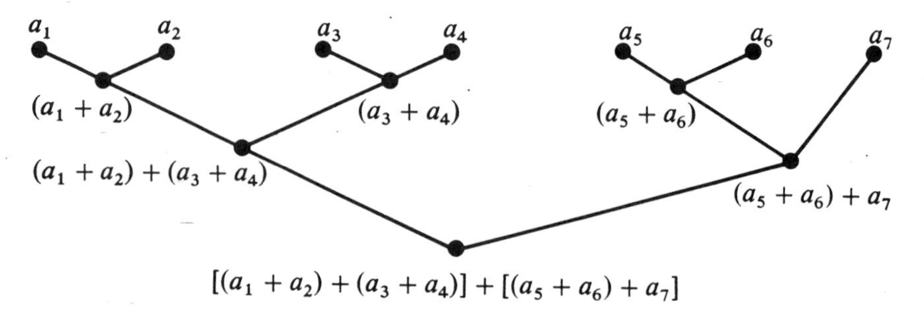

Method C. Instead of accumulating the sums by twos as in Method B, accumulate the sums at any level on the diagram, e.g., by sixteens or by any convenient power of 2.

Method D.

$$t_1 = a_1, \qquad t_{k+1} = t_k + a_{k+1}, \qquad k = 1, 2, \ldots, p - 1$$

$$\eta_{k+1} = t_{k+1} - t_k, \qquad \varepsilon_{k+1} = \eta_{k+1} - a_{k+1}$$

$$w_1 = 0, \qquad w_{k+1} = w_k - \varepsilon_{k+1}$$

$$S = t_p + w_p \tag{4.2.1.5}$$

What is behind this work? When a_{k+1} is added to t_k, only its high accuracy part affects the sum. This high accuracy part is $\eta_{k+1} = t_{k+1} - t_k$. The low accuracy part of a_{k+1} is therefore $a_{k+1} - \eta_{k+1} = -\varepsilon_{k+1}$. The last two lines of the method accumulate the low accuracy parts as w_p and add them to t_p.

Method E. A similar scheme is given by the recurrence

$$t_0 = 0, \qquad \varepsilon_0 = 0$$

$$\bar{a}_k = a_k + \varepsilon_{k-1}$$

$$u_k = t_{k-1} + \bar{a}_k$$

$$\varepsilon_k = (t_{k-1} - u_k) + \bar{a}_k$$

$$S = u_p$$

It should be observed that while methods A, B, and C require the same number of additions, B has a larger storage requirement than A. The storage requirements of method C are correspondingly less. Methods D and E require about four times the number of additions as A, B, and C. It is possible to define a measure w of the amount of accuracy lost through roundoff, and it can be shown that in method A, $w \sim p$. In method B, $w \sim \lceil \log_2(p) \rceil + 1$.

Here $\lceil x \rceil$ designates the "ceiling" of x, i.e., the smallest integer $\geq x$. In methods D and E, w is independent of p. (In D, $w \leq 3$.)

Stummel has applied to methods A and B a new rounding error analysis using the linearization method and new condition numbers. He shows that method B, called cascade summation, can be analyzed easily and concisely by his means.

The programmer who contemplates using roundoff alleviators of type D or E should weigh the advantages chalked up by an almost imperceptible and bounded roundoff against the advantages of merely accumulating the sums in double precision by the simple method A. Method E has been built into "QUAD" (Appendix 2). Fairweather has built method C into a Romberg integration algorithm.

Comparisons of methods A, B, D on the sum $\sum_{k=1}^{p+1} 1/k = S_p = S$ performed on the ICT 1905 by Babuška are revealing. We tabulate $S_p(\text{computed}) - S_p$.

p	Method A	Method B	Method D
64	4.8×10^{-11}	-1.2×10^{-10}	5.8×10^{-11}
2048	4.0×10^{-11}	-1.2×10^{-10}	1.2×10^{-10}
16384	-3.9×10^{-9}	-2.3×10^{-10}	1.2×10^{-10}
262,144	-3.8×10^{-8}	-3.5×10^{-10}	1.2×10^{-10}
524,288	-4.7×10^{-8}	-4.7×10^{-10}	—

Comparisons between methods A and C are given by Wallick. See also Fairweather and Keast.

References. Babuška [1]; Caprani [1]; Dunkl [1]; Fairweather [F1]; Fairweather and Keast [1, 2]; Gram [A1]; Hutchinson [1]; Kahan [1]; Linz [1]; Malcolm [1]; Møller [1, 2]; Nickel [2]; Rutishauser [3]; Stummel [1, 2]; Wallick [1].

4.2.2 *Equally Weighted Integration Rules*

When working to low accuracy, either because of a short word length or because of noise in the function values, an integration rule with equal weights, usually called a Chebyshev-type rule, has certain advantages. Such a rule takes the form

$$If \equiv \int_a^b w(x)f(x)\,dx \simeq C_n \sum_{i=1}^{n} f(x_i) \tag{4.2.2.1}$$

rather than the usual form $If \simeq \sum_{i=1}^{n} w_i f(x_i)$. Using (4.2.2.1), we see that we have replaced an inner product by a sum so that a bound on the generated roundoff error is smaller. Since we generally have $\sum_{i=1}^{n} w_i = C_n$, the bound on the propagated error caused by errors in the function values $f(x_i)$ is also minimized when all the weights w_i are equal.

We have already come across equally weighted rules, namely the rectangular and midpoint rules as well as the Gauss–Chebyshev rule. Additional examples are the composite Gauss 2-point rule, the trapezoidal rule for periodic functions, and the rules of sampling type for multiple integrals (Section 5.9).

The construction of "best" Chebyshev-type rules as well as almost-Chebyshev-type rules in which all function values except a few are multiplied by the same weight has been the subject of many papers. The recent work has been surveyed by Gautschi who also gives many references to the classical results.

Reference. Gautschi [7].

4.2.3 Computational Pitfalls

In addition to problems of roundoff error, one must be aware of additional computational pitfalls in the numerical evaluation of integrals and related computations. These include overflow, underflow, and loss of significance.

a. Overflow

In the computation of integrals with singularities, it may occur that the evaluation of the integrand $f(x)$ at an integration point x_i close to a singularity leads to an overflow. However, if the integration rule is such that the sum $\sum_{j=1}^{n} w_j f(x_j)$ is a good approximation to the value of the integral which is assumed to be within range, then $w_i f(x_i)$ will be small in magnitude. Hence, instead of writing a function routine, FUNCTION $F(X)$, to evaluate and return $f(x)$, one should write the function routine in the form FUNCTION $F(X, W)$ which evaluates $wf(x)$ in such a way as to avoid overflow. Note that this may be difficult to implement in some adaptive routines (Section 6.2) since the value of w_j multiplying $f(x_j)$ is usually determined after $f(x_j)$ has been evaluated.

b. Underflow

When using the IMT rule and similar rules (Section 2.9.2), the integration points tend to cluster around the endpoints of the integration interval. Hence, if the integrand in an integral on [0, 1] has a term involving say x^2, then the evaluation of the integrand at a point close to zero can result in underflow when x^2 is computed. If the computer considers underflow to be an error and does not automatically replace every quantity that underflows by zero, then the programmer must anticipate these underflows and program around them. This was shown by Robinson and de Doncker. On the other hand, it may be important to know the underflowed value since it may subsequently be multiplied by a large quantity. In this case, the programmer must rearrange his computation to avoid the underflow in the first place.

c. *Loss of Significance*

Since in many rules generated by nonlinear transformations of the independent variable such as those mentioned in Sections 2.9.2, 3.4.5, and 3.4.6, the integration points tend to cluster around the endpoints of the integration interval, one must be careful not to lose significance if the integrand, say over $[-1, 1]$, involves a factor of the form $(1 - x)^{\alpha}(1 + x)^{\beta}$. In this case, the direct computation of $1 \pm x_i$ for $|x_i|$ close to 1 would be disastrous. One compensation for this loss of significance is the fact that the weight associated with a point close to an endpoint is very small; nevertheless it can cause considerable trouble especially if α or β are negative. The following example shows this.

Example. (Stenger) Evaluate $I = \int_{-1}^{1} (1 - x^2)^{-1/2}\, dx$ in 16 significant figure floating point arithmetic using (3.4.5.2). Take $h = \log 2$, and truncate the sum at $m = k$. If we set $k = 54$, then $z_k = .99999\ 99999\ 99999\ 0$ where

$$z_k = z_k(h) = (e^{kh} - 1)/(e^{kh} + 1).$$

Hence the computed value of $(1 - z_k^2)^{-1/2}$ is $.707\ldots \times 10^8$, whereas the true value computed by the formula $(1 - z_k^2)^{-1/2} = (1 + e^{kh})/2e^{kh/2}$ is $.671\ldots \times 10^8$. Thus, the term

$$h\left\{\frac{2e^{kh}}{(1 + e^{kh})^2}[1 - z_k^2]^{-1/2} + \frac{2e^{-kh}}{(1 + e^{-kh})^2}[1 - z_k^2]^{-1/2}\right\}$$

contributes an error $.554\ldots \times 10^{-10}$ to the numerical approximation

$$\log 2 \sum_{m=-54}^{54} \frac{2^{m+1}}{(1 + 2^m)^2}\left[1 - \left(\frac{2^m - 1}{2^m + 1}\right)^2\right]^{-1/2}$$

of I. This means that we can achieve only 10 significant figures of accuracy. If we had taken $k = 58$, the situation would have been worse. In that case z_k is computed to be $1.00000\ 00000\ 00000$ so that an error message results when the computer tries to evaluate $(1.0 - 1.0)^{-1/2}$.

In this case, the difficulty can be easily remedied by computing $(1 - z_m)^{-1/2}$ by means of the expression $(1 + e^{mh})/2e^{mh/2}$. See also Takahasi and Mori for a similar substitution when using the DE rule (3.4.5.10).

We illustrate two other cases where there is a loss of significance and show how alternative expressions can be used to overcome the problem. A third example occurs in the computation of the coefficients in Filon's method as discussed in Section 2.10.2.

Example. (Stenger) In formula (3.4.6.21), the expression for the abscissa $z_k = \log(e^{kh} + \sqrt{1 + e^{2kh}})$ is not accurate enough when $e^{kh} \le .1$. In that case, the formula

$$z_k = e^{kh} - \tfrac{1}{6}e^{3kh} + \tfrac{3}{40}e^{5kh} - \tfrac{5}{112}e^{7kh} + \cdots$$

is preferable.

Example. (Lether) In the computation of the weights of the Gauss integration rules using (2.7.3.8), significant digits are lost when the abscissas are close to the endpoints ± 1. Thus, the relative error in the computation of w_{1n} for $n = 256$ was $.29 \times 10^{-10}$ in the IBM 370 computer which has an accuracy of close to 16 significant digits. Using the alternative representation

$$w_{kn} = 2 \Bigg/ \left\{1 + 3x_{kn}^2 + \sum_{j=2}^{n-1} (2j + 1)[P_j(x_{kn})]^2\right\}$$

(cf. (2.7.5.9)), the relative error for the same weight was less than $.20 \times 10^{-13}$. This improvement in accuracy is achieved at a cost of about n^2 additional multiplications.

References. Lether [9]; Robinson and de Doncker [1]; Stenger [9]; Takahasi and Mori [4]; Toda and Ono [1].

4.2.4 Rounding Error Detection in Automatic Integration

In most automatic integration schemes (see Chapter 6), the program generates a sequence of approximations $\{Q_n f\}$ to an integral If together with a sequence of error estimates $\{E_n f\}$. It exits when one of the error estimates, say $E_m f$, satisfies some exit criterion. In such a case success is indicated. Or it exits when the computer resources allocated to this program have been exhausted. In such a case failure is indicated. Since, in general, the sequence $\{Q_n f\}$ converges to If, the theoretical values of $\{E_n f\}$ should tend to zero if they are realistic.

Eventually, in theory, there will occur an element $E_m f$ which satisfies the exit criterion. However, it may occur that the values of $Q_n f$ and $E_n f$ are affected by noise originating from sources such as roundoff error in the evaluation of $f(x)$, loss of significance, use of empirical data, etc., or that the exit criterion is too strict relative to machine accuracy. In this case, the error estimates may never satisfy the exit criterion and the machine will exit when the resources are exhausted. Furthermore, the approximations $\{Q_n f\}$ may get worse with increasing $n > n_0$ for some small value of n_0. Hence, it is important that the program detect this phenomenon as soon as possible, and that it exit with an indication that noise had been detected, providing the best possible approximation to If under the circumstances.

The first to deal with this problem was Lyness in his program SQUANK (Lyness [F1]). However, since this program was designed for functions $f \in C^4[a, b]$ which were not highly oscillatory, it tended to throw out the baby with the bathwater by attributing noisy behavior to any function for which it was not designed instead of to those functions which were polluted by roundoff error. Another approach was proposed by Piessens and implemented in most of the QUADPACK routines (Piessens *et al.*) as well as other programs developed by his Leuven group. In these automatic integration schemes, $Q_{n+1}f = Q_n f - qf + q_1 f + q_2 f$, where qf is an approximation to the integral of f over some subinterval $[c, d]$ of $[a, b]$ and $q_1 f$, $q_2 f$ are approximations to the integral over say, $[c, m]$ and $[m, d]$, respectively, where m is some point in $[c, d]$, usually $(c + d)/2$. For qf, $q_1 f$, and $q_2 f$, error estimates ef, $e_1 f$, and $e_2 f$, respectively, are available. It is expected that if qf, $q_1 f$, and $q_2 f$ are good estimates of the corresponding integrals, then $qf \simeq q_1 f + q_2 f$. Furthermore, since an integration interval is subdivided in order to reduce the truncation error, we expect that $|e_1 f| + |e_2 f| < |ef|$. However, in the presence of roundoff error, this inequality will not necessarily hold. Hence, Piessens checks if

$$.9999 < \frac{qf}{q_1 f + q_2 f} < 1.0001$$

and $|e_1 f| + |e_2 f| > .99|ef|$. If this occurs in ten steps of the algorithm, the computations are terminated with an error indication.

For another approach, see Blue.

References. Blue [2]; de Doncker and Piessens [1]; Lyness [7, 8, F1]; Piessens [13]; Piessens, de Doncker, Überhuber, and Kahaner [1]; Robinson and de Doncker [1]; Roose and de Doncker [1].

4.3 *Truncation Error*

It should be pointed out that if the analysis of truncation error can be carried out completely, then what we are really dealing with is not an approximate or numerical integration but a complete expansion of a particular integral. Therefore, in most of what follows, we strive for approximate or upper bounds for truncation errors.

Truncation Error through Peano's Theorem

In discussing the truncation error of approximate integration, it is frequently useful to regard the error

$$E = E(f) = \int_a^b f(x)\,dx - \sum_{i=1}^{n} w_i f(x_i) \tag{4.3.1}$$

as a *linear functional* defined over a certain class of functions. Linear functionals E have the property that

$$E(cf(x)) = cE(f(x)) \tag{4.3.2}$$

for any constant c, and

$$E(f(x) + g(x)) = E(f(x)) + E(g(x)). \tag{4.3.3}$$

In the theorem that follows, we consider integrands of class $C^{n+1}[a, b]$ and linear functionals of the following structure:

$$\begin{aligned} E(f) = \int_a^b [a_0(x)f(x) + a_1(x)f'(x) + \cdots + a_n(x)f^{(n)}(x)]\, dx \\ - \sum_{i=1}^{j_0} b_{i0} f(x_{i0}) - \sum_{i=1}^{j_1} b_{i1} f'(x_{i1}) - \cdots - \sum_{i=1}^{j_n} b_{in} f^{(n)}(x_{in}). \end{aligned} \tag{4.3.4}$$

Here it is assumed that the functions $a_i(x)$ are piecewise continuous over $[a, b]$ and that the points x_{ij} lie in $[a, b]$.

For the validity of Peano's theorem, it suffices, however, if f has an nth derivative that is absolutely continuous (see Section 4.3.5) on $[a, b]$. Furthermore, we may take the functional E to be of the form

$$E(f) = \sum_{i=0}^{n} \int_a^b f^{(i)}(x)\, d\mu_i(x)$$

where $\mu_i(x)$ are functions of bounded variation on $[a, b]$.

PEANO'S THEOREM. *Let* $E(p(x)) = 0$ *whenever* $p(x) \in \mathcal{P}_n$. *Then, for all* $f(x) \in C^{n+1}[a, b]$,

$$E(f) = \int_a^b f^{(n+1)}(t)K(t)\, dt, \tag{4.3.5}$$

where

$$K(t) \equiv K_n(t) = \frac{1}{n!} E_x[(x - t)_+^n] \tag{4.3.6}$$

and

$$(x - t)_+^n = \begin{cases} (x - t)^n, & x \geq t, \\ 0, & x < t. \end{cases} \tag{4.3.7}$$

The notation E_x means the linear functional E is applied to the x variable in $(x - t)_+^n$. The function $K(t)$ is called the *Peano kernel* for the linear functional E. It is also called an *influence function* for E.

Proof. By Taylor's theorem with exact remainder,

$$f(x) = f(a) + f'(a)(x - a) + \cdots + \frac{f^{(n)}(a)(x - a)^n}{n!}$$
$$+ \frac{1}{n!}\int_a^x f^{(n+1)}(t)(x - t)^n \, dt. \tag{4.3.8}$$

The integral remainder may be written as

$$\frac{1}{n!}\int_a^b f^{(n+1)}(t)(x - t)_+^n \, dt.$$

We apply E to both sides of equation (4.3.8) and use the fact that E vanishes on polynomials of class $\mathscr{P}_n$:

$$E(f(x)) = \frac{1}{n!} E_x \int_a^b f^{(n+1)}(t)(x - t)_+^n \, dt. \tag{4.3.9}$$

Now the type of functional we are working with allows the interchange of E and the integral; hence,

$$E(f) = \frac{1}{n!}\int_a^b f^{(n+1)}(t) E_x(x - t)_+^n \, dt. \tag{4.3.10}$$

COROLLARY

$$|E(f)| \leq \max_{a \leq x \leq b} |f^{(n+1)}(x)| \int_a^b |K(t)| \, dt. \tag{4.3.11}$$

If the Peano kernel does not change its sign over the interval $[a, b]$, then E may be expressed essentially as $f^{(n+1)}$ evaluated at an intermediate point.

COROLLARY. *If, in addition, $K(t)$ does not change its sign on $[a, b]$, then*

$$E(f) = \frac{f^{(n+1)}(\xi)}{(n + 1)!} E(x^{n+1}), \qquad a < \xi < b. \tag{4.3.12}$$

Proof. Under the additional hypotheses we may use the mean-value theorem for integrals and obtain

$$E(f) = f^{(n+1)}(\xi) \int_a^b K(t) \, dt. \tag{4.3.13}$$

Now we insert $f(x) = x^{n+1}$ in (4.3.13) and obtain

$$E(x^{n+1}) = (n + 1)! \int_a^b K(t) \, dt. \tag{4.3.14}$$

This yields (4.3.12).

Example. If we set

$$E(f) = \int_a^b f(x)\,dx - \frac{b-a}{2}[f(a) + f(b)],$$

then $E(f)$ is the truncation error in the trapezoidal rule. We may select $n = 1$ in Peano's theorem and obtain

$$\begin{aligned} K(t) &= \int_a^b (x-t)_+\,dx - \frac{b-a}{2}[(a-t)_+ + (b-t)_+] \\ &= \int_t^b (x-t)\,dx - \frac{b-a}{2}(b-t) = \frac{1}{2}(a-t)(b-t). \end{aligned} \tag{4.3.15}$$

Therefore

$$\int_a^b f(x)\,dx - \frac{b-a}{2}[f(a) + f(b)] = \frac{1}{2}\int_a^b f''(t)(a-t)(b-t)\,dt. \tag{4.3.16}$$

The kernel $K(t)$ is nonpositive throughout $[a, b]$; hence we may apply the second corollary. Now

$$E(x^2) = \int_a^b x^2\,dx - \frac{b-a}{2}(a^2 + b^2) = -\frac{1}{6}(b-a)^3; \tag{4.3.17}$$

thus

$$\int_a^b f(x)\,dx = \frac{b-a}{2}[f(a) + f(b)] - \frac{1}{12}(b-a)^3 f''(\xi), \qquad a < \xi < b. \tag{4.3.18}$$

Example. Let

$$E(f) = \int_{-1}^{1} f(x)\,dx - \tfrac{1}{3}f(-1) - \tfrac{4}{3}f(0) - \tfrac{1}{3}f(1).$$

Now $E(p(x)) = 0$ if $p(x) \in \mathscr{P}_3$. Hence, we may take $n = 3$ in Peano's theorem. We find that $K(t) = \frac{1}{6}E_x[(x-t)_+^3]$. When this expression is computed out, we find that

$$K(t) = \begin{cases} -\frac{1}{72}(1-t)^3(3t+1), & 0 \le t \le 1, \\ K(-t), & -1 \le t \le 0. \end{cases} \tag{4.3.19}$$

Now $K(t) \le 0$ in $[-1, 1]$ so that the second corollary is applicable. Since $E(x^4) = -\frac{4}{15}$, this leads to the error in Simpson's rule:

$$\int_{-1}^{1} f(x)\,dx = \frac{1}{3}f(-1) + \frac{4}{3}f(0) + \frac{1}{3}f(1) - \frac{f^{(4)}(\xi)}{90}, \qquad -1 < \xi < 1. \tag{4.3.20}$$

For an arbitrary rule of approximate integration

$$\int_a^b f(x)\,dx \approx \sum_{k=1}^{n} w_k f(x_k),$$

there is no reason for the Peano kernel to have one sign.

Example. Consider the error in the Clenshaw–Curtis rule with five abscissas. This is an interpolatory rule with abscissas at $\cos k\pi/4$, $k = 0, 1, \ldots, 4$. See Section 2.5.5. In this case, we have

$$E(f) = \int_{-1}^{1} K(t) f^{(6)}(t)\,dt$$

where

$$K(t) = \frac{1}{6}(1-t)^6 - \frac{1}{15}(1-t)^5, \qquad \frac{\sqrt{2}}{2} \le t \le 1$$

$$= \frac{1}{6}(1-t)^6 - \frac{1}{15}(1-t)^5 - \frac{8}{15}\left(\frac{\sqrt{2}}{2} - t\right)^5, \qquad 0 \le t \le \frac{\sqrt{2}}{2}$$

and

$$K(-t) = K(t) \quad \text{for} \quad -1 \le t \le 0.$$

The kernel $K(t)$ changes sign. See Locher.

If we define $K^+(t) = \max(K(t), 0)$ and $K^-(t) = \max(-K(t), 0)$ so that $K(t) = K^+(t) - K^-(t)$, with $K^+(t), K^-(t) \ge 0$, then we have that

$$E(f) = \int_a^b f^{(n+1)}(t)K^+(t)\,dt - \int_a^b f^{(n+1)}(t)K^-(t)\,dt$$

$$= \lambda_1 f^{(n+1)}(\xi_1) - \lambda_2 f^{(n+1)}(\xi_2)$$

where $\lambda_1 = \int_a^b K^+(t)\,dt$ and $\lambda_2 = \int_a^b K^-(t)\,dt$ are nonnegative, and $\lambda_1 - \lambda_2 = E(x^{n+1})/(n+1)!$. This expression may be useful if $f^{(n+1)}(x)$ does not change sign in $[a, b]$ and $0 < m \le |f^{(n+1)}(x)| \le M$. Other values for λ_1, λ_2 in the expression

$$E(f) = \lambda_1 f^{(n+1)}(\xi_1) - \lambda_2 f^{(n+1)}(\xi_2)$$

are given by Allasia and Allasia. For other ways to deal with the case when the Peano kernel changes sign, see Meek.

For integration rules of Newton–Cotes type, it can be shown that the kernel does have one sign. (See Steffensen.) Similarly for the second Fejér rule (2.5.5.6).

COROLLARY. *If a rule R has a Peano kernel that does not change sign and if R is compounded, then the compound rule has a kernel that does not change sign.*

Proof. For simplicity, we work with two intervals $[a, a+h]$, and $[a+h, a+2h]$. We have

$$\int_a^{a+h} f(x)\,dx = \sum w_k f(x_k) + \int_a^{a+h} f^{(n+1)}(t)K(t)\,dt,$$

$$\int_{a+h}^{a+2h} f(x)\,dx = \sum w_k f(x_k') + \int_{a+h}^{a+2h} f^{(n+1)}(t)K(t-h)\,dt,$$

where $x'_k = x_k + h$. Hence

$$\int_a^{a+2h} f(x)\,dx = \sum w_k f(x_k) + \sum w_k f(x'_k) + \int_a^{a+2h} f^{(n+1)}(t)K_1(t)\,dt,$$

where

$$K_1(t) = \begin{cases} K(t), & a \le t \le a+h, \\ K(t-h), & a+h \le t \le a+2h. \end{cases}$$

The kernel $K_1(t)$ clearly does not change its sign on $[a, a+2h]$. Application of the previous two corollaries leads to error terms for compound rules.

A function such as $(x-t)_+^n$, which is a piecewise polynomial and which has continuous derivatives up to a certain order, is known as a *spline function.* The kernel $K(t)$ is a *monospline,* a function that is not strictly a spline but is intimately related to spline functions. The theory of spline functions has many intimate connections with approximate integration, some of which are indicated in Sections 2.3.2, 3.4, 3.9, and 4.7.3.

The constant

$$c = \int_a^b |K(t)|\,dt \tag{4.3.21}$$

plays an important role in certain theories of error. With this constant, we may write the error estimate (4.3.11) in the form

$$|E(f)| \le c \max_{a \le x \le b} |f^{(n+1)}(x)|. \tag{4.3.22}$$

THEOREM. *The error estimate* (4.3.22) *is best possible in the sense that there exists a function* $f(x)$ *with a bounded, piecewise continuous* $(n+1)$th *derivative for which the inequality is an equality.*

Proof. Define the function sign(x) by

$$\operatorname{sign}(x) = \begin{cases} 1 & \text{if} \quad x > 0, \\ 0 & \text{if} \quad x = 0, \\ -1 & \text{is} \quad x < 0. \end{cases}$$

Since $K(x)$ is a monospline, the function sign $K(x)$ is bounded and piecewise continuous in $[a, b]$. Now, solve the differential equation

$$f^{(n+1)}(x) = \operatorname{sign} K(x) \tag{4.3.23}$$

by integrating sign $K(x)$ $n+1$ times using arbitrary constants of integration. Note that since $K(x) \not\equiv 0$, $\max_{a \le x \le b} |f^{(n+1)}(x)| = 1$. From (4.3.5),

$$E(f) = \int_a^b f^{(n+1)}(x)K(x)\,dx = \int_a^b (\text{sign } K(x))K(x)\,dx$$

$$= \int_a^b |K(x)|\,dx = c \max_{a \le x \le b} |f^{(n+1)}(x)|.$$

A second version of Peano's theorem and concomitant corollary appears in Brass. It states that if $E(p(x)) = 0$ whenever $p(x) \in \mathscr{P}_n$, if $f^{(n)}(x) \in BV[a, b]$ and if, when $n = 0$, $f \in C[a, b]$, then

$$E(f) = \int_a^b K_{n+1}(t)\,df^{(n)}(t). \tag{4.3.24}$$

Furthermore,

$$|E(f)| \le V_a^b(f^{(n)}) \max_{a \le t \le b} |K_{n+1}(t)| \tag{4.3.25}$$

where the bound in (4.3.25) is sharp. (See Section 1.6.5 for the notation.)

For additional properties of the Peano kernel, see Stroud [19], Brass, and Engels.

For the development of Peano's theorem in several dimensions, the reader is referred to the work of Sard, some of which appears also in Stroud [18] and Engels. See also Barnhill and Pilcher.

References. Allasia and Allasia [1]; Allasia and Giordano [1]; Allasia and Patrucco [1]; Barnhill and Pilcher [1]; Brass [1, pp. 39–46]; Brass and Schmeisser [1]; Davis [6, pp. 69–75]; Engels [13, pp. 93–116]; Krylov [1, pp. 82–92]; Locher [2]; Meek [1]; Milne [1]; Nikolskii [1]; Sard [2, Chap. 4]; Schoenberg [1]; Steffensen [1, pp. 154–165]; Stroud [18, Chap. 5]; Stroud [19, 163–186].

4.3.1 Peano's Kernels for Gauss Rules

If we set $E(f) = \int_{-1}^{1} w(x)f(x)\,dx - \sum_{k=1}^{n} w_k f(x_k)$, $-1 \le x_k \le 1$, and if we assume that $E(f) = 0$ for $f \in \mathscr{P}_r$, then by Peano's theorem in Section 4.3, the Peano kernel for E is given explicitly by

$$K_r(t) = \frac{1}{r!} E_x[(x - t)_+^r]$$

or

$$r!\,K_r(t) = \int_{-1}^{1} w(x)(x - t)_+^r\,dx - \sum_{x_k > t} w_k(x_k - t)^r. \tag{4.3.1.1}$$

In the special case $w(x) = 1$,

$$r!\,K_r(t) = \frac{(1-t)^{r+1}}{r+1} - \sum_{x_k > t} w_k(x_k - t)^r. \tag{4.3.1.2}$$

Let now $\{x_k\}$ and $\{w_k\}$ be the abscissas and weights for the Gauss rule of order n with respect to $w(x)$ (Section 2.7). This rule is exact for $f \in \mathscr{P}_{2n-1}$. Therefore for each r, $r = 0, 1, \ldots, 2n-1$, we may find a kernel of order r and an error estimate of type (4.3.11). The constants

$$e_r = \int_{-1}^{1} |K_{r-1}(t)|\,dt, \qquad r = 1, 2, 2n \tag{4.3.1.3}$$

have been computed and tabulated by Stroud and Secrest.

Example. For $n = 10$, $r = 2$, it is found that $e_2 = .003859$. This implies the error estimate

$$|E_{G_{10}}(f)| \le .003859 \sup_{-1 \le x \le 1} |f''(x)|$$

for all functions f with bounded piecewise continuous second derivatives.

Stroud recommends the use of the error estimate (4.3.11) with appropriately low n for functions with low order continuity. He uses this estimate to show that Gauss rules of high order are effective for such functions despite the fact that the "standard" error term (2.7.11) involves a high order derivative.

References. Gautschi [9]; Roghi [1]; Stroud [8]; Stroud and Secrest [2].

4.3.2 Peano Kernels for Other Rules

For the Peano kernel for Filon's rule, see Luke, Fosdick, Chase and Fosdick. For Romberg rules, see Stroud. For Laguerre rules, see Stroud and Chen.

References. Fosdick [1]; Chase and Fosdick [1]; Luke [2]; Stroud [7]; Stround and Chen [1].

4.3.5 Rules and Error through the Theory of Linear Differential Equations

The method expounded in this section originated in work by von Mises and by Radon. It consists essentially of a glorified form of integration by parts and was subsequently brought to a high degree of perfection in the work of Ghizzetti and Ossicini. For simplicity, we limit ourselves to the case of the finite interval $-\infty < a \le x \le b < \infty$. $L_1[a, b]$ designates the class of Lebesgue integrable functions on $[a, b]$.

DEFINITION. A function $f(x)$ is said to be *absolutely continuous* in $[a, b]$ ($f \in AC[a, b]$) if $f(x) = \int_a^x g(t)\,dt$ where $g(x) \in L_1[a, b]$. A function $f(x) \in AC^k[a, b]$ if $f^{(k)}(x) \in AC[a, b]$, $k = 0, 1, \ldots$.

If a function satisfies a Lipschitz condition on $[a, b]$, then it is in $AC[a, b]$ and this case is sufficient in practical applications.

We consider a differential operator of nth order

$$L \equiv D^n + a_1(x)D^{n-1} + \cdots + a_{n-1}(x)D + a_n(x). \tag{4.3.5.1}$$

Here $D = d/dx$ and we assume that $a_k(x) \in AC^{n-k-1}[a, b]$, $k = 1, 2, \ldots, n-1$, $a_n(x) \in L_1[a, b]$. The *adjoint* differential operator L^* is defined by

$$L^* \equiv (-1)^n D^n + (-1)^{n-1} D^{n-1}[a_1(x)\cdot\,] + \cdots - D[a_{n-1}(x)\cdot\,] + a_n(x). \tag{4.3.5.2}$$

The operators L and L^* may be applied to functions of class $AC^{n-1}[a, b]$ and yield functions of class $L_1[a, b]$.

The operators L_r will be defined by

$$L_r = D^r + a_1(x)D^{r-1} + \cdots + a_{r-1}(x)D + a_r(x), \qquad r = 0, 1, \ldots, n-1, \tag{4.3.5.3}$$

while L_r^* will be the adjoint of L_r.

We now recall *Lagrange's identity*: Let $u(x)$, $v(x) \in AC^{n-1}[a, b]$, then (almost everywhere in $[a, b]$)

$$vL(u) - uL^*(v) = \frac{d}{dx}\sum_{k=0}^{n-1} u^{(k)}(x)L_{n-k-1}^*(v). \tag{4.3.5.4}$$

The sum on the right is often called the *bilinear concomitant* for L. Integrating the Lagrange identity from α to β, we obtain *Green's formula*:

$$\int_\alpha^\beta uL^*(v)\,dx = \int_\alpha^\beta vL(u)\,dx - \left[\sum_{k=0}^{n-1} u^{(k)}(x)L_{n-k-1}^*(v(x))\right]_{x=\alpha}^{x=\beta}. \tag{4.3.5.5}$$

We shall be interested in developing rules of the type

$$\int_a^b w(x)f(x)\,dx = \sum_{i=1}^{m}\sum_{k=0}^{n-1} w_{ki} f^{(k)}(x_i) + E(f). \tag{4.3.5.6}$$

We assume that $w(x) \in L_1[a, b]$, $f(x) \in AC^{n-1}[a, b]$ and $a \le x_1 < x_2 < \cdots < x_m \le b$. We will be interested only in rules (4.3.5.6) for which

$$E(f) = 0 \quad \text{wherever} \quad L(f) = 0. \tag{4.3.5.7}$$

In other words, the right-hand sum is required to integrate exactly all solutions of the linear differential equation $L(f) = 0$. As examples, if we select $L = D^n$, the rule is exact for $\mathscr{P}_{n-1}$. If $L = D\prod_{k=1}^{r}(D^2 + k^2)$, the rule is exact for trigonometric polynomials of order $\leq r$.

Now let $\phi_i(x)$ be *any* solution of $L^*(\phi) = w(x)$. Then, setting $u = f$, $v = \phi_i$, $\alpha = x_i$, $\beta = x_{i+1}$ in (4.3.5.5) we get

$$\int_{x_i}^{x_{i+1}} w(x)f(x)\,dx = \int_{x_i}^{x_{i+1}} \phi_i(x)L(f)\,dx - \left[\sum_{k=0}^{n-1} f^{(k)}(x)L_{n-k-1}^*(\phi_i(x))\right]_{x_i}^{x_{i+1}} \tag{4.3.5.8}$$

Now, write $x_0 = a$, $x_{m+1} = b$ and sum (4.3.5.8) on i. This yields

$$\int_a^b w(x)f(x)\,dx = -\sum_{i=0}^{m}\left[\sum_{k=0}^{n-1} f^{(k)}(x)L_{n-k-1}^*(\phi_i(x))\right]_{x_i}^{x_{i+1}} + \sum_{i=0}^{m}\int_{x_i}^{x_{i+1}} \phi_i(x)L(f)\,dx. \tag{4.3.5.9}$$

Suppose now that $\phi_0(x)$ and $\phi_m(x)$ are particular solutions of the differential equation $L^*(\phi) = w(x)$ determined so that

$$\phi_0^{(k)}(a) = 0, \qquad \phi_m^{(k)}(b) = 0, \qquad k = 0, 1, \ldots, n-1. \tag{4.3.5.10}$$

Then,

$$L_{n-k-1}^*(\phi_0)\,|_{x=x_0=a} = 0, \qquad L_{n-k-1}^*(\phi_m)\,|_{x=x_{m+1}=b} = 0, \qquad k = 0, 1, \ldots, n-1. \tag{4.3.5.11}$$

This allows us to transform (4.3.5.9) into

$$\int_a^b w(x)f(x)\,dx = \sum_{i=1}^{m}\sum_{k=0}^{n-1}[L_{n-k-1}^*(\phi_i(x) - \phi_{i-1}(x))]_{x=x_i} f^{(k)}(x_i) + \sum_{i=0}^{m}\int_{x_i}^{x_{i+1}} \phi_i(x)L(f)\,dx. \tag{4.3.5.12}$$

If we set

$$K(x) = \phi_i(x) \quad \text{for} \quad x_i \leq x \leq x_{i+1}, \quad i = 0, 1, \ldots, m, \tag{4.3.5.13}$$

and

$$w_{ki} = [L_{n-k-1}^*(\phi_i(x) - \phi_{i-1}(x))]_{x=x_i}, \qquad k = 0, 1, \ldots, n-1, \quad i = 1, 2, \ldots, m. \tag{4.3.5.14}$$

we can write (4.3.5.12) as

$$\int_a^b w(x)f(x)\,dx = \sum_{i=1}^{m}\sum_{k=0}^{n-1} w_{ki} f^{(k)}(x_i) + E(f) \tag{4.3.5.15}$$

with

$$E(f) = \int_a^b K(x)L(f)\,dx. \tag{4.3.5.16}$$

It is clear from (4.3.5.16) that $E(f) = 0$ whenever f satisfies the differential equation $L(f) = 0$; more generally, $E(f) = 0$ whenever $L(f)$ is orthogonal to the kernel $K(x)$.

Summarizing, let $\phi_0(x)$ and $\phi_m(x)$ be solutions of $L^*(\phi) = w(x)$ that satisfy the initial conditions (4.3.5.11). Let $\phi_1(x), \ldots, \phi_{m-1}(x)$ be any $m - 1$ solutions of $L^*(f) = w(x)$. Let constants w_{ki} be determined by (4.3.5.14). Then (4.3.5.15) holds for any $f \in AC^{n-1}[a, b]$. It can be shown, conversely, that any rule of type (4.3.5.15), exact for solutions of $L(f) = 0$, can be obtained in this manner.

The book of Ghizzetti and Ossicini develops this method systematically. Extensions are made to the semi-infinite and infinite intervals. All the "usual" rules are obtained and studies are made of the sign of the kernel $K(x)$. Rules are obtained which are exact for trigonometric and hyperbolic polynomials.

References. Ghizzetti and Ossicini [1]; Radon [1]; von Mises [1, 2].

4.4 Special Devices

Peano's theorem is of great generality, but the price one pays for this is that the kernel $K(t) = E_x[(x - t)_+^n]$ may be difficult to convert to a neat, easily handled form. Furthermore, it is difficult to compute the constants $\int |K(t)|\,dt$ for high-order rules. Other devices are available. We shall exhibit the method used to obtain the error in the Gaussian integration formula.

Let $w(x)$ be an admissible weight function defined on $[a, b]$, let $p_n(x)$ be the corresponding orthogonal polynomials, and let

$$p_n^*(x) = k_n x^n + \cdots, \qquad k_n > 0,$$

be the orthonormal polynomials. The numbers x_k and w_k are the abscissas and weights of the Gauss rule of order n, related to $p_n(x)$ as explained in Section 2.7.

THEOREM. *If* $f(x) \in C^{2n}[a, b]$, *then*

$$E_{G_n}(f) = \int_a^b w(x)f(x)\,dx - \sum_{k=1}^{n} w_k f(x_k) = \frac{f^{(2n)}(\eta)}{(2n)!\,k_n^2}, \qquad a < \eta < b. \tag{4.4.1}$$

Proof. Let $h_{2n-1}(x)$ be that unique polynomial of class $\mathscr{P}_{2n-1}$ for which $h_{2n-1}(x_k) = f(x_k)$, $h'_{2n-1}(x_k) = f'(x_k)$, $k = 1, 2, \ldots, n$. Then, according to the remainder theorem for polynomial interpolation (see, for example, Davis), we have

$$f(x) = h_{2n-1}(x) + \frac{f^{(2n)}(\xi(x))}{(2n)!}(x - x_1)^2(x - x_2)^2 \cdots (x - x_n)^2 \tag{4.4.2}$$

for $a \leq x \leq b$ and $a < \xi(x) < b$. The function

$$\frac{(2n)!\,(f(x) - h_{2n-1}(x))}{(x - x_1)^2 \cdots (x - x_n)^2} = f^{(2n)}(\xi(x))$$

is continuous in $a \leq x \leq b$. Now, multiply (4.4.2) by $w(x)$:

$$w(x)f(x) = w(x)h_{2n-1}(x) + \frac{f^{(2n)}(\xi(x))}{(2n)!}\,w(x)\,\frac{p_n^*(x)^2}{k_n^2}. \tag{4.4.3}$$

Integrate (4.4.3) and employ the general mean-value theorem for integrals:

$$\begin{aligned}\int_a^b w(x)f(x)\,dx &= \int_a^b w(x)h_{2n-1}(x)\,dx + \frac{1}{(2n)!\,k_n^2}\int_a^b f^{(2n)}(\xi(x))w(x)[p_n^*(x)]^2\,dx \\ &= \int_a^b w(x)h_{2n-1}(x)\,dx + \frac{f^{(2n)}(\eta)}{(2n)!\,k_n^2}\int_a^b w(x)[p_n^*(x)]^2\,dx. \end{aligned} \tag{4.4.4}$$

Now, since G_n integrates members of $\mathscr{P}_{2n-1}$ exactly, then

$$\int_a^b w(x)h_{2n-1}(x)\,dx = \sum_{k=1}^{n} w_k h_{2n-1}(x_k) = \sum_{k=1}^{n} w_k f(x_k).$$

Since p_n^* is orthonormal, then $\int_a^b w(x)[p_n^*(x)]^2\,dx = 1$. These equalities reduce (4.4.4) to (4.4.1).

Explicit forms of the Gauss remainder have already been given in Sections 2.7 and 3.6 for the most commonly employed cases.

In a similar fashion, using a mixed Lagrange–Hermite interpolating polynomial which interpolates to the function and its first derivative at interior points and to the function only at one or both endpoints of the integration interval, we get the error expressions given in Section 2.7.1 for Radau and Lobatto integration, respectively.

Another application of the mixed Lagrange–Hermite interpolation formula is in the determination of the error term in Hunter's method for evaluating the Cauchy principal value integral

$$I(f; \lambda) = P \int_a^b \frac{w(x) f(x)}{x - \lambda}\, dx, \qquad a < \lambda < b \tag{4.4.5}$$

using a Gauss based rule. (See Section 2.12.8.) If $\{x_k: k = 1, \ldots, n\}$ are the zeros of $p_n(x)$ defined above and $\lambda \neq x_k$, $k = 1, \ldots, n$, then we can approximate $f(x)$ by the interpolation polynomial $q_{2n}(x) \in \mathscr{P}_{2n}$ such that $q_{2n}(x_k) = f(x_k)$, $q'_{2n}(x_k) = f'(x_k)$, $k = 1, \ldots, n$, and $q_{2n}(\lambda) = f(\lambda)$. The error in this approximation is given by

$$f(x) - q_{2n}(x) = \frac{(x - \lambda) p_n^*(x)^2}{(2n+1)!\, k_n^2} f^{(2n+1)}(\xi(x)), \qquad a < \xi(x) < b \tag{4.4.6}$$

where, as above, $f^{(2n+1)}(\xi(x))$ is continuous in $a \leq x \leq b$. If we now multiply (4.4.6) by $w(x)/(x - \lambda)$ and integrate in the sense of the principal value, we get

$$I(f; \lambda) - I(q_{2n}; \lambda) = \int_a^b \frac{w(x) p_n^*(x)^2}{(2n+1)!\, k_n^2} f^{(2n+1)}(\xi(x))\, dx. \tag{4.4.7}$$

The error term in this case becomes

$$f^{(2n+1)}(\eta)/(2n+1)!\, k_n^2, \qquad a < \eta < b. \tag{4.4.8}$$

The form of $I(q_{2n}; \lambda)$ is given in Section 2.12.8. If $\lambda = x_j$ for some j, the form of $I(q_{2n}; \lambda)$ will be different but the error term will still be given by (4.4.8).

References. Davis [6, p. 67]; Engels [13, Chap. 7]; Hunter [5]; Stroud and Secrest [2].

4.5 Error Estimates through Differences

If $\int_a^b f(x)\, dx \approx \sum_{k=1}^n w_k f(x_k) = R(f)$ is an integration rule of interpolatory type, it is equivalent to passing a polynomial through the points $(x_i, f(x_i))$, $i = 1, 2, \ldots, n$ and integrating this polynomial. Hence, we may obtain error expressions by integrating remainder formulas for polynomial interpolation.

If $p_{n-1}(f; x) \in \mathscr{P}_{n-1}$ and satisfies $p_{n-1}(f; x_i) = f(x_i)$, $i = 1, 2, \ldots, n$, it is known that

$$f(x) - p_{n-1}(f; x) = [f(x), f(x_1), f(x_2), \ldots, f(x_n)] \times (x - x_1)(x - x_2) \cdots (x - x_n). \tag{4.5.1}$$

where $[f(x), f(x_1), \ldots, f(x_n)]$ is the divided difference (Section 1.11.8). Hence, integration yields

$$E(f) = \int_a^b f(x)\,dx - \int_a^b p_{n-1}(f; x)\,dx = \int_a^b [f(x), f(x_1), \ldots, f(x_n)] \times (x - x_1)\cdots(x - x_n)\,dx. \quad (4.5.2)$$

Under smoothness conditions, the divided difference may be replaced by an nth derivative and we obtain the following theorem.

THEOREM. *Let $f(x) \in C^n[a, b]$ and $a \le x_1 < x_2 < \cdots < x_n \le b$. Then,*

$$E(f) = \frac{1}{n!}\int_a^b f^{(n)}(\xi(x))(x - x_1)\cdots(x - x_n)\,dx \quad (4.5.3)$$

for some $\xi(x)$ with $a \le \xi(x) \le b$. If

$$|f^{(n)}(x)| \le M \quad \text{on} \quad a \le x \le b, \quad (4.5.4)$$

then

$$|E(f)| \le \frac{M}{n!}\int_a^b |(x - x_1)\cdots(x - x_n)|\,dx. \quad (4.5.5)$$

Formulas similar to (4.5.2), (4.5.3), and (4.5.5) hold for weighted or product integration as well as for Cauchy principal value integrals.

Error expressions of the usual sort involve high-order derivatives, $f^{(M)}(\xi)$, of the integrand. In the case of Simpson's rule, $E = cf^{(4)}(\xi)$. In the case of a Gauss 96-point rule (which has actually been used in machine calculation), we should have to compute $f^{(192)}(\xi)$. If the integrand is sufficiently elementary, it may be possible to obtain the appropriate derivative $f^{(M)}(x)$ and from it to estimate $\max_{a \le x \le b} |f^{(M)}(x)|$. For example, we can obtain the nth derivative of certain types of rational functions quite easily through interpolation formulas.

Let $z_0, z_1, \ldots, z_p$ be $p + 1$ distinct real or complex numbers. Set

$$w(z) = (z - z_0)(z - z_1)\cdots(z - z_p). \quad (4.5.6)$$

Then, for a given $f(z)$, the polynomial

$$p(f; z) = \sum_{i=0}^{p} f(z_i)\frac{w(z)}{(z - z_i)w'(z_i)} \quad (4.5.7)$$

interpolates to f at $z_0, \ldots, z_p$. [See (1.11.6–1.11.13).] If $f(z)$ is itself a polynomial of degree $\leq p$, then $f(z)$ and $p(f; z)$ coincide at the $p+1$ points $z_0, \ldots, z_p$ and hence must be identical. Therefore

$$f(z) = \sum_{i=0}^{p} f(z_i) \frac{w(z)}{(z - z_i)w'(z_i)} \tag{4.5.8}$$

and

$$\frac{f(z)}{w(z)} = \sum_{i=0}^{p} \frac{f(z_i)}{w'(z_i)} \frac{1}{(z - z_i)}. \tag{4.5.9}$$

Thus, finally,

$$\frac{d^n}{dz^n}\left(\frac{f(z)}{w(z)}\right) = (-1)^n n! \sum_{i=0}^{p} \frac{f(z_i)}{w'(z_i)} \frac{1}{(z - z_i)^{n+1}}. \tag{4.5.10}$$

Example. Estimate

$$\max_{0 \leq x \leq 1} \left| \frac{d^n}{dx^n}\left(\frac{1}{1 + x^4}\right) \right|.$$

Here

$$f(z) \equiv 1, \qquad w(z) = z^4 + 1 = (z - z_0)(z - z_1)(z - z_2)(z - z_3),$$

where $z_0 = e^{\pi i/4}$, $z_1 = e^{3\pi i/4}$, $z_2 = e^{5\pi i/4}$, $z_3 = e^{7\pi i/4}$, $w'(z) = 4z^3$. Hence

$$\frac{d^n}{dz^n}\left(\frac{1}{1 + z^4}\right) = \frac{(-1)^n n!}{4}\left[\frac{1}{z_1(z - z_0)^{n+1}} + \frac{1}{z_0(z - z_1)^{n+1}} + \frac{1}{z_3(z - z_2)^{n+1}} + \frac{1}{z_2(z - z_3)^{n+1}}\right].$$

Since the minimum distance from z_0 or z_3 to $[0, 1]$ is $\frac{1}{2}\sqrt{2}$, and the minimum distance from z_1 or z_2 to $[0, 1]$ is 1, we have

$$\max_{0 \leq x \leq 1} \left| \frac{d^n}{dx^n}\left(\frac{1}{1 + x^4}\right) \right| \leq \frac{n!}{4}[(\sqrt{2})^{n+1} + (1)^{n+1} + (1)^{n+1} + (\sqrt{2})^{n+1}]$$

$$= \frac{n!}{2}[2^{(1/2)(n+1)} + 1].$$

For some additional algebraic devices that are useful in computing formal high-order derivatives, the reader is referred to Steffensen. Also, computer programs have been written for the formal calculation of higher derivatives (see Section 1.2). But even with simple integrands, high-order derivatives can lead to expressions of enormous complexity. We are, of course, interested in ways of avoiding this formidable computation.

References. Davis [6, pp. 56, 64]; Rabinowitz [13]; Steffensen [1, pp. 231–241].

An nth derivative may be approximated by an nth difference. Let $a \leq x_0 < x_1 < x_2 < \cdots < x_n \leq b$ be $n+1$ equally spaced points of distance h apart. If $f(x) \in C^n[a, b]$, then

$$\Delta^n f(x_0) = h^n f^{(n)}(\xi), \qquad x_0 < \xi < x_n. \tag{4.5.11}$$

Thus, $f^{(n)}(x)$ may be approximated by $(1/h^n)\Delta^n f(x_0)$. *Markoff's formulas*, which are Newton forward-difference formulas differentiated, provide complete expansions of derivatives in terms of differences. The first few expansions are

$$\begin{aligned}
hf'(x_0) &= \Delta_0 - \tfrac{1}{2}\Delta_0^2 + \tfrac{1}{3}\Delta_0^3 - \tfrac{1}{4}\Delta_0^4 + \cdots,\\
h^2 f''(x_0) &= \Delta_0^2 - \Delta_0^3 + \tfrac{11}{12}\Delta_0^4 - \cdots,\\
h^3 f'''(x_0) &= \Delta_0^3 - \tfrac{3}{2}\Delta_0^4 + \tfrac{7}{4}\Delta_0^5 - \cdots,\\
h^4 f^{(4)}(x_0) &= \Delta_0^4 - 2\Delta_0^5 + \tfrac{17}{6}\Delta_0^6 - \tfrac{7}{2}\Delta_0^7 + \cdots,\\
&\vdots
\end{aligned}$$

Example. $\int_{1.5}^{1.6} \Gamma(x)\,dx$ has been evaluated by use of a 3-point Simpson rule. Estimate the error E. It involves the fourth derivative of the Gamma function $\Gamma(x)$. The following tabulation of values is from the NBS Handbook of Mathematical Functions.

x	$\Gamma(x)$	Δ	Δ^2	Δ^3	Δ^4
1.50	.8862 269				
1.52	.8870 388	.0008 119			
1.54	.8881 777	.0011 389	.0003 270		
1.56	.8896 392	.0014 615	.0003 226	−.0000 044	
1.58	.8914 196	.0017 804	.0003 189	−.0000 037	−.0000 007
1.60	.8935 154	.0020 958	.0003 154	−.0000 035	−.0000 002

We have $|\Delta^4\Gamma| \approx 7.0 \times 10^{-7}$, so that $|\Gamma^{(4)}| \approx (7.0 \times 10^{-7})/(.02)^4 \approx 5.0$. Hence

$$|E| \approx \frac{(b-a)^5}{2880}\,|\Gamma^{(4)}| \approx \frac{5(10^{-1})^5}{2880} \approx 2 \times 10^{-8}.$$

Reference. NBS Handbook [1].

4.6 Error Estimates through the Theory of Analytic Functions

If the integrand is an *analytic* function, then the high-order derivatives relevant to error estimation may be bounded by use of Cauchy's theorem. This theorem is as follows.

Let B be a simply connected region in the complex z $(= x + iy)$ plane.

Suppose that $f(z)$ is analytic in B. Let z_0 lie in B. Suppose that C is a simple contour which lies in B and goes around z_0 in the positive sense; then

$$f^{(n)}(z_0) = \frac{n!}{2\pi i}\int_C \frac{f(z)}{(z - z_0)^{n+1}}\,dz. \tag{4.6.1}$$

Suppose next that the real-line segment $a \le x \le b$ is contained in B. Select a C that lies in B and that contains this segment in its interior. Introduce this notation: $L(C)$ = length of C, $M_C = \max_{z \in C} |f(z)|$, δ = minimum distance from points of C to points of the segment $a \le x \le b$. We now have

$$|f^{(n)}(x)| \le \frac{n!}{2\pi}\int_C \frac{|f(z)|\,|dz|}{|z - x|^{n+1}} \le \frac{n!\,L(C)M_C}{2\pi\delta^{n+1}}, \qquad a \le x \le b. \tag{4.6.2}$$

Therefore

$$\max_{a \le x \le b} |f^{(n)}(x)| \le \frac{n!\,L(C)M_C}{2\pi\delta^{n+1}}, \tag{4.6.3}$$

and the right-hand number provides a bound for $|f^{(n)}(x)|$ in terms of the maximum modulus of f along C. Note that the terms $L(C)$ and M_C, as well as δ, depend on C. As C becomes larger, $L(C)$, M_C, and δ all increase in value, and the minimum (or nearly minimum) value of the bound can be found if we vary C.

Example. Estimate $\max_{-1 \le x \le 1} |f^{(10)}(x)|$, where $f(x) = \exp[e^x]$. Take C as the circle $|z| = R$, $R > 1$. Now

$$|f(z)| = |\exp[e^z]| = |\exp[e^x \cos y + ie^x \sin y]| = |\exp[e^x \cos y]| \le \exp[e^R].$$

Hence, since $\delta = R - 1$,

$$\max_{-1 \le x \le 1} |f^{(10)}(x)| \le \frac{(10)!\,2\pi R \exp[e^R]}{2\pi(R - 1)^{11}}.$$

This inequality is valid for all $R > 1$. A coarse calculation shows that $R \exp[e^R] \times (R - 1)^{-11}$ has a minimum of about 2.46×10^3 at $R = 2.2$. Hence,

$$\max_{-1 \le x \le 1} |f^{(10)}(x)| \le (10)! \times 2.46 \times 10^3.$$

A particularly good region B to use in (4.6.1) is the union of two semicircles of radius δ and centers at a and b respectively, and the rectangle with vertices at $(a, \pm\delta)$, $(b, \pm\delta)$. For a fixed x in $[a, b]$, the circle $|z - x| \le \delta$ is contained in B and we have from (4.6.1)

$$|f^{(n)}(x)| \le \frac{n!\,2\pi\delta M_x}{2\pi\delta^{n+1}} = \frac{n!\,M_x}{\delta^n} \qquad \text{where} \quad M_x = \max_{|z - x| \le \delta} |f(z)|.$$

Hence,

$$\max_{a \le x \le b} |f^{(n)}(x)| \le n!M/\delta^n \tag{4.6.3a}$$

where $M = \max |f(z)|$, the maximum being taken over the boundary of B.

It is also possible to obtain derivative-free error estimates without first going through the real variable work that leads to expressions containing $f^{(n)}(\xi)$. If $f(z)$ is a regular analytic function in a simply connected region B, then we can write

$$f(z) = \frac{1}{2\pi i}\int_C \frac{f(t)}{t - z}\,dt, \tag{4.6.4}$$

where C is a simple contour lying in B and containing z in its interior. Suppose that E is a linear functional. Then

$$E(f(z)) = \frac{1}{2\pi i}\,E\int_C \frac{f(t)}{t - z}\,dt. \tag{4.6.5}$$

Under certain mild restrictions as to the nature of E, it will be possible to interchange the operator E with $\int_C$ and obtain

$$E(f(z)) = \frac{1}{2\pi i}\int_C E_z\left(\frac{1}{t - z}\right)f(t)\,dt. \tag{4.6.6}$$

Thus

$$E(f(z)) = \frac{1}{2\pi i}\int_C k(t)f(t)\,dt, \tag{4.6.7}$$

where

$$k(t) = E_z\left(\frac{1}{t - z}\right). \tag{4.6.8}$$

Formula (4.6.7) can be regarded as the complex variable analog of Peano's formula. We have

$$|E(f(z))| \le (2\pi)^{-1}\int_C |k(t)|\,|f(t)|\,|dt|, \tag{4.6.9}$$

so that an error estimate may be found in terms of the size of $|f(t)|$ on the contour C. The problem now is how to obtain good estimates for

$$|k(t)| = \left|E_z\left(\frac{1}{t - z}\right)\right|.$$

For approximate integration of interpolatory type, interpolatory formulas in the complex plane may be used to good effect. Let $z_0, z_1, \ldots, z_n$ be $n + 1$ distinct points lying in the complex plane. Suppose that $f(z)$ is a regular analytic function in a simply connected region B that contains these points in its interior. Let

$$w(z) = (z - z_0)(z - z_1) \cdots (z - z_n). \tag{4.6.10}$$

Then

$$p_n(z) = \sum_{k=0}^{n} f(z_k) \frac{w(z)}{(z - z_k)w'(z_k)} \tag{4.6.11}$$

is the polynomial of class $\mathscr{P}_n$, which interpolates to $f(z)$ at z_k:

$$p_n(z_k) = f(z_k), \qquad k = 0, 1, \ldots, n. \tag{4.6.12}$$

According to a theorem of Hermite (see, for example, Davis) the error in polynomial interpolation can be expressed as a contour integral:

$$f(z) = p_n(z) + \frac{1}{2\pi i} \int_C \frac{w(z)f(t)}{w(t)(t - z)}\, dt, \tag{4.6.13}$$

where C is a simple contour contained in B and containing the points $z_0, \ldots, z_n$ in its interior. If now the points $z = a$ and $z = b$ also lie inside C, then, integrating (4.6.13), we obtain

$$\int_a^b f(z)\, dz = \sum_{k=0}^{n} f(z_k) \int_a^b \frac{w(z)}{(z - z_k)w'(z_k)}\, dz + \int_a^b dz \frac{1}{2\pi i} \int_C \frac{w(z)f(t)}{w(t)(t - z)}\, dt. \tag{4.6.14}$$

Hence, interchanging the order of integration, we have

$$\int_a^b f(z)\, dz = \sum_{k=0}^{n} a_k f(z_k) + \frac{1}{2\pi i} \int_C \frac{f(t)}{w(t)} u(t)\, dt, \tag{4.6.15}$$

where $a_0, \ldots, a_n$ are the weights for interpolatory integration at $z_0, \ldots, z_n$ and $u(t)$ designates the function

$$u(t) = \int_a^b \frac{w(z)\, dz}{t - z}. \tag{4.6.16}$$

If we now write

$$E(f) = \int_a^b f(z)\, dz - \sum_{k=0}^{n} a_k f(z_k) = \frac{1}{2\pi i} \int_C \frac{f(t)u(t)}{w(t)}\, dt, \tag{4.6.17}$$

then

$$E(f) = \frac{1}{2\pi i} \int_C \frac{f(t)}{w(t)} \left(\int_a^b \frac{w(z)\, dz}{t - z} \right) dt, \tag{4.6.18}$$

and we obtain an upper bound for E that does not involve derivatives:

$$|E(f)| \leq \frac{1}{2\pi}\int_C \frac{|f(t)|}{|t-z_0|\,|t-z_1|\cdots|t-z_n|} \times \left(\int_a^b \frac{|z-z_0|\,|z-z_1|\cdots|z-z_n|}{|t-z|}\,|dz|\right)|dt|. \quad (4.6.19)$$

The following notations will be used: length of C, $L(C)$; minimum distance from $[a, b]$ to C, δ_C; maximum distance from $z_0, z_1, \ldots, z_n$ to $\{a, b\}$, D; minimum distance from $z_0, z_1, \ldots, z_n$ to C, d. Then, from (4.6.19), we have

$$|E(f)| \leq \frac{L(C)}{2\pi} \max_{t\in C} |f(t)| \frac{1}{d^{n+1}} \frac{(b-a)D^{n+1}}{\delta_C} = \text{const} \max_{t\in C} |f(t)|, \quad (4.6.20)$$

where

$$\text{const} = \frac{L(C)(b-a)}{2\pi\delta_C}\left(\frac{D}{d}\right)^{n+1} \quad (4.6.21)$$

These identities have been applied by McNamee to Gaussian integration. They are also applicable to Gauss rules with respect to any admissible weight function, using the corresponding orthogonal polynomials and functions of the second kind. Let $z_0, z_1, \ldots, z_{n-1}$ be the n abscissas of the G_n rule. Then

$$w(z) = (z-z_0)\cdots(z-z_{n-1}) = c_n P_n(z), \quad (4.6.22)$$

where $P_n(z)$ is the Legendre polynomial of order n and c_n is a normalizing constant. Then the remainder is given by

$$\int_{-1}^{1} f(x)\,dx - G_n(f) = \frac{1}{2\pi i}\int_C \frac{f(t)}{P_n(t)}\left(\int_{-1}^{1}\frac{P_n(z)}{t-z}\,dz\right)dt. \quad (4.6.23)$$

The functions

$$Q_n(t) = \int_{-1}^{1} \frac{P_n(z)}{t-z}\,dz \quad (4.6.24)$$

are *Legendre functions of the second kind* (see Section 1.12). They are regular, single-valued functions in the whole plane with the segment $[-1 \leq t \leq 1]$ deleted. Therefore

$$\int_{-1}^{1} f(x)\,dx - G_n(f) = \frac{1}{2\pi i}\int_C \frac{f(t)Q_n(t)}{P_n(t)}\,dt. \quad (4.6.25)$$

An upper bound for the integral on the right can now be obtained in terms of $\max_{t \in C} |f(t)|$, as was worked out before. An alternative method is to use an asymptotic expression for $Q_n(t)/P_n(t)$ and to employ a very large contour C. It is known that, as $t \to \infty$,

$$\frac{Q_n(t)}{P_n(t)} = c_n t^{-2n-1}\left[1 + \frac{2n^3 + 3n^2 - n - 1}{(2n+3)(2n-1)t^2} + \cdots\right], \tag{4.6.26a}$$

where

$$c_n = \frac{2^{2n+1}(n!)^4}{(2n)!\,(2n+1)!}. \tag{4.6.26b}$$

Example. (McNamee). Take $f(t) = e^t t^3$ and select $n = 5$. On the circle $|t| = R$, the modulus of the integrand in (4.6.25) is dominated by

$$c_5 \exp[-[7 \log R - R]].$$

We therefore choose $R = 7$. This leads to

$$\left|\int_{-1}^{1} e^t t^3\,dt - G_5(e^t t^3)\right| \le 4 \times 10^{-6}.$$

Exact error: 1.0×10^{-7}.

Whereas the approximation $c_n t^{-2n-1}$ to $K_n(t) = Q_n(t)/P_n(t)$ is good for $t \to \infty$, it is not too accurate if $f(t)$ has a singularity close to the interval $[-1, 1]$. Lether has improved on the approximation by adding several terms to the expansion based on the following theorem.

THEOREM. *Let $T_j(x)$ denote the jth Tschebyscheff polynomial of the first kind and define a_v by $a_v = R_n(T_{2n+2v})/R_n(T_n)$, where $R_n(f)$ is the error in the integration of $f(x)$ using the n-point Gaussian rule G_n (see Section 4.8.1). Then, for t bounded away from $[-1, 1]$,*

$$K_n(t) = \frac{c_n 2^{2n+2}}{z^{2n+1}}\left[1 + \frac{(1+a_1)}{z^2 - 1} + \frac{1}{z^2 - 1}\sum_{v=2}^{\infty} \frac{a_v}{z^{2v-2}}\right] \tag{4.6.27}$$

where $z = t + (t^2 - 1)^{1/2}$.

Lether gives exact expressions for $(1 + a_1)$, a_2, and a_3.

Example. (Lether) If $f(t)$ is a meromorphic function having simple poles $z_j \notin [-1, 1]$ with residues A_j, $j = 1, \ldots, m$, then

$$R_n(f) = -\sum_{j=1}^{m} A_j K_n(z_j).$$

Let $f(t) = (t-2)^{-1}$ so that $m = 1$, $z_1 = 2$, and $A_1 = 1$. Then $R_n(f) = -K_n(2)$. For $n = 5$, the exact value of $R_n(f) = -.3046856 \times 10^{-5}$. Using approximations to $K_n(t)$ consisting of 1, 2, and 4 terms in the brackets of (4.6.27), we get the values $-.307 \times 10^{-5}$, $-.30470 \times 10^{-5}$, and $-.3046863 \times 10^{-5}$, respectively.

Chawla has analyzed the error associated with rules of Fejér type by the methods of this section. Let

$$E_C(f) = \int_{-1}^{1} f(x)\,dx - \sum_{k=0}^{n} w_k f(x_k), \qquad x_k = \cos\left(\frac{2k+1}{2n+2}\pi\right) \tag{4.6.28}$$

(weights w_k appropriate to the condition $E_C(p) = 0$, $p \in \mathscr{P}_n$). Let

$$E_P(f) = \int_{-1}^{1} f(x)\,dx - \sum_{k=0}^{n} w_k f(x_k), \qquad x_k = \cos\left(\frac{k\pi}{n}\right) \tag{4.6.29}$$

(weights w_k appropriate to the condition $E_P(p) = 0$, $p \in \mathscr{P}_n$).

Let $f(z)$ be a regular and single-valued analytic function in the closure of the ellipse $\mathscr{E}_\rho$, $\rho > 1$ with foci at ± 1 and semiaxis sum ρ. Let $M = M(\rho) = \max_{z \in \mathscr{E}_\rho} |f(z)|$. Then, for n even,

$$|E_C(f)| \leq \left(\frac{4n}{2n+1}\right)\left(\frac{\rho + \rho^{-1}}{\rho^2 - 1}\right)\frac{M}{\rho^{n+1} - \rho^{-n-1}}, \tag{4.6.30}$$

$$|E_P(f)| \leq \left(\frac{16n^2}{4n^2 - 1}\right)\frac{M}{(\rho^2 - 1)(\rho^n - \rho^{-n})}. \tag{4.6.31}$$

The subscripts C and P refer to the "classical" (Fejér's first rule) and "practical" (Clenshaw–Curtis rule) abscissas, respectively.

A similar result has been derived by Basu for Fejér's second rule using the "Filippi" abscissas $\cos(k\pi/(n+1))$, $k = 1, \ldots, n$, n odd.

Taking C to be the ellipse $\mathscr{E}_\rho$ and setting $t(\theta) = \frac{1}{2}(\rho e^{2\pi i\theta} + \rho^{-1}e^{-2\pi i\theta})$, Smith transforms the integral

$$\frac{1}{2\pi i}\int_{\mathscr{E}_\rho} \frac{f(t)Q_n(t)}{P_n(t)}\,dt$$

to the form

$$\int_0^1 \frac{f(t(\theta))Q_n(t(\theta))}{P_n(t(\theta))}(\rho e^{2\pi i\theta} - \rho^{-1}e^{-2\pi i\theta})\,d\theta = \int_0^1 g(\theta)\,d\theta.$$

$g(\theta)$ is a complex-valued analytic function of the real variable θ and is periodic of period 1. Hence, one can approximate the integral of $g(\theta)$ by the trapezoidal rule (Section 2.9). Both P_n and Q_n can be evaluated using the three-term recurrence (1.12.8). Smith gives some examples and shows how one could in theory iterate this process since the error in the trapezoidal rule can again be expressed as the integral of a periodic function. This is a very tedious proposition.

Takahasi and Mori have given graphs of the level lines of $|k(t)|$ for a variety of integration rules. They also describe a method of estimating the integration error based on choosing contours C that pass through the saddle points of $|k(t)f(t)|$. In addition, they show how to derive integration rules by working backward from the form of $k(t)$. That is, writing

$$k(t) = \int_a^b \frac{w(x)}{t-x}\,dx - \sum_{i=1}^{n} \frac{a_i}{t-x_i}, \tag{4.6.32}$$

they choose the a_i and x_i (or the a_i in case the x_i are given) so as to make $k(t)$ "small."

For the particular class of weight functions

$$w(x) = x^{\beta}(1-x)^{\alpha}(-\log x)^{\nu}, \qquad \alpha+\nu > -1, \quad \beta > -1 \tag{4.6.33}$$

on $[0, 1]$, Sidi [2] has derived a set of integration rules that make $k(t)$ "small." In these rules, the abscissas x_i, $i = 1, \ldots, n$, are the zeros of the polynomial $\sum_{j=0}^{n} \lambda_j z^j$, where

$$\lambda_j = (-1)^j \binom{n}{j} (j+1)^{n+\alpha+\nu-s}, \qquad j = 0, 1, \ldots, n, \tag{4.6.34}$$

with s a small nonnegative integer, say 0, 1, or 2. The weights w_i, which are interpolatory, are given by

$$w_i = \sum_{j=1}^{n} \lambda_j x_i^j \left(\sum_{m=1}^{j} \mu_m / x_i^m \right) \Bigg/ \sum_{j=1}^{n} \lambda_j j x_i^{j-1}, \qquad i = 1, \ldots, n \tag{4.6.35}$$

where the μ_m are the moments of $w(x)$. These rules are remarkable in that the abscissas x_i do not depend on the value of β at all but only on $\alpha + \nu$. Furthermore, for values of $\alpha + \nu$ differing by a small integer, we can still use the same abscissas.

Sidi [4] has done something similar for weighted integrals over $[0, \infty)$.

References. Basu [1]; Chawla [4]; Davis [6, p. 68]; Lether [13]; Locher [2]; J. McNamee [1]; Rabinowitz [8]; Sidi [2, 4]; H. V. Smith [2]; Takahasi and Mori [1, 2].

Freud has established a different expression for the kernel (4.6.8), which we now write $E_n((t-z)^{-1})$ to indicate its dependence on the n-point rule which we take to be of Gauss type over $[-1, 1]$. As previously, the functional E_n operates on $(t-z)^{-1}$ as a function of z. The following derivation of Freud's formula is due to von Sydow.

We are considering the Gauss rule

$$\int_{-1}^{1} w(x) f(x)\,dx = \sum_{i=1}^{n} w_i f(x_i) + E_n f = Q_n f + E_n f \tag{4.6.36}$$

where the x_i are the zeros of the *orthonormal* polynomial

$$p_n^*(x) = k_n x^n + \cdots, \qquad k_n > 0.$$

Since $Q_n f$ is exact for $f \in \mathscr{P}_{2n-1}$, we have, for large $|t|$,

$$E_n((t-z)^{-1}) - E_{n+1}((t-z)^{-1}) = \frac{1}{t^{2n+1}}[E_n(z^{2n}) + O(t^{-1})]. \tag{4.6.37}$$

Furthermore, $E(z^{2n}) = k_n^{-2}E(p_n^{*2}) = k_n^{-2}$ since $Q_n(p_n^{*2}) = 0$. Now

$$E_n((t-z)^{-1}) - E_{n+1}((t-z)^{-1}) = Q_{n+1}((t-z)^{-1}) - Q_n((t-z)^{-1}) \tag{4.6.38}$$

which can be written as $\pi(t)/p_n^*(t)p_{n+1}^*(t)$ for some $\pi(t) \in \mathscr{P}_{2n}$. However, by (4.6.37), $\pi(t) = k_{n+1}/k_n$ from which follows Freud's formula

$$E_n((t-z)^{-1}) = \sum_{j=n}^{\infty} \frac{k_{j+1}}{k_j} \frac{1}{p_j^*(t)p_{j+1}^*(t)}. \tag{4.6.39}$$

We now restrict our attention to weight functions which satisfy the condition that $|\int_{-\pi}^{\pi} \log w(\cos\theta)\, d\theta| < \infty$ and define

$$D(z) = \exp\left\{\frac{1}{4\pi}\int_{-\pi}^{\pi} \log w(\cos\theta)|\sin\theta| \frac{1+ze^{-i\theta}}{1-ze^{-i\theta}}\, d\theta\right\}. \tag{4.6.40}$$

We consider for each $R > 1$, the ellipse C_R with foci at ± 1 and semiaxes $(R \pm R^{-1})/2$. Then, if $f(z)$ is analytic in the interior of C_R and continuous on the boundary, we have

$$|E_n f| = \left|\frac{1}{2\pi i}\int_{C_R} f(t)E_n((t-z)^{-1})\, dt\right| \le \| E_n \| \max_{z \in C_R} |f(z)|. \tag{4.6.41}$$

An estimate for $\| E_n \|$ is given by

$$\| E_n \| \le R^{-2n}(1+\varepsilon_n)\int_{-\pi}^{\pi} |D^2(R^{-1}e^{i\theta})|\, d\theta \tag{4.6.42}$$

with $\varepsilon_n \to 0$ as $n \to \infty$. A bound for the integral in this estimate is given by $2\int_{-1}^{1} w(x)\, dx$.

References. Chen [1]; Freud [2, 3]; von Sydow [1].

4.6.1 Errors through Contour Integration; Asymptotic Analysis

An extensive analysis of integration error for analytic integrands has been given by Donaldson and Elliott, following upon the work of Barrett. The idea of this work is first, as in (4.6.18) or (4.6.25), to express the error

$$E_n(f) = \int_a^b w(x)f(x)\, dx - \sum_{k=1}^{n} a_k f(x_k) \tag{4.6.1.1}$$

as a contour integral

$$E_n(f) = (2\pi i)^{-1} \int_C K_n(z) f(z)\, dz \tag{4.6.1.2}$$

for an appropriate kernel $K_n(z)$ and contour C. Then, instead of merely estimating the error by means of (4.6.9), one evaluates the contour integral with more care and precision with a view toward obtaining an asymptotic expansion as $n \to \infty$. There is first the possibility of deforming the contour C to another contour C^*. Assuming that $K_n(z)$ is analytic in the whole complex plane with the interval $[a, b]$ deleted, the manner in which C can be deformed depends upon the singularity structure of $f(z)$: whether it is entire, meromorphic, has branch points, etc. One next replaces $K_n(z)$ by an asymptotic expansion $K_n^*(z)$ for large n. Hopefully, at this stage we can evaluate or estimate $(2\pi i)^{-1} \int_{C^*} K_n^*(z) f(z)\, dz$.

We begin by considering quite general functions of the second kind. First, we must define an admissible class of functions which can serve as weight functions.

Definition. The function $g(x)$ satisfies a Lipschitz condition of order α on $[a, b]$ ($g(x) \in \mathrm{Lip}_\alpha$, $\alpha > 0$) if there exists a constant M such that

$$|g(x_1) - g(x_2)| \le M |x_1 - x_2|^\alpha, \qquad x_1, x_2 \in [a, b].$$

Definition. The function $g(x) \in H^*$ on $[a, b]$ if for some fixed β, $0 \le \beta < 1$, the functions

$$g^*(x) = (x - c)^\beta g(x)$$

where c is either a or b, satisfy a Lipschitz condition on $[a, b]$.

Example. $x^{-1/2} \in H^*$ on $[0, 1]$.

Theorem. *If $w(x)\phi(x) \in H^*$ on $[a, b]$, the function of the second kind*

$$\psi(z) = \int_a^b \frac{w(t)\phi(t)}{z - t}\, dt, \qquad z \notin [a, b] \tag{4.6.1.3}$$

is analytic in the entire complex plane with the interval $[a, b]$ deleted. Moreover, the limits

$$\lim_{y \to 0+} \psi(x + iy) = \psi(x + i0) \quad \textit{and} \quad \lim_{y \to 0+} \psi(x - iy) = \psi(x - i0)$$

both exist for $a < x < b$, are continuous there, and

$$\psi(x - i0) - \psi(x + i0) = 2\pi i w(x)\phi(x). \tag{4.6.1.4}$$

This relationship is known as the *Sochozki–Plemelj formula*; and for a proof of these things, we refer to Muskhelishvili.

Let now $-\infty \leq a < b \leq \infty$ and D be a domain that contains (a, b). Let $w(x)$ be a weight function (not necessarily positive) defined on (a, b) and such that $\int_a^b w(x)f(x)\,dx$ exists in the sense of Riemann whenever $f(z)$ is analytic in D.

Let $\phi(z)$ be analytic in D and have simple zeros at $a < x_1 < x_2 < \cdots < x_n < b$. Let the only singularities of $1/\phi(z)$ in D be the simple poles at $z = x_i$.

Let $\psi(z)$ be an analytic function in the plane cut along an interval M containing (a, b) and such that

$$\psi(x - 0i) - \psi(x + 0i) = 2\pi i w(x)\phi(x) \quad \text{for} \quad a < x < b. \tag{4.6.1.5}$$

Define constants λ_k, $k = 1, 2, \ldots, n$ by means of

$$\lambda_k = -\psi(x_k)/\phi'(x_k). \tag{4.6.1.6}$$

Observe first that since $\phi(x_k) = 0$,

$$\lim_{\varepsilon \to 0^+} \psi(x_k - \varepsilon i) = \lim_{\varepsilon \to 0^+} \psi(x_k + \varepsilon i)$$

and this common value is designated by $\psi(x_k)$. Secondly, since x_k is a simple zero of $\phi(z)$, $\phi'(x_k) \neq 0$.

Let C^+ be a simple smooth contour lying in the common part of the upper half-plane and D, extending from $x = b$ to $x = a$. Similarly for C^- and the lower half-plane. Set $C = C^+ \cup C^-$.

THEOREM. *Under the above conditions, and if $f(z)$ is analytic in D, then*

$$E(f) = \int_a^b w(x)f(x)\,dx - \sum_{k=1}^{n} \lambda_k f(x_k) = \frac{1}{2\pi i}\int_C \frac{\psi(z)}{\phi(z)} f(z)\,dz. \tag{4.6.1.7}$$

Proof. Call the right-hand integral I. The function $(\psi(z)/\phi(z))f(z)$ has no singularities in $D - M$. Hence we may deform C^+ into the upper edge of the interval (a, b) indented by a small semicircle δ_k^+ of radius δ centered at each x_k. Similarly for C^-. Hence,

$$I = \frac{1}{2\pi i}\left\{\int_a^{x_1-\delta} + \sum_{k=1}^{n-1}\int_{x_k+\delta}^{x_{k+1}-\delta} + \int_{x_n+\delta}^{b}\right\}\frac{\psi(x - i0) - \psi(x + i0)}{\phi(x)} f(x)\,dx$$

$$+ \frac{1}{2\pi i}\sum_{k=1}^{n}\left\{\int_{\delta_k^-} + \int_{\delta_k^+}\right\}\frac{\psi(z)}{\phi(z)} f(z)\,dz.$$

Now, as $\delta \to 0$, the first part of the right-hand side approaches

$$\frac{1}{2\pi i}\int_a^b \frac{\psi(x - i0) - \psi(x + i0)}{\phi(x)} f(x)\,dx,$$

and by (4.6.1.4) this is

$$\int_a^b w(x)f(x)\,dx.$$

Consider next,

$$\lim_{\delta \to 0} \int_{\delta_k^-} \frac{\psi(z)}{\phi(z)} f(z)\, dz.$$

Since $\psi(x - i0)$ exists and is continuous in $a < x < b$, it follows that the function $(\psi(z)/\phi(z))f(z)$ is continuous on and within the semicircle bounded by δ_k^- and the lower edge of the x axis, except possibly at $x = x_k$. Furthermore,

$$\lim_{z \to x_k} \frac{(z - x_k)}{\phi(z)} \psi(z) f(z) = \frac{\psi(x_k) f(x_k)}{\phi'(x_k)}.$$

Hence, by an extension of the Residue theorem (see for example Phillips)

$$\lim_{\delta \to 0} \int_{\delta_k^-} \frac{\psi(z)}{\phi(z)} f(z)\, dz = \frac{\pi i \psi(x_k) f(x_k)}{\phi'(x_k)}.$$

Similarly for δ_k^+, and (4.6.1.7) follows.

Modifications can be made to allow multiple zeros at x_k and $a \le x_1 < \cdots < x_n \le b$ (an endpoint can be an abscissa).

If we start with any analytic function $\phi(z)$ with zeros at x_k, and a weighting function $w(x)$ such that $w\phi \in H^*$, then, by the formula of Sochozki and Plemelj, the function of the second kind defined by (4.6.1.3) satisfies the conditions of the last theorem. Thus, any such ϕ leads to an integration rule and error term

$$\int_a^b w(x) f(x)\, dx = \sum_{k=1}^{n} \lambda_k f(x_k) + \frac{1}{2\pi i} \int_C \frac{\psi(z)}{\phi(z)} f(z)\, dz,$$

$$\lambda_k = \frac{1}{\phi'(x_k)} \int_a^b \frac{w(t)\phi(t)\, dt}{t - x_k}. \tag{4.6.1.8}$$

Familiar rules can be recovered by selecting $\phi(z)$ appropriately. If one takes $\phi(z) = (z - x_1)(z - x_2) \cdots (z - x_n)$, $a < x_1 < \cdots < x_n < b$, we get interpolatory integration rules. If we select $\phi(z) = \phi_n(z) = n$th orthogonal polynomial with respect to a weight $w(x)$, we get the Gauss rules. The selection $\phi(z) = P_{n+1}(z) + P_n(z) = (1 + z)P_n^{(0,1)}(z)$ leads to a Radau rule. The selection

$$\phi(z) = P_{n+2}(z) - P_n(z) = -\frac{2n + 3}{2n + 2}(1 - z^2)P_n^{(1,1)}(z)$$

leads to Lobatto, while the selection

$$\phi(z) = T_{n+1}(z) - T_{n-1}(z) = -\frac{\Gamma(\frac{1}{2})\Gamma(n + 1)}{\Gamma(n + \frac{1}{2})}(1 - z^2)P_{n-1}^{(1/2,1/2)}(z)$$

leads to Clenshaw–Curtis.

One can also generate rules of noninterpolatory type, for example, trapezoidal, midpoint, Euler–Maclaurin, and Romberg rules by appropriate selection of $\phi(z)$.

It is beyond the scope of this book to give the derivations of asymptotic expansions for $K_n(z)$, $n \to \infty$. Such derivations may require the full panoply of methods in asymptotic analysis: saddle-point methods, Laplace methods, etc. We shall content ourselves with quoting some typical results.

Asymptotic expressions for $K_n(z)$ have been derived systematically by Elliott. In the case of the classical orthogonal polynomials, these expressions have been listed in Section 1.13. Now, the more that is known about the singularity structure of the integrand—the location of the singularity, its type, residues, etc.—the more accurately can the asymptotic error be obtained.

Let GJ designate the Gauss–Jacobi rule on $[-1, 1]$ with weight $w(x) = (1-x)^\alpha(1+x)^\beta$, $\alpha, \beta > -1$. Set $N = n + \frac{1}{2}(\alpha + \beta + 1)$. Let GL be the Gauss–Laguerre rule on $[0, \infty)$ with weight $w(x) = x^\alpha e^{-x}$, $\alpha > -1$. Set $k = 4n + 2\alpha + 2$. Let GH be the Gauss–Hermite rule on $(-\infty, \infty)$ with weight $w(x) = e^{-x^2}$ and set $k = 2(n+1)$.

If the integrand $f(x)$ is analytic in a region that contains $[-1, 1]$ in its interior, then, as can be shown from (4.6.19), $E_{GJ_n}(f)$ converges to 0 geometrically. More precisely, if $f(z)$ is analytic in the closed ellipse $\mathscr{E}_\rho$, $\rho > 1$, with foci at ± 1 and semiaxis sum ρ, then

$$|E_{GJ_n}(f)| \leq \text{const}/\rho^{2n}. \tag{4.6.1.9}$$

Still more precisely, we have for large n, and, say, for the ordinary Gauss rule G_n

$$K_n(z) = \frac{Q_n(z)}{P_n(z)} \cong \frac{2\pi}{\{z + (z^2-1)^{1/2}\}^{2n+1}}. \tag{4.6.1.10}$$

The symbol $\cong$ means that the ratio of the two sides approaches 1 as $n \to \infty$. Now on $\mathscr{E}_\rho$ we have $|z + (z^2-1)^{1/2}| = \rho$ so that $|K_n(z)| \cong 2\pi/\rho^{2n+1}$. Since

$$E_{G_n}(f) = (2\pi i)^{-1} \int_{\mathscr{E}_\rho} K_n(z) f(z)\, dz,$$

$$|E_{G_n}(f)| \leq \frac{l(\mathscr{E}_\rho)}{\rho^{2n+1}} \max_{z \in \mathscr{E}_\rho} |f(z)|(1 + o(1)),$$

where $l(\mathscr{E}_\rho) =$ length of $\mathscr{E}_\rho$. Since $l(\mathscr{E}_\rho) < 2\pi\frac{1}{2}(\rho + 1/\rho)$,

$$|E_{G_n}(f)| \leq \frac{\pi(\rho + \rho^{-1})}{\rho^{2n+1}} \max_{z \in \mathscr{E}_\rho} |f(z)|, \qquad n \text{ large}. \tag{4.6.1.11}$$

Assume now that $f(z)$ is regular in the interior of $\mathscr{E}_\rho$ and on the ellipse has only a pair of simple poles at z_0 and $\bar{z}_0$. Let the residues at the poles of the function $(z-1)^\alpha(z+1)^\beta f(z)$ be r and $\bar{r}$. Then

$$E_{GJ_n}(f) \cong -4\pi \operatorname{Re}\{r[z_0 + (z_0^2 - 1)^{1/2}]^{-2n}\}. \tag{4.6.1.12}$$

Choose the root so that $|z_0 + (z_0^2 - 1)^{1/2}| > 1$.

If $f(z)$ has a power singularity at $z = 1$ such that $f(z)w(z) \cong (1-z)^\sigma$, $\sigma > -1$, as $z \to 1$, and if $f(z)$ is regular elsewhere in a region that contains $[-1, 1]$, then

$$E_{GJ_n}(f) \cong \frac{\sin \pi(\sigma - \alpha)}{2^{\sigma-1}\pi N^{2\sigma+2}} \int_0^\infty \frac{t^{2\sigma+1}K_\alpha(t)}{I_\alpha(t)}\,dt. \tag{4.6.1.13}$$

Here $\sigma > -1$, $\sigma - \alpha > -1$, $\sigma - \alpha \neq$ integer, and I and K are the modified Bessel functions. Note that this implies convergence with the rapidity $n^{-2\sigma-2}$.

If $\sigma - \alpha =$ integer in the above, we have

$$E_{GJ_n}(f) = o(N^{-2\sigma-2}). \tag{4.6.1.14}$$

A prototype function with an internal singularity of low continuity order is $|x|^\mu$, $\mu > -1$. The Gaussian error for this integrand has the following asymptotic expansion:

$$E_{G_{2m}}(|x|^\mu) \cong -\frac{4\sin(\pi\mu/2)\Gamma(1+\mu)(1-2^{-\mu})\zeta(1+\mu)}{(4m+1)^{\mu+1}}. \tag{4.6.1.15}$$

For $-1 < \mu < 0$, the integrand is unbounded over $[-1, 1]$, and G_{2m} which "ignores the singularity" converges to the proper answer with slow speed $\cong m^{-\mu-1}$.

If $f(z)$ has no singularities on or within any parabola $\operatorname{Re}(-z)^{1/2} = R =$ const, except for a pair of simple poles at z_0, $\bar{z}_0$ where $w(z)f(z)$ has residues r, $\bar{r}$, then

$$E_{GL_n}(f) \cong -4\pi \operatorname{Re}\{re^{-i\alpha\pi}[\exp(-z_0)^{1/2}]^{-2k^{1/2}}\}. \tag{4.6.1.16}$$

Note that this implies convergence like $\text{const}^{-n^{1/2}}$.

If $f(z)w(z) \cong z^\sigma$ as $z \to 0$, $\sigma > -1$, $\sigma - \alpha > -1$, then

$$E_{GL_n}(f) \cong \frac{4 \sin \pi(\sigma - \alpha)}{\pi k^{\sigma+1}} \int_0^\infty \frac{t^{2\sigma+1}K_\alpha(t)}{I_\alpha(t)}\,dt, \tag{4.6.1.17}$$

so that convergence is like $n^{-\sigma-1}$.

If $f(z)$ has no singularities between the lines $\operatorname{Im} z = \pm R$, $R =$ const, and on them has only a pair of simple poles z_0, $\bar{z}_0$, $\operatorname{Im} z_0 > 0$ with $f(z)w(z)$ having residues r, $\bar{r}$ at these points, then

$$E_{GH_n}(f) \cong -4\pi \operatorname{Re}\{r[\exp(-iz_0)]^{-2k^{1/2}}\}. \tag{4.6.1.18}$$

In the cases GL and GH, certain size conditions at $z = \infty$, not mentioned here, must also be met by the integrand.

The reader who wishes to see other types of integrands handled should consult Donaldson and Elliott.

Example. Select $f(z) = 1/(1 + z^2)$. This function has simple poles at $\pm i$ with residues $\pm i/2$. The first column lists the error obtained by applying G_n to f. The second column lists the error estimate obtained by using the asymptotic expression on the right-hand side of (4.6.1 12).

G_2	.0707 96327	.0766 12819
G_4	.0021 68876	.0022 55271
G_6	.0000 64620	.0000 66389
G_8	.0000 01914	.0000 01954
G_{10}	.0000 00057	.0000 00058
G_{12}	.0000 00002	.0000 00002

References. Barrett [1]; Chawla [5]; Chawla and Jain [1]; Donaldson and Elliott [1–3]; Elliott [4]; Lether [12]; Muskhelishvili [1, Chap. 2, Par. 16–17]; E. G. Phillips [1, p. 117, Theorem 2].

4.6.5 Periodic Analytic Functions; Trapezoidal Error

In this section we obtain an error estimate when the trapezoidal rule is applied to functions that are periodic and analytic. We shall assume that $f(z)$ is regular in the strip

$$|\operatorname{Im} z| < \sigma,$$

has period p, and is real for real values of z. Under these circumstances, we can write

$$f(z) = \sum_{k=-\infty}^{\infty} a_k \exp(2\pi ikz/p), \qquad a_k = \bar{a}_{-k} \tag{4.6.5.1}$$

and this series is uniformly and absolutely convergent in any substrip $|\operatorname{Im} z| \le \sigma'$, $0 \le \sigma' < \sigma$. Along the x axis we have

$$f(x) = \frac{\alpha_0}{2} + \sum_{k=1}^{\infty} \alpha_k \cos\frac{2\pi kx}{p} + \beta_k \sin\frac{2\pi kx}{p},$$

$$a_k = \frac{1}{2}(\alpha_k - i\beta_k). \tag{4.6.5.2}$$

As in (2.9.22) we find

$$E_{T_n}(f(x)) = p\sum_{k=1}^{\infty}\alpha_{kn} = p\sum_{k=-\infty}^{\infty}{}' a_{kn} = 2p \operatorname{Re}\sum_{k=1}^{\infty} a_{kn} \tag{4.6.5.3}$$

where $\sum'$ omits the term $k = 0$.

If we make the change of variable

$$z = \frac{p}{2\pi i}\log w, \qquad w = \exp\left(\frac{2\pi i z}{p}\right),$$

then

$$g(w) = f\left(\frac{p}{2\pi i}\log w\right) = \sum_{k=-\infty}^{\infty} a_k w^k, \tag{4.6.5.4}$$

and in view of the periodicity of $f(z)$, $g(w)$ will be regular and single valued in the annulus

$$A\colon\quad \exp(-2\pi\sigma/p) < |w| < \exp(2\pi\sigma/p).$$

Therefore,

$$a_k = \frac{1}{2\pi i}\int_C \frac{g(w)}{w^{k+1}}\,dw \tag{4.6.5.5}$$

where C lies in A and surrounds $w = 0$.

Now, assuming that C also contains $|w| = 1$ in its interior,

$$\sum_{k=1}^{\infty} a_{kn} = \sum_{k=1}^{\infty}\frac{1}{2\pi i}\int_C \frac{g(w)}{w^{kn+1}}\,dw = \frac{1}{2\pi i}\int_C \frac{g(w)}{w(w^n - 1)}\,dw.$$

Thus,

$$E_{T_n}(f(x)) = \frac{p}{\pi}\operatorname{Im}\int_C \frac{g(w)}{w(w^n - 1)}\,dw. \tag{4.6.5.6}$$

We now have

THEOREM. *Let l_t designate the line $x + it$, $-\infty < x < \infty$. Let $f(z)$ be real on the real axis, have period p, and be regular in $|\operatorname{Im} z| < \sigma$, then*

$$|E_{T_n}(f)| \le 2p \max_{z\in l_t}|f(z)|\cdot\frac{\exp(-2\pi tn/p)}{1 - \exp(-2\pi tn/p)}, \tag{4.6.5.7}$$

for all $0 < t < \sigma$.

Proof. Let r satisfy $1 < r < \exp(2\pi\sigma/p)$ and let C: $|w| = r$. Then from the identity (4.6.5.6) we have

$$|E_{T_n}(f)| \leq \frac{p}{\pi} \int_C \frac{|g(w)|\, ds}{|w|\,|w^n - 1|}$$

$$\leq \frac{p}{\pi} \frac{2\pi r}{r(r^n - 1)} \max_{w \in C} |g(w)|.$$

This follows since $|w^n - 1| \geq r^n - 1$ on C. Now

$$\max_{w \in C} |g(w)| = \max_{z \in l_t} |f(z)|$$

where $t = (p/2\pi) \log r$ and the inequality follows.

We may conclude from this that as $n \to \infty$, the trapezoidal rule converges to the integral with geometric rapidity; the larger the strip of regularity, the more rapid is the convergence.

Example. Compare the example on p. 139. For a deeper analysis of the error in this specific example, see Donaldson and Elliott.

Care must be taken in practical situations to include rounding effects. See Rabinowitz.

For functions with period 2π which satisfy the conditions of the theorem, Schönhage and Kress [9] have established that

$$|E_{T_n}(f)| \leq \gamma_{n\sigma} \sup_{|\mathrm{Im}\, z| \leq \sigma} |\mathrm{Re}\, f(z)|,$$

$$|E_{T_n}(f)| \leq \gamma_{n\sigma} \sup_{|\mathrm{Im}\, z| \leq \sigma} |\mathrm{Im}\, f(z)|$$

where

$$\gamma_{n\sigma} = \frac{8}{1 - e^{-2n\sigma}} \log \frac{1 + e^{-n\sigma}}{1 - e^{-n\sigma}} = 16e^{-n\sigma} + O(e^{-2n\sigma}).$$

For error bounds involving $|f(z) + f(\bar{z})|$, see Knauff.

Hunter has generalized the treatment of periodic analytic functions to include functions meromorphic within a strip, while Martensen has generalized the results to periodic analytic functions of several variables.

References. Brass [1, Chap. 6]; Davis [4]; Donaldson and Elliott [1]; Hunter [4]; Knauff [1]; Knauff and Kress [1]; Kress [2, 4, 8, 9]; Martensen [1]; Rabinowitz [7]; Schönhage [2].

4.7 Application of Functional Analysis to Numerical Integration

Beginning in the late 1940s, applications of the ideas of functional analysis were made to numerical integration. The activity in this area has been considerable and has given rise to mathematics of great intrinsic interest and with connections to many topics in analysis. It is appropriate to devote some space to a sketch of some of its features.

In the functional analysis approach to the errors of numerical integration, one works within a normed or seminormed linear space X of functions. One considers the error functional as a member of the dual space of linear functionals and estimates its norm.

Some aspects of this approach can be exhibited without the use of the heavy machinery of functional analysis. Witness the following simple example.

Let $f(z)$ be a regular analytic function in $|z| < 1$ and suppose that $f(z)$ is also continuous on $|z| = 1$. Let $-1 < a < b < 1$ and set

$$E(f) = \int_a^b f(x)\,dx - \sum_{k=1}^{n} w_k f(x_k). \tag{4.7.1}$$

For simplicity, assume that $-1 < x_k < 1$. By Cauchy's theorem, we have

$$f(z) = \frac{1}{2\pi i}\int_C \frac{f(t)}{t - z}\,dt = \frac{1}{2\pi}\int_C \frac{f(t)}{1 - z\bar{t}}\,ds. \tag{4.7.2}$$

In this last integral, $\bar{t}$ is the complex conjugate of t and ds is the element of length along the unit circle C. Applying E to both sides of (4.7.2) and interchanging E and $\int$, we obtain

$$E(f) = \frac{1}{2\pi}\int_C E\left(\frac{1}{1 - z\bar{t}}\right) f(t)\,ds. \tag{4.7.3}$$

From the Schwarz inequality, we have

$$|E(f)|^2 \leq \frac{1}{4\pi^2}\int_C \left| E\left(\frac{1}{1 - z\bar{t}}\right)\right|^2 ds \int_C |f(t)|^2\,ds. \tag{4.7.4}$$

Now

$$\frac{1}{1 - z\bar{t}} = \sum_{n=0}^{\infty} z^n \bar{t}^n, \quad \text{so that} \quad E\left(\frac{1}{1 - z\bar{t}}\right) = \sum_{n=0}^{\infty} E(z^n)\bar{t}^n.$$

Hence

$$\int_C \left| E\left(\frac{1}{1 - z\bar{t}}\right) \right|^2 ds = \int_C E\left(\frac{1}{1 - z\bar{t}}\right) \overline{E\left(\frac{1}{1 - z\bar{t}}\right)} ds$$

$$= \int_C \sum_{n=0}^{\infty} E(z^n)\bar{t}^n \sum_{m=0}^{\infty} \overline{E(z^m)} t^m \, ds.$$

In view of the fact that $\int_C \bar{t}^n t^m \, ds = 0$ if $m \neq n$ and 2π if $m = n$, we conclude that

$$\frac{1}{4\pi^2} \int_C \left| E\left(\frac{1}{1 - z\bar{t}}\right) \right|^2 ds = \frac{1}{2\pi} \sum_{n=0}^{\infty} | E(z^n) |^2. \tag{4.7.5}$$

If the rule of approximation is exact for polynomials of class $\mathscr{P}_N$, then this sum simplifies to

$$\frac{1}{2\pi} \sum_{n=N+1}^{\infty} | E(z^n) |^2.$$

The error estimate (4.7.4) may be written as

$$| E(f) | \leq \sigma \| f \|, \tag{4.7.6}$$

where

$$\sigma^2 = \frac{1}{2\pi} \sum_{n=0}^{\infty} | E(z^n) |^2 \tag{4.7.7}$$

and

$$\| f \|^2 = \int_C | f(z) |^2 \, ds. \tag{4.7.8}$$

The quantity $\| f \|$ is called the *norm of* f, and σ is the *norm of* E. This estimate is *derivative-free* insofar as it requires only a knowledge of the values of $f(z)$ along $|z| = 1$. The norm $\| f \|$ may be estimated from

$$\| f \|^2 = \int_C | f(z) |^2 \, ds \leq 2\pi \max_{|z|=1} | f(z) |^2, \tag{4.7.9}$$

so that (4.7.6) leads to

$$| E(f) | \leq (2\pi)^{1/2} \sigma \max_{|z|=1} | f(z) |. \tag{4.7.10}$$

The constant σ is independent of f and depends merely on the rule of approximate integration. It may be computed directly from (4.7.7) by machine or it may be handled analytically.

On the basis of a number of summation identities, Hämmerlin has computed the following bounds for σ.

Interval	Rule	Error norm
$[-\frac{1}{2}, \frac{1}{2}]$	$h^{-1} \times T$	$(2\pi)^{1/2}\sigma < .189h^2$
$[-\frac{1}{2}, \frac{1}{2}]$	$h^{-1} \times S$	$(2\pi)^{1/2}\sigma < .253h^4$
$[-\frac{1}{2}, \frac{1}{2}]$	$h^{-1} \times N$	$(2\pi)^{1/2}\sigma < .569h^4$

Note: T = trapezoidal rule, S = Simpson's rule, N = Simpson's $\frac{3}{8}$ rule.

Using the somewhat more general norm

$$\|f\|^2 = (r)^{-1} \int_{C_r} |f(z)|^2 \, ds, \qquad C_r\colon |z| = r > 1,$$

Stetter studies the error in the formula of Gauss type (2.7.13):

$$E(f) = \int_{-1}^{1} \frac{f(x)}{(1 - x^2)^{1/2}} \, dx - \frac{\pi}{n} \sum_{k=1}^{n} f\left(\cos\left(\frac{2k-1}{2n}\pi\right)\right),$$

and derives the inequality

$$|\sigma| \le \frac{1.05}{r^{2n-2}(2n)^{1/2}} \left(\frac{1}{r^4 - 1}\right)^{1/2}. \tag{4.7.11}$$

Suppose that $E(p) = 0$ whenever $p \in \mathscr{P}_m$. Then, if it is convenient to do so, we can replace (4.7.6) by

$$|E(f)| \le \sigma \|f - p\| \tag{4.7.12}$$

for any $p \in \mathscr{P}_m$. For, $|E(f - p)| \le \sigma\|f - p\|$. But, $E(f - p) = E(f) - E(p) = E(f)$.

References. Brass [1, Sect. V.4]; Chai and Wertz [1]; Davis [1, 5]; Davis and Rabinowitz [1]; Hämmerlin [2–4]; Lo, Lee, and Sun [1]; Rabinowitz and Richter [2]; F. Stetter [1, 2]; Stroud and Secrest [2].

The broader aspects of the theory require a knowledge of the elementary portions of the theory of normed spaces and Hilbert spaces. This knowledge is assumed in what follows. The reader is referred to the book by Davis [6] or to any book on elementary functional analysis for details.

A linear space X is called a *normed linear space* if for each element $x \in X$ there is defined a real number designated by $\|x\|$ with the properties

(1) $\|x\| \geq 0$,
(2) $\|x\| = 0$ if and only if $x = 0$,
(3) $\|\alpha x\| = |\alpha| \, \|x\|$ for scalar α,
(4) $\|x + y\| \leq \|x\| + \|y\|$.

The quantity $\|x\|$ is called the *norm of* x. If all conditions except (2) are fulfilled, then $\| \ \|$ is called a *seminorm* and X is a seminormed space.

While the number of seminormed function spaces is endless, only a few have been put to work in theories of numerical integration. Among them, we cite:

(a) Let $q \geq 1$ and an integer $r \geq 0$ be fixed. Let X designate the set of all functions that have an absolutely continuous derivative of order $r - 1$ on $[a, b]$ and whose rth derivative is qth power integrable there. Set

$$\|f\| = \left(\int_a^b |f^{(r)}(x)|^q \, dx \right)^{1/q}. \tag{4.7.13}$$

A weighting function is often included in the integral defining $\|f\|$. If $r = 0$, X is a normed space, while if $r \geq 1$, it is a seminormed space. In this case, if $\|f\| = 0$, then $f \in \mathscr{P}_{r-1}$. If a normed space is required, one way of producing it is to throw into one class all functions of X which differ by a polynomial in $\mathscr{P}_{r-1}$. Another way is to add to the above norm any functional which serves as a norm for $\mathscr{P}_{r-1}$, e.g.,

$$\sum_{i=1}^{r} k_i |f(x_i)|, \qquad k_i > 0, \qquad a \leq x_1 < \cdots < x_r \leq b.$$

(b) Let X be the class $C^r[a, b]$ and set

$$\|f\| = \max_{a \leq x \leq b} |f^{(r)}(x)|. \tag{4.7.14}$$

This is a seminormed space if $r \geq 1$.

(c) Let B designate a region in a Euclidean space of d dimensions. Given a nonnegative integer l and a number $p \geq 1$, let X designate the set of all functions defined on B such that the indicated derivatives below are pth power integrable.

Define

$$\|f\| = \left(\int_B \sum_{|\mathbf{i}|=l} |f^{(\mathbf{i})}|^p \, dx_1 \, dx_2 \cdots dx_d \right)^{1/p}. \tag{4.7.15}$$

X is sometimes called a *Sobolev space*. For the notation, see Section 1.11.7.

(d) Let B be a region in the complex plane. Let $w(z) \geq 0$ in B with $\int_B w(z)\,dx\,dy < \infty$. Fix a $p \geq 1$. Let X be the set of functions that are analytic and single valued in B and such that

$$\|f\| = \left(\int_B w(z)\,|f(z)|^p\,dx\,dy\right)^{1/p} < \infty. \tag{4.7.16}$$

Common selections are $w(z) \equiv 1$, $p = 2$, $B =$ a circle, ellipse, or infinite strip.

(e) A similar definition where the integral is taken over the boundary of B.

The set of single valued analytic functions $f(z)$ in $|z| < 1$ such that

$$\int_{|z|=1} |f|^p\,ds = \lim_{r\to 1^-} \int_{|z|=r} |f(re^{i\theta})|^p\,ds < \infty, \qquad 1 \leq p < \infty,$$

is often called the *Hardy space* H_p. These spaces are important in dealing with integrals over $[-1, 1]$ of functions that have singularities at the endpoints. The space H_2 is of special interest since it is a Hilbert space. It is also connected with the optimal integration rules of Wilf.

The error $E(f) = \int_a^b f(x)\,dx - \sum_{k=1}^n w_k f(x_k)$, as observed in Section 4.3, is a *linear functional* over X. In numerous instances E is also a *bounded* linear functional. This means that

$$\sup_{f\in X} \frac{|E(f)|}{\|f\|} = \sigma < \infty \tag{4.7.17}$$

and if we then define a norm of E by

$$\|E\| = \sigma \tag{4.7.18}$$

we may write

$$|E(f)| \leq \|E\|\,\|f\| \tag{4.7.19}$$

as the fundamental inequality of the error.

It is perhaps easier computationally to work with spaces X of functions that are inner product or Hilbert spaces, for then the whole linear apparatus of least square approximations, orthogonality, etc. becomes available. When the spaces are real, there are intimate connections to the theory of spline approximations, and Green's functions; when they are complex analytic, there are connections to the theory of analytic reproducing kernel functions.

Numerous authors have produced tables of error norms. Hämmerlin works with the Hardy space H_2 and derives norms for the trapezoidal and Simpson's rule as functions of the spacing h. O'Hara and Smith have norms for Clenshaw–Curtis rules. Stroud and Secrest cover Gaussian rules, and Stroud has Romberg rules. Barnhill and his students have worked with multiple integrals. For a fixed X, we may now consider a series of related questions

(a) How does one arrive at $\|f\|$ or at an estimate for it?
(b) For a fixed rule R, how does one compute or estimate from above or below its error norm $\|E\|$? What are asymptotic expressions for $\|E\|$ as $n \to \infty$?
(c) For fixed abscissas $x_1, \ldots, x_n$, how does one select weights $w_1, \ldots, w_n$ so that $\|E\|$ is minimized?
    This program leads to "relatively minimum norm" or "relatively best" rules.
(d) If both x_i and w_i are free to vary, how does one select them so that $\|E\|$ is minimized? This leads to "minimum norm" or "best" rules. The term "optimal" integration rule is also used in this context. However, one must always investigate further to determine with respect to what the particular rule is optimal.

Nor does this end the variety of interesting extremal problems that can be posed. We may require e.g., the minimum of $\|E\|$ under any of the side conditions

(e) abscissas are confined to some given subset
(f) $w_i \geq 0$
(g) the resulting rule required to be precise for a given number of prescribed functions.

In their book on optimal algorithms, Traub and Wózniakowski give nearly 100 titles in a partial list of papers in which optimal integration rules are studied.

The monograph by Levin and Girshovich on optimal integration rules deals mainly with cases in which the abscissas and weights of one- and two-dimensional optimal rules can be given in closed form. They also discuss some asymptotically optimal rules.

References. Barnhill [4]; Barnhill and Nielson [1]; Brass [1, Chap. VII]; Davis [1, 6]; Davis and Rabinowitz [1]; Ghizzetti [1]; Golomb and Weinberger [1]; Levin and Girshovich [1]; Lo, Lee, and Sun [1]; Meyers and Sard [1]; Nikolskii [1]; O'Hara and Smith [1]; Rabinowitz and Richter [2]; Sard [1]; Sikorski [1]; Stroud [7]; Stroud and Secrest [2]; Traub and Wózniakowski [1]; Valentin [1]; Weinberger [1]; Wilf [1, 2].

4.7.1 *Approximate Integration and the Hilbert Space $L^2(B)$.*

Let B be a bounded region in the complex plane. We shall designate by $L^2(B)$, the set of functions that are single valued and analytic in B and such that

$$\|f\|^2 = \int_B\int |f(z)|^2\,dx\,dy < \infty. \tag{4.7.1.1}$$

This norm has the monotonicity property that if $G \subseteq B$, then $\|f\|_G \leq \|f\|_B$, which is occasionally of use in norm estimation. Introducing

$$(f, g) = \int_B \int f(z)\overline{g(z)}\, dx\, dy \tag{4.7.1.2}$$

as an inner product, $L^2(B)$ becomes a Hilbert space. If z_0 is an interior point of B then for any $n = 0, 1, 2, \ldots$ the point functional $L(f) = f^{(n)}(z_0)$ is *bounded* over $L^2(B)$. That is, there exists a constant $C = C(z_0, n)$ such that $|f^{(n)}(z_0)| \leq C\|f\|$ for all $f \in L^2(B)$. In particular, $L(f) = f(z_0)$ is bounded, and as a consequence, $L^2(B)$ has a *reproducing kernel function*

$$K_B(z, \bar{w}) = \sum_{n=0}^{\infty} p_n^*(z)\overline{p_n^*(w)}$$

where $\{p_n^*(z)\}$ is *any* complete orthonormal system for $L^2(B)$. The kernel has the property that $(f(z), K_B(z, \bar{w})) = f(w)$ for any $f \in L^2(B)$. If the interval $a \leq x \leq b$ lies in the interior of B and if $\alpha(x)$ is a function of bounded variation on $[a, b]$, then the functional $L(f) = \int_a^b f(x)\, d\alpha(x)$ is also bounded over $L^2(B)$ and hence also any error expression of the form

$$E(f) = \int_a^b f(x)\, d\alpha(x) - \sum_{k=1}^{n} w_k f^{(n_k)}(x_k) \tag{4.7.1.3}$$

where $a \leq x_i \leq b$. Now for any linear functional L bounded over $L^2(B)$, its *representer* $r(z)$ is given by

$$r(z) = L_{\bar{w}} K_B(z, \bar{w}). \tag{4.7.1.4}$$

This means that

$$L(f) = (f, r), \qquad f \in L^2(B) \tag{4.7.1.5}$$

and that

$$\|L\|^2 = \|r\|^2 = \|L_{\bar{w}} K_B(z, \bar{w})\|^2$$

$$= \sum_{k=0}^{\infty} |L(p_k^*)|^2. \tag{4.7.1.6}$$

It follows that

$$|E(f)| \leq \|E\|\, \|f\| \tag{4.7.1.7}$$

and this estimate is sharp in $L^2(B)$. The right-hand term is derivative free. The constant $\|E\|$ is independent of f and may be tabulated assuming that $\{p_k^*\}$ is known.

Let B be a bounded, simply connected region with boundary C. It is known that if the complement of $B \cup C$ is a simple region whose boundary is exactly C, then the set of powers $1, z, z^2, \ldots$ is complete in $L^2(B)$. Hence there exists a complete orthonormal set of polynomials for $L^2(B)$: $p_0^*(z), p_1^*(z), \ldots$.

For such regions B, if E is a bounded linear functional that vanishes on all functions of $\mathscr{P}_m$, then

$$\|E\|^2 = \sum_{k=m+1}^{\infty} |E(p_k^*)|^2. \tag{4.7.1.8}$$

For any $p_m(z) \in \mathscr{P}_m$ and $f \in L^2(B)$, we have $E(f) = E(f - p_m)$. Therefore one has a sharper estimate of the form

$$|E(f)|^2 = |E(f - p_m)|^2 \leq \left(\sum_{k=m+1}^{\infty} |E(p_k^*)|^2\right)(\|f - p_m\|)^2 \tag{4.7.1.9}$$

for any $p_m \in \mathscr{P}_m$.

If $L, L_1, L_2, \ldots, L_n$ are fixed linear functionals on $L^2(B)$, and if we set

$$E = L - (w_1 L_1 + \cdots + w_n L_n), \tag{4.7.1.10}$$

the problem of finding

$$\min_{w_i} \|E\| = \min_{w_i} \|L - (w_1 L_1 + \cdots + w_n L_n)\|$$

is that of the best approximation of L by linear combinations of $L_1, \ldots, L_n$ in the dual space of $L^2(B)$. The problem has a unique solution L^{RMN} and is the "relatively minimum norm" rule for L. The solution can be described as a Fourier expansion in the following way. Let $r(z), r_1(z), \ldots, r_n(z)$ be the representers of $L, L_1, \ldots, L_n$ respectively. Assuming that $L_1, \ldots, L_n$ and hence $r_1(z), \ldots, r_n(z)$ are independent, orthonormalize them, yielding the functions $r_1^*(z), \ldots, r_n^*(z)$. Set

$$s(z) = \sum_{k=1}^{n} (r, r_k^*) r_k^*(z). \tag{4.7.1.11}$$

Then

$$L^{\text{RMN}}(f) = (f, s) = \sum_{k=1}^{n} (r_k^*, r)(f, r_k^*) \tag{4.7.1.12}$$

and

$$\|L - L^{\text{RMN}}\|^2 = \|r\|^2 - \sum_{k=1}^{n} |(r, r_k^*)|^2. \tag{4.7.1.13}$$

As a particular instance, if we set $f = r_i^*$ in (4.7.1.12), we see that

$$L^{\text{RMN}}(r_i^*) = (r_i^*, s) = (r_i^*, r) = L(r_i^*).$$

By taking appropriate linear combinations,

$$L^{\text{RMN}}(r_i) = L(r_i) \quad \text{where} \quad r_i = L_{i,\bar{w}} K_B(z, \bar{w}). \tag{4.7.1.14}$$

Thus, the approximate rule L^{RMN} is exact for the n functions r_i, and can therefore be regarded as a rule of interpolatory type for these functions. If $L = \int_a^b f(x)\,d\alpha(x)$ and $L_i = f(x_i)$, $i = 1, 2, \ldots, n$, then the RMN rule for L based on abscissas x_i integrates the n functions $K_B(z, x_i)$ exactly. In general, a RMN rule will not integrate the constant function exactly—a fact that has caused eyebrows to be raised for it runs counter to a few needs and many prejudices.

To overcome this objection to RMN rules, Knauff [2] and Chawla and Kaul have derived rules using a fixed set of n abscissas which have a minimal error norm in the class of rules which are exact for all polynomials in $\mathscr{P}_m$ for a given $m \le n - 1$.

Knauff [3] has also determined the weights in RMN rules for certain spaces of periodic analytic functions when equidistant abscissas are used. It turns out that the RMN rule has the form

$$\int_0^{2\pi} f(x)\,dx \approx \frac{2\pi a_n}{n} \sum_{k=1}^{n} f\left(\frac{2\pi k}{n}\right)$$

with different values of a_n depending on the particular space involved. The same weight a_n appears in the RMN rule

$$\int_{-1}^{1} (1 - x^2)^{1/2} f(x)\,dx \approx \frac{\pi a_n}{n} \sum_{k=1}^{n} f\left(\cos \frac{2k-1}{2n} \pi\right)$$

when $f(x)$ belongs to the appropriate space.

References. Chawla and Kaul [1]; Knauff [2, 3]; Knauff and Paulik [1].

If the finite orthonormal system $r_1^*, \ldots, r_n^*$ is augmented to form a complete orthonormal system $r_1^*, r_2^*, \ldots,$ then for any $f \in L^2(B)$, we have $f = \sum_{k=1}^{\infty} (f, r_k^*) r_k^*$ so that $(f, r) = L(f) = \sum_{k=1}^{\infty} (f, r_k^*) L(r_k^*) = \sum_{k=1}^{\infty} (f, r_k^*) \times (r_k^*, r)$. From (4.7.1.12),

$$L(f) - L^{\text{RMN}}(f) = \sum_{k=n+1}^{\infty} (f, r_k^*)(r_k^*, r)$$

so that by the Schwarz inequality

$$\begin{aligned} |L(f) - L^{\text{RMN}}(f)|^2 &\le \sum_{k=n+1}^{\infty} |(f, r_k^*)|^2 \sum_{k=n+1}^{\infty} |(r_k^*, r)|^2 \\ &= \left(\sum_{k=n+1}^{\infty} |(f, r_k^*)|^2\right)\left(\|r\|^2 - \sum_{k=1}^{n} |(r_k^*, r)|^2\right) \\ &= \left(\sum_{k=n+1}^{\infty} |(f, r_k^*)|^2\right) \|L - L^{\text{RMN}}\|^2. \end{aligned} \tag{4.7.1.15}$$

This inequality is known as the *hypercircle inequality* and was originally derived by Prager and Synge in a different context.

Here is a further interpretation. Consider the set of all functions f such that

$$L_1(f) = (f, r_1) = c_1, \ldots, L_n(f) = (f, r_n) = c_n. \qquad (4.7.1.16)$$

This set is called a *hyperplane* H of codimension n. The intersection of a hyperplane H and a ball $\|f\| \leq M$ in the Hilbert space is called a *hypercircle*. If we pass over to the orthonormalized functions r_i^*, then by taking linear combinations, we may write the hyperplane in the equivalent form

$$H: L_i^*(f) = (f, r_i^*) = a_i, \qquad i = 1, 2, \ldots, n, \qquad (4.7.1.17)$$

for appropriately determined constants a_i. H contains a unique function closest to the 0 function. It is given by

$$w = \sum_{k=1}^{n} a_k r_k^* = \sum_{k=1}^{n} (f, r_k^*) r_k^* \qquad (4.7.1.18)$$

and its norm is

$$\|w\|^2 = \sum_{k=1}^{n} |(f, r_k^*)|^2.$$

Hence the hypercircle inequality can be written as

$$|L(f) - L^{\text{RMN}}(f)| \leq (\|f\|^2 - \|w\|^2)^{1/2} \|L - L^{\text{RMN}}\|. \qquad (4.7.1.19)$$

This inequality is satisfied for any function lying in the hypercircle with $M = \|f\|$ and it can be shown that the bounds are attained.

Generally speaking, error estimates in approximate integration are given in the form $|E(f)| \leq C_R \|f\|$ where C_R is a constant which depends upon the integration rule and is independent of the particular function integrated and where $\|f\|$ depends upon f and is independent of the rule. An estimate of $\|f\|$ makes no use of the *specific* functional values obtained in the process of computing the specific integral by the rule. It seems a pity to waste this information and (4.7.1.19) gives improved estimates which make use of the functional values. They occur in the term $\|w\|^2 = \sum_{k=1}^{n} |(f, r_k^*)|^2 = \sum_{k=1}^{n} |a_i|^2$ and the a_i are related to the c_i linearly. (4.7.1.19) may be said to provide a *value-dependent* error estimate. For high values of n, they become increasingly difficult to compute.

For analogs of the hypercircle inequality in normed linear spaces see Meinguet.

For the application of the hypercircle inequality in several dimensions, see Barnhill and Stroud.

References. Barnhill [4]; Golomb and Weinberger [1]; Meinguet [1]; Stroud [18]; Synge [1]; Valentin [1, 2].

Let $L_i(f) = f(z_i)$, $i = 1, 2, \ldots, n$, $z_i \in B$. The problem of minimizing

$$\|E\| = \left\| L - \sum_{i=1}^{n} w_i L_i \right\| \tag{4.7.1.20}$$

as z_i vary in B and w_i vary over the complex numbers is that of determining *minimal norm or MN rules*.

Several authors have attempted to determine minimum norm rules numerically by solving the set of $3n$ nonlinear equations

$$\partial\|E\|/\partial w_i = 0, \qquad \partial\|E\|/\partial x_i = 0, \qquad \partial\|E\|/\partial y_i = 0 \tag{4.7.1.21}$$

$(x_i + \sqrt{-1}y_i = z_i)$.

The evidence seems to be that for moderate values of n these equations can be extremely ill conditioned, and Newton's method does not work. Minimizing $\|E\|$ by means of the Davidon–Fletcher–Powell method appears to be more promising. The subroutine FMFP in the IBM Scientific Subroutine Package can be used for this purpose. See also Chapter E04 in the NAG library and Chapter Z in the IMSL library.

On the theoretical side, the existence of MN rules for $L_2(B)$ was first established by Barrar *et al.* A simpler proof for H_2 was given by Knauff. The question of uniqueness is trickier; in fact, for certain spaces of functions, Brombeer and Macher have shown the existence of several distinct MN rules. Assuming existence, one can show from the equations (4.7.1.21) that a minimal norm rule must be exact not only for the functions $K_B(z, \bar{z}_i)$, $i = 1, 2, \ldots, n$, but also for the derivatives

$$\left.\frac{\partial}{\partial \bar{z}} K_B(z, \bar{z})\right|_{\bar{z}=\bar{z}_i}, \qquad i = 1, 2, \ldots, n.$$

Thus, minimal norm rules integrate these $2n$ functions exactly and in a certain sense therefore are analogs of Gauss rules. Severe numerical problems are encountered when the $2n$ nonlinear equations expressing this fact are used to determine the abscissas and weights.

Suppose that the region B is symmetric with respect to the x axis and contains the interval $[-1, 1]$. It has been established that among all rules with abscissas on $[-1, 1]$, the one with minimum error norm has its abscissas interior to $[-1, 1]$ and has *positive* weights.

A parallel theory can be worked out for spaces employing inner products that are line integrals over the boundary of a region.

Müller-Bausch has proved the existence in certain Hilbert spaces of MN rules of the form

$$\int_{-1}^{1} w(x)f(x)\,dx \simeq \sum_{i=1}^{n} a_i f(x_i) + \sum_{j=0}^{k} (b_j f^{(j)}(-1) + c_j f^{(j)}(1)),$$

$$\sum_{j=0}^{k} (|b_j| + |c_j|) \neq 0.$$

For $k = 0$, the resulting formula is of Lobatto type.

For MN rules in the Hardy spaces H_p, $1 < p < \infty$, it has been shown by Loeb and Werner and by Newman that the error goes to zero as $\exp(-c\sqrt{n})$. Stenger has shown that these rules are intimately related to those based on the Whittaker cardinal function (Section 3.4.6).

For the numerical computations of abscissas and weights of MN rules in Hardy H_2 space (Wilf rules), see Schrader and Engels and Eckhardt.

Example. (Duc-Jacquet [3]) Consider the Hilbert space H of functions f such that $f(0) = 0$, $f, f' \in L_2[0, 1]$ with $(f, g) = \int_0^1 f'(x)g'(x)\,dx$. Then the abscissas and weights in the MN rule for the integral

$$\int_0^1 \frac{f(x)}{x}\,dx \simeq \sum_{i=1}^{n} w_{in} f(x_{in}), \qquad f \in H$$

are given by the following construction. Set $u_1 = 2$, $u_{i+1} = 2 - \lambda_i u_i$, $i = 1, 2, \ldots$ where λ_i is the unique zero of $u_i - \log x/(x - 1)$. Then $w_{in} = 2u_i - 2 > 0$, $x_{nn} = \exp(1 - u_n)$, and $x_{in} = (2 - u_{i+1})x_{i+1}/u_i$, $i = n - 1, \ldots, 1$. Note that the weights w_{in} are independent of n. For $n = 50$, the value of the norm of the error functional is about 0.3.

References. Barrar, Loeb, and Werner [1]; Brombeer and Macher [1]; Davis [5]; Duc-Jacquet [1–3]; Engels and Eckhardt [1]; IBM [2]; IMSL [1]; Larkin [1]; Loeb and Werner [1]; Müller-Bausch [1, 2]; NAG [1]; Newman [1, 2]; Paulik [4]; Rabinowitz and Richter [3]; Richter [1–3]; Schrader [1]; Stenger [8, 9].

4.7.2 *Explicit Formulas*

The ease of obtaining explicit formulas of any sort is related to the simplicity of a complete orthonormal system for $L^2(B)$. The two simplest cases are

(a) The family of circles $|z| \leq R$, with a complete orthonormal system of polynomials

$$p_n^*(z) = \left(\frac{n+1}{\pi}\right)^{1/2} \frac{z^n}{R^{n+1}}, \qquad n = 0, 1, \ldots. \tag{4.7.2.1}$$

(b) The family of ellipses $\mathscr{E}_\rho$, foci at $z = \pm 1$, semiaxis sum $= \rho$, with a complete orthonormal system of polynomials

$$p_n^*(z) = 2(n+1)^{1/2}[\pi(\rho^{2n+2} - \rho^{-2n-2})]^{-1/2} U_n(z) \tag{4.7.2.2}$$

where the $U_n(z)$ are the Tschebyscheff polynomials of the second kind defined by

$$U_n(z) = (1 - z^2)^{-1/2} \sin((n + 1) \arccos z). \tag{4.7.2.3}$$

The advantage of using family (b) rather than family (a) when discussing integration over $[-1, 1]$ is that for $\rho \to 1^+$, the ellipses $\mathscr{E}_\rho$ collapse to the segment $[-1, 1]$ so that *all* analytic functions on $[-1, 1]$ fall within the compass of some $L^2(\mathscr{E}_\rho)$.

Writing $E(f) = \int_{-1}^{+1} f(x)\,dx - \sum_{i=1}^{n} w_i f(x_i)$, and applying (4.7.1.6) leads to

$$\|E\|_{\mathscr{E}_\rho}^2 = \frac{4}{\pi} \sum_{k=0}^{\infty} \frac{(k + 1)[\tau_k - \sum_{i=1}^{n} w_i U_k(x_i)]^2}{\rho^{2k+2} - \rho^{-2k-2}} \tag{4.7.2.4}$$

where

$$\tau_k = \frac{1 + (-1)^k}{k + 1} \tag{4.7.2.5}$$

and $\|E\|_{\mathscr{E}_\rho}$ designate the norm of E over $L^2(\mathscr{E}_\rho)$. Lo, Lee, and Sun give tabulations of $\|E\|_{\mathscr{E}_\rho}$ for a variety of familiar rules while Stroud and Secrest give tabulations and graphs for various Gauss-type rules.

Useful approximations to error norms have also been obtained. Thus, if E_{G_n} is the error in an n-point Gauss rule,

$$\|E_{G_n}\|_{\mathscr{E}_\rho} \sim \frac{2((2n + 1)\pi)^{1/2}}{\rho^{2n+1}}. \tag{4.7.2.6}$$

The agreement is very close except for ρ near 1 and n quite small.

The family of ellipses $\mathscr{E}_\rho$ admits another orthonormal set of polynomials $\tilde{p}_n(z)$ when the following inner product is used:

$$(f, g) = \int_{\mathscr{E}_\rho} f(z)\overline{g(z)}\, |1 - z^2|^{-1/2}\, |dz|. \tag{4.7.2.7}$$

We have

$$\tilde{p}_n(z) = (2/\pi)^{1/2}(\rho^{2n} + \rho^{-2n})^{-1/2} T_n(z). \tag{4.7.2.8}$$

The formula corresponding to (4.7.2.4) is

$$\|E\|_{\mathscr{E}_\rho}^2 = \frac{2}{\pi} \sum_{k=0}^{\infty} (\rho^{2k} + \rho^{-2k})^{-1} \left[\frac{1 + (-1)^k}{1 - k^2} - \sum_{i=1}^{n} w_i T_k(x_i)\right]^2, \tag{4.7.2.9}$$

while an estimate for G_n is given by

$$\|E_{G_n}\|_{\mathscr{E}_\rho} \sim \left(\frac{\pi}{2}\right)^{1/2} \frac{1}{\rho^{2n}} \left[1 + \left(1 + \frac{1}{2n}\right)^2 \Big/ 2\rho^4\right]. \tag{4.7.2.10}$$

A variation of (4.7.2.4) has been applied by de Angelis *et al.* to the computation of the error norm in the integration of the trigonometric integrals $\int_0^T f(t) \cos 2\pi\omega t \, dt$ and $\int_0^T f(t) \sin 2\pi\omega t \, dt$ where $f(t) \in L_2(\mathscr{E}_\rho)$.

References. Davis and Rabinowitz [1]; de Angelis, Murli, and Pirozzi [1]; Lo, Lee, and Sun [1]; Rabinowitz [6]; Rabinowitz and Richter [2]; Stroud and Secrest [2].

4.7.3 *Optimal Rules of Sard Type*

Let r be a fixed integer ≥ 2. Let X designate the set of all functions $f(x) \in C^r[0, 1]$ with the seminorm

$$\|f\|^2 = \int_0^1 |f^{(r)}(x)|^2 \, dx. \tag{4.7.3.1}$$

Designate by $E(f)$ the error functional

$$E(f) = \int_0^1 f(x) \, dx - \sum_{k=1}^n w_k f(x_k) \tag{4.7.3.2}$$

where the w_i and $0 < x_1 < \cdots < x_n < 1$ are fixed. Suppose now that $n \geq r$, and E has been restricted so that $E(f) = 0$ whenever $f \in \mathscr{P}_{r-1}$. By (4.3.10),

$$E(f) = \frac{1}{(r-1)!} \int_0^1 f^{(r)}(t) K(t) \, dt \tag{4.7.3.3}$$

where

$$K(t) = E_x(x-t)_+^{r-1} = \int_0^1 (x-t)_+^{r-1} \, dx - \sum_{k=1}^n w_k (x_k - t)_+^{r-1}$$

$$= \frac{(1-t)^r}{r} - \sum_{k=1}^n w_k (x_k - t)_+^{r-1}. \tag{4.7.3.4}$$

This formula is valid for all $f \in C^r[0, 1]$. The kernel $K(t)$ is evidently of class $C^{r-2}[0, 1]$ and is a monospline.

By the Schwarz inequality,

$$|E(f)| \leq \frac{1}{(r-1)!} \left(\int_0^1 |f^{(r)}(t)|^2 \, dt\right)^{1/2} \left(\int_0^1 |K(t)|^2 \, dt\right)^{1/2}$$

$$= \|f\| \frac{1}{(r-1)!} \left(\int_0^1 |K(t)|^2 \, dt\right)^{1/2}. \tag{4.7.3.5}$$

Hence,

$$\sup_{\substack{f \in X \\ \|f\| \le 1}} |E(f)| \le \frac{1}{(r-1)!}\left(\int_0^1 |K(t)|^2\,dt\right)^{1/2}. \tag{4.7.3.6}$$

On the other hand, if we select a function g such that $g^{(r)}(t) = K(t)$, then g is surely of class $C^r[0, 1]$, $\|g\|^2 = \int_0^1 |K(t)|^2\,dt$, and

$$E(g) = \frac{1}{(r-1)!}\int_0^1 (K^2(t))\,dt, \quad \text{so that} \quad |E(g)| = \frac{1}{(r-1)!}\int_0^1 |K(t)|^2\,dt.$$

Thus,

$$\sup_{\substack{f \in X \\ \|f\| \le 1}} |E(f)| = \frac{1}{(r-1)!}\left(\int_0^1 |K(t)|^2\,dt\right)^{1/2} \equiv \sigma_E(w_k; x_k). \tag{4.7.3.7}$$

Let now $\{x_k\}$ be fixed and the w_k vary over the real numbers. Then, the rule $\sum_{k=1}^n w_k^* f(x_k)$ with error

$$E^*(f) = \int_0^1 f(x)\,dx - \sum_{k=1}^n w_k^* f(x_k)$$

will be called a *relatively minimum norm rule in the sense of Sard* if

$$\sigma_E(w_k^*; x_k) = \min_{w_k} \sigma_E(w_k; x_k). \tag{4.7.3.8}$$

From (4.7.3.4) it follows that the problem of determining such rules is equivalent to that of the best approximation in the sense of least squares of the function $(1-t)^r$ by means of splines of the form $\sum_{k=1}^n w_k(x_k - t)_+^{r-1}$.

If $0 < x_1 < \cdots < x_r < 1$ are allowed to vary in position, we arrive at *minimum norm rules.*

For existence, uniqueness, characterization of these rules see Schoenberg, Karlin. For numerical results, see Meyers and Sard, Powell, Sard and Stern.

Wang and Klein give an algorithm for computing minimum norm rules in the sense of Sard, which also has the possibility of constraining such rules to include the endpoints.

References. Davis [1]; el-Tom [1]; Karlin [1–3]; Meyers and Sard [1]; Nikolskii [1]; Powell [1]; Richter [1, 3]; Sard [1]; Sarma [1]; Schoenberg [2, 4]; Secrest [1, 2]; Stern [1]; Wang and Klein [1].

4.7.4 *The Role of Minimal Norm Rules in Computing*

There are numerous reasons why minimal norm rules (of all varieties) are not used in computing practice and remain, after almost three decades of investigation, a plaything of the theoretician.

(a) The additional accuracy provided either in the values of the integral or in error estimates is only marginal.

(b) The rules are difficult to compute. Many "irrational" numbers must be stored and these numbers are presently available for only low values of n.

(c) Proceeding from low n to high n implies much recomputation of functional values.

(d) There are an infinite variety of normed spaces, each with its own competitive minimal norm rules. The run-of-the mill integrand probably belongs to most of these spaces. Which to choose?

(e) In numerous cases minimal norm rules approach classical rules asymptotically as some parameter becomes large.

(f) Finally, the underlying theory is more sophisticated than that of conventional rules and hence is less appealing to the novice or the worker in a different field. The "meaning" of the rule is not intuitively obvious.

See also the remarks by de Boor and Brass.

References. Brass [1, p. 23]; de Boor [3].

4.8 *Errors for Integrands with Low Continuity*

The errors incurred in approximate integration formulas are conventionally expressed in terms of the higher derivatives of the integrand and are valid only if the integrand is sufficiently smooth. What is the error when a high-order rule is applied to a function with low-order continuity? One method of obtaining this is to use Peano's theorem with a low value of n. This is valid even when the rule is exact for polynomials of high degree. The error is given by (4.3.5) and a bound in terms of max $|f^{(n)}(x)|$ is given by (4.3.11).

We have already seen exact asymptotic estimates obtained for the errors in the trapezoidal and Gauss rules applied to integrands with specified types of singularity.

Still another method of error estimation is through the use of approximation theory.

THEOREM. *Let $f(x) \in C[a, b]$, and suppose that there exists a polynomial of degree $\leq n$, $p_n(x)$, such that*

$$|f(x) - p_n(x)| \leq \varepsilon, \qquad a \leq x \leq b. \tag{4.8.1}$$

Let

$$R(f) = \sum_{k=1}^{m} w_k f(x_k) \approx \int_a^b f(x)\, dx, \qquad a \leq x_k \leq b$$

be a rule of approximate integration which is exact for polynomials of class $\mathscr{P}_n$. *Then if*

$$E(f) = \int_a^b f(x)\,dx - \sum_{k=1}^{m} w_k f(x_k), \tag{4.8.2}$$

we have

$$|E(f)| \le \left[(b-a) + \sum_{k=1}^{m} |w_k|\right]\varepsilon. \tag{4.8.3}$$

Proof.

$$E(f) = \int_a^b f(x)\,dx - R(f) = \int_a^b (f(x) - p_n(x))\,dx + \int_a^b p_n(x)\,dx - R(f)$$

$$= \int_a^b (f(x) - p_n(x))\,dx + R(p_n - f)$$

inasmuch as $\int_a^b p_n(x)\,dx = R(p_n)$. Hence

$$|E(f)| \le \int_a^b |f(x) - p_n(x)|\,dx + \sum_{k=1}^{m} |w_k|\,|p_n(x_k) - f(x_k)|$$

$$\le (b-a)\varepsilon + \varepsilon \sum_{k=1}^{m} |w_k|.$$

COROLLARY. *If* $w_k > 0$, *then*

$$|E(f)| \le 2(b-a)\varepsilon. \tag{4.8.4}$$

Proof. In this case, we have

$$\sum_{k=1}^{m} |w_k| = \sum_{k=1}^{m} w_k = \int_a^b dx = b - a.$$

This estimate of error can now be coupled with an estimate of how close a given continuous function may be approximated by a polynomial. The following theorem due to Jackson is relevant.

THEOREM. *Let* $f(x)$ *be of class* $C[a, b]$ *and have* $w(\delta)$ *as its modulus of continuity there. Then for each* $n = 1, 2, 3, \ldots$ *there exists a polynomial of degree* $\le n$, $p_n(x)$, *such that*

$$|f(x) - p_n(x)| \le 6w\left(\frac{b-a}{2n}\right), \qquad a \le x \le b. \tag{4.8.5}$$

If $f(x)$ is of class $C[a, b]$ and has a bounded derivative there,

$$|f'(x)| \leq M, \qquad a \leq x \leq b, \tag{4.8.6}$$

then for each $n = 1, 2, \ldots$ there exists a polynomial $p_n(x)$ of degree $\leq n$ such that

$$|f(x) - p_n(x)| \leq \frac{3(b-a)M}{n}, \qquad a \leq x \leq b. \tag{4.8.7}$$

If $f(x) \in C^p[a, b]$ $(p > 1)$ and has a bounded $(p+1)$th derivative there:

$$|f^{(p+1)}(x)| \leq M_{p+1}, \qquad a \leq x \leq b, \tag{4.8.8}$$

then for each $n = 1, 2, \ldots$ there exists a polynomial $p_n(x)$ of degree $\leq n$ such that

$$|f(x) - p_n(x)| \leq \frac{6^{p+1}p^p}{p!\,n^{p+1}}(p+1)(b-a)^{p+1}M_{p+1}. \tag{4.8.9}$$

Example. The function $f(x) = x^{3/2}$ is in $C^1[0, 1]$ and satisfies $|f'(x)| \leq \frac{3}{2}$ there. It is not in class $C^2[0, 1]$. Hence, by (4.8.7), there exists a $p_n(x) \in \mathscr{P}_n$ such that

$$|f(x) - p_n(x)| \leq 3 \cdot 1 \cdot \tfrac{3}{2} \cdot n^{-1} = 9/2n = \varepsilon.$$

The n-point Gauss rule is exact for polynomials of degree $\leq 2n - 1$. Hence, by (4.8.4), we have

$$|E_{G_n}(f)| \leq \frac{2 \cdot 9}{2(2n-1)} = \frac{9}{2n-1}.$$

The table in Section 2.7 displays an error much smaller than this bound.

For a further discussion of results of this type including a sharpening of the constants in the approximation theorems, see Baker and Ehersman.

In the same direction and frequently useful although it does not pertain to functions of *low* continuity is the following result of S. Bernstein.

THEOREM. *Let $\mathscr{E}_\rho$ designate the ellipse with foci at ± 1 and ρ = sum of semiaxes. If $f(z)$ is analytic in $\mathscr{E}_\rho$ and is real for z real, and if $|f(z)| \leq M$ in $\mathscr{E}_\rho$, then we can find a polynomial $p_n(x) \in \mathscr{P}_n$ such that*

$$|f(x) - p_n(x)| \leq 2M/\rho^n(\rho - 1), \qquad -1 \leq x \leq 1. \tag{4.8.10}$$

Another application of approximation theory to the estimation of the error in the integration of analytic functions by the Gauss rules (4.6.36) is given by von Sydow. Since the best approximation by a polynomial $p_n(t; z)$ to $(t - z)^{-1}$ satisfies

$$|(t-z)^{-1} - p_n(t; z)| \leq \frac{2|z - \sqrt{z^2 - 1}|^{n+1}}{|\sqrt{z^2 - 1}|(1 - |z - \sqrt{z^2 - 1}|^2)} \tag{4.8.11}$$

(Achieser [1, p. 62]), it follows from (4.6.41) that if f is analytic in the ellipse C_R with foci ± 1 and semiaxes $(R \pm R^{-1})/2$, then

$$\| E_n \| \leq 4 \int_{-1}^{1} w(x)\, dx / R^{2n}(1 - R^{-2}). \tag{4.8.12}$$

References. C. T. H. Baker [1]; Ehersman [1]; Korovkin [1, pp. 86–88]; Locher [1]; Locher and Zeller [1]; Natanson [1, p. 129]; Rabinowitz [5]; Stroud [8]; von Sydow [1].

4.8.1 Application of Tschebyscheff Expansions

Any function continuous and of bounded variation on $I = [-1, 1]$ may be expanded in the uniformly and absolutely convergent series

$$f(x) = \sum_{k=0}^{\infty}{}' a_k T_k(x) \tag{4.8.1.1}$$

where

$$a_k = \frac{2}{\pi} \int_{-1}^{1} \frac{f(x) T_k(x)}{(1 - x^2)^{1/2}}\, dx = \frac{2}{\pi} \int_0^{\pi} g(\theta) \cos k\theta \, d\theta, \qquad k = 0, 1, 2, \ldots \tag{4.8.1.2}$$

and $g(\theta) \equiv f(\cos \theta)$. The coefficients a_k can be bounded as follows:

(a) Define $F_1(x) \equiv (1 - x^2)^{1/2} f'(x)$; if $F_1(x)$ is of bounded variation in I with $|F_1(x)| \leq P_1$ and if c_1 is the number of intervals of monotonicity of $F_1(x)$ in I, then

$$|a_k| \leq 4 c_1 P_1 / \pi k^2 \quad \text{for} \quad k \geq 1. \tag{4.8.1.3}$$

(b) Define $F_2(x) \equiv (1 - x^2) f''(x) - x f'(x)$; if $F_2(x)$ is of bounded variation in I with $|F_2(x)| \leq P_2$, if c_2 is the number of intervals of monotonicity of $F_2(x)$, and if $\lim_{x \to \pm 1} F_1(x) = 0$, then

$$|a_k| \leq 4 c_2 P_2 / \pi k^3 \quad \text{for} \quad k \geq 1. \tag{4.8.1.4}$$

Further bounds for functions of higher continuity may be derived by integrating the right-hand integral in (4.8.1.2) by parts.

If we apply the error operator E_R corresponding to a particular rule $R(f)$ to (4.8.1.1) we have

$$E_R(f) = \sum{}' a_k E_R(T_k).$$

In certain situations, where we have some qualitative knowledge of the behavior of the a_k, this equation may help us in estimating the error provided we know something about $E_R(T_k)$. These values may be precomputed for certain commonly used rules while for the Gauss rules G_n

much is known about their asymptotic behavior. We have the following results:

$$E_{G_n}(T_{2n}) = \frac{2 \cdot 2 \cdot 4 \cdot 4 \cdots 2n \cdot 2n}{1 \cdot 3 \cdot 3 \cdot 5 \cdots (2n-1) \cdot (2n+1)} \to \frac{\pi}{2},$$

$$E_{G_n}(T_{2n+2}) = -E_{G_n}(T_{2n})\left(1 + \frac{2n+1}{(2n-1)(2n+3)}\right),$$

$$E_{G_n}(T_{j(4n+2)\pm 2l}) \simeq \begin{cases} (-1)^j \dfrac{2}{4l^2-1}, & l = 0, 1, \ldots, n-1, \\ (-1)^j \dfrac{\pi}{2}, & l = n, \end{cases}$$

$$E_{G_n}(T_{2p+1}) = 0, \qquad \text{all } p. \tag{4.8.1.5}$$

In other situations, we only have estimates for $|a_k|$ such as those given in (4.8.1.3) and (4.8.1.4). In these cases we have

$$|E_R(f)| \le \sum{}' |a_k| \, |E_R(T_k)|. \tag{4.8.1.6}$$

If f is as in (a); then, e.g.,

$$|E_{G_n}(f)| \le \frac{4c_1 P_1}{\pi} \sum_{k=n}^{\infty} \frac{|E_{G_n}(T_{2k})|}{(2k)^2} \tag{4.8.1.7}$$

and this bound can be approximated by $c_1 P_1/n^2$.

References. Curtis and Rabinowitz [1]; Elliott [1]; Nicholson, Rabinowitz, Richter, and Zeilberger [1]; Rabinowitz [5]; Riess and Johnson [1, 2].

4.9 Practical Error Estimation

While the methods of the preceding sections are suitable for error estimation when evaluating integrals manually, they are not suitable in general for automatic computation. One exception is the work of Gray and Rall. This combines formal differentiation and interval analysis and is applicable in certain situations but is quite expensive. Usually the computer has available only a finite number of values of the integrand and possibly of some of its derivatives and has no information at all about bounds on high-order derivatives, norms, etc. which enter into the standard error estimates. Since the value of the Riemann integral of a function is unchanged when the values of the integrand are modified at a finite number of points, it follows that the knowledge of a finite number of function values is inadequate to estimate the value of the integral and even more so to provide an error estimate. On the other hand, since in fact we do estimate the value of the integral using a finite amount of information, we can also attempt to use such information, incremented possibly with additional values, to provide an error estimate. There is no guarantee of success but in practice most attempts have succeeded since the integrands arising in real life situations are usually well behaved and not pathological.

In practical error estimation, one must navigate carefully between the Scylla of underestimation and the Charybdis of overestimation. The danger of underestimation is quite clear in that it leads to accepting an inaccurate result as accurate, a very undesirable situation. This possibility may be reduced at considerable expense by requesting an accuracy greater than that actually needed so that if the underestimate is not too gross, the accuracy achieved will be satisfactory. If the final error estimate is the sum of many *local* error estimates, i.e., error estimates over subintervals of the integration interval, then underestimation of a few local error estimates may not be catastrophic, provided again that these underestimates are not way off. The reasons are that the underestimates over some subintervals may be compensated by overestimates over other subintervals. In addition, the global error estimate is usually taken to be the sum of the *magnitudes* of the local error estimates so that there is a possibility that the local error estimates may cancel each other to some extent. In any event, a numerical integration program is said to be unreliable if it underestimates the error too often.

To achieve reliability, most programs overestimate the error, usually at the cost of efficiency. This is usually not too high a price to pay to avoid incorrect results, especially if we are only evaluating a few one-dimensional integrals. However, the price may become prohibitive if many integrals are to be evaluated, especially in the multidimensional case. More seriously, a program may exit with an indication of failure when in fact the accuracy requested has indeed been achieved. This can be quite frustrating to the user.

Example. Approximate the ten-dimensional integral

$$I = \int_0^1 \cdots \int_0^1 \left(\sum_{i=1}^{10} x_i \right)^{-2.5} dx_1 \cdots dx_{10} \doteq .021203$$

to relative accuracy ε using at most 160000 function evaluations in the program ADAPT by Genz and Malik.

ε	.1	.01	.001
Computed value	.021226	.021226	.021205
Relative error	.00108	.00108	.00009
Computed relative error	.00821	.00821	.00135
Number of function evaluations	1245	1245	158115

For $\varepsilon = .001$, the program exited with an indication of failure since the next iteration would have required more than the maximum number of function evaluations allowed. The computed error estimate overestimated the true error by a factor of 15. In fact, the result achieved with 1245 function evaluations was almost good enough but the program did not know this.

Laurie has dealt with the problem of overestimation as follows: Assume there are two integration rules A and B where A is "better" than B. Consider now the integral of $f(x)$ over some interval $[c, d]$, $I \equiv If = \int_c^d f(x)\,dx$. Denote by A_1 and A_2, the rules A and $(2 \times A)$, respectively, applied to $f(x)$ over $[c, d]$ with a similar definition for B_1 and B_2. If

$$\alpha = \frac{I - A_2}{I - A_1} \approx \beta = \frac{I - B_2}{I - B_1},$$

then it is easy to show that an improvement over the error estimate $A_2 - B_2$ for $I - A_2$ is given by

$$\varepsilon = \mu(A_2 - A_1)/(1 - \mu)$$

where $\mu = (A_2 - B_2)/(A_1 - B_1)$. However, ε is a reliable estimate for $I - A_2$ under more general circumstances as the following theorem shows:

THEOREM *Assume the following conditions:*

(i) $|A_2 - A_1| < |B_2 - B_1|$, *i.e., rule A is more accurate than rule B when applied to* $f(x)$ *over* $[c, d]$,

(ii) $0 < \alpha \le \beta < 1$, *i.e., both rules converge monotonically when compounded and rule A converges at least as fast as rule B,*

then A_2 *and* $A_2 + \varepsilon$ *bracket* I.

Examples (Laurie).

Let A be the three-point Gauss rule and B, the two-point Gauss rule. We give three cases. In each case $[c, d] = [-1, 1]$.

Case 1. $I = \int_{-1}^{1} e^x\,dx = e - e^{-1} = 2.350402387$

$$\begin{aligned} A_1 &= 2.350336929 & A_2 &= 2.350401261 \\ B_1 &= 2.342696088 & B_2 &= 2.349875138 \\ \alpha &= .0172 & \beta &= .0684 \\ |A_2 - B_2| &= .000526123 & \varepsilon &= .000004757 \\ I - A_2 &= .000001126 & & \end{aligned}$$

The crude estimate $|A_2 - B_2|$ is an overestimate by a factor of 467. The conditions of the theorem hold so that the improved estimate ε is reliable. It overestimates the true error by a factor of 4.2.

Case 2. $I = \int_{-1}^{1} (1 + x)^{-1/2}\, dx = 2\sqrt{2} = 2.828427125$

$$\begin{aligned} A_1 &= 2.476094452 & A_2 &= 2.579283441 \\ B_1 &= 2.334414218 & B_2 &= 2.478832862 \\ \alpha &= .707126256 & \beta &= .707662205 \\ |A_2 - B_2| &= .100450579 & \varepsilon &= .251406268 \\ I - A_2 &= .249143684 \end{aligned}$$

For this improper integral, both rules have the same rate of convergence when compounded, $\alpha \approx \beta$. $|A_2 - B_2|$ underestimates the error by a factor of .4. Since the conditions of the theorem hold, the estimate ε is reliable and in this case, is a very good approximation to the true error. This to be expected since $\alpha \approx \beta$.

Case 3. $I = \int_{-1}^{1} |x + .2|\, dx = 1.04$

$$\begin{aligned} A_1 &= 1.038440743 & A_2 &= 1.048499075 \\ B_1 &= 1.154700538 & B_2 &= 1.000000001 \\ \alpha &= -5.450721081 & \beta &= -.348734188 \\ |A_2 - B_2| &= .048499074 & \varepsilon &= -.002960810 \\ I - A_2 &= -.008499075 \end{aligned}$$

Here, the integrand is so bad that conditions (i) and (ii) are not satisfied and $|\varepsilon|$ underestimates the error.

In practical error analysis the most obvious approach to estimating the error in approximating

$$If \equiv I[c, d]f = \int_c^d w(x)f(x)\, dx \tag{4.9.1}$$

by $Q_n f = \sum_{i=1}^{n} w_{in} f(x_{in})$ is by comparing $Q_n f$ with $Q_m f$ for some $m \neq n$. Here $Q_n f$ and $Q_m f$ belong to some family of rules converging to If and $Q_n f$ is "more accurate" than $Q_m f$ if $n > m$. The interval $[c, d]$ may be either the entire integration interval $[a, b]$ or some subinterval so that we may be dealing with either a global or a local error estimate. We then take $|Q_n f - Q_m f|$ as an estimate of

$$E_n f = |If - Q_n f|.$$

If $m < n$, as is generally the case, then $Q_n f$ will probably be a better approximation to If than $Q_m f$ and indeed $|Q_n f - Q_m f|$ will be a good approximation not to $E_n f$ but to $E_m f$. However, since one is reluctant not to use the more accurate approximation $Q_n f$ once the investment has been made to compute it, one ends up by overestimating the error. This is a typical situation in numerical analysis. We have an approximation A and an error estimate E. On the one hand, we can accept the value A and then the error estimate E will be realistic. On the other hand, we can add the value E to the approximation A to get a more accurate approximation

A', but then we will have no error estimate for A'. The typical way out of this dilemma is to have one's cake and eat it too; i.e., to accept the more accurate value A' and to use the error estimate E which is appropriate to the less accurate value A. This leads to overestimation and its problems. Such an approach has been used in various numerical integration programs. Among them are many Romberg schemes (Section 6.3), the Patterson program where $\{Q_n f\}$ is a sequence of Gauss–Kronrod–Patterson rules of increasing accuracy, some of the QUADPACK routines (Piessens *et al.*), the ADAPT program for multidimensional integration by Genz and Malik, and so forth.

A particular instance of this situation has been discussed by Fritsch *et al.* Let us assume a basic rule satisfying

$$Q^{[1]}f \equiv Q^{[1]}[a, a+h]f = I[a, a+h]f + ch^{m+3}f^{(m+2)}(\xi), \qquad a < \xi < a+h \tag{4.9.2}$$

and define

$$Q^{[2]}f \equiv Q^{[2]}[a, a+h]f = Q^{[1]}[a, a+h/2]f + Q^{[1]}[a+h/2, a+h]f, \tag{4.9.3}$$

$$Ef \equiv E[a, a+h]f = (Q^{[1]}f - Q^{[2]}f)/(2^{m+2} - 1), \tag{4.9.4}$$

$$Q^{[1,2]}f = Q^{[2]}f - Ef. \tag{4.9.5}$$

Then, if $f \in C^{(m+2)}(a, a+h)$,

$$Q^{[2]}f = If + 2c(h/2)^{m+3}f^{(m+2)}(\xi_1) = If + ch^{m+3}f^{(m+2)}(\xi_1)/2^{m+2}, \qquad a < \xi_1 < a+h \tag{4.9.6}$$

and

$$Q^{[1]}f - Q^{[2]}f = (1 - 2^{m+2})ch^{m+3}f^{(m+2)}(\xi_2), \qquad a < \xi_2 < a+h. \tag{4.9.7}$$

The use of Ef as an error estimate for the approximation $Q^{[1,2]}f$ to If is the case described above. The use of Ef to estimate the error in $Q^{[2]}f$ is more reasonable. The use of $(2^{m+2} - 1)Ef$ as the error estimate for $Q^{[2]}f$ is also an instance of the situation described previously. Each of these procedures has been used in practice. A program by Shampine and Allen takes $Q^{[1]}f$ to be Simpson's rule so that $m = 2$ and it uses $(15Ef)^{-l/2}$ as the error estimate for $Q^{[1,2]}f$ where l is the level of subdivision of the initial integration interval $[a, b]$ so that the length of the subinterval under consideration is $(b-a)2^{-l}$. This error estimate is more stringent at shallow levels (small l) and less stringent at deeper levels (large l). Another program uses a seven-point Newton–Cotes rule, $m = 6$, and the error estimate

$\min(Ef, (255Ef)2^{-1/2})$ for the approximation $Q^{[2]}f$. We remark in passing that for the Newton–Cotes rules with $m = 2, 4, 6, 8$ Fritsch *et al.* tabulate the coefficients c_k, d_k, e_k in the expressions

$$Q^{[1]}f = h\sum_{k=0}^{m} c_k f(a + kh/m), \tag{4.9.8}$$

$$Q^{[1,2]}f = h\sum_{k=0}^{2m} d_k f(a + kh/2m), \tag{4.9.9}$$

$$Ef = \frac{1}{2^{m+2} - 1}\sum_{k=0}^{2m} e_k f(a + kh/2m). \tag{4.9.10}$$

A different estimate for the error in Newton–Cotes rules, applicable for any rule of the form (4.9.2), was suggested by Ninomiya. He approximates the error term using a numerical differentiation formula involving two additional function evaluations at suitably chosen points taken midway between the first pair and last pair of abscissas. For example, for the five-point Newton–Cotes rule

$$\int_0^{4h} f(x)\,dx = \frac{h}{90}(7(f_0 + f_4) + 32(f_1 + f_3) + 12f_2) - \frac{h^7 f^{(6)}(\xi)}{7741440}, \qquad a < \xi < 4h \tag{4.9.11}$$

where $f_i = f(ih)$, the error estimate uses the approximation

$$\frac{h^7 f^{(6)}(\xi)}{7741440} \simeq \frac{16h}{6615}(15(f_0 + f_4) + 84(f_1 + f_3) - 70f_2 - 64[f(h/2) + f(7h/2)]). \tag{4.9.12}$$

Ninomiya prefers a nine-point Newton–Cotes rule since the ratio of additional points needed to estimate the derivative to the number of points in the rule is smaller. Notice that in case the integration interval is bisected, the additional points become integration points on the next level so that these functional values can be reused.

Locher suggests the following estimate for the error Ef in Simpson's rule

$$\int_{-1}^{1} f(x)\,dx = \tfrac{1}{3}(f(-1) + 4f(0) + f(1)) + Ef, \tag{4.9.13}$$

$$|Ef| \simeq \tfrac{4}{15}|f(-1) - 2f(-\sqrt{2}/2) + 2f(0) - 2f(\sqrt{2}/2) + f(1)|. \tag{4.9.14}$$

He also gives the following error estimate for the Clenshaw–Curtis rule $(CC_n f)$:

$$\int_{-1}^{1} f(x)\,dx = \sum_{i=0}^{n} w_i f(x_i) + E_n f, \qquad x_i = \cos\left(\frac{n-i}{n}\pi\right), \qquad i = 0, \ldots, n. \tag{4.9.15}$$

Define the divided-difference functional $\lambda_{n-1} f$ by

$$\lambda_{n-1} f = \sum_{i=0}^{n} \mu_i f(x_i), \qquad \mu_i = \prod_{j=0,\, j\neq i}^{n} (x_i - x_j)^{-1}, \qquad i = 0, \ldots, n. \tag{4.9.16}$$

Then

$$|E_n f| \simeq 4|\lambda_{n-1} f| \Big/ \sum_{i=0}^{n} |\mu_i|. \tag{4.9.17}$$

Other estimates for the error in the Clenshaw–Curtis and related rules are based on their formulation as integrals of a Tschebyscheff series (cf. Section 2.13.1). Thus Piessens and his co-workers approximate the integral $\int_{-1}^{1} w(x)f(x)\,dx$ by the series $\sum_{j=0}^{\prime n} a_j M_j$ where

$$a_j = \frac{2}{n} \sum_{k=0}^{n}{}'' f\left(\cos\frac{\pi k}{n}\right) \cos\frac{\pi jk}{n}, \qquad j = 0, \ldots, n, \tag{4.9.18}$$

$$M_j = \int_{-1}^{1} w(x) T_j(x)\,dx, \qquad j = 0, \ldots, n, \tag{4.9.19}$$

the modified moments of $w(x)$, and usually $n = 12$. As an error estimate, they take $\sum_{j=9}^{12} |a_j| \max(|M_j|, 1)$. O'Hara and Smith followed by Oliver have developed more sophisticated error estimates for Clenshaw–Curtis integration based on an analysis of the behavior of the coefficients a_j. Thus, Oliver considers the $(n+1)$-point rule $CC_n f$ over the interval $[\alpha - h, \alpha + h]$ with n even. He then defines

$$a_r = \frac{2}{n} \sum_{i=0}^{n}{}'' f(\alpha + hx_i) T_r(x_i), \qquad r = 0, \ldots, n, \qquad x_i = \cos\frac{i\pi}{n} \tag{4.9.20}$$

and sets

$$K = \max\left\{\left|\frac{a_n}{2a_{n-2}}\right|, \left|\frac{a_{n-2}}{a_{n-4}}\right|, \left|\frac{a_{n-4}}{a_{n-6}}\right|\right\}. \tag{4.9.21}$$

If $K \le K_n(16)$ where $K_n(2^m)$ is tabulated for $m = 1, \ldots, 4$ and $n = 2^p$, $p = 2, \ldots, 7$, let $\sigma = \min(2^m)$ such that $K \le K_n(2^m)$. Then an error estimate for $CC_n f$ is given by

$$\frac{16h\sigma n}{(n^2-1)(n^2-9)} K^3 |a_{n-4}|. \tag{4.9.22}$$

For additional approaches to practical error estimation, see the descriptions of the automatic integration algorithms, a partial list of which is given in the References.

References. Blue [2]; Bunton, Diethelm, and Haigler [1]; Chase and Fosdick [1]; Clenshaw and Curtis [1]; de Boor [1, 2]; Forsythe, Malcolm, and Moler [1]; Fritsch, Kahaner, and Lyness [1]; Garribba, Quartapelle, and Reina [1]; Genz and Malik [1]; Gray and Rall [1–3]; Hausner [1]; Hausner and Hutchison [1]; Hazlewood [1]; Hopp [1]; Kahan [2]; Kahaner [2]; Laurie [3]; Linz [F1]; Locher [4]; Ninomiya [1]; O'Hara and Smith [1, 2]; Oliver [2, 3]; Patterson [F1]; Piessens and Branders [4]; Piessens, de Doncker, Überhuber, and Kahaner [1]; Piessens, Mertens, and Branders [1]; Robinson [5, 6]; Rowland and Varol [1]; Shampine and Allen [1].

Chapter 5

Approximate Integration in Two or More Dimensions

5.1 Introduction

The object of our investigation can be described in the following general way. Let B designate a fixed closed region in d-dimensional Euclidean space and let dV designate the d-dimensional volume element. Find fixed points $P_1, P_2, \ldots, P_n$ (preferably in B) and fixed weights $w_1, w_2, \ldots, w_n$ such that

$$\int_B w(P)f(P)\,dV \approx \sum_{k=1}^{n} w_k f(P_k) \tag{5.1.1}$$

is a useful approximation to the integral on the left for a reasonably large class of functions of d variables defined on B. There is also the possibility of using information about the derivatives of f to form the approximation, but this modification will not be considered.

In passing from one dimension to several dimensions, the diversity of integrals and the difficulty in handling them is greatly increased. In the first place, in one dimension we can restrict our attention to three different types of regions of integration; the finite interval, the singly infinite interval, and the doubly infinite interval, whereas in several dimensions, there are potentially an infinite number of different types of regions to contend with. In the second place, the behavior of functions of several variables can be considerably more complicated than that of functions of one variable (for example, we may have singular behavior on a manifold) and our experience and intuition with them is much more limited. Furthermore, it usually takes much more time to evaluate a function of several variables. Then again, as the dimension becomes higher, more and more points are necessary for successful approximation, and even with current computing speeds the number of functional evaluations may be an important consideration. The necessity for economization has, in fact, led to approximate integration by Monte Carlo methods.

In principle, as in the case of one dimension, we should like to discuss integration over a variety of regions, sequences of rules and their convergence, error estimates, "automatic" integration, how to handle singularities, how to handle highly oscillatory integrands, etc. But these matters are only moderately developed or are currently under development, so that our treatment of this subject will be somewhat superficial.

References. Engels [13]; Haber [7]; Mysovskikh [10, 15, 17]; Niederreiter [6]; Sobolev [4]; Stroud [18].

5.1.5 A Challenge

Staff programmers in computer centers get used to the monsters—exhibiting all manner of difficulties—that enter through the laboratory doors. Here is a double integral arising in the theory of powers of multivariate statistical tests that presented itself:

$$F = -(2\pi)^{-2} \int_{-\infty}^{\infty} \int_{-\infty}^{\infty} \operatorname{Re}\{\phi_1 \phi_2 \phi_3\}\, dt_1\, dt_2,$$

$$\phi_1 = \frac{1}{t_1 t_2 [1 - 2i(t_1 + t_2) - 4(1 - \rho_1^2) t_1 t_2]^{2.5}},$$

$$\phi_2 = \left[1 - \frac{1}{(1 + 2ivt_1)^{12.5}} - \frac{1}{(1 + 2ivt_2)^{12.5}} + \frac{1}{[1 + 2iv(t_1 + t_2) - 4(1 - \rho_1^2)v^2 t_1 t_2]^{12.5}}\right],$$

$$\phi_3 = \exp\left[\delta \frac{i(t_1 + t_2) + 4(1 - \rho_1 \rho_2) t_1 t_2}{1 - 2i(t_1 + t_2) - 4(1 - \rho_1^2) t_1 t_2}\right].$$

Here $i = \sqrt{-1}$ and the ranges of the parameters are $0 \le v \le 1$, $0 \le \rho_1$, $\rho_2 \le 1$, $0 \le \delta \le 10$.

PROBLEM. Compute F with demonstrable six figure accuracy.

For the particular choice of parameters $v = .5$, $\rho_1 = .25$, $\rho_2 = .75$ and $\delta = 5$, F was computed accurately by the programs DITAMO by Robinson and de Doncker and ADAPT by Genz and Malik. In the latter case, the integral was first transformed to the unit square C_2 by two applications of the transformation $t = u/(1 - u^2)$. The following sequence of values was computed:

Integrand calls	Estimated value	Estimated relative accuracy
969	4.805831	$.15 \times 10^{-2}$
1989	4.806415	$.22 \times 10^{-3}$
3995	4.806402	$.25 \times 10^{-4}$
7973	4.806408	$.30 \times 10^{-5}$
11373	4.806408	$.99 \times 10^{-6}$

When the transformation $t = u/(1 - u^4)$ was used, a similar behavior was observed.

References. Genz and Malik [F1]; Robinson and de Doncker [F1].

5.2 *Some Elementary Multiple Integrals over Standard Regions*

In this chapter, the letter d designates the number of dimensions of the space. The standard bounded regions are the d-dimensional cube, the d-dimensional solid sphere (or ball), the surface of the d-dimensional sphere, and the d-dimensional simplex (the generalization of the triangle and the tetrahedron).

1. Let S_d denote the simplex

$$x_1 \geq 0, x_2 \geq 0, \ldots, x_d \geq 0, \qquad x_1 + x_2 + \cdots + x_d \leq 1.$$

Then

$$\int_{S_d} x_1^{p_1-1} x_2^{p_2-1} \cdots x_d^{p_d-1}\, dV = \frac{\Gamma(p_1)\Gamma(p_2)\cdots\Gamma(p_d)}{\Gamma(p_1 + p_2 + \cdots + p_d + 1)}. \tag{5.2.1}$$

The notation $\Gamma(x)$ designates the Gamma function. In particular, the volume V of S_d is given by

$$V = \int_{S_d} dV = \frac{1}{d!}. \tag{5.2.2}$$

2. Let θ_d denote the portion of the ball $x_1^2 + x_2^2 + \cdots + x_d^2 \leq 1$ with positive coordinates: $x_1 \geq 0, x_2 \geq 0, \ldots, x_d \geq 0$. Then

$$\int_{\theta_d} x_1^{p_1-1} x_2^{p_2-1} \cdots x_d^{p_d-1}\, dV = \frac{1}{2^d} \frac{\Gamma(p_1/2)\Gamma(p_2/2)\cdots\Gamma(p_d/2)}{\Gamma(\frac{1}{2}(p_1 + p_2 + \cdots + p_d) + 1)}. \quad (5.2.3)$$

In particular, the volume V of the ball B_d: $x_1^2 + x_2^2 + \cdots + x_d^2 \leq 1$ is

$$V = 2^d \int_{\theta_d} dV = \frac{\pi^{d/2}}{\Gamma(d/2 + 1)}. \quad (5.2.4)$$

3. Let U_d designate the surface of the d-dimensional sphere. U_d: $x_1^2 + x_2^2 + \cdots + x_d^2 = 1$. Let dS designate the differential element of surface area. Then,

$$S = \int_{U_d} dS = \frac{2\pi^{d/2}}{\Gamma(d/2)}. \quad (5.2.5)$$

$$\int_{U_d} x_1^{p_1-1} x_2^{p_2-1} \cdots x_d^{p_d-1}\, dS = \frac{2\Gamma(p_1/2)\cdots\Gamma(p_d/2)}{\Gamma((p_1 + p_2 + \cdots + p_d)/2)} \quad (5.2.6)$$

if all p_i are odd. Otherwise the value is 0.

4. Let $Q(x_1, x_2, \ldots, x_d)$ be a positive-definite symmetric quadratic form. Then

$$V = \int \cdots \int_{Q \leq c^2} dx_1\, dx_2 \cdots dx_d = \frac{\pi^{d/2} c^d}{\Gamma(d/2 + 1)(\det Q)^{1/2}}, \quad (5.2.7)$$

$$\int_{-\infty}^{\infty} \cdots \int_{-\infty}^{\infty} e^{-Q/2}\, dx_1 \cdots dx_d = \frac{(2\pi)^{d/2}}{(\det Q)^{1/2}}. \quad (5.2.8)$$

5. Let $\alpha_1, \alpha_2, \ldots, \alpha_d$ be nonnegative even integers. Then

$$\int_{-\infty}^{\infty} \cdots \int_{-\infty}^{\infty} \exp[-x_1^2 - \cdots - x_d^2] x_1^{\alpha_1} \cdots x_d^{\alpha_d}\, dx_1 \cdots dx_d$$

$$= \Gamma\left(\frac{\alpha_1 + 1}{2}\right)\Gamma\left(\frac{\alpha_2 + 1}{2}\right) \cdots \Gamma\left(\frac{\alpha_d + 1}{2}\right). \quad (5.2.9)$$

$$\int_{-\infty}^{\infty} \cdots \int_{-\infty}^{\infty} \exp[-(x_1^2 + \cdots + x_d^2)^{1/2}] x_1^{\alpha_1} \cdots x_d^{\alpha_d}\, dx_1 \cdots dx_d$$

$$= 2(\alpha_1 + \cdots + \alpha_d + d - 1)! \frac{\Gamma((\alpha_1 + 1)/2) \cdots \Gamma((\alpha_d + 1)/2)}{\Gamma((\alpha_1 + \cdots + \alpha_d + d)/2)}. \quad (5.2.10)$$

6. Let $0 \le a \le b$. Let n be a positive integer. Let $t > -n$ and not be a negative integer. Then,

$$\int_a^b \cdots \int_a^b \left(\sum_{j=1}^n x_j\right)^t dx_1 \cdots dx_n = \frac{\sum_{i=0}^n (-1)^i \binom{n}{i} ((n-i)b + ia)^{n+t}}{(t+1)(t+2)\cdots(t+n)}. \tag{5.2.11}$$

Integrals of monomials over other interesting regions are given by Stroud and Engels.

References. Engels [13]; Gradshteyn and Ryzhik [1]; Jarosch and Rabinowitz [1]; Stroud [18].

5.3 *Change of Order of Integration*

Higher-dimensional integrals can exhibit a variety of peculiarities. The reader is no doubt familiar with integrals that cannot be evaluated in elementary terms in their original form but can be so evaluated if the order of integration is changed. For example: $\int_0^1 dx \int_x^1 e^{y^2}\, dy$ leads to a nonelementary indefinite integral; but in the reverse order

$$\int_0^1 dx \int_x^1 e^{y^2}\, dy = \int_0^1 e^{y^2}\, dy \int_0^y dx = \tfrac{1}{2}(e-1).$$

There is a numerical phenomenon that is reminiscent of this. If a multiple integral is expressed as an iterated integral, the order in which it is so expressed may affect the numerical accuracy of the result.

Example

$$I = \int_0^1 dy \int_0^{(1-y)^{1/2}} dx = \int_0^1 dx \int_0^{1-x^2} dy = \tfrac{2}{3}.$$

If the first iterated integral is computed numerically by use of ordinary rules, for example, Simpson's rule, it is found that, due to the square-root singularity, the answer is rather less accurate than that obtained from the second iterated integral.

Reference. Price [2].

5.4 *Change of Variables*

If a given region B can be transformed into a *standard* region B' for which a rule of approximate integration is available, then this rule may be transformed to provide a rule for B. We shall show how this works in the case of two dimensions. The device is perfectly general.

Let B be a region in the xy plane and let B' (which will be our standard region) be in the uv plane. Let the regions B and B' be related to each other by means of the transformation

$$x = \phi(u, v),$$
$$y = \psi(u, v). \tag{5.4.1}$$

It will be assumed that ϕ and ψ have continuous partial derivatives and that the Jacobian

$$J(u, v) = \begin{vmatrix} \partial\phi/\partial u & \partial\phi/\partial v \\ \partial\psi/\partial u & \partial\psi/\partial v \end{vmatrix} \tag{5.4.2}$$

does not vanish in B'. Suppose, further, that

$$\iint_{B'} h(u, v)\, du\, dv \approx \sum_{k=1}^{n} w_k h(u_k, v_k), \qquad (u_k, v_k) \in B' \tag{5.4.3}$$

is a rule of approximate integration over B'. Now we have

$$\begin{aligned} \iint_B f(x, y)\, dx\, dy &= \iint_{B'} f(\phi(u, v), \psi(u, v))\, |J(u, v)|\, du\, dv \\ &\approx \sum_{k=1}^{n} w_k f(\phi(u_k, v_k), \psi(u_k, v_k))\, |J(u_k, v_k)| \\ &= \sum_{k=1}^{n} W_k f(x_k, y_k), \end{aligned} \tag{5.4.4}$$

where

$$x_k = \phi(u_k, v_k), \qquad y_k = \psi(u_k, v_k), \qquad \text{and} \qquad W_k = w_k\, |J(u_k, v_k)|. \tag{5.4.5}$$

Thus

$$\iint_B f(x, y)\, dx\, dy \approx \sum_{k=1}^{n} W_k f(x_k, y_k), \tag{5.4.6}$$

with abscissas and weights as above, constitutes a rule of approximate integration for B.

An important special case occurs when B and B' are related by a nonsingular affine transformation:

$$x = au + bv + c,$$
$$y = du + ev + f. \tag{5.4.7}$$

In this case, the Jacobian is constant:

$$|J| = \left|\det\begin{pmatrix} a & b \\ d & e \end{pmatrix}\right| = |ae - bd| \neq 0. \tag{5.4.8}$$

Integration rules over parallelograms, ellipses, etc., and their higher dimensional analogs are therefore easy to develop.

Here is another useful change of variable. The integral

$$I = \int_{L_0}^{U_0} \int_{L_1(x_1)}^{U_1(x_1)} \cdots \int_{L_{d-1}(x_1, x_2, \ldots, x_{d-1})}^{U_{d-1}(x_1, x_2, \ldots, x_{d-1})} f(x_1, x_2, \ldots, x_d)\, dx_1\, dx_2 \cdots dx_d \tag{5.4.9}$$

can be transformed into an integral over a hypercube by means of the transformation

$$x_i = \frac{U_{i-1} + L_{i-1}}{2} + y_i \frac{U_{i-1} - L_{i-1}}{2}, \qquad i = 1, 2, \ldots, d. \tag{5.4.10}$$

The Jacobian of this transformation is

$$J = J(x_1, \ldots, x_{d-1}) = \prod_{i=1}^{d} \left(\frac{U_{i-1} - L_{i-1}}{2}\right). \tag{5.4.11}$$

Example

$$I = \int_{-1}^{+1} \int_{-(1-x_1^2)^{1/2}}^{(1-x_1^2)^{1/2}} f(x_1, x_2)\, dx_1\, dx_2;$$

$$x_1 = \frac{U_0 + L_0}{2} + y_1 \frac{U_0 - L_0}{2} = y_1,$$

$$x_2 = \frac{U_1 + L_1}{2} + y_2 \frac{U_1 - L_1}{2} = y_2(1 - x_1^2)^{1/2}.$$

Therefore

$$I = \int_{-1}^{1} \int_{-1}^{1} f(y_1, y_2(1 - y_1^2)^{1/2})(1 - y_1^2)^{1/2}\, dy_1\, dy_2\,.$$

Reference. McNamee and Stenger [1].

5.5 Decomposition into Elementary Regions

Let B designate a bounded region with volume V. Suppose that the whole space is divided up by a cubical grid whose side is h. Designate by N the number of cubes contained entirely in B. Then N (which is the measure of difficulty of obtaining V by the primitive process of counting) satisfies

$$V \sim Nh^d. \tag{5.5.1}$$

The error E made in counting only those cubes that lie in B equals approximately the volume of those cubes which pass through the boundary of B; hence

$$E \sim hS, \tag{5.5.2}$$

where S designates the surface measure of the boundary of B. Now we have

$$S \sim V^{(d-1)/d}. \tag{5.5.3}$$

Combining these approximations, we obtain

$$E/V \sim 1/N^{1/d}. \tag{5.5.4}$$

This is known as the *dimensional effect* and is a very pessimistic (particularly for high d) estimate of the number of functional evaluations necessary to achieve a given accuracy.

The dimensional effect is exhibited in error estimates for multidimensional Riemann sums. We shall consider only integrals on the unit hypercube of dimension d, $H_d = \{\mathbf{x}\colon 0 \le x_i \le 1,\ 1 \le i \le d\}$. We first define the total variation of an integrand proceeding in two stages.

Consider d partitions π_i of $[0, 1]$: $0 = x_{i0} < x_{i1} < \cdots < x_{i, m_i} = 1$. For any function $\phi(\mathbf{t})$ of d variables, define

$$\begin{aligned}\delta_{i, j_i} \phi(\mathbf{t}) = \phi(t_1, \ldots, t_{i-1}, x_{i, j_i}, t_{i+1}, \ldots, t_d) \\ - \phi(t_1, \ldots, t_{i-1}, x_{i, j_i - 1}, t_{i+1}, \ldots, t_d).\end{aligned}$$

Then the d-dimensional variation of f in the sense of Vitali $V^d f$ is defined by

$$V^d f = \sup_{\pi_1, \ldots, \pi_d} \sum_{j_1 = 1}^{m_1} \cdots \sum_{j_d = 1}^{m_d} |\delta_{1, j_1} \delta_{2, j_2} \cdots \delta_{d, j_d} f(\mathbf{x})|.$$

If $V^d f$ is finite, then f is said to be of bounded variation in the sense of Vitali. If $f(\mathbf{x})$ is sufficiently smooth in that the partial derivative below exists and is continuous, then

$$V^d f = \int_{H_d} \left| \frac{\partial^n f}{\partial x_1 \cdots \partial x_d} \right| dV.$$

On the other hand, for the following function in two dimensions

$$f(x, y) = \begin{cases} 0, & x \le y, \\ 1, & x > y, \end{cases}$$

$V^2 f = \infty$ so that the requirement of bounded variation in the sense of Vitali is quite restrictive. However, it is not suitable for error estimation in numerical integration since if f depends on fewer than d variables, then $V^d f = 0$. To deal with such functions, we introduce the following definition.

DEFINITION. f is said to be of *bounded variation in the sense of Hardy and Krause*, $f \in BVHK$, if $V^k f_{\sigma(k)} < \infty$ for $k \le d$ and all $\sigma(k)$, where $\sigma(k)$ is any choice of k of the x_i, $1 \le i \le d$, and $f_{\sigma(k)}$ is the function of the chosen k variables obtained by fixing the remaining d-k variables x_i to be 1. The *total variation* Vf is then given by

$$Vf = \sum_{k=1}^{d} \sum_{\sigma(k)} V^k f_{\sigma(k)}.$$

We now have the following result.

THEOREM. *If $f \in BVHK$ on the hypercube H_d with total variation Vf and $N = n^d$ for some integer $n \ge 1$, then*

$$\left| \int_0^1 \cdots \int_0^1 f \, dx_1 \, dx_2 \cdots dx_d - \frac{1}{N} \sum_{k_1, k_2, \ldots, k_d = 0}^{n-1} f\left(\frac{k_1}{n}, \frac{k_2}{n}, \ldots, \frac{k_d}{n}\right) \right| \le \frac{dVf}{N^{1/d}}. \tag{5.5.5}$$

In order to judge the limits of accuracy that can be achieved with averages formed arbitrarily, we introduce the concept of discrepancy of a finite set of points in H_d. There are several definitions of discrepancy which are useful in error estimation. We shall restrict ourselves to the so-called extreme discrepancy. For any set $X_N = (\mathbf{x}_1, \mathbf{x}_2, \ldots, \mathbf{x}_N)$ of N points in H_d and any point $\mathbf{t} \in H_d$ define the counting function $A(\mathbf{t}; X_N)$ to be equal to the number of points of X_N belonging to the hyperrectangle $\times_{j=1}^{d} [0, t_j]$ where $\mathbf{t} = (t_1, \ldots, t_d)$. Then the *extreme discrepancy* $D_N(X_N)$ is given by

$$D_N(X_N) = \sup_{\mathbf{t} \in H_d} \left| N^{-1} A(\mathbf{t}; X_N) - \prod_{i=1}^{d} t_i \right|. \tag{5.5.6}$$

For an infinite sequence $X = \{\mathbf{x}_1, \mathbf{x}_2, \ldots\}$, the discrepancy $D_N(X)$ is defined to be the discrepancy of the finite set $(\mathbf{x}_1, \mathbf{x}_2, \ldots, \mathbf{x}_N)$. We can now state the following result:

If $f \in BVHK$ on H_d with total variation Vf and X is any finite or infinite sequence of points in H_d, $\mathbf{x}_1, \mathbf{x}_2, \ldots$, then

$$\left| \int_{H_d} f \, dV - \frac{1}{N} \sum_{i=1}^{N} f(\mathbf{x}_i) \right| \le Vf \cdot D_N(X). \tag{5.5.7}$$

In view of this result, it makes sense to try to find low-discrepancy sequences. Now Roth has shown that for any finite set X_N,

$$D_N(X_N) \ge C_d (\log N)^{(d-1)/2} / N, \tag{5.5.8}$$

where we may take $C_d = 2^{-4d}((d-1)\log 2)^{(1-d)/2}$. Furthermore, for any infinite sequence $X = \{\mathbf{x}_1, \mathbf{x}_2, \ldots\}$, $ND_N(X) > C_{d+1}(\log N)^{d/2} - 1$ for infinitely many N. This implies that the best rate of convergence of the error estimate is given by $(\log N)^{(d-1)/2}/N$. Unfortunately, no sequence with such a small discrepancy has been found. However, there do exist finite sets X_N such that $ND_N(X_N) = O((\log N)^{d-1})$ and infinite sequences X such that $ND_N(X) = O((\log N)^d)$.

Example. Let r be a fixed integer >1. Then any positive integer k can be written in the form

$$k = c_0 + c_1 r + \cdots + c_q r^q, \qquad 0 \le c_i < r, \quad c_q \ne 0, \quad q = [\log_r k].$$

The *radical-inverse* function $\phi_r(k)$ is then given by

$$\phi_r(k) = c_0 r^{-1} + c_1 r^{-2} + \cdots + c_q r^{-q-1}. \tag{5.5.9}$$

Let $p(j)$ be the jth prime so that $p(1) = 2$, $p(2) = 3, \ldots$. Then the *Halton sequence* in d dimensions is given by $X = \{\mathbf{x}_i\}$ where

$$\mathbf{x}_i = (\phi_{p(1)}(i), \phi_{p(2)}(i), \ldots, \phi_{p(d)}(i)), \qquad i = 1, 2, \ldots \tag{5.5.10}$$

and the *Hammersley set* $X_N = \{\mathbf{x}_1, \mathbf{x}_2, \ldots, \mathbf{x}_N\}$, by

$$\mathbf{x}_i = (i/N, \phi_{p(1)}(i), \ldots, \phi_{p(d-1)}(i)), \qquad i = 1, \ldots, N. \tag{5.5.11}$$

For the Halton sequence, we have

$$ND_N(X) \le A_d(\log N)^d + O((\log N)^{d-1})$$

where

$$A_d = \prod_{j=1}^{d} \frac{p(j) - 1}{2 \log p(j)}$$

while for the Hammersley set,

$$ND_N(X_N) \le A_{d-1}(\log N)^{d-1} + O((\log N)^{d-2}).$$

Other low-discrepancy sequences include the Zaremba modification of the Halton sequence, the so-called LP_τ-sequences of Sobol and the Faure sequence.

We note that just as the Hammersley set was derived from the Halton sequence, so one can derive a finite set in $d+1$ dimensions from any infinite set in d dimensions $X^{(d)} = \{\mathbf{x}_i^{(d)}\}$ by writing

$$X_N^{(d+1)} = (\mathbf{x}_1^{(d+1)}, \ldots, \mathbf{x}_N^{(d+1)}) \qquad \text{with} \quad \mathbf{x}_j^{(d+1)} = (j/N, \mathbf{x}_j^{(d)}),$$

and the discrepancy $D_N(X_N^{(d+1)})$ is of the same order of magnitude as $D_N(X^{(d)})$. Thus, if we fix the number N of integration points in advance, we can get a set of points with lower discrepancy than if we use the initial segment of an infinite sequence. However, the latter option has the advantage that it enables us to continue the computation without discarding any previously computed function values.

The use of low-discrepancy sequences for multidimensional numerical integration is indicated for integrands which are not too smooth. However, they should be smooth enough to ensure that the total variation is finite.

References. Antonov and Soleev [1]; Faure [1]; Haber [3]; Halton [1, 3]; Halton and Smith [A1]; Halton and Zaremba [1]; Hlawka [1]; Kuipers and Nierderreiter [1]; Niederreiter [2, 6]; Roos and Arnold [1]; Roth [2]; Sobol [1–6]; Stroud [18]; Tocher [1]; Warnock [1]; Zaremba [3].

That the dimensional effect is to some extent inherent in the problem can be concluded from a theorem of Bahvalov which sets lower bounds for what can be accomplished with N-point rules. Designate by C_M^r the class of functions of d variables, all of whose rth order partial derivatives $\partial^r f/\partial x_1^{\alpha_1}\,\partial x_2^{\alpha_2}\cdots\partial x_d^{\alpha_d}$, $\alpha_1+\alpha_2+\cdots+\alpha_d=r$ exist, are continuous in a hypercube H_d and are $\le M$ in absolute value there.

THEOREM. *There is a constant K depending upon M and r such that for any N-point rule R there is a function $f\in C_M^r$ with*

$$\left|\int_{H_d} f\,dV - R(f)\right| > K/N^{r/d}. \tag{5.5.12}$$

References. Bahvalov [1]; Haber [7].

5.6 Cartesian Products and Product Rules

Let B be a region in r-dimensional Euclidean space with points $(x_1, x_2, \ldots, x_r)$, and let G be a region in s-dimensional Euclidean space with points $(y_1, y_2, \ldots, y_s)$. The symbol $B\times G$ designates the *Cartesian product of B and G*, and by this we mean the region in the Euclidean space of $r+s$ dimensions with points $(x_1, \ldots, x_r, y_1, \ldots, y_s)$ that satisfy $(x_1, \ldots, x_r)\in B$, $(y_1, \ldots, y_s)\in G$.

Examples. $B: 0\le x\le 1$, $G: 0\le y\le 1$; then $B\times G$ is the unit square, $0\le x, y\le 1$. If B is the disk $x^2+y^2\le 1$ and G is the segment $0\le z\le 1$, then $B\times G$ is the right cylinder of height 1 whose base is B.

It is convenient to use vector notation and write $\mathbf{x}=(x_1,\ldots,x_r)$ and $\mathbf{y}=(y_1,\ldots,y_s)$. Suppose now that R is an m-point rule of integration over B,

$$R(f)=\sum_{j=1}^{m} w_j f(\mathbf{x}_j)\approx\int_B f(\mathbf{x})\,dV, \qquad \mathbf{x}_j\in B, \tag{5.6.1}$$

and S is an n-point rule of integration over G,

$$S(f)=\sum_{k=1}^{n} v_k f(\mathbf{y}_k)\approx\int_G f(\mathbf{y})\,dV, \qquad \mathbf{y}_k\in G. \tag{5.6.2}$$

Then by the *product rule of R and S* we mean the *mn*-point rule applicable to $B \times G$ and defined by

$$R \times S(f) = \sum_{j=1}^{m} \sum_{k=1}^{n} w_j v_k f(\mathbf{x}_j, \mathbf{y}_k) \approx \int_{B \times G} f(\mathbf{x}, \mathbf{y})\, dV. \tag{5.6.3}$$

An obvious extension of the notion of the product of two rules can be made so as to define the product of three or more rules, and similar definitions can be made for weighted integrals. The following theorem holds for errors.

THEOREM. *If R integrates $f(\mathbf{x})$ exactly over B, if S integrates $g(\mathbf{y})$ exactly over G, and if $h(\mathbf{x}, \mathbf{y}) = f(\mathbf{x})g(\mathbf{y})$, then $R \times S$ will integrate $h(\mathbf{x}, \mathbf{y})$ exactly over $B \times G$.*

Proof

$$\begin{aligned}\int_{B \times G} h(\mathbf{x}, \mathbf{y})\, dV &= \int_{B \times G} f(\mathbf{x})g(\mathbf{y})\, dV = \int_B f(\mathbf{x})\, dV_B \int_G g(\mathbf{y})\, dV_G \\ &= \sum_{j=1}^{m} w_j f(\mathbf{x}_j) \sum_{k=1}^{n} v_k g(\mathbf{y}_k) = \sum_{j=1}^{m} \sum_{k=1}^{n} w_j v_k f(\mathbf{x}_j) g(\mathbf{y}_k) \\ &= R \times S(h).\end{aligned}$$

Examples. Let R designate the rectangle $a \le x \le b$, $c \le y \le d$. Set $m = \frac{1}{2}(a + b)$, $n = \frac{1}{2}(c + d)$, $b - a = h$, $d - c = k$. The evaluation of $\iint_R f(x, y)\, dx\, dy$ by the product of two Simpson's rules yields

$$\begin{aligned}\iint_R f(x, y)\, dx\, dy \approx \frac{hk}{36}[&f(a, c) + f(a, d) + f(b, c) + f(b, d) \\ &+ 4(f(a, n) + f(m, c) + f(b, n) + f(m, d)) + 16f(m, n)]. \end{aligned} \tag{5.6.4}$$

This 9-point rule will integrate exactly all linear combinations of the 16 monomials $x^i y^j$ ($0 \le i$, $j \le 3$). The appropriate weights for the product of two compound Simpson's rules can be found by the multiplication table following.

	1	4	2	4	2	4	⋯	2	4	1
1	1	4	2	4	2	4	⋯	2	4	1
4	4	16	8	16	8	16	⋯	8	16	4
2	2	8	4	8	4	8	⋯	4	8	2
4	4	16	8	16	8	16	⋯	8	16	4
2										
4										
⋮	⋮									⋮
2										
4	4	16	8	16	8	16	⋯	8	16	4
1	1	4	2	4	2	4	⋯	2	4	1

As a second example of a product rule, consider the integral $I = \int_C f(x, y)\,dx\,dy$, where C is the unit circle $x^2 + y^2 \le 1$. Expressing this integral in polar coordinates, we have

$$I = \int_0^{2\pi} d\theta \int_0^1 fr\,dr. \tag{5.6.5}$$

In the r variable, we will integrate approximately, using a Gaussian rule of order n corresponding to the weight function $w(r) = r$ with abscissas $r_1, r_2, \ldots, r_n$ and weights $w_1, w_2, \ldots, w_n$. In the θ variable (since f is periodic in θ) we use a rectangular (trapezoidal) rule with abscissas $\theta_k = (2\pi k/N)$, $k = 0, 1, \ldots, N-1$, and weights all $1/N$. This yields

$$I \approx \frac{1}{N} \sum_{k=0}^{N-1} \sum_{i=1}^{n} w_i f\left(r_i \cos\frac{2\pi k}{N}, r_i \sin\frac{2\pi k}{N}\right). \tag{5.6.6}$$

In this example we have "tucked" the Jacobian into the rule.

As a further instance of this, consider integration over the three-dimensional spherical shell $R^2 \le x^2 + y^2 + z^2 \le 1$. This leads to

$$I = \int_R^1 r^2\,dr \int_0^{\pi} \sin\phi\,d\phi \int_0^{2\pi} F(\theta, \phi, r)\,d\theta. \tag{5.6.7}$$

We wish to approximate this by a product rule of the form

$$I \approx \sum_{i,j,k} A_i B_j C_k F(\theta_i, \phi_j, r_k) \tag{5.6.8}$$

Let $\theta_i = 2\pi i/(m+1)$, $i = 1, 2, \ldots, m+1$. Let $\cos\phi_j$, $j = 1, 2, \ldots, 2m+2$, be the zeros of the Legendre polynomial of degree $2m+2$ on $[-1, 1]$. Let r_k^2 be the zeros of the polynomial in r^2 of degree $m+1$, $Q_{m+1}(r^2)$, where $\int_R^1 r^2 Q_{m+1}(r^2)p(r^2)\,dr = 0$ for any $p \in \mathscr{P}_m$.

Let $A_i = 2\pi/(m+1)$, B_j be the Gauss weights of order $2m+2$, and C_k be the Gauss weights appropriate to Q_{m+1}, i.e.,

$$C_k = \frac{1}{Q'_{m+1}(r_k^2)} \int_R^1 \frac{r^2 Q_{m+1}(r^2)\,dr}{r^2 - r_k^2}.$$

With this selection of constants, the rule (5.6.8) will be of accuracy $4m+3$ in x, y, z.

Extensions to the n-dimensional shells are in Hetherington and Mustard, while the abscissas and weights for the 3-sphere are in Feuchter. See Stroud and Secrest for the infinite sphere (the whole three-dimensional space).

The last two examples above are instances of spherical product Gauss rules. In dimension d, these involve Gaussian abscissas and weights with respect to the weight function $|r|^{d-1}$ over some interval and with respect to the weight functions $(1 - x^2)^{(k-2)/2}$, $k = 1, 2, \ldots, n-1$, over $[-1, 1]$. For details, including a discussion of the more efficient spherical product Lobatto rules, see Stroud.

There is also a product rule, the conical product rule, for cones. Let R_{d-1} be a region in $(d-1)$-dimensional space in the variables $u_1, \ldots, u_{d-1}$. Then the cone Z_d with base R_{d-1} and vertex $(0, \ldots, 0, 1)$ is given by

$$Z_d = \{(x_1, \ldots, x_d)\colon x_i = u_i(1-\lambda),\ i = 1, \ldots, d-1;\ x_d = \lambda,\ 0 \le \lambda \le 1,$$
$$(u_1, \ldots, u_{d-1}) \in R_{d-1}\}.$$

If we are given an integration rule of degree m for R_{d-1}

$$\int_{R_{d-1}} \cdots \int f(x_1, \ldots, x_{d-1})\, dx_1 \cdots dx_{d-1} \approx \sum_{j=1}^{N} B_j f(y_{j1}, \ldots, y_{j,d-1}) \quad (5.6.9)$$

and an integration rule of degree m over $[0, 1]$ of the form

$$\int_0^1 (1-t)^{d-1} g(t)\, dt \approx \sum_{j=1}^{M} A_i g(z_i), \quad (5.6.10)$$

then the rule

$$\int_{Z_d} f\, dV \approx \sum_{i=1}^{M} \sum_{j=1}^{N} w_{ij} f(\mathbf{x}_{ij}) \quad (5.6.11)$$

is an integration rule of degree m for Z_d where $w_{ij} = A_i B_j$ and $\mathbf{x}_{ij} = (y_{j1}(1-z_i), \ldots, y_{j,d-1}(1-z_i), z_i)$, $i = 1, \ldots, M$; $j = 1, \ldots, N$. Note that if (5.6.10) is a rule of Radau or Lobatto type or indeed any rule of degree m with $z_M = 1$, then the N points corresponding to $i = M$ "collapse" into the vertex of the cone, thus saving $N - 1$ function evaluations. (For the definition of the degree of an integration rule in d dimensions, see Section 5.7.)

References. Feuchter [1]; Fishman [1]; Hammer, Marlow and Stroud [1]; R. G. Hetherington [1]; Krylov [1, pp. 122–125]; Mustard [1]; Peirce [1, 2]; Stroud [18, Chap. 2]; Stroud and Secrest [1, 2].

Example. Let G_5 be the Gaussian 5-point rule over $[0, 1]$. Then $G_5 \times G_5 \times G_5 \times G_5 (=(G_5)^4)$ is a product rule of 625 points applicable to the unit 4-dimensional cube. It integrates exactly the 10^4 monomials of the form $x_1^{m_1} x_2^{m_2} x_3^{m_3} x_4^{m_4}$ with $0 \le m_1, m_2, m_3, m_4 \le 9$. The following numerical example is from Thacher.

$$I(k) = \int_0^1 \int_0^1 \int_0^1 \int_0^1 (k \cos w - 7kw \sin w - 6kw^2 \cos w + kw^3 \sin w)\, dx_1\, dx_2\, dx_3\, dx_4$$

$$= \sin k, \qquad w = kx_1 x_2 x_3 x_4 .$$

k	.1	.25	1.0	$\pi/2$	$3\pi/2$
Exact value	.0998 334	.2474 04	.8414 71	1.0000 0	− 1.0000 0
Error in $(G_5)^4$	.0000 0018	.0000 005	.0000 011	.0000 010	.0000 282

As k becomes larger, the integrand becomes less polynomial-like in its behavior and the error begins to increase.

Reference. Thacher [3].

Example

$$I(b) = \frac{1}{\pi^3} \int_0^\pi \int_0^\pi \int_0^\pi \frac{dx\,dy\,dz}{3b - \cos x - \cos y - \cos z}.$$

b^{-1}	Rule	Approximate value	Exact value
.7	A	.25767	.25794
.7	B	.25792	.25794
.8	A	.30786	.30781
.8	B	.30780	.30781
.9	A	.37231	.36993
.9	B	.36990	.36993
1.0	A	.49849	.50546
1.0	B	.50305	.50546

Rule A is $(G_3)^3$, a 27-point rule. Rule B is $(3 \times G_3)^3$, a 729-point rule. Note that the integrand is singular for $b^{-1} = 1.0$.

References. Gradshteyn and Ryzhik [1, p. 619]; Tikson [1].

Example

$$\int_0^1 \int_0^1 \frac{dx\,dy}{1 - xy} = \frac{\pi^2}{6}.$$

Rule	Approximate value
R_1	1.58123
R_2	1.61213
R_3	1.62283
R_4	1.62827
R_5	1.63156
R_6	1.63376
Exact value	1.644934

The rule $R_k = (k \times G_3)^2$ and therefore has $9k^2$ points. Note the singularity of the integrand at $x = y = 1$ and also the difficulty of treating the integral as an iterated simple integral. However, the integrand is positive and monotonic in each variable, and the process of ignoring the singularity is working out.

Example

$$J = 2^{-6} \int_{-1}^{1} \cdots \int_{-1}^{1} \cos[3(1 - x_1)x_2 x_3 x_4 x_5 x_6 + \tfrac{1}{2}]\, dx_1 \cdots dx_6.$$

Rule	Approximate Value
$(G_3)^6$	.85875 262
$(G_4)^6$	.85851 236
$(G_5)^6$	.85852 519
$(G_6)^6$	.85852 470
Exact value	.85852 471

Example

$$J = \int_0^1 \cdots \int_0^1 [\log(x_1 x_2 x_3/x_4 x_5 x_6)]^2 x_1 x_2 x_3 x_4 x_5 x_6\, dx_1 \cdots dx_6.$$

Rule	Approximate Value
$(G_4)^6$	.02407 639
$(G_6)^6$	.02362 196
$(G_8)^6$	.02350 998
Exact value	.02343 75

Note the infinite derivatives of the integrand at the lower limit.

Reference. Cranley and Patterson [2].

The number of points N in a product rule exhibits a dimensional effect. That is to say, if R is a p-point rule on $[0, 1]$, then the corresponding product rule R^d over the d-dimensional cube has p^d points. For a modest value of p, say 5, and for a multiple integral of dimension 18, this would amount to $5^{18} \approx 10^{13}$ functional evaluations. Even with present speeds of computers, this amount of computation would be quite out of the question. On the other hand, for multiple integrals of dimension at most 4 or 5, we could take R up to 10 points and compute a product rule of 10^4 or 10^5 points without excessive expenditure of time. The method adopted for approximate multiple integration will therefore depend strongly on the dimension number as well as on the complexity of the integrand.

References. Hammer [2]; Hammer and Wymore [1].

To circumvent the dimensional effect of product rules in hypercubes while taking advantage of their simplicity and efficiency, integration rules have been proposed that use a subset of the lattice of integration points in the product rule. This has been done in two ways, deterministic and probabilistic. In the former case, Keast, following up the work of Stenger and McNamee, developed an algorithm to generate points and weights for integration rules of degree $2m - 1$ using subsets of the lattice of the product rule $(G_m)^d$. These rules use many fewer than m^d points but some of the weights are negative and increase in magnitude as d and m grow. There may therefore be some loss of significance in their use. Of course, these rules do not have the full integrating power of $(G_m)^d$ since the latter integrate correctly all monomials $x_1^{k_1} \cdots x_d^{k_d}$ such that $k_i \leq 2m - 1, i = 1, \ldots, d$ while the Keast rules only integrate correctly those monomials for which $k_1 + \cdots + k_d \leq 2m - 1$. Nevertheless, in the absence of anything better these rules may be useful.

The probabilistic scheme was first suggested by Hammersley and subsequently by Tsuda. Let us assume that we have a Gauss rule G_m for the unit interval

$$\int_0^1 g(x)\, dx = \sum_{j=1}^{m} w_j g(v_j) + c_m g^{(2m-1)}(\xi), \qquad 0 < \xi < 1.$$

Then we shall approximate the integral of $f(\mathbf{x})$ over the unit hypercube $H_d = [0, 1]^d$ by the average

$$\int_{H_d} f\, dV \approx \frac{1}{N} \sum_{i=1}^{N} f(\mathbf{x}_i)$$

where each point $\mathbf{x}_i$ belongs to the lattice of $(G_m)^d$ and is chosen in a random fashion as follows:

Define $z_0 = 0$, $z_k = \sum_{j=1}^{k} w_j$, $k = 1, \ldots, m$, so that $z_m = 1$. For each dimension l, $1 \leq l \leq d$, choose a random number ξ_l, $0 < \xi_l \leq 1$, and determine the index p_l such that $z_{p_l - 1} < \xi_l \leq z_{p_l}$. Then

$$\mathbf{x}_i = (v_{p_1}, \ldots, v_{p_d}).$$

Each point $\mathbf{x}_i$ is of course determined by a different d-tuple of random numbers. For further discussion and examples, see Tsuda who used the Fejér rule (2.5.5.5) as the basic integration rule instead of the Gauss rule.

References. Hammersley [1]; Keast [4, 5]; Lyness and Keast [1]; McNamee and Stenger [1]; Stenger [7]; Tsuda [1].

A simple error estimate for product rules has been given by Haber. Consider

$$\int_{G_d} f \equiv \int_0^1 \cdots \int_0^1 f(x_1, x_2, \ldots, x_d)\, dx_1\, dx_2 \cdots dx_d$$

$$\simeq \sum_{i_1=1}^{n_1} \sum_{i_2=1}^{n_2} \cdots \sum_{i_d=1}^{n_d} a_{i_1,1} a_{i_2,2}, \ldots, a_{i_d,d} f(x_{i_1,1}, x_{i_2,2}, \ldots, x_{i_d,d})$$

$$\equiv \left(\prod_{i=1}^{d} Q_i\right) f$$

where

$$Q_i(g) = \sum_{j=1}^{n_i} a_{j,i}\, g(x_{j,i}), \qquad i = 1, 2, \ldots, d.$$

Assume that

$$\left| \int_0^1 f(x_1, \ldots, x_d)\, dx_i - Q_i(f; x_i) \right| \le E_i$$

for all values of $x_1, x_2, \ldots, x_{i-1}, x_{i+1}, \ldots, x_d$ lying between 0 and 1. If we set $A_i = \sum_{j=1}^{n_i} |a_{j,i}|$, then

$$\left| \int_{G_d} f - \left(\prod_{i=1}^{d} Q_i\right) f \right| \le E_1 + A_1 E_2 + A_1 A_2 E_3 + \cdots + A_1 A_2 \cdots A_{d-1} E_d.$$

Other error estimates for product rules, in particular for product Gauss rules, are given by Stenger and Lether.

References. Haber [7]; Lether [1–4]; Stenger [2]; Stroud and Secrest [2].

5.6.1 *Generalized Product Rules*

If the region B is sufficiently simple, the multiple integral $I = \int \cdots \int f\, dx_1\, dx_2 \cdots dx_d$ may be expressed as an iterated integral of the form

$$I = \int_{L_0}^{U_0} dx_1 \int_{L_1(x_1)}^{U_1(x_1)} dx_2 \cdots \int_{L_{d-1}(x_1, x_2, \ldots, x_{d-1})}^{U_{d-1}(x_1, x_2, \ldots, x_{d-1})} f(x_1, x_2, \ldots, x_d)\, dx_d, \tag{5.6.1.1}$$

or as a sum of such integrals. The notion of a product rule may be extended to integrals of this form. For simplicity, we work in two variables, x, y, and write

$$I = \int_a^b dx \int_{\psi(x)}^{\phi(x)} f(x, y)\, dy = \int_a^b g(x)\, dx, \tag{5.6.1.2}$$

where

$$g(x) = \int_{\psi(x)}^{\phi(x)} f(x, y)\, dy. \tag{5.6.1.3}$$

We use an N-point rule R_1,

$$\int_a^b g(x)\, dx \approx \sum_{k=1}^{N} w_k g(x_k), \tag{5.6.1.4}$$

to integrate in the x variable. Thus

$$I \approx \sum_{k=1}^{N} w_k \int_{\psi(x_k)}^{\phi(x_k)} f(x_k, y)\, dy. \tag{5.6.1.5}$$

For each of the N integrals, we use an M-point rule R_2 for its evaluation:

$$\int_{\psi(x_k)}^{\phi(x_k)} f(x_k, y)\, dy \approx \sum_{j=1}^{M} v_{jk} f(x_k, y_{jk}). \tag{5.6.1.6}$$

The double subscripts on the right reflect the fact that the abscissas and weights of R_2 must be adjusted for each value of k to the interval $\psi(x_k) \le y \le \phi(x_k)$. Thus

$$I \approx \sum_{k=1}^{N} w_k \sum_{j=1}^{M} v_{jk} f(x_k, y_{jk}) \tag{5.6.1.7}$$

is an MN-point rule for I.

Note that this is the same formula that results from applying the product rule $R_1 \times R_2$ to the transformed integral over the square $[a, b] \times [a, b]$ with Jacobian

$$J(x) = [\phi(x) - \psi(x)]/(b - a).$$

Example. (Thacher)

$$I = \int_{-1}^{1} \int_{-(1-x^2)^{1/2}}^{(1-x^2)^{1/2}} \int_{-(1-x^2-y^2)^{1/2}}^{(1-x^2-y^2)^{1/2}} \frac{dz\, dy\, dx}{x^2 + y^2 + (z-k)^2} = \pi\left(2 + \left(\frac{1}{k} - k\right) \log\left|\frac{1+k}{1-k}\right|\right).$$

An $s \times G_p$ (G = Gauss) rule was employed in each variable, and the following values were obtained.

k	$\frac{1}{2}$		2	
Exact	11.4602 7375		1.1060 9686	
s	1	2	1	2
$p = 2$	5.454466	9.361670	1.0368787	1.1184317
$p = 3$	11.838664	12.408983	1.1343568	1.1094294

References. Freeman [A1]; Lether [14]; Thacher [2].

5.7 Rules Exact for Monomials

In one dimension we have seen that the program of obtaining rules that are exact for monomials $1, x, x^2, \ldots, x^n$ coincides with the program of integrating an interpolation polynomial. This has no exact generalization in more than one dimension for the simple reason that polynomial interpolation is not always possible in two and more dimensions. For example, given n points $P_1, P_2, \ldots, P_n$ in the plane, and given n monomials in two variables $1, x, y, x^2, xy, \ldots$, it is not always possible to find a linear combination of the monomials which takes on prescribed values at the points P_i.

If the points P_i constitute a product lattice, then there exists a generalization of Lagrange's formula. We set

$$u(x) = \prod_{i=0}^{m} (x - x_i), \qquad u_i(x) = \frac{u(x)}{x - x_i},$$
$$v(y) = \prod_{j=0}^{n} (y - y_j), \qquad v_j(y) = \frac{v(y)}{y - y_j}. \tag{5.7.1}$$

Then

$$p_{mn}(x, y) = \sum_{i=0}^{m} \sum_{j=0}^{n} f(x_i, y_j) \frac{u_i(x)v_j(y)}{u_i(x_i)v_j(y_j)} \tag{5.7.2}$$

is a polynomial in x and y which takes on the values $f(x_i, y_j)$ at the $(m+1)(n+1)$ points (x_i, y_j) $(i = 0, 1, \ldots, m; j = 0, 1, \ldots, n)$, which align themselves in a rectangular lattice. This device carries over to higher dimensions. This formula may be integrated formally to produce integration rules of the type

$$\iint_B f(x, y)\, dx\, dy \approx \sum_{i=0}^{m} \sum_{j=0}^{n} w_{ij} f(x_i, y_j) \tag{5.7.3}$$

with

$$w_{ij} = \iint_B \frac{u_i(x)v_j(y)}{u_i(x_i)v_j(y_j)}\, dx\, dy.$$

There is an interpolation formula due to Biermann that makes use of functional values on a triangular lattice of points

$$\begin{array}{lllll} (x_0, y_0), & (x_1, y_0), & \ldots, & & (x_m, y_0) \\ (x_0, y_1), & (x_1, y_1), & \ldots, & (x_{m-1}, y_1) & \\ \vdots & & & & \\ (x_0, y_m) & & & & \end{array}$$

Biermann's interpolation formula can also be integrated.

Additional configurations of points in the plane for which interpolation formulas always exist and can be integrated are given by Salzer.

Genz discusses the construction of a general set of points in d dimensions which can be used for multivariate interpolation by polynomials of degree p. When such a polynomial is integrated, it yields an integration rule of degree p. The set X_p is given by

$$X_p = \{\mathbf{g_k}: |\mathbf{k}| \leq p\}$$

where

$$\mathbf{g_k} = (g_{1, k_1}, g_{2, k_2}, \ldots, g_{d, k_d})$$

with integers $k_i \geq 0$ and $|\mathbf{k}| = \sum_{i=1}^{d} k_i$ and where the elements in each of the sequences $\{g_{n, i}: i = 0, 1, \ldots\}$ are distinct.

Example. For $d = 3$, the points in X_2 are as follows:

$$(g_{10}, g_{20}, g_{30}), \quad (g_{11}, g_{20}, g_{30}), \quad (g_{10}, g_{21}, g_{30}), \quad (g_{10}, g_{20}, g_{31}), \quad (g_{11}, g_{21}, g_{30}),$$
$$(g_{11}, g_{20}, g_{31}), \quad (g_{10}, g_{21}, g_{31}), \quad (g_{12}, g_{20}, g_{30}), \quad (g_{10}, g_{22}, g_{30}), \quad (g_{10}, g_{20}, g_{32})$$

where $\{g_{n0}, g_{n1}, g_{n2}\}$ are distinct for $n = 1, 2, 3$. However, g_{10} need not be distinct from any g_{2i} or g_{3i}, and so forth.

Note that if $q > p$, $X_p \subset X_q$.

For integrals over the hypercube $[-1, 1]^d$, Genz shows how to choose the sets X_p so that many of the weights in the corresponding integration rule vanish thus saving function evaluations. Furthermore, no function evaluations are lost in going from an integration rule of degree p to one of higher degree.

References. Genz [8]; Salzer [10]; Stancu [2]; Steffensen [1, p. 215]; Thacher [6, 7].

In order to obtain formulas that involve fewer points than required for a product layout, we abandon the interpolatory point of view and ask for points $P_1, P_2, \ldots, P_n$ and corresponding weights $w_1, w_2, \ldots, w_n$ such that

$$\int_B f\,dV \approx \sum_{k=1}^{n} w_k f(P_k)$$

is exact for some m monomials. If m is equal to the number of free parameters, then such a rule is called a *minimal point rule.* The determination of such points and weights leads to systems of simultaneous nonlinear equations.

We shall exhibit some of the special rules that have been obtained. Particular emphasis has been laid on the development of rules that are precise for monomials up to degree p, that is, monomials of the form $\prod_{i=1}^{d} x_i^{a_i}$ where $a_i \geq 0$ and $a_1 + a_2 + \cdots + a_d \leq p$. Such a rule is said to be of *precision* or of *degree p.* It is customary to add the requirement that there exist at least one monomial of degree $p + 1$ for which the rule is not exact.

There are $\binom{d+n}{n} = (d+n)!/n!\,d!$ distinct monomials of degree $\leq n$ in d variables.

If rule R_i is of precision p_i in the variable x_i, then the product rule $R = R_1 \times R_2 \times \cdots \times R_d$ is of precision $p = \min\{p_1, p_2, \ldots, p_d\}$ over the d-dimensional hypercube. However, R will usually be much more accurate than other rules of precision p inasmuch as it will integrate exactly all monomials $\prod_{i=1}^{d} x_i^{a_i}$ such that $a_i \leq p_i$ and not only those monomials for which $a_1 + \cdots + a_d \leq p$.

We display the relevant equations for achieving precision. Let d be the dimension, $x_1, x_2, \ldots, x_d$, the variables, and n, the number of points. The point P_k is $(x_{1k}, x_{2k}, \ldots, x_{dk})$.

Degree 0: 1 equation,

$$\sum_{k=1}^{n} w_k = \int_B dV$$

Degree 1: d equations,

$$\sum_{k=1}^{n} w_k x_{rk} = \int_B x_r\,dV, \qquad r = 1, 2, \ldots, d$$

Degree 2: $d(d+1)/2!$ equations,

$$\sum_{k=1}^{n} w_k x_{rk} x_{sk} = \int_B x_r x_s\,dV, \qquad r = 1, 2, \ldots, d, \qquad r \leq s \leq d$$

and so on.

To achieve precision p there are, all in all, $\binom{d+p}{p}$ equations in $n + nd = n(d + 1)$ variables to be fulfilled. If therefore $n \geq (1/(d + 1))\binom{d+p}{p}$, the expectation would be that there exists a solution to this system. However, since the system is nonlinear, there is no a priori guarantee that it has a real solution. Furthermore, there are situations in which these equations can be solved with far fewer points.

A rule such that $n = N \equiv \lceil \binom{d+p}{p}/(d + 1)\rceil$ is called an *efficient* rule. Here $\lceil x \rceil$ denotes the "ceiling" function defined to be the smallest integer $\geq x$. A rule of precision p containing more than N points is called *subefficient* while one containing less is said to be *hyperefficient*. Examples of hyperefficient rules are the $2d$-point rules of precision 3 with $d \geq 7$ and the $(2d^2 + 1)$-point rules of precision 5 with $d \geq 8$ derived below.

Given a required precision p, it is possible to set crude upper and lower limits for n, the number of points required.

THEOREM.

$$\binom{d + [p/2]}{[p/2]} \leq n \leq \binom{d + p}{p}. \tag{5.7.4}$$

Proof. Consider all the monomials of degree $\leq [p/2]$. There are $m = \binom{d+[p/2]}{[p/2]}$ of them. Designate them by $M_1, M_2, \ldots, M_m$. Suppose now that $R(f) = \sum_{k=1}^{n} w_k f(P_k)$ is a rule of precision p for a region B. The linear system of n equations in m unknowns $b_1, b_2, \ldots, b_m$,

$$\sum_{t=1}^{m} b_t M_t(P_k) = 0, \qquad k = 1, 2, \ldots, n$$

will have a nontrivial solution if $m > n$. But then, $g = b_1 M_1 + b_2 M_2 + \cdots + b_m M_m$ will be a polynomial of degree $\leq [p/2]$, not identically zero. Hence g^2 is a polynomial of degree $\leq p$ and $g^2 \not\equiv 0$. Furthermore $g(P_k) = 0$, $k = 1, 2, \ldots, n$. Hence $0 \not\equiv \int_B g^2\, dV = R(g^2) = \sum_{k=1}^{n} w_k g^2(P_k) = 0$. Contradiction.

The upper bound follows by fixing the points P_k in such a way that the determinant of the resulting linear system is not zero, and therefore can be solved. It also follows trivially from Tchakaloff's theorem which will be given shortly, and which provides much more under the circumstances.

For centrally symmetric regions, that is regions Ω such that if $\mathbf{x} \in \Omega$, then $-\mathbf{x} \in \Omega$, lower bounds greater than

$$\kappa \equiv \binom{d + [p/2]}{[p/2]}$$

exist for integration rules of odd degree, $p = 2k + 1$. Denoting by v the number of odd monomials in d variables of degree $\leq k$, we have that if an n-point integration rule of degree $2k + 1$ contains the origin as one of its points, then n must satisfy the inequality

$$n \geq \begin{cases} 2(\kappa - \nu) & \text{for } k \text{ even} \\ 2\nu & \text{for } k \text{ odd} \end{cases}$$

and if the origin is not one of its points,

$$n \geq \begin{cases} 2(\kappa - \nu) - 1 & \text{for } k \text{ even} \\ 2\nu + 1 & \text{for } k \text{ odd.} \end{cases}$$

References. Davis [8]; Möller [1, 4]; Mysovskikh [15]; Schmid [1]; Stroud [2].

We now describe some specific monomial rules.

1 2d-Point Rule of Precision 3 for a Hypercube

Designate the hypercube $-1 \leq x_i \leq 1$, $i = 1, 2, \ldots, d$, by H_d. Let $dV = dx_1\, dx_2 \cdots dx_d$. We shall look for an integration formula of the form

$$\int_{H_d} f\, dV \approx w \sum^{2d} f(\pm u, 0, 0, \ldots, 0). \tag{5.7.5}$$

The sum on the right is extended over all $2d$ possibilities of $\pm u$ in each of the d argument positions. That is to say, $f(0, \pm u, 0, \ldots, 0)$, etc.

Inserting $f \equiv 1$ into (5.7.5), we find that the

$$\text{volume of } H_d = 2^d = 2dw, \tag{5.7.6}$$

or

$$w = \frac{2^{d-1}}{d}. \tag{5.7.7}$$

Formula (5.7.5) is automatically fulfilled for $f = x_i$, $i = 1, \ldots, d$ by symmetry.

Inserting $f = x_1^2$ into (5.7.5), we obtain

$$\begin{aligned} \int_{H_d} x_1^2\, dV &= \int_{-1}^{1} x_1^2\, dx_1 \int_{-1}^{+1} \cdots \int_{-1}^{+1} dx_2\, dx_3 \cdots dx_d \\ &= \tfrac{2}{3} \cdot 2^{d-1} = w(2u^2), \end{aligned} \tag{5.7.8}$$

or

$$u = \left(\frac{2^{d-1}}{3w}\right)^{1/2} = \left(\frac{d}{3}\right)^{1/2}.$$

The formula will be exact for $f = x_i x_j$, $i \neq j$, by symmetry, and for $f = x_2^2$, $x_3^2, \ldots, x_d^2$, on the basis of the previous equation. Furthermore, by symmetry, it will be exact for monomials of degree three. Since it is not exact for x_1^4, it is a rule of precision 3.

This method will work for any region that is fully symmetric. A region B is said to be a fully symmetric (FS) region if $\mathbf{x} = (x_1, x_2, \ldots, x_d) \in B$ implies that $(\pm x_{p(1)}, \pm x_{p(2)}, \ldots, \pm x_{p(d)}) \in B$ where the set $\{p(1), \ldots, p(d)\}$ is any permutation of $\{1, 2, \ldots, d\}$. The set of all such points will be denoted by $\mathbf{x}_{\text{FS}}$ or $(x_1, \ldots, x_d)_{\text{FS}}$. The most important FS regions are the hypercube $[-a, a]^d$, the hypersphere, and the entire space.

We may also add a weight function $g(x_1, \ldots, x_d)$ to the integrand providing that it is also *fully symmetric*. By this is meant that $g(x_1, x_2, \ldots, x_d) = g(\pm x_{p(1)}, \pm x_{p(2)}, \ldots, \pm x_{p(d)})$. We may, therefore, obtain rules, for example, for the infinite space with the weights e^{-r} and e^{-r^2} where $r^2 = x_1^2 + x_2^2 + \cdots + x_d^2$.

For $d > 3$, we have $u > 1$, and the evaluation points are exterior to the hypercube. This is not a desirable feature of an integration formula. However, Stroud gives the following $2d$-point rule of precision 3 with all points inside the hypercube. Let P_k $(k = 1, \ldots, d)$ denote the point $(x_{1k}, x_{2k}, \ldots, x_{dk})$, where

$$x_{2r-1,k} = \left(\frac{2}{3}\right)^{1/2} \cos\frac{(2r-1)k\pi}{d}, \qquad x_{2r,k} = \left(\frac{2}{3}\right)^{1/2} \sin\frac{(2r-1)k\pi}{d},$$

$$r = 1, 2, \ldots, [d/2],$$

and if d is odd, $x_{dk} = (-1)^k/\sqrt{3}$. Let $P_{k+d} = -P_k$. Then

$$\int_{H_d} f\,dV \approx \frac{2^{d-1}}{d} \sum_{k=1}^{2d} f(P_k) \tag{5.7.9}$$

is an integration rule of precision 3 over the hypercube H_d with all points inside H_d and with equal weights.

Stroud also gives a similar $(d+1)$-point rule of precision 2. Let P_k, $k = 0, \ldots, d$, denote the point $(x_{1k}, x_{2k}, \ldots, x_{dk})$, where

$$x_{2r-1,k} = \left(\frac{2}{3}\right)^{1/2} \cos\frac{2rk\pi}{d+1}, \qquad x_{2r,k} = \left(\frac{2}{3}\right)^{1/2} \sin\frac{2rk\pi}{d+1},$$

$$r = 1, 2, \ldots, \left[\frac{d}{2}\right],$$

and if d is odd, $x_{dk} = (-1)^k/\sqrt{3}$. Then

$$\int_{H_d} f\,dV \approx \frac{2^d}{d+1} \sum_{k=0}^{d} f(P_k) \tag{5.7.10}$$

is an integration rule of precision 2 over the hypercube H_d with all points inside H_d and with equal weights.

Reference. Stroud [1].

2 *$(2d^2 + 1)$-Point Rule of Precision 5 for a Hypercube*

We shall look for an integration formula of the form

$$\int_{H_d} f\,dV \approx w_1 f(0, 0, \ldots, 0) + w_2 \sum^{2d} f(\pm u, 0, \ldots, 0)$$

$$+ w_3 \sum^{2d(d-1)} f(\pm u, \pm u, 0, \ldots, 0). \tag{5.7.11}$$

The first sum on the right is extended over the $2d$ possibilities of $\pm u$ in d positions and the second sum is extended over the $2(d)(d-1) = 4\binom{d}{2}$ possibilities of $\pm u$, $\pm u$ in pairs of positions.

As before, if a monomial contains an odd power, it will satisfy (5.7.11) automatically. Inserting, successively, $f = 1$, x_1^2, x_1^4, $x_1^2 x_2^2$ into (5.7.11), we obtain

$$w_1 + 2dw_2 + 2d(d-1)w_3 = \int_{H_d} dV = I_0,$$

$$2u^2 w_2 + 4(d-1)u^2 w_3 = \int_{H_d} x_1^2\, dV = I_2,$$

$$2u^4 w_2 + 4(d-1)u^4 w_3 = \int_{H_d} x_1^4\, dV = I_4,$$

$$4u^4 w_3 = \int_{H_d} x_1^2 x_2^2\, dV = I_{22}. \tag{5.7.12}$$

If we divide the third equation by the second, we obtain u^2. The values of w_i then follow immediately:

$$w_1 = I_0 - d\left(\frac{I_2}{I_4}\right)^2 \left(I_4 - \frac{d-1}{2} I_{22}\right) = \frac{2^d}{162}(25d^2 - 115d + 162),$$

$$w_2 = \frac{1}{2}\left(\frac{I_2}{I_4}\right)^2 (I_4 - (d-1)I_{22}) = \frac{2^d}{162}(70 - 25d),$$

$$w_3 = \frac{1}{4} I_{22}\left(\frac{I_2}{I_4}\right)^2 = \frac{25}{324} 2^d,$$

$$u = \left(\frac{I_4}{I_2}\right)^{1/2} = \left(\frac{3}{5}\right)^{1/2}. \tag{5.7.13}$$

If $d \geq 3$, the weights will be of mixed sign, a not too desirable situation. Note from the derivation of this rule that similar rules may be derived for any fully symmetric region B and any fully symmetric weight function for which the integrals $I_0 = \int_B w\,dV$, $I_2 = \int_B wx_1^2\,dV$, $I_4 = \int_B wx_1^4\,dV$, $I_{22} = \int_B wx_1^2 x_2^2\,dV$ are finite.

References. Hammer and Stroud [2]; J. C. P. Miller [2]; Morrison [1]; Stenger [1].

Example. $I(b)$ as in Section 5.6.

b^{-1}	Rule A	Rule B	Exact value
1.0[a]	.42573	.46375	.50546
0.9	.35960	.36929	.36993
0.8	.30458	.30772	.30781
0.7	.25676	.25792	.25794

Rule A: 5th-degree rule = 19 points.
Rule B: $2^3 \times$ Rule A = 152 points.

[a] Integrand singular.

Examples

d	Exact	P_2	P_3	P_5
	$I_d^4 = \int_{-1}^{1} \cdots \int_{-1}^{1} \left(\sum_{i=1}^{d} x_i\right)^4 dx_1 \cdots dx_d = 2^d[\frac{1}{3}d^2 - \frac{2}{15}d]$			
8	.5188 2667 (4)	.5234 (4)	.5006 (4)	.5188 26 (4)
11	.7959 8933 (5)	.6550 (5)	.7422 (5)	.7959 88 (5)
14	.1039 8379 (7)	.1708 (7)	.1682 (7)	.1039 83 (7)
17	.1232 9506 (8)	.1684 (8)	.1773 (8)	.1232 94 (8)
20	.1370 1393 (9)	.3146 (9)	.3122 (9)	.1370 12 (9)
23	.1453 4661 (10)	.2866 (10)	.2944 (10)	.1453 44 (10)
	$I_d^6 = \int_{-1}^{1} \cdots \int_{-1}^{1} \left(\sum_{i=1}^{d} x_i\right)^6 dx_1 \cdots dx_d = 2^d d[(d-1)(5d-1)/9 + \frac{1}{7}]$			
8	.6241 5237 (5)	.4756 (5)	.4490 (5)	.1458 17 (5)
11	.1354 8982 (7)	.7498 (6)	.9471 (6)	.2279 83 (6)
14	.2289 3908 (8)	.4686 (8)	.4652 (8)	.3009 40 (7)
17	.3330 6640 (9)	.5031 (9)	.5474 (9)	.3591 88 (8)
20	.4386 0434 (10)	.1763 (11)	.1766 (11)	.4009 73 (9)
23	.5379 2942 (11)	.1683 (12)	.1756 (12)	.4267 76 (10)
	$I_d^5 = \int_{-1}^{1} \cdots \int_{-1}^{1} \left(\sum_{i=1}^{d} x_i\right)^5 dx_1 \cdots dx_d = 0$			
8				−.34 (−2)
11				−.70 (−1)
14				−.10 (1)
17				−.34 (2)
20				−.71 (3)
23				−.98 (4)

The rules P_2, P_3, and P_5 are the 2nd-, 3rd-, and 5th-degree rules of the text; the parentheses, for example, in .5188 (4), means $.5188 \times 10^4$.

We have here striking examples of the difficulties in using rules with negative weights. This shows up in applying P_5 to I_d^5. Furthermore, the results for I_d^6, where P_3 and even P_2 are closer to the exact answer than P_5, would seem to indicate that rules which choose their points along distinguished subspaces of low dimension may lead to poor answers.

McNamee and Stenger have given similar monomial rules of precision 7, 9, and 11 for the hypercube of general dimension d. The form assumed for the degree 7 rule is

$$\begin{aligned}\int_{H_d} f\,dV \approx{}& A_0 f(0,\ldots,0) + A_1 \sum f(\pm u, 0,\ldots,0)\\ &+ A_2 \sum f(\pm v, 0,\ldots,0) + A_{11} \sum f(\pm u, \pm u, 0,\ldots,0)\\ &+ A_{22} \sum f(\pm v, \pm v, 0,\ldots,0) + A_{111} \sum f(\pm u, \pm u, \pm u, 0,\ldots,0)\end{aligned} \tag{5.7.14}$$

The sums are extended over all possibilities of $\pm$ and all permutations of the nonzero arguments in one, two, and three positions.

As the coefficients appear to be highly oscillatory and the practicability of the rules has not been established, we merely provide some references. The method generalizes to fully symmetric regions and weight functions.

References. McNamee and Stenger [1]; Stenger [1, 7]; Stroud [18, Chap. 8].

Integration rules (5.7.5), (5.7.11), and (5.7.14) are fully symmetric (FS) rules. An integration rule over an FS region B is said to be *fully symmetric* if it is of the form

$$\int_B w(\mathbf{x}) f(\mathbf{x})\,d\mathbf{x} \approx \sum_{i=1}^{n} w_i \sum_{\mathbf{y} \in (\mathbf{x}_i)_{\mathrm{FS}}} f(\mathbf{y}) \tag{5.7.15}$$

where $w(\mathbf{x})$ is an FS weight function. Such a rule is determined by n weights w_i and n *generators* $\mathbf{x}_i$. Clearly, if such a rule is of degree $2s$, then it is of degree $2s + 1$.

Since any point $\mathbf{y} \in \mathbf{x}_{\mathrm{FS}}$ determines the set $\mathbf{x}_{\mathrm{FS}}$, we arbitrarily call the generator of the set that point $\mathbf{x} = (x_1, x_2, \ldots, x_d)$ for which $x_1 \geq x_2 \geq \cdots \geq x_d \geq 0$. In such a generator, there will be p distinct nonzero components z_i, $0 \leq p \leq d$, each one with a multiplicity $n_i \geq 1$. The *class* of a generator $\mathbf{x}$ is the set of positive integers $\mathcal{N} = \{n_1, n_2, \ldots, n_p\}$ where $\sum_{i=1}^{p} n_i \leq d$. The empty set $\mathcal{N} = \phi$ corresponds to the generator $(0, 0, \ldots, 0)$. If we define $p(r)$ to be the number of partitions of the positive integer r into a sum of positive integers, then the number of classes of generators in dimension d is given by $M = 1 + \sum_{r=1}^{p} p(r)$. If $\mathbf{x}$ is of class $\mathcal{N} = \{n_1, n_2, \ldots, n_p\}$, then the number of points in the set $\mathbf{x}_{FS}$ is given by

$$|\mathcal{N}| = \frac{2^{d-n_0} d!}{(d-n_0)!\, n_1! \cdots n_p!}$$

where $n_0 = d - \sum_{i=1}^{p} n_i$ is the number of zeros in $\mathbf{x}$.

We shall now assume that we have assigned an ordering to the various classes of generators so that they are indexed by the integers $1, \ldots, M$. For example, in three dimensions, we could order the seven classes of generators as follows: $\mathcal{N}_1 = \phi$, $\mathcal{N}_2 = \{1\}$, $\mathcal{N}_3 = \{2\}$, $\mathcal{N}_4 = \{1, 1\}$, $\mathcal{N}_5 = \{3\}$, $\mathcal{N}_6 = \{2, 1\}$, $\mathcal{N}_7 = \{1, 1, 1\}$. The *structure* of an FS integration rule is determined by an M-vector of nonnegative integers $\mathbf{K} = (K_1, \ldots, K_M)$ which indicates how many generators of each class appear in the rule. Once we have decided on a particular structure, we can write down a system of nonlinear moment equations similar to (5.7.12) to determine the weights w_i and generators $\mathbf{x}_i$ of the rule. However, not all structures $\mathbf{K}$ lead to consistent systems of equations. A necessary condition for consistency is linear consistency which means that no linear combination of equations in the system can yield the equation $0 = 1$. This imposes certain linear constraints on the elements of $\mathbf{K}$. Such constraints were determined by Mantel and Rabinowitz in the two- and three-dimensional case. Subsequently, using a different approach, Keast and Lyness rederived the sets of constraints for these two cases as well as for the four-dimensional case.

Once we have these constraints, we can derive structures leading to rules with a minimal number of points by solving the integer programming problem

$$\text{minimize} \quad \sum_{i=1}^{M} K_i |\mathcal{N}_i|$$

subject to the consistency constraints mentioned above and the nonnegativity constraints $K_i \geq 0$, $i = 1, \ldots, M$. In addition to optimal solutions to this problem, one can also find suboptimal solutions. These are necessary since a consistent structure does not always yield an integration rule because the nonlinear system of moment equations may not have a real solution. Furthermore, a real solution may not be satisfactory in that some of the points may lie outside the integration region and the weights may have too large a figure of demerit as defined below (page 377). Thus if a particular structure does not lead to an acceptable solution, we proceed to the next best structure.

This procedure was followed by Mantel and Rabinowitz who derived three-dimensional FS monomial rules of moderate degree for the standard FS regions. Previously, Rabinowitz and Richter, following the same procedure on an informal basis, had obtained FS monomial rules of high degree with a minimal number of points for the corresponding two-dimensional regions.

This procedure can also be carried out for other types of symmetry such as central symmetry ($\mathbf{x} \in B \Rightarrow -\mathbf{x} \in B$), ordinary symmetry ($(x_1, x_2, \ldots, x_d) \in B \Rightarrow (\pm x_1, \pm x_2, \ldots, \pm x_d) \in B$), and rotational symmetry. However, this has been limited to two-dimensional regions and rules of low degree in d dimensions.

References. Collatz [1]; Dobrodeev [1]; Haegemans [1]; Keast [6]; Keast and Diaz [1]; Keast and Lyness [1]; Lyness and Jespersen [1]; Lyness and Monegato [1]; Mantel [1]; Mantel and Rabinowitz [1]; Mysovskikh [1]; Piessens and Haegemans [3, 4]; Rabinowitz and Richter [1].

Examples (Rabinowitz and Richter). We give below the errors in approximating the double integral

$$\int_{-1}^{1} \int_{-1}^{1} f(x, y)\, dx\, dy$$

for four functions using minimal-point integration rules and product Gauss rules. The notation M_n indicates a rule using n points. The degree of M_{20} is 9, of M_{25} and M_{28}, 11, of M_{37}, 13, and of M_{44} and M_{48}, 15. Two rules are given for degrees 11 and 15 since the ones with the smaller number of points were not acceptable inasmuch as they included points outside the cube. These examples seem to indicate that not only is a product Gauss rule of given precision superior to other formulas of that precision, but also that in many cases, a Gauss rule of lower precision is equal if not superior to a minimal rule of higher precision. This is due to the fact that a product Gauss rule of precision p integrates exactly *all* monomials $x^i y^j$ for which $i, j \le p$ while the two-dimensional rules of precision p integrate exactly only those monomials for which $i + j \le p$. As Cranley and Patterson point out, this situation gets much worse as the dimension d grows in that monomials with relatively large coefficients in the Taylor expansion of the function of d variables are not integrated exactly.

Rule	e^{xy}	$(1 - xy)^{-1}$	$\sin(\frac{1}{2}\pi(x + 2y))$	$((x + 1)^2 + (y + 2)^3)^{-1}$
M_{20}	2.3 (−8)	1.1 (−2)	2.6 (−7)	1.1 (−9)
G_5^2	1.1 (−13)	2.6 (−2)	2.2 (−8)	6.8 (−11)
M_{25}	1.9 (−10)	1.0 (−2)	1.3 (−9)	2.3 (−12)
M_{28}	3.0 (−10)	2.9 (−2)	2.7 (−9)	6.3 (−12)
G_6^2	$< 10^{-15}$	1.9 (−2)	1.0 (−10)	1.1 (−13)
M_{37}	1.4 (−12)	2.3 (−2)	1.8 (−11)	1.9 (−13)
G_7^2	$< 10^{-15}$	1.3 (−2)	3.6 (−13)	1.0 (−14)
M_{44}	3.8 (−14)	1.8 (−2)	1.1 (−13)	4.5 (−14)
M_{48}	1.8 (−14)	1.8 (−2)	9.4 (−14)	4.9 (−14)
G_8^2	$< 10^{-15}$	1.1 (−2)	8.0 (−16)	8.0 (−17)

Note the singularities of $(1 - xy)^{-1}$ at the points (1, 1), and (−1, −1).

References. Cranley and Patterson [1]; Rabinowitz and Richter [1].

Integration rules over two- and three-dimensional simplexes (triangles and tetrahedra) are of particular interest in the application of the *finite element method* (Zienkiewicz [1]). The emphasis here is on rules of moderate degree with a small number of integration points (cf. Strang and Fix [1, Sect. 4.3]). Many such rules have been derived and in more than one way. The minimal symmetric rules for the triangle developed by Lyness and Jespersen and independently by Laursen and Gellert are especially useful. Attempts have also been made to reduce the number of points by relaxing to some extent the requirement that the rule be of a certain precision. Thus, Reddy has computed a three-point rule for the triangle which integrates exactly all polynomials of degree 3 of the form

$$p_3(x, y) = a + bx + cy + dx^2 + exy + fy^2 + rx^3 + s(x^2y + xy^2) + ty^3.$$

For additional work along this line as well as for rules for functions with singularities on the boundary, see Cowper, Cristecu and Loubignac, Hillion, Lannoy, Laursen and Gellert, Majorana *et al.*, Moan, and Reddy and Shippy.

Finite element literature is also concerned with numerical integration over quadrilaterals and corresponding three-dimensional regions. We list a few references to this topic: see Gray and van Genuchten, Hellen, Irons, Pina *et al.*, and Rodin.

Several programs have been written for automatic integration over triangles. Various pairs of rules are used as the basic rule pair required for local error estimation. These include Gaussian conical product rules and minimal symmetric monomial rules. An interesting pair of rules used by Laurie consists of a seven-point degree five Radon rule similar to (5.7.18) and a 19-point degree eight Kronrod-type extension of this rule.

References. Bartholemew [1]; Cowper [1]; Cristecu and Loubignac [1]; de Doncker and Robinson [1]; Gray and van Genuchten [1]; Haegemans [2]; Hellen [1]; Hemker [1]; Hillion [1, 2]; Irons [1, 2]; Lannoy [1]; Lauffer [1]; Laurie [2]; Laursen and Gellert [1]; Lyness and Jespersen [1]; Majorana, Odorizzi, and Vitaliani [1]; Moan [1]; Nicolaides [1]; G. M. Phillips [1]; Pina, Fernandes, and Brebbia [1]; Reddy [1]; Reddy and Shippy [1]; Rodin [1]; Silvester [1]; Stroud [18].

Grundmann and Möller have derived a family of integration rules for the simplex

$$S_d = \{(x_1, \ldots, x_d): x_i \geq 0, i = 1, \ldots, d; x_1 + \cdots + x_d \leq 1\}$$

which are invariant under all affine transformations of S_d onto itself. Let B be a region and G, a finite group of transformations T of B onto B. An integration rule R over B is said to be invariant with respect to G if it is of the form

$$Rf = \sum_{j=1}^{n} w_j \sum_{T \in G} f(T\mathbf{x}_j), \qquad \mathbf{x}_j \in B.$$

For any odd $p = 2s + 1$, the rule

$$\int_{S_d} f\,dV \simeq \sum_{i=0}^{s} (-1)^i 2^{-2s} \frac{(p+d-2i)^p}{i!(p+d-i)!} \times \sum_{\substack{|\boldsymbol{\beta}|=s-i \\ \beta_0 \geq \cdots \geq \beta_d}} f\left(\left(\frac{2\beta_0+1}{p+d-2i}, \ldots, \frac{2\beta_d+1}{p+d-2i}\right)_p\right) \tag{5.7.16}$$

is of degree p. Here, $\boldsymbol{\beta} = (\beta_0, \ldots, \beta_d)$ is a $(d+1)$-tuple of nonnegative integers with $|\boldsymbol{\beta}| = \beta_0 + \cdots + \beta_d$ and the notation $(y_0, y_1, \ldots, y_d)_p$ for the $(d+1)$-tuple $\mathbf{y} = (y_0, \ldots, y_d)$ denotes the set of all d-tuples that are the last d components of all $(d+1)$-tuples derived from $\mathbf{y}$ by any permutation of the $d+1$ components of $\mathbf{y}$. The number of integration points in (5.7.16) is $\binom{d+s+1}{s}$ which is close to the lower bound (5.7.4) for large d. However, a significant proportion of the weights are negative and the ratio of the sum of the magnitudes of the weights to the sum of the weights equals $(1/s!)(d/2)^s + O(d^{s-1})$ as $d \to \infty$. On the other hand, for a fixed d, the sequence of rules is an embedded sequence in that all points used in a rule of degree p are used in the $(p+2)$-degree rule. In this way no function evaluations are wasted in going to rules of higher degree. In addition, the points and weights are given by explicit arithmetic formulas so that they can be evaluated by a program when they are required and need not be tabulated in advance. This makes (5.7.16) well suited for automatic integration over simplexes in any dimension. But we must be aware of the possible loss of significance and work to greater precision than required in the final result. A program implementing (5.7.16) appears in the NAG library.

For other approaches to numerical integration over simplexes, see the works of de Doncker and of Lyness and his co-workers.

References. de Doncker [2, 3]; Grundmann and Möller [1]; Lyness [18]; Lyness and Genz [1]; Lyness and Monegato [2]; Lyness and Puri [1]; NAG [1]; van der Sluis [2].

An imbedded family of fully symmetric integration rules has also been derived for the hypercube $[-1, 1]^d$ by Genz and Malik. If we denote the degree of the rule by $2m+1$, then these rules are useful in the range $3 \leq m \leq 6$, $3 \leq d \leq 3m$. The rule of degree $2m+1$ is of the form

$$R^{(m,d)}f = \sum_{\mathbf{p} \in P^{(m,d)}} W_{\mathbf{p}}^{(m,d)} \sum_{\mathbf{y} \in (\boldsymbol{\lambda}_{\mathbf{p}})_{FS}} f(\mathbf{y}) + W^{(m,d)} \sum_{\mathbf{y} \in (\boldsymbol{\delta})_{FS}} f(\mathbf{y})$$

where $\boldsymbol{\delta} = (\delta, \delta, \ldots, \delta)$, the W's are appropriate weights, $P^{(m,d)}$ is the set of partitions of all nonnegative integers $<m$ into d nonnegative integers and $\boldsymbol{\lambda}_{\mathbf{p}}$ is a vector determined by a particular partition $\mathbf{p}$ as follows: We have a fixed sequence of distinct real numbers, $\lambda_0 = 0, \lambda_1, \ldots, \lambda_5$, $0 < \lambda_i < 1$, $i = 1, \ldots, 5$, and $R^{(m,d)}f$ is based on the first m elements in this sequence. If we determine the partition $\mathbf{p} = (p_1, \ldots, p_d)$ uniquely by the requirement $m - 1 \geq p_1 \geq p_2 \geq \cdots \geq p_d \geq 0$, $\sum_{i=1}^{d} p_i \leq m - 1$, then $\boldsymbol{\lambda}_{\mathbf{p}}$ will be given by $\boldsymbol{\lambda}_{\mathbf{p}} = (\lambda_{p_1}, \lambda_{p_2}, \ldots, \lambda_{p_d})$.

Example. For $m = 4$ and $d \geq 3$, the partitions involved are $(0, 0, \ldots, 0)$, $(1, 0, \ldots, 0)$, $(2, 0, \ldots, 0)$, $(1, 1, 0, \ldots, 0)$, $(3, 0, \ldots, 0)$, $(2, 1, \ldots, 0)$, and $(1, 1, 1, 0, \ldots, 0)$, and the rule $R^{(4,d)}$ of degree 9 has 8 weights and uses the generators λ_1, λ_2, and λ_3. The additional function evaluations not used in the rule $R^{(3,d)}$ of degree 7 come from the last three partitions in the list.

The reason the sequence ends with λ_5 is that λ_6 is greater than 1 so that it leads to points outside the hypercube. This is not desirable.

These rules have formed the basis for a variable precision multidimensional adaptive algorithm.

References. Genz and Malik [2]; Malik [1].

There appears to be no systematic theoretical approach to monomial rules nor any systematic evaluation of their practical effectiveness.

Two properties of approximate integration rules are considered particularly desirable: the abscissas should lie in the region and the weights should be positive. When ad hoc methods are applied to develop monomial rules, there is, as we have seen, little control over these conditions, and they may not be fulfilled. Yet it is theoretically possible to find such rules. This is guaranteed by a theorem of Tchakaloff. We formulate it in the case of two dimensions, but it is perfectly general. Note that the number of monomials $x^p y^q$ of total degree $\leq n$ is $N = \frac{1}{2}(n + 1)(n + 2)$.

THEOREM. *Let B be a closed, bounded set in the plane with positive area. Then there exist $N = \frac{1}{2}(n + 1)(n + 2)$ points in B, $P_1, \ldots, P_N$, and N positive weights $w_1, \ldots, w_N$ such that*

$$\iint_B f(x, y)\, dx\, dy = \sum_{i=1}^{N} w_i f(P_i) \tag{5.7.17}$$

whenever f is a polynomial in x, y of degree $\leq n$.

This theorem has been generalized to unbounded regions, to positive functionals and to sets of linearly independent functions other than polynomials. The most elegant proof is existential in nature and makes use of the theory of convex bodies. A more "elementary" and constructive proof can be found in Davis [8]. See also Wilson, Stroud, and Engels. Efficient numerical procedures for obtaining such rules corresponding to an arbitrary B and high N are yet to be devised.

A related question is that of obtaining nonnegative interpolation formulas for harmonic functions in several variables. See Davis [12], and Davis and Wilson.

References. Davis [8, 9, 12]; Davis and Wilson [1]; Dennis and Goldstein [1]; Engels [13, pp. 286–290]; Mysovskikh [11]; Rogosinski [1, 2]; Stroud [18, pp. 58–67]; V. Tchakaloff [1]; Wilhelmsen [1]; Wilson [1];.

Following the analysis of Section 4.2, we may use the quantity $V^{-1}\sum_{i=1}^{n} |w_i|$ as a figure of demerit of a rule with respect to roundoff error or with respect to errors in measurement in $f(x_i)$ when these have been determined by experimentation. (Certain theories suggest the use of the quantity $V^{-1}(\sum_{i=1}^{n} |w_i|^2)^{1/2}$.)

In numerous families of monomial rules that have been obtained by ad hoc methods, this figure of demerit appears to grow rapidly with d and p.

Orthogonal Polynomials

As in the one-dimensional case, there is a close connection between orthogonal polynomials in several variables and integration rules. The classic example is the Radon 7-point formula of precision 5 for the square S: $-1 \leq x, y \leq 1$:

$$\int_S f(x, y)\,dx\,dy \approx w_1 f(P_1) + w_2[f(P_2) + f(P_3) + f(P_4) + f(P_5)] + w_3[f(P_6) + f(P_7)] \tag{5.7.18}$$

where

$$P_1 = (0, 0); \qquad P_2, P_3, P_4, P_5 = (\pm s, \pm t), \quad s = \sqrt{\tfrac{1}{3}},\ t = \sqrt{\tfrac{3}{5}};$$

$$P_6, P_7 = (\pm r, 0), \qquad r = \sqrt{\tfrac{14}{15}}; \qquad w_1 = \tfrac{8}{7}, \qquad w_2 = \tfrac{5}{9}, \qquad w_3 = \tfrac{20}{63}.$$

The points P_i are common zeros of the polynomials $y^3 - \frac{3}{5}y$ and $x^3 + xy^2 - \frac{14}{15}x$ which are orthogonal polynomials over S.

The satisfactory theory of one-dimensional integration using zeros of orthogonal polynomials (points inside the interval, weights positive) does not appear to carry over to several dimensions. However, we can make the following statements whose precise formulation and proofs can be found in the work of Stroud.

(1) Given a set of orthogonal polynomials over a region and their common zeros; it is possible to determine if these zeros (or a subset of them)

can be used as the points in an efficient integration formula. This can be done by inspecting only the polynomials and their zeros.

(2) Given an n-point rule of precision p. If $n < \binom{p+d}{p}$, it is possible to find a set of polynomials with certain orthogonality properties and having the points of the rule as their common zeros.

The theory of orthogonal polynomials in multidimensional numerical integration was deepened with the introduction of polynomial ideals by Möller. It was further refined by Schmid who introduced real polynomial ideals with the desirable property that common zeros are real and distinct although they do not necessarily lie in the region of integration. These developments have made possible the application of ideal theory to the problem of finding minimal rules. Unfortunately, practical results have only materialized in the two-dimensional case where new minimal integration rules of moderate degree have been discovered. It appears that the extension of these results to higher degrees and more dimensions involves a considerable computational effort. Nevertheless, this work has produced many beautiful theoretical results.

References. Engels [13]; Franke [1, 2]; Günther [1–3]; Haegemans [3]; Haegemans and Piessens [1, 2]; Hirsch [1, 2]; Linden, Kroll, and Schmid [1]; Möller [1–5]; Morrow and Patterson [1]; Mysovskikh [2, 8, 9, 14, 17]; Radon [2]; Schmid [1–6]; Stroud [14, 17, 18].

We note in this context that the theory and practice of numerical integration in two dimensions is much more developed than in $d > 2$ dimensions. Thus, in addition to the many integration rules for squares, circles, and triangles, there are several papers on the numerical evaluation of two-dimensional Cauchy principal value integrals. There are also papers by M. Levin and Neumann based on the concept of blending functions. The latter has succeeded in deriving Kronrod-type extensions of the Gauss product rules G_n^2 which use much fewer points than the Gauss–Kronrod product rule. Another example is the extension by D. Levin to two dimensions of his work on highly oscillatory integrals. Finally almost all work on error estimation has been confined to the two-dimensional case.

References. Ahlin [1]; Barnhill [4]; Barnhill and Nielson [1]; Chawla [3]; Chawla and Jayarajan [2]; El-Tom [3]; Lether [2–4, 6]; D. Levin [2]; M. Levin [3]; Monegato and Lyness [1]; G. Neumann [1, 2]; T. E. Price [1]; Sard [2, Chap. 4]; Squire [9]; Stroud [18, Chap. 5]; Theocaris, Ioakimidis, and Kazantzakis [1].

Another approach to monomial rules is via invariant formulas. We have seen one such example in the family of integration rules for the simplex. Much more developed is the case of the surface of the d-dimensional sphere. We list some pertinent references.

Once we have an integration rule for the surface of a sphere, we can generate rules for the solid sphere and for spherical shells by following the procedure in Stroud [18, Sect. 2.8].

References. Goethals and Seidel [1]; Konjaev [1, 2]; V. I. Lebedev [1–4]; McLaren [1]; Mysovskikh [15]; Salihov [1]; Sharipkhodzhaeva [1]; Sobolev [1].

We close this section by noting that there are many rules that are exact for monomials over a variety of regions. Rules can be developed for integrals with continuous weight functions or for weight functions with integrable singularities. A rather complete coverage of such rules until 1971 appears in the comprehensive monograph by Stroud [18]. Additional rules appear in the references of this section. Other papers containing such rules are listed below.

References. Albrecht and Engels [1]; Dobrodeev [1, 2]; Engels [13]; Franke [3]; Fritsch [2]; Guenther and Roetman [1]; Huelsman [1]; Lether [14]; G. M. Phillips [3–6]; J. F. Price [2].

5.8 Compound Rules

Suppose that we are integrating over a hypercube. We may subdivide the hypercube into a sum of smaller hypercubes by means of a rectangular grid. If we apply a fixed rule R to each of the smaller hypercubes, the rule that emerges from this process will be called a *compound rule*. If we work in d dimensions and divide the side of the cube into m parts, applying R to each of the m^d smaller cubes, the total rule will be designated by $m^d \times R$.

In one dimension, compound rules are frequently used to good advantage. In higher dimensions, difficulties intervene. The number of points in a compound rule acts like the number of points in a product rule. Both exhibit the dimensional effect m^d. For high d, this strictly limits the number of subdivisions that can be made.

If compound rules are desired, it is in the interest of economy to use a base rule R whose points are so disposed that they will overlap when the R's of contiguous hypercubes are combined.

Let H_d designate the hypercube $-1 \le x_i \le 1$, $i = 1, 2, \ldots, d$. Lyness has proposed the following rule of degree 5, which has evaluation points at the vertices of the hypercube:

$$\int_{H_d} f\,dV = \frac{2^d}{9}\left[(8 - 5d)f(0, 0, \ldots, 0) + 2^{-d}\sum^{2^d} f(\pm 1, \pm 1, \ldots, \pm 1) + \frac{5}{2}\sum^{2d} f\left(\pm\sqrt{\frac{2}{5}}, 0, 0, \ldots, 0\right)\right]. \tag{5.8.1}$$

In the first $\sum$ on the right, we sum over the 2^d possibilities of ± 1 in all the argument positions. In the second $\sum$, we sum over the $2d$ possibilities of $\pm\sqrt{\frac{2}{5}}$ in each of the argument positions and 0 elsewhere.

Thacher has adapted to higher dimensions a one-dimensional integration rule of Ralston that has weights of opposite sign and equal magnitude at the end points of the interval. When compounded, the internal points cancel out. Thacher's rule is of degree 2.

Thacher's idea has been extended to rules of higher degree by Engels.

Programming difficulties come into play here. We can write a program for any particular dimension, but to write a FORTRAN program with the dimension d as a parameter in such a way that we take advantage of the overlapping points, is rather difficult.

There is also the possibility of using compound rules together with a Richardson speed-up method. For example, we may subdivide the side of the hypercube into two and three equal parts and use the results for $h = 1$, $h = \frac{1}{2}$, $h = \frac{1}{3}$, as the basis for an extrapolation to the limit. A program is given in Appendix 2 for $d < 10$.

The Richardson extrapolation process (Section 1.15) is based on the Euler–Maclaurin expansion for approximate integration over a hypercube. Lyness [14, 15] has generalized this expansion to include the case of an integrand with a specific kind of singularity at a vertex of the hypercube. We assume that the hypercube $C_d = [0, 1]^d$ and that the vertex is at the origin. The intergrand in question has the form

$$f(\mathbf{x}) = r^\alpha \varphi(\boldsymbol{\theta}) h(r) g(\mathbf{x}), \qquad \alpha > -d \tag{5.8.2}$$

where $r^2 = x_1^2 + \cdots + x_d^2$ and $(r, \boldsymbol{\theta})$ are the hyperspherical coordinates of $\mathbf{x}$ (cf. Section 1.11.3).

We start with an integration rule R over C_d given by

$$Rt = \sum_{j=1}^{n} w_j t(\mathbf{x}_j), \qquad \mathbf{x}_j \in C_d, \qquad \sum_{j=1}^{n} w_j = 1 \tag{5.8.3}$$

where, for simplicity, we assume that the origin is not an integration point. We define the m^d copy of this rule to be the rule

$$(m^d \times R)t = m^{-d} \sum_{k_1=0}^{m-1} \sum_{k_2=0}^{m-1} \cdots \sum_{k_d=0}^{m-1} \sum_{j=1}^{n} w_j t\left(\frac{\mathbf{x}_j + \mathbf{k}}{m}\right) \tag{5.8.4}$$

where $\mathbf{k} = (k_1, k_2, \ldots, k_d)$. The rule R is said to be *symmetric* if $Rt_i = Rs$, $i = 1, \ldots, d$ where

$$t_i(x_1, \ldots, x_i, \ldots, x_d) = s(x_1, \ldots, x_{i-1}, 1 - x_i, x_{i+1}, \ldots, x_d).$$

Thus the product midpoint rule $Rt = t(\frac{1}{2}, \ldots, \frac{1}{2})$ is symmetric.

If the functions $\varphi(\boldsymbol{\theta})$, $h(r)$, and $g(\mathbf{x})$ in (5.8.2) are analytic in their stated variables at all points in C_d, then for any integration rule R of the form (5.8.3), the error

$$E_m f = \int_{C_d} f\, dV - (m^d \times R)f$$

has the asymptotic expansion

$$E_m f \sim \sum_{i=0} \frac{A_{\alpha+d+i} + C_{\alpha+d+i} \log m}{m^{\alpha+d+i}} + \sum_{s=1} \frac{B_s}{m^s} \tag{5.8.5}$$

where $C_{\alpha+d+i} = 0$ if α is not an integer. In addition, if R is symmetric, $B_s = C_s = 0$ for s odd. If R is an integration rule of precision p, then $B_s = C_s = 0$, $s \le p$. If $h(r)g(\mathbf{x}) = h(-r)g(-\mathbf{x})$, then $A_{\alpha+d+i} = C_{\alpha+d+i} = 0$ for i odd while if $h(r)g(\mathbf{x}) = -h(-r)g(-\mathbf{x})$, $A_{\alpha+d+i} = C_{\alpha+d+i} = 0$ for i even.

If $f(\mathbf{x})$ is analytic in all variables in C_d, $A_k = C_k = 0$, and if $f(\mathbf{x})$ is a polynomial of degree q, then, in addition, $B_s = 0$, $s > q$. If $f(\mathbf{x})$ and all its partial derivatives of total order $\le r$ are integrable over C_d, then

$$E_m f = \sum_{s=1}^{r-1} \frac{B_s}{m^s} + O(m^{-r}). \tag{5.8.6}$$

This last results yields the standard Euler–Maclaurin expansion for the compound product midpoint rule inasmuch as B_s vanishes for s odd, $s > 1$, and the midpoint rule is of precision 1.

Example. (Lyness [15]) Let $f(x, y) = r^{-3/2} e^{-r^2} e^{-x^2}$, then if R is a symmetric rule,

$$E_m f \sim \frac{A_{1/2}}{m^{1/2}} + \frac{B_2}{m^2} + \frac{A_{5/2}}{m^{5/2}} + \frac{B_4}{m^4} + \frac{A_{9/2}}{m^{9/2}} + \cdots.$$

If, in addition, R is of precision 7, then

$$E_m f \sim \frac{A_{1/2}}{m^{1/2}} + \frac{A_{5/2}}{m^{5/2}} + \frac{A_{9/2}}{m^{9/2}} + \frac{A_{13/2}}{m^{13/2}} + \frac{B_8}{m^8} + \cdots.$$

For the extension to functions of the form $(\log r)^q f(\mathbf{x})$, where q is a positive integer, and other theoretical aspects, see Lyness [14]. For practical aspects of evaluating such integrals, see the companion paper (Lyness [15]).

As noted above, the use of an m^d copy rule, $m^d \times R$, runs up against the dimension barrier so that expansions (5.8.5) and (5.8.6) are only useful for low dimensions if high accuracy is required and for higher dimensions if one is satisfied with low accuracy. To circumvent this problem, Genz has suggested that one drop the requirement of using the same compound rule in each dimension. This makes things much more complicated but results in a gain in efficiency. Thus, taking as the basic rule R over [0, 1], the midpoint rule $Mf = f(\frac{1}{2})$, Genz generates the sequence

$$M_{\mathbf{m}} f = \prod_{i=1}^{d} (m_i + 1)^{-1} \sum_{k_1=0}^{m_1} \sum_{k_2=0}^{m_2} \cdots \sum_{k_d=0}^{m_d} f\left(\frac{2k_1 + 1}{2m_1 + 2}, \ldots, \frac{2k_d + 1}{2m_d + 2}\right)$$

where $\mathbf{m} = (m_1, \ldots, m_d)$ is a d-tuple of nonnegative integers and the sequence $\{\mathbf{m}\}$ is ordered as follows: If $m_2 \neq 0$, then the successor $S(\mathbf{m})$ of $\mathbf{m}$ is given by $(m_1 + 1, m_2 - 1, m_3, \ldots, m_d)$. Otherwise, $S(\mathbf{m}) = (0, 0, \ldots, 0, m_1 + 1, m_{j+1} - 1, m_{j+2}, \ldots, m_d)$ if $m_2 = m_3 = \cdots = m_j = 0$. The error expansion for the approximation $M_{\mathbf{m}} f$ for sufficiently smooth functions is given by

$$\int_{C_d} f\, dV - M_{\mathbf{m}} f = \sum_{1 \le |\mathbf{k}| \le p} C_{\mathbf{k}} \mathbf{h}_{\mathbf{m}}^{\mathbf{k}} + \sum_{|\mathbf{k}| = p+1} O(\mathbf{h}_{\mathbf{m}}^{\mathbf{k}})$$

where $\mathbf{h}_{\mathbf{m}} = (h_{m_1}, h_{m_2}, \ldots, h_{m_d})$, $h_j = (1 + j)^{-2}$,

$$\mathbf{k} = (k_1, \ldots, k_d), \qquad |\mathbf{k}| = k_1 + \cdots + k_d, \qquad \mathbf{h}_{\mathbf{m}}^{\mathbf{k}} = \prod_{i=1}^{d} h_{m_i}^{k_i}$$

and where the $C_{\mathbf{k}}$ depend on f but not on the sequence $\{\mathbf{m}\}$.

Once we have computed several elements of the sequence $\{M_{\mathbf{m}} f\}$, we must devise an appropriate extrapolation procedure. Genz has given several algorithms for this nontrivial problem, each with its advantages and disadvantages. His program INTLAG implements one of these algorithms which is based on multivariable Lagrange interpolation.

References. Baker and Hodgson [1]; Engels [2]; Genz [1, 3–5, F2]; Kuntzmann [1]; Lyness [1, 14, 15]; Lyness and McHugh [1]; Ralston [1]; Stroud [18]; Thacher [3].

Compound rules also exist for simplexes. See the work of Lyness and his co-workers and of de Doncker.

Simpson and Yazici have described a way to organize the computation in the extrapolation method for triangles which is suitable for vector processing computer architectures.

References. de Doncker [2, 3]; Duffy [1]; Lyness [18]; Lyness and Genz [1]; Lyness and Monegato [2]; Lyness and Puri [1]; Sidi [7]; Simpson and Yazici [1].

5.8.5 The Method of Sag and Szekeres

This method proceeds in several steps. In the first step, the domain of integration is transformed to the unit hypersphere in such a way that the integrand and all its derivatives are reduced to zero on the surface of the hypersphere. Then a product trapezoidal rule is employed in the interior of the sphere. As in the one-dimensional case, the accuracy of the rule is enhanced by the vanishing of the derivatives at the boundary. Since the Jacobian of the transformation is found to be small near the surface of the hypersphere, the mesh points are scanned only in a concentric hypersphere of radius $r_0 = \frac{8}{10}$. This reduces the number of functional evaluations considerably.

Sag and Szekeres give transformations for the unit hypercube and simplex. We reproduce the first.

The transformation

$$\xi_i = \tanh \frac{ux_i}{1-r}, \qquad i = 1, 2, \ldots, d,$$

$$r^2 = x_1^2 + \cdots + x_d^2 \tag{5.8.5.1}$$

where u is a constant with a recommended value of $u = 1.5$, transforms the hypercube H_d: $-1 \le \xi_i \le 1$ into S_d: $\sum_{i=1}^d x_i^2 \le 1$. The Jacobian is

$$J(x_1, \ldots, x_d) = \partial(\xi)/\partial(x) = u^d(1-r)^{-d-1} \prod_{i=1}^{d} (1 - \xi_i^2). \tag{5.8.5.2}$$

With a mesh size of h, the integration rule is now

$$\int_{S_d} f(x_1, \ldots, x_d)\, dV \approx h^d \sum_i f(x_{i1}, \ldots, x_{id}) \tag{5.8.5.3}$$

where $\mathbf{x}_i = (x_{i1}, \ldots, x_{id})$ are the mesh points inside the sphere of radius r_0. A scheme for generating these mesh points is suggested.

While this method performs well with many integrals, it can be quite inefficient with others. There is some feeling that it is effective in dealing with integrands with singularities on the surface of the hypersphere. Several criticisms leveled at this method are that it does not integrate constants properly (2.5% error at $d = 15$) and that for high dimensions, by far the larger fraction of the volume of the hypersphere lies beyond $r_0 = \frac{8}{10}$ so that the mesh points there are not sampled.

A program implementing this method appears in the NAG library.

The method of Sag and Szekeres has been refined and extended by Roose and de Doncker who gave an algorithm for the automatic integration over the d-dimensional solid sphere of functions which may have singularities either on the surface or at the center of the sphere.

References. Cranley and Patterson [2]; NAG [1]; Roose and de Doncker [1]; Sag and Szekeres [1].

5.8.6 *Methods of Approximation*

The Bernstein polynomial approximation in d variables can be integrated. If $f(x_1, x_2, \ldots, x_d) = f(\mathbf{x})$ is defined and continuous in the hypercube H_d: $0 \le x_i \le 1$, $i = 1, 2, \ldots, d$, then its Bernstein polynomial is defined by

$$B(f;\mathbf{x}) = \sum_{v_1=0}^{m_1} \sum_{v_2=0}^{m_2} \cdots \sum_{v_d=0}^{m_d} f\left(\frac{v_1}{m_1}, \frac{v_2}{m_2}, \ldots, \frac{v_d}{m_d}\right) \prod_{k=1}^{d} p_{v_k, m_k}(x_k)$$

where

$$p_{v,m}(x) = \binom{m}{v} x^v (1-x)^{m-v}. \tag{5.8.6.1}$$

As $m_1, m_2, \ldots, m_d \to \infty$, $B(f;\mathbf{x}) \to f(\mathbf{x})$ uniformly in H_d. Now,

$$\int_0^1 \cdots \int_0^1 B(f;\mathbf{x})\, dx_1 \cdots dx_d = \sum_{v_1=0}^{m_1} \cdots \sum_{v_d=0}^{m_d} f\left(\frac{v_1}{m_1}, \ldots, \frac{v_d}{m_d}\right) \Big/ \prod_{k=1}^{d} (m_k + 1). \tag{5.8.6.2}$$

Hence this method is equivalent to averaging over the grid points.

For another method of approximation, see Schwartz.

Reference. C. Schwartz [1].

5.9 *Multiple Integration by Sampling*

The method of sampling in numerical analysis can be described quite generally as follows. We wish to compute a number I that arises in a mathematical problem. Suppose it turns out that I is also the *expected value* of a certain *stochastic* (that is, chance) *process*. The expected value of the process is estimated by sampling, and the estimate is used as an approximation to I.

These generalities are illustrated by the following simple problem. Suppose we would like to compute

$$I = \int_a^b f(x)\, dx. \tag{5.9.1}$$

The mean value of $f(x)$ over the interval $[a, b]$ is $I/(b-a)$. Let $x_1, x_2, \ldots, x_n$ be n points selected at random in $[a, b]$. We then sample the height of $f(x)$ by computing $f(x_1), f(x_2), \ldots, f(x_n)$ and forming the sample average:

$$\hat{f}_n = \frac{1}{n} \sum_{i=1}^{n} f(x_i). \tag{5.9.2}$$

We would expect that $\hat{f}_n \approx I/(b-a)$ and, hence, that

$$\int_a^b f(x)\,dx \approx \frac{b-a}{n}[f(x_1) + \cdots + f(x_n)]. \tag{5.9.3}$$

When random values are used for x_i, this method is also known as the *Monte Carlo method*. As we shall see, various modifications must be used in practice.

Example. Compute $I = \int_0^1 x\,dx = \frac{1}{2}$ by simple Monte Carlo. Select $n = 50$ and choose 50 two-decimal numbers at random from a uniform distribution in the interval [0, 1].

.37	.68	.05	.63	.77
.40	.57	.86	.93	.41
.16	.54	.23	.74	.07
.26	.47	.38	.55	.11
.74	.97	.77	.53	.80
.09	.28	.52	.90	.20
.89	.16	.68	.63	.87
.74	.56	.88	.92	.42
.81	.02	.15	.64	.24
.39	.99	.53	.26	.71

(These numbers were obtained from a table of 2500 random 5-digit numbers in NBS Handbook [1, pp. 991–995].)

This selection of numbers yields the estimate

$$\int_0^1 x\,dx \approx .539. \tag{5.9.4}$$

A more extensive computation with $n = 100$ yields

$$\int_0^1 x\,dx \approx .527. \tag{5.9.5}$$

Behind this computation is the *Law of Large Numbers. Let $x_1, x_2, \ldots, x_n, \ldots$ be random variables selected according to a probability density function $\mu(x)$,*

$$\int_{-\infty}^{\infty} \mu(x)\,dx = 1.$$

Let $I = \int_{-\infty}^{\infty} f(x)\mu(x)\,dx$ exist. Then, given an $\varepsilon > 0$, we have

$$\lim_{n\to\infty} \text{probability}\left(I - \varepsilon \le \frac{1}{n}\sum_{i=1}^{n} f(x_i) \le I + \varepsilon\right) = 1. \tag{5.9.6}$$

This tells us that the probability that the sample average is close to the mean value I can be made arbitrarily close to 1 if we take sufficiently large samples.

In Monte Carlo computations we are not inclined to consider numerous samples of the same size but, rather, a single sample of infinite length. The *Strong Law of Large Numbers* is therefore particularly relevant:

$$\text{Probability}\left(\lim_{n\to\infty}\frac{1}{n}\sum_{i=1}^{n} f(x_i) = I\right) = 1. \tag{5.9.7}$$

In the numerical example above, we select $f(x) = x$ and

$$\mu(x) = \begin{cases} 1, & 0 \le x \le 1, \\ 0, & \text{otherwise.} \end{cases}$$

The workings of the Monte Carlo method raise two questions: (1) How can we obtain random sequences of numbers? (2) How can we estimate the error incurred in our approximations? It is possible to use sequences of numbers prepared in advance or to connect up the computer to a physical process with a random feature to it. In practice, however, this is not done. The first requires too much storage. The second has the poor feature (at least from the standpoint of numerical analysis) that the numbers cannot be duplicated easily and, hence, a computation cannot be readily checked. Furthermore, the physical processes involved are generally much too slow relative to the speed of current computers and are not as reliable as computers. What is done in practice is to generate a so-called *pseudorandom sequence*† of numbers and to use it instead. A pseudorandom sequence of numbers is a deterministic sequence of numbers defined mathematically and, according to a much quoted definition of D. H. Lehmer, is "a vague notion embodying the idea of a sequence in which each item is unpredictable to the uninitiated, and whose digits pass a certain number of tests traditional with statisticians and depending somewhat on the uses to which the sequence is to be put." We shall return to the question of pseudorandom sequences in a later section.

If we adopt the statistical point of view, estimates of error may be obtained through the use of the *Central Limit Theorem*. As above, let

$$I = \int_{-\infty}^{\infty} f(x)\mu(x)\,dx \tag{5.9.8}$$

† Many authors use the term *pseudorandom* for *equidistributed* and the term *random* for *pseudorandom*. Note this usage in Section 5.9.2.5. The current trend is to use the term Monte Carlo or pseudo-Monte Carlo for methods using pseudorandom sequences and quasi-Monte Carlo for those using equidistributed sequences.

be the *mean* (or expected value) of f, and let

$$\sigma^2 = \int_{-\infty}^{\infty} (f(x) - I)^2 \mu(x)\, dx = \int_{-\infty}^{\infty} f^2(x)\mu(x)\, dx - I^2 \qquad (5.9.9)$$

designate the *variance* of $f(x)$. Then the Central Limit Theorem tells us that

$$\text{prob}\left(\left| \frac{1}{n} \sum_{i=1}^{n} f(x_i) - I \right| \le \frac{\lambda\sigma}{\sqrt{n}} \right) = \frac{1}{\sqrt{2\pi}} \int_{-\lambda}^{\lambda} e^{-x^2/2}\, dx + O\left(\frac{1}{\sqrt{n}} \right). \qquad (5.9.10)$$

A similar formula holds for multiple integrals.

A table of the probability integral

$$PI = \frac{1}{\sqrt{2\pi}} \int_{-\lambda}^{\lambda} e^{-x^2/2}\, dx$$

yields the following typical values.

λ	PI
.6745	.50
1.645	.90
1.960	.95
2.576	.99
3.291	.999
3.891	.9999

If we designate the error $(1/n)\sum_{i=1}^{n} f(x_i) - I$ by E, then, using these numerical values, we may convert (5.9.10) into the following statements.

$$\text{Error } E \text{ at } 50\% \text{ level of probability} \le \frac{.6745\sigma}{\sqrt{n}}.$$

$$\text{Error } E \text{ at } 90\% \text{ level of probability} \le \frac{1.645\sigma}{\sqrt{n}}.$$

$$\text{Error } E \text{ at } 95\% \text{ level of probability} \le \frac{1.960\sigma}{\sqrt{n}}. \qquad (5.9.11)$$

For a fixed level of confidence (that is, for $\lambda = \text{const}$), the error bound $\lambda\sigma/\sqrt{n}$ varies directly as σ and inversely as $n^{1/2}$. This is known as the "$n^{-1/2}$ law," and this rapidity of convergence is typical of Monte Carlo work. While this rate of convergence is slow, the advantage of Monte Carlo is that it is independent of the dimension, except that σ seems to increase with d. Furthermore, the rate of convergence is independent of the smoothness of the integrand. This is especially useful in integrating functions over irregular bounded regions in several dimensions. All we need do is to embed the region in a hypercube and define the integrand to vanish outside the region. A program implementing this idea is given by Engels.

Note also that the variance is defined only for integrands which are square-integrable even though this is not required for the convergence (in probability) of the Monte Carlo method.

Example. Referring to the previous integration, we have

$$I = \int_0^1 x\,dx = \frac{1}{2}, \qquad \sigma^2 = \int_0^1 x^2\,dx - \frac{1}{4} = \frac{1}{12}.$$

Error at 95% level $\leq 1.96\sigma/\sqrt{n} = 1.96/\sqrt{50}\,2\sqrt{3} = .08$ ($n = 50$). Observed error = .039.
Error at 95% level $\leq 1.96/\sqrt{100}\,2\sqrt{3} = .057$ ($n = 100$). Observed error = .027.

In a "practical" case, we need an estimate of σ in order to compute an error estimate. But the standard deviation σ will be unknown. We can, of course, use the sample variance

$$V_s = \frac{1}{n}\sum_{i=1}^{n} f^2(x_i) - \left(\frac{1}{n}\sum_{i=1}^{n} f(x_i)\right)^2 \tag{5.9.12}$$

to estimate σ^2. It is perhaps better to use $[n/(n-1)]V_s$ to estimate the variance. For, suppose we sample from a parent population of N items, taking all the $\binom{N}{n}$ samples of size n. If $\bar{V}$ is the mean of all the sample variances, and if V is the variance of the whole population, it can be shown that

$$\bar{V} = \frac{n-1}{n} V\left(1 + \frac{1}{N-1}\right). \tag{5.9.13}$$

Hence, if N is very large, we have

$$\bar{V} \approx \frac{n-1}{n} V \tag{5.9.14}$$

and, therefore,

$$V \approx \frac{n}{n-1}\bar{V}. \tag{5.9.15}$$

However, in Monte Carlo computations, n is ordinarily so large that $n/(n-1) \approx 1$ and the correction is negligible.

References. Engels [F1]; Fraser [1, pp. 114–123]; Kahn [1].

Let us examine the error bound $\lambda\sigma/\sqrt{n}$ more carefully. In order to reduce this error (for a fixed level of confidence) we can do several things; we can increase n, the number of functional evaluations, or we can reduce the variance σ^2. If we increase n by a factor of 100, we increase the accuracy of the answer by a factor of only 10. Even though this variation is discouraging, with present computer speeds, we may easily do 10^6 functional evaluations, and the resulting accuracy of 1 part in 10^2 or 10^3 may very well meet the requirements. In some cases we can replace the original problem by a modified problem, which has the effect of reducing σ sharply. This is known as a *variance-reducing* scheme and will be discussed subsequently.

There is another device that can be used for increasing accuracy: to abandon the idea of random sequences of points and to employ sequences of points specifically tailored for integration. One such class of sequences are known as *equidistributed sequences*, and their theory can be discussed outside the framework of statistics. The use of such sequences leads to error estimates that are roughly of the order of $1/n$ and, hence, converge more rapidly.

5.9.1 *Variance Reduction*

We begin with a simple sort of variance reduction. Suppose we are interested in computing $I = \int_0^1 f(x)\,dx$, and suppose we can find a $g(x)$ such that

$$|f(x) - g(x)| \leq \varepsilon, \qquad 0 \leq x \leq 1, \tag{5.9.1.1}$$

and such that

$$\int_0^1 g(x)\,dx = J \tag{5.9.1.2}$$

is a known quantity. Such a $g(x)$ is called a *control variate* for $f(x)$. Compute

$$I_1 = \int_0^1 (f(x) - g(x))\,dx \tag{5.9.1.3}$$

by Monte Carlo. The variance in this modified problem equals

$$\sigma_1^2 = \int_0^1 (f - g)^2\,dx - \left(\int_0^1 (f - g)\,dx\right)^2 \tag{5.9.1.4}$$

and, hence, $\sigma_1 \leq \varepsilon$.

We therefore take

$$\frac{1}{n}\sum_{i=1}^{n} (f(x_i) - g(x_i)) + J \tag{5.9.1.5}$$

as an estimator for I.

It should be observed that the variance has been reduced at the cost of doubling the number of functional evaluations. If one and the same problem is done two ways, having variances σ_1^2, σ_2^2 and taking times proportional to N_1, N_2, respectively, then the relative effectiveness of the variance reduction is

$$\frac{\lambda\sigma_1}{n^{1/2}(N_2/N_1)^{1/2}}\bigg/\frac{\lambda\sigma_2}{n^{1/2}} = \frac{N_1^{1/2}\sigma_1}{N_2^{1/2}\sigma_2}. \tag{5.9.1.6}$$

Example. Let $f(x) = e^x$ and $g(x) = 1 + x$. The variance of e^x is $\frac{1}{2}(e^2 - 1) - (e - 1)^2 = .242$, and the variance of $e^x - 1 - x$ is $\frac{1}{2}(e - 1)(5 - e) - \frac{23}{12} = .044$. If it takes 20% more time to compute $e^x - 1 - x$ as opposed to e^x, then the relative effectiveness of the reduction is 2.6 : 1.

A second method of variance reduction is that of *importance sampling*. Write $I = \int_0^1 f(x)\,dx$ as

$$I = \int_0^1 \frac{f(x)}{p(x)}\,p(x)\,dx, \tag{5.9.1.7}$$

where $p(x) > 0$ and

$$\int_0^1 p(x)\,dx = 1. \tag{5.9.1.8}$$

Now use as an estimator for I

$$\hat{f}_n = \frac{1}{n}\sum_{i=1}^{n}\frac{f(x_i)}{p(x_i)}, \tag{5.9.1.9}$$

where x_i is a random variable that has been extracted from the probability density distribution of $p(x)$ on $0 \le x \le 1$. The relevant variance is now

$$\sigma^2 = \int_0^1 \frac{f^2(x)}{p^2(x)}\,p(x)\,dx - \left(\int_0^1 \frac{f(x)}{p(x)}\,p(x)\,dx\right)^2. \tag{5.9.1.10}$$

Now assuming that $f(x) > 0$ (if it is not, we add a constant), we select

$$p(x) \approx f(x)\Big/\int_0^1 f(x)\,dx. \tag{5.9.1.11}$$

Then we have

$$\sigma^2 \approx 0. \tag{5.9.1.12}$$

Ideally, then, we should sample in proportion to the value of the function with constant of proportionality equal to the value of the integral. This would yield a 0 variance, but we should then need to know the solution to our problem beforehand. Furthermore, it is very difficult to sample with respect to the probability density function $cf(x)$. At any rate, we know that, by using a $p(x)$ whose integral is known and whose behavior approximates that of $f(x)$, we should expect to reduce the variance. Again, this gain must be measured against the loss of time involved in computing with random variables selected from a nonconstant probability density $p(x)$.

Example. $f(x) = e^x$. Take $p(x) = \frac{2}{3}(1 + x)$. The estimator is

$$\hat{f}_n = \frac{3}{2n} \sum_{i=1}^{n} \frac{e^{x_i}}{1 + x_i},$$

where the x_i have a probability density distribution $\frac{2}{3}(1 + x)$. The relevant variance is

$$\sigma^2 = \frac{3}{2}\int_0^1 \frac{e^{2x}\,dx}{1 + x} - \left(\int_0^1 e^x\,dx\right)^2 = \frac{3}{2}e^{-2}[Ei(4) - Ei(2)] - (e - 1)^2 = .0269,$$

$$Ei(t) = \int_1^t \frac{e^t}{t}\,dt.$$

If we approximate $cf(x)$ by a piecewise constant function $p(x)$, then sampling with respect to this $p(x)$ is no longer a problem. In d dimensions, the situation is much more complicated. What one can attempt is to approximate $f(\mathbf{x})/\int_{C_d} f\,dV$ by the product of d one-dimensional piecewise constant functions $\prod_{i=1}^{d} p_i(x_i)$. Here C_d is the unit hypercube $[0, 1]^d$. The functions $p_i(x_i)$ are determined using an iterative scheme in which the variability of $f(\mathbf{x})$ in each dimension is evaluated along with the approximation to the integral. The $p_i(x_i)$ are adjusted after each iteration until they converge. Then one samples with respect to the final density distribution to get an approximation with a small variance. This idea has been implemented by Lepage in the program VEGAS.

Sasaki has proposed functions of the form

$$h_1(x_1, x_2)h_2(x_2, x_3) \cdots h_{d-1}(x_{d-1}, x_d)$$

either as control variates or as bases for importance sampling. They are useful in cases where the variation of the integrand is not too large.

References. Lepage [1, 2]; Sasaki [1].

Other variance reduction methods are *stratified sampling* and the use of *antithetic variates*. In stratified sampling, we subdivide the unit hypercube C_d into hyperrectangles and choose, in a random fashion, a fixed number k of samples in each hyperrectangle. In Section 5.9.5, we give Haber's implementation of this method in which each hyperrectangle has the same volume. Another implementation by Sheppy and Lautrup uses an approach similar to that of Lepage. Here the end result of each iteration is a new subdivision of C_d into a set of hyperrectangles R_j, each of which is the product of d one-dimensional intervals on which the piecewise constant functions $p_i(x_i)$ are constant. This method forms the basis for a program in the NAG library. A third subdivision scheme has been proposed by Friedman and Wright and implemented in the program DIVONNE. Here the aim is to subdivide C_d into a set of hyperrectangles $\{R_i: i = 1, \ldots, n\}$ in such a way that the values

$$(\max_{\mathbf{x} \in R_i} f(\mathbf{x}) - \min_{\mathbf{x} \in R_i} f(\mathbf{x}))\, \mathrm{vol}(R_i)$$

are approximately constant over the set $\{R_i\}$.

See also Halton and Zeidman.

References. Friedman and Wright [1, 2]; Halton and Zeidman [1]; Hammersley and Morton [1]; Lautrup [1]; Morton [2]; NAG [1].

Variation reduction techniques appear to be equivalent to the approximation of the integrand by functions that can be handled analytically and, hence, may not be available when the dimension is high or the integrand is intractable. To be effective, they must exhibit the dimensional effect, and so this limits their use to problems of moderate dimension.

Ermakov and Zolotukhin have put forward a method of variance reduction that employs expansions in orthogonal functions. The method was extended by Handscomb who pointed out that it involves sampling from a quite involved frequency distribution—a process which may be attended by considerable difficulty.

A computer implementation of this method has been proposed by Bogues *et al.* The procedure appears to be quite complicated.

Rosenberg uses Bernstein polynomials of f in d variables (see Section 5.8.6) as control variates and as importance sampling functions. This, again, involves sampling from a complicated frequency distribution.

These ideas have been pursued in a somewhat different vein by Cranley and Patterson who look for estimators that will be perfectly correlated for all monomials of degree $\leq 2m - 1$. Writing the multidimensional integral in the form

$$I = 2^{-d} \int_{-1}^{1} \cdots \int_{-1}^{1} f(x_1, \ldots, x_d)\, dV,$$

an estimator for I is sought in the form

$$D = \sum_{i=1}^{m} a_i c_i(\mathbf{x}) + \sum_{i=1}^{m} b_i c_i(-\mathbf{x})$$

where

$$c_i(\mathbf{x}) = \frac{1}{i^d} \sum_{j_1=1}^{i} \cdots \sum_{j_d=1}^{i} f\left(\frac{x_1 - i + 2j_1 - 1}{i}, \ldots, \frac{x_d - i + 2j_d - 1}{i}\right),$$

$$c_i(-\mathbf{x}) = \frac{1}{i^d} \sum_{j_1=1}^{i} \cdots \sum_{j_d=1}^{i} f\left(\frac{-x_1 - i + 2j_1 - 1}{i}, \ldots, \frac{-x_d - i + 2j_d - 1}{i}\right). \tag{5.9.1.13}$$

The requirement of perfect correlation for monomials leads to a system of linear equations for the numerical coefficients a_i, b_i. Insofar as the scheme requires $2\sum_{j=1}^{m} j^d$ functional evaluations for the same point, it exhibits the dimensional effect, and is probably useful for $d \leq$ around 6. For details and many examples, see Cranley and Patterson.

References. Bogues, Morrow, and Patterson [1]; Cranley and Patterson [2]; Ermakov and Zolotukhin [1]; Granovskii [1]; Hammersley and Handscomb [1]; Handscomb [1]; Kahn [1, 2]; Powell and Swann [1]; Rosenberg [1]; Wilf [2].

5.9.2 Pseudorandom Sequences; Number Theoretic Methods

Pseudorandom sequences have been most commonly generated by the power-residue or linear congruential method. The integers $x_1, x_2, \ldots$ are defined recursively by means of

$$x_{n+1} = ax_n + c \pmod{m}, \qquad x_0 = \text{a starting value}. \tag{5.9.2.1}$$

Here a, c, and m are certain integers and this notation means that x_{n+1} is the remainder when $ax_n + c$ is divided by m. Since division by m can produce at most m different remainders, it is clear that the sequence $x_0, x_1, \ldots$ will be a periodic sequence whose period cannot exceed m. Therefore, what is wanted is a selection of a, c, and m which will produce a period that is large relative to the number of random numbers required in a computation. To go into the number theory of (5.9.2.1) would take us away from our main interests, and we shall merely describe a procedure for a binary computer.

The computer is assumed to have a word size of b bits. Arithmetic is carried out with the binary point to the extreme right. For the number a, choose an integer of the form $8t \pm 3$ and close to $2^{b/2}$. Choose $m = 2^b$ and $c = 0$. For x_0, choose any odd integer. The multiplication ax_0 now produces a product of $2b$ bits. The b high-order bits are discarded and the b low-order bits are the residue x_1. The process is now iterated. To obtain numbers in the unit interval, the binary point is considered to be at the far left. The period of the sequence $x_0, x_1, \ldots$ is 2^{b-2}, so that with a 35-bit machine the sequence has a period of $2^{33} \approx 8.5 \times 10^9$.

In order for a sequence to qualify for the title "pseudorandom," its beginning portions must pass a number of statistical tests. These tests include, for example, tests for uniform distribution, for independence of successive numbers, for autocorrelation between x_n and x_{n+p}, for runs up and down, "poker" tests on combinations of digits, etc. For work in multiple integration, one of these tests should obviously be the random quality of points in higher-dimensional space generated by the pseudorandom sequence $\{x_i\}$. There is some evidence that the points $(x_i, x_{i+1}, \ldots, x_{i+d-1})$ delivered by linear congruential methods may be poorly distributed and that the distribution is sensitive to a, c, m, and to the dimensionality d of the space. For a discussion of this, see Marsaglia who also lists "relatively good" congruential generators for the moduli $m = 2^{32}, 2^{35}, 2^{36}$. We remark that the fact that such sequences do not have properties of random sequences does not mean that they are inappropriate for numerical multiple integration. It just means that we cannot apply probability theory to interpret the results. If we can show that such sequences of points have low discrepancy, then we can apply the theory of Section 5.5. Dieter suggests that some of these shortcomings may be avoided by the use of a "generalized Fibonacci generator"

$$x_{i+n} = a_1 x_{i+n-1} + a_2 x_{i+n-2} + \cdots + a_n x_i \qquad (\text{mod } p)$$

p prime, but these matters have not yet been worked out fully.

For further discussion of pseudorandom sequences including their generation, testing, and philosophy, see Jansson, Knuth, and Freiberger and Grenander.

References. Dieter [1]; J. N. Franklin [1]; Freiberger and Grenander [1]; Hammersley and Handscomb [1]; Hull and Dobell [1]; IBM [1]; Jansson [1]; Knuth [1]; Marsaglia [1–3]; Niederreiter [4, 6]; Shreider [1]; Taussky and Todd [1].

Example. Estimate $I = \int_0^1 x\,dx = \frac{1}{2}$, using a pseudorandom sequence. ($\sigma^2 = \int_0^1 x^2\,dx - I^2 = \frac{1}{12}$.) The sequence x_n was generated according to the recurrence $x_{n+1} = 5^{17} x_n \bmod(2^{42})$, $x_0 = 1$.

Number of points	Approximate value of I
2^1	.37
2^2	.52
2^4	.45
2^8	.507
2^{16}	.502
2^{18}	.5008

Error at 95% level with 2^{18} points is $(1.96\sigma)2^{-9} = .0011$; exact error $= .0008$.

Example. Estimate $I = \int_0^1 \cdots \int_0^1 e^{x_1 x_2 x_3 x_4} dx_1 dx_2 dx_3 dx_4$. Using a pseudorandom sequence, we computed $I_N = 1/N \sum_{k=1}^{N} e^{x_{1k}x_{2k}x_{3k}x_{4k}}$. Designate the numbers $x_{11}, x_{21}, x_{31}, x_{41}, x_{12}, x_{22}, x_{32}, x_{42}, \ldots$ by $y_1, y_2, \ldots$. The y's were generated by the recurrence

$$y_{n+1} = ay_n \bmod(10^{10}), \qquad a = 101203, \qquad y_0 = 9876543211.$$

N	I_N
2	1.2523 816
4	1.0397 712
8	1.0332 334
16	1.0548 477
32	1.0693 787
64	1.0705 580
128	1.0573 892
256	1.0739 589
512	1.0706 032
1024	1.0741 810
2048	1.0714 488
4096	1.0677 460
8192	1.0695 889
Exact value	1.0693 9761

In this example $\sigma = .09$ and, with $8192 = 2^{13}$ points, the error at the 95% level is $(1.96\sigma)2^{-13/2} = .002$; exact error $= .0002$.

Example. Estimate

$$I = 2^{-7} \int_0^1 \cdots \int_0^1 (x_1 + \cdots + x_8)^2 \, dx_1 \cdots dx_8 = 25/192.$$

Pseudorandom numbers were chosen as in the previous example.

N	I_N
2	.1118 1381
4	.0978 4566
8	.1363 5815
16	.1279 8534
32	.1361 6714
64	.1222 1112
128	.1290 1039
256	.1292 9553
512	.1342 6877
1024	.1304 3647
2048	.1292 0104
4096	.1298 9805
8192	.1310 6422
Exact value	.1302 0833

5.9.2.5 "Built In" Randomizers

Many computer languages have randomizers built in. Thus, FORTRAN has the subroutine RAND, APL has the randomizer designated by ?N, etc. Most of these are based upon linear congruential methods and are used largely for the many noncomputational randomizations that occur in computer work. The serious employer of Monte Carlo methods should carry out experiments to satisfy himself of the quality of the results in the specific type of problem with which he is concerned.

References. Dudewicz and Ralley [1]; Forsythe, Malcolm, and Moler [F1]; Schrage [1].

5.9.3 Equidistributed Sequences

DEFINITION. A (deterministic) sequence of points $x_1, x_2, \ldots$ in $[a, b]$ is called equidistributed or uniformly distributed in $[a, b]$ if

$$\lim_{n\to\infty} \frac{b-a}{n} \sum_{i=1}^{n} f(x_i) = \int_a^b f(x)\, dx \tag{5.9.3.1}$$

for all bounded, Riemann-integrable functions $f(x)$.

The term "equidistributed" comes from the following consideration. Let $[\tau_1, \tau_2]$ be a subinterval of $[a, b]$. Let $N_n[\tau_1, \tau_2]$ = the number of points of $x_1, x_2, \ldots, x_n$ that lie in $[\tau_1, \tau_2]$. Then

$$\lim_{n\to\infty} \frac{N_n}{n} = \frac{\tau_2 - \tau_1}{b - a}. \tag{5.9.3.2}$$

To show (5.9.3.2) merely select $f(x) \equiv 1$ on $[\tau_1, \tau_2]$ and $f(x) \equiv 0$ elsewhere in the interval $[a, b]$ and use (5.9.3.1). Thus, the fraction of points of an equidistributed sequence that lie in any interval is asymptotically proportional to the length of the interval. This property may also be used as a definition of equidistribution.

We next exhibit a sequence that is equidistributed. Let (ξ) designate the fractional part of ξ, that is,

$$(\xi) = \xi - [\xi], \tag{5.9.3.3}$$

where $[\xi]$ is the largest integer $\leq \xi$.

THEOREM. *If θ is an irrational number, then the sequence*

$$x_n = (n\theta), \qquad n = 1, 2, \ldots, \tag{5.9.3.4}$$

is equidistributed in $[0, 1]$.

Proof. We give only an indication of a proof. The details can be found in Davis [6, p. 357]. Since

$$e^{2\pi ikx_j} = e^{2\pi ik(j\theta - [j\theta])} = e^{2\pi ikj\theta},$$

$$\frac{1}{n}[e^{2\pi ikx_1} + e^{2\pi ikx_2} + \cdots + e^{2\pi ikx_n}]$$

$$= \frac{1}{n}[e^{2\pi ik\theta} + (e^{2\pi ik\theta})^2 + \cdots + (e^{2\pi ik\theta})^n] = \frac{1}{n}e^{2\pi ik\theta}\frac{e^{2\pi ikn\theta} - 1}{e^{2\pi ik\theta} - 1}.$$

Hence for $k = \pm 1, \pm 2, \ldots,$

$$\lim_{n\to\infty}\frac{1}{n}[e^{2\pi ikx_1} + \cdots + e^{2\pi ikx_n}] = \int_0^1 e^{2\pi ikx}\,dx. \tag{5.9.3.5}$$

For $k = 0$, (5.9.3.5) holds trivially. The proof is completed by approximating a Riemann-integrable function by the functions $e^{2\pi ikx}$, $k = 0, \pm 1, \ldots.$

While equidistribution is a property of random sequences, there are certain statistical features of random sequences that the sequence (5.9.3.4) does not have. For example, designate by $\text{prob}(x_n > x_{n+1})$ the limit (if it exists)

$$\text{prob}(x_n > x_{n+1}) = \lim_{N\to\infty}\frac{1}{N}\sum_{\substack{x_n > x_{n+1},\\ 1\le n\le N}} 1. \tag{5.9.3.6}$$

For a random sequence, we would require that $\text{prob}(x_n > x_{n+1}) = \frac{1}{2}$. Now, if $0 < \theta < 1$, then $x_n > x_{n+1}$ if and only if $1 - \theta \le x_n < 1$. The probability of this last event is, by equidistribution, θ. Hence, $\text{prob}(x_n > x_{n+1}) \ne \frac{1}{2}$.

The interest in the sequences $(n\theta)$ for multidimensional integration lies in two facts. In the first place, suppose that $\theta_1, \theta_2, \ldots, \theta_d$ are d irrational numbers such that $1, \theta_1, \ldots, \theta_d$ are *linearly independent over the rational numbers*, that is, $\alpha_0 + \alpha_1\theta_1 + \cdots + \alpha_d\theta_d \ne 0$ for rational α_i not all vanishing. Then it can be shown that the points

$$P_n\colon ((n\theta_1), (n\theta_2), \ldots, (n\theta_d)), \qquad n = 1, 2, \ldots, \tag{5.9.3.7}$$

are equidistributed over the hypercube $0 \le x_i \le 1, i = 1, 2, \ldots, d$. This means that

$$\lim_{N\to\infty} I_N = \lim_{N\to\infty}\frac{1}{N}\sum_{n=1}^{N} f((n\theta_1), \ldots, (n\theta_d))$$
$$= \int_0^1 \cdots \int_0^1 f\,dx_1\,dx_2 \cdots dx_d \tag{5.9.3.8}$$

for any bounded Riemann-integrable function f.

In the second place, truncation error estimates can be obtained which are better than the $N^{-1/2}$ error provided by statistical theory. We shall show how this comes about in the one-dimensional case.

Let $f(x)$ be defined in $0 \le x \le 1$, $f(0) = f(1)$, and be sufficiently smooth there. Then we have the absolutely and uniformly convergent Fourier series

$$f(x) = \sum_{k=-\infty}^{\infty} a_k e^{2\pi ikx}, \qquad a_{-k} = \overline{a_k}, \qquad a_0 = \int_0^1 f(x)\,dx. \tag{5.9.3.9}$$

Therefore, by periodicity, we have

$$f((n\theta)) = \sum_{k=-\infty}^{\infty} a_k e^{2\pi ik(n\theta)} = \sum_{k=-\infty}^{\infty} a_k e^{2\pi ikn\theta}. \tag{5.9.3.10}$$

Summing, we obtain

$$\begin{aligned}\frac{1}{N}\sum_{n=1}^{N} f((n\theta)) &= \sum_{k=-\infty}^{\infty} a_k \frac{1}{N}\sum_{n=1}^{N} e^{2\pi ikn\theta} \\ &= \sum_{k=-\infty}^{\infty} a_k \frac{1}{N}\frac{e^{2\pi iNk\theta}-1}{e^{2\pi ik\theta}-1}.\end{aligned} \tag{5.9.3.11}$$

Therefore,

$$\frac{1}{N}\sum_{n=1}^{N} f((n\theta)) - \int_0^1 f(x)\,dx = \frac{1}{N}\sum_{k=-\infty}^{\infty}{}' a_k \frac{e^{2\pi iNk\theta}-1}{e^{2\pi ik\theta}-1}. \tag{5.9.3.12}$$

The prime on the summation sign indicates that the term corresponding to $k = 0$ is omitted from the sum. Hence

$$\left|\frac{1}{N}\sum_{n=1}^{N} f((n\theta)) - \int_0^1 f(x)\,dx\right| \le \frac{1}{N}\sum_{k=-\infty}^{\infty}{}' |a_k| \left|\frac{e^{2\pi iNk\theta}-1}{e^{2\pi ik\theta}-1}\right|. \tag{5.9.3.13}$$

Since

$$|e^{iu}-1| = |\cos u + i\sin u - 1| = (2(1-\cos u))^{1/2} = 2\left|\sin\frac{u}{2}\right|, \tag{5.9.3.14}$$

it follows that

$$\left|\frac{e^{2\pi iNk\theta}-1}{e^{2\pi ik\theta}-1}\right| = \left|\frac{\sin \pi Nk\theta}{\sin \pi k\theta}\right|. \tag{5.9.3.15}$$

Now, $|\sin \pi Nk\theta| \le 1$. To obtain a lower bound for the denominator, observe the inequalities $\sin \pi x \ge 2x$ in $0 \le x \le \frac{1}{2}$ and $\sin \pi x \ge 2(1-x)$ in $\frac{1}{2} \le x \le 1$. Hence, in general, we have

$$|\sin \pi x| \ge 2|x - \langle x\rangle|, \tag{5.9.3.16}$$

where $\langle x \rangle$ designates the nearest integer to x. Combining these inequalities, we obtain

$$\left| \frac{1}{N} \sum_{n=1}^{N} f((n\theta)) - \int_0^1 f(x)\,dx \right| \leq \frac{1}{2N} \sum_{k=-\infty}^{\infty}{}' |a_k| \frac{1}{|k\theta - \langle k\theta \rangle|}. \tag{5.9.3.17}$$

LEMMA (Liouville). Let α be the root of a polynomial of degree n that has integer coefficients and is irreducible over the rationals. Then there exists a constant $c > 0$ such that, for every pair of integers p, q with $q > 0$,

$$\left| \alpha - \frac{p}{q} \right| \geq \frac{c}{q^n}. \tag{5.9.3.18}$$

Proof. Let

$$f(x) = a_0 x^n + a_1 x^{n-1} + \cdots + a_n,$$

where a_i are integers and $a_0 > 0$. Write

$$f(x) = a_0(x - \alpha)(x - \alpha_2) \cdots (x - \alpha_n).$$

Then,

$$q^n f(p/q) = a_0 p^n + a_1 p^{n-1} q + \cdots + a_n q^n.$$

Therefore, $q^n f(p/q)$ is an integer. It is not 0, for otherwise f would have a factor $x - (p/q)$ and would be reducible. Hence, it follows that

$$|q^n f(p/q)| \geq 1. \tag{5.9.3.19}$$

Now,

$$\frac{f(x)}{a_0(x - \alpha)} = \prod_{k=2}^{n} (x - \alpha_k).$$

Therefore, we have

$$\frac{p}{q} - \alpha = \frac{q^n f(p/q)}{a_0 q^n \prod_{k=2}^{n} (p/q - \alpha_k)}$$

and

$$\left| \frac{p}{q} - \alpha \right| = \frac{|q^n f(p/q)|}{a_0 q^n \prod_{k=2}^{n} |p/q - \alpha_k|} \geq \frac{1}{a_0 q^n \prod_{k=2}^{n} |p/q - \alpha_k|}. \tag{5.9.3.20}$$

Set

$$\beta = \max(|\alpha|, |\alpha_2|, \ldots, |\alpha_n|), \tag{5.9.3.21}$$

and consider the following two cases.

Case I. $|p/q| > 2\beta$. Now

$$\left|\frac{p}{q} - \alpha\right| \geq \left|\frac{p}{q}\right| - |\alpha| > 2\beta - |\alpha| = \beta + (\beta - |\alpha|) \geq \frac{\beta}{q^n}$$

since $q \geq 1$.

Case II. $|p/q| \leq 2\beta$. Then

$$\left|\frac{p}{q} - \alpha_k\right| \leq \left|\frac{p}{q}\right| + |\alpha_k| \leq 2\beta + \beta = 3\beta.$$

Therefore, by (5.9.3.20), we have

$$\left|\frac{p}{q} - \alpha\right| \geq \frac{1}{a_0 q^n \prod_{k=2}^{n} (3\beta)} = \frac{1}{a_0 q^n (3\beta)^{n-1}}.$$

If we now set

$$c = \min\left(\beta, \frac{1}{a_0(3\beta)^{n-1}}\right), \tag{5.9.3.22}$$

then (5.9.3.18) follows.

COROLLARY. *If θ is a quadratic irrational number, then we can find a constant $c > 0$ such that for all integers k,*

$$\frac{1}{|k\theta - \langle k\theta\rangle|} \leq \frac{|k|}{c}. \tag{5.9.3.23}$$

Proof

$$|k\theta - \langle k\theta\rangle| = |k| \left|\theta - \frac{\langle k\theta\rangle}{k}\right| \geq |k| \frac{c}{|k|^2}.$$

This follows from (5.9.3.18), with $n = 2$. Taking reciprocals, we obtain (5.9.3.23).

THEOREM. *Let $f(x)$ be periodic in $[0, 1]$ and be of class $C^3[0, 1]$ so that we have $f(0) = f(1)$, $f'(0) = f'(1)$, $f''(0) = f''(1)$. Let θ be a quadratic irrational number. Then*

$$\left|\frac{1}{N} \sum_{n=1}^{N} f((n\theta)) - \int_0^1 f(x)\, dx\right| \leq \frac{c}{N}. \tag{5.9.3.24}$$

Proof. We have

$$f(x) = \sum_{k=-\infty}^{\infty} a_k e^{2\pi i k x} \quad \text{with} \quad a_k = \int_0^1 f(x) e^{-2\pi i k x}\, dx.$$

Integrating by parts, we have, for $k \neq 0$,

$$a_k = \frac{f(x)e^{-2\pi ikx}}{-2\pi ik}\bigg|_0^1 + \frac{1}{2\pi ik}\int_0^1 f'e^{-2\pi ikx}\,dx = \frac{1}{2\pi ik}\int_0^1 f'e^{-2\pi ikx}\,dx.$$

Integrating by parts twice more, we have

$$a_k = \frac{1}{(2\pi ik)^3}\int_0^1 f'''e^{-2\pi ikx}\,dx.$$

Therefore

$$|a_k| \le \frac{c_1}{|k|^3},$$

where

$$c_1 = \frac{1}{(2\pi)^3}\int_0^1 |f'''(x)|\,dx.$$

From (5.9.3.23), we have

$$\frac{1}{|k\theta - \langle k\theta\rangle|} \le \frac{|k|}{c_2}$$

and, hence, by (5.9.3.17)

$$\left|\frac{1}{N}\sum_{k=1}^{N} f((n\theta)) - \int_0^1 f(x)\,dx\right| \le \frac{1}{2Nc_2}\sum_{k=-\infty}^{\infty}{}' |a_k|\,|k|$$

$$\le \frac{c_1}{2Nc_2}\sum_{k=-\infty}^{\infty}{}' \frac{|k|}{|k|^3} = \frac{c}{N}.$$

Even if $f(x)$ does not fulfill the stringent conditions of this theorem, the convergence may be more rapid than $N^{-1/2}$. There is, for example, the Hardy–Littlewood–Ostrowski Theorem: *Let θ be a positive irrational number whose continued fraction expansion*

$$\theta = a_0 + \cfrac{1}{a_1 + \cfrac{1}{a_2 + \cdots}}$$

satisfies $0 < a_i \le A$ (this will be the case if θ is, for example, a quadratic irrationality). Then

$$\left|\frac{1}{N}\sum_{n=1}^{N} (n\theta) - \int_0^1 x\,dx\right| \le \frac{3}{2}A\frac{\log N}{N}. \tag{5.9.3.25}$$

Thus, we have a rate of convergence which is more rapid than $N^{-1/2}$ but somewhat less than N^{-1}.

More generally, if $f(x)$ has total variation V, then, it can be shown that there exists a constant A such that

$$|R_N(f)| = \left|\frac{1}{N}\sum_{n=1}^{N} f((n\theta)) - \int_0^1 f(x)\,dx\right| \le \frac{4AV\log((A+1)N)}{(\log(A+1))N}$$

In particular, if $\theta = \frac{1}{2}(\sqrt{5}-1)$,

$$|R_N(f)| \le \frac{(\frac{7}{6})V\log(6N)}{N}.$$

Richtmyer has extended this theorem to higher dimensions. Let $\theta_1, \theta_2, \ldots, \theta_d$ be d irrational numbers such that $1, \theta_1, \ldots, \theta_d$ are linearly independent over the rationals. Define the class of functions $\mathscr{E}^k$, $k > 1$, to consist of those functions f which are periodic with period 1 in each variable and such that the coefficients $c_{\mathbf{h}}$ in the Fourier series

$$f(\mathbf{t}) = \sum_{h_1=-\infty}^{\infty} \cdots \sum_{h_d=-\infty}^{\infty} c_{\mathbf{h}} \exp(2\pi i(h_1 t_1 + \cdots + h_d t_d), \qquad \mathbf{h} = (h_1, \ldots, h_d)$$

satisfy the inequality $|c_{\mathbf{h}}| \le Cr(\mathbf{h})^{-k}$ for all $\mathbf{h} \ne \mathbf{0}$ and some $C > 0$ where $r(\mathbf{h}) = \prod_{j=1}^{d} \max(1, |h_j|)$. Then if $f \in \mathscr{E}^k$,

$$\left|\int_0^1 \cdots \int_0^1 f\,dV - \frac{1}{N}\sum_{n=1}^{N} f((n\theta_1), \ldots, (n\theta_d))\right| \le \frac{c}{N}. \tag{5.9.3.26}$$

If $k > 2$, we can achieve higher rates of convergence by replacing the simple average of functional values in (5.9.3.26) by a weighted average. Let q be a positive integer and define the weights $a_{Nn}^{(q)}$ from the identity

$$\left(\sum_{j=0}^{N-1} z^j\right)^q = \sum_{n=0}^{q(N-1)} a_{Nn}^{(q)} z^n.$$

Then, for every $f \in \mathscr{E}^k$ with $k > q$ we have

$$\int_0^1 \cdots \int_0^1 f\,dV - \frac{1}{N^q}\sum_{n=0}^{q(N-1)} a_{Nn}^{(q)} f((n\theta_1), \ldots, (n\theta_d)) = O(N^{-q}).$$

Halve has shown how to derive explicit formulas for the $a_{Nn}^{(q)}$. Thus, for $q = 2$

$$a_{Nn}^{(2)} = 2N - 1 - n, \qquad n = 0, \ldots, 2N-2,$$

while for $q = 3$

$$a_{Nn}^{(3)} = \begin{cases} 3N(N-1) + 1 - n^2, & 0 \le n \le N-1, \\ \frac{1}{2}(3N-2-n)(3N-1-n), & N \le n \le 3N-3. \end{cases}$$

For an alternate set of weights, see Sugihara and Murota.

References. Bertrandias [1]; Davis [6, pp. 354–357]; J. N. Franklin [1]; Halve [1]; Haselgrove [1]; Hlawka [2]; Hua and Wang [1]; Koksma [1]; Korobov [1]; Kuipers and Niederreiter [1]; LeVeque [1]; Niederreiter [3, 6]; Ostrowski [1]; Peck [1]; Richtmyer [1, 2]; Richtmyer, Devaney, and Metropolis [1]; Roth [1]; Sugihara and Murota [1].

Example. Compute $\int_0^1 \cdots \int_0^1 e^{x_1 x_2 x_3 x_4}\, dx_1\, dx_2\, dx_3\, dx_4$, using an equidistributed sequence of points. Select $\theta_1 = \sqrt{2}$, $\theta_2 = \sqrt{3}$, $\theta_3 = \frac{1}{3}\sqrt{6}$, $\theta_4 = \sqrt{10}$.

N	I_N
2	1.0556 385
4	1.0646 192
8	1.0592 766
16	1.0615 566
32	1.0626 119
64	1.0586 261
128	1.0657 314
256	1.0673 119
512	1.0668 403
1024	1.0681 500
2048	1.0685 418
4096	1.0685 545
8192	1.0688 021
Exact value	1.0693 9761

One sequence of equidistributed points, composed of linearly independent irrational numbers, that has been used successfully in integration problems in d dimensions is

$$P_n: \left(\left(\frac{n(n+1)}{2} p_1^{1/2} \right), \left(\frac{n(n+1)}{2} p_2^{1/2} \right), \ldots, \left(\frac{n(n+1)}{2} p_d^{1/2} \right) \right) \tag{5.9.3.27}$$

where $p_1, p_2, \ldots$ is the sequence of *prime* numbers 2, 3, 5, 7, 11, ... and where (ξ) designates the fractional part of ξ. This sequence has the property that for any integer $n \geq 1$, the sequences (P_1, P_{n+1}), (P_2, P_{n+2}), ... are equidistributed over the unit cube of dimension $2d$. Another derived from the work of Baker is based on the set $e^{r_1}, \ldots, e^{r_d}$ with distinct nonzero rationals $r_1, \ldots, r_d$. A third, proposed by Niederreiter, selects $\theta_j = 2^{j/(d+1)}$, $1 \leq j \leq d$. If $2d + 3$ is a prime, the set

$$2 \cos \frac{2\pi}{2d+3}, \quad 2 \cos \frac{4\pi}{2d+3}, \ldots, 2 \cos \frac{2\pi d}{2d+3}$$

has been recommended.

References. A. Baker [1]; Bass and Guilloud [1]; Haber [4, 7]; Hua and Wang [1]; Niederreiter [1, 3, 6]; Zaremba [1, 3, 6]; Zinterhof [1].

Examples. Using the sequence (5.9.3.27) in four dimensions, Haber reports the following experiments.

(a)

$$I = \int_0^1 \cdots \int_0^1 f_1 \, dx_1 \, dx_2 \, dx_3 \, dx_4$$

$$f_1 = \exp(x_1 x_2 x_3 x_4) - 1, \qquad I = .0693976,$$

E = error. This integrand is analytic.

N	E
16	−.0067
81	−.0040
256	−.0010
625	.0026
1,296	−.0003
2,401	−.0015
4,096	−.0009
6,561	−.0014
10,000	.0012
65,536	−.00021

(b)

$$I = \int_0^1 \cdots \int_0^1 f_2 \, dx_1 \, dx_2 \, dx_3 \, dx_4$$

$$f_2 = \begin{cases} 1 & \text{if } x_1^2 + x_2^2 + x_3^2 + x_4^2 \le 1 \\ 0 & \text{otherwise.} \end{cases}$$

This integrand is discontinuous.

N	E
16	.0584
81	−.0002
256	−.0119
625	−.0180
1,296	−.0087
2,401	−.0085
4,096	−.0075
6,561	−.0064
10,000	−.0011
65,536	.00052

For alternative methods for generating equidistributed sequences of d-tuples, see Section 5.5. In the one-dimensional case, the sequence $X = \{\phi_2(i): i = 1, 2, \ldots\}$ where $\phi_r(i)$ is given by (5.5.9) is known as the van der

Corput sequence and the corresponding finite two-dimensional sequence $\{(i/N, \phi_2(i)): i = 1, \ldots, N\}$ is called the Roth sequence. For the van der Corput sequence, Sobol has shown that if $f(x) = x^{\beta}$, $\beta > -1$ or $f(x) = x^{-1} \log^{\gamma} x$, $\gamma > 1$, then

$$\lim_{N\to\infty} \frac{1}{N} \sum_{i=1}^{N} f(x_i) = \int_0^1 f(x)\, dx \equiv If \tag{5.9.3.28}$$

so that we can ignore the singularity. This result is a special case of the following theorem.

Let $f(x)$ be differentiable in $(0, 1]$ and unbounded as $x \to 0$ such that If exists. Let $X = \{x_i\}$ be a sequence in $[0, 1]$ with discrepancy $D_N(X)$ and define $a_N = \min_{1 \le i \le N} x_i$. Then if

$$D_N(X) \int_{a_N}^1 |f'(x)|\, dx \to 0 \quad \text{with} \quad N,$$

(5.9.3.28) holds. Sobol has extended these results to interior singularities in $(0, 1)$ and to integrals over the hypercube $[0, 1]^d$ with singularities at the origin.

References. Roth [1]; Sobol [4]; van der Corput [1].

Examples (Roos and Arnold). We give here some results of numerical multiple integration using three sampling procedures:

(A) pseudorandom numbers generated by

$$10^{-13} x_{j+1} = x_j a (\text{mod } 10^{-13}), \qquad j = 0, 1, 2, \ldots$$
$$x_0 = 10^{-13}, \qquad a = .2541865828329$$

(B) the sequence (5.9.3.7) with $\theta_j = p_j^{1/2}$, $j = 1, \ldots, d$, where p_j is the jth prime

(C) the sequence (5.5.10).

(1) $$\int_0^1 \cdots \int_0^1 \frac{1}{d} \sum_{i=1}^{d} |4x_i - 2|\, dx_1 \cdots dx_d = 1.$$

	$d = 20$			$d = 25$		
n	A	B	C	A	B	C
1000	1.03518	.99958	1.00048	.96171	.99962	1.00063
2000	1.02274	.99972	.99946	.99151	.99973	.99978
3000	1.04525	.99986	1.00007	.97229	.99988	.99997
4000	1.05118	.99986	1.00026	.93753	.99986	1.00017
5000	1.03695	.99986	1.00054	.92273	.99988	1.00058
6000	1.01259	.99989	1.00038	.94954	.99988	1.00034
7000	1.01116	.99993	1.00030	.94950	.99993	1.00020

$$(2) \quad \int_0^1 \cdots \int_0^1 \prod_{i=1}^{d} |4x_i - 2| \, dx_1 \cdots dx_d = 1$$

	$d = 20$			$d = 25$		
n	A	B	C	A	B	C
1000	.13016	1.00620	2.33065	.02861	.58274	34.4708
2000	.06914	.99920	1.44952	.16614	1.21436	17.4312
3000	.05657	.82612	1.36950	.11080	1.01688	12.1670
4000	.23207	.86309	1.20404	.08311	.95756	9.60404
5000	.18694	.79677	1.15070	.08345	.86182	7.81476
6000	.15578	.90277	1.10199	.07212	.94265	6.68951
7000	.23956	.94012	1.00672	.06183	.94438	5.78880

$$(3) \quad \int_0^1 \cdots \int_0^1 \prod_{i=1}^{d} \left(\frac{\pi}{2} \sin \pi x_i\right) dx_1 \cdots dx_d = 1$$

	$d = 20$			$d = 25$		
n	A	B	C	A	B	C
1000	.10480	.86967	.62829	1.38715	.56125	.26959
2000	.07652	.87048	.79619	.70158	.74389	.58298
3000	.05105	.82744	.83804	.47505	.67361	1.01796
4000	.04966	.85621	1.13006	.37217	.76168	1.04671
5000	.04237	1.22634	1.02751	1.32295	1.23185	.94256
6000	.03531	1.21321	1.00647	1.10249	1.11632	.97415
7000	.03225	1.19119	1.01634	1.02024	1.13249	.94017

It is very difficult to draw any conclusions about the efficacy of these methods based on the numbers given here. It is clear that 7000 function evaluations are not enough to give a clear picture of what is happening when the dimension is high. Monte Carlo (A) comes off worst as expected although the poor performance in case 2 and in case 3 with $d = 20$ is rather surprising. Overall method B seems best. However, it is possible that method C would fare better if the sequences started with a large value for i rather than with $i = 1$.

Reference. Roos and Arnold [1].

Much effort is being spent on finding sequences of points in multidimensional space that are "good" for multiple integration. These are based on recent advances in number theory. One direction is that of low discrepancy sequences such as (5.5.10) since (5.5.7) gives a bound on integration error which depends on the discrepancy. These rules are good for functions f that need not be too smooth since they only require that the variation Vf be finite. However, the lower bound on the discrepancy given by (5.5.8) limits the accuracy that can be achieved by this method. For smooth functions, two alternative classes of points are available. We have already mentioned the first class, namely, the sequences $(n\boldsymbol{\alpha})$ where $\boldsymbol{\alpha}$ is the point $(\alpha_1, \alpha_2, \ldots, \alpha_d)$ and the numbers $1, \alpha_1, \ldots, \alpha_d$ are linearly independent over the rationals. The other class are the *good lattice points* or *optimal coefficients* developed by Korobov and Hlawka which we describe below. The theory behind both classes of methods is based on properties of periodic functions, but as we shall see, these methods can also be applied to sufficiently smooth non-periodic functions.

Let C_d be the unit hypercube $[0, 1]^d$ and let $f \in \mathscr{E}^k$, $k > 1$, with constant $C = C(f)$ so that

$$f(\mathbf{x}) = \sum_{\mathbf{m}} c_{\mathbf{m}} \exp(2\pi i \, \mathbf{m} \cdot \mathbf{x})$$

with $|c_{\mathbf{m}}| \le C r(\mathbf{m})^{-k}$, $\mathbf{m} \ne \mathbf{0}$, where $r(\mathbf{m}) = \prod_{j=1}^{d} \max(1, |m_j|)$. Here $\mathbf{m} \cdot \mathbf{x} = m_1 x_1 + \cdots + m_d x_d$. Then for any lattice point $\mathbf{g} = (g_1, \ldots, g_d)$, g_i integer, and any integer N,

$$E_N(f; \mathbf{g}) \equiv \frac{1}{N} \sum_{n=1}^{N} f\left(\frac{1}{N}\mathbf{g}\right) - \int_{C_d} f \, dV = \sum_{\mathbf{m} \ne \mathbf{0}} \delta_N(\mathbf{m} \cdot \mathbf{g}) c_{\mathbf{m}}$$

where

$$\delta_N(\mathbf{m} \cdot \mathbf{g}) = \begin{cases} 1 & \text{if } \mathbf{m} \cdot \mathbf{g} \equiv 0 \pmod{N}, \\ 0 & \text{otherwise.} \end{cases}$$

Hence $|E_N(f; g)| \le C P^{(k)}(\mathbf{g}, N)$ where

$$P^{(k)}(\mathbf{g}, N) = \sum_{\mathbf{m} \ne \mathbf{0}} r(\mathbf{m})^{-k}, \qquad \text{for } \mathbf{m} \cdot \mathbf{g} \equiv 0 \pmod{N} \quad (5.9.3.29)$$

If we could find a lattice point $\mathbf{g}$ such that $P^{(k)}(\mathbf{g}, N)$ were small, then we would have a good rule for $f \in \mathscr{E}^k$. Since $P^{(k)}(\mathbf{g}, N)$ will be small if the smallest $r(\mathbf{m})$ in the sum (5.9.3.29) is large, we can define good lattice points independent of k as follows.

DEFINITION. Let $N_1, N_2, \ldots$ be an increasing sequence of positive integers. A sequence $\mathbf{g}(N_1)$, $\mathbf{g}(N_2)$, $\ldots$ of d-tuples of integers is a good lattice point sequence if there exist $A > 0$ and β such that for all i, every nonzero solution $\mathbf{m}$ of the congruence $\mathbf{m} \cdot \mathbf{g}(N_i) \equiv 0 \pmod{N_i}$ satisfies

$$r(\mathbf{m}) > A N_i / \log^{\beta} N_i. \qquad (5.9.3.30)$$

The smallest β for which (5.9.3.30) holds is called the index β^* and depends on the dimension d. It follows that the integration rule

$$Q(f, N_i, \mathbf{g}(N_i)) \equiv N_i^{-1} \sum_{h=1}^{N_i} f\left(\frac{h}{N_i} \mathbf{g}(N_i)\right)$$

is exact for all trigonometric polynomials of degree $p < AN_i/\log^{\beta^*} N_i$. For any good lattice point $\mathbf{g}$ and any $f \in \mathscr{E}^k$, the error $E_N(f; \mathbf{g})$ is $O((\log N)^{k\beta^*}/N^k)$. Good lattice point sequences exist for any increasing sequence of positive integers but it is relatively difficult to find them. It has been shown that there exist sequences for which the index $\beta^* \leq d - 1$. A lower bound on β^* is given by $\beta^* \geq (d-1)/2$. This follows from a result by Sharygin [2] who showed that for any N points $\mathbf{x}_j, j = 1, \ldots, N$, in C_d and any integer $k > 1$, there exists a function $f \in \mathscr{E}^k$ such that $f(\mathbf{x}_j) = 0$, $j = 1, \ldots, N$, and $\int_{C_d} f \, dV \geq C(k, d) \log^{(d-1)/2} N/N^k$.

Tables of good lattice points have been published by various authors using different methods. In general, the computation of such points involves a considerable amount of computation. Haber has proposed the following method which gives rather good lattice points at a reasonable cost. We first define the function

$$F_k(x) = \sum_{\mathbf{m}} r(\mathbf{m})^{-k} \exp(2\pi i \, \mathbf{m} \cdot \mathbf{x}), \qquad k > 1.$$

Then, for $f \in \mathscr{E}^k$ with constant C, $E_N(f; \mathbf{g}) \leq CE_N(F_k; \mathbf{g})$ for any lattice point $\mathbf{g}$. Since good lattice points are defined independently of the exponent k, it follows that a point $\mathbf{g}$ which makes $E_N(F_2; \mathbf{g})$ small is a good lattice point. Now $E_N(F_2; \mathbf{g})$ is easy to compute since

$$F_2(\mathbf{x}) = \prod_{i=1}^{d} (1 + 2\pi^2 B_2^*(x_i))$$

where B_2^* is the periodized Bernoulli polynomial $B_2(x) = x^2 - x + \frac{1}{6}$. Haber computed $E_N(F_2; \mathbf{g})$ for a set of about 50 random lattice points and tabulated the minimizing point $\hat{\mathbf{g}}$ for a sequence of N's and in various dimensions. Since it has been shown that about 50% of all lattice points, all of whose components are relatively prime to N, are good lattice points, we are reasonable certain that the $\hat{\mathbf{g}}$ are good. They are not the best in the sense, say, of minimizing the quantity $E_N(F_2; \mathbf{g})$ but they are close. On the basis of numerical evidence, Haber concludes that if

$$E_N(F_2; \hat{\mathbf{g}}) \sim K(d) N^{-2} \log^{\beta(d)} N,$$

then $K(d)$ decreases with d and the following are approximate values for $\beta(d)$:

d	2	3	4	5	6	7	8
$\beta(d)$	1.2	2.8	4.3	5.8	6.9	8.3	9.6

Thus $\beta(d)$ is closer to the lower bound $d - 1$ than to the upper bound $2d - 2$.

To apply the methods of good lattice points or of irrational numbers to a nonperiodic function $f(\mathbf{x})$, we must periodize it, i.e., find a function $h \in \mathscr{E}^k$ for some $k > 1$ such that

$$\int_{C_d} f\,dV = \int_{C_d} h\,dV. \tag{5.9.3.31}$$

For f sufficiently smooth this is possible by virtue of the following theorem of Korobov.

THEOREM. *If, for an integer $r \geq 1$,*
(i) *all partial derivatives*

$$\frac{\partial^{j_1+\cdots+j_d}}{\partial x^{j_1} \cdots \partial x_d^{j_d}} f \qquad (0 \leq j_i \leq r; \quad i = 1, \ldots, d)$$

are of bounded variation in the sense of Hardy and Krause on C_d,

$$\text{(ii)} \quad \Delta_i \frac{\partial^{j_1+\cdots+j_d}}{\partial x^{j_1} \cdots \partial x_d^{j_d}} f = 0 \qquad (0 \leq j_i \leq r - 1; \quad i = 1, \ldots, d) \tag{5.9.3.32}$$

where $\Delta_i \phi(x_1, \ldots, x_d) = \phi(x_1, \ldots, x_{i-1}, 1, x_{i+1}, \ldots, x_d) - \phi(x_1, \ldots, x_{i-1}, 0, x_{i+1}, \ldots, x_d)$,

then the Fourier coefficients $c_\mathbf{m}$ of f satisfy for all $\mathbf{m} \neq \mathbf{0}$,

$$|c_\mathbf{m}| \leq 3W^{(r)}/2\pi r(\mathbf{m})^{r+1}$$

where $W^{(r)}$ depends on the smoothness of $\partial^{rd} f/\partial x_1^r \cdots \partial x_d^r$.

As a first example of a practical periodization, define

$$h(x_1, \ldots, x_d) = f(1 - 2|x_1 - \tfrac{1}{2}|, \ldots, 1 - 2|x_d - \tfrac{1}{2}|). \tag{5.9.3.33}$$

Then (5.9.3.31) holds; however, (5.9.3.32) holds only for $h(\mathbf{x})$ but nor for any of its partial derivatives, so that we can only achieve $h \in \mathscr{E}^2$. On the other hand, we can save 50% of the computational effort when applying the method of good lattice points to h. Thus, for N odd, we have

$$\int_{C_d} h\,dV \simeq \frac{1}{N} f(\mathbf{0}) + \frac{2}{N} \sum_{n=1}^{(N-1)/2} f\left(\left\{\frac{ng_1}{N}\right\}, \ldots, \left\{\frac{ng_d}{N}\right\}\right)$$

where $\{a\}$ denotes the fractional part of a.

A second periodization is achieved by a change of variables. Let

$$\psi_p(u) = (2p-1)\binom{2p-2}{p-1}\int_0^u t^{p-1}(1-t)^{p-1}\,dt$$

so that $\psi_2(u) = 3u^2 - 2u^3$, $\psi_3(u) = 10u^3 - 15u^4 + 6u^5$, etc. We have $\psi_p'(u) = \psi_p''(u) = \cdots = \psi_p^{(p-1)}(u) = 0$, $u = 0, 1$. Hence, if (i) in the theorem holds for $f(\mathbf{x})$ with $r = p - 1$, then (i) and (ii) hold for

$$h(\mathbf{x}) = f(\psi_p(x_1), \ldots, \psi_p(x_d))\psi_p'(x_1)\cdots\psi_p'(x_d)$$

with $|c_{\mathbf{m}}| \leq K_p r(\mathbf{m})^{-p}$, and (5.9.3.31) holds. Since K_p grows rapidly with p, it is not recommended to take p greater than 4 even for very smooth functions.

For a comprehensive coverage of number-theoretic methods for multiple numerical integration, including extensive tables of good lattice points, see the monograph by Hua and Wang. Other tables appear in some of the works listed in the references.

References. Cheng, Suzukawa, and Wolfsberg [1]; Conroy [1]; Cranley and Patterson [4]; Daudey, Diner, and Savinelli [1]; Foglia [1]; Haber [7, 10, 14]; Halve [1]; Hlawka [1, 2]; Hlawka, Ferneis, and Zinterhoff [1]; Hua and Wang [1]; Keast [1, 2]; Korobov [1]; LaBudde and Warnock [1]; Maisonneuve [1]; Moon [1]; Niederreiter [5, 6]; Saltykov [1]; Sharygin [1, 2]; Solodov [1]; Stroud [18, Chap. 6]; Zakrzewska, Dudek, and Nazarewicz [1]; Zaremba [1–9].

5.9.4 *Method of Averaging*

The intimate relationship that exists between the theory of equidistribution and the theory of Fourier series suggests that it might be possible to apply summability methods (that is, averaging) to speed up the convergence of results obtained from equidistributed sequences. This lies at the basis of a theory which was worked out by Haselgrove and of which we can give only the barest description.

Let us assume that $f(x_1, x_2, \ldots, x_d)$ is a function of period 2 in each variable and that its Fourier expansion is

$$f(x_1, x_2, \ldots, x_d) = \sum_{-\infty}^{\infty} a_{n_1, n_2, \ldots, n_d} \exp[i\pi(n_1 x_1 + n_2 x_2 + \cdots + n_d x_d)]. \tag{5.9.4.1}$$

Assume, further, that the Fourier coefficients satisfy an inequality of the form

$$|a_{n_1, n_2, \ldots, n_d}| \leq M\,|n_1 n_2 \cdots n_d|^{-2}. \tag{5.9.4.2}$$

In (5.9.4.2), M is a constant and if any $n_i = 0$, the inequality is to hold with the zero factor omitted from the denominator on the right-hand side. This condition is equivalent to a smoothness condition on the derivatives of f.

Let $\alpha_1, \alpha_2, \ldots, \alpha_d$ be d irrational numbers such that $1, \alpha_1, \ldots, \alpha_d$ are linearly independent over the rationals. Set

$$S_1 = S_1(N) = \sum_{m=-N}^{N} f(2m\alpha_1, 2m\alpha_2, \ldots, 2m\alpha_d), \tag{5.9.4.3}$$

$$S_2 = S_2(N) = \sum_{p=0}^{N} S_1(p), \tag{5.9.4.4}$$

$$s_1(N) = \frac{1}{2N+1} S_1(N), \tag{5.9.4.5}$$

$$s_2(N) = \frac{1}{(N+1)^2} S_2(N). \tag{5.9.4.6}$$

The quantity $s_2(N)$ is the Cesàro mean of the sequence of values

$$f(2m\alpha_1, \ldots, 2m\alpha_d).$$

If

$$I = 2^{-d} \int_{-1}^{1} \cdots \int_{-1}^{1} f(x_1, x_2, \ldots, x_d)\, dx_1\, dx_2 \cdots dx_d, \tag{5.9.4.7}$$

then it can be shown by an analysis similar to the one given in Section 5.9.3 that under the above conditions there exist numbers $\alpha_1, \ldots, \alpha_d$ such that

$$|I - s_1(N)| \leq \frac{\text{const}}{N} \tag{5.9.4.8}$$

and

$$|I - s_2(N)| \leq \frac{\text{const}}{N^{2-\varepsilon}}, \qquad \varepsilon > 0. \tag{5.9.4.9}$$

If the integrand is not periodic, it must be reduced to the periodic case. Consider, for example,

$$J = \int_{0}^{1} \cdots \int_{0}^{1} F(x_1, x_2, \ldots, x_d)\, dx_1, dx_2 \cdots dx_d. \tag{5.9.4.10}$$

Then it follows that

$$J = 2^{-d} \int_{-1}^{1} \cdots \int_{-1}^{1} F(|x_1|, |x_2|, \ldots, |x_d|)\, dx_1\, dx_2 \cdots dx_d. \tag{5.9.4.11}$$

The function

$$f(x_1, x_2, \ldots, x_d) = F(|x_1|, |x_2|, \ldots, |x_d|)$$

will then be periodic in $x_1, \ldots, x_d$ but not very smooth.

Haselgrove tabulates good values of α to use for $d \leq 8$.

Example (Haselgrove)

$$I = \int_0^1 \cdots \int_0^1 e^{-x_1 x_2 \cdots x_5}\, dx_1\, dx_2 \cdots dx_5 = .9706\ 5719.$$

N	$s_1(N)$	$s_2(N)$	Monte Carlo
1000	.9706 2392	.9706 2580	.9676 3166
2000	.9708 2902	.9706 3927	.9687 0265
3000	.9705 4070	.9706 6765	.9688 5258
4000	.9706 8153	.9706 6383	.9694 4396
5000	.9706 5925	.9706 5630	.9695 0137
6000	.9706 1983	.9706 5761	.9699 0269
7000	.9706 8925	.9706 5639	.9701 8578
8000	.9706 4881	.9706 5632	.9703 0504
9000	.9706 3833	.9706 5706	.9703 8771
10000	.9706 6307	.9706 5854	.9703 2729
11000	.9706 5947	.9706 5860	.9702 9480
12000	.9706 7426	.9706 5744	.9704 8290

The "Monte Carlo" column refers to computations carried out with pseudo-random points. Notice that at $N = 12{,}000$, s_2 achieves twice the number of significant figures as the Monte Carlo computation.

Reference. Haselgrove [1].

Here are some further examples in higher dimensions, computed at the Weizmann Institute. The irrational numbers α_i were selected to be

$$\alpha_i = .5(p_i^{1/2} - [p_i^{1/2}])$$

where $p_1 = 2$, $p_2 = 3$, $p_3 = 5, \ldots$ are the successive prime numbers.

$$\int_0^1 \cdots \int_0^1 \left(\sum_{i=1}^{10} x_i\right)^{-9.5} dV = 1.044\,(-5)$$

N	$s_1(N)$	$s_2(N)$
10,000	3.338 (−6)	3.545 (−6)
20,000	3.448 (−6)	3.478 (−6)
30,000	3.482 (−6)	3.430 (−6)

$$\int_0^1 \cdots \int_0^1 \left(\sum_{i=1}^{10} x_i\right)^{-2.5} dV = 2.1203\,(-2)$$

2,000	2.1196 (−2)	2.1275 (−2)
5,000	2.1176 (−2)	2.1126 (−2)
10,000	2.1183 (−2)	2.1203 (−2)

$$\int_0^1 \cdots \int_0^1 \left(\sum_{i=1}^{10} x_i\right)^{1.5} dV = 1.132097\,(1)$$

2,000	1.13223 (1)	1.132048 (1)
5,000	1.13215 (1)	1.132137 (1)
10,000	1.13212 (1)	1.132112 (1)

$$\int_0^1 \cdots \int_0^1 \left(\sum_{i=1}^{20} x_i\right)^{1/2} dV = 3.15558$$

N	$s_1(N)$	$s_2(N)$
1,000	3.15426	3.15483
2,000	3.15495	3.15519
3,000	3.15517	3.15528

$$\int_0^1 \cdots \int_0^1 \left(\sum_{i=1}^{20} x_i\right)^{-1/2} dV = .31828$$

1,000	.31772	.31802
2,000	.31799	.31814
3,000	.31809	.31819

Note the singularities of these integrands.

5.9.5 *Haber's Method of Stratified Sampling*

This method appears to have some utility in the dimensional range $6 \le d \le 10$, approximately. Designate the hypercube $0 \le x_i \le 1$, $i = 1, 2, \ldots, d$ by C_d and $I(f) = \int_{C_d} f\,dV$. For a given integer M, divide C_d into $N = M^d$ equal subcubes of side $1/M$. Designate these subcubes by $R_1, R_2, \ldots, R_N$ and in each R_i select a point P_i at random. Write

$$Q_{1,N}(f) = N^{-1} \sum_{i=1}^{N} f(P_i). \tag{5.9.5.1}$$

$Q_{1,N}$ is a random variable with mean $I(f)$ and standard deviation $\sigma_{1,N}(f)$. If all the derivatives $\partial f/\partial x_i$, $i = 1, 2, \ldots, d$ are continuous in C_d, then Haber has shown that

$$\sigma_{1,N}(f) = (\gamma_1 + \varepsilon_N)/(N^{1/2+1/d}) \tag{5.9.5.2}$$

where $\varepsilon_N \to 0$ as $N \to \infty$ and

$$\gamma_1^2 = \int_{C_d} \left[\left(\frac{\partial f}{\partial x_1}\right)^2 + \cdots + \left(\frac{\partial f}{\partial x_d}\right)^2\right] dx_1 \cdots dx_d. \tag{5.9.5.3}$$

This represents a gain in accuracy of the factor $N^{-1/d}$ as opposed to simple Monte Carlo in C_d. Of course, this is of diminishing importance as d becomes large.

Haber also uses antithetic variates to arrive at other estimates. Let P_i^* be such that $(P_i + P_i^*)/2$ is the centroid of R_i. Define

$$Q_{2,N}(f) = (2N)^{-1} \sum_{i=1}^{N} (f(P_i) + f(P_i^*)). \tag{5.9.5.4}$$

If all the derivatives $\partial^2 f/\partial x_i \partial x_j$, $i, j = 1, 2, \ldots, d$ are continuous in C_d, then

$$\sigma_{2,N}(f) = (\gamma_2 + \varepsilon_N)/(N^{1/2+2/d}). \tag{5.9.5.5}$$

where γ_2 involves integrals of second-order partial derivatives.

For points P_i and π_i both selected at random in R_i, write

$$Q_{3,N}(f) = (2N)^{-1} \sum_{i=1}^{N} (f(P_i) + f(\pi_i)) \tag{5.9.5.6}$$

$$Q_{4,N}(f) = (4N)^{-1} \sum_{i=1}^{N} (f(P_i) + f(P_i^*) + f(\pi_i) + f(\pi_i^*)) \tag{5.9.5.7}$$

$Q_{3,N}$ and $Q_{4,N}$ are seen to be the averages of two different values $Q_{1,N}$ and $Q_{2,N}$. The quantity

$$\delta_{3,N}^2 = (4N^2)^{-1} \sum_{i=1}^{N} [f(P_i) - f(\pi_i)]^2 \tag{5.9.5.8}$$

which can be computed as an adjunct to (5.9.5.6) gives us an estimate for $\sigma_{3,N}^2$.

The quantity

$$\delta_{4,N}^2 = 1/(4N)^2 \sum_{i=1}^{N} [f(P_i) + f(P_i^*) - f(\pi_i) - f(\pi_i^*)]^2 \tag{5.9.5.9}$$

gives us an estimate for $\sigma_{4,N}^2$.

A variant of this method has been proposed by Okamoto and Takahasi. It computes a quantity similar to $Q_{2,N}(f)$ which approximates $I(f)$ while only requiring that f have continuous first partial derivatives in C_d. In addition, one can estimate the variance without doubling the number of integration points.

References. Haber [4–7]; Okamoto [1]; Okamoto and Takahasi [1].

Example. Compute $I(E) = \int_0^1 \cdots \int_0^1 (\sum_{i=1}^{10} x_i)^E \, dx_1 \cdots dx_{10}$.

		Q_3	$3\delta_3$	Q_4	$3\delta_4$
$E = -9.5$	$M = 2$	3.851 (−6)	1.618 (−6)	3.107 (−6)	8.353 (−7)
	$M = 3$	4.465 (−6)	1.446 (−6)	4.532 (−6)	9.700 (−7)
Exact value: 1.044 (−5)					

Note that the integrand is not square-integrable.

$E = -2.5$	$M = 2$	2.1289 (−2)	4.725 (−4)	2.122 (−2)	1.311 (−4)
	$M = 3$	2.1207 (−2)	4.601 (−4)	2.1201 (−2)	1.087 (−5)
Exact value: 2.1203 (−2)					

$E = 1.5$	$M = 2$	11.35693	9.9165 (−2)	11.32107	3.4054 (−3)
	$M = 3$	11.31959	8.8664 (−3)	11.32085	1.8743 (−4)
Exact value: 11.32097					

Number of functional evaluations $Q_3 : 2 \cdot M^{10}$, $Q_4 : 4 \cdot M^{10}$. These results were computed using the subroutine HABER given in Appendix 2.

5.9.6 *Sampling on the Surface of a Sphere*

For this topic the reader is referred to the works below.

References. Arnol'd and Krylov [1]; Gerl [1]; Hlawka [3]; Kac [1]; Tashiro [1]; Watson [1].

5.10 The Present State of the Art

As a general rule, it is not possible to obtain as much accuracy with higher-dimensional integrals as it is with one-dimensional integrals for reasonable computing times. One must naturally allow more computing time for higher dimensions, but we are not prepared to stay on the machine for a month with just one integral.

Given the present state of theory and hardware (1983), one can distinguish three dimensional ranges in addition to the two-dimensional case which deserves separate treatment:

Range I: integrals of dimension 3 to about 6 or 7,
Range II: dimension 7 or 8 to about 15,
Range III: dimension greater than 15.

In two dimensions, the situation is satisfactory. There is the possibility of using the different varieties of product rules and the monomial rules of quite high degree which exist for squares, triangles, circular disks, and the entire plane. Neumann has derived an efficient Kronrod-like extension for product Gauss rules over the square. de Doncker and Robinson have used an adaptive extrapolation procedure for integration over a triangle which takes into account singularities on its boundary, and to a lesser extent in its interior. They have also written an automatic program based on a product IMT rule for integrals of the form

$$\int_a^b \int_c^{g(x)} f(x, y)\, dy\, dx$$

where c and b may be infinite, $g(x)$ may exhibit singular behavior, and $f(x, y)$ may have singularities on the boundary of the region of integration. More general automatic programs such as that by Roose and de Doncker for the hypersphere and by Genz and Malik for the hypercube perform very well in two dimensions. An additional possibility is to use back-to-back one-dimensional high-quality automatic integration rules provided one takes the proper precautions as indicated by Fritsch *et al.*

Two-dimensional surface fitting is an active research area at the present and as the subject develops, integration of irregularly spaced data in the plane will become feasible. At the moment, Dierckx has a program to fit a bicubic-spline surface to data defined on a rectilinear mesh over a rectangle and to evaluate its integral.

In Range I, the dimensional effect is not yet sufficient to rule out the use of (generalized) product rules (assuming that we have a product region). Monomial rules for product or special regions are feasible and the compounding of rules is possible. In dimensions 3 and 4 we may use automatic product rules obtained by forming the product of automatic one-dimensional rules. However, it is preferable to use three- or four-dimensional rules that have been automatized. For difficult integrals, a sequence of compound rules generated in an adaptive manner has been combined with the epsilon-algorithm to produce respectable results with a reasonable amount of computation. Sampling methods or methods based upon equidistribution may also be useful. Adaptive programs based on monomial rules and on sampling have proven effective for integration over hypercubes. The former are more accurate and efficient for smooth integrands. The latter come into their own for problematic integrands but give less accuracy. All in all, assuming that the integral has no particular difficulties such as infinite ranges, singularities, or highly oscillatory integrands, moderate accuracy (5–6 figures) is possible.

Range II is the borderline range. The dimensional effect becomes more severe. Compounding becomes impossible. Monomial rules may be used, but unless the weights are essentially positive, roundoff may destroy accuracy with single precision arithmetic so that double precision arithmetic is called for when using such rules. Effective procedure have been given by Keast and Genz [8] for generating such monomial rules. In this range, sampling methods and their more expensive variants, for example, the adaptive importance sampling program by Lepage as well as equidistribution methods are definitely competitive. Low accuracy (three or four figures) is achievable.

The integrals of Range III have "really high" dimensionality. Sampling or equidistribution methods are indicated. They are time consuming, but with some care, are reliable. Sophisticated methods of variance reduction appear to exhibit a dimensional effect and are probably ruled out in this range. Some authors feel that the dimensional effect may even play a role in crude methods inasmuch as it may occur in the constant in the asymptotic error term. Experience with monomial rules in this range is difficult to interpret. Order of magnitude accuracy (one or two significant figures) is possible.

In all ranges, crude Monte Carlo exists as a last resort, especially for integrals defined over nonstandard regions and for integrands of low-order continuity. It is usually quite reliable but very expensive and not very accurate.

We conclude this section with the observation that the Mark 10 version of the NAG library contains five new programs for multiple integration, of which one supersedes one of the three multiple integration programs in Mark 9. Since mathematical software for numerical integration in one dimension, as presented in QUADPACK and NAG, is of high quality, the efforts of the numerical integration community are now directed towards producing software for multiple integration. Let us hope that these efforts will be crowned with success and that it will not be too long before we see MULTIQUADPACK.

References. de Doncker and Robinson [1, F1]; Dierckx [2]; Fritsch, Kahaner, and Lyness [1]; Genz [1, 4, 8]; Genz and Malik [1]; Keast [4, 5]; Lepage [1, 2]; NAG [1]; G. Neumann [1]; QUADPACK [F1]; Robinson and de Doncker [1]; Roose and de Doncker [1].

Chapter 6

Automatic Integration

6.1 The Goals of Automatic Integration

The aim of an automatic integration scheme is to relieve the person who has to compute an integral of any need to think. By an *automatic integrator* we mean a program for numerical integration with the following features. The user inserts (1) the limits of integration, (2) a routine for computing $f(x)$, (3) a tolerance ε, and (4) an upper bound N on the number of functional evaluations. The program then exits either

(a) with a value I of the integral, which is allegedly correct to within the tolerance, that is,

$$\left| I - \int_a^b f(x)\, dx \right| \leq \varepsilon \quad \text{or} \quad \frac{|I - \int_a^b f(x)\, dx|}{\int_a^b |f(x)|\, dx} \leq \varepsilon,$$

or (b) with a statement that the upper bound N has been achieved but not the tolerance ε, and the statement may include the "best" value of the integral determined in the process.

A more sophisticated requirement than those given in (a) was suggested by de Boor and used in most of the QUADPACK programs as well as in de Boor's own program, CADRE (Section 6.3). It requires that

$$\left| I - \int_a^b f(x)\,dx \right| \leq \max\left(\varepsilon_{\text{abs}}, \varepsilon_{\text{rel}} \left| \int_a^b f(x)\,dx \right| \right)$$

where ε_{abs} and ε_{rel} are absolute and relative tolerances, respectively.

In any case, the program may print out the "theoretical" error which has been achieved and which is used to control the exit. It should be observed that this "theoretical" error is really only a "figure of merit" for the computation.

An automatic integrator is a program intended to be used "blind," that is, without any theoretical analysis of the situation either before the problem is placed on the machine or after it comes off the machine. The results of the automatic integration are often utilized or combined in further computation.

The following two general exit criteria may serve as prototypes.

(1) A sequence of rules, $R_1(f)$, $R_2(f)$, ..., which is known to be convergent (for a wide class of functions) is applied to the integrand f. The question is then asked: Is $|R_{n+1}(f) - R_n(f)| \leq \varepsilon$? If the answer is yes, the program exits and prints out the value $R_{n+1}(f)$. If the upper bound N is achieved without a "yes" to the question, the program selects the value of n for which

$$|R_{n+1}(f) - R_n(f)| = d_n$$

is minimum, and prints out both $R_n(f)$ and the difference d_n.

(2) A sequence $\{R_n\}$ of convergent rules is prescribed. Simultaneously with R_n one computes an estimate of error E_n based principally on the functional information generated in computing R_n. The program exits when $|E_n| \leq \varepsilon$.

While it is possible to use any sequence of rules which is convergent, in practice many rules are chosen in such a way that all or almost all the information gathered at the nth stage is not discarded but is used in forming the $(n + 1)$th stage. Examples of this are the trapezoidal rules using $2^k + 1$ points, $k = 0, 1, \ldots,$ the Gauss–Kronrod–Patterson sequence (Section 2.7.1.1), the Clenshaw–Curtis rules using $2^k N + 1$ points, $k = 0, 1, \ldots,$ and the generalized composite Newton–Cotes rules, i.e., rules in which a particular Newton–Cotes integration rule (usually closed and containing an odd number of points) is used in each subinterval of a general partition of the interval of integration.

The attractiveness of an automatic integrator should be plain to anyone who has ever been confronted with the job of carrying out many integrations of complicated integrands. Furthermore, experience has shown that if the tolerance (ε) requirements are modest, such a program can do a reasonable job.

But it is important for a proper understanding of the situation to formulate a number of objections and cautionary remarks. On the theoretical side, no estimate of the accuracy of an approximate integral based on a finite amount of functional information has any validity *in the absence of theoretical information about the function such as theoretically accurate bounds on derivatives, statements about monotonicity, convexity, etc.*

To clarify this, we can say that no condition of the form

$$\left| \sum_{k=1}^{m} b_k f(y_k) \right| \leq \delta \tag{6.1.1}$$

can imply that

$$\left| \int_a^b f(x)\,dx - \sum_{k=1}^{n} w_k f(x_k) \right| \leq \varepsilon, \tag{6.1.2}$$

where the w_k and the b_k are *constants that are independent of* f. Note that this covers the automatic integrators just discussed. It is possible to find a continuous function (in fact even a polynomial function) for which (6.1.1) will hold and for which

$$R(f) = \sum_{k=1}^{n} w_k f(x_k)$$

differs from $\int_a^b f(x)$ by any preassigned quantity. To obtain such a function, merely define f to be 0 on $\{x_k\}$ and $\{y_k\}$ and adjust its values elsewhere so that $\int_a^b f(x)\,dx$ is large.

In an attempt to provide a theoretical foundation to automatic integration, Rice has introduced the concept of characteristic length. This is defined as follows: Assume that we have a generalized composite integration rule based on the integration rule Qf and that with each application $Q_i f$ of Qf on a subinterval Δ_i there exists a process which yields a local error estimate $\hat{E}_i f$ for the error

$$E_i f \equiv \int_{\Delta_i} f(x)\,dx - Q_i f.$$

Then the *characteristic length* $\lambda = \lambda(f)$ of a function f relative to the integration rule Qf is that number λ such that if the length of a subinterval Δ_i is less than λ, then $|\hat{E}_i| \leq |E_i|$. A typical result is the following: Assume that f is singular at a finite set of points $S = \{s_i \mid i = 1, 2, \ldots, R\}$

and that if $x_0 \notin S$ then $f^{(4)}(x)$ is continuous in a neighborhood of x_0. Assume further that there exist positive constants K and α such that

$$|f^{(4)}(x)| \leq K \left| \prod_{i=1}^{R} (x - s_i) \right|^{\alpha - 4}.$$

Consider now a modified version of CADRE (Section 6.3) in which the local error estimate is used only when a subinterval has length less than the characteristic length $\lambda(f)$. Then if $\lambda(f) > 0$,

$$\left| \int_a^b f(x)\,dx - Q_N f \right| \leq O(N^{-4})$$

as $N \to \infty$, where $Q_N f$ is the approximation generated by CADRE using N function evaluations, $Q_N f = \sum_{i=1}^{N} w_i f(x_i)$.

The characteristic length λ depends on f and on the process yielding $\hat{E}_i f$; λ may be zero for some functions, e.g., $x^t \sin(1/x)$, $t > 0$, even though they are quite smooth ($t > 2$ in our example). In general, λ is as intractable a quantity as the norm of f in some complex space or the bound of a high-order derivative.

As mentioned above, many automatic integrators print out the tolerance or "theoretical" error used to provide shutoff. Note that this "theoretical" error is not really the error (if it were, we could add it to the computed value to obtain a perfect result!) but only a "figure of merit." As we have seen, it is quite possible for this "theoretical" error to be many times smaller than the true error; but in many or even most cases of interest, this figure of merit is useful because it gives some indication of the amount of accuracy achieved. In conservative routines which try to achieve reliability at the cost of efficiency, the error estimate usually overestimates the true error by several orders of magnitude. Consequently, these routines may indicate failure when the actual error is well within the requested tolerance.

There are several pitfalls as far as the practical working of such programs is concerned. The tolerance ε may be chosen so small that it is found impossible to meet. More and more points are then called for and the program exits having reached the maximum allowable number of functional evaluations. In the meantime, roundoff may enter to worsen progressively the result. Or, ε may be chosen too large, and spurious agreement may lead the program to accept a value which is quite unsatisfactory. Some striking examples of the latter, with rather innocent-looking integrands, were given by Clenshaw and Curtis.

The integral

$$I = \int_{-1}^{+1} \left(\tfrac{23}{25} \cosh x - \cos x\right) dx$$

leads to the following numerical evaluation (S = Simpson):

$$1 \times S: \quad .4795546,$$
$$2 \times S: \quad .4795551.$$

The difference is

$$.0000005.$$

But we know that

$$I = .4794282,$$

so that accepting $2 \times S$ would have led to an error of

$$.0001269,$$

which is more than 200 times the difference.

The integral

$$J = \int_{-1}^{1} \frac{dx}{x^4 + x^2 + .9}$$

leads to the following evaluations (G = Gauss):

$$G_3: \quad 1.585026,$$
$$G_4: \quad 1.585060.$$

The difference is

$$.000034.$$

But we know that

$$J = 1.582233;$$

thus accepting G_4 would lead to an error of .0028, or about 100 times the difference.

In working with the Romberg–Håvie automatic integrator, we encountered this example:

$$I = \int_{0}^{1} \frac{2\,dx}{2 + \sin(31.4159x)}.$$

Specified error controlling automatic shutoff: 1.4×10^{-5}. True error of exit value: 1.5×10^{-1}.

Thus, if we specify $\varepsilon \geq 1.4 \times 10^{-5}$, a quite unsatisfactory answer will be obtained.

A fourth example has been given by Cranley and Patterson as follows: When the 7- and 15-point formulas of Patterson (see Section 6.2.0) are applied to the integral

$$\int_{-1}^{1} \exp(-6.793x^2)/(1.000001 - x^2)\, dx,$$

the results agree to six digits. The actual error is 1 in the second digit. See also Casaletto *et al.* and Kahaner.

Lyness has studied this phenomenon in some detail and has shown why it is almost inevitable. The introduction of the notion of characteristic length $\lambda(f)$ of a function f, which is designed to avoid such situations, only shifts the problem to that of determining $\lambda(f)$ for a particular algorithm.

Yet we should realize that, in principle, an automatic integrator does nothing that might not reasonably be done by a programmer forced to work in the absence of any theoretical analysis of his integral. A series of printouts, monitored either by eye or by the machine, leads him squarely and surely to the same dilemma: What faith can he put in the alleged value? There is no answer other than to build up experience and set modest goals.

Nevertheless, one should not be too pessimistic. Indeed, as far as one-dimensional integration over finite intervals is concerned, the performance of the most recent automatic integrators such as QAGS in QUADPACK is quite impressive. They are very reliable and at the same time, reasonably efficient.

Many papers have appeared in which various automatic integration programs have been compared. Almost every author who writes a new automatic integrator publishes a table comparing the performance of his program with those of the competition to demonstrate the superiority of his algorithm. Others do these computations to help decide which program should be incorporated in some subroutine library. The qualities required of an automatic integrator are efficiency, reliability, and robustness. Efficiency is usually measured by either the amount of computer time or the number of integrand evaluations required to calculate a set of integrals. Each measure has its disadvantage; the former is machine dependent while the latter does not take into account the auxiliary computations included in the program. A reliable program is one that if it exits successfully, then we are reasonably certain that the magnitude of the actual error does not exceed the requested tolerance. However, the catch is that a program may be very reliable but relatively useless inasmuch as it exits frequently with an indication of failure to achieve the desired accuracy. Thus we define a third quality robustness which means that the program will integrate correctly a broad range of integrals with an occasional failure. Robustness is usually achieved at the expense of efficiency and reliability but some of the latest automatic integrators such as those in QUADPACK have proven to be quite efficient, reliable, and robust.

For the much more difficult problem of comparing multiple integration programs, a program has been written by Genz which will check the performance of such programs on six families of functions which have the following attributes: oscillatory, product peak, corner peak, Gaussian, discontinuous derivative, and discontinuity.

The question arises: suppose that an automatic integration scheme is applied to a function $f(x)$ over an interval $[a, b]$ with a given error tolerance ε and exits with an indication of failure. What is one to do next? The following are possibilities.

(1) Increase, if possible, the number of functional evaluations allowed.

(2) Raise the error tolerance ε.

(3) Subdivide the interval $[a, b]$ into two or more subintervals, preferably in a random manner, and apply the integration scheme separately on each subinterval.

(4) Try a different automatic integration scheme, perhaps one with special features suitable for the integrand in question.

(5) Try to locate the interior singularities of the integrand and integrate between them, thus converting interior singularities to endpoint singularities which are much more tractable. Alternatively, use a program such as QAGP in QUADPACK (=D01ALF in the NAG library) into which one inputs a list of singular points and which then integrates automatically between them.

(6) Treat the integral analytically, using the techniques of Appendix 1.

A bibliography on automatic integration covering the period 1958–1975 has been published by de Doncker and Piessens.

References. Blue [1]; Casaletto, Pickett, and Rice [1]; Clenshaw and Curtis [1]; Cranley and Patterson [3]; de Boor [1, F1]; de Doncker and Piessens [2, 3]; Einarsson [4]; Gander [1]; Garribba, Quartapelle, and Reina [1]; Genz [7]; Håvie [6]; Hazlewood [1]; Hillstrom [1, 2]; Kahaner [1, 2]; Keast [3]; Lyness [20]; Lyness and Kaganove [1, 2]; NAG [1]; Ninomiya [1]; Oliver [3]; Piessens and de Doncker [1, 2]; QUADPACK [F1]; J. R. Rice [3, 6]; Robinson [3, 7, 8]; Roothart and Fiolet [1]; Stetter and Ueberhuber [1].

6.1.1 Automatic Integration as a Functional

The integral $I(f) = \int_a^b f(x)\,dx$ has numerous fundamental properties, the most important of which are listed in Section 1.5. We may think of automatic integration as a process whereby a program for an integrand f and two limits a, b is converted into a number $\mathrm{AI}(f; a, b; \varepsilon, N)$. For a fixed computing environment and for fixed ε and N, we may regard AI as a functional operating on $f; a, b$. The reader should be cautioned that depending upon the way in which the automatic integrator has been written, this functional may or may not have the simplest properties enjoyed by $I(f)$. It may or may not be *strictly* linear, homogeneous, or additive ((1.5.5)–(1.5.8)). The rule $\int_a^b f(x)\,dx = -\int_b^a f(x)\,dx$ may fail. Neglecting roundoff, a fixed rule $R(f) = \sum_{k=1}^n a_k f(x_k)$ will have these properties, but in automatic integrators, the quantities n, a_k, x_k used in forming this sum may all depend upon f, a, b, ε, N.

Lyness has pointed out another distinction between fixed rules $R(f)$ and automatic integrators $AI(f)$. If we consider the function $I(f;\lambda) = \int_a^b f(x;\lambda)\,dx$ where f varies smoothly with λ, then $I(f;\lambda)$ will be a smooth function of λ as will $R(f;\lambda)$. However, $AI(f;\lambda)$ will in general be a jagged function of λ.

Reference. Lyness [20].

6.2 *Some Automatic Integrators*

The writing of automatic integrators has become a favorite pastime of many programmers and since, say, 1960 scores of such schemes—published and unpublished—have been around. In fact, there is a feeling in some circles that the one-dimensional finite interval case has been subjected to overkill and that one should concentrate one's efforts on automatic multiple integration.

Automatic integration schemes can be classified as *adaptive* or *nonadaptive* and *iterative* or *noniterative*. In adaptive integration, the points at which the integrand is evaluated are chosen in a way that depends on the nature of the integrand. In nonadaptive integration, the sequence of integration points is chosen according to a fixed scheme independent of the nature of the integrand; the number of such points, however, will usually depend on the integrand. In an iterative scheme, successive approximations to the integral are computed until there is agreement to within the given tolerance. In a noniterative scheme, information from a first approximation is used to generate a second approximation, which is then taken as the final answer.

An automatic integrator is generally written with a certain class F of integrands in mind. The error estimates on which the shutoff is based come from a theory appropriate to this class of integrands. It is possible to splice together into one program n automatic integrators each of which is based upon a different class of functions $F_1, F_2, \ldots, F_n$. After a preliminary scan of certain functional values and based upon certain characteristic behavior within each class F_i, the program makes a decision as to which class F_j the integrand is most likely to be in and then the control is shifted to the F_j portion of the program. Such an automatic integrator will be called *class conscious.*

A first step in this direction was taken by Lyness who asked that the user make the choice among various automatic integrators based on the class of the integrand.

References. Lyness [11, 16].

A simple nonadaptive iterative scheme has been widely published. It compares an $n \times S$ with a $2n \times S$, $n = 1, 2, \ldots$, until the difference is less than a fixed ε times the absolute value of $2n \times S$, where S denotes Simpson's rule (2.2.2). We shall now give some more complicated approaches.

References. Higman [1]; McCormick and Salvadori [1, pp. 312–315]; Noble [1, p. 239]; Pennington [1, pp. 204–205].

6.2.0 *An Iterative Nonadaptive Scheme Based on Kronrod Formulas*

Patterson has generalized the Kronrod rules (see Section 2.7.1.1) and used them to construct an automatic integrator. He takes a sequence of rules R_n, $n = 3, 7, 15, 31, 63, 127, 255$, with $R_3 = G_3$. The n-point rule R_n integrates polynomials of degree $\leq \frac{1}{2}(3n + 1)$ exactly. The abscissas of R_{2n+1} include those of R_n so that no functional evaluations are lost. The weights, furthermore, are all positive. The shut off criterion is

$$|R_{2m+1}(f) - R_m(f)| \leq \varepsilon |R_{2m+1}(f)|.$$

It obviously requires quite a bit of key punching as well as storage space to insert 128 abscissas and 254 weights into the computer program.

Patterson embeds this scheme in a nonadaptive control algorithm with the following strategy. At step n the range of integration is subdivided into 2^n equal panels. The panels are scanned in a nonsequential manner and the scheme is applied to each. A scan is terminated and the delinquent panel placed at the head of the queue for examination on the next scan should both of the following conditions occur:

(a) The scheme fails to converge on a panel p, i.e., $|R_{2m+1}(f; p) - R_{2m}(f; p)| > \varepsilon |R_{2m+1}(f; p)|$ for all applicable m where $R_m(f; p)$ denotes the application of the appropriate transformation of R_m to the panel p.

(b) The estimated absolute error committed on this panel exceeds the tolerance times the estimated absolute value of the integral over the original range, i.e., $|R_{255}(f;\ p) - R_{127}(f;\ p)| > \varepsilon|R_{255}(f)|$. Note that the same value of ε is used throughout.

This leads to step $n + 1$. The process terminates when no convergence failure occurs on any subinterval, or when $n = n_{\max} = 12$.

Patterson has also embedded this scheme in an adaptive control algorithm which he claims is generally more efficient but is likely to be less reliable.

References. Cranley and Patterson [3]; Patterson [1, F1].

Patterson's work can serve as a prototype in which an automatic integration scheme is embedded in a higher-level program. This program applies the scheme to the original interval and checks for success or failure. In the case of failure, the interval is subdivided and the scheme applied to the resultant subintervals according to some strategy. A discussion of various possible strategies in adaptive integration is given by Rice.

Reference. J. R. Rice [6].

The two major strategies in adaptive integration are the local and the global. Both assume the existence of a scheme which, for any subinterval Δ, computes an approximation $Q(\Delta)$ to the integral over Δ and an estimate $E(\Delta)$ of the magnitude of the error in this approximation. The global strategy proceeds as follows: Given an initial interval Δ_0, usually specified by the endpoints a, b, and a tolerance ε, which may be either absolute or relative, $Q(\Delta_0)$ and $E(\Delta_0)$ are evaluated. If $E(\Delta_0) \le \varepsilon$ (absolute tolerance) or $E(\Delta_0) \le \varepsilon|Q(\Delta_0)|$ (relative tolerance), then we are through. Otherwise, we subdivide Δ_0 into p subintervals, $\Delta_{11}, \ldots, \Delta_{1p}$, compute $Q(\Delta_{1i})$ and $E(\Delta_{1i})$, $i = 1, \ldots, p$, and proceed to stage 1. Current practice favors subdivision by bisection. However, Hanke has produced theoretical arguments backed by some experimental evidence suggesting that trisection is preferable.

At stage n, we have a list of subintervals Δ_{ni}, $i = 1, \ldots, m(n) \equiv n(p-1) + 1$. For each subinterval Δ_{ni}, we have available $Q(\Delta_{ni})$ and $E(\Delta_{ni})$, where some of these values are carried over from the previous stage and some have been computed for the subintervals just generated. We now check if $\sum E(\Delta_{ni}) \le \varepsilon$ or $\sum E(\Delta_{ni}) \le \varepsilon|\sum Q(\Delta_{ni})|$. If so, we are through. If not, we subdivide that interval Δ_{nk} such that $E(\Delta_{nk}) = \max_{1 \le i \le m(n)} E(\Delta_{ni})$. This leads to stage $n + 1$ where $\Delta_{n+1,i} = \Delta_{ni}$ for $i = 1, \ldots, m(n)$, $i \ne k$, while $\Delta_{n+1,k}$ is one of the subintervals of $\Delta_{n,k}$ and $\Delta_{n+1,i}$, $i = m(n) + 1, \ldots, m(n) + p - 1$, are the other $p - 1$ subintervals of Δ_{nk}. If the exit criterion is not satisfied after a maximum number of stages, the routine exits with an error indication together with the latest approximation to the integral and its error estimate.

The advantages of the global strategy are that it generally uses fewer function evaluations than the local strategy and that it always returns an approximation to the integral. It also has a restart feature in that after it exits, either with success or failure, one can reenter and continue the computation to achieve more accuracy by reducing the tolerance or to achieve the initial accuracy by increasing the maximum number of stages allowed. On the other hand, the global strategy requires much computer storage to hold the information for each of the many subintervals, which may also include values of the integrand at points in the subinterval for use if the subinterval is subsequently subdivided. It also requires either a search through the list of subintervals to find Δ_{nk} or a reordering of the list at each stage. This can be done in various ways; see, for example, Malcolm and Simpson or Kahaner. A global strategy is used in most of the QUADPACK adaptive programmes as well as in the multidimensional adaptive program ADAPT by Genz and Malik.

The local strategy was the one first used in adaptive integration and is described in several textbooks on numerical analysis such as Forsythe *et al.* and Shampine and Allen. These also contain programs implementing this strategy. As in the global case, we start with Δ_0, compute $Q(\Delta_0)$ and $E(\Delta_0)$ and check for success. In case of failure, we subdivide Δ_0 into p subintervals $\Delta_{11}, \ldots, \Delta_{1p}$ and put them on a list L in some predetermined order and enter stage 1. At stage n, there exists a list L of subintervals which have not yet been treated at all plus the value of the integral and its error estimate over those subintervals for which an error criterion has been satisfied. Each such subinterval is discarded as soon as the error criterion has been satisfied for it and is no longer available to the program. In general, these discarded subintervals are adjacent to each other and start from an endpoint but this is not strictly necessary for the operation of the algorithm. If the list L is empty, we are through. Otherwise, we choose a subinterval Δ from L according to some criterion, delete Δ from L, and evaluate $Q(\Delta)$ and $E(\Delta)$. If $E(\Delta)$ is satisfactory, we add $Q(\Delta)$ and $E(\Delta)$ to the running integral and error sums, respectively. Otherwise, Δ is subdivided into p subintervals which we place on L in some order. Note that in both global and local strategies, the p subintervals of a particular interval need not be of equal length although this is usually the case. As in the global case, additional information may be stored with each subinterval. The above process is continued until either L is exhausted, the maximum number of stages is reached, or the maximum number of function evaluations has been made.

The local strategy can be implemented quite easily using a programming language allowing recursion such as ALGOL, PL/1, or PASCAL. It requires less storage space than the global strategy but generally is less efficient as far as function evaluations and computer time are concerned. It also has the disadvantage that if it fails, it does not provide an approximation to the integral over the entire interval.

Kahaner and Wyman have combined both strategies in their program GLAQ, Global Local Adaptive Quadrature, for use on computers with a limited amount of storage such as microcomputers.

References. Forsythe, Malcolm, and Moler [1, F1]; Gander [1]; Genz and Malik [1, F1]; Hanke [1]; Kahaner [4]; Kahaner and Wells [1, 2]; Kahaner and Wyman [1]; Malcolm and Simpson [1]; McKeeman [A1]; QUADPACK [F1]; J. R. Rice [3, 6]; Shampine and Allen [1, F1].

6.2.1 *An Adaptive Noniterative Scheme for Automatic Integration Based on the Midpoint Rule*

The input to this scheme is a, b, ε, n, f. The output is $I \approx \int_a^b f(x)\,dx$, with the error hopefully less than ε; n is the number of intervals initially chosen. The method used is as follows: Let $h = (b - a)/n$. Compute $f(x)$ at

$$x_i = a + [(2i - 1)/2n]h, \qquad i = 1, \ldots, n,$$

and use as first approximation to I the sum $h \sum_{i=1}^{n} f_i$, where $f_i = f(x_i)$. Then D_i is computed for $i = 1, \ldots, n$, where

$$D_1 = f_1 - 2f_2 + f_3, \qquad D_n = f_n - 2f_{n-1} + f_{n-2},$$
$$D_i = f_{i-1} - 2f_i + f_{i+1}, \qquad i = 2, \ldots, n - 1.$$

If $|D_i|\,h/24 < \varepsilon/n$, we accept the value hf_i for that panel and go on to $i + 1$; otherwise

$$\bar{h} = h\Big/\left(\frac{|D_i|\,h}{24}\frac{n}{\varepsilon}\right)^{1/2}$$

is computed and h_i is chosen such that $h_i < \bar{h}$ and h_i divides h: $h = n_i h_i$. The value for that panel is then $h_i \sum_{j=1}^{n_i} f(x_{ij})$, where

$$x_{ij} = a + (i - 1)h + \frac{2j - 1}{2n_i} h_i, \qquad j = 1, \ldots, n_i;$$

I is then taken as the sum of the values for the n panels.

The theoretical background for this method is the fact that the error term in the midpoint rule

$$\int_{x_0}^{x_0+h} f(x)\,dx \approx hf(x_0 + h/2)$$

is $h^3 f''(\xi)/24$, $x_0 < \xi < x_0 + h$. Hence, if we assume that $D_i \approx h^2 f''(\xi_i)$ and that we want the error term in each panel to be less than ε/n, we choose h_i as above. The total error will then be less than ε.

This method is a noniterative adaptive scheme. This means that the number of functional evaluations in each interval depends on the behavior of the integrand in that interval. It is not iterative because, after choosing the h_i, it does not check whether the new approximation satisfies the error criteria. The method was tested on various functions and gave the results listed in the table that follows.

Reference. Russell [1].

Examples

Function	$\int_0^1 f(x)\,dx$	ε	n	Computed value	Number of evaluations
$x^{1/2}$	.6666 6667	.001	5	.6678 3588	11
			20	.6668 9109	24
		.0000 01	5	.6666 7952	205
			20	.6666 6964	276
$x^{3/2}$	.4000 0000	.001	5	.3997 4105	15
			20	.3998 5290	20
		.0000 01	5	.3999 9823	219
			20	.3999 9848	276
$x^{-1/2}$	2.0000 0000	.001	5	1.8975 480	23
			20	1.9546 326	48
		.0000 01	5	1.9807 263	599
			20	1.9919 003	1134
$1/(1 + x^4)$	.8669 7299	.001	5	.8671 5831	15
			20	.8670 7717	20
		.0000 01	5	.8669 7346	257
			20	.8669 7378	212
$2/(2 + \sin 10\pi x)$	1.1547 005	.001	5	.9999 9999	5
			20	1.1547 004	140
		.0000 01	5	.9999 9999	5
			20	1.1546 989	3700

These results demonstrate several weaknesses of this method. In the first place there is no assurance that the final answer will be correct to within the tolerance ε. In the second place, the number of evaluations of the integrand tends to be excessive for small ε. Finally, the last example is a case where the initial value of n was too small and there was no further computation of the integrand because the error test was satisfied at all the integration points. The method is demonstrably inferior to other schemes.

Reference. Casaletto, Pickett, and Rice [1].

6.2.2 *An Example of an Adaptive Iterative Integration Based on Simpson's Rule*

A routine of this type has been given by Villars. It has an input a, b, ε, and possibly h_0; otherwise $h_0 = (b - a)/4$. It computes over the interval $[x_i, x_i + 4h_i]$ with $2 \times S$. The error corresponding to the 4th difference is compared with the pro-rata part of the total allowable error ε. If the 4th difference error is too large, the working interval h_i is halved and the calculation is repeated. If it is too small, that is, if it is less than $\frac{1}{16}$ of the pro-rata error, the interval size h_i is doubled for the next increment. Whenever a value for a given interval is accepted, the error calculated from the 4th difference is subtracted out. If h is reduced too much, the program exits with an error indication. When h is reduced, previously computed values of the integrand go to waste.

The approach of Villars is similar to that used in some algorithms for the numerical solution of initial value problems of ordinary differential equations. A more elaborate program based on these ideas has been written by Garribba *et al.*

A similar routine, which is more sophisticated, has been given by Kuncir. Its input includes a parameter N, which limits the step size to $\geq (b - a)/2^N$. In case the step size has to be reduced further, the routine exits with two numbers i and I, where i is the location on the x-axis at which the integration was terminated and I is the integral from a to i. In the case where the integration is successful, $i = b$ and I is the integral desired. In this routine there are no superfluous evaluations of the integrand. On the other hand, it does not subtract out the 4th difference error.

Kuncir's approach uses a local strategy. Lyness has given a theoretical treatment of adaptive Simpson schemes using local strategies. He also discusses various slight modifications which do not affect the basic idea, and gives a program that embodies his suggestions. Malcolm and Simpson give a global version of the adaptive Simpson scheme.

References. Gander [1]; Garribba, Quartapelle, and Reina [1, F1]; Henriksson [A1]; Kuncir [A1]; Lyness [7, 8, F1]; Malcolm and Simpson [1, F1]; Villars [1, 2].

6.2.3 *Adaptive Newton–Cotes Integration*

This scheme is a generalized version of the one by Kuncir with the following different features. The algorithms are written in recursive form. However, the special adaptive Simpson's rule integrator has been given in nonrecursive form, and a FORTRAN version of it is given in Appendix 2. This latter routine has been tested extensively and some results are given in the next table. The adaptive Simpson's rule has also been used successfully in integrating $|x|^{-1/2}$ through zero, ignoring the singularity. The general algorithm is written to work with the first seven Newton–Cotes rules, using two to eight points. The input is f, x, a, b, $n - 1$, ε. The function f may be a function of several variables; x is the variable of integration and is included to facilitate iterated integration. For a fixed n-point rule, the interval is divided into n subintervals and an n-point Newton–Cotes integration is performed over each subinterval and compared with the value over the entire interval. That is, $n \times NC_n$ is compared with NC_n. If

$$|NC_n - n \times NC_n| < \varepsilon(n \times NC_n(|f|)),$$

we accept the answer. Otherwise, we treat each subinterval as the original interval, using as comparison $\varepsilon/\sqrt{n}$ rather than ε/n since the latter proves too strict in practice. Each subinterval is subdivided as many times as necessary for agreement, with the number of subdivisions bounded by $100/n$. If there is no agreement at this stage, the last result is nevertheless accepted for that small subinterval, and the integration proceeds. Thus, there is no error exit to this procedure.

Hillstrom has carried out some numerical investigations which indicate that if the integrand has a high sharp peak or if great accuracy is required, then an adaptive high-degree rule is most efficient.

References. Hillstrom [1, 2]; McKeeman [A1, A3, 1]; McKeeman and Tesler [A1].

Examples

Function $f(x)$	Exact value of $\int_0^1 f(x)\,dx$	Romberg value for $\varepsilon = 10^{-3}, 10^{-6}$	Number of points[a]	Adaptive Simpson value for $\varepsilon = 10^{-3}, 10^{-6}$	Number of points[b]
$x^{1/2}$	.6666 6667	.6665 3263	65	.6666 5866	55
		.6666 6633	4097	.6666 6655	199
$x^{3/2}$	.4000 0000	.4000 0854	17	.4000 1016	19
		.3999 9995	129	.3999 9992	91
$\dfrac{1}{1+x}$	.6931 4718	.6931 4739	9	.6931 4743	19
		.6931 4706	33	.6931 4711	55
$\dfrac{1}{1+x^4}$	.8669 7299	.8669 7292	17	.8669 7326	19
		.8669 7300	65	.8669 7293	67

Examples *(continued)*

Function $f(x)$	Exact value of $\int_0^1 f(x)\,dx$	Romberg value for $\varepsilon = 10^{-3}, 10^{-6}$	Number of points[a]	Adaptive Simpson value for $\varepsilon = 10^{-3}, 10^{-6}$	Number of points[b]
$\frac{1}{1+e^x}$	.3798 8551	.3798 8544	9	.3798 8543	19
		.3798 8546	17	.3798 8543	19
$\frac{x}{e^x-1}$	.7775 0463	.7775 0448	9	.7775 0459	19
		.7775 0453	17	.7775 0459	19
$\frac{2}{2+\sin 10\pi x}$	1.1547 005	1.1547 003	65	1.1546 288	163
		1.1547 004	257	1.1547 002	883

[a] The minimum number of points is 9 and the number of points is always of the form $2^m + 1$.

[b] The minimum number of points is 19 and the number of points is always of the form $7 + 12k$.

6.2.4 *Other Adaptive Integration Schemes*

Rice has studied the subject of adaptive integration in depth and has come up with a meta-algorithm which consists of the following groups of components: interval processor components, bound estimator components, special behavior components, and interval collection management components. He arrives at the conclusion that there are at least one million adaptive quadrature algorithms which are potentially reasonable and interesting. We give here two such schemes, one by O'Hara and Smith and one by Cranley and Patterson. In subsequent sections we shall give several others.

References. Amble [1]; de Boor [2]; Delves [2]; Hausner [1]; Hausner and Hutchison [1]; J. R. Rice [3, 4, 6]; Robinson [2, 6].

The Adaptive Scheme of O'Hara and Smith

This scheme uses a local strategy and proceeds as follows:

Select a fixed rule of integration R, and suppose that an approximate method of error estimation has also been selected. (For example, comparing R against another rule.) Designate the rule R applied to the subinterval $p \le x \le q$ by R_{pq} and the corresponding error estimate by E_{pq}. We are interested in computing $\int_a^b f(x)\,dx$ to within a prescribed accuracy ε. Compute R_{ab}. If $|E_{ab}| \le \varepsilon$, accept R_{ab}. If not, let $c = \frac{1}{2}(a+b)$. Compute R_{ac}. If $|E_{ac}| \le k_{ac}\varepsilon$ (a policy for selecting k_{ac} will be mentioned later), accept R_{ac} as the value of $\int_a^c$. If not, set $d = \frac{1}{2}(a + c)$ and ask if $|E_{ad}| \leq k_{ad}\varepsilon$. Proceed in this manner. Ultimately, we reach a point e with $|E_{ae}| \leq k_{ae}\varepsilon$. R_{ae} is now accepted as $\int_a^e$. The integral $\int_e^b$ is now worked on and with an assigned tolerance of $\varepsilon_e = \varepsilon - |E_{ae}|$. This is begun by considering the interval $e \leq x \leq d$, and asking if $|E_{ed}| \leq k_{ed}\varepsilon_e$, etc. The suggested policy for the fractions k is to take $k_{pq} = \frac{1}{10}$ if $q \neq b$, otherwise $k_{pq} = 1$.

O'Hara and Smith further suggest that R be selected as a 9-point Romberg rule and that for an error estimate E, it be compared against $2 \times NC_5$ which uses the same abscissas. Further checks are incorporated into their program.

Reference. O'Hara and Smith [2].

The Cranley–Patterson Scheme

This scheme uses a global strategy and proceeds as follows:

Given the integration rule $Q(f; a, b)$ which has an absolute error estimate $E(f; a, b)$ associated with it and given a function $g(x)$, endpoints c, d and an absolute error tolerance ε. At step n, assume that $[c, d]$ is divided into n panels, for each of which the result of applying Q and E is known. If $\sum E \leq \varepsilon$, we terminate with $\sum Q$ as our approximation to $I = \int_c^d g(x)\,dx$. Otherwise, the panel on which E is greatest is halved and Q and E applied on each half. This takes us to step $n + 1$ unless $n = n_{\max}$ at which point we terminate with failure. This is done even though we have an estimate for I, namely $\sum Q$; unfortunately our error estimate $\sum E$ is greater than our desired tolerance.

Reference. Cranley and Patterson [3].

Lemme and Rice have investigated the use of parallel processors in adaptive integration. They show that using M processors, one can achieve a speed up of at least a constant times $M/\log M$ if one organizes the computation properly.

References. Lemme and Rice [1, 2]; J. R. Rice [5, 7, 8].

6.3 Romberg Integration

Romberg integration is essentially an application of Richardson's extrapolation procedure to the Euler–Maclaurin sum formula. Romberg's name has been attached to it because he was the first to describe the algorithm in recursive form. A nonrecursive treatment of Richardson extrapolation is given in Henrici. A complete discussion of the theory of Romberg integration is given in the paper by Bauer, Rutishauser, and Stiefel. We shall describe the main features of the classical algorithm and give some of its modifications.

Let $f(x)$ be a bounded, Riemann-integrable function on [0, 1] and set $I = \int_0^1 f(x)\,dx$. Let

$$T_0^{(k)} = h\sum_{j=0}^{2^k}{}'' f(hj), \qquad h = 2^{-k}, \tag{6.3.1}$$

be trapezoidal sums. The double prime indicates that the first and last terms are to be multiplied by $\frac{1}{2}$. Then, from the Euler–Maclaurin sum formula (2.9.15), assuming that $f(x)$ is sufficiently differentiable, we obtain

$$I - T_0^{(k)} = C_1 h^2 + C_2 h^4 + \cdots = C_1 2^{-2k} + C_2 2^{-4k} + \cdots, \tag{6.3.2}$$

where the constants C_i depend on f but are independent of h. Now let

$$T_m^{(k)} = \frac{4^m T_{m-1}^{(k+1)} - T_{m-1}^{(k)}}{4^m - 1} = T_{m-1}^{(k+1)} + \frac{T_{m-1}^{(k+1)} - T_{m-1}^{(k)}}{4^m - 1}. \tag{6.3.3}$$

This enables us to construct the following triangular array, which we call *the T-table.*

$$\begin{array}{llll} T_0^{(0)} & & & \\ T_0^{(1)} & T_1^{(0)} & & \\ T_0^{(2)} & T_1^{(1)} & T_2^{(0)} & \\ \vdots & & & \ddots \end{array} \tag{6.3.4}$$

Each application of $T_m^{(k)}$ knocks out the term with h^{2m} in the expansion of the error. The values $T_1^{(k)}$ turn out to be exactly those obtained by use of Simpson's rule. The following properties of the T-table have been proved.

(0)

$$I - T_k^{(i)} \sim (-1)^{k-1} \sum_{q=k}^{\infty} \frac{(-1)^q \gamma_{q,k-1} B_{2k} h_i^{2q}}{(2q)!} \cdot [f^{(2q-1)}(1) - f^{(2q-1)}(0)], \qquad i \geq k,$$

where

$$h_i = 2^{-i} \qquad \text{and} \qquad \gamma_{q,k} = \prod_{j=1}^{k} \frac{2^{2q} - 2^{2j}}{2^{2j} - 1}, \qquad k = 1, 2, \ldots.$$

(1) The columns and diagonals of the T-table converge to I.

(2) If $f \in C^{2m+2}$, then

$$|T_m^{(k)} - I| \leq \left| \frac{4^{-k(m+1)} B_{2m+2} f^{(2m+2)}(\xi)}{2^{m(m+1)}(2m+2)!} \right|, \qquad 0 \leq \xi \leq 1.$$

(3) If $f(z)$ is analytic in a region containing [0, 1], then any diagonal converges *superlinearly*, that is, asymptotically faster than any geometric series.

This may be seen by combining the estimate in (2) with that given in (4.6.3). We obtain an inequality of the form

$$|T_m^{(k)} - I| \le \frac{\text{const} \cdot 4^{-k(m+1)} B_{2m+2}}{2^{m(m+1)} \delta^{2m}}$$

where the const is independent of k and m.

(4) We can write

$$T_m^{(k)} = h \sum_{j=0}^{2^{m+k}}{}'' d_j^{(m)} f(jh), \qquad h = 2^{-(m+k)}.$$

The coefficients $d_j^{(m)}$ are given by the formulas

$$d_j^{(m)} = c_{m0} + 2c_{m1} + 4c_{m2} + \cdots + 2^p c_{mp}$$

where 2^p is the highest power of 2 which divides j. Moreover,

$$c_{mk} = \frac{(-1)^k c_{m-k,0}}{3 \cdot 15 \cdot 63 \cdots (4^k - 1)}$$

and

$$c_{m0} = 1 / \prod_{k=1}^{m} \left(1 - \frac{1}{4^k}\right).$$

Most programs do not work with the coefficients directly through these formulas; they work recursively. However, there has been some tendency recently to compute and discuss individual rules of the T-table. An algorithm for computing the coefficients $d_j^{(m)}$ has been published by Welsch.

The coefficients $d_j^{(m)}$ satisfy the inequalities

$$.48 < d_j^{(m)} < 1.46.$$

ALGOL procedures for classical Romberg integration are given by Bauer and by others listed in Appendix 3. A FORTRAN program QUAD by Dunkl appears in Appendix 2. This program was used to integrate the functions given in the table that follows as well as the table in Section 6.2.3.

The classical Romberg scheme proceeds by successively halving the interval. The advantage of this is that it uses all functional values computed at each stage. On the other hand, the number of values computed goes up exponentially. Hence, variations of the Romberg method have been proposed, based on the following result. Let $\{n_k\}$ be a sequence of integers such that $n_{k+1}/n_k > c > 1$. Let

$$T_0^{(k)} = h_k \sum_{j=0}^{n_k}{}'' f(jh_k), \qquad h_k = \frac{1}{n_k}.$$

Form a T-table with

$$T_m^{(k)} = \frac{h_k^2 T_{m-1}^{(k+1)} - h_{k+m}^2 T_{m-1}^{(k)}}{h_k^2 - h_{k+m}^2} = T_{m-1}^{(k+1)} + \frac{T_{m-1}^{(k+1)} - T_{m-1}^{(k)}}{(h_k^2/h_{k+m}^2) - 1}; \tag{6.3.5}$$

then the columns and diagonals converge to I. Bauer *et al.* have worked out the formulas where $\{n_k\} = \{1, 2, 3, 6, 9, 18, 27, 54, \ldots\}$. In this case, the $T_m^{(k)}$ are given by

$$T_m^{(k)} = \begin{cases} \dfrac{3^m T_{m-1}^{(k+1)} - T_{m-1}^{(k)}}{3^m - 1}, & m \text{ even}, \\[2ex] \dfrac{4 \cdot 3^m T_{m-1}^{(k+1)} - 3T_{m-1}^{(k)}}{4 \cdot 3^m - 3}, & m \text{ odd}, k \text{ even}, \\[2ex] \dfrac{3 \cdot 3^m T_{m-1}^{(k+1)} - 4T_{m-1}^{(k)}}{3 \cdot 3^m - 4}, & m \text{ odd}, k \text{ odd}. \end{cases}$$

Example. The convergence (that is, shutoff) criterion used in computing the values below was that three successive values along the main diagonal of the T-table agree within $\varepsilon = .001$; n is the number of functional evaluations needed to achieve convergence.

Function	$\int_0^1 f(x)\,dx$	Classical Romberg	n	Modified Romberg	n
$x^{1/2}$	.6666 6667	.6665 3263	65	.6667 6597	163
$x^{3/2}$	.4000 0000	.4000 0854	17	.3999 0517	55
$1/(1 + x)$	.6931 4718	.6931 4739	9	.6932 5864	37
$1/(1 + x^4)$	.8669 7299	.8669 7292	17	.8670 3917	55
$1/(1 + e^x)$	.3798 8551	.3798 8544	9	.3796 2453	7
$x/(e^x - 1)$	.7775 0463	.7775 0448	9	.7774 0864	19
$2/(2 + \sin 10\pi x)$	1.1547 005	1.1547 003	65	1.1547 015	37

Two other sequences have been suggested. One is $n_k = k + 1$, which does not satisfy the condition mentioned above but which nevertheless yields a T-table that converges, provided that $m \leq 15$. Here there may be numerical instability due to roundoff. However, Laurie has shown that this instability is mild so that in case only moderate accuracy is required, the use of this sequence is reasonable. The other sequence is $\{n_k\} = \{1, 2, 3, 4, 6, 8, 12, \ldots\}$, which has been recommended highly by Bulirsch and by Oliver as optimal in the sense of giving the best accuracy with the least amount of roundoff for a fixed amount of computation. The extensive experimental work of Fairweather and Keast has confirmed this recommendation. For an efficient way to organize the computation in this case so as to avoid the necessity for reevaluating integrands, see Oliver.

The following "practical" bracketing criterion for Romberg integration has been given by Bulirsch and Rutishauser.

Given a sequence $\{h_i\}$ such that $h_{i+1}/h_i = q < 1$, $i = 0, 1, \ldots$, define the trapezoidal sum

$$T_i = T(h_i) = h_i \sum_{j=0}^{m_i}{}'' f(a + jh_i)$$

where $m_i = (b - a)/h_i$. Define

$$T_{ik} = T_{i,k-1} - \frac{T_{i,k-1} - T_{i-1,k-1}}{(h_{i-k}^2/h_i^2) - 1}, \qquad i = 0, 1, 2, \ldots; \qquad k = 1, 2, \ldots i$$

and

$$U_{ik} = (1 + \alpha_{ik})T_{i+1,k} - \alpha_{ik} T_{ik}$$

with

$$\alpha_{ik} = 1 + \frac{2}{(h_{i-k}^2/h_{i+1}^2) - 1}.$$

For some fixed $k \geq 4$, set $J_m^i = \min\{T_{ik}, U_{ik}\}$, $J_M^i = \max\{T_{ik}, U_{ik}\}$. If starting from some I, we have that

$$J_m^I < J_m^{I+1} < \cdots < J_m^i < \cdots < J_M^i < \cdots < J_M^{I+1} < J_M^I,$$

then with "practical" certainty

$$J_m^i < \int_a^b f(x)\,dx < J_M^i.$$

The midpoint rule has an error expansion similar to that of the trapezoidal rule. If we have

$$M_0^{(k)} = h \sum_{j=1}^{2^k} f((j - \tfrac{1}{2})h), \qquad h = 2^{-k}, \tag{6.3.6}$$

then it follows that

$$I - M_0^{(k)} = -\tfrac{1}{2}C_1 h^2 - \tfrac{7}{8}C_2 h^4 - \tfrac{31}{32}C_2 h^6 - \cdots, \tag{6.3.7}$$

where the C_i are the same as in (6.3.2).

Hence we can build up a corresponding M-table.

$$\begin{array}{lll} M_0^{(0)} & & \\ M_0^{(1)} & M_1^{(0)} & \\ M_0^{(2)} & M_1^{(1)} & M_2^{(0)} \\ \vdots & & \ddots \end{array}$$

Its properties are similar to those of the T-table.

Property (0) becomes

(0′)

$$I - M_k^{(i)} \sim -(-)^{k-1} \sum_{q=k}^{\infty} \frac{(-1)^q \gamma_{q,k-1} \beta_q B_{2q} h_i^{2q}}{(2q)!} [f^{(2q-1)}(1) - f^{(2q-1)}(0)],$$

$$\beta_q = 1 - 2^{-(2q-1)}.$$

Since, in general, $T_0^{(k+1)}$ is computed as $\frac{1}{2}(T_0^{(k)} + M_0^{(k)})$, there is very little extra work involved in generating the M-table. Furthermore, we have

$$T_m^{(k)} = \tfrac{1}{2}(T_m^{(k-1)} + M_m^{(k-1)}) \tag{6.3.8}$$

and, in addition,

$$T_m^{(k)} = M_{m-1}^{(k)} + \frac{(2 \cdot 4^{m-1} - 1)(T_{m-1}^{(k)} - M_{m-1}^{(k)})}{4^m - 1}. \tag{6.3.9}$$

These relations enable us to generate the values of $T_m^{(k)}$ from $T_0^{(0)}$ and the values $M_0^{(0)}, M_1^{(0)}, \ldots$ without forming the $T_k^{(0)}$ explicitly; this saves a roundoff, which can be troublesome in Romberg integration.

In the case where the T-table is computed, the criterion of convergence is agreement to within ε of three successive values along the main diagonal. In the extended case, the criterion is agreement to within ε of $T_m^{(0)}$ and $M_m^{(0)}$. This usually occurs sooner than in the previous case and, when it does, the final value of the integral is taken as $\frac{1}{2}(T_m^{(0)} + M_m^{(0)}) = T_m^{(1)}$.

For many integrands, "bracketing" of the integral by the T-table and the M-table may be anticipated. From inspecting the first term of the series in properties (0) and (0′), we can "expect" that $I - T_k^{(i)}$ and $I - M_k^{(i)}$ have opposite signs. Moreover, one expects that $T_n^{(i)}$ is a better approximation to I than $T_n^{(i-1)}$ since the step size of the former is half that of the latter. From (6.3.8) therefore, we can expect that I lies in the interval spanned by $T_k^{(i)}$ and $M_k^{(i)}$: $T_k^{(i)} < I < M_k^{(i)}$ or $M_k^{(i)} < I < T_k^{(i)}$. Write $\varepsilon = \frac{1}{2} | T_k^{(i)} - M_k^{(i)} |$. Then, from (6.3.8), we expect the bracketing

$$T_k^{(i+1)} - \varepsilon < I < T_k^{(i+1)} + \varepsilon.$$

A considerable literature exists giving precise theoretical conditions (frequently asymptotic) under which bracketing occurs. See Håvie, Ström,

Bulirsch and Stoer, Albrecht, and Förster. The program RMINSP in Appendix 2 incorporates some of these ideas.

Although the Romberg scheme converges whenever the trapezoidal rule does, nevertheless, for functions of the form

$$f(x) = x^{\beta}\phi(x), \qquad \phi(x)\text{ analytic}, \qquad 0 < \beta < 1,$$

convergence will be rather slow (see $x^{1/2}$ in the table in Section 6.2.3). Bulirsch gives two devices by which convergence can be speeded up. First, the $T_0^{(k)}$ column is modified to yield

$$\hat{T}_0^{(k)} = T_0^{(k)} + G(\beta)f\left(\frac{h_k G(1+\beta)}{G(\beta)}\right)h_k^{1+\beta} \tag{6.3.10}$$

with

$$G(\beta) = \sin\frac{\pi}{2}\frac{\beta\Gamma(1+\beta)\zeta(1+\beta)}{\pi^{1+\beta}2^{\beta}} = -\zeta(-\beta), \tag{6.3.11}$$

where ζ is the Riemann zeta function (see Section 3.4). Then the $T_m^{(k)}$ are computed by

$$T_m^{(k)} = \frac{h_k^{1+\beta}T_{m-1}^{(k+1)} - h_{k+m}^{1+\beta}T_{m-1}^{(k)}}{h_k^{1+\beta} - h_{k+m}^{1+\beta}}. \tag{6.3.12}$$

In a similar fashion, Rutishauser gives the following scheme for

$$f(x) = x^{-1/2}\phi(x),$$

$\phi(x)$ analytic. Let

$$T_2^{(k)} = h\left\{\alpha_1 f(\alpha_2 h) + \sum_{j=1}^{2^k-1} f(jh) + \frac{1}{2}f(1)\right\}, \qquad h = 2^{-k}, \tag{6.3.13}$$

where

$$\alpha_1 = (\zeta(-\tfrac{1}{2})\zeta(\tfrac{1}{2}))^{1/2} = .5509\ 8782\ 8,$$

$$\alpha_2 = \frac{\zeta(-\frac{1}{2})}{\zeta(\frac{1}{2})} = .1423\ 5326\ 0. \tag{6.3.14}$$

Then

$$T_{3m}^{(k)} = \frac{4^m T_{3m-1}^{(k+1)} - T_{3m-1}^{(k)}}{4^m - 1},$$

$$T_{3m+1}^{(k)} = \frac{4^{m+1}T_{3m}^{(k+1)} - \sqrt{8}\,T_{3m}^{(k)}}{4^{m+1} - \sqrt{8}}, \qquad m = 1, 2, 3, \ldots,$$

$$T_{3m+2}^{(k)} = \frac{4^{m+1}T_{3m+1}^{(k+1)} - \sqrt{2}\,T_{3m+1}^{(k)}}{4^{m+1} - \sqrt{2}} \tag{6.3.15}$$

yield the desired T-table.

Further examples of the application of Romberg integration to singular integrals are given by Fox, Fox and Hayes, Shanks, and Cohen. In the case of nonpolynomial behavior of the integrand, Håvie applies a modified Romberg integration to the transformed integral

$$I = \int_{-1}^{1} f(x)\,dx = \int_{0}^{\pi} f(\cos\phi)\sin\phi\,d\phi$$

with considerable success.

Romberg integration, an iterative nonadaptive scheme for automatic integration, has proved successful in practice and, as can be seen from the table in Section 6.2.3, is competitive with the adaptive Simpson's rule. It is automatic in that the number of functional evaluations depends on the behavior of the integrand over the entire interval of integration. It is nonadaptive in that it evaluates the integrand at a fixed set of points, for in adaptive schemes, the set of points at which the integrand is evaluated depends on the integrand itself. In general, this leads to greater efficiency for the adaptive schemes in terms of functional evaluations. However, the fixed set of points used in Romberg integration is advantageous when we wish to integrate a function $f(\alpha, x)$ depending on a parameter α for various values of the parameter. In many cases, $f(\alpha, x)$ can be written as $g(x)h(\alpha, x)$ or it contains subexpressions depending only on x. In these cases, the parts depending only on x can be computed as they are needed, stored in a proper sequence, and used again and again.

The classical Romberg integration scheme based on (6.3.3) is very economical as far as computer memory is concerned in that the program needed to implement it is quite simple. Furthermore, one needs to retain at each stage only the last row of the T-table plus a few additional quantities. Hence, Romberg integration is ideally suited for situations in which memory is scarce. This is the case with programmable hand calculators. In fact, one such calculator has a built-in integration program which is activated by an integral function key. The program designed by Kahan applies Romberg integration to the transformed integral

$$\int_{a}^{b} f(x)\,dx = \int_{-1}^{1} g(y)\,dy$$

where $g(y) = \frac{3}{4}(b-a)f(\frac{1}{4}(b-a)y(3-y^2) + \frac{1}{2}(b+a))(1-y^2)$. This transformation avoids the even spacing of points in the trapezoidal rule which can lead to early incorrect termination of the process for certain trigonometric functions. It also alleviates singularities at the endpoints of the interval. A novel feature of Kahan's program is his termination criterion. The user does not specify an error tolerance ε; instead, he indicates the accuracy of the integrand evaluation program. Then, if $f(x)$ is evaluated with an uncertainty $\Delta f(x)$ which may be a bound on the magnitude of either the absolute or relative error, the program terminates when three successive diagonal elements of the T-table agree to within an approximate value of $\int_a^b \Delta f(x)\,dx$.

Prager has given an adaptive Romberg scheme which works as follows. The input consists of a, b, h, and n. A T-table is set up for the interval $[a, a + h]$ with k not greater than 4. If $T_k^{(0)}$ agrees with $T_{k-1}^{(0)}$ to n figures, the build-up of the T-table ceases and $T_k^{(0)}$ is accepted as the integral over the interval. If agreement occurs with $k = 1$, h is increased to $1.5h$. If it occurs with $k = 4$, h is decreased to $.6h$; otherwise h remains the same. In all cases, integration then proceeds over the next interval. If $T_4^{(0)}$ does not agree with $T_3^{(0)}$ to n figures, h is decreased to $.6h$, and integration over the shortened interval is attempted. There is an error exit when the number of intervals over which integration is performed exceeds $8(b - a)/h$. Examples of the use of this scheme are given in Sections 2.12.1 and 3.2.

Further adaptive schemes have been proposed by E. K. Miller and others. A cautious adaptive Romberg extrapolation scheme has been implemented by de Boor in the FORTRAN program CADRE given in Appendix 2. This program may also be found in the IMSL library.

The essence of "cautious extrapolation," first put forward by Lynch, seems to come down to this: Assume that we have a generalized Euler–Maclaurin expansion of the form

$$I \equiv \int_a^b f(x)\,dx = T(h) + \sum_{j=1}^{n} a_j h^{\gamma_j} + O(h^{\gamma_n}), \qquad \gamma_1 < \gamma_2 < \cdots < \gamma_n .$$

Then, if we knew the values γ_j we could eliminate successive terms in the sum by extrapolation. However, it may occur that the coefficient a_j of a later term is very much larger than that of an early term and the error in the extrapolated value may be large. Of course, as $h \to 0$, the first term of the sum dominates, but for what value of h? To determine this, while simultaneously finding the value of γ_1, we compute the ratio

$$R(h) = (T(2h) - T(4h))/(T(h) - T(2h)).$$

It is easy to show that as $h \to 0$, $R(h) \to 2^{\gamma_1}$. Hence, if $R(h) \approx c$, h small, we may be reasonably certain that $\gamma_1 \approx \log_2 c$ and that $a_1 h^{\gamma_1}$ is the dominant term of the error. We then eliminate this error, generating a new column of approximations. We continue this process on the new column until the error is less than a given tolerance.

For an integrand that is sufficiently differentiable in the interval, we have $\gamma_1 = 2$, so that $\lim_{h \to 0} R(h) = 4$. The ratio $R(h)$ is also used as an indicator that the integrand may have an algebraic endpoint singularity (f has the form $(x-a)^\alpha g$ or $(b-x)^\alpha g$ with $-1<\alpha<1$, $\alpha \neq 0$, and g sufficiently differentiable). In this case, it can be shown that $\lim_{h\to 0} R(h) = 2^{1+\alpha}$. In the presence of a jump discontinuity in the integrand, $\lim_{h\to 0} |R(h)| = 2$.

CADRE is written to take distinct action appropriate to these possibilities. It exits with one of five possible flags. Upon seeing the flag and after having decided whether he is "cautious," "reasonable" or "adventurous," the user can accept or reject the result accordingly.

CADRE is a program that exhibits rudimentary signs of class consciousness, i.e., the computer makes an educated guess as to what class of functions the integrand is in and then calls in an appropriate theory. Automation is at a higher level than formerly and the necessity to think mathematically about the integrand has been replaced by psychoanalytic introspection about the user's personality.

For additional discussions of caution in the context of Romberg integration, see Blue, Cohen, and Fairweather and Keast. An alternative strategy for caution in the context of adaptive integration is given by Hazlewood. Here, "caution" means that one makes an independent check of the assumptions about the properties of the integrand made in the course of the integration procedure.

A different method for discovering and treating singularities has been proposed by Ninomiya in his adaptive Newton–Cotes integration algorithm.

References. Albrecht [2]; Bauer [1, A1]; Bauer, Rutishauser, and Stiefel [1]; Brass [1]; Bulirsch [1]; Bulirsch and Ruthishauser [1]; Bulirsch and Stoer [1]; Bunton, Diethelm, and Haigler [1]; Bunton, Diethelm, and Winje [1]; Cohen [1]; de Boor [1, 2, F1]; Dunkl [1]; Fairweather [F1]; Fairweather and Keast [1, 2]; Filippi [2]; Förster [1]; Fox [2]; Fox and Hayes [1]; Gram [A1]; Håvie [1-3, 6, 9, 11]; Hazlewood [1, A1]; Henrici [2]; IMSL [1]; Joyce [1]; Kahan [2]; Krasun and Prager [1]; Kubik [A1]; Laurie [1]; Lynch [1]; E. K. Miller [1]; J. C. P. Miller [3]; NAG [1]; H. Neumann [1]; Ninomiya [1]; Oliver [1]; Prager [1]; Rabinowitz [2]; Robinson [5]; Romberg [1]; Rutishauser [2]; J. A. Shanks [1]; Stiefel [1, 2]; Stiefel and Rutishauser [1]; Ström [1]; Stroud [7]; Thacher [4]; Wallick [F1]; Welsch [A2].

6.3.1 *Further Variants of Romberg*

Bulirsch and Stoer compute trapezoidal sums for several values of h, fit a rational function of h^2 to this data, and then extrapolate to $h=0$. Here are the details.

Let $n_0, n_1, \ldots$ be a sequence of integers satisfying $n_i/n_{i+1} \le \beta < 1$, $i = 0, 1, \ldots$ for some constant β. The sequence $\{1, 2, 3, 4, 6, 8, 12, \ldots\}$ is particularly recommended. Designate by $T(h)$ the trapezoidal sum

$$T(h) = T_n(h) = h[\tfrac{1}{2}f(a) + f(a+h) + \cdots + f(b-h) + \tfrac{1}{2}f(b)]$$

where $h = (b - a)/n$. Designate by $\hat{T}_k^i(h)$ the rational function of h^2

$$\hat{T}_k^i(h) = \frac{c_0 + c_1 h^2 + \cdots + c_\mu h^{2\mu}}{d_0 + d_1 h^2 + \cdots + d_\nu h^{2\nu}}$$

$$\mu = \left[\frac{k}{2}\right], \qquad \nu = k - \mu$$

which interpolates to the trapezoidal sums as follows:

$$\hat{T}_k^i(h_j) = T(h_j), \qquad j = i, i + 1, \ldots, i + k$$

where $h_j = (b - a)/n_j$. Now extrapolate to 0 by writing

$$T_k^i = \hat{T}_k^i(0).$$

Then, the elements in the triangle

$$\begin{array}{lllll}
T(h_0) = T_0^0 & & & & \\
 & T_1^0 & & & \\
T(h_1) = T_0^1 & & T_2^0 & & \\
 & T_1^1 & \cdot & \cdots & \\
T(h_2) = T_0^2 & \cdot & \cdot & & T_m^0 \\
\cdot \quad \cdot & \cdot & \cdot & \cdots & \\
\cdot \quad \cdot & \cdot & T_2^{m-2} & & \\
\cdot \quad \cdot & \cdot & & & \\
\cdot \quad \cdot & T_1^{m-1} & & & \\
T(h_m) = T_0^m & & & &
\end{array}$$

can be calculated from the first column on the left by means of the formulas

$$\Delta_0^m = T(h_m), \qquad \nabla_0^m = T(h_m)$$

and for $k = 1, \ldots, m$,

$$\Delta_k^m = \frac{\nabla_{k-1}^m \, \delta_{k-1}^m}{(h_{m-k}/h_m)^2 \, \Delta_{k-1}^{m-1} - \nabla_{k-1}^m},$$

$$\nabla_k^m = \frac{(h_{m-k}/h_m)^2 \, \Delta_{k-1}^{m-1} \, \delta_{k-1}^m}{(h_{m-k}/h_m)^2 \, \Delta_{k-1}^{m-1} - \nabla_{k-1}^m}, \tag{6.3.1.1}$$

$$T_k^{m-k} = \sum_{l=0}^{k} \Delta_l^m$$

where we have written

$$\delta_{k-1}^m = \nabla_{k-1}^m - \Delta_{k-1}^{m-1}.$$

If the denominator of (6.3.1.1) vanishes, Δ_k^m and ∇_k^m are set to zero. Stoer and Bulirsch give the following more compact formula for generating the triangle (6.3.1.1):

$$T_k^j = T_{k-1}^j + \frac{T_{k-1}^{j+1} - T_{k-1}^j}{\left[\dfrac{h_j}{h_{j+k}}\right]^2 \left[1 - \dfrac{T_{k-1}^{j+1} - T_{k-1}^j}{T_{k-1}^{j+1} - T_{k-2}^{j+1}}\right] - 1}, \qquad \begin{array}{l} j = 0, 1, \ldots, m-k, \\ k = 1, 2, \ldots, m \end{array}$$

where $T_{-1}^j = 0$, $T_0^j = T(h_j)$, $j = 0, 1, \ldots, m$.

If we set

$$U_k^i = 2T_k^{i+1} - T_k^i,$$

then for sufficiently small h and for k fixed and $i \to \infty$, T_k^i and U_k^i converge to $\int_a^b f(x)\,dx$ monotonically and from opposite sides.

Difficulties may be anticipated if f is not 3 or 4 times differentiable, or if f has period $b - a$. Furthermore, its use is not recommended if f is too simple, if $b - a$ is too small, or if not much accuracy is needed, as the method tends to be time consuming.

A FORTRAN program IRATEX implementing this method is given in Appendix 2.

For the use of the epsilon algorithm in the context of Romberg integration, see Kahaner, Genz, and Chisholm *et al.*

References. Bulirsch and Stoer [2]; Chisholm, Genz, and Rowlands [1]; Genz [2]; Kahaner [3, F3]; Stoer and Bulirsch [1, p. 135].

A sophisticated algorithm has been proposed by de Doncker combining extrapolation with adaptive integration, a possibility suggested previously by Genz. It is based on the following ideas. Assume that we have a basic integration rule

$$If \equiv \int_0^1 f(x)\,dx \simeq Qf = \sum_{i=1}^m w_i f(x_i), \qquad \sum_{i=1}^m w_i = 1, \quad 0 \le x_i \le 1,$$

and that the integrand g can be expressed in the form

$$g(x) = u(x) \sum_{i=1}^s \sigma_i(x)$$

where $\sigma_i(x) = \operatorname{sgn}^{\gamma_i}|x - t_i|\ |x - t_i|^{\delta_i} \log^{\eta_i}|x - t_i|$, $i = 1, \ldots, s$, with $t_i \in [0, 1]$, $\gamma_i = 0$ or 1, $\delta_i > -1$, η_i nonnegative integers, and $u(x) \in C^r[0, 1]$. Then, we have the following asymptotic expansion for the error in the composite rule $S_j g \equiv (2^j \times Q)g$:

$$Eg = Ig - S_j g = \sum_{k=0}^{r-1} \sum_{i=1}^{s} 2^{-j(\delta_i + k + 1)} \sum_{l=0}^{L_{ik}} \alpha_{ikl} j^l + O(2^{-jr}) \qquad (6.3.1.2)$$

where the L_{ik} depend on the t_i and η_i and the α_{ikl} are independent of j (see Lyness and Ninham). This expansion is valid provided the t_i are rational numbers and this is the case in digital computation. Now Genz has shown that one can apply the epsilon algorithm (Section 1.15) to a sequence of approximations $\{S_j g\}$ to Ig to eliminate successive terms in the error expansion (6.3.1.2). However, when g has singularities, convergence may be quite slow and since the number of points grows exponentially with j, we may have a very long computation. Note that the sequence $\{S_j g\}$ corresponds to the classical Romberg sequence, so that one could try to slow down the rapid growth of integration points by using a different sequence of composite rules. However, this will not work since the asymptotic expansion of the error will not be amenable to extrapolation by the epsilon algorithm. The specific form $S_j g = (2^j \times Q)g$ is crucial.

To counteract the rapid growth of integration points, an adaptive approach is taken and a modified sequence $\hat{S}_j g$ is formed as follows: On each subinterval $I_{jk} = [k2^{-j}, (k+1)2^{-j}]$ under consideration, the error in the approximation to $\int_{I_{jk}} g$ by $Q_{jk} g$ is monitored, where $Q_{jk} g$ denotes the rule Q transformed to the interval I_{jk}. If the error is sufficiently small, then in the next stage, the interval I_{jk} is not subdivided. This means that $S_{j+1} g$ is not computed and is replaced by a very good approximation $\hat{S}_{j+1} g$. Thus, we are "deceiving" the epsilon algorithm since strictly speaking, it works only for the sequence $\{S_j g\}$. However, since the sequence $\{\hat{S}_j g\}$ is numerically close to $\{S_j g\}$, it is expected that the epsilon algorithm will also converge for the sequence $\{\hat{S}_j g\}$ with a substantial saving in function evaluations. The computer implementation of this idea is quite complicated but the resulting program is very efficient and reliable as shown by Robinson. This program, QAGS, appears in QUADPACK and is also available in the NAG library (D01AJF). The idea carries over to multiple integration and forms the basis for the program for automatic integration over a triangle by de Doncker and Robinson.

References. de Doncker [1, 3]; de Doncker and Robinson [1, F1]; Genz [1, 4]; Kahaner and Stoer [1]; Lyness and Ninham [1]; NAG [1]; QUADPACK [F1]; Robinson [8].

6.4 Automatic Integration Using Tschebyscheff Polynomials

Clenshaw and Curtis suggest approximating the integrand $f(x)$ by a series of Tschebyscheff polynomials $\sum_{i=0}^{N} a_i T_i(x)$ and then integrating this expansion to get a second Tschebyscheff expansion $\sum_{i=0}^{N+1} b_i T_i(x)$ for the indefinite integral $F(x) = \int_{-1}^{x} f(t)\,dt$ as well as a value for the definite integral $F(1) = 2(b_1 + b_3 + b_5 + \cdots)$. If the last three values of b_i in the expansion of either $F(x)$ or $F(1)$ are small, the sum is accepted. Otherwise, N is replaced by $2N$

and the computations are repeated, using previously computed results, until the error criterion is satisfied. Theoretically, we do not have to double the number of terms each time but can use any ascending sequence of integers for successive approximation. However, doubling has computational advantages. Since we shall discuss a modification of this method which is similar to it but more accurate, we refer the reader to the original paper for further details about the original method (see Sections 2.5.5 and 2.13.1).

Quite a large literature has arisen around the Clenshaw–Curtis method. O'Hara and Smith give various criteria enabling one to determine whether enough terms have been taken in the Tschebyscheff expansion. Based on their work, Oliver has prepared an adaptive algorithm in which decisions are made at each stage as to whether to double the number of terms in the expansion or halve the particular interval under consideration and start anew in each subinterval. Gentleman has also proposed a criterion for determining whether enough terms have been taken in the expansion. He has incorporated it in a program whose main innovation is the use of the FFT (Section 3.9.5) to compute the sums $\sum a_i T_i(x)$ needed.

Piessens and his co-workers have applied a modified Clenshaw–Curtis algorithm to evaluate integrals of the form

$$I(f, k) = \int_a^b k(x) f(x)\, dx$$

where $k(x)$ may be one of the singular functions $(x - a)^\alpha$, $(x - a)^\alpha \log(x - a)$, $(b - x)^\alpha$, $(b - x)^\alpha \log(b - x)$, $\alpha > -1$, or one of the trigonometric functions $\sin \omega x$, $\cos \omega x$, $\omega > 0$. They do this by first approximating $f(x)$ by a Tschebyscheff series

$$f(x) \simeq \sum_{i=0}^{n} c_i T_i(x).$$

Then $I(f, k) \simeq \sum_{i=0}^{n} c_i I(T_i, k)$ where the $I(T_i, k)$ are the modified moments of k with respect to the Tschebyscheff polynomials and can be computed by recurrence formulas. A similar approach is used for the Cauchy principal value integral

$$P \int_a^b \frac{f(x)}{x - c}\, dx, \qquad a < c < b.$$

These ideas have been incorporated in some of the QUADPACK programs. These programs can also be found in the NAG library.

Håvie has modified the Clenshaw–Curtis method to generate two sequences of approximations which hopefully bracket the integral. He recommends that the integration be carried out by using prestored abscissas and weights and evaluating the sum $\sum w_i f(x_i)$ rather than by computing the coefficients in the Tschebyscheff expansions. However, in the interest of saving storage at the expense of efficiency, we give in Appendix 2 the program CHINSP by Håvie using the usual approach.

A modification to the Clenshaw–Curtis idea which is more accurate for the indefinite integral has been proposed by Filippi. His approach is to approximate not $f(x)$ but $F(x)$ as a Tschebyscheff series. Let

$$F(x) = \frac{A_0}{2} + A_1 T_1(x) + \cdots, \tag{6.4.1}$$

where

$$A_n = \frac{2}{\pi}\int_{-1}^{1} F(x)T_n(x)(1 - x^2)^{-1/2}\,dx, \qquad n = 0, 1, 2, \ldots. \tag{6.4.2}$$

Now

$$T_n(x)(1 - x^2)^{-1/2} = -\frac{1}{n^2}\frac{d}{dx}((1 - x^2)^{1/2}T'_n(x)) \qquad (n > 1), \tag{6.4.3}$$

so that making use of $F'(x) = f(x)$, we obtain

$$A_n = \frac{2}{\pi n^2}\int_{-1}^{1} f(x)T'_n(x)(1 - x^2)^{1/2}\,dx \qquad (n \geq 1). \tag{6.4.4}$$

The polynomials

$$T'_n(x) = n\frac{\sin(n \arccos x)}{\sin(\arccos x)}$$

form an orthogonal set with respect to the weight $(1 - x^2)^{1/2}$. Hence the coefficients A_n are the Fourier coefficients in the expansion

$$f(x) = A_1 + A_2 T'_2(x) + \cdots + A_n T'_n(x) + \cdots. \tag{6.4.5}$$

The partial sums

$$q_{N-1}(x;f) \equiv q_{N-1}(x) = A_1 + A_2 T'_2(x) + \cdots + A_N T'_N(x)$$

are not necessarily the best approximations to $f(x)$. They are characterized instead by the condition

$$\int_{-1}^{1} [f(x) - q_{N-1}(x)]^2(1 - x^2)^{1/2}\,dx = \min \tag{6.4.6}$$

for any Nth-degree polynomial f. The Lagrange interpolation polynomial $\bar{q}_{N-1}(x)$ to $f(x)$ with interpolation points $t_r = r\pi/(N + 1)$, the zeros of T'_{N+1}, is an optimal approximation to $q_{N-1}(x)$. Hence, if we write

$$\bar{q}_{N-1}(x) = a_1 + a_2 T'_2(x) + \cdots + a_N T'_N(x), \tag{6.4.7}$$

then we obtain

$$F(x) \sim \bar{Q}_N(x) = \frac{a_0}{2} + a_1 T_1(x) + \cdots + a_N T_N(x), \qquad (6.4.8)$$

where

$$a_n = \frac{2}{n(N+1)} \sum_{r=1}^{N} f(\cos t_r) \sin t_r \sin nt_r, \qquad n = 1, 2, \ldots, N \quad (6.4.9)$$

and

$$\frac{a_0}{2} = a_1 - a_2 + a_3 - \cdots + (-1)^{N+1} a_N . \qquad (6.4.10)$$

The value of the definite integral is given by

$$\int_{-1}^{+1} f(x)\, dx = 2(a_1 + a_3 + a_5 \cdots). \qquad (6.4.11)$$

The computation of a_n proceeds as follows: Let $z_{N+1} = z_{N+2} = 0$. Compute $z_N, z_{N-1}, \ldots, z_1$ by the recurrence relation

$$z_r = 2z_{r+1} \cos t_n - z_{r+2} + f(\cos t_r) \sin t_r .$$

Then $a_n = 2z_1 \sin t_n / n(N+1)$.

For automatic integration we start with $N = N_1$ and check the last three coefficients a_N, a_{N-1}, a_{N-2}. If they are less than a given ε, we have finished. Otherwise, we take $N = 2N_1 + 1$, which enables us to use the previously computed values of $f(\cos t_r) \sin t_r$ and proceed as before, stopping when we reach $N_{\max}$.

References. Basu [1]; Branders and Piessens [1, F1]; Brass [1]; Brass and Schmeisser [1, 2]; Chawla [4]; Clenshaw and Curtis [1]; Elliott [3]; Engels [13]; Filippi [1, 4]; Fraser and Wilson [1]; Gentleman [2, F1]; Haussman and Zeller [1]; Håvie [4, 6, 8]; Hopgood and Litherland [A1]; Imhof [2]; Locher [2, 4]; NAG [1]; O'Hara [1]; O'Hara and Smith [1, 2]; Oliver [2, 3]; Piessens and Branders [4, 7, F2]; Piessens, de Doncker, Überhuber, and Kahaner [1]; Piessens, Mertens, and Branders [1, F1]; Piessens, van Roy-Branders, and Mertens [1, F1]; QUADPACK [F1]; Riess and Johnson [2]; F. J. Smith [1]; Wagner [1]; Wright [1].

6.4.5 *Automatic Integration of Oscillatory Functions*

Chase and Fosdick have written an automatic routine for the integration of finite Fourier transforms based on Filon's method (Section 2.10.2). Linz has written an algorithm for the adaptive calculation of the Fourier cosine integral.

Piessens and his co-workers have written several automatic programs for finite Fourier transforms using various approaches. One of these programs has subsequently been incorporated by Piessens into a program for the automatic computation of Bessel function integrals.

The following integrals

$$\int_0^{2\pi} f(x) \sin nx \, dx$$

were computed using the Chase and Fosdick routine and we list here the number of function evaluations needed when the tolerance requested was 10^{-6}. The maximum number of functional evaluations allowed was 4096. For comparison, the number of function evaluations needed when CADRE (Section 6.3) was used is given in parenthesis. Wherever the accuracy specified was not reached, the number is in italic.

n	$x \cos x$	$x \cos 50x$	$(1 - x^2/4\pi^2)^{1/2}$	$\log x$
1	256(129)	2048(2041)	4096(97)	2048(197)
2	256(153)	2048(2081)	*4096*(145)	4096(145)
4	128(241)	4096(2129)	*4096*(225)	4096(289)
10	128(521)	4096(2561)	*4096*(497)	4096(497)
20	128(937)	4096(3081)	*4096*(961)	4096(*473*)
30	256(1201)	4096(3569)	*4096*(1361)	*4096*(1193)

We see that CADRE is superior for functions with singularities and is quite competitive otherwise except for smooth functions and large n. Note one case where CADRE gave an incorrect answer.

References. Chase and Fosdick [F1]; de Boor [F1]; Einarsson [F1]; Haegemans and Piessens [A1]; Hopp [1]; Linz [F1]; Piessens [14, 15]; Piessens and Branders [4, 7]; Piessens, de Doncker, Überhuber, and Kahaner [1]; Piessens and Haegemans [2]; Teijelo [A1].

6.5 Automatic Integration in Several Variables

In a certain sense, multiple integration by sampling is an automatic integration procedure. The input consists of the integrand, the region, a tolerance ε, and possibly an upper bound N on the number of functional evaluations permitted. The output consists of the value of the integral, hopefully correct to within ε, unless convergence was not achieved within N evaluations. In such a case, the routine will indicate this. Convergence is taken to occur when many successive approximations to the integral agree to within ε. Alternatively, we may stop when the estimated variance is small. A method based on a combination of number theoretic and probabilistic reasoning, suggested by Cranley and Patterson, has been implemented in the automatic program D01GCF on the NAG library.

Other attempts at automatic integration in several variables include a generalization of Romberg integration to two dimensions, proposed by Laurent, and the previously mentioned work by Lyness and McHugh on Richardson extrapolation.

The program NDIMRI which impliments a generalization of modified Romberg integration to n dimensions, $n \leq 9$, is given in Appendix 2.

Robinson and de Doncker have written a versatile automatic nonadaptive program for two-dimensional integrals of the form

$$\int_a^b dx \int_c^{g(x)} f(x, y)\, dy$$

where c and either a or b may be infinite. This program is based on a product IMT rule (Section 2.9.2) and hence can deal with singular behavior of $g(x)$ and with boundary singularities of $f(x, y)$.

A program based on the method of Sag and Szekeres (Section 5.8.5) appears in the NAG library. Another automatic program based on a modification of this method has been written by Roose and de Doncker for integration over hyperspheres of functions which may have singularities on the surface or at the center of the hypersphere. These programs are nonadaptive.

The following adaptive integration programs for integration over triangles have appeared recently: TRIADA by Haegemans, CUBTRI by Laurie, TRIEX by de Doncker and Robinson, and ADPCUB by Barnhill and Little.

An adaptive multidimensional integration scheme based on the epsilon-algorithm has been successfully applied by Genz to functions with singularities at the boundary.

Genz and Malik, following in the footsteps of van Dooren and de Ridder, have written the adaptive integration program ADAPT, for the d-dimensional hyperrectangle, $2 \leq d \leq 15$. ADAPT, which also appears in the NAG library, uses a fifth-degree fully symmetric integration rule $Q_5 f$ and a seventh-degree extension $Q_7 f$ as the basic integration rules over the hypercube $[-1, 1]^d$. Here

$$Q_5 f = w_0 f(0, 0, \ldots, 0) + w_1 \sum f(\lambda_1, 0, \ldots, 0) + w_2 \sum f(\lambda_2, 0, \ldots, 0) + w_3 \sum f(\lambda_2, \lambda_2, 0, \ldots, 0) \tag{6.5.1}$$

and

$$Q_7 f = \hat{w}_0 f(0, 0, \ldots, 0) + \hat{w}_1 \sum f(\lambda_1, 0, \ldots, 0) + \hat{w}_2 \sum f(\lambda_2, 0, \ldots, 0) + \hat{w}_3 \sum f(\lambda_2, \lambda_2, 0, \ldots, 0) + \hat{w}_4 \sum f(\lambda_3, \lambda_3, \ldots, \lambda_3). \tag{6.5.2}$$

All sums in (6.5.1) and (6.5.2) are fully symmetric sums over all permutations of coordinates, including sign changes. The λ_i are independent of the dimension d and satisfy $0 < \lambda_i < 1$. The weights w_j and $\hat{w}_j$ are given by simple functions of d and are not always positive. ADAPT uses a global strategy and at each stage, bisects the hyperrectangle with the largest error estimate $|Q_7 f - Q_5 f|$ in the direction in which the local fourth difference of f is largest.

A rather expensive automatic integration procedure results when we apply one of the automatic integration procedures in one dimension to the iterated form of a multiple integral. Such integration procedures are usually written in recursive form in that the integrand can be an integral and, hence, the procedure calls upon itself. Thus the adaptive Newton–Cotes integration procedure mentioned in Section 6.2.3 can be used for the integration of an iterated integral.

Another example of the recursive algorithm is "Multiple Integral" of McKeeman. This assumes the availability of a procedure for integrating in one dimension. Although the procedure requires that the region be a hyper-rectangle, Cadwell has shown that this is not necessary.

An analysis of this so-called back-to-back integration process for two-dimensional integration based on possibly different one-dimensional adaptive integration routines has been carried out by Fritsch *et al.* They show that this process may be unstable in certain situations depending on the strategies adopted in the one-dimensional routines.

Back-to-back programs for two-dimensional integrals appear in the IMSL and NAG libraries.

Example. (Cadwell). The one-dimensional integration rule used in this example is an iterative nonadaptive Simpson's rule employing at least 5 points. The symbol ε designates the one-dimensional tolerance.

$$\int_V e^{-x_1^2-x_2^2-x_3^2-x_4^2}\,dV = \frac{\pi^2}{16}(1 - 2e^{-1}) = .16300,$$

where V is the region $x_i \geq 0$, $\sum x_i^2 \leq 1$.

ε	Number of points	Approximate integral
.01	625	.16125
.001	3345	.16281
.0001	69113	.16298

Examples. Here are some two and three dimensional Romberg results. The selection of integrands with a variety of singularities is deliberate; but since such integrands often occur in practice, the number of points that are required is realistic. In higher dimensions it becomes increasingly difficult to deal with singularities by special programming devices. The entries below are the diagonals of the T-table.

$$f(x, y) = |x^2 + y^2 - .25|$$

$$\int_{-1}^{1}\int_{-1}^{1} f(x, y)\, dx\, dy = \frac{5}{3} + \frac{\pi}{16} = 1.8630162.$$

Number of points	Approximate integral
3^2	1.8230453
$2^2 \cdot 3^2$	1.8757202
$2^4 \cdot 3^2$	1.8623424
$2^6 \cdot 3^2$	1.8629020
$2^8 \cdot 3^2$	1.8630223
$2^{10} \cdot 3^2$	1.8630197
$2^{12} \cdot 3^2$	1.8630155

Note the singularity along the circle of radius $\frac{1}{2}$.

$$f(x, y) = \frac{1}{1 - xy}$$

$$\int_{0}^{1}\int_{0}^{1} f(x, y)\, dx\, dy = \pi^2/6 = 1.644934$$

Number of points	Approximate integral
2^0	1.493529
2^2	1.570612
2^4	1.607527
2^6	1.626093
2^8	1.635469
2^{10}	1.640189
2^{12}	1.642558
2^{14}	1.643745
2^{16}	1.644339
2^{18}	1.644624

Compare this against the same example in Section 5.6.

$$\int_{0}^{1}\int_{0}^{1} |x - y|^{1/2}\, dx\, dy = \frac{8}{15} = .53333333$$

Number of points	Approximate integral
2^0	.471405
2^2	.514908
2^4	.527116
2^6	.531168
2^8	.532572
2^{10}	.533065
2^{12}	.533239
2^{14}	.533300

Note the singularity along $x = y$.

$$f(x, y, z) = \frac{1}{3 - \cos \pi x - \cos \pi y - \cos \pi z}$$

$$\int_0^1 \int_0^1 \int_0^1 f(x, y, z)\, dx\, dy\, dz = .50546$$

Number of points	Romberg	Product midpoint rule
5^3	.47741	.47741
10^3	.49623	.49153
15^3	.50037	.49618
20^3	.50213	.49850
30^3	.50324	.50083
45^3	.50398	.50237
65^3	.50445	.50332

Compare against the same example in Section 5.6.

$$f(x, y, z) = \frac{1}{3.75 - \cos \pi x - \cos \pi y - \cos \pi z}$$

$$\int_0^1 \int_0^1 \int_0^1 f(x, y, z)\, dx\, dy\, dz = .30781$$

Number of points	Romberg	Product midpoint rule
1^3	.26667	.26667
2^3	.31833	.30541
3^3	.30832	.30765
4^3	.30745	.30780
6^3	.30780	.30781
9^3	.30781	.30781

Here the integrand is not singular. Moreover, it is periodic and Romberg is worse than the product midpoint rule.

$$f(x, y, z) = |x^2 + y^2 + z^2 - .125|$$

$$\int_{-1}^{1}\int_{-1}^{1}\int_{-1}^{1} f(x, y, z)\, dx\, dy\, dz = 7 + \pi/120 = 7.0261799$$

Number of points	Romberg	Product midpoint
5^3	6.69600	6.69600
10^3	7.01856	6.93792
15^3	7.02038	6.98362
20^3	7.01621	6.99840
30^3	7.01956	7.00968
45^3	7.01827	7.01457
65^3	7.01853	7.01662

Note the singularity along the sphere of radius $\frac{1}{2}$, and the difficulty of arriving at an answer correct to 4S.

References. Anders [1]; Antes [1]; Barnhill and Little [1]; Cadwell [1]; Cranley and Patterson [4]; de Doncker and Robinson [1, F1]; Freeman [A1]; Fritsch, Kahaner and Lyness [1]; Gallaher [A2]; Genz [1, 3, 4, 5]; Genz and Malik [1]; Haegemans [2]; Halton and Zeidman [1]; IMSL [1]; Kahaner and Wells [1, 2]; Kuntzmann [1]; Laurent [1]; Laurie [2]; Lewellen [1]; Lyness [20]; Lyness and McHugh [1]; McKeeman [A2, A3]; NAG [1]; Olynk [A1]; QUADPACK [F1]; Robinson [4]; Robinson and de Doncker [1]; Roose and de Doncker [1]; Sag and Szekeres [1]; Stroud [18]; Thacher [2]; van Dooren and de Ridder [1].

Examples. We give here an extensive set of functions defined on $C_d = [0, 1]^d$ which were integrated using ADAPT. Some of these integrals were evaluated by other methods in Chapter 5. With each example, we list one or more values of RELTOL, the required relative tolerance given to the program, RELEST, the relative accuracy of the result estimated by the program, RELACC, the actual relative accuracy of the results and N, the number of function evaluations required. If the program exited with a failure indication because N reached the maximum allowed number of function evaluations and RELEST was still larger than RELTOL, we add F to the number N. Note that this maximum varied with each group of functions. One feature which stands out is that, with the exception of the case CP4 where early incorrect termination occurred, RELEST is much larger than RELACC. As a result, in many cases failure is indicated whereas RELACC is actually less than RELTOL.

The first set of examples was suggested by Tsuda. The functions $f_j(x)$ have been normalized so that the value of the integral is 1. The parameter j determines the value of f at the origin, $f_j(\mathbf{0}) = 10^j$.

$$f_j(x) = \frac{c^{2d}10^2}{\prod_{i=1}^{d}(c + x_i)}, \qquad c = \frac{1}{10^{j/d} - 1}, \quad j = 1, \ldots, 4.$$

	RELTOL	RELEST	RELACC	N
$\int_{C_2} f_1\,dV$	1(−2)	2(−3)	7(−6)	119
	1(−3)	8(−4)	8(−7)	153
	1(−4)	9(−5)	3(−7)	289
$\int_{C_2} f_2\,dV$	1(−2)	9(−3)	5(−5)	323
	1(−3)	8(−4)	7(−7)	595
	1(−4)	1(−4)	2(−7)	1139
$\int_{C_2} f_3\,dV$	1(−2)	8(−3)	7(−5)	765
	1(−3)	9(−4)	0	1173
	1(−4)	1(−4)	6(−8)	2329
$\int_{C_2} f_4\,dV$	1(−2)	9(−3)	5(−5)	1309
	1(−3)	1(−3)	1(−6)	1921
	1(−4)	1(−4)	5(−7)	3519
$\int_{C_8} f_1\,dV$	1(−2)-1(−3)	2(−4)	5(−6)	401
	1(−4)	9(−5)	4(−6)	2807
$\int_{C_8} f_2\,dV$	1(−2)	7(−3)	3(−4)	1203
	1(−3)	1(−3)	1(−4)	14837
	1(−4)	1.4(−4)	1(−5)	160000*F*
$\int_{C_8} f_3\,dV$	1(−2)	1(−2)	2(−3)	14035
	1(−3)	1.5(−3)	3(−4)	160000*F*
$\int_{C_8} f_4\,dV$	1(−2)	1(−2)	3(−3)	119097
	1(−3)	7(−3)	2(−3)	160000*F*

The following three examples are ten-dimensional integrals with singularities of various degrees of strength:

$$I(E) = \int_{C_{10}} \left(\sum_{i=1}^{10} x_i\right)^E d\mathbf{x}.$$

	RELTOL	RELEST	RELACC	N
$I(1.5) = 1.13209(1)$	1(−1)-1(−4)	2(−6)	5(−8)	1245
$I(-2.5) = 2.1203(-2)$	1(−1)-1(−2)	8(−3)	1(−4)	1245
	1(−3)	1.4(−3)	1(−5)	158115*F*
$I(-9.5) = 1.044(-5)$	1(−1)	6(−1)	1.2(−1)	158115*F*

The following two four-dimensional Gaussian integrals were discussed by Lepage. We define

$$G(c) = \int_{C_4} \exp(-100 \sum_{i=1}^{4} (x_i - c)^2)\, dV \simeq \pi^2/10000.$$

	RELTOL	RELEST	RELACC	N
$G(1/2)$	1(−2)	1(−2)	3(−3)	23883
	1(−3)	3(−3)	3(−4)	50000F
$G(1/3) + G(2/3)$	1(−2)	1(−2)	2(−3)	27189
	1(−3)	1(−3)	7(−5)	143697
	1(−4)	8(−4)	8(−5)	160000F

The next set of integrals was suggested by Thacher.

$$T(k) = \int_{C_4} k((1 - 6w^2) \cos w - (7 - w^2)w \sin w)\, dV = \sin k, \qquad w = kx_1x_2x_3x_4.$$

	RELTOL	RELEST	RELACC	N
$T(.1) =$	1(−2)-1(−4)	9(−5)	3(−6)	57
.998334166(−1)	1(−5)	8(−6)	1(−7)	627
	1(−6)	9(−7)	7(−9)	2565
$T(.25) =$	1(−2)-1(−3)	5(−4)	2(−5)	57
.247403959	1(−4)	9(−5)	1(−6)	399
	1(−5)	1(−5)	1(−7)	1653
	1(−6)	1(−6)	4(−9)	6897
$T(1) =$	1(−2)	4(−3)	2(−4)	57
.8414709848	1(−3)	7(−4)	6(−6)	285
	1(−4)	9(−5)	4(−7)	2565
	1(−5)	1(−5)	1(−8)	11571
	1(−6)	1(−6)	5(−10)	48393
$T(\pi/2) = 1$	1(−2)	1(−2)	1(−4)	285
	1(−3)	1(−3)	3(−6)	1083
	1(−4)	1(−4)	2(−6)	3933
	1(−5)	1(−5)	2(−8)	29355
	1(−6)	4(−5)	2(−9)	50000F
$T(3\pi/2) = -1$	1(−2)	1(−2)	1(−4)	6555
	1(−3)			
	1(−4)			

Cranley and Patterson have suggested these six-dimensional integrals as good tests of a multiple integration program, and ADAPT did not fare too well on these. In case CP4, an incorrect answer was returned as allegedly correct, the only time this occurred. C_6 designates the hypercube $-1 \le x_i \le 1$, $i = 1, \ldots, 6$.

$$CP1 = \int_{C_6} \cos(3(1 - x_1)x_2 x_3 x_4 x_5 x_6 + \tfrac{1}{2})\, dV = .54945(2),$$

$$CP2 = \int_{C_6} \cos\left(\sum_{i=1}^{6} x_i + \tfrac{1}{2}\right) dV = .199389(2),$$

$$CP3 = \int_{C_6} dV \bigg/ \left(2 + \sin \pi\sqrt{87} \sum_{i=1}^{6} x_i\right) = .369504(2),$$

$$CP4 = \int_{C_6} dV \bigg/ \left(2 + \sin \pi\sqrt{87} \prod_{i=1}^{6} x_i\right) = .33448(2).$$

	RELTOL	RELEST	RELACC	N
$CP1$	1(−2)	4(−2)	1.6(−2)	160000F
$CP2$	1(−2)	9(−3)	1(−5)	3427
	1(−3)	1(−3)	5(−6)	40081
	1(−4)	3(−4)	8(−7)	160000F
$CP3$	1(−2)	3(0)	5(−2)	160000F
$CP4$	1(−2)-1(−4)	6(−5)	3(−2)	149
	1(−5)	5(−2)	6(−3)	160000F

We close with a set of six four-dimensional integrals of varying degrees of difficulty.

$$g_1(\mathbf{x}) = \exp(x_1 x_2 x_3 x_4), \qquad \text{very smooth,}$$

$$g_2(\mathbf{x}) = \begin{cases} 1, & \sum_{i=1}^{4} x_i^2 \le 1, \\ 0, & \sum_{i=1}^{4} x_i^2 > 1, \end{cases} \qquad \text{discontinuous,}$$

$$g_3(\mathbf{x}) = \prod_{i=1}^{4} |x_i - \tfrac{1}{2}|, \qquad \text{discontinuous derivative,}$$

$$g_4(\mathbf{x}) = \prod_{i=1}^{4} \sin \pi x_i, \qquad \text{smooth,}$$

$$g_5(\mathbf{x}) = \sum_{i=1}^{4} |x_i - \tfrac{1}{2}|, \qquad \text{discontinuous derivative,}$$

$$g_6(\mathbf{x}) = \sin(100\pi(x_1 + x_2 - x_3 + 2x_4)) + 1, \qquad \text{highly oscillatory.}$$

	RELTOL	RELEST	RELACC	N
$\int_{C_4} g_1\, dV =$ 1.06939761	1(−2)-1(−3)	6(−4)	3(−5)	57
	1(−4)	9(−5)	3(−6)	399
	1(−5)	1(−5)	1(−8)	1767
	1(−6)	1(−6)	1(−9)	7467
$\int_{C_4} g_2\, dV = \pi^2/32$	1(−2)	8(−2)	3(−3)	50000F
$\int_{C_4} g_3\, dV = 2^{-8}$	1(−2)	1	9(−2)	50000F
$\int_{C_4} g_4\, dV = 16/\pi^4$	1(−2)	9(−3)	2(−3)	627
	1(−3)	9(−4)	3(−5)	3591
	1(−4)	1(−4)	4(−6)	9063
	1(−5)	1.3(−5)	1(−7)	50000F
$\int_{C_4} g_5\, dV = 1$	1(−2)	6(−3)	1(−3)	1653
	1(−3)-1(−6)	3(−13)	0	1767
$\int_{C_4} g_6\, dV = 1$	1(−2)-1(−6)	2(−11)	6(−12)	57

Note that the minimum number of points used by ADAPT is $2^d + 2d + 1$ and that this is the amount actually used in several of the above examples.

The Thacher integrals $T(k)$, the Cranley–Patterson group CPj, and the integrals of the six functions $g_i(\mathbf{x})$ were also computed using the NAG program D01FAF, which is based on adaptive stratified sampling (Section 5.9.1). The results indicate that ADAPT and D01FAF complement each other to some extent and that the latter program should be used for difficult integrands with a low accuracy tolerance. Note, that in these results, RELACC is closer to RELEST than in the ADAPT results but that there are several cases where RELACC exceeds RELEST. This indicates that D01FAF is not as reliable as ADAPT.

	RELTOL	RELEST	RELACC	N
$T(.1)$	1(−2)	8(−4)	1(−4)	96
	1(−3)	4(−4)	4(−4)	144
	1(−4)	1(−4)	2(−5)	924
	1(−5)	1(−5)	5(−7)	14706
$T(.25)$	1(−2)	7(−3)	1(−3)	96
	1(−3)	1(−3)	1.6(−3)	576
	1(−4)	7(−5)	6(−5)	12132
	1(−5)	2(−5)	2(−5)	56172F

	RELTOL	RELEST	RELACC	N
$T(1)$	1(−2)	8(−3)	9(−3)	1008
	1(−3)	8(−4)	9(−5)	22071
	1(−4)	4(−4)	2(−4)	61245*F*
$T(\pi/2)$	1(−2)	8(−3)	5(−3)	4764
	1(−3)	1.3(−3)	3(−4)	48867*F*
$T(3\pi/2)$	1(−2)	3(−2)	3(−3)	51615*F*

	RELTOL	RELEST	RELACC	N
$CP1$	1(−2)	7(−3)	9(−3)	672
	1(−3)	8(−4)	8(−4)	49824
	1(−4)	5(−4)	1.1(−4)	161376*F*
$CP2$	1(−2)	7(−3)	6(−3)	44160
	1(−3)	2.3(−3)	2.5(−3)	195360*F*
$CP3$	1(−2)	9(−3)	5(−3)	5430
	1(−3)	2(−3)	4(−4)	126984*F*
$CP4$	1(−2)	9(−3)	2(−3)	1872
	1(−3)	1(−3)	1(−3)	90912
	1(−4)	8(−4)	5(−4)	165152*F*

	RELTOL	RELEST	RELACC	N
$\int_{C_4} g_1\, dV$	1(−2)	6(−3)	3(−3)	252
	1(−3)	8(−4)	1(−4)	3606
	1(−4)	1.3(−4)	9(−5)	40941*F*
$\int_{C_4} g_2\, dV$	1(−2)	1(−2)	8(−3)	12516
	1(−3)	3(−3)	2(−3)	81084*F*
$\int_{C_4} g_3\, dV$	1(−2)	8(−3)	5(−3)	36540
	1(−3)	8(−3)	5(−3)	55260*F*
$\int_{C_4} g_4\, dV$	1(−2)	9(−3)	9(−3)	20424
	1(−3)	7(−3)	8(−4)	36664*F*
$\int_{C_4} g_5\, dV$	1(−2)	9(−3)	9(−3)	2088
	1(−3)	2(−3)	2(−3)	40152*F*
$\int_{C_4} g_6\, dV$	1(−2)	9(−3)	6(−3)	17595
	1(−3)	5(−3)	4(−3)	64023*F*

References. Cranley and Patterson [2]; Genz and Malik [1, F1]; Lepage [1]; NAG [1]; Thacher [3]; Tsuda [1].

6.6 Concluding Remarks

At the present stage of computer technology and practice, automatic integrators are generally recommended for integrating functions of one variable over a finite interval. If the programmer is confronted with an isolated (that is, a one-shot) integral, an adaptive program seems best for it appears to sacrifice some economy for reliability. CADRE and QAGS in QUADPACK appear to be particularly robust.

For indefinite integration, orthogonal polynomials—in particular Tschebyscheff or Legendre polynomials—may be used. If vast amounts of computation are required, automatic integration can become expensive, and preliminary analysis leading to the use of classical procedures may result in considerable savings. This would be the case, for example, if a multiple integral of dimension $d \geq 3$ is computed as an iterated automated integral.

Automatic integrations are not useful in the following situations: integration of data, solution of integral equations, or other situations where N functional evaluations lead to a system of equations in N unknowns. Nor is it useful for the integration of a function depending on a parameter λ and where it is desired that the resulting function of λ be smooth. Integrals with special difficulties such as infinite ranges of integration, singularities, highly oscillatory integrands should be treated in a special way, and there is some tendency to develop automatic routines tailored for each sort of difficulty. An illustration of this is the set of QUADPACK programs with the accompanying decision trees. A similar situation occurs in Chapter D01-Quadrature in the NAG library. The programmer will therefore be in the position of the sportsman who has given up worms but now must know which fly to use to catch his fish.

One can discern two opposing tendencies in computing. On the one hand, there is the desire to provide the consumer with full automation and decision avoidance. The writing of polyalgorithms has this as its goal. On the other hand, there is a movement toward interactive computation wherein the consumer monitors his output at critical points and makes decisions that are guided by intuition and knowledge of his problem.

Conservative, but realistic practice tries to combine these two possibilities in harmonious proportions.

References. Lyness [11, 20]; NAG [1]; QUADPACK [F1].

Appendix 1

On the Practical Evaluation of Integrals[†]

Milton Abramowitz

Someone has recently defined an applied mathematician as an individual enclosed in a small office and engaged in the study of mathematical problems which interest him personally; he waits for someone to stick his head in the door and introduce himself by saying, "I've got a problem." Usually the person coming for help may be a physicist, engineer, meteorologist, statistician, or chemist who has suddenly reached a point in his investigation where he encounters a mathematical problem calling for an unusual or nonstandard technique for its solution. It is of considerable importance for the mathematician to be able to provide practical answers to such questions. By a practical answer one does not mean a result which is obtained after months of detailed analysis but, rather, a solution or explanation obtained in a minimum of time. The reply to questions presented to him may require analysis or merely the ability to furnish a reference to where one can find the particular topic discussed. The range of mathematical topics from which queries may arise is all-inclusive. However, a topic which arises frequently enough to merit some discussion is one which particularizes the statement "I've got a problem," to "I've got an integral." We propose to discuss here some typical questions along this line which have arisen in the experience of the writer. It is difficult to give a general classification of the integrals which might be encountered, and for the purposes of this article it is best to give illustrations by selected examples.

Before proceeding to the special problem it is important to mention other aspects of providing practical assistance. There is no doubt that mathematically rigorous or precisely derived results are both desirable and necessary. However, in trying to provide useful results to research workers in other fields the mathematician should avoid such niceties at the outset. Furthermore, one should not try at first to answer questions in all completeness. When one is asked for a solution with special characteristics as required by the physical problem, the course of the investigation should be guided by these requirements. A rough-and-ready reply is often of much more value than a solution obtained with precise attention to rigor. The epsilons and deltas can come later.

The integrals to be discussed will in general be functions of one or more parameters. In attempting to evaluate a particular integral one should try as soon as possible to decide whether to obtain an analytic solution or settle for a numerical quadrature. However, prior to undertaking any work it is advisable to study the integrals from the following points of view:

(1) Confirm the existence of the integral,
(2) Ascertain the important ranges of the parameters involved,
(3) Reduce the integral to its simplest form,
(4) Determine the essential parameters which are involved,
(5) Determine the accuracy to which numerical values (if desired) are to be given.

The fourth item mentioned above is of considerable importance. The integral will arise in a physical problem where many parameters are often involved. The research worker will very often overlook the fact that these quantities may be combined in some way to yield a smaller number of parameters. In analyzing the integral such a reduction is highly desirable. Furthermore, the manner in which the parameters combine is usually of some physical significance so that the mathematical analysis may bring to light additional important information. As an example consider the integral

$$\int_0^\infty e^{-(ax^2+bx)}\,dx,$$

whose value is known. Here it would appear that from the point of view of evaluating the integral, a and b are essentially distinct. However, we write successively

$$\int_0^\infty e^{-(ax^2+bx)}\,dx = e^{b^2/4a}\int_0^\infty e^{-(ax^2+bx+b^2/4a)}\,dx$$

$$= e^{b^2/4a}\int_0^\infty e^{-(a^{1/2}x+b/2a^{1/2})^2}\,dx$$

$$= \frac{e^{b^2/4a}}{\sqrt{a}}\int_{b/2a^{1/2}}^\infty e^{-u^2}\,du = \frac{e^{b^2/4a}}{\sqrt{a}}\operatorname{erfc}\frac{b}{2\sqrt{a}},$$

and thus the essential quantity in evaluating the integral is $b/2\sqrt{a}$.

If the particular integrand in question involves functions which have been adequately tabulated, a numerical integration may not only furnish an adequate solution but may be the simplest method of computation. Naturally, if one wishes to exhibit a specific property of the integral representation of a transcendental function which describes the behavior of a physical system, a numerical table may not be satisfactory and analysis will be necessary.

Although the kinds of integrals which one encounters are too varied to classify, the questions one might ask about a particular integral are predominantly of the following type.

(1) What are the numerical values of the integral when the parameters involved are assigned definite values? Here one may wish a particular value, a graph, or a table of values.

(2) What are the series expansions for the integral in terms of the parameters? One need not necessarily be limited to power series. Furthermore, since power series usually have limited ranges of utility, other representations may be desirable.

(3) What is the asymptotic behavior of the integral in terms of the parameter involved? (For example, $\Gamma(x+1) = \int_0^\infty e^{-t}t^x\,dt \sim (2\pi x)^{1/2}x^x e^{-x}$, when x is large.)

(4) What is the complete asymptotic expansion of the integral in terms of the parameters involved?

$$(\Gamma(x+1) \sim (2\pi x)^{1/2}x^x e^{-x}\{1 + 1/12x + 1/288x^2 - \cdots\}).$$

Let us now illustrate some procedures which have been employed in specific instances and which one might follow. The most natural course, if one cannot find the integral in a table, is to identify it in terms of integrals which are already known.

Example 1. Reduction to a Known Form. The following integral arose in the theory of radiation:

$$f(a) = \int_{-1}^{+1} \frac{u^3}{(1 - u^2)^{1/2}} \sin au \, du.$$

The presence of the factor $(\sin au)/(1 - u^2)^{1/2}$ in the integrand suggests the possibility of a transformation which will yield a representation in terms of Bessel functions. This suggests the transformation $u = \sin t$, and one gets

$$f(a) = \int_{-\pi/2}^{+\pi/2} \sin^3 t \sin(a \sin t) \, dt$$

$$= 2 \int_{0}^{+\pi/2} \sin^3 t \sin(a \sin t) \, dt.$$

Now with the help of the formula $4 \sin^3 t = 3 \sin t - \sin 3t$ and the known integral representation for the Bessel functions,

$$J_{2n+1}(a) = (2/\pi) \int_{0}^{\pi/2} \sin(a \sin t) \sin(2n + 1)t \, dt,$$

we find

$$f(a) = (\pi/4)\{3J_1(a) - J_3(a)\}.$$

Example 2. Evaluation by a Limiting Procedure. Sometimes the integral under consideration may not be in a form which can be found in the tables, and some modification may be required to make use of known results. For example, it may be possible to introduce a factor depending on a parameter and then obtain the desired evaluation by a limiting process with respect to the parameter. A case in point is the integral

$$I = \int_{0}^{\infty} J_0(t) \, dt.$$

From the known result for the Laplace transform of $J_0(t)$ we have

$$\int_{0}^{\infty} e^{-pt} J_0(t) \, dt = 1/(1 + p^2)^{1/2}.$$

This integral is defined for $R(p) > 0$, and in the limit as $p \to 0$ it may be shown to be defined. Thus if we let $p \to 0$ in the right member, we get $I = 1$. In using this technique one must always make certain of the existence of the limiting integral. For if one starts with the integral

$$\int_{0}^{\infty} e^{-pt} \sin t \, dt = 1/(p^2 + 1),$$

although the limit of the right member as $p \to 0$ is unity, the integral does not exist for $p = 0$.

Example 3. Use of Functional Relationships. Sometimes one is able to make use of functional relationships involving the higher transcendental functions to obtain not an approximation to the particular integral but the complete expansion. The usefulness of such results is dependent on the adequate tabulation of the functions involved. An example of this technique provided by the integral which arises in an electrical circuit problem with time-varying capacitance is [1]

$$E(a, x) = \int_{-\infty}^{x} e^{(t + a\cos t)}\, dt,$$

where $-1 < a < 1$. Here one makes use of the known inversion of the relation

$$y = \tau - a \sin \tau$$

in terms of the Bessel functions $J_n(y)$, namely

$$\tau = y + \sum_{n=1}^{\infty} \frac{2}{n} J_n(na) \sin ny.$$

After making the substitution $t = (\pi/2) + \tau$, the integral becomes

$$e^{\pi/2} \int_{-\infty}^{x-\pi/2} e^{\tau - a\sin\tau}\, d\tau = e^{\pi/2} \int_{-\infty}^{Y} e^{y} \frac{d\tau}{dy}\, dy,$$

where $Y = x - \pi/2 + a\cos x$. Differentiating the preceding expansion and substituting for $d\tau/dy$, we then obtain, by termwise integration of the series, the expansion

$$E(a, x) = e^{(x + a\cos x)} \left\{ 1 + 2 \sum_{n=1}^{\infty} J_n(na) \frac{n \sin nY \cos nY}{n^2 + 1} \right\}.$$

Thus the procedure has reduced the problem to the evaluation of a rapidly convergent series of adequately tabulated functions.

Example 4. The Combined Use of Functional Relationships and Termwise Integration. An example of an integral [2] which arises in the theory of cooperative phenomena and where functional relationships turn out to be particularly useful is given by

$$I(b) = \frac{1}{\pi^3} \int_0^{\pi} \int_0^{\pi} \int_0^{\pi} \frac{dx\, dy\, dz}{3b - (\cos x + \cos y + \cos z)}.$$

Integrating with respect to z, we get for $b > 1$,

$$I(b) = \frac{1}{\pi^2}\int_0^{\pi}\int_0^{\pi}\frac{dx\,dy}{[(3b - \cos x - \cos y)^2 - 1]^{1/2}}.$$

To determine a series expansion for this integral, let us write

$$u = \frac{2}{3b - \cos x},$$

$$t = \cos y,$$

and obtain

$$I(b) = \frac{1}{\pi^2}\int_0^{\pi}\int_{-1}^{+1}\frac{dt\,dx}{\{[(2/u - t)^2 - 1][1 - t^2]\}^{1/2}}.$$

The integral with respect to t may now be transformed into the normal form of the elliptic integral $K(u)$, namely

$$K(u) = \int_0^1 \frac{ds}{\{(1 - s^2)(1 - u^2s^2)\}^{1/2}}.$$

To achieve this, we make the change of variable

$$t = \frac{As^2 + B}{Cs^2 + D},$$

and, since there are only three essential constants in this transformation, we may choose $D = 1$. In order that we have the correspondence $t = 1 \rightarrow s = 1$, $t = -1 \rightarrow s = 0$, we find $B = -1$ and $A = C + 2$, so that

$$t = \frac{(C + 2)s^2 - 1}{Cs^2 + 1}.$$

Carrying out the indicated substitution, we obtain

$$I(b) = \frac{1}{\pi^2}\int_0^{\pi}\int_0^1 \frac{2(C + 1)^{1/2}\,ds\,dx}{[(1 - s^2)R_4(s)]^{1/2}},$$

where $R_4(s) = \{2/u(s + 1) - [(C + 2)s^2 - 1]\}^2 - (Cs^2 + 1)^2$.

Now in order that $R_4(s)$ shall be a quadratic, the coefficient of s^4 must vanish, and we get the condition for C as

$$\frac{4}{u^2}C^2 + (C + 2)^2 - \frac{4}{u}C(C + 2) - C^2 = 0.$$

Solving for C, we get $C = u$ and $C = u/(1 - u)$. It can be shown that the condition $b > 1$ leads to rejection of the second value of C. Substituting $C = u$, we find

$$I(b) = \frac{1}{\pi^2}\int_0^\pi \int_0^1 \frac{u\, ds\, dx}{[(1 - s^2)(1 - u^2 s^2)]^{1/2}}$$

$$= \frac{1}{\pi^2}\int_0^\pi uK(u)\, dx.$$

This form can now be used to obtain an expansion for $I(b)$ which converges rapidly except in the neighborhood of $b = 1$. To this end one makes use of the series expansion for $K(u)$, namely

$$K(u) = \frac{\pi}{2}\left\{1 + \sum_{m=1}^{\infty} \left[\frac{1 \cdot 3 \cdot 5 \cdots (2m - 1)}{2 \cdot 4 \cdot 6 \cdots 2m}\right]^2 u^{2m}\right\}.$$

Substituting this series for the integral and integrating termwise, we obtain

$$I(b) = \frac{1}{\sqrt{9b^2 - 1}} + \sum_{m=1}^{\infty} \left[\frac{1 \cdot 3 \cdot 5 \cdots (2m - 1)}{2 \cdot 4 \cdot 6 \cdots 2m}\right]^2 \frac{2^{2m} g_{2m}}{(2m)!},$$

where

$$g_n = \left[\frac{d^n}{d(3b)^n}\{9b^2 - 1\}^{-1/2}\right],$$

as shown below, and satisfies the recurrence relation

$$(9b^2 - 1)g_{n+1} + 3b(2n + 1)g_n + n^2 g_{n-1} = 0.$$

The evaluation of the quantities g_{2n} is of interest and we may show† that they may be expressed in terms of the Legendre polynomial $P_n(z)$. We have

$$\int_0^\pi u^{2n+1}\, dx = 2^{2n+1}\int_0^\pi \frac{dx}{(3b - \cos x)^{2n+1}} = 2^{2n+1} g_{2n}\,.$$

For $n = 0$ the integral may be evaluated by elementary methods (or reference to a table of integrals) and the result is $\pi(9b^2 - 1)^{-1/2}$. For $n > 0$ the integrals are obtained by differentiation with respect to the parameter $3b$. However, from the known integral [3] for $P_n(z)$,

$$P_n(z) = \frac{1}{\pi}\int_0^\pi \{z - (z^2 - 1)^{1/2} \cos\phi\}^n\, d\phi,$$

† The author wishes to thank Dr. P. Henrici for pointing this out.

and from the relation $P_{-n-1}(z) = P_n(z)$ we get for $z = 3b/(9b^2 - 1)^{1/2}$,

$$g_{2n} = \frac{(2n)!}{\pi}(9b^2 - 1)^{-(n+1)/2} P_{2n}\left(\frac{3b}{(9b^2 - 1)^{1/2}}\right).$$

The recurrence relation for the g_n can now be found readily from the known recurrence relation for the Legendre polynomials,

$$(n + 1)P_{n+1} - (2n + 1)zP_n + nP_{n-1} = 0.$$

We note that, when $b = 1$, the integral is improper due to the singularity at $x = 0$. In the region of $b = 1$ the integral can be evaluated efficiently by extracting the contributions from the singularity while for $b = 1$ the value is known [4]. The technique involved in extracting the singularity is complicated and we illustrate it by the following simpler example.

Example 5. Extraction of Singular Part. Consider

$$I(q) = \int_0^q \frac{e^{-x}\,dx}{1 - x}, \qquad 0 \le q \le 1.$$

We note that the integrand has a pole at $x = 1$ and $I(1) = \infty$. However, in the neighborhood of $x = 1$, $e^{-x}(1 - x)^{-1}$ behaves like $e^{-1}(1 - x)^{-1}$, so that we may write $I(q)$ as

$$\begin{aligned} I(q) &= e^{-1}\int_0^q \frac{dx}{1 - x} + \int_0^q \left(\frac{e^{-x}}{1 - x} - \frac{e^{-1}}{1 - x}\right) dx \\ &= -e^{-1}\log(1 - q) + \int_0^q \left(\frac{e^{-x}}{1 - x} - \frac{e^{-1}}{1 - x}\right) dx. \end{aligned}$$

Now the second integral has no singularity in the neighborhood of $q = 1$ and may be evaluated quite easily by quadratures. Thus, the fundamental notion here is to modify the integrand by subtracting from it an expression (integrable in closed form) which eliminates the singularity and yields a form which can be integrated numerically. This matter has been discussed in some detail in [5].

Example 6. Reduction to a Differential Equation. The infinite integral [6] which arises in the determination of the response of a detector to a random noise voltage having a narrow spectrum,

$$f(x) = \int_0^\infty \frac{e^{-u^2}\,du}{x + u},$$

provides an interesting illustration of some methods available for evaluating integrals. To obtain an expression useful for large values of x is simple. One merely expresses $1/(x + u)$ as a geometric progression in (x/u) and integrates term by term. The result is an asymptotic series of the form

$$f(x) = \frac{1}{2} \sum_{r=0}^{n-1} \frac{(-1)^r}{x^{r+1}} \Gamma\left(\frac{r+1}{2}\right) + R_n,$$

where the remainder can be shown to be smaller than the next term neglected. This is an asymptotic expansion which actually diverges, but by keeping n fixed and letting x increase we can make the error arbitrarily small. However, if x is fixed, the terms in the series will decrease up to a certain value, from which they will start to increase. Thus, to obtain the smallest error one must stop the calculation just before the smallest term. To obtain an expansion valid for x in the neighborhood of $x = 0$, Goodwin and Staton [6] show that the integral satisfies the differential equation

$$\frac{df}{dx} + 2xf = \sqrt{\pi} - \frac{1}{x}.$$

From this equation one sees that f behaves like $-\log x$ as $x \to 0$ so that the limit as $x \to 0$ of $(f + \log x)$ must be found. By considering the limit

$$\lim_{x \to 0} \left\{ \int_0^\infty \frac{e^{-u^2}}{u + x} du - \int_0^\infty \frac{du}{(u^2 + 1)(u + x)} \right\} = \int_0^\infty \left[\frac{e^{-u^2}}{u} - \frac{1}{u(1 + u^2)} \right] du,$$

where the second integral of the left member has the value $[\log x - \frac{1}{2}\pi x]/(1 + x^2)$ and the right member has the value $-\gamma/2$ (γ being Euler's constant), it then follows immediately that $f(x) + \log x \to -\frac{1}{2}\gamma$. The ascending series can now be obtained from the differential equation above. It is to be noted that there is no standard procedure which can be used to form the second integral of the left member.

If we write $y = f(x) + e^{-x^2} \log x$, the differential equation becomes

$$y' + 2xy = \sqrt{\pi} - \frac{1 - e^{-x^2}}{x} = \sqrt{\pi} + \sum_{n=1}^{\infty} \frac{(-1)^n x^{2n-1}}{n!},$$

and now from the theory of differential equations we know that a series for y can be determined in the form

$$y = \sum_{n=0}^{\infty} a_n x^n$$

with

$$a_0 = -\tfrac{1}{2}\gamma, \qquad \gamma = \text{Euler's constant}.$$

If a table of the function $f(x)$ were required, alternative methods of calculation with the series, which are effective, are either a numerical integration of the differential equation or a direct evaluation by numerical quadrature of the integral.

Example 7. Method of Laplace Transformation. Instead of employing the foregoing method for determining the ascending series, it is possible to use the Laplace transform to obtain the result in more direct fashion. The Laplace transform is defined as

$$L\{g(t)\} = \int_0^\infty e^{-tp} g(t)\, dt.$$

If we make the substitution $u = xv$ in the integral for $f(x)$, we get

$$g(t) = \int_0^\infty \frac{e^{-tv^2}}{1+v}\, dv. \qquad t = x^2.$$

The transform of $g(t)$ with respect to t is

$$\int_0^\infty \frac{dv}{(1+v)(p+v^2)} = \frac{\pi}{2\sqrt{p}(p+1)} + \frac{1}{2}\frac{\log p}{(p+1)}.$$

Now from the known transforms [16]

$$L^{-1}\left\{\frac{1}{\sqrt{p}(p+1)}\right\} = \frac{2e^{-t}}{\sqrt{\pi}} \int_0^{t^{1/2}} e^{u^2}\, du, \qquad L^{-1}\left\{\frac{\log p}{(p+1)}\right\} = -e^{-t} Ei(t),$$

where $Ei(t)$ is the exponential integral, we obtain by inversion of the Laplace transform (here this merely involves reference to a table of Laplace transforms) the following expression for $f(x)$:

$$f(x) = \sqrt{\pi}\, e^{-x^2} \int_0^x e^{u^2}\, du - \tfrac{1}{2} e^{-x^2} Ei(x^2).$$

Actually one could also obtain the power series for $f(x)$ by expanding the transform in a series of $1/p$ and inverting the transform termwise. However, the above method [7] gives us a result which expresses $f(x)$ in terms of a function which has already been tabulated.

Example 8. Saddle-Point Approximation, Improvement by Method of Differential Equations. Another method of particular value is the saddle-point method. As an example of this technique we take the integral [8]

$$f(x) = \int_0^\infty e^{-u^2 - x/u}\, du,$$

arising in the theory of absorption coefficients for thermal neutrons. We shall find an approximation to this integral useful for large values of x. The fundamental idea to be employed here is the assumption that the principal contribution to the integral comes in the neighborhood of $u = u_0$, where $g(u) = u^2 + x/u$ has a minimum value. The point u_0 which is the saddle-point is determined from the condition that $g'(u_0) = 0$. However, it is convenient to make the following change of variable,

$$u = vt, \qquad t = (x/2)^{1/3},$$

and thus obtain

$$t \int_0^\infty e^{-t^2(v^2 + 2/v)}\, dv.$$

Now $h(v) = v^2 + 2/v$ has a minimum at $v = 1$, and from Taylor's theorem we have

$$h(v) = h(1) + (v-1)h'(1) + \frac{(v-1)^2}{2!} h''(1) \cdots$$

in the neighborhood of $v = 1$. Since $h(1) = 3$, $h'(1) = 0$, $h''(1) = 6$, the integral may be approximated by

$$te^{-3t^2} \int_0^\infty e^{-3t^2(v-1)^2}\, dv.$$

To evaluate the integral we now set $\sqrt{3}\, t(v-1) = s$ and obtain

$$f(x) \cong \frac{e^{-3t^2}}{\sqrt{3}} \int_{-3t^{1/2}}^\infty e^{-s^2}\, ds.$$

Now, since x and therefore t is large, we can assume that the lower limit in the integral can be extended to infinity without introducing any appreciable error. We thus have

$$f(x) \cong \frac{e^{-3t^2}}{\sqrt{3}} \int_{-\infty}^\infty e^{-s^2}\, ds = \left(\frac{\pi}{3}\right)^{1/2} e^{-3t^2}, \qquad t = \left(\frac{x}{2}\right)^{1/3}$$

This represents an asymptotic approximation to the integral and for most purposes would prove adequate. However, should it be necessary to obtain an improvement to this result, the following technique is recommended. First, it can be shown that the original integral satisfies the differential equation

$$xf''' + f'' + 2f = 0.$$

Secondly, in the asymptotic result obtained earlier the variable x appears in the form $3(x/2)^{2/3} = z$, say. This suggests introducing z as a new variable on the assumption that the complete asymptotic expansion is a function of z. If we then carry out this substitution, we get the differential equation

$$\frac{d^3f}{dz^3} + \frac{1}{4z^2}\frac{df}{dz} + f = 0.$$

Now, on the basis of the previous result we can write $f(z) = (\pi/3)^{1/2}e^{-z}U(z)$, where $U(\infty) = 1$, and obtain a differential equation for U:

$$\frac{d^3U}{dz^3} - 3\frac{d^2U}{dz^2} + \left[3 + \frac{1}{4z^2}\right]\frac{dU}{dz} - \left(\frac{1}{4z^2}\right)U = 0.$$

From the theory of ordinary differential equations we can now obtain a solution to this equation in the form

$$U = 1 + \frac{a_1}{z} + \frac{a_2}{z^2}\cdots,$$

where the coefficients a_i can be obtained by substituting this expression in the previous equation. We stress the fact that all the steps mentioned here are of an elementary character. From the practical point of view, the determination of the differential equation for $f(x)$ may call for some ingenuity, but all the subsequent steps are straightforward.

Example 9. Inversion of Order of Integration using Contour Integrals. As a final example let us consider the modified Airy integral

$$A(x) = \int_0^\infty e^{-t^3 - xt}\,dt.$$

The fundamental idea to be demonstrated is the replacement of some part or all of the integrand by an equivalent definite integral and inversion of the order of integration. In addition, we shall employ the notion of the Mellin–Barnes type integral to obtain both the power series in x and an asymptotic expansion in $1/x$. Although these results can be obtained here from the integral directly, the method serves to illustrate the technique.

It can be shown [9] that

$$e^{-s} = \frac{1}{2\pi i}\int_{-a-i\infty}^{-a+i\infty} \Gamma(-z)(s)^z\,dz,$$

where a is real and positive and the path of integration is a straight line parallel to the imaginary axis. Thus, if we substitute for e^{-xt} in the integral for $A(x)$, we get

$$\begin{aligned} A(x) &= \int_0^\infty e^{-t^3}\left[\frac{1}{2\pi i}\int_{-a-i\infty}^{-a+i\infty} \Gamma(-z)(xt)^z\,dz\right] dt \\ &= \frac{1}{2\pi i}\int_{-a-i\infty}^{-a+i\infty} \Gamma(-z)(x)^z\left[\int_0^\infty e^{-t^3}t^z\,dt\right] dz \\ &= \frac{1}{3}\frac{1}{2\pi i}\int_{-a-i\infty}^{-a+i\infty} \Gamma(-z)\Gamma\left(\frac{z+1}{3}\right)(x)^z\,dz. \end{aligned}$$

Now we consider the paths of integration composed of the line $R(z) = -a$, where $0 < a < 1$, and that part of the circle $|z| = m + \frac{1}{2}$ to the left or right of this line where m is an integer. The value of a must be chosen so as to separate the poles of $\Gamma(-z)$ from those of $\Gamma[(z+1)/3]$. It can then be shown that the contribution from the line-integral around either circular arc vanishes as $m \to \infty$, and the value of the integral will be obtained as the sum of the residues resulting from the poles enclosed by the closed contour.

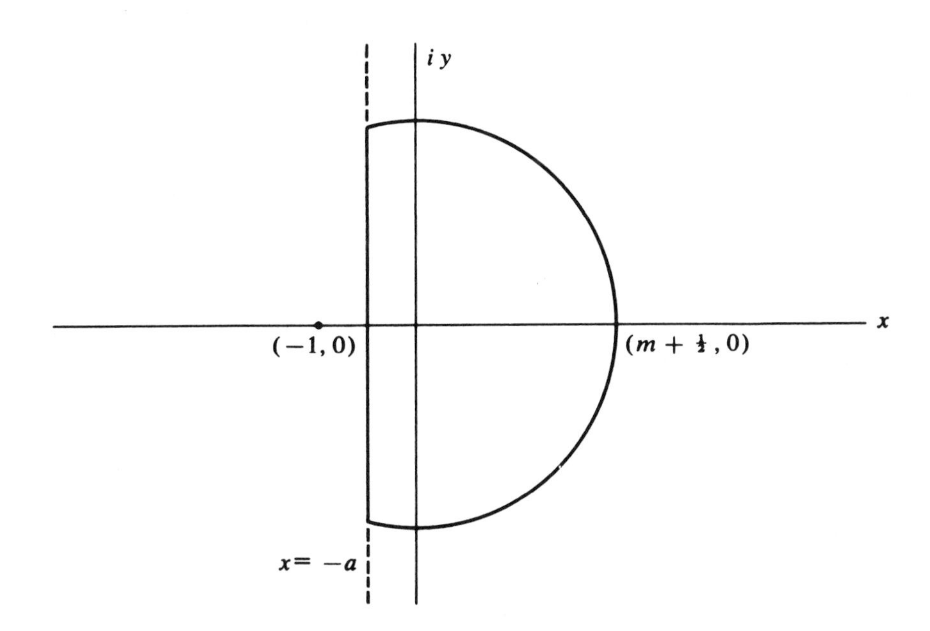

Contour of integration

Furthermore, we note that two different expansions can be derived, depending on whether the circular arc is drawn to the right or the left of the imaginary axis. Let us first consider the region as fixed by the circle to the right. The poles of the integrand are those arising from $\Gamma(-z)$, namely $z = n$ (n a positive integer). The residue of $\Gamma(-z)$ is $(-1)^n/n!$; thus, if $m \to \infty$, the ascending series is

$$A(x) = \frac{1}{3} \sum_{n=0}^{\infty} \frac{\Gamma[(n+1)/3](-x)^n}{n!}.$$

If we now take the circle to the left of the imaginary axis, the poles are now those resulting from $\Gamma[(z+1)/3]$, namely when $(z+1)/3$ is zero or a negative integer, that is, when $z = -3n - 1$. The residue of $\Gamma[(z+1)/3]$ is thus $3(-1)^n/\Gamma(n+1)$, and the contribution to the integral at each pole is $(x)^{-3n-1}\Gamma(3n+1)/\Gamma(n+1)$. It can be shown that the series so obtained actually diverges but that, if one keeps only a finite number of terms, the remainder is smaller than the first term neglected. We thus obtain

$$A(x) \sim \sum_{n=0}^{\infty} \frac{(-1)^n\Gamma(3n+1)}{\Gamma(n+1)x^{3n+1}},$$

where the symbol $\sim$ is written to indicate the fact that the series is an asymptotic expansion. We have omitted a discussion of the convergence and the remainder terms in these expansions for reasons of space.

It seems worthwhile to summarize some suggested methods which might be tried in making an investigation of a particular integral:

(1) Try to recognize the integral or find it listed in existing tables of integrals. Here one should also refer to tables of transforms such as Laplace transforms, Fourier transforms, Mellin transforms, and Hankel transforms. For the convenience of the reader a short list of tables is given in references [13] to [19]. A useful device is to differentiate with respect to one or more parameters and thereby obtain a known integral. One may also replace an appropriate constant by a parameter and then apply this technique.

(2) Try the method of integration by parts. This procedure is relatively simple and in very many cases yields at least one expansion for the integral. For example, in the case of the exponential integral $Ei(t) = \int_0^\infty e^{-ut}\, du/u$ one can derive the asymptotic expansion for large values of t immediately. The technique of exponentializing the integrand [10], namely writing the integral in the form $\int_a^b e^{p(x)}\, dx$ and integrating by parts successively, is very useful.

(3) Modify the integrand by subtracting out a singular function which can be integrated and reduces the integrand to a regular function.

(4) Evaluate the integral by numerical quadrature to produce a table. The integrand may possess a singularity and therefore require some modification

before the quadrature can be effected. Even in those cases where the integral can be evaluated in closed form it may be worth using numerical integration in preference to any of the techniques suggested. In fact one may know the value of the integral and still use quadratures rather than the explicit value.

(5) Substitute for part or all of the integrand the equivalent power series expansion and integrate term by term. In the case of the Airy integral discussed earlier, expansion of the exponent e^{-xt} and termwise integration yield the ascending power series while expansion of the part e^{-t^3} gives the descending series.

(6) Use some functional relation satisfied by a part or all of the integrand. As an illustration of this we may cite the integral discussed by Henrici, in which he uses an expansion in terms of a series of Bessel functions.

(7) Substitute an equivalent definite integral for a part or all of the integrand and invert the order of integration.

(8) Find the transform (e.g., Laplace transform) of the particular integral. The resulting transformed integral may be simplified to the extent where the integration may be carried out explicitly and the desired expression may be obtained by inversion.

(9) Determine the differential equation satisfied by the integral or a modification of the integral in order to employ the methods made available by the theory of differential equations.

(10) Employ an appropriate complex integral taken along a suitably chosen contour. A well-known example of this technique is the integral

$$\int_0^\infty \frac{\cos mx}{x^2 + a^2}\, dx,$$

where one evaluates the line integral

$$\int \frac{\exp(imz)}{z^2 + a^2}\, dz$$

along an appropriate path. This subject is discussed in all of the standard works on complex variables.

(11) Approximate the integrand in the region where the principal contribution to the integral is obtained by functions whose integration can then be effected. Approximation in terms of orthogonal functions may be useful.

(12) Use the saddle-point method to find an asymptotic representation for the integral. When the differential equation satisfied by the integral is known, the technique discussed earlier may be used.

(13) Use the method of steepest descent to determine an asymptotic representation for the integral. A comprehensive discussion of this subject may be found in the report [11] of the British Admiralty.

It should be emphasized again that the list of methods outlined can not be considered exhaustive. In addition to the foregoing types of integrals, which arise mainly in physical applications, mention might also be made of the endless variety that are met in a field such as mathematical statistics. In fact, the essential difficulty in many problems involving probability distributions is simply the evaluation of certain integrals. An example of such an integral, in which a combination of integration by parts and differentiation with respect to a parameter yielded a known integral, may be found in [21].

In closing this discussion we note that it is often possible to handle a particular integral in several ways. In the case of the integral

$$\int_0^\infty \frac{\sin x}{x}\,dx$$

there are a variety of ways [20] in which it can be evaluated. In the article mentioned, five distinct methods are discussed while reference to others is given. The attack made on any particular integral depends on the tools one has at hand. It is hoped that the foregoing discussion will provide the reader with an introduction to some useful techniques. It is not intended to be a comprehensive study, and all questions of mathematical rigor have been omitted. Finally, the author wishes to thank J. Todd, P. Henrici, and H. Antosiewicz for their suggestions during the preparation of this report.

REFERENCES

1. HENRICI, P., *Weitere Bermerkung zu* $\int e^{b(x+a\cos x)}\,dx$, Zeitschrift für angewandte Mathematik und Physik, Vol. III (1952), p. 466.

2. TIKSON, M., *Tabulation of an integral arising in cooperative phenomena*, Journal of Research of National Bureau of Standards, Vol. 50 (1953) March, p. 177.

3. MACROBERT, T. M., *Spherical Harmonics*, Dover (1948), p. 101.

4. OBERHETTINGER, F., and MAGNUS, W., *Anwendung der Elliptischen Funktionen in Physik und Technik*, p. 8.

5. HARTREE, D. R., *Numerical Analysis*, p. 106, 1952.

6. GOODWIN, E. T., and STATON, J., *Table of* $\int_0^\infty e^{-u^2}/(u+x)\,du$, Quarterly Journal of Mechanics and Applied Math. (1) (1948), p. 319.

7. RITCHIE, O. H., *On a Definite Integral*, Mathematical Tables and Other Aids to Computation, (1950), 4, p. 26.

8. ABRAMOWITZ, M., *Evaluation of the Integral*, $\int_0^\infty e^{-u^2-x/u}\,du$, Journal of Math. and Phys., Vol. 32, July (1953), p. 188.

9. MACROBERT, T. M., *Functions of a Complex Variable*, Macmillan, p. 151, 1950.

10. SALZER, H. E., *Coefficients in an Asymptotic Expansion for* $\int_a^b e^{p(u)}\,du$, Mathematical Tables and Other Aids to Computation, Vol. 2, October (1946), p. 188.

11. COPSON, E. T., *The Asymptotic Expansion of a Function Defined by a Definite Integral or Contour Integral*, Admiralty Computing Service, 1946.

12. ANTOSIEWICZ, H. A., *On a Certain Integral Involving Bessel Functions*, NBS Report.

13. ERDÉLYI, A., MAGNUS, W., OBERHETTINGER, F., and TRICOMI, F. G., *Higher Transcendental Functions*, Vol. 1 and 2, McGraw-Hill, 1953.

14. GRÖBNER, W., and HOFREITER, N., *Integraltafel*, Springer, Berlin, 1949.

15. DWIGHT, H. B., *Tables of Integrals*, Macmillan, 1951.

16. DOETSCH, G., *Tabellen zur Laplace-Transformation und Einleitung zum Gebrauch*, Springer, Berlin, 1947.

17. MACLACHLAN, N. W., and HUMBERT, P., *Formulaire pour le calcul symbolique*, Mémorial des Sciences Mathématiques, Fascicule C, Gauthier-Villars, Paris, 1941.

18. DE HAAN, B., *Nouvelles Tables d'Intégrales Définies*, Leiden, 1867.

19. CAMPBELL, G. A., and FOSTER, R. M., *Fourier Integrals for Practical Application*, Van Nostrand, New York, 1951.

20. HARDY, G. H., *The Integral* $\int_0^\infty (\sin x/x)\,dx$, Math. Gazette 5, No. 80 (1909), p. 98.

21. LIEBLEIN, J., *On the Exact Evaluation of the Variances and Covariances of Order Statistics in Samples from the Extreme-Value Distribution*, The Annals of Mathematical Statistics, Vol. 24 (1953), No. 2, p. 282.

NATIONAL BUREAU OF STANDARDS
WASHINGTON, D.C.

A further reference in the spirit of the Abramowitz article is Smith and Lyness who show how the Hilbert transform may be exploited profitably. The importance of asymptotic analysis should lead the reader to Olver's definitive work on the topic and to his later paper on asymptotic approximations and error bounds. For surveys on asymptotic expansions of integrals, see Jones and Wong.

References. Jones [1]; Olver [2, 3]; Smith and Lyness [1]; Wong [1].

Appendix 2

FORTRAN Programs

Goals of Programs

A variety of goals—not all compatible—may be pursued in the construction of an algorithm for numerical integration and in its realization as a computer program. Among these goals are:

(1) Accommodation of special requirements as to the form of the algorithm. (For example: the rule is required to use $f'(0)$.)
(2) Accuracy as regards truncation error.
(3) Ease in estimating truncation error.
(4) Accuracy as regards roundoff error. Roundoff abatement.
(5) Ability to detect roundoff error.
(6) Foolproofness; fail-safe features.
(7) Economy in storage.
(8) Wide applicability.
(9) Flexibility of program.
(10) Ease of programming and card preparation. (For example, use of stored versus generated constants.)
(11) Ease of use.
(12) Computational efficiency.
(13) Transparency of program.

The programs included in this appendix fulfill a variety of purposes. Some illustrate a point in the text, while others should be rather useful. There are utility programs such as GAUSS and SUM which require that the user supply certain numerical data (e.g., abscissas and weights). Most programs assume that the integrand is specified by a FUNCTION program to be supplied by the user; but some, such as AVINT, CUBINT, SPLINT, and FILON integrate functions given in tabular form. All one-dimensional programs of the former class are automatic, while among the multidimensional programs, only NDIMRI is automatic. P5 is included as an example of monomial rules. HABER has much to recommend it for practical use. Among the automatic one-dimensional programs, CADRE is recommended as a robust general-purpose routine which usually gives reliable results for a wide variety of functions.

For "state of the art" programs in numerical integration, the reader is referred to QUADPACK [F1] and NAG [1].

GAUSS

GAUSS (p. 364) is an all-purpose integration program to be used when the number of integration abscissas, their values, and the values of the corresponding weights are known in advance. GAUSS can be used for compound Gauss, Newton–Cotes, Lobatto, Radau, etc. integration over a finite interval and for Laguerre, generalized Laguerre, and Hermite integration over semi-infinite and infinite intervals. It can also be used for any simple integration formula where all the abscissas and weights are given. The parameters are N, X, W, FUN, KEY, A, B, M, SUM, IND; X and W are the names of two arrays containing the abscissas and weights, respectively, of a particular N-point integration rule. KEY indicates the type of rule as explained below, A and B are the endpoints for the case of integration over a finite interval, and M is the number of subintervals into which [A, B] is divided. In the Laguerre case, we permit integrals of the form $\int_a^\infty e^{-x} f(x)\, dx$, and A is the value of a. The result appears in SUM if IND = 1. IND is set to 0 if N < 1, KEY < 1, KEY > 6, or M < 1 when KEY $\leq$ 3.

KEY = 1 for a composite symmetric rule such as that of Gauss. The abscissas and weights are assumed to be normalized over the interval $[-1, 1]$. Only the nonnegative abscissas and their corresponding weights are to be stored in X and W, respectively, and these are to be given in ascending order of the abscissas.

KEY = 2 for a composite nonsymmetric rule such as that of Radau. Abscissas and weights are assumed to be normalized over $[-1, 1]$ and can be given in any order.

KEY = 3 for a composite symmetric rule such as that of closed Newton–Cotes or Lobatto, which includes the endpoints of the interval of integration among the abscissas. Abscissas and weights are to be given as for KEY = 1.

KEY = 4 for a symmetric rule over $(-\infty, \infty)$, such as Hermite, or for any simple symmetric rule where the exact abscissas and weights are given. Abscissas and weights are to be given only for nonnegative abscissas in ascending order.

KEY = 5 for $\int_a^\infty e^{-x} f(x)\, dx$, using Laguerre integration. Abscissas and weights can be given in any order.

KEY = 6 for any approximate integration such as generalized Laguerre integration, where the integral is approximated by $\sum_{i=1}^{N} w_i f(x_i)$ and the N values of w_i and x_i are given.

A program FUNCTION FUN (X) must be supplied by the user and FUN must be declared EXTERNAL in the calling program.

```
      SUBROUTINE GAUSS(N,X,W,FUN,KEY,A,B,M,SUM,IND)
      DIMENSION X(N),W(N)
      IND=0
      IF(N.LT.1 .OR. KEY.LT.1 .OR. KEY.GT.6) RETURN
      IF(KEY.GE.4) IND=1
      L=(N+1)/2
      K=2*L-N+1
      SUM=0.
      IF (KEY-5) 7,5,6
    7 IF (KEY .EQ. 4) GO TO 4
      IF(M.LT.1)RETURN
      IND=1
      H1=(B-A)/M
      H=H1*.5
      AO=A+H
      GO TO (1,2,3),KEY
C     KEY=1,M*SYMMETRIC RULE,E.G. GAUSS
    1 DO 10 M1=1,M
      IF(K .EQ. 2)SUM=SUM+W(1)*FUN(AO)
      DO 15 I=K,L
      H2=H*X(I)
   15 SUM=SUM+W(I)*(FUN(AO+H2)+FUN(AO-H2))
   10 AO=AO+H1
   16 SUM=H*SUM
      RETURN
C     KEY=2,M*NON-SYMMETRIC RULE,E.G. RADAU
    2 DO 20 M1=1,M
      DO.21 I=1,N
   21 SUM=SUM+W(I)*FUN(AO+H*X(I))
   20 AO=AO+H1
      GO TO 16
C     KEY=3,M*SYMMETRIC RULE WITH ENDPOINT 1,E.G. LOBATTO
    3 SUM=W(L)*(FUN(B)-FUN(A))
      W1=2*W(L)
      L1=L-1
      DO 30 M1=1,M
      IF(K .EQ. 2)SUM=SUM+W(1)*FUN(AO)
      IF( K .GT. L1 ) GO TO 33
   32 DO 35 I=K,L1
      H2=H*X(I)
   35 SUM=SUM+W(I)*(FUN(AO+H2)+FUN(AO-H2))
```

```
   33 SUM=SUM+W1*FUN(AO-H)
   30 AO=AO+H1
      GO TO 16
C     KEY=4,INFINITE SYMMETRIC RULE,E.G. HERMITE
    4 IF(K .EQ. 2)SUM=SUM+W(1)*FUN(X(1))
      DO 40 I=K,L
   40 SUM=SUM+W(I)*(FUN(X(I))+FUN(-X(I)))
      RETURN
C     KEY=5,LAGUERRE INTEGRATION FROM A TO INFINITY
    5 DO 50 I=1,N
   50 SUM=SUM+W(I)*FUN(X(I)+A)
      SUM=EXP(-A)*SUM
      RETURN
C     KEY=6 PURE INNER PRODUCT,E.G. GENERALIZED LAGUERRE
    6 DO 60 I=1,N
   60 SUM=SUM+W(I)*FUN(X(I))
      RETURN
      END
```

AVINT (*see Section 2.3*)

This program computes an approximation to $\int_{XLO}^{XUP} Y(x)\,dx$. X, Y are one-dimensional arrays of N elements where $N \geq 3$. The X(I) are usually unequally spaced out and must be distinct and in ascending order. If this is not so or if $N < 3$, the program exits with IND = 0. Otherwise IND = 1. There are no restrictions on XLO and XUP except that of accuracy which requires them to be close to the interval [X(1), X(N)]. AVINT is adapted from Hennion [A1].

```
      FUNCTION AVINT(X,Y,N,XLO,XUP,IND)
      DIMENSION X(N),Y(N)
      IND=0
      IF(N.LT.3) RETURN
      DO 10 I=2,N
      IF(X(I).LE.X(I-1)) RETURN
   10 CONTINUE
      SUM=0.
      IF(XLO.LE.XUP) GO TO 5
      SYL=XUP
      XUP=XLO
      XLO=SYL
      IND=-1
      GO TO 6
    5 IND=1
      SYL=XLO
    6 IB=1
      J=N
      DO 1 I=1,N
      IF(X(I).GE.XLO) GO TO 7
    1 IB=IB+1
    7 IB=MAX0(2,IB)
      IB=MIN0(IB,N-1)
      DO 2 I=1,N
      IF(XUP.GE.X(J)) GO TO 8
    2 J=J-1
    8 J=MIN0(J,N-1)
      J=MAX0(IB,J-1)
      DO 3 JM=IB,J
      X1=X(JM-1)
      X2=X(JM)
```

```
      X3=X(JM+1)
      TERM1=Y(JM-1)/((X1-X2)*(X1-X3))
      TERM2=Y(JM)/((X2-X1)*(X2-X3))
      TERM3=Y(JM+1)/((X3-X1)*(X3-X2))
      A=TERM1+TERM2+TERM3
      B=-(X2+X3)*TERM1-(X1+X3)*TERM2-(X1+X2)*TERM3
      C=X2*X3*TERM1+X1*X3*TERM2+X1*X2*TERM3
      IF(JM.GT.IB) GO TO 14
      CA=A
      CB=B
      CC=C
      GO TO 15
   14 CA=.5*(A+CA)
      CB=.5*(B+CB)
      CC=.5*(C+CC)
   15 SUM=SUM+CA*(X2**3-SYL**3)/3.+CB*.5*(X2**2-SYL**2)+CC*(X2-SYL)
      CA=A
      CB=B
      CC=C
    3 SYL=X2
      AVINT=SUM+CA*(XUP**3-SYL**3)/3.+CB*.5*(XUP**2-SYL**2)+CC*(XUP-SYL)
      IF(IND.EQ.1) RETURN
      IND=1
      SYL=XUP
      XUP=XLO
      XLO=SYL
      AVINT=-AVINT
      RETURN
      END
```

CUBINT (*see Section 2.3*)

This subroutine computes RESULT, an approximation to $\int_{X(IA)}^{X(IB)} F(x)\,dx$, and ERROR, an estimate of the error in the integration. For polynomials of degree ≤ 4, the exact value of the integral (up to rounding error) = RESULT + ERROR. X, F are one-dimensional arrays of N elements, where N must be ≥ 4, while IA and IB must be integers between 1 and N. If any of these conditions does not hold, IND is set to 0 and the subroutine exits without any RESULT. Otherwise, IND = 1. CUBINT is adapted from Gill and Miller [1] where the algorithm is developed and discussed.

```
      SUBROUTINE CUBINT(X,F,N,IA,IB,RESULT,ERROR,IND)
      DIMENSION X(N),F(N)
      IND=0
      IF (N.LT.4 .OR. IA.LT.1 .OR. IB.GT.N) RETURN
      IND=1
      RESULT=0.
      ERROR=0.
      IF (IA.EQ.IB) RETURN
      IF (IA.LT.IB) GO TO 2
      IND=-1
      IT=IB
      IB=IA
      IA=IT
    2 S=0.
      C=0.
      R4=0.
      J=N-2
      IF (IA.LT.N-1 .OR. N.EQ.4) J=MAX0(3,IA)
      K=4
```

```
      IF (IB.GT.2. OR. N.EQ.4) K=MINO(N,IB+2)-1
      DO 1  I=J,K
      IF( I.GT.J) GO TO 5
      H2=X(J-1)-X(J-2)
      D3=(F(J-1)-F(J-2))/H2
      H3=X(J)-X(J-1)
      D1=(F(J)-F(J-1))/H3
      H1=H2+H3
      D2=(D1-D3)/H1
      H4=X(J+1)-X(J)
      R1=(F(J+1)-F(J))/H4
      R2=(R1-D1)/(H4+H3)
      H1 =H1 +H4
      R3=(R2-D2)/H1
      IF (IA .GT. 1) GO TO 8
      RESULT=H2*(F(1)+H2*(.5*D3-H2*(D2/6.-(H2+H3+H3)*R3/12.)))
      S=-H2**3*(H2*(3.*H2+5.*H4)+10.*H3*H1)/60.
      GO TO 8
    5 H4=X(I+1)-X(I)
      R1=(F(I+1)-F(I))/H4
      R4=H4+H3
      R2=(R1-D1)/R4
      R4=R4+H2
      R3=(R2-D2)/R4
      R4=(R3-D3)/(R4+H1)
    8 IF (I.GT.IB .OR. I.LE.IA) GO TO 11
      RESULT=RESULT+H3*((F(I)+F(I-1)) *.5-H3*H3*(D2+R2+(H2-H4)*R3)/12.)
      C=H3**3*(2.*H3*H3+5.*(H3*(H4+H2)+2.*H2*H4))/120.
      ERROR=ERROR+(C+S)*R4
      IF (I.EQ.J) GO TO 14
      S=C
      GO TO 15
   14 S=S+C+C
      GO TO 15
   11 ERROR=ERROR+R4*S
   15 IF (I .LT. K) GO TO 20
      IF (IB .LT. N) GO TO 22
      RESULT=RESULT+H4*(F(N)-H4*(.5*R1+H4*(R2/6.+(H3+H3+H4)*R3/12.)))
      ERROK=ERROR-H4**3*R4*(H4*(3.*H4+5.*H2)+10.*H3*(H2+H3+H4))/60.
   22 IF (IB .GE. N-1) ERROR=ERROR+S*R4
      GO TO 1
   20 H1=H2
      H2=H3
      H3=H4
      D1=R1
      D2=R2
      D3=R3
    1 CONTINUE
      IF (IND.EQ.1) RETURN
      IT=IB
      IB=IA
      IA=IT
      RESULT=-RESULT
      ERROR=-ERROR
      IND=1
      RETURN
      END
```

SPLINT (*see Section 2.3.2*)

This program computes $\int_A^B s(x)\,dx$ where $s(x)$ is the natural spline interpolating the values Y(I) at the nodes X(I), I = 1, ..., N, $3 \le N \le 99$. In addition, it computes the indefinite integral $E(I) = \int_{X(1)}^{X(I)} s(x)\,dx$ and the coefficients Y1(I), Y2(I), and Y3(I), I = 1, ..., N. These coefficients enable one to evaluate $s(x)$ for $X(1) \le x \le X(N)$ according to the formula

$$s(x) = Y(I) + (x - X(I))Y1(I) + (x - X(I))^2 Y2(I) + (x - X(I))^3 Y3(I),$$

$X(I) \le x \le X(I + 1)$, I = 1, ..., N − 1. The X(I) must be distinct and in ascending order. If this is not so or if $N < 3$ or $N > 99$, the program exits with IND = 0. Otherwise IND = 1.

```
      FUNCTION SPLINT(X,Y,N,A,B,Y1,Y2,Y3,E,IND)
      DIMENSION X(N),Y(N),Y1(N),Y2(N),Y3(N),E(N),H(99),T(99)
      IND=0
      IF (N .LT. 3 .OR. N .GT. 99) RETURN
      M2=N-1
      S=0.
      DO 1 I=1,M2
      H(I)=X(I+1)-X(I)
      IF (H(I) .LE. 0.) RETURN
      R=(Y(I+1)-Y(I))/H(I)
      Y2(I)=R-S
    1 S=R
      SPLINT=0.
      S=0.
      R=0.
      Y2(1)=0.
      Y2(N)=0.
      DO 2 I=2,M2
      Y2(I)=Y2(I)+R*Y2(I-1)
      T(I)=2.*(X(I-1)-X(I+1))-R*S
      S=H(I)
    2 R=S/T(I)
      DO 3 J=2,M2
      I=M2+2-J
    3 Y2(I)=(H(I)*Y2(I+1)-Y2(I))/T(I)
      DO 4 I=1,M2
      S=H(I)
      R=Y2(I+1)-Y2(I)
      Y3(I)=R/S
      Y2(I)=3.*Y2(I)
    4 Y1(I)=(Y(I+1)-Y(I))/S-(Y2(I)+R)*S
      E(1)=0.
      DO 5 I=1,M2
      S=X(I+1)-X(I)
    5 E(I+1)=E(I)+(((Y3(I)*.25*S+Y2(I)/3.)*S+Y1(I)*.5)*S+Y(I))*S
      R=A
      U=1.
      DO 6 J=1,2
      IF ( R .GT. X(1) ) GO TO 7
      SPLINT=SPLINT-U*((R-X(1))*Y1(1)*.5+Y(1))*(R-X(1))
      GO TO 8
    7 IF ( R .LT. X(N) ) GO TO 9
      SPLINT=SPLINT-U*(E(N)+(R-X(N))*(Y(N)+.5*(Y(N-1)+(X(N)-X(N-1))
     X *Y1(N-1))*(R-X(N))))
      GO TO 8
    9 DO 10 I=1,M2
      IF (R-X(I+1)) 11,11,10
   11 R=R-X(I)
      SPLINT=SPLINT-U*(E(I)+(((Y3(I)*.25*R+Y2(I)/3.)*R+Y1(I)*.5)*R
     X +Y(I))*R)
      GO TO 8
```

```
10 CONTINUE
 8 U=-1.
 6 R=B
   IND=1
   RETURN
   END
```

GRULE (*see Section 2.7*)

This subroutine computes the [(N + 1)/2] nonnegative abscissas X(I) and corresponding weights W(I) of the N-point Gauss–Legendre integration rule, normalized to the interval [−1, 1]. The abscissas appear in descending order.

```
      SUBROUTINE GRULE(N,X,W)
      DOUBLE PRECISION PKM1,PK,T1,PKP1,DEN,D1,DPN,D2PN,D3PN,D4PN,U,V,H,
     *P,DP,FX
      DOUBLE PRECISION X(N),W(N)
      M=(N+1)/2
      E1=N*(N+1)
      DO 1 I=1,M
      T=(4*I-1)*3.1415926536/(4*N+2)
      X0=(1.-(1.-1./N)/(8.*N*N))*COS(T)
      PKM1=1.
      PK=X0
      DO 3 K=2,N
      T1=X0*PK
      PKP1=T1-PKM1-(T1-PKM1)/K+T1
      PKM1=PK
    3 PK=PKP1
      DEN=1.-X0*X0
      D1=N*(PKM1-X0*PK)
      DPN=D1/DEN
      D2PN=(2.*X0*DPN-E1*PK)/DEN
      D3PN=(4.*X0*D2PN+(2.-E1)*DPN)/DEN
      D4PN=(6.*X0*D3PN+(6.-E1)*D2PN)/DEN
      U=PK/DPN
      V=D2PN/DPN
      H=-U*(1.+.5*U*(V+U*(V*V-D3PN/(3.*DPN))))
      P=PK+H*(DPN+.5*H*(D2PN+H/3.*(D3PN+.25*H*D4PN)))
      DP=DPN+H*(D2PN+.5*H*(D3PN+H*D4PN/3.))
      H=H-P/DP
      X(I)=X0+H
      FX=D1-H*E1*(PK+.5*H*(DPN+H/3.*(D2PN+.25*H*(D3PN+.2*H*D4PN))))
    1 W(I)=2.*(1.-X(I)*X(I))/(FX*FX)
      IF (M+M .GT. N) X(M)=0.
      RETURN
      END
```

FILON (*see Section 2.10.2*)

This program computes

$$\int_A^B F(x)\cos Tx\,dx \qquad \text{if KEY} = 1$$

and

$$\int_A^B F(x)\sin Tx\,dx \qquad \text{if KEY} = 2,$$

using Filon's method. F(x) is assumed to be given as a table of functional values in the array F at M equidistant points from A to B, M odd.

For the case where F(x) can be computed by a subroutine, the use of the program of Chase and Fosdick [F1] is indicated.

```
      FUNCTION FILON (F,T,A,B,M,KEY)
      DIMENSION F(M)
      N=M-1
      H=(B-A)/N
      TH=T*H
      S=SIN(TH)
      C=COS(TH)
      AL=1./TH+S*C/TH**2-2.*S*S/TH**3
      BE=(2.+2.*C-4.*S*C/TH)/TH**2
      GA=4.*(S/TH-C)/TH**2
      SUM=0.0
      F1=F(1)
      F2=F(M)
      S1=SIN(A*T)
      S2=SIN(B*T)
      C1=COS(A*T)
      C2=COS(B*T)
      A1=A+H
      GO TO (1,2),KEY
    1 SU=F2*S2-F1*S1
      SU1=-.5*(F2*C2-F1*C1)
      DO 3 I=2,N,2
      SUM=SUM+F(I)*COS(A1*T)
      A1=A1+H
      SU1=SU1+F(I+1)*COS(A1*T)
    3 A1=A1+H
      GO TO 4
    2 SU=F1*C1-F2*C2
      SU1=-.5*(F2*S2-F1*S1)
      DO 5 I=2,N,2
      SUM=SUM+F(I)*SIN(A1*T)
      A1=A1+H
      SU1=SU1+F(I+1)*SIN(A1*T)
    5 A1=A1+H
    4 FILON=H*(AL*SU+BE*SU1+GA*SUM)
      RETURN
      END
```

SUM (*see Section 5.6*)

This program is used to compute the N-dimensional integral of FUN(x) over a product region B, using a product rule, $1 \leq N \leq 20$. B may be a product of any combination of finite, semi-infinite, or infinite intervals. For each factor in B, it is assumed that an integration rule is given and, hence, B is defined implicitly by the integration rules. The number of points in the Ith rule is given in N1(I). The abscissas and weights of the Ith rule are given in the Ith columns of the two-dimensional arrays X and W, respectively. The dimensions of X and W are (NMAX, N), where NMAX is the maximum number of points in any of the N rules. A program FUNCTION FUN(X, N) must be supplied by the user with X declared by the statement DIMENSION X(N). FUN must be declared EXTERNAL in the calling program. If $N < 1$ or $N > 20$ or $N1(I) < 1$ for any I, the program exits with IND = 0. Otherwise IND = 1.

```
      FUNCTION SUM(X,W,N1,N,NMAX,FUN,IND)
      DIMENSION X(NMAX,N),W(NMAX,N),N1(N),Z(20),M(20)
      IND=0
      IF(N.LT.1 .OR. N.GT.20) RETURN
      DO 1 I=1,N
      IF(N1(I).LT.1) RETURN
    1 M(I)=1
      IND=1
      SUM=0.
    6 K=1
      W1=1.
      DO 2 I=1,N
      M1=M(I)
      Z(I)=X(M1,I)
    2 W1=W1*W(M1,I)
      SUM=SUM+W1*FUN(Z,N)
    8 IF(M(K) .EQ. N1(K)) GO TO 4
      M(K)=M(K)+1
      GO TO 6
    4 M(K)=1
      K=K+1
      IF (K.LE.N) GO TO 8
      RETURN
      END
```

P5 (see Section 5.7)

This program computes a precision 5 approximation to

$$\int_{O(1)-H(1)}^{O(1)+H(1)} \cdots \int_{O(N)-H(N)}^{O(N)+H(N)} F(x_1, \ldots, x_N)\, dx_1 \cdots dx_N, \qquad 2 \le N \le 20.$$

A program FUNCTION F(X, N) must be supplied by the user with X declared by the statement DIMENSION X(N). F must be declared EXTERNAL in the calling program. If $N < 2$ or $N > 20$, the program exits with IND = 0. Otherwise IND = 1.

```
      FUNCTION P5(F,N,O,H,IND)
      DIMENSION O(N),H(N),X(20)
      IND=0
      IF(N.LT.2 .OR. N.GT.20) RETURN
      IND=1
      NN=N-1
      A2=25./324.
      A=SQRT(.6)
      EN=N
      A0=(25.*EN*EN-115.*EN+162.)/162.
      A1=(70.-25.*EN)/162.
      H1=1.
      DO 4 I=1,N
    4 H1=2.*H1*H(I)
      DO 1 I=1,N
```

```
    1 X(I)=O(I)
      S=A0*F(X,N)
      S1=0.
      DO 2 I=1,N
      X(I)=O(I)+A*H(I)
      S1=S1+F(X,N)
      X(I)=O(I)-A*H(I)
      S1=S1+F(X,N)
    2 X(I)=O(I)
      S2=0.
      B=A
    5 DO 7 I=1,NN
      X(I)=O(I)+B*H(I)
      C=A
      I1=I+1
    6 DO 3 J=I1,N
      X(J)=O(J)+C*H(J)
      S2=S2+F(X,N)
    3 X(J)=O(J)
      C=-C
      IF(C .LT. 0.) GO TO 6
    7 X(I)=O(I)
      B=-B
      IF(B .LT. 0.) GO TO 5
      P5=H1*(S+A1*S1+A2*S2)
      RETURN
      END
```

HABER (*see Section 5.9.5*)

This subroutine computes a sequence of estimates EST1(K) and EST2(K), $1 \leq \text{K1} \leq \text{K} \leq \text{K2}$ for the N-dimensional integral

$$\int_0^1 \cdots \int_0^1 \text{FUN}(x_1, \ldots, x_\text{N})\, dx_1 \cdots dx_\text{N}, \qquad 1 \leq \text{N} \leq 10$$

by Haber's method. For each estimate EST1(K), two additional quantities ERR1(K) and DEV1(K) are computed. If the values of DEV1(K) do not vary by more than 10% between consecutive values of K, then ERR1(K) can be taken as a reliable bound on the difference between EST1(K) and the integral. A similar situation holds for EST2(K), DEV2(K), and ERR2(K). The total number of function evaluations is $4(\text{K1}^\text{N} + (\text{K1}+1)^\text{N} + \cdots + \text{K2}^\text{N})$ and K2 should be chosen so as to make this quantity reasonable. If FUN($\mathbf{x}$) is discontinuous, the time may be halved by eliminating the computation of the EST2(K). In other situations, these values are much better than the EST1(K). A program FUNCTION FUN(X, N) must be supplied by the user with X declared by the statement DIMENSION X(N). FUN must be declared EXTERNAL in the calling program. If $\text{N} < 1$ or $\text{N} > 10$ or $\text{K1} < 1$ or $\text{K2} < \text{K1}$, the program exits with IND = 0. Otherwise IND = 1.

```
      SUBROUTINE HABER(K1,K2,N,FUN,EST1,ERR1,DEV1,EST2,ERR2,DEV2,IND)
      DOUBLE PRECISION AL,BE,GA,B,G
      DOUBLE PRECISION S1,D1,S2,D2
      DIMENSION EST1(K2),ERR1(K2),DEV1(K2),
     X          EST2(K2),ERR2(K2),DEV2(K2),
     X          AL(10),BE(10),GA(10),DEX(10),P1(10),
```

```
  X           P2(10),P3(10),P4(10)
   DATA AL/.4142135623730950 ,.7320508075688773 ,.2360679774997897,
  *         .6457513110645906 ,.3166247903553998 ,.6055512754639893,
  *         .1231056256176605 ,.3589989435406736 ,.7958315233127195,
  *         .3851648071345040/
   IND=0
   IF(N.LT.1 .OR. N.GT.10 .OR. K1.LT.1 .OR. K1.GT.K2) RETURN
   IND=1
   DO 1 I=1,N
   BE(I)=AL(I)
   GA(I)=AL(I)
 1 DEX(I)=0.
   DO 2 K=K1,K2
   AK=K
   KEY=0
   AK1=AK-1.1
   S1=0.
   D1=0.
   S2=0.
   D2=0.
   AKN=AK**N
   T=SQRT(AKN)*AK
   BK=1./AK
 5 KEY=KEY+1
   IF (KEY .EQ. 1) GO TO 6
   KEY=KEY-1
   J=1
 4 IF(DEX(J).GT.AK1) GO TO 8
   DEX(J)=DEX(J)+1.
   GO TO 6
 8 DEX(J)=0.
   J=J+1
   IF(J.LE.N) GO TO 4
   GO TO 3
 6 DO 7 I=1,N
   B=BE(I)+AL(I)
   IF (B .GT. 1.) B=B-1.
   G=GA(I)+B
   IF (G .GT. 1.) G=G-1.
   BE(I)=B+AL(I)
   IF (BE(I).GT.1.) BE(I)=BE(I)-1.
   GA(I)=BE(I)+G
   IF(GA(I).GT.1.) GA(I)=GA(I)-1.
   P1(I)=(DEX(I)+G)*BK
   P2(I)=(DEX(I)+1.-G)*BK
   P3(I)=(DEX(I)+GA(I))*BK
 7 P4(I)=(DEX(I)+1.-GA(I))*BK
   Y1=FUN(P1,N)
   Y3=FUN(P2,N)
   Y2=FUN(P3,N)
   Y4=FUN(P4,N)
   S1=S1+Y1+Y2
   D1=D1+(Y1-Y2)**2
   S2=S2+Y3+Y4
   D2=D2+(Y1+Y3-Y2-Y4)**2
   GO TO 5
 3 EST1(K)= .5*S1/AKN
   ERR1(K)= 1.5*SQRT(D1)/AKN
   DEV1(K)=ERR1(K)*T
   EST2(K)=.25*(S1+S2)/AKN
   ERR2(K)= .75*SQRT(D2)/AKN
 2 DEV2(K)=ERR2(K)*T*AK
   RETURN
   END
```

SIMP (*see Section 6.2.3*)

This program is a FORTRAN adaptation from McKeeman and Tesler [A1]. It computes an approximation to $\int_{A1}^{B}$ FUN(x) dx to within the tolerance EP except for exceptional situations. A program FUNCTION FUN(X) must be supplied by the user and FUN declared EXTERNAL in the calling program. An alternative program incorporating several improvements is SQUANK by Lyness [F1].

```
      FUNCTION SIMP(A1,B,EP,FUN)
C     NONRECURSIVE ADAPTIVE INTEGRATION
C     ALGORITHM 182 CACM 6 (1963) 315
      DIMENSION DX(30),EPSP(30),X2(30),X3(30),F2(30),F3(30),F4(30),
     1FMP(30),FBP(30),EST2(30),EST3(30),PVAL(30,3),NRTR(30)
      COMMON PVAL,SUM,LVL,L1
C     THE PARAMETER SETUP FOR THE INITIAL CALL
      A=A1
      EPS=EP
      LVL=0
      ABSAR  =0.
      EST=0.
      DA=B-A
      FA=FUN(A)
      FM=4.*FUN ((A+B)*.5)
      FB=FUN(B)
C     1=RECUR
1     LVL=LVL+1
      DX(LVL)=DA/3.
      SX=DX(LVL)/6.
      F1=4.*FUN(.5*DX(LVL)+A)
      X2(LVL)=A+DX(LVL)
      F2(LVL)=FUN(X2(LVL))
      X3(LVL)=X2(LVL)+DX(LVL)
      F3(LVL)=FUN(X3(LVL))
      EPSP(LVL)=EPS
      F4(LVL)=4.*FUN      ( DX(LVL)*.5+X3(LVL))
      FMP(LVL)=FM
      EST1=SX*(FA+F1+F2(LVL))
      FBP(LVL)=FB
      EST2(LVL)=SX*(F2(LVL)+F3(LVL)+FM)
      EST3(LVL)=SX*(F3(LVL)+F4(LVL)+FB)
      SUM=EST1+EST2(LVL)+EST3(LVL)
      ABSAR  =ABSAR  -ABSF(EST)+ABSF(EST1)+ABSF(EST2(LVL))+ABSF(EST3(LVL
     1))
      IF(ABSF(EST-SUM)-EPSP(LVL)*ABSAR  ) 2,2,3
3     IF(LVL-30) 4,2,2
C     2=UP
2     LVL=LVL-1
      L=NRTR(LVL)
      PVAL(LVL,L)=SUM
      GO TO (11,12,13)L
C     11=R1,12=R2,13=R3
4     NRTR(LVL)=1
      EST=EST1
      FM=F1
      FB=F2(LVL)
7     EPS=EPSP(LVL)/1.7
      DA=DX(LVL)
      GO TO 1
11    NRTR(LVL)=2
      FA=F2(LVL)
      FM=FMP(LVL)
      FB=F3(LVL)
      EST=EST2(LVL)
```

```
      A=X2(LVL)
      GO TO 7
12    NRTR(LVL)=3
      FA=F3(LVL)
      FM=F4(LVL)
      FB=FBP(LVL)
      EST=EST3(LVL)
      A=X3(LVL)
      GO TO 7
13    SUM=PVAL(LVL,1)+PVAL(LVL,2)+PVAL(LVL,3)
      IF(LVL-1) 5,5,2
5     SIMP   =SUM
      RETURN
```

QUAD (*see Section 6.3*)

This program is used to approximate $I = \int_A^B \mathrm{FUN}(x)\,dx$ by Romberg integration and was written by Dunkl [1] (courtesy SHARE Installation TY, University of Toronto, Toronto, Ontario, Canada). The maximum number of steps taken is fifteen and, if convergence is not achieved by then, the final value computed is taken as the approximation to the integral. The parameters EPS and ETA are error tolerances and the result QUAD hopefully satisfies the less restrictive of the following two conditions:

(i) $|\mathrm{QUAD} - I| < \mathrm{EPS}$ (absolute error),
(ii) $\mathrm{QUAD} = \int_A^B \{1 + y(x)\}\mathrm{FUN}(x)\,dx$, where $|y(x)| < \mathrm{ETA} + 2^{-26}$ (relative error).

In case (ii), if FUN(x) does not change sign for any x between A and B, then $|\mathrm{QUAD} - I|/|I| < \mathrm{ETA} + 2^{-26}$.

The parameter MIN indicates at what step in the integration procedure we start to check for convergence; $3 < \mathrm{MIN} < 15$. A program FUNCTION FUN(X) must be supplied by the user and FUN declared EXTERNAL in the calling program.

```
      FUNCTION QUAD(A,B,FUN,EPS,ETA,MIN)
C     ROMBERG INTEGRATION
      DIMENSION Q(16)
1     H=B-A
      FCNA=FUN(A)
      FCNB=FUN(B)
      TABS=ABSF(H)*(ABSF(FCNA)+ABSF(FCNB))/2.
      T=H*(FCNA+FCNB)/2.
      NX=1
      DO 12 N=1,15
      H=H/2.
      SUM=0
      SCORR=0
      SUMABS=0
      DO 2 I=1,NX
      XI=2.*FLOATF(I)-1.
      FCNXI=FUN(A+XI*H)
      SUMABS=SUMABS+ABSF(FCNXI)
      FCNXI=FCNXI+SCORR
```

```
      SS=SUM+FCNXI
      SCORR=(SUM-SS)+FCNXI
2     SUM=SS
      T=T/2.+H*SUM
      TABS=TABS/2.+ABSF(H)*SUMABS
      Q(N)=2.*(T+H*SUM)/3.
      IF(N-2) 10,3,3
3     F=4.
      DO 4 J=2,N
      I=N+1-J
      F=F*4.
4     Q(I)=Q(I+1)+(Q(I+1)-Q(I))/(F-1.)
      IF(N-3) 9,5,5
5     IF(N-MIN) 9,6,6
6     X=ABSF(Q(1)-QX2)+ABSF(QX2-QX1)
      IF(TABS) 7,8,7
7     IF(X/TABS-3.*(ABSF(ETA)+0.14901161E-7))11,11,8
8     IF(X-3.*ABSF(EPS))11,11,9
9     QX1=QX2
10    QX2=Q(1)
12    NX=NX*2
11    QUAD=Q(1)
      RETURN
      END
```

RMINSP (*see Section 6.3*)

This program computes $\int_{A}^{B}$ FUNC(x) dx, hopefully to a relative accuracy of EPSIN. EPSOUT is a good estimate for the relative error. IOP = 1 or 2. If IOP = 1, the ordinary modified Romberg algorithm is used. If IOP = 2, the modified Romberg algorithm is applied to a transformed integral. This option should be used if FUNC(x) has singularities at the endpoints of the integration interval. A program FUNCTION FUNC(X) must be supplied by the user and FUNC must be declared EXTERNAL in the calling program. RNDERR in the DATA statement is a machine-dependent parameter which should be set to the relative machine accuracy. For further information, see the COMMENTS in the program.

RMINSP was supplied by its author Prof. T. Håvie who developed it at CERN and is discussed in Håvie [1, 2, 6].

```
      FUNCTION RMINSP(A,B,EPSIN,EPSOUT,FUNC,IOP)
C
C     ROMBERG/HAAVIE EXTENDED INTEGRATION THIS ALGORITHM IS AN EXTENDED
C     VERSION OF ALGORITHM NO 257 FROM COMM.ACM. (SEE ALSO BIT 6 (1966)
C     24-30 AND BIT 7 (1967),103-113)
C     PARAMETERS
C     A        = LOWER BOUNDARY
C     B        = UPPER BOUNDARY
C     EPSIN    = ACCURACY REQUIRED FOR THE APPROXINATION
C     EPSOUT   = IMPROVED ERROR ESTIMATE FOR THE APPROXIMATION
C     FUNC     = FUNCTION ROUTINE FOR THE FUNCTION FUNC(X).TO BE DE-
C                CLARED EXTERNAL IN THE CALLING ROUTINE
C     IOP      = OPTION PARAMETER , IOP=1 , MODIFIED ROMBERG ALGORITHM,
C                                           ORDINARY CASE
C                                   IOP=2 , MODIFIED ROMBERG ALGORITHM,
C                                           COSINE TRANSFORMED CASE
```

```
C      INTEGRATION PARAMETERS
C      NUPPER   = 9 , CORRESPONDS TO 1024 SUB-INTERVALS FOR THE UNFOLDED
C                 INTEGRAL THE MAX. NO OF FUNCTION EVALUATIONS THUS BEING
C                 1025 THE HIGHEST END-POINT APPROXIMATION IS THUS USING
C                 1024 INTERVALS WHILE THE HIGHEST MID-POINT APPROXIMA-
C                 TION IS USING 512 INTERVALS
C      RNDERR   = 1.E-14 , THE RELATIVE MACHINE ACCURACY IN SINGLE PRECI-
C                 SION (CDC-6000 SERIES).
C
       DIMENSION ACOF(11) , BCOF(10)
       DOUBLE PRECISION PI,FAC1,FAC2
       DATA ZERO,FOURTH,HALF,ONE,TWO,FOUR,PI,FAC1,FAC2,NUPPER,RNDERR
      *    /0.,.25,.5,1.,2.,4.,3.14159265358979,.411233516712057,
      *      .822467033441132,9,1.E-14 /
C      SET COEFFICIENTS IN FORMULA FOR ACCUMULATED ROUND OFF ERROR,
C      ROUNDE=RNDERR*(R1+R2*N),WHERE R1,R2 ARE TWO EMPIRICAL CONSTANTS
C      AND N IS THE CURRENT NUMBER OF FUNCTION VALUES USED.
 1000  R1=ONE
       R2=TWO
       IF (IOP.EQ.2) R1=5.E+1
       IF (IOP.EQ.1) R2=1.E-2*R2
       ERROR=EPSIN
C      INITIAL CALCULATIONS
       ALF=HALF*(B-A)
       BET=HALF*(B+A)
       ACOF(1)=FUNC(A)+FUNC(B)
       BCOF(1)=FUNC(BET)
       IF (IOP.EQ 2) GO TO 1010
C      MODIFIED ROMBERG ALGORITHM,ORDINARY CASE
       HNSTEP=TWO
       BCOF(1)=HNSTEP*BCOF(1)
       FACTOR=ONE
       GO TO 1030
C      MODIFIED ROMBERG ALGORITHM,COSINE TRANSFORMED CASE
 1010  HNSTEP=PI
       AR=FAC1
       ENDPTS=ACOF(1)
       ACOF(1)=FAC2*ACOF(1)
       BCOF(1)=HNSTEP*BCOF(1)-AR*ENDPTS
       FACTOR=FOUR
       AR=FOURTH*AR
C      INITIAL PARAMETERS SPECIAL FOR THE MODIFIED ROMBERG ALGORITHM,
C      COSINE TRANSFORMED CASE.
       TRIARG=FOURTH*PI
       ALFNO=-ONE
 1030  HNSTEP=HALF*HNSTEP
       NHALF=1
       N=2
       RN=TWO
 1040  ACOF(1)=HALF*(ACOF(1)+BCOF(1))
       ACOF(2)=ACOF(1)-(ACOF(1)-BCOF(1))/(FOUR*FACTOR-ONE)
C      END OF INITIAL CALCULATIONS
C
C      START ACTUAL CALCULATION
       DO 2140 I=1,NUPPER
       LINE=I+1
       UMID=ZERO
       IF (IOP.EQ.2) GO TO 1060
C      MODIFIED ROMBERG ALGORITHM,ORDINARY CASE
C      COMPUTE FIRST ELEMENT IN MID-POINT FORMULA FOR ORDINARY CASE
       ALFNJ=HALF*HNSTEP
       DO 1050 J=1,NHALF
       XPLUS=ALF*ALFNJ+BET
       XMIN=-ALF*ALFNJ+BET
       UMID=UMID+FUNC(XPLUS)+FUNC(XMIN)
       ALFNJ=ALFNJ+HNSTEP
```

```
 1050 CONTINUE
      UMID=HNSTEP*UMID
      GO TO 2000
C     MODIFIED ROMBERG ALGORITHM,COSINE TRANSFORMED CASE
C     COMPUTE FIRST ELEMENT IN MID-POINT FORMULA FOR COSINE TRANSFORMED
C     ROMBERG ALGORITHM
 1060 CONST1=-SIN(TRIARG)
      CONST2=HALF*ALFNO/CONST1
      IF (IOP.EQ.2) ETANK=CONST2
      ALFNO=CONST1
      BETNO=CONST2
      GAMMAN=ONE-TWO*ALFNO**2
      DELTAN=-TWO*ALFNO*BETNO
      DO 1070 J=1,NHALF
      ALFNJ=GAMMAN*CONST1+DELTAN*CONST2
      BETNJ=GAMMAN*CONST2-DELTAN*CONST1
      XPLUS=ALF*ALFNJ+BET
      XMIN=-ALF*ALFNJ+BET
      UMID=UMID+BETNJ*(FUNC(XPLUS)+FUNC(XMIN))
      CONST1=ALFNJ
      CONST2=BETNJ
 1070 CONTINUE
      UMID=HNSTEP*UMID-AR*ENDPTS
      AR=FOURTH*AR
C     MODIFIED ROMBERG ALGORITHM,CALCULATE (I+1)-TH ROW IN U-TABLE
 2000 CONST1=FOUR*FACTOR
      INDEX=I+1
      DO 2010 J=2,INDEX
      TEND=UMID+(UMID-BCOF(J-1))/(CONST1-ONE)
      BCOF(J-1)=UMID
      UMID=TEND
      CONST1=FOUR*CONST1
 2010 CONTINUE
      BCOF(INDEX)=TEND
      XPLUS=CONST1
C     CALCULATION OF (I+1)-TH ROW IN U-TABLE FINISHED
C
C     TEST IF REQUIRED ACCURACY IS OBTAINED
 2020 EPSOUT=ONE
      IOUT=1
      DO 2040 J=1,INDEX
      CONST1=HALF*(ACOF(J)+BCOF(J))
      CONST2=HALF*ABS((ACOF(J)-BCOF(J))/CONST1)
      IF (CONST2.GT.EPSOUT) GO TO 2030
      EPSOUT=CONST2
      IOUT=J
 2030 ACOF(J)=CONST1
 2040 CONTINUE
C     TESTING ON ACCURACY FINISHED
      IF (IOUT.EQ.INDEX) IOUT=IOUT+1
      ACOF(INDEX+1)=ACOF(INDEX)-(ACOF(INDEX)-BCOF(INDEX))/(XPLUS-ONE)
      LINE=I+2
      ROUNDE=RNDERR*(R1+R2*RN)
      IF (EPSOUT.LT.ROUNDE) EPSOUT=ROUNDE
      IF (ERROR.LT.ROUNDE) ERROR=ROUNDE
      IF (EPSOUT.LE. ERROR) GO TO 2050
 2130 NHALF=N
      N=2*N
      RN=TWO*RN
      HNSTEP=HALF*HNSTEP
      IF (IOP.GT.1) TRIARG=HALF*TRIARG
 2140 CONTINUE
C     ACCURACY NOT REACHED WITH MAX NO OF SUBDIVISIONS
      N   =NHALF
C     CALCULATION FOR MODIFIED ROMBERG ALGORITHM FINISHED
 2050 N=2*N
```

```
2070 INDEX=INDEX-1
2080 N=N+1
     J=IOUT
     IF ((J-1).LT. INDEX) GO TO 2090
     J=INDEX
2090 TEND=ALF*(TWO*ACOF(J)-BCOF(J))
     UMID=ALF*BCOF(J)
     RMINSP=ALF*ACOF(IOUT)
2100 RETURN
     END
```

IRATEX (*see Section 6.3.1*)

This subroutine computes QD, an approximation to $\int_A^B F(x)\,dx$, hopefully to within relative error EPS. ERR is an estimate of the absolute error. If IND = 0, the desired accuracy has not been achieved. If IND = 1, the result is probably correct to the desired accuracy. A program FUNCTION F(X) must be supplied by the user and F must be declared EXTERNAL in the calling program. ETA in the DATA statement is a machine-dependent parameter which should be set to the relative machine accuracy. IRATEX is based on the ALGOL procedure "Quadrature" by Bulirsch and Stoer [A1].

```
    SUBROUTINE IRATEX(F,A,B,EPS,QD,ERR,IND)
    DIMENSION DT(7),D(6)
    LOGICAL BO,BU,ODD
    DATA ETA /1.E-7/
    EPS=AMAX1(EPS,ETA)
    N=2
    NN=3
    BA=B-A
    T1=0.
    GR=0.
    SM=0.
    T2A=(F(A)+F(B))*.5
    T2=T2A
    TB=ABS(T2A)
    C=T2*BA
    DT(1)=C
    DO 1 I=2,7
  1 DT(I)=0.
    ODD=.TRUE.
    BU=.FALSE.
    IND=0
    DO 2 M=1,15
    BO=M.GE.7
    HM=BA/N
    IF (ODD) GO TO 3
  4 DO 7 I=1,N,6
    W=I*HM
  7 T1=T1+F(A+W)+F(B-W)
    ENT=T1+T2A
    T2A=T2
    D(1)=2.25
    D(3)=9.
    D(5)=36.
    GO TO 6
```

```
  3 DO 5 I=1,N,2
    W=F(A+I*HM)
    T2=T2+W
  5 TB=ABS(W)+TB
    ENT=T2
    TAB=TB*ABS(HM)
    D(1)=16./9.
    D(3)=4.*D(1)
    D(5)=4.*D(3)
  6 DDT=DT(1)
    T=ENT*HM
    DT(1)=T
    ENT=T
    IF (BO) GO TO 8
  9 MR=M
    W=N*N
    D(M)=W
    GO TO 10
  8 MR=6
    D(6)=64.
    W=144.
 10 DO 12 I=1,MR
    D1=D(I)*DDT
    DEN=D1-ENT
    E=ENT-DDT
    TNT=ENT
    V=0.
    ENT=0.
    IF (ABS(DEN) .LT. EPS) GO TO 11
    E=E/DEN
    V=TNT*E
    ENT=D1*E
    T=V+T
 11 DDT=DT(I+1)
 12 DT(I+1)=V
    TA=C
    C=T
    QD=C
    IF (.NOT.BO) T=T-V
    V=T-TA
    T=V+T
    ERR=ABS(V)
    IF (TA .GE. T) GO TO 13
    D1=TA
    TA=T
    T=D1
 13 BO=BO .OR. TA .LT. GR .AND. T .GT. SM
    IF (BU .AND. BO .AND. ERR .LT. EPS*W*TAB) GO TO 14
    GR=TA
    SM=T
    ODD= .NOT. ODD
    I=N
    N=NN
    NN=I+I
    BU=BO
    D(2)=4.
  2 D(4)=16.
    BO=.FALSE.
 14 V=ETA*TAB
    IF (ERR .LT.V) ERR=V
    IF (BO) IND=1
    RETURN
    END
```

CADRE (*see Section 6.3*)

This program, a stripped-down version of the program CADRE (de Boor [F1]) computes $\int_A^B F(x)\,dx$, hopefully to a relative accuracy RERR or an absolute accuracy AERR, whichever is larger. ERROR is computed by the program as an estimate of the absolute error. IFLAG gives an indication of the reliability of the result. A cautious man accepts the result only if IFLAG ≤ 2, a reasonable one if also IFLAG $= 3$, while an adventurous one would even accept the result if IFLAG $= 4$ or 5. A program FUNCTION F(X) must be supplied by the user and F must be declared EXTERNAL in the calling program. TOLMCH in the DATA statement is a machine-dependent parameter which depends on the length of the floating-point mantissa.

For further details, refer to de Boor [1] and the full FORTRAN listing contained therein.

```
      FUNCTION CADRE(F,A,B,AERR,RERR,ERROR,IFLAG)
      DIMENSION T(10,10),R(10),AIT(10),DIF(10),RN(4),
     *          TS(2049),IBEGS(30),BEGIN(30),FINIS(30),EST(30)
      REAL LENGTH, JUMPTL
      LOGICAL H2CONV,AITKEN,RIGHT,REGLAR,REGLSV(30)
      DOUBLE PRECISION    ALG402
      DATA TOLMCH,AITLOW,H2TOL,AITTOL,JUMPTL,MAXTS,MAXTBL,MXSTGE
     *  / 2.E-13, 1.1  , .15 ,  .1  , .01  , 2049,  10  ,  30/
      DATA RN/.7142005 ,.3466282 ,.8437510 ,.1263305/
      DATA ALG402 /.3010299956639795/
      CADRE = 0.
      ERROR = 0.
      IFLAG = 1
      LENGTH = ABS(B-A)
      IF (LENGTH .EQ. 0.)                 RETURN
      ERRR = AMIN1(.1,AMAX1(ABS(RERR), 10.*TOLMCH))
      ERRA = ABS(AERR)
      STEPMN = AMAX1(LENGTH/2**MXSTGE,
     *         AMAX1(LENGTH,ABS(A),ABS(B))*TOLMCH)
      STAGE = .5
      ISTAGE = 1
      CUREST = 0.
      FNSIZE = 0.
      PREVER = 0.
      REGLAR=.FALSE.
      BEG = A
      FBEG = F(BEG)/2.
      TS(1) = FBEG
      IBEG = 1
      END=B
      FEND = F(END)/2.
      TS(2) = FEND
      IEND = 2
    5 RIGHT=.FALSE.
    6 STEP = END - BEG
      ASTEP = ABS(STEP)
      IF (ASTEP .LT. STEPMN)              GO TO 950
      T(1,1) = FBEG + FEND
```

```
      TABS = ABS(FBEG) + ABS(FEND)
      L = 1
      N = 1
      H2CONV=.FALSE.
      AITKEN=.FALSE.
                                            GO TO 10
    9 CONTINUE
   10 LM1 = L
      L = L + 1
      N2 = N*2
      FN = N2
      ISTEP = (IEND - IBEG)/N
      IF (ISTEP .GT. 1)                     GO TO 12
      II = IEND
      IEND = IEND + N
      IF (IEND .GT. MAXTS)                  GO TO 900
      HOVN = STEP/FN
      III = IEND
      DO 11 I=1,N2,2
      TS(III) = TS(II)
      TS(III-1) = F(END - I*HOVN)
      III = III-2
   11 II = II-1
      ISTEP = 2
   12 ISTEP2 = IBEG + ISTEP/2
      SUM = 0.
      SUMABS = 0.
      DO 13 I=ISTEP2,IEND,ISTEP
      SUM = SUM + TS(I)
   13 SUMABS = SUMABS + ABS(TS(I))
      T(L,1) = T(L-1,1)/2. + SUM/FN
      TABS = TABS/2. + SUMABS/FN
      ABSI = ASTEP*TABS
      N = N2
      IT = 1
      VINT = STEP*T(L,1)
      TABTLM = TABS*TOLMCH
      FNSIZE = AMAX1(FNSIZE,ABS(T(L,1)))
      ERGOAL = AMAX1(ASTEP*TOLMCH*FNSIZE,
     *          STAGE*AMAX1(ERRA,ERRR*ABS(CUREST+VINT)))
      FEXTRP = 1.
      DO 14 I=1,LM1
      FEXTRP = FEXTRP*4.
      T(I,L) = T(L,I) - T(L-1,I)
   14 T(L,I+1) = T(L,I) + T(I,L)/(FEXTRP-1.)
      ERRER = ASTEP*ABS(T(1,L))
      IF (L .GT. 2)                         GO TO 15
      IF (ABS(T(1,2)) .LE. TABTLM)          GO TO 60
                                            GO TO 10
   15 DO 16 I=2,LM1
      DIFF = 0.
      IF (ABS(T(I-1,L)) .GT. TABTLM) DIFF = T(I-1,LM1)/T(I-1,L)
   16 T(I-1,LM1) = DIFF
      IF (ABS(4.-T(1,LM1)) .LE. H2TOL) GO TO 20
      IF (T(1,LM1) .EQ. 0.)                 GO TO 18
      IF (ABS(2.-ABS(T(1,LM1))) .LT. JUMPTL) GO TO 50
      IF (L .EQ. 3)                         GO TO 9
      H2CONV=.FALSE.
      IF (ABS((T(1,LM1)-T(1,L-2))/T(1,LM1)) .LE. AITTOL)
     *                                      GO TO 30
   17 IF (REGLAR)                           GO TO 18
      IF (L .EQ. 4)                         GO TO 9
   18 IF (ERRER .LE. ERGOAL)                GO TO 70
                                            GO TO 91
   20 IF (H2CONV)                           GO TO 21
```

```
      AITKEN=.FALSE.
      H2CONV=.TRUE.
   21 FEXTRP = 4.
   22 IT = IT + 1
      VINT = STEP*T(L,IT)
      ERRER = ABS(STEP/(FEXTRP-1.)*T(IT-1,L))
      IF (ERRER .LE. ERGOAL)                  GO TO 80
      IF (IT .EQ. LM1)                        GO TO 40
      IF (T(IT,LM1) .EQ. 0.)                  GO TO 22
      IF (T(IT,LM1) .LE. FEXTRP)              GO TO 40
      IF (ABS(T(IT,LM1)/4.-FEXTRP)/FEXTRP .LT. AITTOL)
     *        FEXTRP = FEXTRP*4.
                                              GO TO 22
   30 IF (T(1,LM1) .LT. AITLOW)               GO TO 91
      IF (AITKEN)                             GO TO 31
      H2CONV=.FALSE.
      AITKEN=.TRUE.
   31 FEXTRP = T(L-2,LM1)
      IF (FEXTRP .GT. 4.5)                    GO TO 21
      IF (FEXTRP .LT. AITLOW)                 GO TO 91
      IF (ABS(FEXTRP-T(L-3,LM1))/T(1,LM1) .GT. H2TOL)
     *                                        GO TO 91
      SING = FEXTRP
      FEXTM1 = FEXTRP - 1.
      AIT(1)=0.
      DO 32 I=2,L
      AIT(I) = T(I,1) + (T(I,1)-T(I-1,1))/FEXTM1
      R(I) = T(1,I-1)
   32 DIF(I) = AIT(I) - AIT(I-1)
      IT = 2
   33 VINT = STEP*AIT(L)
  333 ERRER = ERRER/FEXTM1
      IF (ERRER .GT. ERGOAL)                  GO TO 34
      ALPHA = ALOG10(SING)/ALG402 - 1.
      IFLAG = MAX0(IFLAG,2)
                                              GO TO 80
   34 IT = IT + 1
      IF (IT .EQ. LM1)                        GO TO 40
      IF (IT .GT. 3)                          GO TO 35
      H2NEXT = 4.
      SINGNX = 2.*SING
   35 IF (H2NEXT .LT. SINGNX)                 GO TO 36
      FEXTRP = SINGNX
      SINGNX = 2.*SINGNX
                                              GO TO 37
   36 FEXTRP = H2NEXT
      H2NEXT = 4.*H2NEXT
   37 DO 38 I=IT,LM1
      R(I+1) = 0.
   38 IF (ABS(DIF(I+1)) .GT. TABTLM) R(I+1) = DIF(I)/DIF(I+1)
      H2TFEX = -H2TOL*FEXTRP
      IF (R(L) - FEXTRP .LT. H2TFEX)          GO TO 40
      IF (R(L-1)-FEXTRP .LT. H2TFEX)          GO TO 40
      ERRER = ASTEP*ABS(DIF(L))
      FEXTM1 = FEXTRP - 1.
      DO 39 I=IT,L
      AIT(I) = AIT(I) + DIF(I)/FEXTM1
   39 DIF(I) = AIT(I) - AIT(I-1)
                                              GO TO 33
   40 FEXTRP = AMAX1(PREVER/ERRER,AITLOW)
      PREVER = ERRER
      IF (L .LT. 5)                           GO TO 10
      IF (L-IT .GT. 2 .AND. ISTAGE .LT. MXSTGE) GO TO 90
      IF (ERRER/FEXTRP**(MAXTBL-L) .LT. ERGOAL) GO TO 10
                                              GO TO 90
```

```
50 IF (ERRER .GT. ERGOAL)                    GO TO 90
   DIFF = ABS(T(1,L))*2.*FN
                                             GO TO 80
60 SLOPE = (FEND-FBEG)*2.
   FBEG2 = FBEG*2.
   DO 61 I=1,4
   DIFF = ABS(F(BEG+RN(I)*STEP) - FBEG2-RN(I)*SLOPE)
   IF (DIFF .GT. TABTLM)                     GO TO 72
61 CONTINUE
                                             GO TO 80
70 SLOPE = (FEND-FBEG)*2.
   FBEG2 = FBEG*2.
   I = 1
71 DIFF = ABS(F(BEG+RN(I)*STEP) - FBEG2-RN(I)*SLOPE)
72 ERRER = AMAX1(ERRER,ASTEP*DIFF)
   IF (ERRER .GT. ERGOAL)                    GO TO 91
   I = I+1
   IF (I .LE. 4)                             GO TO 71
   IFLAG = 3
80 CADRE = CADRE + VINT
   ERROR = ERROR + ERRER
83 IF (RIGHT)                                GO TO 85
   ISTAGE = ISTAGE - 1
   IF (ISTAGE .EQ. 0)                        RETURN
   REGLAR = REGLSV(ISTAGE)
   BEG = BEGIN(ISTAGE)
   END=FINIS(ISTAGE)
   CUREST = CUREST - EST(ISTAGE+1) + VINT
   IEND = IBEG - 1
   FEND = TS(IEND)
   IBEG = IBEGS(ISTAGE)
                                             GO TO 94
85 CUREST = CUREST + VINT
   STAGE = STAGE*2.
   IEND = IBEG
   IBEG = IBEGS(ISTAGE)
   END=BEG
   BEG = BEGIN(ISTAGE)
   FEND = FBEG
   FBEG = TS(IBEG)
                                             GO TO 5
90 REGLAR=.TRUE.
91 IF (ISTAGE .EQ. MXSTGE)                   GO TO 950
93 IF (RIGHT)                                GO TO 95
   REGLSV(ISTAGE+1) = REGLAR
   BEGIN(ISTAGE) = BEG
   IBEGS(ISTAGE) = IBEG
   STAGE = STAGE/2.
94 RIGHT=.TRUE.
   BEG = (BEG+END)/2.
   IBEG = (IBEG+IEND)/2
   TS(IBEG) = TS(IBEG)/2.
   FBEG = TS(IBEG)
                                             GO TO 6
95 NNLEFT = IBEG - IBEGS(ISTAGE)
   IF (END+NNLEFT .GE. MAXTS)                GO TO 900
   III = IBEGS(ISTAGE)
   II = IEND
   DO 96 I=III,IBEG
   II = II + 1
96 TS(II) = TS(I)
   DO 97 I=IBEG,II
   TS(III) = TS(I)
97 III = III + 1
   IEND = IEND + 1
```

```
      IBEG = IEND - NNLEFT
      FEND = FBEG
      FBEG = TS(IBEG)
      FINIS(ISTAGE)=END
      END=BEG
      BEG = BEGIN(ISTAGE)
      BEGIN(ISTAGE) = END
      REGLSV(ISTAGE) = REGLAR
      ISTAGE = ISTAGE + 1
      REGLAR = REGLSV(ISTAGE)
      EST(ISTAGE) = VINT
      CUREST = CUREST + EST(ISTAGE)
                                          GO TO 5
  900 IFLAG = 4
                                          GO TO 999
  950 IFLAG = 5
  999 CADRE = CUREST + VINT
                                          RETURN
      END
```

CHINSP (*see Section 6.4*)

This program computes $\int_A^B$ FUNC(x) dx, hopefully to a relative accuracy of EPSIN. EPSOUT is a good estimate for the relative error. The method used is a modified Clenshaw–Curtis scheme. A program FUNCTION FUNC(X) must be supplied by the user and FUNC must be declared EXTERNAL in the calling program. RNDERR in the DATA statement is a machine-dependent parameter which should be set to the relative machine accuracy. For further information, see the COMMENTS in the program.

CHINSP was supplied by its author Prof. T. Håvie who developed it at CERN and is discussed in Håvie [4, 6, 8].

```
      FUNCTION CHINSP(A,B,EPSIN,EPSOUT,FUNC)
C
C     MODIFIED CLENSHAW/CURTIS INTEGRATION THIS ALGORITMM IS BASED ON
C     THE METHOD PUBLISHED IN BIT 9 (1969) , 338-350
C     PURPOSE = TO OBTAIN THE VALUE OF A REAL AND DEFINITE INTEGRAL AP-
C               PROXIMATING THE INTEGRAND F(X) BY
C               F(X)= 5*C(1)+C(2)*T(X,1)+...  ...  ...  +C(M)*T(X,M-1)
C               WHERE T(X,K) IS THE CHEBYSHEV POLYNOMIAL OF ORDER K
C               MODIFIED CLENSHAW/CURTIS METHOD
C     PARAMETERS
C     A         = LOWER BOUNDARY
C     B         = UPPER BOUNDARY
C     EPSIN     = ACCURACY REQUIRED FOR THE APPROXINATION
C     EPSOUT    = IMPROVED ERROR ESTIMATE FOR THE APPROXIMATION
C     FUNC      = FUNCTION ROUTINE FOR THE FUNCTION FUNC(X). TO BE DE-
C                 CLARED EXTERNAL IN THE CALLING ROUTINE
C     INTEGRATION PARAMETERS
C     NUPPER    = 9 , CORRESPONDS TO 1024 SUB-INTERVALS FOR THE UNFOLDED
C                 INTEGRAL THE MAX NO OF FUNCTION EVALUATIONS THUS BEEING
C                 1025 THE HIGHEST END-POINT APPROXIMATION IS THUS USING
C                 1024 INTERVALS WHILE THE HIGHEST MID-POINT APPROXIMA-
C                 TION IS USING 512 INTERVALS
C     RNDERR    = 1.E-14 , THE RELATIVE MACHINE ACCURACY IN SINGLE PRECI-
C                 SION (CDC-6000 SERIES)
C
      DIMENSION ACOF(257) , BCOF(257) , CCOF(513)
      REAL KSINK
```

```
      DOUBLE PRECISION PIDIV4
      DATA ZERO,HALF,ONE,TWO,PIDIV4,NUPPER,RNDERR /0.,.5,1.,2.,
     * .785398163397448,9,      1.E-14/
C     ROUNDE=RNDERR*(R1+R2*N),WHERE R1,R2 ARE TWO EMPIRICAL CONSTANTS
C     SET COEFFICIENTS IN FORMULA FOR ACCUMULATED ROUND OFF ERROR,
C     AND N IS THE CURRENT NUMBER OF FUNCTION VALUES USED
 1000 R1=ONE
      R2=TWO
      ERROR=EPSIN
C     INTEGRATION INTERVAL PARAMETERS
      ALF=HALF*(B-A)
      BET=HALF*(B+A)
C     PARAMETERS FOR TRIGONOMETRIC RECURRENCE RELATIONS
      TRIARG=PIDIV4
      ALFNO =-ONE
C     PARAMETERS FOR INTEGRATION STEPSIZE AND LOOPS
      RN    =TWO
      N     =2
      NHALF =1
      HNSTEP=ONE
C     INITIAL CALCULATION FOR THE END-POINT APPROXIMATION
      CONST1=HALF*(FUNC(A)+FUNC(B) )
      CONST2=FUNC(BET)
      ACOF(1)=HALF*(CONST1+CONST2)
      ACOF(2)=HALF*(CONST1-CONST2)
      BCOF(2)=ACOF(2)
      TEND  =TWO*(ACOF(1)-ACOF(2)/(ONE+TWO))
C     START ACTUAL CALCULATION
 1020 DO 1150 I=1,NUPPER
C     COMPUTE FUNCTION VALUES
      CONST1=-SIN(TRIARG)
      CONST2= HALF*ALFNO/CONST1
      ALFNO = CONST1
      BETNO = CONST2
      GAMMAN= ONE-TWO*ALFNO**2
      DELTAN=-TWO*ALFNO*BETNO
      BCOF(1)=ZERO
      DO 1030 J=1,NHALF
      ALFNJ=GAMMAN*CONST1+DELTAN*CONST2
      BETNJ=GAMMAN*CONST2-DELTAN*CONST1
      XPLUS= ALF*ALFNJ+BET
      XMIN =-ALF*ALFNJ+BET
      CCOF(J)=FUNC(XPLUS)+FUNC(XMIN)
      BCOF(1)=BCOF(1)+CCOF(J)
      CONST1 =ALFNJ
      CONST2 =BETNJ
 1030 CONTINUE
      BCOF(1)=HALF*HNSTEP*BCOF(1)
C     CALCULATION OF FIRST B-COEFFICIENT FINISHED COMPUTE THE HIGHER
C     COEFFICIENTS IF NHALF GREATER THAN ONE
      IF (NHALF.EQ.1) GO TO 1070
      CONST1= ONE
      CONST2= ZERO
      NCOF  = NHALF-1
      KSIGN =-1
      DO 1060 K=1,NCOF
C     COMPUTE TRIGONOMETRIC SUM FOR B-COEFFICIENT
      ETANK= GAMMAN*CONST1-DELTAN*CONST2
      KSINK= GAMMAN*CONST2+DELTAN*CONST1
      COF   = TWO*(TWO*ETANK**2-ONE)
      A2    = ZERO
      A1    = ZERO
      A0    = CCOF(NHALF)
      DO 1040 J=1,NCOF
      A2    = A1
      A1    = A0
      INDEX=NHALF-J
```

```
      AO    = CCOF(INDEX)+COF*A1-A2
 1040 CONTINUE
      BCOF(K+1)=HNSTEP*(AO-A1)*ETANK
      BCOF(K+1)=KSIGN*BCOF(K+1)
      KSIGN=-KSIGN
      CONST1= ETANK
      CONST2= KSINK
 1060 CONTINUE
C     CALCULATION OF B-COEFFICIENTS FINISHED
C
C     COMPUTE NEW MODIFIED MID-POINT APPROXIMATION WHEN THE INTERVAL
C     OF INTEGRATION IS DIVIDED IN N EQUAL SUB INTERVALS
 1070 UMID= ZERO
      RK    = RN
      NN    = NHALF+1
      DO 1080 K=1,NN
      INDEX=NN+1-K
      UMID=UMID+BCOF(INDEX)/(RK**2-ONE)
      RK    = RK-TWO
 1080 CONTINUE
      UMID=-TWO*UMID
 1090 CONTINUE
C     COMPUTE NEW C-COEFFICIENTS FOR END-POINT APPROXIMATION AND LARGEST
C     ABSOLUTE VALUE OF COEFFICIENTS
      NN = N+2
      COFMAX=ZERO
      DO 1100 J=1,NHALF
      INDEX=NN-J
      CCOF(J    )=HALF*(ACOF(J)+BCOF(J))
      CCOF(INDEX)=HALF*(ACOF(J)-BCOF(J))
      CONST1= ABS(CCOF(J))
      IF (CONST1 GT COFMAX) COFMAX=CONST1
      CONST1= ABS(CCOF(INDEX))
      IF (CONST1 GT. COFMAX) COFMAX=CONST1
 1100 CONTINUE
      CCOF(NHALF+1)=ACOF(NHALF+1)
C     CALCULATION OF NEW COEFFICIENTS FINISHED
C
C     COMPUTE NEW END-POINT APPROXIMATION WHEN THE INTERVAL OF INTEGRA-
C     TION IS DIVIDED IN 2N EQUAL SUB INTERVALS
      WMEAN =HALF*(TEND+UMID)
      BOUNDS=HALF*(TEND-UMID)
      DELN= ZERO
      RK    = TWO*RN
      DO 1110 J=1,NHALF
      INDEX=N+2-J
      DELN= DELN+CCOF(INDEX)/(RK**2-ONE)
      RK    = RK-TWO
 1110 CONTINUE
      DELN=-TWO*DELN
 1120 TNEW =WMEAN+DELN
      EPSOUT=ABS(BOUNDS/TNEW)
      IF (COFMAX LT.RNDERR) GO TO 1130
      ROUNDE=RNDERR*(R1+R2*RN)
      IF (EPSOUT LT.ROUNDE) EPSOUT=ROUNDE
      IF (ERROR.LT.ROUNDE) ERROR=ROUNDE
      IF (EPSOUT GT ERROR) GO TO 1130
C     REQUIRED ACCURACY OBTAINED OR THE MAXIMUM NUMBER OF FUNCTION VAL-
C     UES USED WITHOUT OBTAINING THE REQUIRED ACCURACY
 1124 N=2*N+1
 1125 TEND=ALF*(TEND+DELN)
      UMID=ALF*(UMID+DELN)
      DELN=ALF*DELN
 1126 CHINSP=ALF*TNEW
 1128 RETURN
C     IF I=NUPPER THEN THE REQUIRED ACCURACY IS NOT OBTAINED.
 1130 IF (I.EQ.NUPPER) GO TO 1124
```

```
      DO 1140 J=1,N
      ACOF(J)=CCOF(J)
 1140 CONTINUE
      ACOF(N+1)=CCOF(N+1)
      BCOF(N+1)=CCOF(N+1)
      TEND=TNEW
      NHALF =N
      N      =2*N
      RN     =TWO*RN
      HNSTEP=HALF*HNSTEP
      TRIARG=HALF*TRIARG
 1150 CONTINUE
      RETURN
      END
```

NDIMRI (*see Section 6.5*)

This subroutine computes AINT, an approximation to the N-dimensional integral over a parallelepiped

$$\int_{A(1)}^{B(1)} \cdots \int_{A(N)}^{B(N)} F(x_1, \ldots, x_N)\, dx_1 \cdots dx_N, \qquad 1 \le N \le 9$$

using a Romberg-type method based on the midpoint rule. Initially, each interval $B(I) - A(I)$ is subdivided into $[1/H(I)]$ subintervals, $0 < H(I) \le 1$, $I = 1, \ldots, N$, to allow for the possibility of taking more points in appropriate variables. M is the maximum number of iterations allowed and should decrease with increasing N since the number of functional evaluations in the Jth iteration is at least J^N. If M is set > 15, it is reset to 15. AL is a constant between 1.5 and 2, preferably 1.5, which regulates the increase in number of points used. If K(J) is the basic number of points used (i.e., assuming $H(I) = 1$) in any direction at iteration J, then $K(1) = 1$, $K(2) = 2$, $K(J) = [AL * K(J\text{-}1)]$, $J \ge 2$. E is a tolerance which should increase as N increases since the expected accuracy depends to a certain extent on the number of iterations allowed. If the estimated relative accuracy attained is less than E, KEY is set to 1. If this is not achieved within M iterations, KEY is set to -1. If $N < 1$ or $N > 9$ or $M < 1$ or $AL < 1.5$ or $AL > 2$ or $H(I) \le 0$ or $H(I) > 1$, the subroutine exits with $KEY = 0$. A program FUNCTION F(X, N) must be supplied by the user with X declared by the statement DIMENSION X(N). F must be declared EXTERNAL in the calling program. TOL in the DATA statement is a machine-dependent parameter which should be set to the relative machine accuracy. If $E < TOL$, it is reset to TOL.

```
      SUBROUTINE NDIMRI(N,A,B,H,AL,M,E,F,AINT,KEY)
C  N=DIMENSION LESS THAN 10
C  A=ARRAY OF LOWER LIMITS
C  B=ARRAY OF UPPER LIMITS
C  H=ARRAY OF FRACTIONS OF FORM 1/N,N POSITIVE INTEGER
```

```
C  AL=NUMBER   BETWEEN 1.5 AND 2. ,PREFERABLY 1.5
C  M=MAXIMUM NUMBER OF ITERATIONS,SHOULD DECREASE AS N INCREASES
C  E=TOLERANCE,SHOULD INCREASE WITH N
C  AINT=RESULT ,F=FUNCTION ROUTINE ,KEY=INDICATOR SET BY SUBROUTINE
      DIMENSION A(N),B(N),C(9),K(15),H(N),G(9),NN(9),P(9),AA(15),V(15)
     X,X(9),D(9)
      DATA K(1),K(2) /1,2/
      DATA TOL /1.E-7/
      KEY=0
      IF (N.LT.1 .OR. N.GT.9 .OR. M.LT.1 .OR. AL.LT.1.5 .OR. AL.GT.2.)
     * RETURN
      DO 2 I=1,N
      IF (H(I).LE.0. .OR. H(I).GT.1.) RETURN
    2 D(I)=A(I)
      EE=AMAX1(E,TOL)
      MM=MINO(M,15)
      DO 1 I=N,8
      P(I+1)=0.
      NN(I+1)=1
    1 D(I+1)=0.
   10 L=1
      DO 20 I=1,N
   20 C(I)=B(I)-D(I)
   21 U=0.
      KT=0
      DO 22 I=1,N
      G(I)=H(I)/K(L)
      NN(I)=1./G(I)+.5
   22 P(I)=C(I)*G(I)
   38 NN9=NN(9)
      DO 30 I9=1,NN9
      X(9)=D(9)+P(9)*(I9-.5)
   37 NN8=NN(8)
      DO 30 I8=1,NN8
      X(8)=D(8)+P(8)*(I8-.5)
   36 NN7=NN(7)
      DO 30 I7=1,NN7
      X(7)=D(7)+P(7)*(I7-.5)
   35 NN6=NN(6)
      DO 30 I6=1,NN6
      X(6)=D(6)+P(6)*(I6-.5)
   34 NN5=NN(5)
      DO 30 I5=1,NN5
      X(5)=D(5)+P(5)*(I5-.5)
   33 NN4=NN(4)
      DO 30 I4=1,NN4
      X(4)=D(4)+P(4)*(I4-.5)
   32 NN3=NN(3)
      DO 30 I3=1,NN3
      X(3)=D(3)+P(3)*(I3-.5)
   31 NN2=NN(2)
      DO 30 I2=1,NN2
      X(2)=D(2)+P(2)*(I2-.5)
      NN1=NN(1)
      DO 30 I1=1,NN1
      X(1)=D(1)+P(1)*(I1-.5)
   30 U=U+F(X,N)
      DO 40 I=1,N
   40 U=U*P(I)
      V(L)=U
      IF (L-1) 43,43,44
   43 AA(1)=V(1)
      L=L+1
      GO TO 21
```

```
44 EN=K(L)
   DO 45 LL=2,L
   I=L+1-LL
45 V(I)=V(I+1)+(V(I+1)-V(I))/((EN/K(I))**2-1.)
   AINT=V(1)
   KEY=1
   IF(ABS(AINT-AA(L-1)) .LT. ABS(AINT*EE)) RETURN
   KEY=-1
   IF (L .EQ. MM) RETURN
   AA(L)=AINT
   L=L+1
   K(L)=AL*K(L-1)
   GO TO 21
   END
```

Appendix 3

Bibliography of ALGOL, FORTRAN, and PL/I Procedures

The symbols A, F, P stand for ALGOL, FORTRAN, and PL/I programs respectively.

ANTES, H.
F1. Four-dimensional Romberg integration, in Antes [1, p. 51].

BAUER, F. L.
A1. Algorithm 60. Romberg integration, *CACM*† **4** (1961) 255. See also *CACM* **5** (1962) 168, 281 and *CACM* **7** (1964) 420.

BERGLAND, G. D., and DOLAN, M. T.
F1. Fast Fourier transform algorithms, in Digital Signal Process Committee [1, 1.2].

BLUE, J. L.
F1. DQUAD, in Blue [1, B1–12].

BOLAND, W. R.
F1. Algorithm 436. Product type trapezoidal integration, *CACM* **15** (1972) 1070.
F2. Algorithm 437. Product type Simpson's integration, *CACM* **15** (1972) 1070–1071.
F3. Algorithm 438. Product type two-point Gauss–Legendre–Simpson's integration, *CACM* **15** (1972) 1071.

† Abbreviations used throughout. *CACM* = Communications of the Association for Computing Machinery. *BIT* = Nordisk Tidskrift for Informationsbehandling. *PAAN* = *Procédures ALGOL en Analyse Numérique*. CNRS, Paris, 1967.

F4. Algorithm 439. Product type three-point Gauss–Legendre–Simpson's integration, *CACM* **15** (1972) 1072.

BRANDERS, M., and PIESSENS, R.
F1. GCCINT, in Branders and Piessens [1, pp. 58–65].

BROUCKE, R.
F1. Algorithm 446. Ten subroutines for the manipulation of Chebyshev Series, *CACM* **16** (1973) 254–256.

BULIRSCH, R.
A1. Romberg–Neville integration, romnevint, in Bulirsch [1, pp. 15–16].

BULIRSCH, R., and STOER, J.
A1. Quadrature, in Bulirsch and Stoer [2, pp. 274–275].
A2. Trapezsummenextrapolation, in Bulirsch and Stoer [3, pp. 425–426].

CHASE, S. M., and FOSDICK, L. D.
F1. Algorithm 353. Filon quadrature, *CACM* **12** (1969) 457–458. See also *CACM* **13** (1970) 263.

CONTE, S. D., and DE BOOR, C.
F1. FFT, in Conte and de Boor [1, pp. 281–282].
F2. RMBERG, in Conte and de Boor [1, pp. 343–344].

DE BOOR, C.
F1. CADRE, Cautious adaptive Romberg extrapolation, in de Boor [1, pp. 430–438].

DE DONCKER, E., and PIESSENS, R.
F1. TAMORI, in de Doncker and Piessens [1, pp. 274–278].

DE DONCKER, E., and ROBINSON, I.
F1. TRIEX, Integration over a TRIangle using non-linear EXtrapolations, TOMS **10** (1984).

DIERCKX, P.
F1. SMOOT, DERIV, and SPLINT, in Dierckx [1, pp. 177–184].
F2. SURFAC, BICUBI, and SPLINT, in Dierckx [2, pp. 117–129].

DURIS, C. S.
F1. PRODNC, in Duris [3, pp. 26–30].
F2. QRCOPY, in Duris [3, pp. 35–39].
F3. QRINT, in Duris [3, pp. 42–43].

EINARSSON, B.
F1. Algorithm 418. Calculation of Fourier integrals, *CACM* **15** (1972) 47–48. See also p. 469 and *CACM* **17** (1974) 324.

ELHAY, S., and KAUTSKY, J.
F1. COWIQ and SIWIQ. Fortran subroutines for the weights of interpolatory quadratures, Technical Rep. TR82-08, Dept. of Computing Science, Univ. of Adelaide, South Australia, 1982.

ENGELS, H.
F1. MCQ, in Engels [13, pp. 81–82].

FAIRWEATHER, G.
F1. Algorithm 351. Modified Romberg quadrature, *CACM* **12** (1969) 324–325. See also *CACM* **13** (1970) 263, 374–375, 449.

Fairweather, G., and Keast, P.
F1. Computer programs, in Fairweather and Keast [1, pp. B1–22].

Ferguson, Jr., W. E.
F1. FFTGM, in Ferguson [1, p. 407].
F2. FFTDB, in Ferguson [1, p. 408].

Forsythe, G. E., Malcolm, M. A., and Moler, C. B.
F1. QUANC8, in Forsythe, Malcolm, and Moler [1, pp. 102–105].

Fraser, D.
F1. Optimized mass storage FFT program, in Digital Signal Processing Committee [1, 1.5].

Freeman, Jr., R. D.
A1. Algorithm 32. Multint, *CACM* **4** (1961) 106. See also *CACM* **6** (1963) 69 and *CACM* **11** (1968) 826.

Gallaher, L. J.
A1. Algorithm 303. An adaptive quadrature procedure with random panel sizes, *CACM* **10** (1967) 373–374.
A2. Algorithm 440. A multidimensional Monte Carlo quadrature with adaptive stratified sampling, *CACM* **16** (1973) 49–50.

Garribba, S., Quartapelle, L., and Reina, G.
F1. SNIFF, in Garribba, Quartapelle, and Reina [1, pp. 373–375].

Gautschi, W.
A1. Algorithm 331. Gaussian quadrature formulas, *CACM* **11** (1968) 432–436. See also *CACM* **12** (1969) 280–281 and *CACM* **13** (1970) 512.

Gentleman, W. M.
F1. Algorithm 424. Clenshaw–Curtis quadrature, *CACM* **15** (1972) 353–355. See also *CACM* **16** (1973) 490.

Genz, A. C.
F1. Computer programs, in Genz [4, Appendix].
F2. INTLAG, in Genz [5, pp. 170–172].

Genz, A. C., and Malik, A. A.
F1. ADAPT, in Genz and Malik [1, pp. 300–302].

Gill, P. E., and Miller, G. F.
A1. Int 4pt, in Gill and Miller [1, pp. 82–83].

Golub, G. H., and Welsch, J. H.
A1. Gaussquadrule, *Math. Comp. Microfiche Suppl.* **23** (1969), No. 106.

Good, I. J., and Gaskins, R. A.
F1. Double integration over any non-overlapping polygonal region, in Good and Gaskins [1, p. 356].
F2. Multiple integration over any simplex, in Good and Gaskins [1, pp. 357–358].

Gram, C.
A1. ALGOL Programming, Contribution No. 8. Definite integral by Romberg's method, ALGOL procedure, *BIT* **4** (1964), 54–60. See also *BIT* **4** (1964) 118–120.

GUSTAFSON, S-A.
A1. Algorithm 417. Rapid computation of weights of interpolatory quadrature rules, *CACM* **14** (1971) 807.

HAEGEMANS, A.
F1. TRIADA, in Haegemans [2, pp. 184–186].

HAEGEMANS, A., and PIESSENS, R.
A1. Algorithm 41. Fourier. Computation of Fourier-transform integrals, *Apl. Mat.* **21** (1976) 229–236.

HALTON, J. H., and SMITH, G. B.
A1. Algorithm 247. Radical-inverse quasi-random point sequence, CACM **7** (1964) 701–702.

HALTON, J. H., and ZEIDMAN, E. A.
F1. MCSS, in Halton and Zeidman [1, pp. 144–158].

HAUSNER, A.
F1. NL9, in Hausner [1, pp. 396–410].

HAUSNER, A., and HUTCHISON, J. D.
F1. FOGIE, in Hausner and Hutchison [1, pp. 152–166].

HÅVIE, T.
A1. Integrator, in Håvie [2, pp. 110–113].

HAZLEWOOD, L. J.
A1. Algorithm 101. An algorithm for cautious adaptive quadrature, *Comput. J.* **21** (1978) 180–183.

HENNION, P. E.
A1. Algorithm 77. Interpolation, differentiation, and integration, *CACM* **5** (1962) 96. See also *CACM* **5** (1962) 348 and *CACM* **6** (1963) 446–447, 663.
A2. Algorithm 84. Simpson's integration, *CACM* **5** (1962) 208. See also *CACM* **5** (1962) 392, 440, 557.

HENRIKSSON, S.
A1. ALGOL Programming, Contribution No. 2. Simpson numerical integration with variable length of step, *BIT* **1** (1961) 290.

HERBOLD, R. J.
A1. Algorithm 1. Quad I, *CACM* **3** (1960) 74.

HOPGOOD, F. R. A., and LITHERLAND, G.
A1. Algorithm 279. Chebyshev quadrature, *CACM* **9** (1966) 270. See also *CACM* **9** (1966) 434 and *CACM* **10** (1967) 294, 666.

HOPP, T. H.
F1. IFEIX, in Hopp [1, pp. 203–221].

IBM SYSTEM/360 SCIENTIFIC SUBROUTINE PACKAGE (PL/I) (360A-CM-07X) H20-0586-0, 1968.
P1. QTFG, Integration of a monotonically tabulated function by trapezoidal rule;
QTFE, Integration of an equidistantly tabulated function by trapezoidal rule, pp. 92–93.
P2. QSF, Integration of an equidistantly tabulated function by Simpson's rule, pp. 93–94.
P3. QHFG, Integration of a monotonically tabulated function with first derivative by a Hermitian formula of first order;

QHSG, Integration of a monotonically tabulated function with first and second derivatives by a Hermitian formula of second order;
QHFE, Integration of an equidistantly tabulated function with first derivative by a Hermitian formula of first order;
QHSE, Integration of an equidistantly tabulated function with first and second derivatives by a Hermitian formula of second order, pp. 94–96.

P4. QATR, Integration of a given function by the trapezoidal rule together with Romberg's extrapolation method, pp. 97–99.

P5. QG_n ($n = 2, 4, 8, 16, 24, 32, 48$), Integration of a given function by n-point Gaussian quadrature formula, pp. 99–101.

P6. QL_n ($n = 2, 4, 8, 12, 16, 24$), Integration of a given function by n-point Gaussian–Laguerre quadrature formula, pp. 101–103

P7. QH_n ($n = 2, 4, 8, 16, 24, 32, 48$), Integration of a given function by n-point Gaussian–Hermite quadrature formula, pp. 103-105.

P8. QA_n ($n = 2, 4, 8, 12, 16, 24$), Integration of a given function by associated n-point Gaussian–Laguerre quadrature formula, pp. 105–106.

IBM System/360 Scientific Subroutine Package (360A-CM-03X) Version III, H20-0205-3, 1968.

F1. (D)QTFG, Integration of monotonically tabulated function by trapezoidal rule, p. 289.

F2. (D)QTFE, Integration of equidistantly tabulated function by trapezoidal rule, p. 290.

F3. (D)QSF, Integration of equidistantly tabulated function by Simpson's rule, pp. 291–292.

F4. (D)QHFG, Integration of monotonically tabulated function with first derivative by Hermitian formula of first order, p. 293.

F5. (D)QHFE, Integration of equidistantly tabulated function with first derivative by Hermitian formula of first order, p. 294.

F6. (D)QHSG, Integration of monotonically tabulated function with first and second derivatives by Hermitian formula of second order, p. 295.

F7. (D)QHSE, Integration of equidistantly tabulated function with first and second derivatives by Hermitian formula of second order, p. 296.

F8. (D)QATR, Integration of a given function by trapezoidal rule together with Romberg's extrapolation method, pp. 297–298.

F9. QG_n ($n = 2(1)10$), DQG_n ($n = 4, 8, 12, 16, 24, 32$), Integration of a given function by Gaussian quadrature formulas, pp. 299–303.

F10. QL_n ($n = 2(1)10$), DQL_n ($n = 4, 8, 12, 16, 24, 32$), Integration of a given function by Gaussian–Laguerre quadrature formulas, pp. 303–307.

F11. QH_n ($n = 2(1)10$), DQH_n ($n = 8, 16, 24, 32, 48, 64$), Integration of a given function by Gaussian–Hermite quadrature formulas, pp. 308–313.

F12. QA_n ($n = 2(1)10$), DAQ_n ($n = 4, 8, 12, 16, 24, 32$), Integration of a given function by associated Gaussian–Laguerre quadrature formulas, pp. 314–318.

Kahaner, D.

F1. QNC7, Numerical quadrature based on seven-point Newton–Cotes rule used adaptively, in Kahaner [2, pp. 238–239].

F2. QUAD, Numerical quadrature based on ten-point Newton–Cotes rule used adaptively, in Kahaner [2, pp. 240–241].

F3. SHNK, Epsilon algorithm, in Kahaner [2, p. 241].

Keast, P.

F1. Computer programs, in Keast [4, pp. 32–52].

KUBIK, R. N.
A1. Algorithm 257. Håvie integrator, *CACM* **8** (1965) 381. See also *CACM* **9** (1966) 795, 871.

KUNCIR, G. F.
A1. Algorithm 103. Simpson's rule integrator, *CACM* **5** (1962) 347.

LABOUR.
A1. Procédures d'intégration avec majoration de l'erreur commise. Cas d'une fonction régulière, *PAAN* pp. 224–228.
A2. Procédure d'intégration avec majoration de l'erreur commise. Intégrales de la forme $\int_a^b f(x) \cos \lambda x \, dx$ et $\int_a^b f(x) \sin \lambda x \, dx$, *PAAN*, pp. 229–231.
A3. Intégrale d'une fonction de deux variables sur un rectangle. Méthode de Cotes, *PAAN* pp. 235–236.
A4. Intégrale d'une fonction de deux variables sur un rectangle. Méthode composite de degré 3, *PAAN* pp. 237–238.

LAURENT, J.-P.
A1. Quadrature sur un intervalle borné (Méthode de Romberg), *PAAN* pp. 239–240.
A2. Quadrature sur un domaine rectangulaire (Méthode de Romberg), *PAAN* pp. 241–242.

LAURIE, D. P., and ROLFES, L.
F1. Algorithm 015. Computation of Gaussian quadrature rules from modified moments, *J. Comput. Appl. Math.* **5** (1979) 235–243.

LEPAGE, G. P.
F1. VEGAS, in Lepage [2, pp. 21–25].

LINZ, P.
F1. Algorithm 427. Fourier cosine integral, *CACM* **15** (1972) 358–360.

LYNESS, J. N.
F1. Algorithm 379. SQUANK (Simpson quadrature used adaptively—Noise killed), *CACM* **13** (1970) 260–263. See also *CACM* **15** (1972) 1073.

LYNESS, J. N. and SANDE, G.
F1. Algorithm 413. ENTCAF and ENTCRF: Evaluation of normalized Taylor coefficients of an analytic function, *CACM* **14** (1971) 669–675.

MCCLELLAN, J. H., and NAWAB, H.
F1. Complex general-N Winograd Fourier transform algorithm, in Digital Signal Processing Committee [1, 1.7].

MCKEEMAN, W. M.
A1. Algorithm 145. Adaptive numerical integration by Simpson's rule, *CACM* **5** (1962) 604. See also *CACM* **6** (1963) 167 and *CACM* **8** (1965) 171.
A2. Algorithm 146. Multiple integration, *CACM* **5** (1962) 604–605. See also *CACM* **7** (1964) 296.
A3. Algorithm 198. Adaptive integration and multiple integration, *CACM* **6** (1963) 443–444.

MCKEEMAN, W. M., and TESLER, L.
A1. Algorithm 182. Nonrecursive adaptive integration, *CACM* **6** (1963) 315. See also *CACM* **7** (1964) 244.

McNAMEE, J. M.
F1. GENINT, in McNamee, [1, pp. 277–278].

MALCOLM, M. A., and SIMPSON, R. B.
F1. SQUAGE, in Malcolm and Simpson [1, pp. 136–138].

MALIK, A. A.
F1. Computer programs, in Malik [1, Chapter 5].

MELENDEZ, J., and KAHANER, D.
F1. GAUSS, Numerical quadrature based on five- and seven-point Gauss quadrature used adaptively, in Kahaner [2, p. 239].

MONRO, D. M.
F1. Algorithm AS 83. Complex discrete fast Fourier transform, *Applied Statistics* **24** (1975) 153–160.
F2. Algorithm AS 97. Real discrete fast Fourier transform, *Applied Statistics* **25** (1976) 166–172.

MONRO, D. M., and BRANCH, J. L.
F1. Algorithm AS 117. The Chirp discrete Fourier transform of general length, *Applied Statistics* **26** (1977) 351–361.

MORRIS, L. R.
F1. Time-efficient radix-4 fast Fourier transform, in Digital Signal Processing Committee [1, 1.8].

OLIVER, J.
A1. Adapquad, in Oliver [3, pp. 145–147].

OLYNYK, F.
A1. Algorithm 233. Simpson's rule for multiple integration, *CACM* **7** (1964) 348–349. See also *CACM* **13** (1970) 512.

PATTERSON, T. N. L.
F1. Algorithm 468. Algorithm for automatic numerical integration over a finite interval, *CACM* **16** (1973) 694–699.

PFALZ, J. L.
A1. Algorithm 98. Evaluation of definite complex line integrals, *CACM* **5** (1962) 345.

PIESSENS, R.
F1. Algorithm 453. Gaussian quadrature formulas for Bromwich's integral, *CACM* **16** (1973) 486–487.
F2. AIND*8, in Piessens [10, p. 401].
F3. Algorithm 473. Computation of Legendre series coefficients, CACM **17** (1974) 25.
F4. AIND*8, in Piessens [13, Appendix].
F5. DEFOS, in Piessens [14, Appendix].
A6. Algorithm 113. Inversion of the Laplace transform, *Comput. J.* **25** (1982) 278–281.

PIESSENS, R., and BRANDERS, M.
F1. Kronro, in Piessens and Branders [3, pp. 344–347]
F2. AINOS, in Piessens and Branders [4, pp. 157–164].

PIESSENS, R., and DE DONCKER, E.
F1. DEFIN, in Piessens and de Doncker [1, Appendix A].

PIESSENS, R., and HAEGEMANS, A.
F1. OSCIN, in Piessens and Haegemans [2, pp. 187–192].

PIESSENS, R., MERTENS, I., and BRANDERS, M.
F1. AINAB, in Piessens, Mertens, and Branders [1, pp. 67–68].

PIESSENS, R., VAN ROY-BRANDERS, M., and MERTERNS, I.
F1. AINC, in Piessens, van Roy-Branders, and Mertens [1, pp. 34–35].

QUADPACK
F1. Computer programs, in Piessens, de Doncker, Überhuber, and Kahaner [1, pp. 130–294].

RABINER, L. R.
F1. FFT subroutines for sequences with special properties, in Digital Signal Processing Committee [1, 1.3].
F.2 Chirp z-transform algorithm program, in Digital Signal Processing Committee [1, 1.6].

RADER, C. M.
F1. FOUREA-A short demonstration version of the FFT, in Digital Signal Processing Committee [1, 1.1].

ROBINSON, I.
F1. Fortran programs, in Robinson [3, pp. 248–299].
A2. multiquad, in Robinson [4, pp. 12–19].
F3. AGM, in Robinson [6, pp. 111–115].

ROBINSON, I., and DE DONCKER, E.
F1. DITAMO, in Robinson and de Doncker [1, pp. 269–281].

ROOSE, D., and DE DONCKER, E.
F1. SFERIN, in Roose and de Doncker [1, pp. 211–224].

ROOTHART, C. J., and FIOLET, H.
A1. Five procedures, in Roothart and Fiolet [1, pp. 2–9].

RUTISHAUSER, H.
A1. Algorithm 125. Weightcoeff, *CACM* **5** (1962) 510–511.

SACK, R. A., and DONOVAN, A. F.
A1. Terminal matrix in Sack and Donovan [1, pp. 476–477].

SHAMPINE, L. F., and ALLEN, JR., R. C.
F1. SIMP, in Shampine and Allen [1, pp. 240–242].

SIKORSKI, K., STENGER, F., and SCHWING, J.
F1. A Fortran program for integration in H_p. To appear in TOMS **10** (1984).

SINGLETON, R. C.
A1. Algorithm 338. Algol procedures for the fast Fourier transform, CACM **11** (1968) 773–776.
A2. Algorithm 339. An Algol procedure for the fast Fourier transform with arbitrary factors, CACM **11** (1968) 776–779. See also CACM **12** (1969) 187.
A3. Algorithm 345. An Algol convolution procedure based on the fast Fourier transform, CACM **12** (1969) 179–184. See also CACM **12** (1969) 566.
F4. Mixed radix fast Fourier transforms, in Digital Signal Processing Committee [1, 1.4].
F5. Two-dimensional mixed radix mass storage Fourier transform, in Digital Signal Processing Committee [1, 1.9].

SQUIRE, W.
F1. Integration functions and subroutines, *Integration for Engineers and Scientists*, Appendix 2, pp. 273–286. Amer. Elsevier, New York, 1970.
F2. QINF, in Squire [6, pp. 480–481].
F3. QFIN, in Squire [8, p. 710].

STEHFEST, H.
A1. Algorithm 368. Numerical inversion of Laplace transform, CACM **13** (1970) 47–49.

STROUD, A. H.
F1. Computer programs, *Approximate Calculation of Multiple Integrals*, Chapter 10. Prentice-Hall, Englewood Cliffs, New Jersey, 1971.
F2. ROMBERG, in Stroud [19, p. 157].
F3. QSTOER, in Stroud [19, pp. 197–198].

STROUD, A. H., and SECREST, D.
F1. Subroutine Jacobi, in Stroud and Secrest [2, pp. 29–31].
F2. Subroutine Laguer, in Stroud and Secrest [2, pp. 32–34].
F3. Subroutine Hermit, in Stroud and Secrest [2, pp. 35–36].

TEIJELO, L.
A1. Algorithm 255. Computation of Fourier coefficients, *CACM* **8** (1965) 279. See also *CACM* **12** (1969) 636.

VAN DOOREN, P., and DE RIDDER, L.
F1. HALF, in van Dooren and de Ridder [1, pp. 211–217].

VASILEVA, L. G., BRUSLENŠKAJA, O. V., and LAPŠIN, E. A.
A1. *Programs for Numerical Integration in ALGOL*, Izdat. Moskov. Gas. Univ., Moscow, 1972 (Russian).

VERJUS-CHEIN.
A1. Quadrature sur un domaine parallélépipédique (Méthode de Romberg), *PAAN* pp. 243–244.

WALLICK, G. C.
F1. Algorithm 400. Modified Håvie integration, *CACM* **13** (1970) 622–624. See also *CACM* **17** (1974) 324.

WEILLON, F.
A1. Algorithm 486. Numerical inversion of Laplace transform, CACM **17** (1974) 587–589.

WELSCH, J. H.
A1. Algorithm 280. Abscissas and weights for Gregory quadrature, *CACM* **9** (1966) 271.
A2. Algorithm 281. Abscissas and weights for Romberg quadrature, *CACM* **9** (1966) 271–272. See also *CACM* **10** (1967) 188.

WERNER, H., and ZWICK, D.
F1. QUAD1, in Werner and Zwick [1, pp. 18–20].
F2. QUAD2, in Werner and Zwick [1, pp. 21–23].

Appendix 4

Bibliography of Tables

Of making many tables there is no end. This is true even when the tables are restricted to the weights and abscissas of integration rules and even after several authors (Hamming, Rutishauser, Golub and Welsch, Gautschi, etc.) have published algorithms for generating integration rules to suit one's needs. Tables continue to appear in large numbers and a bibliography of such tables would be obsolete almost immediately. We shall, therefore, content ourselves with the following.

(1) A listing of the section headings in Fletcher, Miller, Rosenhead, and Comrie's "Index of Mathematical Tables", which deal with numerical integration.
(2) A description of the tables and formulas on integration in the NBS Handbook of Mathematical Functions.
(3) A description of the tables in the comprehensive book of Gaussian integration rules by Stroud and Secrest and some other sources of Gauss formulas.
(4) A description of some tables of unusual interest.

For the rest we refer the reader to the pages of the journal "Mathematics of Computation" and its predecessor "Mathematical Tables and Other Aids to Computation" (MTAC) both for countless tables of integration rules and for reviews of other such tables, published and unpublished, as well as for errata in published tables.

4.1 Section Headings in the "Index of Mathematical Tables"

(See Fletcher, Miller, Rosenhead, and Comrie, *An Index of Mathematical Tables*, Addison-Wesley, Reading, Mass.)

23.5 Numerical Integration, using Ordinates (pp. 584–591)
.51 Cotes' Formulae
.512 Sard's Formulae
.515 Other Formulae with Equal Intervals
.518 Integration of Linear Sums of Exponential Functions
.52 Partial-Range Formulae, etc.
.53 Formulae for Forward Integration
.54 Steffensen's Formulae
.55 Maclaurin's Formulae
.56 Filon's Formulae
.57 Chebyshev's Formulae
.575 Other Formulae with Equal Coefficients
.58 Formulae with Unequal Intervals and Coefficients
.581 Gaussian Quadrature Formulae
.5815 Radau's Quadrature Formulae
.582 Gauss–Laguerre Quadrature Formulae
.583 Generalized Gauss–Laguerre Quadrature Formulae
.584 Gauss–Hermite Quadrature Formulae
.585 Other Quadrature Formulae with Unequal Intervals and Coefficients
.588 A Formula for Repeated Integration
.59 Lagrangian Integration Polynomials
23.6 Integration and Summation, using Differences (pp. 591–593)
.61 Single Integrals Anywhere in the Interval
.62 Repeated Integrals Anywhere in the Interval
.67 Osculatory Quadrature Formulae
.68 $J = 1/(nw) \int_0^{nw} f(x)\,dx$ in Terms of Differences
.69 Lubbock Coefficients
.695 Summation of Slowly Convergent Series
23.7 Double Integrals, etc. (p. 593)

4.2 Tables and Formulas for Integration in the NBS Handbook

1. Newton–Cotes Formulas (Closed Type), $n = 2(1)11$, pp. 886–887.
2. Newton–Cotes Formulas (Open Type), $n = 2(1)7$, p. 887.
3. Lagrangian Integration Coefficients,

$$\int_{x_m}^{x_{m+1}} f(x)\,dx \approx \frac{1}{h}\sum_k A_k(m) f(x_k), \qquad n = 3(1)10,$$ p. 915.

4. Abscissas and Weight Factors for Gaussian Integration, $n = 2(1)12(4)24(8)48(16)96$, pp. 916–919.

5. Abscissas for Equal Weight Chebyshev Integration

$$\int_{-1}^{1} f(x)\,dx \approx \frac{2}{n}\sum_{i=1}^{n} f(x_i), \qquad n = 2(1)7, 9,$$ p. 920.

6. Abscissas and Weight Factors for Lobatto Integration, $n = 3(1)10$, p. 920.
7. Abscissas and Weight Factors for Gaussian Integration for Integrands with a Logarithmic Singularity, $n = 2(1)4$, p. 920.
8. Abscissas and Weight Factors for Gaussian Integration of Moments

$$\int_{0}^{1} x^k f(x)\,dx \approx \sum_{i=1}^{n} w_i f(x_i), \qquad k = 0(1)5, \qquad n = 1(1)8,$$ pp. 921–922.

9. Abscissas and Weight Factors for Laguerre Integration, $n = 2(1)10, 12, 15$, p. 923.
10. Abscissas and Weight Factors for Hermite Integration, $n = 2(1)10, 12, 16, 20$, p. 924.
11. Coefficients for Filon's Quadrature Formula, n = number of points in formula. p. 924.

4.3 *Tables of Gaussian Type Integration Formulas in Stroud and Secrest "Gaussian Quadrature Formulas"*

All integration formulas to thirty significant figures.

1. $\int_{-1}^{1} f(x)\,dx \approx \sum_{i=1}^{N} A_i f(x_i)$ (Gauss), $N = 2(1)64(4)96(8)168, 256, 384, 512$

2. $\int_{-1}^{1} (1 - x^2)^{\alpha} f(x)\,dx \approx \sum_{i=1}^{N} A_i f(x_i)$, $N = 2(1)20$, $\alpha = -\frac{1}{2}, \frac{1}{2}, 1, \frac{3}{2}$

3. $\int_{-1}^{1} (1 + x)^{\beta} f(x)\,dx \approx \sum_{i=1}^{N} A_i f(x_i)$. $\beta = 1$, $N = 2(1)30$; $\beta = 2, 3, 4$, $N = 2(1)20$

4. $\int_{-1}^{1} |x|^{\alpha} f(x)\,dx \approx \sum_{i=1}^{N} A_i f(x_i)$, $N = 2(1)20$, $\alpha = 1, 2, 3, 4$

5. $\int_{-\infty}^{\infty} e^{-x^2} f(x)\,dx \approx \sum_{i=1}^{N} A_i f(x_i)$ (Hermite), $N = 2(1)64(4)96(8)136$

6. $\int_{0}^{\infty} e^{-x} f(x)\,dx \approx \sum_{i=1}^{N} A_i f(x_i)$ (Laguerre), $N = 2(1)32(4)68$

7. $\int_{-\infty}^{\infty} |x|^{\alpha} e^{-x^2} f(x)\,dx \approx \sum_{i=1}^{N} A_i f(x_i)$, $N = 2(1)20$, $\alpha = 1, 2, 3$

8. $\int_{-\infty}^{\infty} |x|^{\alpha} e^{-|x|} f(x)\,dx \approx \sum_{i=1}^{N} A_i f(x_i)$, $N = 2(1)20$, $\alpha = 1, 2, 3$

9. $\int_0^1 \ln\left(\frac{1}{x}\right) f(x)\,dx \approx \sum_{i=1}^{N} A_i f(x_i), \qquad N = 2(1)16$

10. $\frac{1}{2\pi i}\int_{c-i\infty}^{c+i\infty} p^{-1}e^{p}F(p)\,dp \approx \sum_{i=1}^{N} A_i F(p_i), \qquad N = 2(1)24.$

11. $\int_{-1}^{1} f(x)\,dx \approx Af(-1) + \sum_{i=1}^{N} A_i f(x_i) + Af(+1)$ (Lobatto),

$N = 2(1)32(4)96.$

12. $\int_{-1}^{1} f(x)\,dx \approx Af(-1) + \sum_{i=1}^{N} A_i f(x_i)$ (Radau), $N = 2(1)19(4)47.$

13. $\int_{-1}^{1} f(x)\,dx \approx \sum_{i=1}^{N} A_i f(x_i) + \sum_{k=0}^{M} B_{2k} f^{(2k)}(0), \qquad N = 2(2)16,$

$M = 1, 2, 3.$

14. $\int_{-\infty}^{\infty} e^{-x^2} f(x)\,dx \approx \sum_{i=1}^{N} A_i f(x_i) + \sum_{k=0}^{M} B_{2k} f^{(2k)}(0), \qquad N = 2(2)16,$

$M = 1, 2, 3.$

4.3.1 Tables of Other Gauss Formulas

Boujot and Maroni tabulate Gauss formulas $N = 2(1)12$, 12S for the following weights: $x^{\alpha-1}(1-x)^{\beta-1}$, $\alpha, \beta > 0$, over $[0, 1]$. $x^{\alpha-1}(\log 1/x)^{\beta-1}$, $\alpha, \beta > 0$ over $[0, 1]$. $e^{-x}x^{\alpha-1}$, $\alpha > 0$ over $[0, \infty)$. Common values of α and β are used. Also $w(x)/x$ and $w(x)/x(1-x)$ where

$$w(x) = \frac{1}{\pi^2 + \log^2\left(\frac{1-x}{x}\right)}$$

over $[0, 1]$.

Berger has tabulated zeros and weights for Gauss–Laguerre and Gauss–Hermite quadrature to over 23S for selected values of N between 100 and 2000.

Krylov and Pal'tsev have given Gauss formulas $N = 1(1)10$, 15S for the following weights: $x^\alpha \log e/x$, $\alpha \doteq -.9(.01)0(.1)5$, $x^\beta \log e/x \log(e/(1-x))$, $\beta = 0(1)5$, $\log(1/x)$, over $[0, 1]$ and $x^\beta e^{-x}\log(1 + 1/x)$, $\beta = 0(1)5$ over $[0, \infty)$.

Piessens and Branders have given Gauss formulas for selected values of N to 25S for the following weights: $x^\alpha e^{-ax}$, $a = 1, 2, 5$, $\alpha = -\frac{1}{2}, 0, \frac{1}{2}$; e^{-ax^2}, $a = 1, 2, 5, 10$; $x^{-\alpha}\log(1/x)$, $\alpha = -\frac{1}{2}, -\frac{1}{3}, \frac{1}{5}, \frac{1}{4}, \frac{1}{3}, \frac{1}{2}$; $(1-x)^{-\alpha}x^{-\beta}\log(1/x)$, $\alpha, \beta = -\frac{1}{2}, -\frac{1}{3}, \frac{1}{5}, \frac{1}{4}, \frac{1}{3}, \frac{1}{2}$, over $[0, 1]$, $\cos x$ and $\sin x$ over $[-\pi, \pi]$, and e^{-ax^2}, $a = 1, 2, 5, 10$; $(x + a)^{-\alpha}$, $a = 1.001, 1.01, 1.1$, $\alpha = \frac{1}{2}, 1, 2$; $(x^2 + a^2)^{-\alpha}$, $a = .001, .01, .1, 1$, $\alpha = \frac{1}{2}$, 1, 2 over $[-1, 1]$.

4.4 Tables of Unusual Interest

1. J. C. P. Miller, *Quadrature in Terms of Equally Spaced Function Values* (91 pages) contains almost every formula ever developed which expresses an integral as a weighted sum of equally spaced values of the integrand. Miller gives both finite

difference formulas and formulas in terms of equally spaced function values. He also gives a number of special formulas. In many formulas, the coefficients are given both as exact fractions and in decimal form. Error coefficients are given both for truncation errors and rounding errors.

2. Krylov, Lugin, and Ianovich, *Tables for the Numerical Integration of Functions with Power Singularities* (434 pages) gives abscissas and weights to 8S (significant figures) for the integration

$$\int_0^1 x^\beta(1-x)^\alpha f(x)\,dx, \qquad \alpha, \beta = -.9(.1)3, \qquad \beta < \alpha, \qquad n = 1\text{–}8.$$

3. Aizenshtat, Krylov, and Metleskii, *Tables for Calculating Laplace Transforms and Integrals of the Form* $\int_0^\infty x^s e^{-x} f(x)\,dx$ (378 pages) gives x_i, w_i, and $w_i e^{x_i}$ for $s = -.9(.02)0, .55(.05)3, -\frac{3}{4}, -\frac{1}{4}, m + k/3, m = -1(1)2, k = 1, 2, n = 1(1)15$, 8S.

4. Shao, Chen, and Frank, *Tables of Zeros and Gaussian Weights of Certain Associated Laguerre Polynomials and the Related Generalized Hermite Polynomials* (311 pages) gives x_i, w_i for generalized Laguerre integration for $\alpha = -.5(.5)10$, $n = 4, 8(8)32(16)64(32)128$ and for generalized Hermite integration for $\lambda = 0(1)10$, $n = 8(8)32(16)64(32)128(64)256$ to 25S.

5. Krylov and Skoblya, *Handbook of Numerical Inversion of Laplace Transforms* (293 pages) gives the p_k and A_k occurring in the integration formula

$$\frac{1}{2\pi i}\int_{c-i\infty}^{c+i\infty} e^p p^{-s}\phi(p)\,dp \approx \sum_{k=1}^{n} A_k \phi(p_k)$$

which is exact for $\phi(p) = p^{-j}, j = 0, 1, 2, \ldots, 2n-1$. Table 1: $s = 1(1)5; n = 1(1)15$, 20S. Table 2: $s = .01(.01)3, s \neq 1, 2, 3; n = 1(1)10$, 7–8S.

6. Kruglikova, *Tables for the Numerical Fourier Transform* (31 pages) gives w_k and x_k occurring in the integration formulas

$$\int_0^\infty \left(1 + \begin{matrix}\sin x\\ \cos x\end{matrix}\right) f(x)\,dx \approx \sum_{k=1}^{n} w_k f(x_k),$$

which is exact for $f(x) = (1+x)^{-s-i}$, $i = 0, 1, \ldots, 2n-1$. Here x_k are given to 10–14S and w_k to 5–14S for $s = r + \frac{1}{4}, r + \frac{1}{3}, r + \frac{1}{2}, r + \frac{2}{3}, r + \frac{3}{4}, r + 1$ $(r = 1, 2, 3)$ and for $n = 1(1)8$.

7. Piessens, R., *Tables of Use in the Inversion of the Laplace Transform.* Rep. TW1, TW2, TW3. Appl. Math. Sect., Univ. of Leuven, Heverlee, Belgium, 1969.

8. The reader might profitably consult Krylov and Shulgina which contains a wealth of tables.

9. The book *Approximate Calculation of Multiple Integrals* by A. H. Stroud includes a compendium of tables of formulas for numerical integration over various regions in several dimensions that is virtually complete as of 1971.

10. Kutt, *Quadrature Formulae for Finite Part Integrals* gives weights a_i to 30S for the equispaced rule

$$⨍_0^1 \frac{f(x)}{x^\lambda}\,dx \approx \sum_{i=1}^{N} a_i f(i/N), \qquad N = 3(1)20$$

and abscissas x_i and weights w_i to 30S for the Gauss-type integration rule

$$\int_0^1 \frac{f(x)}{x^\lambda}\,dx \approx \sum_{i=1}^{N} w_i f(x_i), \qquad N = 2(1)20$$

for $\lambda = 1, \frac{4}{3}, \frac{3}{2}, \frac{5}{3}, 2, 3, 4, 5$. For the equispaced case, auxiliary material is included to allow evaluation of finite part integrals of the form

$$\int_s^r \frac{f(x)}{(x-s)^\lambda}\,dx, \qquad \lambda = 2, 3, 4, 5.$$

11. Among the many tabulations of good lattice points, the most extensive are those by Hua and Wang, LaBudde and Warnock, and Haber.

References. Aizenshtat, Krylov, and Metleskii [1]; Berger [2, 3]; Boujot and Maroni [1]; Fletcher, Miller, Rosenhead, and Comrie [1]; Gautschi [A1]; Golub and Welsch [A1]; Haber [14]; Hamming [1, Chap. 10]; Hua and Wang [1]; Kruglikova [1]; Krylov, Lugin, and Ianovich [1]; Krylov and Pal'tsev [1]; Krylov and Shulgina [1]; Krylov and Skoblya [1]; Kutt [1]; LaBudde and Warnock [1]; J. C. P. Miller [1]; NBS Handbook [1]; Piessens and Branders [5]; Rutishauser [A1]; Shao, Chen, and Frank [1]; Skoblya [1]; Stroud [18]; Stroud and Secrest [2].

Appendix 5

Bibliography of Books and Articles

ABD-ELALL, L. F., DELVES, L. M., and REID J. K.

1. A numerical method for locating the zeros and poles of a meromorphic function, in *Numerical Methods for Nonlinear Algebraic Equations* (Rabinowitz, P., ed.), pp. 47–59. Gordon & Breach, London, 1970.

ABRAMOVICI, F.

1. The accurate calculation of Fourier integrals by the fast Fourier transform technique, *J. Comp. Phys.* **11** (1973) 28–37. See also *J. Comp. Phys.* **17** (1975) 446–449.

ABRAMOWITZ, M.

1. On the practical evaluation of integrals. *SIAM J. Appl. Math.* **2** (1954) 20–35.

ABRAMOWITZ, M., and STEGUN I. A. (ed.)

1. *Handbook of Mathematical Functions*, Nat. Bur. Stand. Appl. Math. Ser. No. 55. U.S. Govt. Printing Office, Washington, D.C., 1964.

ACHARYA, B. P., and DAS, R. N.

1. Numerical determination of Cauchy principal value integrals, *Computing* **27** (1981) 373–378.

ACHIESER, N. I.

1. *Vorlesungen über Approximationstheorie*. **2**. Aufl. Akademie-Verlag, Berlin, 1967.

ACM

1. *Collected Algorithms from CACM* (Communications of Association for Computing Machinery), ACM, New York.

AHLBERG, J. H., NILSON, E. N., and WALSH J. L.
1. *The Theory of Splines and Their Applications.* Academic Press, New York, 1967.

AHLIN, A. C.
1. On error bounds for Gaussian cubature. *SIAM Rev.* **4** (1962) 25–39.

AIZENSHTAT, V. S., KRYLOV, V. I., and METLESKII, A. S.
1. *Tables for Calculating Laplace Transforms and Integrals of the Form* $\int_0^\infty x^s e^{-x} f(x)\,dx$. Izdat. Akad. Nauk BSSR, Minsk, 1962.

AKSEN, M. B.
1. Upper bounds for approximations by quadrature formulae for certain classes of functions, *U.S.S.R. Computational Math. and Math. Phys.* **4**, No. 3 (1964) 220–226.

ALAYLIOGLU, A.
1. Numerical evaluation of finite Fourier integrals, *J. Comput. Appl. Math.* **9** (1983) 305–313.

ALAYLIOGLU, A., EVANS, G. A., and HYSLOP, J.
1. The evaluation of oscillatory integrals with infinite limits, *J. Comp. Phys.* **13** (1973) 433–438.
2. Automatic generation of quadrature formulae for oscillatory integrals, *Comput. J.* **18** (1975) 173–176.
3. The use of Chebyshev series for the evaluation of oscillatory integrals, *Comput. J.* **19** (1976) 258–267.

ALBRECHT, J.
1. Formeln zur numerischen Integration über Kreisbereiche, *ZAMM*† **40** (1960) 514–517.
2. Intervallschachtelungen beim Romberg-Verfahren, *ZAMM* **52** (1972) 433–435.

ALBRECHT, J., and COLLATZ, L.
1. Zur numerischen Auswertung mehrdimensionaler Integrale, *ZAMM* **38** (1958) 1–15.

ALBRECHT, J., and ENGELS, H.
1. Zur numerischen Integration über Kreisbereichen, in *Constructive Theory of Functions of Several Variables* (Schempp, W., and Zeller, K., eds.), pp. 1–5, Lecture Notes in Mathematics 571. Springer-Verlag, Berlin, 1977.

ALDER, B., FERNBACH, S., and ROTENBERG, M. (eds.).
1. *Methods in Computational Physics*, Vol. 2, Academic Press, New York, 1963.

ALLASIA, G., and ALLASIA, M.
1. Una rappresentazione del resto delle formule di quadratura, *Atti Accad. Sci. Torino I*, **110** (1975–1976) 353–358.

† Abbreviations used throughout
ZAMM = Zeitschrift für Angewandte Mathematik und Mechanik.
BIT = Nordisk Tidskrift for Informationbehandling.
CACM = Communications of the Association for Computing Machinery.
JACM = Journal of the Association for Computing Machinery.
ZAMP = Zeitschrift für Angewandte Mathematik und Physik.
MTAC = Mathematical Tables and Other Aids to Computation.
JIMA = Journal of the Institute of Mathematics and Its Applications.
ISNM = International Series of Numerical Mathematics.
TOMS = ACM Transactions on Mathematical Software.
IJNME = International Journal for Numerical Methods in Engineering.

ALLASIA, G., and GIORDANO, C.

1. Rappresentazione e valutazione del resto nelle formule di integrazione numerica generalizzate, *Calcolo* **17** (1980) 257–269.

ALLASIA, G., and PATRUCCO, P.

1. Valutazione del resto delle formule di quadrature, *Rend. Sem. Mat. Univ. Politec. Torino* **34** (1975–1976) 263–274.

AMBLE, O.

1. An autoadaptive method for numerical integration, *Norske Vid. Selsk. Forh. (Trondheim)* **42** (1969) 43–50.

ANDERS, E. B.

1. An extension of Romberg integration procedures to N variables, *JACM* **13** (1966) 505–510.

ANDERSEN, D. G.

1. Gaussian quadrature formulae for $\int_0^1 -\ln(x)f(x)\,dx$, *Math. Comp.* **19** (1965) 477–481.

ANDERSON, W. L.

1. Fast Hankel transforms using related and lagged convolutions, *TOMS* **8** (1982) 344–368. See also *TOMS* **8** (1982) 369–370.

ANDREWS, H.

1. A high-speed algorithm for the computer generation of Fourier transforms, *IEEE Trans. Comput.* **C-17** (1968) 373–375.

ANDREWS, H. C., and PRATT, W. K.

1. Transform image coding, *Symp. Comput. Process. Commun., Polytech. Inst. of Brooklyn, April 1969*, pp. 63–84.

ANSELONE, P. M., and OPFER, G.

1. Numerical integration of weakly singular functions, in Hämmerlin [5, pp. 11–43].

ANTES, H.

1. Über die vierdimensionale Romberg-Integration mit Schranken, *Computing* **9** (1972) 45–52.

ANTONOV, I. A., and SOLEEV, V. M.

1. An economic method of computing LP_τ-sequences, U.S.S.R. *Comput. Math. Math. Phys.* **19** (1979) 252–256.

ANUTA, P. E.

1. Spatial registration of multispectral and multitemporal digital imagery using fast Fourier transform techniques, *IEEE Trans. Geosci. Electron.* **GE-8** (1970) 353–368.

ARNOL'D, V. I., and KRYLOV, A. L.

1. Uniform distribution of points on a sphere and some ergodic properties of solutions of linear ordinary differential equations in a complex region, *Soviet Math. Dokl.* **4** (1963) 1–5.

ASCHER, U., and WEISS, R.

1. Collocation for singular perturbation problems. I: First order systems with constant coefficients, *SIAM J. Numer. Anal.* **20** (1983) 537–557.

ASKEY, R.

1. Positivity of the Cotes numbers for some Jacobi abscissas, *Numer. Math.* **19** (1972) 46–48.
2. Positivity of the Cotes numbers for some Jacobi abscissas—II, *JIMA* **24** (1979) 95–98.

ASKEY, R., and FITCH, J.

1. Positivity of the Cotes number for some ultraspherical abscissas, *SIAM J. Numer. Anal.* **5** (1968) 199–201.

ATCHISON, T. A.

1. Selecting nonlinear transformations for the evaluation of improper integrals, *J. Res. Nat. Bur. Standards Sect. B* **74** (1970) 183–185.

ATCHISON, T. A., and GRAY, H. L.

1. Nonlinear transformations related to the evaluation of improper integrals II, *SIAM J. Numer. Anal.* **5** (1968) 451–459.

ATKINSON, K.

1. The numerical evaluation of the Cauchy transform on simple closed curves, *SIAM J. Numer. Anal.* **9** (1972) 284–299.
2. *A Survey of Numerical Methods for the Solution of Fredholm Integral Equations of the Second Kind.* SIAM Publ., Philadelphia, 1976.
3. An automatic program for linear Fredholm integral equations of the second kind, TOMS **2** (1976) 154–171.
4. *An Introduction to Numerical Analysis.* Wiley, New York, 1978.
5. Numerical integration on the sphere, *J. Austral. Math. Soc. Ser. B* **23** (1981–1982) 332–347.

ATKINSON, M. P., and LANG, S. R.

1. A comparison of some inverse Laplace transform techniques for use in circuit design, *Comput. J.* **15** (1972) 138–139.

AXELSSON, O.

1. Global integration of differential equations through Lobatto quadrature, *BIT* **4** (1964) 69–86.

BABURIN, O. V., and LEBEDEV, V. I.

1. Computation of tables of the roots and weights of Hermitian and Laguerre polynomials for $n = 1(1)101$, *U.S.S.R. Comput. Math. Math. Phys.* **7**(5) (1967) 76–88.

BABUŠKA, I.

1. Numerical stability in mathematical analysis, in *Information Processing 68*, (Morrell, A. J. H., ed.), Vol. 1, pp. 11–23. North-Holland Publ., Amsterdam, 1969.

BABUŠKA, I., PRÁGER, M., and VITÁSEK, E.

1. *Numerical Processes in Differential Equations.* Wiley, New York, 1966.

BAKER, A.

1. On some diophantine inequalities involving the exponential function, *Canad. J. Math.* **17** (1965) 616–626.

BAKER, C. T. H.

1. On the nature of certain quadrature formulas and their errors, *SIAM J. Numer. Anal.* **5** (1968) 783–804.
2. *The Numerical Treatment of Integral Equations.* Clarendon Press, Oxford, 1977.

BAKER, C. T. H., and HODGSON, G. S.

1. Asymptotic expansions for integration formulas in one or more dimensions, *SIAM J. Numer. Anal.* **8** (1971) 473–480.

BAKER, C. T. H., and RADCLIFFE, P. A.
1. Error bounds for some Chebyshev methods of approximation and integration, *SIAM J. Numer. Anal.* **7** (1970) 317–327.

BAKHVALOV, N. S.
1. Approximate computation of multiple integrals, *Vestnik Moskov. Univ. Ser. Mat. Mek. Astr. Fiz. Him.* **4** (1959) 3–18.
2. On the hypothesis of the independence of round-off errors in numerical integration, *U.S.S.R. Computational Math. and Math. Phys.* **4**, No. 3 (1964) 1–8.
3. *Numerical Methods.* Mir Publishers, Moscow, 1977.

BAKHVALOV, N. S., and VASILEVA, L. G.
1. Evaluation of the integrals of oscillating functions by interpolation at nodes of Gaussian quadratures, *U.S.S.R. Computational Math. and Math. Phys.* **8**, No. 1 (1968) 241–249.

BAKKER, M.
1. A note on C° Galerkin methods for two-point boundary problems, *Numer. Math.* **38** (1982) 447–453.

BAKUN, W. H., and EISENBERG, A.
1. Fourier integrals and quadrature-introduced aliasing, *Bull. Seismological Soc. Amer.* **60** (1970) 1291–1296.

BANDEMER, H.
1. Mechanische Quadraturverfahren durch sukzessive Intervallteilung, *ZAMM* **47** (1967) 115–120.

BARATELLA, P.
1. Un'estensione ottimale della formula di quadratura di Radau, *Rend. Sem. Mat. Univ. Politec. Torino* **37** (1979) 147–158.

BARATELLA, P., GARETTO, M., and VINARDI, G.
1. Approximation of the Bessel function $J_\nu(x)$ by numerical integration, *J. Comput. Appl. Math.* **7** (1981) 87–91.

BARATELLA, P., and VINARDI, G.
1. An iterative method for zeros of classical Laguerre polynomials, Atti Accad. Sci. Torino Cl. Sci. Fis. Mat. Natur. **114** (1980/81) 147–160.

BAREISS, E. H., and NEUMAN, C. P.
1. Singular integrals and singular integral equations with a Cauchy kernel and the method of symmetric pairing, Rep. 6988. Argonne Nat. Lab., Argonne, Illinois, 1965.

BARNEA, D. I., and SILVERMAN, H. F.
1. A class of algorithms for fast digital image registration, *IEEE Trans. Computers* **C-21** (1972) 179–186.

BARNHILL, R. E.
1. The convergence of quadratures on complex contours, *SIAM J. Numer. Anal.* **2** (1965) 321–336.
2. Complex quadratures with remainders of minimum norm, *Numer. Math.* **7** (1965) 384–390.
3. Optimal quadratures in $L^2(E_\rho)$, *SIAM J. Numer. Anal.* **4** (1967) 390–397, 534–541.
4. An error analysis for numerical multiple integration, *Math. Comp.* **22** (1968) 98–109, 286–292.

5. Asymptotic properties of minimum norm and optimal quadratures, *Numer. Math.* **12** (1968) 384–393.
6. The convergence of complex cubatures, *SIAM J. Numer. Anal.* **6** (1969) 82–89.

Barnhill, R. E., and Little, F. F.
1. Adaptive triangular cubatures, CAGD Rep. 80/3, Dept. of Mathematics, Univ. of Utah, Salt Lake City, UT, 1980.

Barnhill, R. E., and Nielson, G. M.
1. An error analysis for numerical multiple integration, III, *Math. Comp.* **24** (1970) 301–314.

Barnhill, R. E., and Pilcher, D. T.
1. Sard kernels for certain bivariate cubatures, *CACM* **16** (1973) 567–570.

Barnhill, R. E., and Wixom, J. A.
1. Quadratures with remainders of minimum norm, *Math. Comp.* **21** (1967) 66–75, 382–387.

Barrar, R. B., Loeb, H. L., and Werner, H.
1. On the existence of optimal integration formulas for analytic functions, *Numer. Math.* **23** (1974) 105–117.

Barrett, W.
1. Convergence properties of Gaussian quadrature formulae, *Comput. J.* **3** (1960–1961) 272–277.
2. On the convergence of Cotes' quadrature formulae, *J. London Math. Soc.* **39** (1964) 296–302.
3. On the convergence of sequences of rational approximations to analytic functions of a certain class, *JIMA* **7** (1971) 308–323.

Barrucand, P.
1. Quadratures elliptiques, un nouveau procédé d'intégration numérique, *C. R. Acad. Sci. Paris* **260** (1965) 6015–6018.
2. Integration numérique, abscisse de Kronrod–Patterson et polynômes de Szegö, *C. R. Acad. Sci. Paris Ser. A* **270** (1970) 336–338.
3. Une application de la théorie des nombres à l'intégration numérique, *C. R. Acad. Sci. Paris Ser. A* **270** (1970) 633–636.

Barth, W.
1. Quadraturfehlerabschätzung bei mehrfachen Integralen, *ZAMP* **18** (1967) 760–762.

Bartholemew, G. E.
1. Numerical integration over a triangle, *MTAC* **13** (1959) 295–298.

Bass, J., and Guilloud, J.
1. Méthode de Monte Carlo et suites uniformement denses, *Chiffres* **1** (1958) 149–155.

Basu, N. K.
1. Error estimates for a Chebyshev quadrature method, *Math. Comp.* **24** (1970) 863–867.
2. Evaluation of a definite integral using Tschebyscheff approximation, *Mathematica* **13** (1971) 13–23.

Basu, N. K., and Kundu, M. C.
1. Polynomial approximation and the quadrature problem over a semi-infinite interval, *Apl. Math.* **20** (1975) 216–221.
2. Some methods of numerical integration over a semi-infinite interval, *Apl. Math.* **22** (1977) 237–243.

BAUER, F. L.
1. La méthode d'intégration numérique de Romberg, *Colloq. Anal. Numér.* (Mons, 1961), pp. 119–129. Librairie Universitaire, Louvain, 1961.

BAUER, F. L., RUTISHAUSER, H., and STIEFEL, E. L.
1. New aspects in numerical quadrature, in *Experimental Arithmetic, High Speed Computing, and Mathematics*, pp. 199–218. Amer. Math. Soc., Providence, Rhode Island, 1963.

BAUER, F. L., and STETTER, H. J.
1. Zur numerischen Fourier-Transformation, *Numer. Math.* **1** (1959) 208–220.

BEARD, R. E.
1. Some notes on approximate product integration, *J. Inst. Actu,* **73** (1947) 356–416.

BEHFOROOZ, G. H., and PAPAMICHAEL, N.
1. End conditions for interpolatory cubic splines with unequally spaced knots, *J. Comput. Appl. Math.* **6** (1980) 59–65.

BEIGHTON, S., and NOBLE, B.
1. An error estimate for Stenger's quadrature formula, *Math. Comp.* **38** (1982) 539–545.

BELL, D. A.
1. Approximations in Fourier transforms, *Comput. J.* **6** (1963–1964) 244–247.

BELLMAN, R. E., KALABA, R. E., and LOCKETT, J. A.
1. *Numerical Inversion of the Laplace Transform. Applications to Biology, Economics, Engineering and Physics.* Amer. Elsevier, New York, 1966.

BEREANU, B.
1. On the convergence of Cartesian multidimensional quadrature formulae, *Numer. Math.* **19** (1972) 348–350.

BEREZIN, I. S., and ZHIDKOV, N. P.
1. *Computing Methods*, Vol. 1. Pergamon, Oxford, 1965.

BERGER, B. S.
1. Dynamic response of an infinite cylindrical shell in an acoustic medium, *J. Appl. Mech.* **36** (1969) 342–345.
2. *Tables of Zeros and Weights for Gauss–Laguerre Quadrature to 23–24S for $N = 100,150,200(100)900$.* Dept. of Mech. Eng., Univ. of Maryland, College Park.
3. *Tables of Zeros and Weights for Gauss–Hermite Quadrature to 26–27S for $N = 200(200)1000,2000$.* Dept. of Mech. Eng., Univ. of Maryland, College Park.

BERGMAN, S.
1. *The Kernel Function and Conformal Mapping.* Mathematical Surveys, Vol. 5, Amer. Math. Soc., New York, 1950.

BERTRANDIAS, J. P.
1. Calcul d'une intégrale au moyen de la suite $X_n = An$. Evaluation de l'erreur, *Publ. Inst. Statist. Univ. Paris* **9** (1960) 335–357.

BICKLEY, W. G.
1. Formulae for numerical integration, *Math. Gaz.* **23** (1939) 352–359.

BILLAUER, A.
1. On Gaussian quadrature by divided differences of a modified function, *BIT* **14** (1974) 359–361.

BINGHAM, C., GODFREY, M. D., and TUKEY, J. D.
1. Modern techniques of power spectrum estimation, *IEEE Trans. Audio Electroacoust.* **AU-15** (1967) 56–66.

BIRKHOFF, G., and DE BOOR, C.
1. Error bounds for spline interpolation, *J. Math. Mech.* **13** (1964) 827–835.

BIRKHOFF, G., and YOUNG, D.
1. Numerical quadrature of analytic and harmonic functions, *J. Math. and Phys.* **29** (1950) 217–221.

BLAKEMORE, M., EVANS, G. A., and HYSLOP, J.
1. Comparison of some methods for evaluating infinite range oscillatory integrals, *J. Comp. Phys.* **22** (1976) 352–376.

BLANC, C., and LINIGER, W.
1. Stochastische Fehlerauswertung bei numerischen Methoden, *ZAMM* **35** (1955) 121–130.

BLOOMFIELD, P.
1. *Fourier Analysis of Time Series.* Wiley, New York, 1976.

BLUE, J. L.
1. Automatic numerical quadrature—DQUAD, Computer Science Tech. Rep. 25, Bell Labs., Murray Hill, NJ, 1975.
2. Automatic numerical quadrature, *Bell System Tech. J.* **56** (1977) 1651–1678.

BOAS, JR., R. P.
1. Generalized Taylor series, quadrature formulas and a formula by Kronecker, *SIAM Rev.* **12** (1970) 116–119.

BOAS, JR., R. P., and SCHOENFELD, L.
1. Indefinite integration by residues, *SIAM Rev.* **8** (1966) 173–183.

BOGUES, K., MORROW, C. R., and PATTERSON, T. N. L.
1. An implementation of the method of Ermakov and Zolotukin for multidimensional integration and interpolation, *Numer. Math.* **37** (1981) 49–60.

BOLAND, W. R.
1. The convergence of product-type quadrature formulas, *SIAM J. Numer. Anal.* **9** (1972) 6–13.
2. Properties of product-type quadrature formulas, *BIT* **13** (1973) 287–291.
3. The numerical solution of Fredholm integral equations using product-type quadrature formulas, *BIT* **12** (1972) 5–16.

BOLAND, W. R., and DURIS, C. S.
1. Product type quadrature formulas, *BIT* **11** (1971) 139–158.

BOLAND, W. R., and HAYMOND, R. E.
1. The numerical solution of Fredholm integral equations with rapidly varying kernels, *Appl. Math. Comp.* **3** (1977) 25–38.

BOYKO, E. R., NASSIFF, P. J., and PEZZULLO, J. C.
1. Integration weights for experimental data, *J. Comp. Phys.* **27** (1978) 443–446.

BOUJOT, J.-P.
1. Calcul de moments appliqué à l'evaluation numérique d'intégrales, Publ. No. VB/8.10.5/AI. Inst. Blaise Pascal, Paris, 1965.

BOUJOT, J.-P., and MARONI, P.
1. Algorithme général de construction de tables de Gauss pour les problèmes de quadratures, Publ. No. MMX/8.1.8/AI. Inst. Blaise Pascal, Paris, 1968.

BOUJOT, J.-P., RAPHALEN, A., and SOULÉ, J. L.
1. Présentation des methodes et utilisation des sous-programmes de la rubrique quadratures, Rapport D.C.E. Fontenay No. 55. Commissariat à L'Energie Atomique, 1968.

BOUZITAT, J.
1. Sur l'intégration numérique approchée par la méthode de Gauss généralisée et sur une extension de cette méthode, *C. R. Acad. Sci. Paris*, **229** (1949) 1201–1203.

BRACEWELL, R.
1. *The Fourier Transform and its Applications.* McGraw-Hill, New York, 1965.

BRANDERS, M., and PIESSENS, R.
1. Algorithm 001—An extension of Clenshaw–Curtis quadrature, *J. Comput. Appl. Math.* **1** (1975) 55–65.

BRASS, H.
1. *Quadraturverfahren.* Vandenhoeck and Ruprecht, Göttingen, 1977.

BRASS, H., and SCHMEISSER, G.
1. The definiteness of Filippi's quadrature formulae and related problems, in Hämmerlin [5, pp. 109–119].
2. Error estimates for interpolatory quadrature formulae, *Numer. Math.* **37** (1981) 371–386.

BREZINSKI, C.
1. *Accélération de la convergence en analyse numérique*, Lecture Notes in Math. 584. Springer-Verlag, Berlin, 1977.
2. *Algorithmes d'accélération de la convergence.* Editions Technip, Paris, 1978.
3. A general extrapolation algorithm, *Numer. Math.* **35** (1980) 175–187.

BRIGHAM, E. O.
1. *The Fast Fourier Transform.* Prentice Hall, Englewood Cliffs, N.J., 1974.

BRIGHAM, E. O., and MORROW, R. E.
1. The fast Fourier transform, *IEEE Spectrum* **4**, No. 12 (1967) 63–70.

BROMBEER, R., and MACHER, R.
1. Zum Eindeutigkeitsproblem bei optimalen Quadraturformeln mit freien Knoten, *ZAMM* **56** (1976) T280–281.

BUCK, R. C.
1. *Advanced Calculus.* McGraw-Hill, New York, 1956.

BULIRSCH, R.
1. Bemerkungen zur Romberg-Integration, *Numer. Math.* **6** (1964) 6–16.

BULIRSCH, R., and RUTISHAUSER, H.

1. Interpolation und genäherte Quadratur, in *Mathematische Hilfsmittel des Ingenieurs* (Sauer, R., and Szabó, I., ed.), Pt. III, pp. 232–319. Springer-Verlag, Berlin and New York, 1968.

BULIRSCH, R., and STOER, J.

1. Asymptotic upper and lower bounds for results of extrapolation methods, *Numer. Math.* **8** (1966) 93–104.
2. Handbook series numerical integration: Numerical quadrature by extrapolation, *Numer. Math.* **9** (1967) 271–278.
3. Fehlerabschätzungen und Extrapolation mit rationalen Funktionen bei Verfahren vom Richardson-Typus, *Numer. Math.* **6** (1964) 413–427.

BUNSE, W.

1. Blending product-type quadrature rules, *BIT* **22** (1982) 477–486.

BUNTON, W. R., DIETHELM, M., and HAIGLER, K.

1. Romberg quadrature subroutines for single and multiple integrals, Rep. TM 314-221. Jet Propulsion Lab., Pasadena, California, 1969.

BUNTON, W. R., DIETHELM, M., and WINJE, G. L.

1. Modified Romberg quadrature: A subroutine to support general scientific computing, Rep. TM314-258. Jet Propulsion Lab., Pasadena, California, 1970.

BURGOYNE, F. D.

1. Quadrature formulas over infinite intervals in terms of differences, *Math. Comp.* **17** (1963) 298–301.

BURNETT, D. S., and SOROKA, W. W.

1. An efficient numerical technique for evaluating large quantities of highly oscillatory integrals, *JIMA* **10** (1972) 325–332.

BURROWS, B. L.

1. A new approach to numerical integration, *JIMA* **26** (1980) 151–173.

BUTCHER, J. C.

1. Integration processes based on Radau quadrature formulas, *Math. Comp.* **18** (1964) 233–244.

BUYST, L., and SCHOTSMANS, L.

1. A method of Gaussian type for the numerical integration of oscillating functions, *ICC Bull.* **3** (1964) 210–214.

BYKOVA, T. M.

1. A cubature formula for calculating triple integrals, which is exact for polynomials of fourth degree, and has eleven nodes, *Izvest. Akad. Nauk BSSR Ser. Fiz.-Mat. Nauk* **1** (1970) 51–54.

BYKOVA, T. M., and MYSOVSKIKH, I. P.

1. The Radon cubature formula for a region with central symmetry, *U.S.S.R. Computational Math. and Math. Phys.* **9**, No. 3 (1969) 253–259.

CADETE, M. O. R.

1. *Cálculo Automatico de Integrais Definidos* (Portugese). Instituto Gulbenkian de Ciência, Oeiras, Portugal, 1975.

CADWELL, J. H.
1. A recursive program for the general n-dimensional integral, *CACM* **6** (1963) 35–36.

CAPOVANI, M., GHELARDONI, G., and LOMBARDI, G.
1. Algoritmi per il calcolo di formule Gaussiane di quadratura, Noto Interna B75–28. Istituto di Elaborazione della Informazione del C.N.R., Pisa, 1975.
2. Un methodo per il calcolo dei nodi e dei pesi delle formule di quadratura di Gauss–Hermite, *Calcolo* **13** (1976), 441–452.

CAPRANI, O.
1. Implementation of a low round-off summation method, *BIT* **11** (1971) 271–275.

CASALETTO, J., PICKETT, M., and RICE, J. R.
1. A comparison of some numerical integration programs, *SIGNUM Newsletter* **4**, No. 3 (1969) 30–40.

CESCHINO, F., and LETIN, M.
1. Généralisation de la méthode de Havie au calcul des integrales multiples, *C. R. Acad. Sci. Paris Ser. A* **275** (1972) 505–508.

CHAI, A. S., and WERTZ, H. J.
1. Some considerations in practical computation, Tech. Summary Rep. 469. MRC, Madison, Wisconsin, 1964.

CHAKRAVARTI, P. C.
1. *Integrals and Sums: Some New Formulae for their Numerical Evaluation.* Athlone Press, London, 1970.

CHASE, S. M., and FOSDICK, L. D.
1. An algorithm for Filon quadrature, *CACM* **12** (1969) 453–457.

CHAWLA, M. M.
1. On the Chebyshev polynomials of the second kind, *SIAM Rev.* **9** (1967) 729–733.
2. Convergence of Newton-Cotes quadratures for analytic functions, *BIT* **11** (1971) 159–167.
3. On the estimation of errors of Gaussian cubature formulas, *SIAM J. Numer. Anal.* **5** (1968) 172–181.
4. Error estimates for the Clenshaw–Curtis quadrature, *Math. Comp.* **22** (1968) 651–656.
5. Asymptotic estimates for the error of the Gauss–Legendre quadrature formula, *Comput. J.* **11** (1968) 339–340.
6. Error bounds for the Gauss–Chebyshev quadrature formula of the closed type, *Math. Comp.* **22** (1968) 889–891.
7. On Davis's method for the estimation of errors of Gauss–Chebyshev quadratures, *SIAM J. Numer. Anal.* **6** (1969) 108–117.
8. Estimation of errors of Gauss–Chebyshev quadratures, *Comput. J.* **13** (1970) 107–109.
9. Estimating errors of numerical approximations for analytic functions, *Numer. Math.* **16** (1971) 370–374.
10. Hilbert spaces for estimating errors of quadratures for analytic functions, *BIT* **10** (1970) 145–155.

CHAWLA, M. M., and JAIN, M. K.
1. Asymptotic error estimates for the Gauss quadrature formula, *Math. Comp.* **22** (1968) 91–97.
2. Error estimates for Gauss quadrature formulas for analytic functions, *Math. Comp.* **22** (1968) 82–90.

CHAWLA, M. M., and JAYARAJAN, N.
1. Quadrature formulas for Cauchy principal value integrals, Computing **15** (1975) 347–355.
2. Bivariate cardinal interpolation and cubature formulas, *SIAM J. Numer. Anal.* **12** (1975) 605–616.

CHAWLA, M. M., and KAUL, V.
1. Optimal rules with polynomial precision for Hilbert spaces possessing reproducing kernel functions, *Numer. Math.* **22** (1974) 207–218; **25** (1976) 379–382.

CHAWLA, M. M., and KUMAR, S.
1. Convergence of quadratures for Cauchy principal value integrals, *Computing* **23** (1979) 67–72.

CHAWLA, M. M., and RAMAKRISHNAN, T. R.
1. Modified Gauss–Jacobi quadrature formulas for the numerical evaluation of Cauchy type singular integrals, *BIT* **14** (1974) 14–21.

CHEBYSHEV, P. L. (Tchebyscheff)
1. Sur les quadratures, *J. Math. Pures Appl.* [2] **19** (1874) 19–34.

CHEN, T. H. C.
1. Asymptotic error estimates for Gaussian quadrature formulas, *Math. Comp.* **38** (1982) 143–152.

CHENG, V., SUZUKAWA, H., and WOLFSBERG, M.
1. Investigations of a non-random numerical method for multidimensional integration, *J. Chem. Phys.* **59** (1973) 3992–3999.

CHISHOLM, J. S. R., GENZ, A., and ROWLANDS, G. E.
1. Accelerated convergence of sequences of quadrature approximations, *J. Comp. Phys.* **10** (1972) 284–307.

CHRISTOFFEL, E. B.
1. Über die Gaussische Quadratur und eine Verallgemeinerung derselben, *J. Reine Angew. Math.* **55** (1858) 61–82.

CICENIA, R. A.
1. Numerical integration formulae involving derivatives, *JIMA* **18** (1976) 79–85.
2. Remainders in numerical integration formulae involving derivatives, *JIMA* **24** (1979) 347–352.

CLARK, M., Jr., and HANSEN, K. F.
1. *Numerical Methods of Reactor Analysis*. Academic Press, New York, 1964.

CLENDENIN, W. W.
1. A method for numerical calculation of Fourier integrals, *Numer. Math.* **8** (1966) 422–436.
2. A method for numerical quadrature over an infinite interval using systematically increasing mesh size, *Numer. Math.* **16** (1971) 397–407.

CLENSHAW, C. W.
1. A note on the summation of Chebyshev series, *MTAC* **9** (1955) 118–120.
2. *Chebyshev Series for Mathematical Functions*, Math. Tables, Vol. 5, Nat. Phys. Lab., HM Stationery Office, London, 1962.

CLENSHAW, C. W., and CURTIS, A. R.
1. A method for numerical integration on an automatic computer, *Numer. Math.* **2** (1960) 197–205.

COCHRAN, W. T. *et al.*
1. What is the fast Fourier transform, *Proc. IEEE* **55**, (1967) 1664–1674.

COHEN, A. M.
1. Cautious Romberg extrapolation, *Internat. J. Comput. Math.* **8** (1980) 137–147.

COLLATZ, L.
1. Applications of nonlinear optimization to approximation problems, in *Integer and Nonlinear Programming*, (Abadie, J., ed.), pp. 285–308. North-Holland Publ., Amsterdam, 1970.

CONCUS, P., CASSATT, D., JAEHNIG, G., and MELBY, E.
1. Tables for the evaluation of $\int_0^\infty x^\beta e^{-x} f(x)\,dx$ by Gauss–Laguerre quadrature, *Math. Comp.* **17** (1963) 245–256.

CONROY, H.
1. Molecular Schrödinger equation VIII. A new method for the evaluation of multidimensional integrals, *J. Chem. Phys.* **47** (1967) 5307–5318; see also *J. Chem. Phys.* **50** (1969) 565–566.

CONTE, S. D., and DE BOOR, C.
1. *Elementary Numerical Analysis.* An Algorithmic Approach. Third edition. McGraw-Hill, New York, 1980.

COOLEY, J. W., LEWIS, P. A. W., and WELCH, P. D.
1. Application of the fast Fourier transform to computation of Fourier integrals, Fourier series and convolution integrals, *IEEE Trans. Audio Electroacoust.* **AU-15** (1967) 79–85.
2. Historical notes on the fast Fourier transform, *Proc. IEEE* **55**, (1967) 1675–1677.
3. The fast Fourier transform and its application to time series analysis, in *Statistical Methods for Digital Computers* (Enslein, K., Ralston, A., and Wilf, H. S., eds.), pp. 377–423. Wiley, New York, 1977.

COOLEY, J. W., and TUKEY, J. W.
1. An algorithm for the machine calculation of complex Fourier series, *Math. Comp.* **19** (1965) 297–301.

CORNILLE, P.
1. Computation of Hankel transforms, *SIAM Rev.* **14** (1972) 278–285.

COURANT, R., and JOHN, F.
1. *Introduction to Calculus and Analysis*, Vol. 2. Wiley-Interscience, New York, 1974.

COWDREY, D. R., and REEVES, C. M.
1. An application of the Monte Carlo method to the evaluation of some molecular integrals, *Comput. J.* **6** (1963–1964) 277–286.

COWPER, G. R.
1. Gaussian quadrature formulas for triangles, *IJNME* **7** (1973) 405–408.

COX, M. G.
1. Practical spline approximation, in *Topics in Numerical Analysis* (Turner, P. R., ed.), pp. 79–112, Lecture Notes in Mathematics 965. Springer-Verlag, Berlin, 1982.

CRANLEY, R., and PATTERSON, T. N. L.

1. The evaluation of multidimensional integrals, *Comput. J.* **11** (1968) 102–110.

2. A regression method for the Monte Carlo evaluation of multidimensional integrals, *Numer. Math.* **16** (1970) 58–72.

3. On the automatic numerical evaluation of definite integrals, *Comput. J.* **14** (1971) 189–198.

4. Randomization of number theoretic methods for multiple integration, *SIAM J. Numer. Anal.* **13** (1976) 904–914.

CRISTECU, M., and LOUBIGNAC, G.

1. Quadratures de Gauss pour des functions avec singularité en $1/R$ sur des carrés ou des triangles, in *Méthodes numériques dans les sciences de l'ingénieur* (Absi, E., and Glowinski, R., eds.), pp. 21–31. GAMNI, Dunod, Paris, 1979.

CURTIS, A. R., and RABINOWITZ, P.

1. On the Gaussian integration of Chebyshev polynomials, *Math. Comp.* **26** (1972) 207–211.

DANIELL, P. J.

1. Remainders in interpolation and quadrature formulae, *Math. Gaz.* **24** (1940) 238–244.

DAUDEY, J. P., DINES, S., and SAVINELLI, R.

1. Numerical integration techniques for quantum chemistry. The role of periodization in the calculation of electronic integrals, *Theoret. Chim. Acta* **37** (1975) 275–283.

DAVENPORT, J. H.

1. *On the Integration of Algebraic Functions*, Lecture Notes in Computer Science 102. Springer-Verlag, Berlin, 1981.

DAVIES, B., and MARTIN, B.

1. Numerical inversion of the Laplace transform: A survey and comparison of methods, *J. Comp. Phys.* **33** (1979) 1–32.

DAVIS, P. J.

1. Errors of numerical approximation for analytic functions, *J. Rational Mech. Anal.* **2** (1953) 303–313.

2. On simple quadratures, *Proc. Amer. Math. Soc.* **4** (1953) 127–136.

3. On a problem in the theory of mechanical quadratures, *Pacific J. Math.* **5** (1955) 669–674.

4. On the numerical integration of periodic analytic functions, in *On Numerical Approximation* (Langer, R. E., ed.), pp. 45–59. Univ. of Wisconsin Press, Madison, 1959.

5. Errors of numerical approximation for analytic functions, in *Survey of Numerical Analysis* (Todd, J., ed.). pp. 468–484. McGraw-Hill, New York, 1962.

6. *Interpolation and Approximation.* Ginn (Blaisdell), Boston, Massachusetts, 1963.

7. Triangle formulas in the complex plane, *Math. Comp.* **18** (1964) 569–577.

8. A construction of nonnegative approximate quadrates, *Math. Comp.* **21** (1967) 578–582.

9. Approximate integration rules with nonnegative weights, in *Lecture Series in Differential Equations* (Aziz, A. K., ed.), Vol. II. pp. 233–256. Van Nostrand Reinhold, Princeton, New Jersey, 1969.

10. Additional simple quadratures in the complex plane, *Aequationes Math.* **3** (1969) 149–155.

11. Applying diverse areas of mathematics to the approximate computation of integrals, in *Jeffery-Williams Lectures 1968–1972*, pp. 91–123. Canad. Math. Congr., Montreal, Quebec, 1972.

12. Non-negative interpolation formulas for harmonic and analytic functions, in *Approximation Theory* (Talbot, A., ed.), pp. 83–100. Academic Press, New York, 1970.

13. *Circulant Matrices*. Wiley, New York, 1979.

DAVIS, P. J., and POLANSKY, I.

1. Numerical interpolation, differentiation, and integration, *Handbook of Mathematical Functions*, Chapter 25, pp. 875–924. Nat. Bur. Standards, Washington, 1964.

DAVIS, P. J., and RABINOWITZ, P.

1. On the estimation of quadrature errors for analytic functions, *MTAC* **8** (1954) 193–203.
2. Some Monte Carlo experiments in computing multiple integrals, *MTAC* **10** (1956) 1–8.
3. Abscissas and weights for Gaussian quadratures of high order, *J. Res. Nat. Bur. Standards* **56** (1956) 35–37.
4. Additional abscissas and weights for Gaussian quadratures of high order: Values for $n = 64$, 80, and 96, *J. Res. Nat. Bur. Standards* **60** (1958) 613–614.
5. Advances in orthonormalizing computation, *Advances in Computers* **2** (1961) 55–133.
6. Some geometrical theorems for abscissas and weights of Gauss type, *J. Math. Anal. Appl.* **2** (1961) 428–437.
7. Ignoring the singularity in approximate integration, *SIAM J. Numer. Anal.* **2** (1965) 367–383.
8. A multiple purpose orthonormalizing code and its uses, *JACM* **1** (1964) 183–191.
9. On the nonexistence of simplex integration rules for infinite integrals, *Math. Comp.* **26** (1972) 687–688.
10. *Numerical Integration*. Ginn (Blaisdell), Boston, Massachusetts, 1967.

DAVIS, P. J., and WILSON, M. W.

1. Non-negative interpolation formulas for uniformly elliptic equations, *J. Approx. Theory* **1** (1968) 374–380.

DAY, J. T.

1. Quadrature methods of arbitrary order for solving linear ordinary differential equations, *BIT* **6** (1966) 181–190.

DEÁK, I.

1. Comparison of methods for generating uniformly distributed random points in and on a hypersphere, *Problems of Control and Information Theory* **8** (1979) 105–113.

DEANGELIS, P. L., MURLI, A., and PIROZZI, M. A.

1. Sulla evalutazione dell'errore nel calcolo numerico degli integrali trigonometrici, *Calcolo* **13** (1976) 421–440.

DE BALBINE, G., and FRANKLIN, J. N.

1. The calculation of Fourier integrals, *Math. Comp.* **20** (1966) 570–589.

DE BOOR, C.

1. CADRE: An algorithm for numerical quadrature, in J. R. Rice [2, pp. 417–449].
2. On writing an automatic integration algorithm, in J. R. Rice [2, pp. 201–209].
3. The approximation of functions and linear functionals: Best vs. good approximation, in *Numerical Analysis* (Golub, G. H., and Oliger, J., eds.), pp. 53–70, Proc. Symp. Appl. Math. XXII. Amer. Math. Soc., Providence, RI, 1978.
4. *A Practical Guide to Splines*. Springer-Verlag, New York, 1978.

DE BOOR, C., and RICE, J. R.

1. Least squares cubic spline approximation, I—Fixed Knots, CSD TR 20; II—Variable Knots, CSD TR 21. Purdue Univ., Lafayette, Indiana, 1968.

DECELL, JR., H. P., and LEA, R. N.
1. Numerical integration and Riemann–Stieltjes sums, *SIAM Rev.* **8** (1966) 196–200.

DE DONCKER, E.
1. An adaptive extrapolation algorithm for automatic integration, *SIGNUM Newsletter* **13**(2) (1978) 12–18.
2. New Euler–Maclaurin expansions and their application to quadrature over the s-dimensional simplex, *Math. Comp.* **33** (1979) 1003–1018.
3. Numerical Integration and Asymptotic Expansions, Ph.D. Thesis, Dept. of Mathematics, Catholic University of Leuven, 1980.

DE DONCKER, E., and PIESSENS, R.
1. Algorithm 32. Automatic computation of integrals with singular integrand, over a finite or an infinite range, *Computing* **17** (1976) 265–279.
2. A bibliography on automatic integration, *J. Comput. Appl. Math.* **2** (1976) 273–280.
3. Measurement of quality of quadrature routines, Tech. Rep., App. Maths. and Prog. Div., Kath. Univ. Leuven, 1976.

DE DONCKER, E., and ROBINSON, I.
1. An algorithm for automatic integration over a triangle using non-linear extrapolation, *TOMS* **10** (1984).

DE HOOG, F., and WEISS, R.
1. Asymptotic expansions for product integration, *Math. Comp.* **27** (1973) 295–306.

DELVES, L. M.
1. The numerical evaluation of principal value integrals, *Comput. J.* **10** (1967–1968) 389–391.
2. On adaptive variable order quadrature routines, *Proc. Fifth Manitoba Conf. Numer. Math.* (1975) 257–271.
3. A fast method for the solution of Fredholm integral equations, *JIMA* **20** (1977) 173–182.

DELVES, L. M., and ABD-ELAL, L. F.
1. A fast Galerkin scheme for singular integral equations, *JIMA* **23** (1979) 139–166.

DELVES, L. M., and PHILIPS, C.
1. A fast implementation of the global element method, *JIMA* **25** (1980) 177–197.

DELVES, L. M., and WALSH, J. (eds.)
1. *Numerical Solution of Integral Equations.* Clarendon Press, Oxford, 1974.

DENNIS, J. E., and GOLDSTEIN, A. A.
1. Cubature and the Tchakaloff cone, *J. Comput. System Sci.* **3** (1969) 218–220.

DE VORE, R. A., and SCHERER, K. (eds.)
1. *Quantitative Approximation.* Academic Press, New York, 1980.

DIERCKX, P.
1. Algorithm 003. An algorithm for smoothing, differentiation and integration of experimental data using spline functions, *J. Comput. Appl. Math.* **1** (1975) 165–184.
2. Algorithm 008. An algorithm for least-squares fitting of cubic spline surfaces to functions on a rectilinear mesh over a rectangle, *J. Comput. Appl. Math.* **3** (1977) 113–129.

DIERCKX, P., and PIESSENS, R.
1. Calculation of Fourier coefficients of discrete functions using cubic splines, *J. Comput. Appl. Math.* **3** (1977) 207–209.

DIETER, U.

1. Statistical interdependence of pseudo-random numbers generated by the linear congruential method, in Zaremba [5, pp. 287–317].

DIGITAL SIGNAL PROCESSING COMMITTEE (eds.).

1. *Programs for Digital Signal Processing*, IEEE Press, New York, 1979.

DITKIN, V. A.

1. On certain approximate formulas for the calculation of triple integrals, *Dokl. Akad. Nauk SSSR* **62** (1948) 445–447.

DIXON, V. A.

1. Numerical quadrature: A survey of the available algorithms, in Evans [1, pp. 105–137].

DOBRODEEV, L. N.

1. Cubature formulas of the seventh order of accuracy for a hypersphere and a hypercube, *U.S.S.R. Computational Math. and Math. Phys.* **10**, No. 1 (1970) 252–253.
2. Cubature rules with equal coefficients for integrating functions with respect to symmetric domains, *U.S.S.R. Comput. Math. Math. Phys.* **18**(4) (1978) 27–34.

DONALDSON, J. D.

1. Asymptotic estimates of the errors in the numerical integration of analytic functions, Ph.D. Thesis, Dept. of Math., Univ. of Tasmania, Hobart, 1967.
2. Estimates of upper bounds for quadrature errors, *SIAM J. Numer. Anal.* **10** (1973) 13–22.

DONALDSON, J. D., and ELLIOTT, D.

1. QUADRATURE I: A unified approach to the development of quadrature rules, Tech. Rep. 23. Dept. of Math., Univ. of Tasmania, Hobart, 1970.
2. QUADRATURE II: The estimation of remainders in certain quadrature rules, Tech. Rep. 24. Dept. of Math., Univ. of Tasmania, Hobart, 1970.
3. A unified approach to quadrature rules with asymptotic estimates of their remainders, *SIAM J. Numer. Anal.* **9** (1972) 573–602.

DORONIN, G. Y.

1. On the question of formulas of mechanical quadrature, *Sb. Nauch. Tr. Dnepropetrovsk. Inzh.-Stroit. Inst.* **1–2** (1955) 210–217.

DUBNER, H., and ABATE, J.

1. Numerical inversion of Laplace transforms by relating them to the finite Fourier cosine transform, *JACM* **15** (1968) 115–123.

DUC-JACQUET, M.

1. Meilleures formules d'intégration dans certains espaces de Hilbert de fonctions, *C.R. Acad. Sci. Paris* **271** (1970) 795–797.
2. Sur l'approximation d'une fonctionnelle par des fonctionnelles qui dépendent d'un paramètre, dans des espaces de Hilbert, *C.R. Acad. Sci. Paris* **272** (1971) 469–472.
3. Approximation des fonctionnelles lineaires sur les espaces Hilbertiens autoreproduisants, Ph.D. Thesis, Univ. Grenoble, 1973.

DUDEWICZ, E. J., and RALLEY, T. G.

1. *The Handbook of Random Number Generation and Testing with TESTRAND Computer Code*. American Sciences Press, Columbus, OH., 1981.

DUFFY, M. G.

1. Quadrature over a pyramid or cube of integrands with a singularity at a vertex, *SIAM J. Numer. Anal.* **19** (1982) 1260–1262.

DUNKL, C. F.
1. Romberg quadrature to prescribed accuracy, SHARE File Number 7090-1481 TYQUAD.

DURIS, C. S.
1. A simplex sufficiency condition for quadrature formulas, *Math. Comp.* **20** (1966) 68–78.
2. Optimal quadrature formulas using generalized inverses. Pt. I: General theory and minimum variance formulas, *Math. Comp.* **25** (1971) 495–504.
3. The (q, r) copy of product-type Newton–Cotes quadrature formulas, Math. Rep. 74–1, Drexel Univ., Philadelphia, 1974.
4. Generating and compounding product-type Newton–Cotes quadrature formulas, *TOMS* **2** (1976) 50–58.

DURIS, C. S., and LYNESS, J. N.
1. Compound quadrature rules for the product of two functions, *SIAM J. Numer. Anal.* **12** (1975) 681–697.

DZJADYK, V. K., and PANASOVIC, V. A.
1. Estimate of the remainder for certain cubature formulas, *Ukrain. Mat. Ž.* **20** (1968) 147–155.

EBERLEIN, W. F.
1. A new method for numerical evaluation of the Fourier transform, *J. Math. Anal. Appl.* **65** (1978) 80–84.

ECKHARDT, U.
1. Einige Eigenschaften Wilfscher Quadraturformeln, *Numer. Math.* **12** (1968) 1–7.

EHLICH, H.
1. Untersuchungen zur numerischen Fourieranalyse, *Math. Z.* **91** (1966) 380–420.

EHRESMANN, D.
1. Zur Abschätzung des Bestapproximationsfehlers bei der Approximation differenzierbare Funktionen durch Polynome, *Numer. Math.* **13** (1969) 94–100.

EHRMAN, J.
1. Double-double accumulation of inner products, Extrinsic Program Library No. C003. Computation Center, Stanford Univ., Stanford, California, 1967.

EINARSSON, B.
1. Numerical calculation of Fourier integrals with cubic splines, *BIT* **8** (1968) 279–286.
2. On the calculation of Fourier integrals, in *Information Processing 71* (Freeman, C. V., ed.) Vol. 2, pp. 1346–1350. North-Holland Publ., Amsterdam, 1972.
3. Use of Richardson extrapolation for the numerical calculation of Fourier transforms, *J. Comp. Phys.* **21** (1976) 365–370.
4. Testing and evaluation of some subroutines for numerical quadrature, in Evans [1, pp. 149–157].

EISNER, E.
1. Numerical integration of a function that has a pole, *CACM* **10** (1967) 239–243; see also *CACM* **10** (1967) 610.

ELLIOTT, D.
1. A Chebyshev series method for the numerical solution of Fredholm integral equations, *Comput. J.* **6** (1963–1964) 102–111.

2. The evaluation and estimation of the coefficients in the Chebyshev series expansion of a function, *Math. Comp.* **18** (1964) 274–284.

3. Truncation errors in two Chebyshev series approximations, *Math. Comp.* **19** (1965) 234–248.

4. Uniform asymptotic expansions of the Jacobi polynomials and an associated function, *Math. Comp.* **25** (1971) 309–316.

ELLIOTT, D., and DONALDSON, J. D.

1. On quadrature rules for ordinary and Cauchy principal value integrals over contours, *SIAM J. Numer. Anal.* **14** (1977) 1078–1087.

ELLIOTT, D., and PAGET, D. F.

1. On the convergence of a quadrature rule for evaluating certain Cauchy principal value integrals, *Numer. Math.* **23** (1975) 311–319; **25** (1976) 287–289.

2. Product-integration rules and their convergence, *BIT* **16** (1976) 32–40.

3. The convergence of product integration rules, *BIT* **18** (1978) 137–141.

4. Gauss type quadrature rules for Cauchy principal value integrals, *Math. Comp.* **33** (1979) 301–309.

ELLIOTT, D., and SZEKERES, G.

1. Some estimates of the coefficients in the Chebyshev series expansion of a function, *Math. Comp.* **19** (1965) 25–32.

ELLIOTT, D. F., and RAO, R.

1. *Fast Transforms. Algorithms, Analyses and Applications.* Academic Press, New York, 1983.

ELSNER, L.

1. Zur Richardson-Extrapolation bei der Simpsonschen Integrationsformel, *ZAMM* **46** (1966) T47–48.

EL-TOM, M. E. A.

1. On optimum quadrature formulae over the square, Rapport No. 28. Semin. Math. Appl. Méc., Univ. Catholique, Louvain, 1969.

2. On ignoring the singularity in approximate integration, *SIAM J. Numer. Anal.* **8** (1971) 412–424.

3. On best cubature formulas and spline interpolation, *Numer. Math.* **32** (1979) 291–306.

ENGELS, H.

1. Neue Ergebnisse zur numerischen Kubatur, Thesis, Math.-Naturwiss. Fak. Tech. Hoch., Aachen, 1966.

2. Kubaturformeln für grosse Rechteckgebiete in n Dimensionen, *Elek. Daten.* **8** (1967) 345–350.

3. Zur iterativen Behandlung der Interpolationsquadratur, *Elek. Daten.* **10** (1969) 262–264.

4. Über eine Klasse von Quadraturformeln mit äquidistanten Stützstellen für grosse Intervalle, *ZAMM* **49** (1969) 751–753.

5. Ein Verfahren zur Verkleinerung des Restbetrages bei der Interpolationsquadratur, *ZAMM* **50** (1970) T40–42.

6. Über einige hermitesche Quadraturverfahren, *Angew. Informatik.* **13** (1971) 529–533.

7. Richardson-Extrapolation mit Hilfe kubischer Splines, KFA Jülich, Jül-791-MA, 1971.

8. Eine Zusammenstellung von Arbeiten über Numerische Quadratur und Kubatur, KFA Jülich, Jül-Bibl-16, 1972.

9. Über gleichgewichtete Kubaturformeln für Kreis- und Sechseck-Gebiet, *Elek. Daten.* **11** (1970) 216–223.

10. Über gleichgewichtete Kubaturformeln für ein Dreiecksgebiet, *Elek. Daten.* **11** (1970) 535–539.

11. Gleichgewichtete Kubaturformeln für ein dreieckiges ebenes Integrationsgebiet, *ZAAM* **51** (1971) T49–51.

12. Asymptotische Entwicklungen aller Nullstellen des Legendre Polynoms $L_n(x)$ nach Potenzen von $1/[n(n+1)]$, in Hämmerlin [5, pp. 120–130].

13. *Numerical Quadrature and Cubature.* Academic Press, London, 1980.

ENGELS, H., and ECKHARDT, U.

1. The determination of nodes and weights in Wilf quadrature formulas, *Abh. Math. Sem. Univ. Hamburg* **48** (1979) 34–41.

ENGELS, H., and MEUER, H-W.

1. Die Entwicklung von Interpolations-quadratur- und Differentiationsformeln zu gegebenen Stützstellen mit PL/I-FORMAC, Bericht der KFA Jülich, Jül-662-MA, 1970.

ERDÉLYI, A.

1. *Asymptotic Expansions.* Dover, New York, 1956.

EPSTEIN, G.

1. Recursive fast Fourier transforms, *Proc. AFIPS Fall Joint Comput. Conf., 1968,* **33** Pt. 1, pp. 141–143.

ERDÉLYI, A., and WYMAN, M.

1. The asymptotic evaluation of certain integrals, *Arch. Rational Mech. Anal.* **14** (1963) 217–260.

ERDÉLYI, A., *et al.*

1. *Higher Transcendental Functions,* 3 Vols. McGraw-Hill, New York, 1953–1955.
2. *Tables of Integral Transforms,* 2 Vols. McGraw-Hill, New York, 1954.

ERMAKOV, S. M., and ZOLOTUKHIN, V. G.

1. Polynomial approximations and the Monte-Carlo method, in *Theory of Probability and Applications, Amer. Math. Soc. Transl.* **5** (1960) 428–431.

ESPINOSA-MALDONADO, R. J., and BYRNE, G. D.

1. On the convergence of quadrature formulas, *SIAM J. Numer. Anal.* **8** (1971) 110–114.

EVANS, D. J. (ed.)

1. *Software for Numerical Mathematics.* Academic Press, London, 1974.

EVANS, G. A., FORBES, R. C., and HYSLOP, J.

1. Polynomial transformations for singular integrals, *Internat. J. Comput. Math.* **14** (1983) 157–170.

EVANS, G. A., HYSLOP, J., and MORGAN, A. P. G.

1. An extrapolation procedure for the evaluation of singular integrals, *Internat. J. Comput. Math.* **12** (1983) 251–265.
2. Iterative solution of Volterra integral equations using Clenshaw–Curtis quadrature. *J. Comp. Phys.* **40** (1981) 64–76.

FAIRWEATHER, G., and KEAST, P.

1. A comparison of non-adaptive Romberg quadrature routines, Tech. Rep. No. 96, Dept. of Computer Science, Univ. of Toronto, 1976.
2. An investigation of Romberg quadrature, *TOMS* **4** (1978) 316–322.

Farrington, C. C.
1. Numerical quadrature of discontinuous functions, *Nat. Meeting ACM, 16th, Los Angeles, September 5–8, 1961*, Preprints. ACM, New York, 1961.

Faure, H.
1. Discrépance de suites associées à un système de numération (en dimension *s*), *Acta Arith.* **41** (1982) 337–351.

Fejér, L.
1. Mechanische Quadraturen mit positiven Cotesschen Zahlen, *Math. Z.* **37** (1933) 287–309.

Feldheim, E.
1. Théorie de la convergence des procédés d'interpolation et de quadrature mécanique, *Mémor. Sci. Math.* **95** (1939) 1–90.

Feldstein, A., and Miller, R. K.
1. Error bounds for compound quadrature of weakly singular integrals, *Math. Comp.* **25** (1971) 505–520.

Ferguson, Jr., W. E.
1. A simple derivation of Glassman's general *N* fast Fourier transform, *Comput. Math. Appl.* **8** (1982) 401–411.

Fettis, H. E.
1. Numerical calculation of certain definite integrals by Poisson's summation formula, *MTAC* **9** (1955) 85–92.
2. Further remarks concerning the relative accuracy of Simpson's and the trapezoidal rule for a certain class of functions, *ZAMM* **38** (1958) 159–160.

Fettis, H. E., and Pexton, R. L.
1. More on the calculation of oscillatory integrals, *J. Comp. Phys.* **47** (1982) 473–476.

Feuchter, C. A.
1. Numerical integration over a sphere, *Math. Comp.* **22** (1968) 293–297.

Filippi, S.
1. Angenäherte Tschebyscheff-Approximation einer Stammfunktion—eine Modifikation des Verfahrens von Clenshaw und Curtis, *Numer. Math.* **6** (1964) 320–328.
2. Das Verfahren von Romberg–Stiefel–Bauer als Spezialfall des allgemeinen Prinzips von Richardson, *Z. Moderne Rechentech. Automat.* (MTW) **11** (1964) Pt. 1, 49–54, Pt. 2, 98–100.
3. Neue Gauss-Typ-Quadraturverfahren, *Z. Moderne Rechentech. Automat.* (MTW) **13** (1966) Pt. 1, 174–180, Pt. 2, 239–245.
4. *Untersuchungen über die Fourier—Tschebyscheff—Approximation von Stammfunktionen.* Vestdeutschen Verlag, Cologne, 1970.

Filippi, S., and Engels, H.
1. Neues über die numerische Behandlung zweidimensionaler Integrale, *Elek. Daten.* **8** (1967) 353–360, 399–402.

Filippi, S., and Esser, H.
1. Eine Reihe neuer Sätze und Ergebnisse zur numerischen Quadratur, *Elek. Daten.* **11** (1969) 166–180.

Filon, L. N. G.
1. On a quadrature formula for trigonometric integrals, *Proc. Roy. Soc. Edinburgh* **49** (1928–1929) 38–47.

FISHMAN, H.
1. Numerical integration constants *MTAC* **11** (1957) 1–9.

FLETCHER, A., MILLER, J. C. P., ROSENHEAD, L., and COMRIE, L. J.
1. *An Index of Mathematical Tables*, 2nd ed. Addison-Wesley, Reading, Massachusetts, 1962.

FLETCHER, R., and REEVES, C. M.
1. A mechanization of algebraic differentiation and the automatic generation of formulae for molecular integrals of Gaussian orbitals, *Comput. J.* **6** (1963–1964) 287–292.

FLINN, E. A.
1. A modification of Filon's method of numerical integration, *JACM* **7** (1960) 181–184.

FOGLIA, C.
1. On number theoretical methods for multidimensional numerical integration, *Comp. Phys. Comm.* **25** (1982) 113–118.

FORNARO, R. J.
1. Numerical evaluation of integrals around simple closed curves, *SIAM J. Numer. Anal.* **10** (1973) 623–634.

FÖRSTER, K.-J.
1. Fehlerschranken bei der Romberg-Quadratur, *ZAMM* **62** (1982) 133–135.

FORSYTHE, G. E.
1. Singularity and near singularity in numerical analysis, *Amer. Math. Monthly* **65** (1958) 229–240.

FORSYTHE, G. E., MALCOLM, M. A., and MOLER, C. B.
1. *Computer Methods for Mathematical Computations*. Prentice-Hall, Englewood Cliffs, N.J., 1977.

FOSDICK, L. D.
1. A special case of the Filon quadrature formula, *Math. Comp.* **22** (1968) 77–81.

FOX, L.
1. Some comments on the accuracy and convergence of finite difference processes in automatic computation. *Congr. Assoc. Française de Calcul, 1st, 1960*, pp. 63–73. Gauthier-Villars, Paris, 1961.
2. Romberg integration for a class of singular integrals, *Comput. J.* **10** (1967) 87–93.

FOX, L., and HAYES, L.
1. On the definite integration of singular integrands, *SIAM Rev.* **12** (1970) 449–457.

FOX, L., and PARKER, I. B.
1. *Chebyshev Polynomials in Numerical Analysis*. Oxford Univ. Press, London and New York, 1968.

FRANKE, R.
1. Orthogonal polynomials and approximate multiple integration, *SIAM J. Numer. Anal.* **8** (1971) 757–766.
2. Obtaining cubatures for rectangles and other planar regions by using orthogonal polynomials, *Math. Comp.* **25**, (1971) 803–817.
3. Minimal point cubatures of precision seven for symmetric planar regions, *SIAM J. Numer. Anal.* **10** (1973) 853–862.

FRANKLIN, J. N.
1. Deterministic simulation of random processes, *Math. Comp.* **17** (1963) 28–59.

FRANKLIN, P.
1. *A Treatise on Advanced Calculus.* Wiley, New York, 1940.

FRASER, D. A. S.
1. *Statistics: An Introduction.* Wiley, New York, 1958.

FRASER, W., and WILSON, M. W.
1. Remarks on the Clenshaw-Curtis quadrature scheme, *SIAM Rev.* **8** (1966) 322–327.

FREIBERGER, W. F., and GRENANDER, U.
1. *A Short Course in Computational Probability and Statistics.* Springer-Verlag, Berlin and New York, 1971.

FREUD, G.
1. *Orthogonale Polynome.* Birkhäuser, Basel, 1969. English Ed., Pergamon, Oxford, 1971.
2. Error estimates for Gauss-Jacobi quadrature formulae, in *Topics in Numerical Analysis* (Miller, J. J. H., ed.), pp. 115–121. Academic Press, London, 1973.
3. Numerical estimates for the error of Gauss-Jacobi quadrature formulae, in *Topics in Numerical Analysis* (Miller, J. J. H., ed.), Vol. II, pp. 43–50. Academic Press, London, 1975.

FRIEDMAN, J. H., and WRIGHT, M. H.
1. A nested partitioning procedure for numerical multiple integration, *TOMS* **7** (1981) 76–92.
2. DIVONNE4, a program for multiple integration and adaptive importance sampling, CGTM No. 193, Computation Research Group, Stanford Linear Accelerator, Stanford, CA, 1981.

FRITSCH, F. N.
1. On the number of nodes in self-contained integration formulas of degree two for compact planar regions, *Numer. Math.* **16** (1970) 224–230.
2. On self-contained numerical integration formulas for symmetric regions, *SIAM J. Numer. Anal.* **8** (1971) 213–221.

FRITSCH, F. N., KAHANER, D. K., and LYNESS, J. N.
1. Double integration using one-dimensional adaptive quadrature routines: A software interface problem, *TOMS* **7** (1981) 46–75.

FRÖBERG, C.-E.
1. *Introduction to Numerical Analysis.* Addison-Wesley, Reading, Massachusetts, 1965.

GAFFNEY, P. W.
1. The calculation of indefinite integrals of *B*-splines, *JIMA* **17** (1976) 37–41.

GAKHOV, F. D.
1. *Boundary Value Problems.* Pergamon Press, Oxford, 1966.

GALANT, D.
1. An implementation of Christoffel's theorem in the theory of orthogonal polynomials, *Math. Comp.* **25** (1971) 111–113.

GANDER, W.
1. A simple adaptive quadrature algorithm, Research Rep. 83–03, Seminar für Angewandte Mathematik, ETH, Zurich, 1983.

GARRIBBA, S., QUARTAPELLE, L., and REINA, G.

1. Algorithm 36. *SNIFF*: Efficient self-tuning algorithm for numerical integration, *Computing* **20** (1978) 363–375.

GATES, JR., L. D.

1. Numerical solution of differential equations by repeated quadratures, *SIAM Rev.* **6** (1964) 134–147.

GATTESCHI, L.

1. On the construction of some Gaussian quadrature rules, in Hämmerlin [5, pp. 138–146].
2. On some orthogonal polynomial integrals, *Math. Comp.* **35** (1980) 1291–1298.

GAUTSCHI, W.

1. Recursive computation of certain integrals, *JACM* **8** (1961) 21–40.
2. Numerical quadrature in the presence of a singularity, *SIAM J. Numer. Anal.* **4** (1967) 357–362.
3. Construction of Gauss–Christoffel quadrature formulas, *Math. Comp.* **22** (1968) 251–270.
4. On the condition of a matrix arising in the numerical inversion of the Laplace transform. *Math. Comp.* **23** (1969) 109–118.
5. On the construction of Gaussian quadrature rules from modified moments, *Math. Comp.* **24** (1970) 245–260. See also the microfiche supplement to Math. Comp., Vol. 24, No. 110.
6. Efficient computation of the complex error function, *SIAM J. Numer. Anal.* **7** (1970) 187–198.
7. Advances in Chebyshev quadrature, in *Numerical Analysis Dundee 1975* (Watson, G. A., ed.), pp. 100–121, Lecture Notes in Mathematics 506. Springer-Verlag, Berlin, 1976.
8. On generating Gaussian quadrature rules, in Hämmerlin [5, pp. 147–154].
9. A survey of Gauss-Christoffel quadrature formulae, in *E. B. Christoffel; The Influence of his Work in Mathematics and the Physical Sciences* (Butzer, P. L., and Feher, F., eds.), pp. 72–147. Birkhäuser, Basel, 1981.
10. On generating orthgonal polynomials, *SIAM J. Sci. Stat. Comput.* **3** (1982) 289–317.
11. How and how not to check Gaussian quadrature formulae, *BIT* **23** (1983) 209–216.

GAUTSCHI, W., and VARGA, R. S.

1. Error bounds for Gaussian quadrature of analytic functions, *SIAM J. Numer. Anal.* **20** (1983) 1170–1186.

GELFAND, I. M., and SHILOV, G. E.

1. *Generalized Functions*, Vol. 1. Academic Press, New York, 1964.

GENTLEMAN, W. M.

1. An error analysis of Goertzel's (Watt's) method for computing Fourier coefficients, *Comput. J.* **12** (1969) 160–165.
2. Implementing Clenshaw–Curtis quadrature, *CACM* **15** (1972) 337–342, 343–346.

GENZ, A.

1. An adaptive multidimensional quadrature procedure, *Computer Phys. Comm.* **4** (1972) 11–15.
2. Applications of the ε-algorithm to quadrature problems in *Padé Approximants and their Applications* (Graves-Morris, P. R., ed.), pp. 105–116. Academic Press, New York, 1973.
3. Some extrapolation methods for the numerical calculation of multidimensional integrals, in Evans [1, pp. 159–172].
4. The approximate calculation of multidimensional integrals using extrapolation methods, Ph.D. Thesis, Dept. of Applied Mathematics, Univ. of Kent at Canterbury, 1975.
5. A Lagrange extrapolation algorithm for sequences of approximations to multiple integrals, *SIAM J. Sci. Stat. Comput.* **3** (1982) 160–172.

6. Parallel methods for the numerical calculation of multiple integrals, *Comp. Phys. Comm.* **26** (1982) 349–352.

7. Testing multidimensional integration routines, Proceedings of INRIA Conference on Software Tools, Methods and Languages. North-Holland Publ., Amsterdam, 1984.

8. Computing the weights for multidimensional integration rules using a multivariable Lagrange interpolation formula, to be published.

Genz, A. C., and Malik, A. A.

1. Algorithm 019. Remarks on algorithm 006: An adaptive algorithm for numerical integration over an N-dimensional rectangular region, *J. Comput. Appl. Math.* **6** (1980) 295–302.

2. An imbedded family of fully symmetric numerical integration rules, *SIAM J. Numer. Anal.* **20** (1983) 580–588.

Gerisch, W., Strogies, W., and Wagner, A.

1. Zur numerischen Laplace-Transformation in der Wärmeleitungstheorie, *ZAMM* **49** (1969) 374–377.

Gerl, P.

1. Gleichverteilung auf der Kugel. Archiv der Math. **24** (1973) 203–207.

Geronimus, Ya. L.

1. *Polynomials Orthogonal on a Circle and Interval.* Pergamon, Oxford, 1960.

Ghizzetti, A.

1. Sulle formule di quadratura, *Rend. Sem. Mat. Fis. Milano* **26** (1954–1955) 1–16.

2. Sulla convergenza dei procedimenti di calcolo, degli integrali definiti, forniti dalle formule di quadratura, *Rend. Sem. Mat. Univ. Padova* **26** (1956) 201–222.

3. Sulla struttura delle formule di quadratura, *Rend. Sem. Mat. Univ. Politec. Torino* **27** (1967–1968) 69–84.

4. Procedure for constructing quadrature formulae on infinite intervals, *Numer. Math.* **12** (1968) 111–119.

Ghizzetti, A., and Ossicini, A.

1. *Quadrature Formulae.* Academic Press, New York, 1970.

Gibb, D.

1. *A Course in Interpolation and Numerical Integration for the Mathematical Laboratory.* Bell, London, 1915.

Gill, P. E., and Miller, G. F.

1. An algorithm for the integration of unequally spaced data, *Comput. J.* **15** (1972), 80–83.

Goethals, J. M., and Seidel, J. J.

1. Cubature formulae, polytopes and spherical designs, in *The Geometric Vein: The Coxeter Festschrift* (Davis, C., Grünbaum, B., and Sherk, F. A., eds.), pp. 203–218. Springer-Verlag, New York, 1981.

Gold, B., and Rader, C. M.

1. *Digital Processing of Signals.* McGraw-Hill, New York, 1969.

Gold, R., Cohen, C. E., and Olson, I.

1. Methods for the numerical evaluation of Fourier integrals, *J. Comp. Phys.* **1** (1966–1967) 536–540.

Goldberg, R. R.

1. *Methods of Real Analysis.* Ginn (Blaisdell), Boston, Massachusetts, 1964.

GOLDBERG, R. R., and VARGA, R. S.
1. Moebius inversion of Fourier transforms, *Duke Math. J.* **23** (1956) 553–559.

GOLOMB, M., and WEINBERGER, H. F.
1. Optimal approximation and error bounds, in Langer [1, pp. 117–190].

GOLUB, G. H.
1. Some modified matrix eigenvalue problems, *SIAM Rev.* **15** (1973) 318–334.

GOLUB, G. H., and KAUTSKY, J.
1. Calculation of Gauss quadratures with multiple free and fixed knots, *Numer. Math.* **41** (1983) 147–163.

GOLUB, G. H., and WELSCH, J. H.
1. Calculation of Gauss quadrature rules, *Math. Comp.* **23** (1969) 221–230.

GOOD, I. J., and GASKINS, R. A.
1. The centroid method of numerical integration, *Numer. Math.* **16** (1971) 343–359.

GOODRICH, R. F., and STENGER, F.
1. Movable singularities and quadrature, *Math. Comp.* **24** (1970) 283–300.

GOODWIN, E. T.
1. The evaluation of integrals of the form $\int_{-\infty}^{\infty} f(x)e^{-x^2}\,dx$, *Proc. Cambridge Philos. Soc.* **45** (1949) 241–245.

GRADSHTEYN, I. S., and RYZHIK, I. M.
1. *Tables of Integrals, Series, and Products.* Academic Press, New York, 1965.

GRANOVSKII, B. L.
1. Random quadratures of the Gaussian type, *U.S.S.R. Computational Math. and Math. Phys.* **8** No. 4 (1968) 244–252.

GRAY, H. L., and ATCHISON, T. A.
1. Nonlinear transformations related to the evaluation of improper integrals I, *SIAM J. Numer. Anal.* **4** (1967) 363–371.
2. The generalized G-transform, *Math. Comp.* **22** (1968) 595–605.
3. Applications of the G and B transforms to the Laplace transform, *Proc. ACM Nat. Conf., 23rd, 1968*, pp. 73–77.

GRAY, H. L., ATCHISON, T. A., and MCWILLIAMS, G. V.
1. Higher order G-transformations, *SIAM J. Numer. Anal.* **8** (1971) 365–381.

GRAY, H. L., and SCHUCANY, W. R.
1. Some limiting cases of the G-transformation, *Math. Comp.* **23** (1969) 849–859.

GRAY, J. H., and RALL, L. B.
1. A computational system for numerical integration with rigorous error estimation, *Proc. 1974 Army Numer. Anal. Conf.*, pp. 341–355.
2. INTE: A Univac 1108/1110 program for numerical integration with rigorous error estimation, Tech. Summary Rep. 1428. MRC, Madison, Wisconsin, 1975.
3. Automatic Euler-Maclaurin integration. *Proc. 1976 Army Numer. Anal. Comp. Conf.*, pp. 431–444.

GRAY, W. G., and VAN GENUCHTEN, M. T.
1. Economical alternatives to Gaussian quadrature over isoparametric quadrilaterals, *IJNME* **12** (1978) 1478–1484.

GREENWOOD, R. E., Jr., CARNAHAN, P. D. M., and NOLLEY, J. W.
1. Numerical integration formulas for use with weight functions x^2 and $x/\sqrt{1 - x^2}$, *Math. Comp.* **13** (1959) 37–40.

GREVILLE, T. N. E.
1. Introduction to spline functions, in *Theory and Applications of Spline Functions* (Greville, T. N. E., ed.), pp. 1–35, Academic Press, New York, 1969.
2. Spline functions, interpolation, and numerical quadrature, in Ralston and Wilf [1, Vol. II. pp. 156–168].

GRIBBLE, J. D.
1. Further properties of inner product quadrature formulas, *BIT* **17** (1977) 392–408. See also *BIT* **20** (1980) 260.
2. Interpolatory inner product quadrature formulas, *BIT* **20** (1980) 466–474.
3. Inner product quadrature formulas exact on maximal product spaces of functions, *J. Comput. Appl. Math.* **8** (1982) 73–79.

GRÖBNER, W., and HOFREITER, N.
1. *Integraltafel*, 2 Vols. Springer-Verlag, Berlin and New York, 1961.

GRUNDMANN, A., and MÖLLER, H. M.
1. Invariant integration formulas for the n-simplex by combinatorial methods, *SIAM J. Numer. Anal.* **15** (1978) 282–290.

GUENTHER, R. B., and ROETMAN, E. L.
1. Newton–Cotes formulae in n-dimensions, *Numer. Math.* **14** (1970) 330–345.

GÜNTHER, C.
1. Zur Konstruktion mehrdimensionaler Integrationsformeln, *ZAMM* **53** (1973) T194–195.
2. Third degree integration formulas with four real points and positive weights in two dimensions, *SIAM J. Numer. Anal.* **11** (1974) 480–493.
3. Zweidimensionale Quadraturformeln vom Grad 7 mit 14 Punkten, *Numer. Math.* **24** (1975) 309–316.

GUSTAFSON, S.-Å.
1. Convergence acceleration by means of numerical quadrature, *BIT* **6** (1966) 117–128.
2. Rapid computation of general interpolation formulae and mechanical quadrature rules, *CACM* **14** (1971) 797–801.
3. Control and estimation of computational errors in the evaluation of interpolation formulae and quadrature rules, *Math. Comp.* **24** (1970) 847–854.
4. A method of computing limit values, *SIAM J. Numer. Anal.* **10** (1973) 1080–1090.
5. Die Berechnung von verallgemeinerten Quadratformeln vom Gausschen Typus, eine Optimierungsaufgabe, in *Numerische Methoden bei Optimierungsaufgaben* (Collatz, L., and Wetterling, W., eds.), pp. 59–71, ISNM17. Birkhäuser, Basel, 1973.

GUSTAFSON, S.-Å., and DALHLQUIST, G.
1. On the computation of slowly convergent Fourier integrals, *Methoden und Verfahren der Math. Phys.* **6** (1972) 93–112.

HABER, S.
1. A note on some quadrature formulas for the interval $(-\infty, \infty)$, *Math. Comp.* **18** (1964) 313–314.
2. Midpoint quadrature formulas, *Math. Comp.* **21** (1967) 719–721.

3. On a sequence of points of interest for numerical quadrature, *J. Res. Nat. Bur. Standards Sect. B* **70** (1966) 127–136.

4. A modified Monte-Carlo quadrature, *Math. Comp.* **20** (1966) 361–368; **21** (1967) 388–397.

5. A combination of Monte Carlo and classical methods for evaluating multiple integrals, *Bull. Amer. Math. Soc.* **74** (1968) 683–686.

6. Stochastic quadrature formulas, *Math. Comp.* **23** (1969) 751–764.

7. Numerical evaluation of multiple integrals, *SIAM Rev.* **12** (1970) 481–526.

8. On certain optimal quadrature formulas, *J. Res. Nat. Bur. Standards Sect. B* **75** (1971) 85–88.

9. The error in numerical integration of analytic functions, *Quart. Appl. Math.* **29** (1971–1972) 411–420.

10. Experiments on optimal coefficients, in Zaremba [5, pp. 11–37].

11. Adaptive integration and improper integrals, *Math. Comp.* **29** (1975) 806–809.

12. The tanh rule for numerical integration, *SIAM J. Numer. Anal.* **14** (1977) 668–685.

13. A number theoretic problem in numerical approximation of integrals, in *Approximation Theory III* (Cheney, E. W., ed.), pp. 473–480. Academic Press, New York, 1980.

14. Parameters for integrating periodic functions of several variables, *Math. Comp.* **37** (1983) 115–129.

HABER, S., and OSGOOD, C. F.

1. On a theorem of Piatetsky-Shapiro and approximation of multiple integrals, *Math. Comp.* **23** (1969) 165–168.

HABER, S., and SHISHA, O.

1. Improper integrals, simple integrals, and numerical quadrature, *J. Approx. Theory* **11** (1974) 1–15.

HAEGEMANS, A.

1. Circularly symmetrical integration formulas for two-dimensional circularly symmetrical regions, *BIT* **16** (1976) 52–59.

2. Algorithm 34. An algorithm for the automatic integration over a triangle, *Computing* **19** (1977) 179–187.

3. Construction of known and new cubature formulas of degree five for three-dimensional symmetric regions, using orthogonal polynomials, in Hämmerlin [6, pp. 119–127].

HAEGEMANS, A., and PIESSENS, R.

1. Construction of cubature formulas of degree eleven for symmetric planar regions, using orthogonal polynomials, *Numer. Math.* **25** (1976) 139–148.

2. Construction of cubature formulas of degree seven and nine over symmetric planar regions, using orthogonal polynomials, *SIAM J. Numer. Anal.* **14** (1977) 492–508.

HALTON, J. H.

1. On the efficiency of certain quasi-random sequences of points in evaluating multi-dimensional integrals, *Numer. Math.* **2** (1960) 84–90.

2. A retrospective and prospective survey of the Monte Carlo method, *SIAM Rev.* **12** (1970) 1–63.

3. Estimating the accuracy of quasi-Monte Carlo integration, in Zaremba [5, pp. 345–360].

HALTON, J. H., and HANDSCOMB, D. C.

1. A method for increasing the efficiency of Monte Carlo integration, *JACM* **4** (1957) 329–340.

HALTON, J. H., and ZAREMBA, S. K.

1. The extreme and L^2 discrepancies of some plane sets, *Monatsh. Math.* **73** (1969) 316–328.

HALTON, J. H., and ZEIDMAN, E. A.

1. The evaluation of multidimensional integrals by the Monte Carlo sequential stratification technique, Tech. Rep. No. 137. Comp. Sci. Dept., Univ. of Wisconsin, Madison, WI, 1971.

HALVE, W. J. M.

1. A number-theoretic approach to numerical multiple integration, Memorandum 1981–06. Dept. of Mathematics, Eindhoven Univ. of Technology, Eindhoven, 1981.

HAMMER, P. C.

1. The midpoint method of numerical integration, *Math. Mag.* **31** (1957–1958) 193–195.
2. Numerical evaluation of multiple integrals, in Langer [1, pp. 99–115].

HAMMER, P. C., MARLOWE, O. J., and STROUD, A. H.

1. Numerical integration over simplexes and cones, *MTAC* **10** (1956) 130–137.

HAMMER, P. C., and STROUD, A. H.

1. Numerical integration over simplexes, *MTAC* **10** (1956) 137–139.
2. Numerical evaluation of multiple integrals II, *MTAC* **12** (1958) 272–280.

HAMMER, P. C., and WICKE, H. H.

1. Quadrature formulas involving derivatives of the integrand, *Math. Comp.* **14** (1960) 3–7.

HAMMER, P. C., and WYMORE, A. W.

1. Numerical evaluation of multiple integrals I, *MTAC* **11** (1957) 59–67.

HÄMMERLIN, G.

1. Zur numerischen Integration periodischer Funktionen, *ZAMM* **39** (1959) 80–82.
2. Über ableitungsfreie Schranken für Quadraturfehler, *Numer. Math.* **5** (1963) 226–233.
3. Über ableitungsfreie Schranken für Quadraturfehler II. Ergänzungen und Möglichkeiten zur Verbesserung, *Numer. Math.* **7** (1965) 232–237.
4. Zur Abschätzung von Quadraturfehlern für analytische Funktionen, *Numer. Math.* **8** (1966) 334–344.
5. (Ed.), *Numerische Integration*, ISNM45. Birkhäuser Verlag, Basel, 1979.
6. (Ed.), *Numerical Integration*, ISNM57. Birkhäuser Verlag, Basel, 1982.

HAMMERSLEY, J. M.

1. Monte-Carlo methods for solving multivariable problems, *Ann. N.Y. Acad. Sci.* **86** (1960) 844–874.

HAMMERSLEY, J. M., and HANDSCOMB, D. C.

1. *Monte Carlo Methods*. Methuen, London, 1964.

HAMMERSLEY, J. M., and MORTON, K. W.

1. A new Monte Carlo technique: antithetic variates, *Proc. Camb. Philos. Soc.* **52** (1956) 449–475.

HAMMING, R. W.

1. *Numerical Methods for Scientists and Engineers*. McGraw-Hill, New York, 1962.
2. A class of integration formulas, *Amer. Math. Monthly* **78** (1971) 518–522.

HAMMING, R. W., and EPSTEIN, M. P.

1. Non-interpolatory quadrature formulas. *SIAM J. Numer. Anal.* **9** (1972) 464–475.

HAMMING, R. W., and PINKHAM, R. S.
1. A class of integration formulas, *JACM* **13** (1966) 430–438.

HANDSCOMB, D. C.
1. Remarks on a Monte Carlo integration method, *Numer. Math.* **6** (1964) 261–268.

HANKE, W.
1. Die Optimale Sektion bei adaptiven Integrationsverfahren mit globaler Strategie, *ZAMM* **62** (1982) T327–329.

HARPER, W. M.
1. Quadrature formulas for infinite integrals, *Math. Comp.* **16** (1962) 170–175.

HARRINGTON, S. J.
1. A new symbolic integration system in REDUCE, *Comput. J.* **22** (1979) 127–131.

HARRIS, C. G., and EVANS, W. A. B.
1. Extension of numerical quadrature formulae to cater for end point singular behaviours over finite intervals, *Internat. J. Comput. Math.* **6** (1977) 219–227.

HART, J. F., *et al.*
1. *Computer Approximations.* Wiley, New York, 1968.

HARTREE, D. R.
1. The evaluation of a diffraction integral, *Proc. Cambridge Philos. Soc.* **50** (1954) 567–574.
2. *Numerical Analysis*, 2nd ed. Oxford Univ. Press, London and New York, 1958.

HASEGAWA, T., TORII, T., and NINOMIYA, I.
1. Generalized Chebyshev interpolation and its application to automatic quadrature, *Math. Comp.* **41** (1983) 537–553.

HASELGROVE, C. B.
1. A method for numerical integration, *Math. Comp.* **15** (1961) 323–337.

HAUSSMANN, W., and ZELLER, K.
1. Quadraturrest, Approximation und Chebyshev-Polynome, in Hämmerlin [6, pp. 128–137].

HAUSNER, A.
1. NL9—An adaptive routine for numerical quadrature, *Proc. 1977 Army Numer. Anal. Comp. Conf.*, pp. 367–410.

HAUSNER, A., and HUTCHISON, J. D.
1. FOGIE: An adaptive code for numerical integrals using Gaussian quadrature, *Proc. 1975 Army Numer. Anal. Comp. Conf.*, pp. 139–177.

HÅVIE, T.
1. On a modification of Romberg's algorithm, *BIT* **6** (1966) 24–30.
2. On the practical application of the modified Romberg algorithm, *BIT* **7** (1967) 103–113.
3. Derivation of explicit expressions for the error terms in the ordinary and the modified Romberg algorithms, *BIT* **9** (1969) 18–29.
4. On a modification of the Clenshaw–Curtis quadrature formula, *BIT* **9** (1969) 338–350.
5. Some algorithms for numerical quadrature using the derivatives of the integrand in the integration interval, *BIT* **10** (1970) 277–294.
6. Some methods for automatic integration and their implementation on the CERN CDC 65/6600 computers, Rep. 71-26. CERN, Geneva, 1971.

7. Remarks on an expansion for integrals of rapidly oscillating functions, *BIT* **13** (1973) 16–29.
8. Further remarks on the modified Clenshaw–Curtis quadrature formula, Rep. 71-21. CERN, Geneva, 1971.
9. Error derivation in Romberg integration, *BIT* **12** (1972) 516–527.
10. Some expansions for integrals with weight functions, *BIT* **14** (1974) 306–313.
11. Romberg integration as a problem in interpolation theory, *BIT* **17** (1977) 418–429.
12. Generalized Neville-type extrapolation schemes, *BIT* **19** (1979) 204–213.

HAYES, D. R., and RUBIN, L.
1. A proof of the Newton–Cotes quadrature formulas with error term, *Amer. Math. Monthly* **77** (1970) 1065–1072. Addendum, *Amer. Math. Monthly* **78** (1971) 988.

HAZLEWOOD, L. J.
1. An alternative strategy for cautious adaptive quadrature, *JIMA* **20** (1977) 505–518.

HEKKER, V.
1. Certain cubature formulas for circle, *Metody Vyčisl.* **5** (1968) 10–14.

HELLEN, T. K.
1. Effective quadrature rules for quadratic solid isoparametric finite elements, *IJNME* **4** (1972) 597–599.

HELMS, H. D.
1. Fast Fourier transform method of computing difference equations and simulating filters. *IEEE Trans. Audio Electroacoust.*, **AU-15** (1967) 85–90.

HEMKER, P. W.
1. A sequence of nested cubature rules, Rep. NW3/75. Mathematisch Centrum, Amsterdam, 1975.

HENRICI, P.
1. *Discrete Variable Methods in Ordinary Differential Equations.* Wiley, New York, 1962.
2. *Elements of Numerical Analysis.* Wiley, New York, 1964.
3. Fast Fourier methods in computational complex analysis, *SIAM Rev.* **21** (1979) 481–527.

HERRERO, F. A.
1. Routine for improper integrals with a programmable calculator, *Amer. J. Phys.* **48** (1980) 679–681.

HETHERINGTON, J. H.
1. An error bound for quadratures, *Math. Comp.* **26** (1972) 695–698.
2. Note on error bounds for numerical integration, *Math. Comp.* **27** (1973) 307–316.

HETHERINGTON, R. G.
1. Numerical integration over hypershells, Ph.D. Thesis, Numer. Anal. Dept., Univ. of Wisconsin, Madison, 1961.

HIGMAN, B.
1. What EVERYBODY should know about ALGOL, *Comput. J.* **6** (1963–1964) 50–56.

HILDEBRAND, F. B.
1. *Introduction to Numerical Analysis.* Second Edition. McGraw-Hill, New York, 1974.

HILLION, P.
1. Numerical integration on a triangle, *IJNME* **11** (1977) 797–815.
2. Numerical integration on a tetrahedron, *Calcolo* **18** (1981) 117–130.

HILLION, P., and NURDIN, G.
1. Integration of highly oscillatory functions, *J. Comp. Phys.* **23** (1977) 74–81.

HILLSTROM, K. E.
1. Comparison of several adaptive Newton–Cotes quadrature routines in evaluating definite integrals with peaked integrands, ANL-7511. Argonne Nat. Lab., Argonne, Illinois, November 1968.
2. Comparison of several adaptive Newton–Cotes quadrature routines in evaluating definite integrals with peaked integrands. II, *CACM* **13** (1970) 362–365.

HIRSCH, P. M.
1. Numerical evaluation and estimation of multiple integrals, Ph.D. Thesis, Univ. of Wisconsin, Madison, 1966.
2. Evaluation of orthogonal polynomials and relationship to evaluating multiple integrals, *Math. Comp.* **22** (1968) 280–285. Corrigenda, *Math. Comp.* **22** (1968) 911.

HLAWKA, E.
1. Zur angenäherten Berechnung mehrfacher Integrale. *Monatsh. Math.* **66** (1962) 140–151.
2. Uniform distribution modulo 1 and numerical analysis. *Compositio Math.* **16** (1964) 92–105.
3. Gleichverteilung auf Produkten von Sphären. *J. Reine Angew. Math.* **330** (1982) 1–43.

HLAWKA, E., FERNEIS, F., and ZINTERHOFF, P.
1. *Zahlentheoretische Methoden in der Numerischen Mathematik.* R. Oldenbourg, Vienna, 1981.

HLAWKA, E., and MÜCK, R.
1. A transformation of equidistributed sequences, in Zaremba [5, pp. 371–388].

HOBSON, E. W.
1. *The Theory of Functions of a Real Variable*, Vol. I. Cambridge Univ. Press, London and New York, 1927.

HOCHSTRASSER, U. W.
1. Orthogonal polynomials. *Handbook of Mathematical Functions*, Chap. 22, pp. 771–802. Nat. Bur. Standards, Washington, D.C., 1964.

HOFSOMMER, D. J.
1. Note on the computation of the zeros of functions satisfying a second order differential equation. *MTAC* **12** (1958) 58–60.

HOLLADAY, J. C.
1. A smoothest curve approximation, *MTAC* **11** (1957) 233–243.

HOPP, T. H.
1. A routine for numerical evaluation of integrals with oscillatory integrands, *Proc. 1979 Army Numer. Anal. Comp. Conf.*, pp. 187–221.

HSU, L. C.
1. Concerning the numerical integration of periodic functions of several variables. *Acta Sci. Math. (Szeged)* **20** (1959) 230–233.

HUA, L. K., and WANG, Y.
1. *Applications of Number Theory to Numerical Analysis.* Springer-Verlag, Berlin, 1981.

HUANG, T. S. (ed.)
1. *Two-Dimensional Digital Signal Processing II*, Topics in Applied Physics Vol. 43. Springer-Verlag, Berlin, 1981.

HUELSMAN, III, C. B.
1. Quadrature formulas over fully symmetric planar regions, *SIAM J. Numer. Anal.* **10** (1973) 539–552.

HULL, T. E., and DOBELL, A. R.
1. Random number generators, *SIAM Rev.* **4** (1962) 230–254.

HUNKINS, D. R.
1. Product-type multiple integration formulas, *BIT* **13** (1973) 408–411.

HUNTER, D. B.
1. The calculation of certain Bessel functions, *Math. Comp.* **18** (1964) 123–128.
2. Romberg's method for certain integrals involving a singularity, *BIT* **7** (1967) 200–205.
3. The evaluation of a class of functions defined by an integral, *Math. Comp.* **22** (1968) 440–444.
4. The evaluation of integrals of periodic analytic functions, *BIT* **11** (1971) 175–180.
5. Some Gauss-type formulae for the evaluation of Cauchy principal values of integrals, *Numer. Math.* **19** (1972) 419–424.
6. The numerical evaluation of Cauchy principal values of integrals by Romberg integration, *Numer. Math.* **21** (1973) 185–192.

HUNTER, D. B., and PARSONS, S. J.
1. A note on the numerical evaluation of finite integrals of oscillatory functions, *BIT* **15** (1975) 221–223.

HURWITZ, JR., H., PFEIFFER, R. A., and ZWEIFEL, P. F.
1. Numerical quadrature of Fourier transform integrals II, *MTAC* **13** (1959) 87–90.

HURWITZ, JR., H., and ZWEIFEL, P. F.
1. Numerical quadrature of Fourier transform integrals, *MTAC* **10** (1956) 140–149.

HUTCHINSON, D.
1. A final word on reducing truncation errors, *CACM* **8** (1965) 262.

IBM
1. Reference Manual C20-8011. Random Number Generation and Testing.
2. System/360 Scientific Subroutine Package, Version III, Programmer's Manual, H20-0205-3 (1968).
3. System/360 Scientific Subroutine Package (PL/1), Program Description and Operations Manual, GH20-0586.

IBRAGIMOV, I. I., and ALLIEV, R. M.
1. Some best cubature formulas, *Izv. Akad. Nauk Azerbaĭdžan SSR Ser. Fiz.-Tehn. Mat. Nauk* **3–4** (1967) 154–161.

ICHIDA, K., and KIYONO, T.
1. Numerical integration in the irregular region, *Computing* **12** (1974) 9–15.

IMHOF, J. P.
1. Remarks on quadrature formulas, *SIAM J. Appl. Math.* **11** (1963) 336–341.
2. On the method for numerical integration of Clenshaw and Curtis, *Numer. Math.* **5** (1963) 138–141.

IMSL
1. *IMSL Library*, edition 8. International Mathematical and Statistical Libraries, Inc., Houston, TX, 1980.

IOAKIMIDIS, N. I.
1. One more approach to the computation of Cauchy principal value integrals, *J. Comput. Appl. Math.* **7** (1981) 289–291.
2. A remark on the application of closed and semi-closed quadrature rules to the direct numerical solution of singular integral equations, *J. Comp. Phys.* **42** (1981) 396–402.

IONESCU, D. V.
1. Formules de cubature, le domaine d'intégration étant un triangle quelconque, *Acad. R. P. Romîne Fil. Cluj. Stud. Cerc. Şti. Ser. I* **6** (1955) 7–49.

IRI, M., MORIGUTI, S., and TAKASAWA, Y.
1. On a certain quadrature formula (in Japanese), No. 91, pp. 82–118, Kokyuroku of Res. Inst. for Math. Sci., Kyoto Univ., 1970.

IRONS, B. M.
1. Engineering applications of numerical integration in stiffness methods, *AIAA J.* **4** (1966) 2035–2037.
2. Quadrature rules for brick based finite elements, *IJNME* **3** (1971) 293–294.

ISAACSON, E., and KELLER, H. B.
1. *Analysis of Numerical Methods*. Wiley, New York, 1966.

IVANOVA, A. N.
1. On convergence of sequences of quadrature formulas of Gauss type on an infinite interval, *Dokl. Akad. Nauk SSSR* **104** (1955) 169–172.

JACKSON, D.
1. *The Theory of Approximation*. Amer. Math. Soc., New York, 1930.

JACOBS, D. (ed.)
1. *The State of the Art in Numerical Analysis*. Academic Press, London, 1977.

JAGERMAN, D.
1. Investigation of a modified mid-point quadrature formula, *Math. Comp.* **20** (1966) 79–89.

JAIN, M. K., and SHARMA, K. D.
1. Numerical solution of linear differential equations and Volterra's integral equation using Lobatto quadrature formula, *Comput. J.* **10** (1967–1968) 101–107.

JANSSON, B.
1. *Random Number Generators*. Almqvist & Wiksell, Stockholm, 1966.

JAROSCH, H., and RABINOWITZ, P.
1. An exact expression for a class of definite multiple integrals, *ICC Bull.* **5** (1966) 95–97.

JONES, D. S.
1. Asymptotic behavior of integrals, *SIAM Rev.* **14** (1972) 286–317.

JORDAN, D. F.
1. Extra-precision accumulating inner product, ANL-F1545-DOTP. Argonne Nat. Lab., Appl. Math. Div. System/360 Library Subroutine, Argonne, Illinois, 1967.

JOYCE, D. C.
1. Survey of extrapolation processes in numerical analysis, *SIAM Rev.* **13** (1971) 435–490.

JUNCOSA, M. L.
1. A general analytic approach for integrals of coverage and more general type using automatic symbol manipulation, in *Information Processing 71* (Freeman, C. V., ed.), Vol. 2, pp. 1356–1362. North-Holland Publ., Amsterdam, 1972.

KAC, I. S.
1. Uniform distribution on a sphere. *Bull. Acad. Polon. Sci. Cl.* **III 5**, (1957) 485–486.

KAHAN, W.
1. Further remarks on reducing truncation errors. *CACM* **8** (1965) 40.
2. Handheld calculator evaluates integrals, *Hewlett-Packard J.* **31**(8) (1980) 23–32.

KAHANER, D. K.
1. Comparison of numerical quadrature formulas, LA-4137. Los Alamos Sci. Lab., Los Alamos, New Mexico, 1969.
2. Comparison of numerical quadrature formulas II, in J. R. Rice [2, pp. 229–259].
3. Numerical quadrature by the ε-algorithm, *Math. Comp.* **26** (1972) 689–693.
4. Algorithm 561. Fortran implementation of heap programs for efficient table maintenance, *TOMS* **6** (1980) 444–449.
5. Sources of information on quadrature software, in *Sources and Development of Mathematical Software* (Cowell, W., ed.), pp. 134–164. Prentice-Hall, Englewood Cliffs, NJ, 1984.

KAHANER, D., and STOER, J.
1. Extrapolated adaptive quadrature, *SIAM J. Sci. Stat. Comput.* **4** (1983) 31–44.

KAHANER, D. K., and Wells, M. B.
1. An algorithm for N-dimensional adaptive quadrature using advanced programming techniques, Rep. LA-UR 76-2310. Los Alamos Sci. Lab., Los Alamos, NM, 1976.
2. An experimental algorithm for N-dimensional adaptive quadrature, *TOMS* **5** (1979) 86–96.

KAHANER, D. K., and WYMAN, W. L.
1. Mathematical software in BASIC: GLAQ, numerical evaluation of definite integrals, *IEEE Micro* **3**, No. 5 (1983) 42–46.
2. Mathematical Software in BASIC. DINT: Data integration (submitted for publication).

KAHN, H.
1. *Applications of Monte-Carlo.* Res. Mem. RM-1237-AEC, Revised ed. Rand Corp., Santa Monica, California, 1956.
2. Multiple quadrature by Monte Carlo methods, in *Mathematical Methods for Digital Computers* (Ralston, A., and Wilf, H. S., eds.), pp. 249–257. Wiley, New York, 1960.

KAMBO, N. S.
1. Error of the Newton–Cotes and Gauss–Legendre quadrature formulas, *Math. Comp.* **24** (1970) 261–269.
2. Error bounds for the Lobatto and Radau quadrature formulas, *Numer. Math.* **16** (1971) 383–388.
3. Error bounds for Clenshaw–Curtiss quadrature formulas, *BIT* **11** (1971) 299–309.
4. Error of certain Gauss quadrature formulas, *JIMA* **7** (1971) 303–307.

KANTOROVITCH, L. V.
1. On approximate calculation of certain types of definite integrals and other applications of the method of selection of the singularities, *Mat. Sb.* **41** (1934) 235–245.

KAPLAN, E. L.
1. Numerical integration near a singularity, *J. Math. and Phys.* **31** (1952) 1–28.

KAPORIN, I. E.
1. A new fast Fourier transform algorithm, *U.S.S.R. Comput. Math. Math. Phys.* **20**(4) (1980) 253–259.

KARLIN, S.
1. The fundamental theorem of algebra for monosplines satisfying certain boundary conditions and applications to optimal quadrature formulas, in *Approximations with Special Emphasis on Spline Functions* (Schoenberg, I. J., ed.), pp. 467–484. Academic Press, New York, 1969.
2. Best quadrature formulas and interpolation by splines satisfying boundary conditions, in *Approximations with Special Emphasis on Spline Functions* (Schoenberg, I. J., ed.), pp. 447–466. Academic Press, New York, 1969.
3. Best quadrature formulas and splines, *J. Approx. Theory* **4** (1971) 59–90.

KARLIN, S., and STUDDEN, W. J.
1. *Tchebycheff Systems with Applications in Analysis and Statistics.* Wiley (Interscience), New York, 1966.

KAUTSKY, J.
1. Matrices related to interpolatory quadratures, *Numer. Math.* **36** (1981) 309–318.

KAUTSKY, J., and ELHAY, S.
1. Calculation of the weights of interpolatory quadratures, *Numer. Math.* **40** (1982) 407–422.
2. Gauss quadrature and Jacobi matrices for weight functions, not of one sign. School of Mathematical Sciences, Flinders Univ. of South Australia, 1983.

KAUTSKY, J., and GOLUB, G. H.
1. On the calculation of Jacobi matrices, Linear Algebra Appl. **52/53** (1983) 439–455.

KEARFOTT, R. B.
1. A sinc approximation for the indefinite integral, *Math. Comp.* **41** (1983) 559–572.

KEAST, P.
1. Multi-dimensional quadrature formulae, Tech. Rep. No. 40. Dept. of Comput. Sci., Univ. of Toronto, Toronto, 1972.
2. Optimal parameters for multidimensional integration, *SIAM J. Numer. Anal.* **10** (1973) 831–838.
3. The evaluation of one-dimensional quadrature routines, Tech. Rep. No. 83. Dept. of Comput. Sci., Univ. of Toronto, Toronto, 1975.

4. Families of s-dimensional, degree $2t + 1$ quadrature rules for product spaces, Tech. Rep. No. 116. Dept. of Comput. Sci., Univ. of Toronto, Toronto, 1978.

5. Some fully symmetric quadrature formulae for product spaces, *JIMA* **23** (1979) 251–264.

6. On the null spaces of fully symmetric basic rules for quadrature formulas in s-dimensions. Tech. Rep. No. 134. Dept. of Comput. Sci., Univ. of Toronto, Toronto, 1979.

KEAST, P., and DIAZ, J. C.

1. Fully symmetric integration formulas for the surface of the sphere in s dimensions, *SIAM J. Numer. Anal.* **20** (1983) 406–419.

KEAST, P., and LYNESS, J. N.

1. On the structure of fully symmetric multidimensional quadrature rules, *SIAM J. Numer. Anal.* **16** (1979) 11–29.

KIEFER, J. E., and WEISS, G. H.

1. A comparison of two methods for accelerating the convergence of Fourier series, *Comput. Math. Appl.* **7** (1981) 527–535.

KING, R.

1. Runge-Kutta methods with constrained minimum error bounds, *Math. Comp.* **20** (1966) 386–391.

KITAHARA, N., and YANO, H.

1. An approximation of real analytic functions based on Cauchy's integral representation, *J. Inform. Process.* **4** (1981) 91–93.

KLERER, M., and GROSSMAN, F.

1. Error rates in tables of indefinite integrals, *Indust. Math.* **18** (1968) 31–62.

2. *A New Table of Indefinite Integrals—Computer Processed.* Dover, New York, 1971.

KNAUFF, W.

1. Fehlernormen zur Quadratur analytischer Funktionen, *Computing* **17** (1977) 309–322.

2. Optimale Approximation mit Nebenbedingungen an lineare Funktionale auf $H^2(E_\rho)$ und $L^2(E_\rho)$, *Computing* **14** (1975) 235–250.

3. Gewichtsoptimale Quadraturformeln bei analytischen Funktionen, *Computing* **18** (1977) 59–66.

KNAUFF, W., and KRESS, R.

1. Optimale Approximation linearer Funktionale auf periodischen Funktionen, *Numer. Math.* **22** (1974) 187–205.

KNAUFF, W., and PAULIK, A.

1. A note on Davis type error bounds, *BIT* **18** (1978) 175–183.

KNESCHKE, A.

1. Über die genäherte Quadratur, *Monatsh. Math. Phys.* **51** (1943) 15–23.

2. Theorie der genäherten Quadratur, *J. Reine Angew. Math.* **187** (1949) 115–128.

KNIGHT, C. J., and NEWBERRY, A. C. R.

1. Trigonometric and Gaussian quadrature, *Math. Comp.* **24** (1970) 575–581.

KNOPP, K.

1. *Theory and Application of Infinite Series.* Blackie, Glasgow and London, 1951.

KNUTH, D. E.
1. *The Art of Computer Programming*, Vol. 2, *Seminumerical Algorithms*. Addison-Wesley, Reading, Massachusetts, 1969.

KOFRON, J.
1. Die ableitungsfreien Fehlerabschätzungen von Quadraturformeln, *Apl. Mat.* **17** (1972) 39–52, 124–136.

KÖLBIG, K. S.
1. A Fortran programme and some numerical test results for integration over a triangle, Rep. 64–32. CERN, Geneva, 1964.

KOKSMA, J. F.
1. The theory of asymptotic distribution modulo one, *Compositio Math.* **16** (1964) 1–22.

KONJAEV, S. I.
1. Ninth-order quadrature formulas invariant with respect to the icosahedral group, *Soviet Math. Dokl.* **18** (1977) 497–501.
2. Quadratures of Gaussian type for a sphere invariant under the icosahedral group with inversion, *Math. Notes* **25** (1979) 326–329.

KOPAL, Z.
1. *Numerical Analysis*, 2nd ed., Wiley, New York, 1961.

KOROBOV, N. M.
1. *Number-Theoretic Methods in Approximate Analysis*. GIFL, Moscow, 1963 (Russian).

KOROVKIN, P. P.
1. *Linear Operators and Approximation Theory*. Hindustan, Delhi, 1960.

KOWALEWSKI, G.
1. *Interpolation und Genäherte Quadratur*. Teubner, Leipzig, 1932.

KRAFT, R., and WENSRICH, C. J.
1. *Monte Carlo Bibliography*, Rep. UCRL-6581. Lawrence Livermore Lab., Univ. of California, Livermore, 1961.

KRASUN, A. M., and PRAGER, W.
1. Remark on Romberg quadrature, *CACM* **8** (1965) 236–237.

KRESS, R.
1. Interpolation auf einem unendlichen Intervall, *Computing* **6** (1970) 274–288.
2. Ein ableitungsfreies Restglied für die trigonometrische Interpolation periodischer analytischer Funktionen, *Numer. Math.* **16** (1971) 389–396.
3. Interpolation auf einem unendlichen Intervall, *ZAMM* **51** (1971) T63–64.
4. Zur numerischen Integration periodischer Funktionen nach der Rechteckregel, *Numer. Math.* **20** (1972) 87–92.
5. Über die numerische Berechnung konjugierter Funktionen, *Computing* **10** (1972) 177–187.
6. On general Hermite trigonometric interpolation, *Numer. Math.* **20** (1972) 125–138.
7. On the general Hermite cardinal interpolation, *Math. Comp.* **26** (1972) 925–933.
8. Zur Quadratur uneigentlicher Integrale bei analytischen Funktionen, *Computing* **13** (1974) 267–277.
9. On error norms of the trapezoidal rule, *SIAM J. Numer. Anal.* **15** (1978) 433–443.

KRESS, R., and MARTENSEN, E.
1. Anwendung der Rechteckregel auf die reele Hilberttransformation mit unendlichem Intervall, *ZAMM* **50** (1970) T61–64.

KRONROD, A. S.
1. *Nodes and Weights of Quadrature Formulas.* Consultants Bureau, New York, 1965.

KRUGLIKOVA, L.
1. *Tables for the Numerical Fourier Transform.* Izdat. Nauka i Tehnika, Minsk, 1964.

KRUMHAAR, H.
1. Error estimates for Luke's approximation formulas for Bessel and Hankel functions, *ZAMM* **45** (1965) 245–255.

KRYLOFF, N.
1. Sur la theorie des quadratures mécaniques et sur certaines questions qui s'y rattachent, *Ann. École Sup. Mines Petersbourg* (1915).
2. Sur quelques formules d'approximation fondées sur la généralisation des quadratures, dites mécaniques, *C. R. Acad. Sci. Paris* **168** (1919) 721–723.
3. Sur quelques recherches dans le domaine de la théorie de l'interpolation et des quadratures, dites mécaniques, *Proc. Internat. Congr., Toronto, 1924*, **1** pp. 651–656.

KRYLOV, V. I.
1. *Approximate Calculation of Integrals*, translated by A. H. Stroud. Macmillan, New York, 1962.

KRYLOV, V. I., and ARLYUK, T. K.
1. On the convergence of quadrature processes containing values of the derivatives of the integrated function, *Dokl. Akad. Nauk BSSR* **7** (1963) 721–723; Engl. Transl., Tech. Summary Rep. 684. Math. Res. Center, Madison, Wisconsin, 1966.

KRYLOV, V. I., and KRUGLIKOVA, L. G.
1. *Handbook of Numerical Harmonic Analysis.* Isr. Program for Sci. Transl., Jerusalem, 1969.

KRYLOV, V. I., LUGIN, V. V., and IANOVITCH, L. A.
1. *Tables for the Numerical Integration of Functions with Power Singularities.* Izdat. Akad. Nauk. BSSR, Minsk, 1963.

KRYLOV, V. I., and PAL'TSEV, A. A.
1. *Tables for Numerical Integration of Functions with Logarithmic and Power Singularities.* Israel Program for Scientific Translations, Jerusalem, 1974.

KRYLOV, V. I., and SHULGINA, L. T.
1. *Spravochnaya Kniga Po Chiślennomu Integrirovaniyu*, Izdat. Nauka, Moscow, 1966.

KRYLOV, V. I., and SKOBLYA, N. S.
1. *Handbook of Numerical Inversion of Laplace Transforms.* Isr. Program for Sci. Transl., Jerusalem, 1969.

KUIPERS, L., and NIEDERREITER, H.
1. *Uniform Distribution of Sequences.* Wiley, New York, 1974.

KUKARKIN, A. B., and NOVIKOVA, E. I.
1. Evaluation of integrals of rapidly oscillating functions by Longman's method, *U.S.S.R. Comput. Math. Math. Phys.* **21** (5) (1981) 16–25.

KUMAR, S.
1. A note on quadrature formulae for Cauchy principal value integrals, *JIMA* **26** (1980) 447–451.

KUNTZMANN, J. (ed.)
1. *Procédures ALGOL en Analyse Numérique*. CNRS, Paris, 1967.

KUNZ, K. S.
1. *Numerical Analysis*. McGraw-Hill, New York, 1957.

KUSSMAUL, R.
1. Clenshaw–Curtis quadrature with a weighting function, *Computing* **9** (1972) 159–164.

KUTT, H. R.
1. Quadrature formulae for finite-part integrals, CSIR Special Report WISK 178. Pretoria, 1975.
2. On the numerical evaluation of finite-part integrals involving an algebraic singularity, CSIR Special Report WISK 179. Pretoria, 1975.
3. The numerical evaluation of principal value integrals by finite-part integration, *Numer. Math.* **24** (1975) 205–210.
4. Gaussian quadrature formulae for improper integrals involving a logarithmic singularity, CSIR Special Report WISK 232. Pretoria, 1976.

LABUDDE, R. A., and WARNOCK, T. T.
1. Numerical multiple integration via number-theoretical lattices: I. Integration of periodic functions, Tech. Rep. TR78-3. Dept. of Mathematical and Computing Sciences, Old Dominion Univ., Norfolk, VA, 1978.

LAMBERT, J. D., and MITCHELL, A. R.
1. The use of higher derivatives in quadrature formulae, *Comput. J.* **5** (1962–1963) 322–327.

LANCE, G. N.
1. *Numerical Methods for High Speed Computers*. Iliffe, London, 1960.

LANCZOS, C.
1. *Tables of Chebyshev Polynomials*. Nat. Bur. Stand. Appl. Math. Ser. No. 9, Introduction, U.S. Govt. Printing Office, Washington, D.C., 1952.
2. *Applied Analysis*. Prentice-Hall, Englewood Cliffs, New Jersey, 1956.
3. *Discourse on Fourier Series*. Hafner, New York, 1966.

LANGER, R. E. (ed.)
1. *On Numerical Approximation*. Univ. of Wisconsin Press, Madison, WI, 1959.

LANNOY, F. G.
1. Triangular finite elements and numerical integration, *Comput. & Structures* **7** (1977) 613.

LAPIDUS, L., and SEINFELD, J. H.
1. *Numerical Solution of Ordinary Differential Equations*. Academic Press, New York, 1971.

LAPSHIN, E. A.
1. The statistical investigation of round-off errors in numerical integration on a fixed point computer, *U.S.S.R. Computational Math. and Math. Phys.* **7**, No. 2 (1967) 1–21.

LARKIN, F. M.
1. Optimal approximation in Hilbert spaces with reproducing kernel functions, *Math. Comp.* **24** (1970) 911–921.

LAUFFER, R.

1. Interpolation mehrfacher Integrale, *Arch. Math.* (*Basel*) **6** (1955) 159–164.

LAURENT, P.-J.

1. Formules de quadrature approchée sur domaines rectangulaires convergents pour toute fonction intégrable Riemann, *C.R. Acad. Sci. Paris* **258** (1964) 798–801.

LAURIE, D. P.

1. Propagation of initial rounding error in Romberg-like quadrature, *BIT* **15** (1975) 277–282.
2. Algorithm 584. CUBTRI: Automatic cubature over a triangle, *TOMS* **8** (1982) 210–218.
3. Sharper error estimates in adaptive quadrature, *BIT*, **23** (1983) 258–261. See also TWISK 259, CSIR, Pretoria, 1982.

LAURSEN, M. E., and GELLERT, M.

1. Some criteria for numerically integrated matrices and quadrature formulas for triangles, *IJNME* **12** (1978) 67–76.

LAUTRUP, B.

1. An adaptive multidimensional integration procedure, in *Proc. 2nd Colloquium on Advanced Computing Methods in Theoretical Physics*, pp. 157–82. Marseille, 1971.

LAX, M., and AGRAWAL, G. P.

1. Evaluation of Fourier integrals using *B*-splines, *Math. Comp.* **39** (1982) 535–548.

LEACH, E. B.

1. The remainder term in numerical integration formulas, *Amer. Math. Monthly* **68** (1961) 273–275.

LEBEDEV, A. V.

1. *Method for the Construction of Cubature Formulae with Precise Estimates.* Inst. Točnoi Meh. Vyčisl. Tehn. Akad. Nauk SSSR, Moscow, 1964.

LEBEDEV, V. I.

1. Values of the nodes and weights of ninth to seventeenth order Gauss–Markov quadrature formulae invariant under the octahedron group with inversion, *U.S.S.R. Comp. Math. Math. Phys.* **15**(1) (1975) 44–51.
2. Quadratures on a sphere, *U.S.S.R. Comp. Math. Math. Phys.* **16**(2) (1976) 10–24.
3. On a type of quadrature formulas of increased algebraic accuracy for a sphere, *Soviet Math. Dokl.* **17** (1976) 1515–1517.
4. Spherical quadrature formulas exact to orders 25–29, *Siberian Math. J.* **18** (1977) 99–106.

LEBEDEV, V. I., and BABURIN, O. V.

1. Calculation of the principal values, weights and nodes of the Gauss quadrature formulae of integrals, *U.S.S.R. Computational Math. and Math. Phys.* **5**, No. 3 (1965) 81–92.

LEHMAN, D. R., PARKE, W. C., and MAXIMON, L. C.

1. Numerical evaluation of integrals containing a spherical Bessel function by product integration, *J. Math. Phys.* **22** (1981) 1399–1413.

LEMME, J. M., and RICE, J. R.

1. Speedup in parallel algorithms for adaptive quadrature, *JACM* **26** (1979) 65–71.
2. Adaptive quadrature algorithms for the ILLIAC IV, *Internat. J. Comput. Inform. Sci.* **9** (1980) 63–72.

LEPAGE, G. P.

1. A new algorithm for adaptive multidimensional integration, *J. Comp. Phys.* **27** (1978) 192–203.

2. VEGAS—An adaptive multidimensional integration program, Report CLNS-80/447. Newman Laboratory of Nuclear Studies, Cornell Univ., Ithaca, N.Y. 1980.

LETHER, F. G.

1. An error representation for product cubature rules, *SIAM J. Numer. Anal.* **7** (1970) 363–365.

2. Cross-product cubature error bounds, *Math. Comp.* **24** (1970) 583–592.

3. Error bounds for fully symmetric cubature rules, *SIAM J. Numer. Anal.* **8** (1971) 49–60.

4. Cubature error bounds for Gauss–Legendre product rules, *SIAM J. Numer. Anal.* **8** (1971) 36–42.

5. Error bounds for fully symmetric quadrature rules, *SIAM J. Numer. Anal.* **11** (1974) 1–9.

6. Cubature error bounds for analytic functions, *Math. Comp.* **27** (1973) 655–668.

7. On Birkhoff–Young quadrature of analytical functions, *J. Comput. Appl. Math.* **2** (1976) 81–84.

8. Modified quadrature formulas for functions with nearby poles, *J. Comput. Appl. Math.* **3** (1977) 3–9.

9. On the construction of Gauss–Legendre quadrature rules, *J. Comput. Appl. Math.* **4** (1978) 47–51.

10. Algorithm 007. Computation of double integrals over a triangle, *J. Comput. Appl. Math.* **2** (1976) 219–224.

11. Subtracting out complex singularities in numerical integration, *Math. Comp.* **31** (1977) 223–229.

12. Error estimates for Gaussian quadrature, *Appl. Math. Comp.* **7** (1980) 237–246.

13. Some observations on Gauss–Legendre quadrature error estimates for analytic functions, *J. Comput. Appl. Math.* **7** (1981) 63–66.

14. A generalized product rule for the circle, *SIAM J. Numer. Anal.* **8** (1971) 249–253.

LETHER, F., WILHELMSEN, D., and FRAZIER, R.

1. On the positivity of Cotes numbers for interpolatory quadratures with Jacobi abscissas, *Computing* **21** (1979) 171–175.

LEVEQUE, W. J.

1. *Topics in Number Theory*, Vol. 2. Addison-Wesley, Reading, Massachusetts, 1956.

LEVIN, D.

1. Development of non-linear transformations for improving convergence of sequences, *Internat. J. Comput. Math.* **3** (1973) 371–388.

2. Procedures for computing one and two dimensional integrals of functions with rapid irregular oscillations, *Math. Comp.* **38** (1982) 531–538.

LEVIN, D., and SIDI, A.

1. Two new classes of nonlinear transformations for accelerating the convergence of infinite integrals and series, *Appl. Math. Comp.* **9** (1981) 175–215.

LEVIN, M.

1. Extremal problems connected with a quadrature formula, *Izv. Akad. Nauk. Est. SSR, Ser. Fiz.-Mat. Tehn. Nauk* **12** (1963) 44–56.

2. An extremal problem for a class of functions, *Izv. Akad. Nauk Est. SSR, Ser. Fiz.-Mat. Tehn. Nauk* **12** (1963) 141–145.

3. On the evaluation of double integrals, *Math. Comp.* **39** (1982) 173–177.
4. Optimal quadrature formula of Markov's and Locher's type with weight function, *Numer. Math.* **40** (1982) 31–37.

LEVIN, M., and GIRSHOVICH, J.
1. *Optimal Quadrature Formulas.* Teubner, Leipzig, 1979.

LEWANOWICZ, S.
1. Construction of a recurrence relation for modified moments, *J. Comput. Appl. Math.* **5** (1979) 193–206.

LEWELLEN, P. C.
1. Two-dimensional adaptive quadrature over rectangular regions, *Comp. Phys. Comm.* **27** (1982) 167–178.

LEWIS, P. A. W.
1. A computer program for the statistical analysis of series of events, *IBM Systems J.* **5**, No. 4 (1966) 202–225.

LIGHTHILL, M. J.
1. *Introduction to Fourier Analysis and Generalized Functions.* Cambridge Univ. Press, London and New York, 1958.

LINDEN, J., KROLL, N., and SCHMID, H. J.
1. Minimale Kubaturformeln für Integrale über dem Einheitsquadrat, Preprint No. 373, Sonderforschungsbereich 72, Approximation und Optimierung. University of Bonn, Bonn, 1980.

LINZ, P.
1. Accurate floating-point summation, *CACM* **13** (1970) 361–362.
2. An adaptive quadrature algorithm for Fourier cosine integrals, Courant Inst., New York.

LIPOW, P., and STENGER, F.,
1. How slowly can quadrature formulas converge, *Math. Comp.* **26** (1972) 917–922.

LITTLEWOOD, R. K., and ZAKIAN, V.
1. Numerical evaluation of Fourier integrals, *JIMA* **18** (1976) 331–339.

LO, Y. T., LEE, S. W., and SUN, B.
1. On Davis' method of estimating quadrature errors, *Math. Comp.* **19** (1965) 133–138.

LOCHER, F.
1. Approximationsverfahren zur Gewinnung von ableitungsfreien Schranken für Quadraturfehler. Thesis, Tübingen, 1968.
2. Fehlerabschätzungen für das Quadraturverfahren von Clenshaw und Curtis, *Computing* **4** (1969) 304–315.
3. Zur Struktur von Quadraturformeln, *Numer. Math.* **20** (1973) 317–326.
4. Fehlerkontrolle bei der numerischen Quadratur, in Hämmerlin [5, pp. 198–210].

LOCHER, F., and ZELLER, K.
1. Approximationsgüte und numerische Integration, *Math. Z.* **104** (1968) 249–251.

LOEB, H. L., and WERNER, H.
1. Optimal numerical quadrature in H_p spaces, *Math. Z.* **138** (1974) 111–117.

LOHMANN, W.
1. Numerische Auswertung von Integralen über eine volle Periode von periodischen Integrandenfunktionen mit der "Rechteckregel", *ZAMM* **36** (1956) 464–465.

LONGMAN, I. M.

1. Note on a method for computing infinite integrals of oscillatory functions, *Proc. Cambridge Philos. Soc.* **52** (1956) 764–768.

2. Tables for the rapid and accurate numerical evaluation of certain infinite integrals involving Bessel functions *MTAC* **11** (1957) 166–180.

3. On the numerical evaluation of Cauchy principal values of integrals, *MTAC* **12** (1958) 205–207.

4. A method for the numerical evaluation of finite integrals of oscillatory functions, *Math. Comp.* **14** (1960) 53–59.

LOTKIN, M.

1. A new integrating procedure of high accuracy, *J. Math. and Phys.* **31** (1952) 29–34.

LÖTZBEYER, W.

1. Asymptotische Eigenschaften linear und nichtlinearer Quadraturformeln, *ZAMM* **52** (1972) T211–214.

LOWAN, A. N., DAVIDS, N., and LEVENSON, A.

1. Table of the zeros of the Legendre polynomials of order 1 to 16 and the weight coefficients for Gauss' mechanical quadrature formula, *Bull. Amer. Math. Soc.* **48** (1942) 739–743.

LUBINSKY, D. S.

1. Geometric convergence of Lagrangian interpolation and numerical integration rules over unbounded contours and intervals, *J. Approx. Theory* **39** (1983) 338–360.

LUBINSKY, D. S., and RABINOWITZ, P.

1. Rates of convergence of Gaussian quadrature for singular integrands, *Math. Comp.* **43** (1984).

LUBINSKY, D. S., and SIDI, A.

1. Convergence of product integration rules for functions with interior and endpoint singularities over bounded and unbounded intervals, Tech. Rep. No. 215. Computer Science Dept., Technion, Haifa, Israel, 1981.

LUBKIN, S.

1. A method of summing infinite series, *J. Res. Nat. Bur. Standards* **48** (1952) 228–254.

LUGANNANI, R., and RICE, S.

1. Use of Gaussian convergence factors in numerical evaluation of slowly convergent integrals, *J. Comp. Phys.* **37** (1980) 264–267.

LUKE, Y. L.

1. Mechanical quadrature near a singularity, *MTAC* **6** (1952) 215–219.

2. On the computation of oscillatory integrals, *Proc. Cambridge Philos. Soc.* **50** (1954) 269–277.

3. Evaluation of an integral arising in numerical integration near a logarithmic singularity, *MTAC* **10** (1956) 14–21.

4. Simple formulas for the evaluation of some higher transcendental functions, *J. Math. and Phys.* **34** (1955) 298–307.

5. *Integrals of Bessel Functions.* McGraw-Hill, New York, 1962.

6. Approximations for elliptic integrals, *Math. Comp.* **22** (1968) 627–634.

7. *The Special Functions and Their Approximations*, Vol. II. Academic Press, New York, 1969.

Lund, J. R.

1. Sinc function quadrature rules for the Fourier integral, *Math. Comp.* **41** (1983) 103–113.

Lynch, R. E.

1. Generalized trapezoid formulas and errors in Romberg Quadrature, in *Blanch Anniv. Vol.* (Mond, B., ed.), pp. 215–229. Office of Aerospace Res., USAF, Washington, D.C. 1967.

Lyness, J. N.

1. Symmetric integration rules for hypercubes, *Math. Comp.* **19** (1965) 260–276, 394–407, 625–637.
2. Limits on the number of function evaluations required by certain high-dimensional integration rules of hypercubic symmetry, *Math. Comp.* **19** (1965) 638–643.
3. Integration rules of hypercubic symmetry over a certain spherically symmetric space, *Math. Comp.* **19** (1965) 471–476.
4. The calculation of Fourier coefficients, *SIAM J. Numer. Anal.* **4** (1967) 301–315.
5. The calculation of Stieltjes' integral, *Numer. Math.* **12** (1968) 252–265.
6. Quadrature methods based on complex function values, *Math. Comp.* **23** (1969) 601–619.
7. The effect of inadequate convergence criteria in automatic routines, *Comput. J.* **12** (1969) 279–281. See also *Comput. J.* **13** (1970) 121.
8. Notes on the adaptive Simpson quadrature routine, *JACM* **16** (1969) 483–495.
9. The calculation of Fourier coefficients by the Möbius inversion of the Poisson summation formula, Pt. I, Functions whose early derivatives are continuous, *Math. Comp.* **24** (1970) 101–135; Pt. II, Piecewise continuous functions and functions with poles near the interval [0, 1], *Math. Comp.* **25** (1971) 59–78; Pt. III, Functions having algebraic singularities, *Math. Comp.* **25** (1971) 483–493.
10. Adjusted forms of the Fourier coefficient asymptotic expansion and applications in numerical quadrature, *Math. Comp.* **25** (1971) 87–104.
11. Guidelines for automatic quadrature routines, in *Information Processing 71* (Freeman, C. V., ed.) Vol. 2, pp. 1351–1355. North-Holland Publ., Amsterdam, 1972.
12. An algorithm for Gauss–Romberg integration, *BIT* **12** (1972) 194–203.
13. Computational techniques based on the Lanczos representation, *Math. Comp.* **28** (1974) 81–123.
14. An error functional expansion for N-dimensional quadrature with an integrand function singular at a point, *Math. Comp.* **30** (1976) 1–23.
15. Applications of extrapolation techniques to multidimensional quadrature of some integrand functions with a singularity, *J. Comp. Phys.* **20** (1976) 346–364.
16. An interface problem in numerical software, *Proc. Sixth Manitoba Conf. on Numer. Math.* (1976) 251–263.
17. Quid, quo, quadrature, in Jacobs [1, pp. 535–560].
18. Quadrature over a simplex: Part 1. A representation for the integrand function. Part 2. A representation for the error functional, *SIAM J. Numer. Anal.* **15** (1978) 122–133, 870–887.
19. QUG1: Trigonometric Fourier coefficients, Tech. Memo. 370. Applied Math. Div., Argonne Nat. Lab., Argonne, 1L, 1981.
20. When not to use an automatic quadrature routine, *SIAM Rev.* **25** (1983) 63–87.
21. QUG2-Integration over a triangle, Tech Memo. 13. Math. and Computer Sci. Div., Argonne Nat Lab., Argonne, IL, 1983.

Lyness, J. N., and Delves, L. M.

1. On numerical contour integration round a closed contour, *Math. Comp.* **21** (1967) 561–577.

LYNESS, J. N., and GATTESCHI, L.
1. A note on cubature over a triangle of a function having specified singularities, in Hämmerlin [6, pp. 164–169].
2. On quasi degree quadrature rules, *Numer. Math.* **39** (1982) 259–267.

LYNESS, J. N., and GENZ, A. C.
1. On simplex trapezoidal rule families, *SIAM J. Numer. Anal.* **17** (1980) 126–147.

LYNESS, J. N., and JESPERSEN, D.
1. Moderate degree symmetric quadrature rules for the triangle, *JIMA* **15** (1975) 19–32.

LYNESS, J. N., and KAGANOVE, J. J.
1. Comments on the nature of automatic quadrature routines, *TOMS* **2** (1976) 65–81.
2. A technique for comparing automatic quadrature routines, *Comput. J.* **20** (1977) 170–177.

LYNESS, J. N., and KEAST, P.
1. On *p*-generator fully symmetric quadrature rules, *Numer. Math.* **30** (1978) 51–63.

LYNESS, J. N., and MCHUGH, B. J. J.
1. Integration over multidimensional hypercubes—I. A progressive procedure, *Comput. J.* **6** (1963–1964) 264–270.
2. On the remainder term in the N-dimensional Euler–Maclaurin expansion, *Numer. Math.* **15** (1970) 333–344.

LYNESS, J. N., and MOLER, C. B.
1. Generalized Romberg methods for integrals of derivatives, *Numer. Math.* **14** (1969) 1–13.

LYNESS, J. N., and MONEGATO, G.
1. Quadrature rules for regions having regular hexagonal symmetry, *SIAM J. Numer. Anal.* **14** (1977) 283–295.
2. Quadrature error functional expansions for the simplex when the integrand function has singularities at vertices, *Math. Comp.* **34** (1980) 213–225.

LYNESS, J. N., and NINHAM, B. W.
1. Numerical quadrature and asymptotic expansions. *Math. Comp.* **21** (1967) 162–178.

LYNESS, J. N., and PURI, K. K.
1. The Euler–Maclaurin expansion for the simplex, *Math. Comp.* **27** (1973) 273–293.

LYUSTERNIK, L. A.
1. Certain cubature formulas for double integrals, *Dokl. Akad. Nauk SSSR* **62** (1948) 449–452.

LYUSTERNIK, L. A., and DITKIN, V. A.
1. Approximate formulas for the calculation of multiple integrals, *Izv. Akad. Nauk SSSR Otd. Tehn. Nauk* (1948) 1163–1168.
2. The construction of approximate formulas for the calculation of multiple integrals, *Dokl. Akad. Nauk SSSR* **61** (1948) 441–444.

MCCORMICK, J. M., and SALVADORI, M. G.
1. *Numerical Methods in FORTRAN*. Prentice-Hall, Englewood Cliffs, New Jersey, 1964.

MCCRACKEN, D. D., and DORN, W. S.
1. *Numerical Methods and FORTRAN Programming*. Wiley, New York, 1964.

McGrath, J. F.

1. Gaussian product-type quadratures, *Appl. Math. Comp.* **5** (1979) 265–280.

2. Matrix evaluation of product-type quadrature coefficients, in *Information Linkage between Applied Mathematics and Industry II* (Schoenstadt, A. L., Faulkner, F. D., Franke, F., and Russak, I. B., eds.), pp. 125–134. Academic Press, New York, 1980.

McKeeman, W. M.

1. Certification of algorithm 145. Adaptive numerical integration by Simpson's rule, *CACM* **6** (1963) 167–168.

McLaren, A. D.

1. Optimal numerical integration on a sphere, *Math. Comp.* **17** (1963) 361–383.

McNamee, J.

1. Error-bounds for the evaluation of integrals by the Euler–Maclaurin formula and by Gauss-type formulae, *Math. Comp.* **18** (1964) 368–381.

McNamee, J., and Stenger, F.

1. Construction of fully symmetric numerical integration formulas, *Numer. Math.* **10** (1967) 327–344.

McNamee, J. M.

1. A program to integrate a function tabulated at unequal intervals, *IJNME* **17** (1981) 271–279.

McWilliams, G. V., and Thompson, R. W.

1. Methods of convergence improvement for some improper integrals, *CACM* **11** (1968) 499–502.

Macon, N.

1. *Numerical Analysis.* Wiley, New York, 1963.

Maisonneuve, D.

1. Recherche et utilisation des "Bons Treillis." Programmation et résultats numériques, in Zaremba [5, pp. 121–200].

Majorana, C., Odorizzi, S., and Vitaliani, R.

1. Shortened quadrature rules for finite elements, *Adv. Eng. Software* **4** (1982) 52–57.

Malcolm, M. A.

1. On accurate floating-point summation, *CACM* **14** (1971) 731–736.

Malcolm, M. A., and Simpson, R. B.

1. Local versus global strategies for adaptive quadrature, *TOMS* **1** (1975) 129–146.

Malik, A. A.

1. Some new fully symmetric rules for multiple integrals with a variable order adaptive algorithm, Ph.D. Thesis, Univ. of Kent, Canterbury, 1980.

Mansfield, L. E.

1. On the optimal approximation of linear functionals in spaces of bivariate functions, *SIAM J. Numer. Anal.* **8** (1971) 115–126.

Mansion, P.

1. Théorème général de Peano sur le reste dans les formules de quadrature, *Mathesis* **34** (1914) 169–174.

MANTEL, F.
1. Structure Analysis of Multidimensional Numerical Cubatures, Ph.D. Thesis, Dept. of Appl. Math., Weizmann Institute of Science, Rehovot, Israel, 1980.

MANTEL, F., and RABINOWITZ, P.
1. The application of integer programming to the computation of fully symmetric integration formulas in two and three dimensions, *SIAM J. Numer. Anal.* **14** (1977) 391–425.

MARIČEV, O. I.
1. *A Method of Calculating Integrals from Special Functions* (Theory and Tables of Formulas) (Gahov, F. D., ed.). 'Nauka i Tehnika', Minsk, 1978 (Russian).

MARKOFF, A.
1. Sur la méthode de Gauss pour le calcul approché des intégrales, *Math. Ann.* **25** (1885) 427–432.

MARONI, P.
1. Equations singulières à noyaux de Cauchy, *Congr. Calcul et Traitement Information, 4th, 1964*, AFIRO, pp. 377–395. Dunod, Paris, 1965.

MARSAGLIA, G.
1. Random numbers fall mainly in the planes, *Proc. Nat. Acad. Sci. U.S.A.* **61** (1968) 25–28.
2. Regularities in congruential random number generators, *Numer. Math.* **16** (1970) 8–10.
3. The structure of linear congruential sequences, in Zaremba [5, pp. 249–285].

MARSAL, D.
1. Eine Verallgemeinerung der Summenformel von Euler und ihre Anwendung auf ausgewählte Probleme der numerischen Integration, *Numer. Math.* **7** (1965) 147–154.
2. Die numerische Quadratur singulärer Funktionalgleichungen. Teil 1: Die Tabellierung uneigentlicher Integralfunktionen, Cauchyscher Hauptwerte und Integraltransformationen mit vorgegebener Genauigkeit, *ZAMM* **50** (1970) 281–294.

MARSDEN, M. J., and TAYLOR, G. D.
1. Numerical evaluation of Fourier integrals, in *Numerische Methoden der Approximationstheorie I* (Collatz, L., and Meinardus, G., eds.), pp. 61–76, ISNM 16. Birkhäuser, Basel, 1972.

MARTENSEN, E.
1. Verallgemeinerung der Quadraturfehlerabschätzung von Davis auf mehrfache Integrale, *ZAMP* **17** (1966) 633–635.
2. Zur numerischen Auswertung uneigentlicher Integrale, *ZAMM* **48** (1968) T83–85.

MARTI, J. T.
1. An algorithm recursively computing the exact Fourier coefficients of *B*-splines with nonequidistant knots, *ZAMP* **29** (1978) 301–305.

MASKELL, S. J., and SACK, R. A.
1. Generalized Lobatto quadrature formulas for contour integrals, in *Studies in Numerical Analysis* (Scaife, B. K. P., ed.), pp. 295–310. Academic Press, London, 1974.

MECHEL, F.
1. Calculation of the modified Bessel functions of the second kind with complex argument, *Math. Comp.* **20** (1966) 407–412.

MEEK, D. S.
1. What to do when the Peano kernel changes sign, *Proc. Ninth Manitoba Conf. Numer. Math. Comp.* (1979) 305–313.

MEINGUET, J.
1. Methods for estimating the remainder in linear rules of approximation. Application to the Romberg algorithm. *Numer. Math.* **8** (1966) 345–366.

MEIR, A., and SHARMA, A.
1. On the method of Romberg quadrature, *SIAM J. Numer. Anal.* **2** (1965) 250–258.

MEISEL, W. S.
1. A numerical integration formula useful in Fourier analysis, *CACM* **11** (1968) 51.

MEISTER, B.
1. On a family of cubature formulae, *Comput. J.* **8** (1965–1966) 368–371.

MEUER, H. W., and ENGELS, H.
1. Zur Konstruktion von Interpolationsquadraturformeln mittels FORMAC, *ZAMM* **51** (1971) T65–66.

MEYERS, L. F., and SARD, A.
1. Best approximate integration formulas, *J. Math. and Phys.* **29** (1950) 118–123.

MICCHELLI, C. A., and RIVLIN, T. J.
1. Quadrature formulae and Hermite–Birkhoff interpolation, *Advances in Math.* **11** (1973) 93–112.
2. The Turán formulae and highest precision quadrature rules for Chebyshev coefficients, IBM Res. Center, Yorktown Heights, New York, 1972.

MICHELS, H. H.
1. Abscissas and weight coefficients for Lobatto quadrature, *Math. Comp.* **17** (1963) 237–244.

MIKHLIN, S. G.
1. *Multidimensional Singular Integrals and Integral Equations.* Pergamon, Oxford, 1965.

MIKLOŠKO, J.
1. Numerical computation of the Fourier coefficients of an analytically given function, *ZAMM* **47** (1967) 470–473.
2. Numerical integration of high speed oscillating functions, *ZAMM* **48** (1968) T87–90.
3. Numerical integration with highly oscillating weight functions, in *Information Processing 68* (Morrell, A. J. H., ed.), Vol. 1, pp. 138–144. North-Holland Publ., Amsterdam, 1969.
4. Numerical integration with highly oscillating weight functions, *Apl. Mat.* **15** (1970) 133–145.
5. Asymptotic properties and the convergence of numerical quadratures, *Numer. Math.* **15** (1970) 234–249.

MILLER, E. K.
1. A variable interval width quadrature technique based on Romberg's method, *J. Comp. Phys.* **5** (1970) 265–279.

MILLER, G. F.
1. On the convergence of the Chebyshev series for functions possessing a singularity in the range of representation, *SIAM J. Numer. Anal.* **3** (1966) 390–409.

MILLER, J. C. P.
1. Quadrature in terms of equally-spaced function values, Tech. Summary Rep. 167. MRC, Madison, Wisconsin, 1960.
2. Numerical quadrature over a rectangular domain in two or more dimensions, *MTAC* **14** (1960), Pt. 1: pp. 13–20. Pt. 2: pp. 130–138. Pt. 3: pp. 240–248.

3. Neville's and Romberg's processes: A fresh appraisal with extensions, *Philos. Trans. Roy. Soc. London Ser. A* **263** (1968–1969) 525–562.

MILLER, M. K., and GUY, JR., W. T.
1. Numerical inversion of the Laplace transform by use of Jacobi polynomials, *SIAM J. Numer. Anal.* **3** (1966) 624–635.

MILLER, R. K.
1. On ignoring the singularity in numerical quadrature, *Math. Comp.* **25** (1971) 521–532.

MILNE, W. E.
1. The remainder in linear methods of approximation, *J. Res. Nat. Bur. Standards* **43** (1949) 501–511.
2. The trapezoidal rule, unpublished, 1953.

MINEUR, H.
1. *Techniques de Calcul Numérique*. Béranger, Paris and Liège, 1952.

MOAN, T.
1. Experiences with orthogonal polynomials and 'best' numerical integration formulas on a triangle; with particular reference to finite element approximations, *ZAMM* **54** (1974) 501–508.

MØLLER, O.
1. Quasi double-precision in floating point addition, *BIT* **5** (1965) 37–50.
2. Notes on quasi double-precision, *BIT* **5** (1965) 251–255.

MÖLLER, H. M.
1. Polynomideale und Kubaturformeln, Ph.D. Thesis, Dept. of Mathematics, Univ. of Dortmund, 1973.
2. Kubaturformeln mit minimaler Knotenzahl, *Numer. Math.* **25** (1976) 185–200.
3. Mehrdimensionale Hermite-Interpolation und numerische Integration, *Math. Z.* **148** (1976) 107–118.
4. Lower bounds for the number of nodes in cubature formulae, in Hämmerlin [5, pp. 221–230].
5. The construction of cubature formulae and ideals of principal classes, in Schempp and Zeller [1, pp. 249–264].

MONEGATO, G.
1. A note on extended Gaussian quadrature rules, *Math. Comp.* **30** (1976) 812–817.
2. Positivity of the weights of extended Gauss–Legendre quadrature rules, *Math. Comp.* **32** (1978) 243–245.
3. Some remarks on the construction of extended Gaussian quadrature rules, *Math. Comp.* **32** (1978) 247–252.
4. An overview of results and questions related to Kronrod schemes, in Hämmerlin [5, pp. 231–240].
5. On polynomials orthogonal with respect to particular variable-signed weight functions, *ZAMP* **31** (1980) 549–555.
6. Stieltjes polynomials and related quadrature rules, *SIAM Rev.* **24** (1982) 137–158.
7. The numerical evaluation of one-dimensional Cauchy principal value integrals, *Computing* **29** (1982) 337–354.

MONEGATO, G., and LYNESS, J. N.
1. On the numerical evaluation of a particular singular two dimensional integral, *Math. Comp.* **33** (1979) 993–1002.

Monro, D. M.
1. Interpolation by fast Fourier and Chebyshev transforms, *IJNME* **14** (1979) 1679–1692.

Moon, Y.-S.
1. Some numerical experiments on number theoretic methods in the approximation of multi-dimensional integrals, Tech. Rep. No. 72. Dept. of Comput. Sci., Univ. of Toronto, Toronto, 1974.

Moore, R. E.
1. *Interval Analysis.* Prentice-Hall, Englewood Cliffs, New Jersey, 1966.
2. The automatic analysis and control of error in digital computing based on the use of interval numbers, in *Error in Digital Computation,* Vol. 1 (Rall, L. B., ed.), pp. 61–130. Wiley, New York, 1965.

Moors, B. P.
1. *Valeur Approximative d'une Intégrale Définie.* Gauthier-Villars, Paris, 1905.

Moran, P. A. P.
1. Approximate relations between series and integrals, *MTAC* **12** (1958) 34–37.

Morawitz, H.
1. A numerical approach to principal value integrals in dispersion relations, *J. Comp. Phys.* **6** (1970) 120–123.

Morelli, A.
1. Formula di quadratura con valori della funzione e delle sue derivate anche in punti fuori dell'intervallo d'integrazione, *Atti Accad. Sci. Torino* **102** (1967–1968) 569–579.

Mori, M.
1. An IMT-type double exponential formula for numerical integration, *Publ. RIMS, Kyoto Univ.* **14** (1978) 713–729.
2. (Ed.), *Numerical Integration and Related Topics.* RIMS Kokyuroka 401, Kyoto Univ. (1980).

Morris, Jr., A. H.
1. Elementary indefinite integration theory for the computer, *J. Comput. System Sci.* **3** (1969) 387–408.

Morrison, D.
1. Numerical quadrature in many dimensions, *JACM* **6** (1959) 219–222.

Morrow, C. R., and Patterson, T. N. L.
1. Construction of algebraic cubature rules using polynomial ideal theory, *SIAM J. Numer. Anal.* **15** (1978) 953–976.

Morton, K. W.
1. On the treatment of Monte Carlo methods in text books, *MTAC* **10** (1956) 223–224.
2. A generalization of the antithetic variate technique for evaluating integrals, *J. Math. Phys.* **36** (1957) 289–293.

Moses, J.
1. Symbolic integration, Rep. MAC-TR-47, Ph.D. Thesis, MIT, Cambridge, Massachusetts, 1967.
2. Symbolic integration: The stormy decade, *CACM* **14** (1971) 548–560.

MÜLLER-BAUSCH, I.
1. Minimale Quadraturformeln mit Randtermen zu Hilberträumen mit reproduzierenden Kernen, Ph.D. Thesis, RWTH Aachen, 1979.
2. Minimale Quadraturformeln mit Randtermen zu Hilbert–Räumen mit reproduzierenden Kernen, *ZAMM* **60** (1980) T304–306.

MUROTA, K., and IRI, M.
1. Parameter tuning and repeated application of the IMT-type transformation in numerical quadrature, *Numer. Math.* **38** (1982) 347–363.

MUSKHELISHVILI, N. I.
1. *Singular Integral Equations*. Noordhoff, Groningen, 1953.

MUSTARD, D.
1. Numerical integration over the n-dimensional spherical shell, *Math. Comp.* **18** (1964) 578–589.

MUSTARD, D., LYNESS, J. N., and BLATT, J. M.
1. Numerical quadrature in n dimensions, *Comput. J.* **6** (1963–1964) 75–87.

MYSOVSKIKH, I. P.
1. On the construction of cubature formulae for very simple domains, *U.S.S.R. Computational Math. and Math. Phys.* **4**, No. 1 (1964) 1–17.
2. The construction of cubature formulae and orthogonal polynomials, *U.S.S.R. Computational Math. and Math. Phys.* **7**, No. 1 (1967) 252–257.
3. Cubature formulas for evaluating integrals over a sphere, *Dokl. Akad. Nauk SSSR* **147** (1962) 552–555.
4. On cubature formulas for circle and sphere, *Metody Vyčisl.* **1** (1963) 3–11.
5. Cubature formulae for evaluating integrals on the surface of a sphere, *Sibirsk. Mat. Z.* **5** (1964) 721–723.
6. On the estimation of the remainder of cubature formulas over hypersphere, *Mat. Zametki* **6**, 5 (1969) 627–632.
7. Proof of the minimality of the number of nodes in the cubature formula for a hypersphere, *U.S.S.R. Computational Math. and Math. Phys.* **6**, No. 4 (1966) 15–27.
8. Cubature formulae and orthogonal polynomials, *U.S.S.R. Comput. Math. Math. Phys.* **9**(2) (1969) 217–228.
9. The application of orthogonal polynomials to cubature formulae, *U.S.S.R. Comput. Math. Math. Phys.* **12** (1972) 228–239.
10. Interpolatory cubature formulae, in *Theory of Cubature Formulae and Applications of Functional Analysis to Problems of Mathematical Physics* (Sobolev, S. L., ed.), pp. 73–90. Novosibirsk, 1973 (Russian).
11. On Chakalov's theorem, *U.S.S.R. Comput. Math. Math. Phys.* **15**(6) (1975) 221–227.
12. Numerical characteristics of orthogonal polynomials in two variables, *Vestnik Leningrad Univ. Math.* **3** (1976) 323–332.
13. On the evaluation of integrals over the surface of a sphere, *Soviet Math. Dokl.* **18** (1977) 925–929.
14. Common roots of polynomials and cubature formulas, *Vestnik Leningrad Univ. Math.* **6** (1979) 155–167.
15. The approximation of multiple integrals by using interpolatory cubature formulae, in De Vore and Scherer [1, pp. 217–243].
16. Cubature formulas for computing integrals over a hypersphere, *Soviet Math. Dokl.* **3** (1962) 1670–1673.
17. *Interpolatory Cubature Formulas*. Izd. 'Nauka', Moscow-Leningrad, 1981 (Russian).

NAG
1. NAG Library, Mark 10. Numerical Analysis Group, NAG Central Office, Oxford, 1983.

NATANSON, I. P.
1. *Constructive Function Theory*, Vol. I. Ungar, New York, 1964.

NAVOT, I.
1. An extension of the Euler–Maclaurin summation formula to functions with a branch singularity, *J. Math. and Phys.* **40** (1961) 271–276.
2. A further extension of the Euler–Maclaurin summation formula, *J. Math. and Phys.* **41** (1962) 155–163.
3. The Euler–Maclaurin functional for functions with a quasi-step discontinuity, *Math. Comp.* **17** (1963) 337–345.
4. The Euler–Maclaurin functional for functions with a complex singularity near the range of integration, *SIAM J. Numer. Anal.* **2** (1965) 259–264.

NBS Handbook
1. *Handbook of Mathematical Functions*, Nat. Bur. Stand. Appl. Math. Ser. No. 55, U.S. Govt. Printing Office, Washington, D.C., 1964.

NEHARI, Z.
1. *Conformal Mapping*. McGraw-Hill, New York, 1952.

NEUMAN, E.
1. Moments and Fourier transforms of *B*-splines, *J. Comput. Appl. Math.* **7** (1981) 51–62.
2. Calculation of complex Fourier coefficients using natural splines, *Computing* **29** (1982) 327–336.

NEUMANN, G.
1. Boolesche interpolatoriesche Kubatur, Ph.D. Thesis, Universität GH Siegen, 1982.
2. Boolean constructed cubature formulas of interpolatory type, in Hämmerlin [6, pp. 177–186].

NEUMANN, H.
1. Über Fehlerabschätzungen zum Romberg–Verfahren, *ZAMM* **46** (1966) 152–153.

NEWBERY, A. C. R.
1. Some extensions of Legendre quadrature, *Math. Comp.* **23** (1969) 173–176.

NEWMAN, D. J.
1. Quadrature formulae for H^p functions, *Math. Z.* **166** (1979) 111–115.
2. *Approximation with Rational Functions*, Regional Conference Series in Mathematics No. 41. Amer. Math. Soc., Providence, RI, 1979.

NG, E. W.
1. (Ed.), *Symbolic and Algebraic Computation*, Lecture Notes in Computer Science 72. Springer-Verlag, Berlin, 1979.
2. Symbolic-numeric interface: a review, in Ng [1, pp. 330–345].

NG, K.-C.
1. On the accuracy of numerical Fourier transforms, *J. Comp. Phys.* **16** (1974) 396–400.

NICHOLSON, D., RABINOWITZ, P., RICHTER, N., and ZEILBERGER, D.
1. On the error in the numerical integration of Chebyshev polynomials, *Math. Comp.* **25** (1971) 79–86.

NICKEL, K.
1. Quadraturverfahren mit Fehlerschranken, *Computing* **3** (1968) 47–64.
2. Das Kahan-Babuškasche Summierungsverfahren in Triplex-ALGOL 60, *ZAMM* **50** (1970) 369–373.

NICOLAIDES, R. A.
1. Interpolation in *n* variables, Tech. Note I.C.S.I. 274. Univ. of London, 1970.

NIEDERREITER, H.
1. On a number-theoretical integration method, *Aequationes Math.* **8** (1972) 304–311.
2. Methods for estimating discrepancy, in Zaremba [5, pp. 203–236].
3. Application of Diophantine approximations to numerical integration, in Osgood [1, 129–199].
4. Pseudo-random numbers and optimal coefficients, *Adv. Math.* **26** (1977) 99–181.
5. Existence of good lattice points in the sense of Hlawka, *Monatsh. Math.* **86** (1978) 203–219.
6. Quasi-Monte Carlo methods and pseudo-random numbers, *Bull. Amer. Math. Soc.* **84** (1978) 957–1041.

NIKOLSKII, S. M.
1. *Quadrature Formulae.* H.M. Stationery Office, London, 1966.

NINHAM, B. W.
1. Generalised functions and divergent integrals, *Numer. Math.* **8** (1966) 444–457.

NINHAM, B. W., and LYNESS, J. N.
1. Further asymptotic expansions for the error functional, *Math. Comp.* **23** (1969) 71–83.

NINOMIYA, I.
1. Improvements of adaptive Newton–Cotes quadrature methods, *J. Inform. Process.* **3** (1980) 162–170.

NOBLE, B.
1. *Numerical Methods*, Vol. 2, *Differences, Integration and Differential Equations.* Wiley (Interscience), New York, 1964.
2. The numerical solution of integral equations, in Jacobs [1, pp. 915–966].

NOBLE, B., and BEIGHTON, S.
1. Error estimates for three methods of evaluating Cauchy principal value integrals, *JIMA* **26** (1980) 431–446.

NORDEN, H. V.
1. Numerical inversion of the Laplace transform, *Acta Acad. Abo. Ser. B* **22** (1961) 3–31.

NORMAN, A. C., and DAVENPORT, J. H.
1. Symbolic integration—the dust settles?, in Ng [1, pp. 398–407].

NUSSBAUMER, H. J.
1. *Fast Fourier Transforms and Convolution Algorithms*, Second Edition. Springer-Verlag, Berlin, 1982.
2. Two dimensional convolution and DFT computation, in Huang [1, pp. 37–88].

NUTFULLIN, S. N.
1. Cubature formulae with rational weights. *Učen. Zap. Kemerovsk. Gos. Ped. Inst.* **10** (1969) 60–62.

OBRESCHKOFF, N.

1. Neue Quadraturformeln, *Abh. Preuss. Akad. Wiss. Math.-Nat. Kl. No.* **4** (1940) 1–20.

O'HARA, H.

1. The automatic evaluation of definite integrals, Ph.D. Thesis, Queen's Univ., Belfast, 1969.

O'HARA, H., and SMITH, F. J.

1. Error estimation in the Clenshaw–Curtis quadrature formula, *Comput. J.* **11** (1968) 213–219.
2. The evaluation of definite integrals by interval subdivision, *Comput. J.* **12** (1969) 179–182.

OHRINGER, L.

1. Newton–Cotes Integration (floating point). Guide General Program Library 9.5.002.

OKAMOTO, M.

1. Asymptotic normality in Monte Carlo integration, *Math. Comp.* **30** (1976) 831–837.

OKAMOTO, M., and TAKAHASI, R.

1. Bisection methods for Monte Carlo integration, *Math. Japon.* **22** (1977) 403–411.

OLIVER, J.

1. The efficiency of extrapolation methods for numerical integration, *Numer. Math.* **17** (1971) 17–32.
2. A practical strategy for the Clenshaw–Curtis quadrature method, *JIMA* **8** (1971) 53–56.
3. A doubly-adaptive Clenshaw–Curtis quadrature method, *Comput. J.* **15** (1972) 141–147.

OLVER, F. W. J.

1. Bessel functions of integer order, *Handbook of Mathematical Functions*, Chapter 9, pp. 355–433. Nat. Bur. of Standards, Washington, D.C., 1964.
2. *Asymptotics and Special Functions*, Academic Press, New York and London, 1974.
3. Asymptotic approximations and error bounds, *SIAM Rev.* **22** (1980) 188–203.

OSBORNE, M. R.

1. Asymptotic formulae for numerical quadrature, *JIMA* **13** (1974) 219–227.

OSGOOD, C. F.

1. (Ed.), *Diophantine Approximation and its Applications*, Academic Press, New York and London.
2. Approximating convolution products better than the DFT while keeping the FFT, *J. Comp. Phys.* **42** (1981) 382–395.

OSGOOD, C. F., and SHISHA, O.

1. The dominated integral, *J. Approx. Theory* **17** (1976) 150–165.
2. Numerical quadrature of improper integrals and the dominated integral, *J. Approx. Theory* **20** (1977) 139–152.

OSSICINI, A.

1. Costruzione di formule di quadratura di tipo gaussiano *Ann. Mat. Pura Appl.* [4] **72** (1966) 213–237.
2. Sulle costanti di Christoffel della formula di quadratura di Gauss–Jacobi, *Ist. Lombardo Accad. Sci. Lett. Rend. A* **101** (1967) 169–180.
3. Le funzione di influenza nel problema di Gauss sulle formule di quadratura, *Matematiche (Cantania)* **23** (1968) 7–30.

OSTROWSKI, A. M.

1. Bermerkungen zur Theorie der Diophantischen Approximationen, *Abh. Math. Sem. Univ. Hamburg* **1** (1922) 77–98.

PAGET, D. F.

1. Generalized Product Integration, Ph.D. Thesis, Univ. Tasmania, Hobart, 1976.
2. A quadrature rule for finite-part integrals, *BIT* **21** (1981) 212–220.
3. The numerical evaluation of Hadamard finite-part integrals, *Numer. Math.* **36** (1981) 447–453.

PAGET, D. F., and ELLIOT, D.

1. An algorithm for the numerical evaluation of certain Cauchy principal value integrals, *Numer. Math.* **19** (1972) 373–385.

PANASOVIC, V. A.

1. Estimate of the remainder of cubature formulae of L. V. Kantorovic, *Visnik Kiev Univ. Ser. Mat.-Meh.* **10** (1968) 72–79.

PANTIS, G.

1. The evaluation of integrals with oscillatory integrands, *J. Comp. Phys.* **17** (1975) 229–233.

PAPOULIS, A.

1. A new method of inversion of the Laplace transform, *Quart. Appl. Math.* **14** (1956–1957) 405–414.

PATTERSON, T. N. L.

1. The optimum addition of points to quadrature formulae, *Math. Comp.* **22** (1968) 847–856. Errata, *Math. Comp.* **23** (1969) 892.
2. On some Gauss and Lobatto based integration formulae, *Math. Comp.* **22** (1968) 877–881.
3. Integration formulae involving derivatives, *Math. Comp.* **23** (1969) 411–412.
4. On high precision methods for the evaluation of Fourier integrals with finite and infinite limits, *Numer. Math.* **24** (1976) 41–52.

PAULIK, A.

1. On the optimal approximation of bounded linear functionals in Hilbert spaces with inner product invariant in rotation or translation, *J. Comput. Appl. Math.* **2** (1976) 267–272.
2. Zur Existenz optimaler Quadraturformeln mit freien Knoten bei Integration analytischer Funktionen, *Numer. Math.* **27** (1977) 177–188.
3. On the determination of the weights in multi-dimensional numerical quadrature, *Computing* **21** (1978) 71–79.
4. Zur Existenz optimaler Quadraturformeln mit freien Knoten bei Integration analytischer Funktionen, *Numer. Math.* **27** (1977) 395–405.

PEANO, G.

1. Resto nelle formule di quadratura espresso con un integrale definito, *Atti Accad. Naz. Lincei. Rend.* **22** (1913) 562–569.
2. Residuo in formulas de quadratura, *Mathesis* **39** (1914) 5–10.

PECK, L. G.

1. On uniform distribution of algebraic numbers, *Proc. Amer. Math. Soc.* **4** (1953) 440–443.

PEIRCE, W. H.

1. Numerical integration over the planar annulus, *SIAM J. Appl. Math.* **5** (1957) 66–73.
2. Numerical integration over the spherical shell, *MTAC* **11** (1957) 244–249.

PENNINGTON, R. H.

1. *Introductory Computer Methods and Numerical Analysis.* Macmillan, New York, 1965.

PEVNYI, A. B.

1. On the best quadrature formula of Markov type in a class of functions, *U.S.S.R. Comput. Math. Math. Phys.* **22** (3) (1982) 65–72.

PHILLIPS, E. G.

1. *Functions of a Complex Variable, with Applications,* 7th ed. Oliver & Boyd, Edinburgh, 1951.

PHILLIPS, G. M.

1. Numerical integration in two and three dimensions, *Comput. J.* **10** (1967–1968) 202–204.
2. Numerical integration over an N-dimensional rectangular region, *Comput. J.* **10** (1967–1968) 297–299.
3. Error estimates for certain integration rules on the triangle, in *Conference on Applications of Numerical Analysis* (Morris, J. L., ed.), Lecture Notes in Math. No. 228, pp. 321–326. Springer-Verlag, Berlin and New York, 1971.
4. Seventh degree integration rules for the cube, *BIT* **19** (1979) 98–103.
5. Seventh degree integration rules for the sphere, *BIT* **20** (1980) 117–119.
6. Seventh degree integration rules for R^3, *BIT* **21** (1981) 126–128.
7. A survey of one-dimensional and multidimensional numerical integration, *Comp. Phys. Comm.* **20** (1980) 17–27.

PICONE, M.

1. Vedute generali sull'interpolazione e qualche loro conseguenza, *Ann. Scuola Norm. Sup. Pisa* [3] **5** (1951) 193–244.
2. Un metodo di Signorini per il calcolo numerico di integrali unidimensionali esteso a quello di integrali pluridimensionali, *Ist. Lombardo Accad. Sci. Lett., Rend. A* **100** (1966) 427–438.
3. Sul calcolo numerico di integrali pluridimensionali per decomposizione in prodotto dell'integrando, *Rev. Roumaine Math. Pures Appl.* **12** (1967) 1349–1363.
4. Sulla relazione fondamentale del calcolo numerico di un integrale pluridimensionale per decomposizione in prodotto dell'integrando, *Atti Acad. Naz. Lincei, Rend., Cl. Sci. Fis. Mat. Natur.* [8] **43** (1967) 3–8.

PIESSENS, R.

1. New quadrature formulas for the numerical inversion of the Laplace transform, *BIT* **9** (1969) 351–361.
2. Gaussian quadrature formulas for the numerical integration of Bromwich's integral and the inversion of the Laplace transform. Rep. TW1, Appl. Math. Sect., Univ. of Leuven, 1969.
3. Numerical inversion of the Laplace transform, *IEEE Trans. Automatic Control* **AC-14**, (1969) 299–301.
4. Gaussian quadrature formulas for the integration of oscillating functions, *ZAMM* **50** (1970) 698–700.
5. Numerical evaluation of Cauchy principal values of integrals, *BIT* **10** (1970) 476–480.
6. Gaussian quadrature formulas for the numerical integration of Bromwich's integral and the inversion of the Laplace transform, *J. Engrg. Math.* **5** (1971) 1–9.
7. On a numerical method for the calculation of transient responses, *J. Franklin Inst.* **292** (1971) 57–64.
8. Some aspects of Gaussian quadrature formulae for the numerical inversion of the Laplace transform, *Comput. J.* **14** (1971) 433–436.
9. Gaussian quadrature formulas for the evaluation of Fourier-cosine coefficients, *ZAMM* **52** (1972) 56–58.

10. An algorithm for automatic integration, *Angew. Informatik* **15** (1973) 399–401.
11. Calculation of Fourier coefficients of a function given at a set of arbitrary points, *Electron. Lett.* **7** (1971) 681–682.
12. Improved method for calculation of Fourier coefficients of a function given at a set of arbitrary points, *Electron. Lett.* **8** (1972) 250–251.
13. A quadrature routine with round-off error guard, Rep. TW 17. App. Maths. and Prog. Div., Kath. Univ. Leuven, 1974.
14. An automatic routine for the integration of oscillating functions, Rep. TW 30. App. Maths. and Prog. Div., Kath. Univ. Leuven, 1975.
15. Automatic computation of Bessel function integrals, *Comp. Phys. Comm.* **25** (1982) 289–295.

PIESSENS, R., and BRANDERS, M.
1. Numerical inversion of the Laplace transform using generalised Laguerre polynomials, *Proc. IEE* **118** (1971) 1517–1522.
2. The evaluation and application of some modified moments, *BIT* **13** (1973) 443–450.
3. A note on the optimal addition of abcissas to quadrature formulas of Gauss and Lobatto type, *Math. Comp.* **28** (1974) 135–140, 344–347.
4. Algorithm 002. Computations of oscillating integrals, *J. Comput. Appl. Math.* **1** (1975) 153–164.
5. *Tables of Gaussian Quadrature Formulas*. Academic, Leuven, 1975.
6. Numerical solution of integral equations of mathematical physics, using Chebyshev polynomials, *J. Comp. Phys.* **21** (1976) 178–196.
7. Modified Clenshaw–Curtis method for the computation of Bessel function integrals, *BIT* **23** (1983) 370–381.

PIESSENS, R., and DE DONCKER, E.
1. Testing a new one-dimensional quadrature routine (preliminary results), Report TW 24. App. Maths. and Prog. Div., Kath. Univ. Leuven, 1975.
2. Testing and comparison of quadrature routines, Tech. Rep. App. Maths. and Prog. Div., Kath. Univ. Leuven, 1976.

PIESSENS, R., DE DONCKER, E., ÜBERHUBER, C., and KAHANER, D.
1. *QUADPACK, A Quadrature Subroutine Package*, Series in Computational Math. 1. Springer-Verlag, Berlin, 1983.

PIESSENS, R., and HAEGEMANS, A.
1. Numerical calculation of Fourier-transform integrals, *Electron. Lett.* **9** (1973) 108–109.
2. Algorithm 22. Algorithm for the automatic integration of highly oscillatory functions, *Computing* **13** (1974) 183–193.
3. Cubature formulas of degree nine for symmetric planar regions, *Math. Comp.* **29** (1975) 810–815.
4. Cubature formulas of degree eleven for symmetric planar regions, *J. Comput. Appl. Math.* **1** (1975) 79–83.

PIESSENS, R., MERTENS, I., and BRANDERS, M.
1. Automatic integration of functions having end point singularities, *Angew. Informatik* **16** (1974) 65–68.

PIESSENS, R., and POLEUNIS, F.
1. A numerical method for the integration of oscillatory functions, *BIT* **11** (1971) 317–327.

PIESSENS, R., VAN ROY-BRANDERS, M., and MERTENS, I.
1. The automatic evaluation of Cauchy principal value integrals, *Angew. Informatik* **18** (1976) 31–35.

PINA H. L. G., FERNANDES, J. L. M., and BREBBIA, C. A.
1. Some numerical integration formulae over triangles and squares with a $1/r$ singularity, *Appl. Math. Modelling* **5** (1981) 209–211.

PÓLYA, G.
1. Über die Konvergenz von Quadraturverfahren, *Math. Z.* **37** (1933) 264–286.

PÓLYA, G., and SZEGÖ, G.
1. *Aufgaben und Lehrsätze aus der Analysis*, 2 Vols. Dover, New York, 1945.

PONOMARENKO, A. K.
1. On certain quadrature formulae, *U.S.S.R. Computational Math. and Math. Phys.* **6**, No. 4 (1966) 210–216.
2. Cubature formulae for some space integrals, *U.S.S.R. Computational Math. and Math. Phys.* **8** No. 6 (1968) 208–213.

POPOVICIU, T.
1. Sur une généralisation de la formule d'intégration numérique de Gauss, *Acad. R. P. Romîne Fil. Iaşi. Stud. Cerc. Şti.* **6** (1955) 29–57.
2. La simplicité du reste dans certaines formules de quadrature, *Mathematica* (*Cluj*) **6** (29) (1964), 1157–1184.

POWELL, M. J. D.
1. On best L_2 spline approximations, in *Numerische Mathematik, Differentialgeichungen, Approximationstheorie*, pp. 317–339. ISNM, Birkhäuser, Basel, 1968.

POWELL, M. J. D., and SWANN, J.
1. Weighted uniform sampling—A Monte Carlo technique for reducing variance, *JIMA* **2** (1966) 228–236.

PRAGER, W.
1. *Introduction to Basic FORTRAN Programming and Numerical Methods.* Ginn (Blaisdell), Boston, Massachusetts, 1965.

PRICE, J. F.
1. Discussion of quadrature formulas for use on digital computers, Rep. D1-82-0052. Boeing Sci. Res. Labs., 1960.
2. Examples and notes on multiple integration, Rep. D1-82-0231. Boeing Sci. Res. Labs., 1963.

PRICE, JR., T. E.
1. Cubature error bounds for a class of analytic functions, *SIAM J. Numer. Anal.* **13** (1976) 227–235.
2. Orthogonal polynomials for nonclassical weight functions, *SIAM J. Numer. Anal.* **16** (1979) 999–1006.

PŘIKRYL, P.
1. On computation of several Fourier coefficients, *ZAMM* **48** (1968) T97–98.

PROTHERO, A., and ROBINSON, A.
1. On the stability and accuracy of one-step methods for solving stiff systems of ordinary differential equations, *Math. Comp.* **28** (1974) 145–162.

PRUESS, S.
1. The approximation of linear functionals and h^2-extrapolation, *SIAM Rev.* **17** (1975) 641–651.

PUKK, R. A.

1. Investigation of algorithms for optimizing the number of nodes of quadrature formulae for a given accuracy of quadrature, *U.S.S.R. Computational Math. and Math. Phys.* **5**, No. 2 (1965) 21–39.

PYKHTEEV, G. N.

1. On the evaluation of certain singular integrals with a kernel of Cauchy type, *J. Appl. Math. Mech.* (PMM) **23** (1959) 1536–1548.

RABINOWITZ, P.

1. Abscissas and weights for Lobatto quadrature of high order, *Math. Comp.* **14** (1960) 47–52.
2. Automatic integration of a function with a parameter, *CACM* **9** (1966) 804–806.
3. Gaussian integration in the presence of a singularity, *SIAM J. Numer. Anal.* **4** (1967) 191–201.
4. Applications of linear programming to numerical analysis, *SIAM Rev.* **10**, No. 2 (1968) 121–159.
5. Error bounds in Gaussian integration of functions of low-order continuity, *Math. Comp.* **22** (1968) 431–434.
6. Practical error coefficients for estimating quadrature errors for analytic functions, *CACM* **11** (1968) 45–46.
7. Practical error coefficients in the integration of periodic analytic functions by the trapezoidal rule, *CACM* **11** (1968) 764–765.
8. Rough and ready error estimates in Gaussian integration of analytic functions, *CACM* **12** (1969) 268–270.
9. Ignoring the singularity in numerical integration, in *Topics in Numerical Analysis III* (Miller, J. J. H., ed.), pp. 361–368. Academic Press, London, 1977.
10. The numerical evaluation of Cauchy principal value integrals, *Symposium on Numerical Mathematics*, Durban, 1978, pp. 54–82.
11. On avoiding the singularity in the numerical integration of proper integrals, *BIT* **19** (1979) 104–110.
12. The exact degree of precision of generalized Gauss Kronrod integration rules, *Math. Comp.* **35** (1980) 1275–1283.
13. Gauss–Kronrod integration rules for Cauchy principal value integrals, *Math. Comp.* **38** (1983) 63–78.
14. Generalized composite integration rules in the presence of a singularity, *Calcolo*.

RABINOWITZ, P., and RICHTER, N.

1. Perfectly symmetric two-dimensional integration formulas with minimal numbers of points, *Math. Comp.* **23** (1969) 765–779.
2. New error coefficients for estimating quadrature errors for analytic functions, *Math. Comp.* **24** (1970) 561–570.
3. Asymptotic properties of minimal integration rules, *Math. Comp.* **24** (1970) 593–609.
4. Chebyshev-type integration rules of minimum norm, *Math. Comp.* **24** (1970) 831–845.
5. New integration rules for functions with singularities on or near the imaginary axis, to be published.

RABINOWITZ, P., and SLOAN, I. H.

1. Product integration in the presence of a singularity, *SIAM J. Numer. Anal.* **21** (1984) 149–166.

RABINOWITZ, P., and WEISS, G.

1. Tables of abscissas and weights for numerical evaluation of integrals of the form $\int_0^\infty e^{-x}x^n f(x)\,dx$, *MTAC* **13** (1959) 285–294.

RABINOWITZ, P., KAUTSKY, J., and ELHAY, S.

1. Some conjectures concerning imbedded sequences of positive interpolatory integration rules (submitted for publication).

RADON, J.

1. Restausdrücke bei Interpolations- und Quadraturformeln durch bestimmte Integrale, *Monatsh. Math. Phys.* **42** (1935) 389–396.
2. Zur mechanischen Kubatur, *Monatsh. Math.* **52** (1948) 286–300.

RALL, L. B.

1. Numerical integration and the solution of integral equations by the use of Riemann sums, *SIAM Rev.* **7** (1965) 55–64.
2. *Computational Solution of Nonlinear Operator Equations.* Wiley, New York, 1969.
3. Applications of software for automatic differentiation in numerical computation, *Computing, Supplement* **2** (1980) 141–156.
4. Optimization of interval computation, in *Interval Mathematics 1980* (Nickel, K., ed.), pp. 489–498. Academic Press, New York, 1980.
5. *Automatic Differentiation: Techniques and Applications,* Lecture Notes in Computer Science 140. Springer-Verlag, Berlin, 1981.

RALSTON, A.

1. A family of quadrature formulas which achieve high accuracy in composite rules, *JACM* **6** (1959) 384–394.
2. Methods for numerical quadrature, in Ralston and Wilf [1, Vol. I, pp. 242–248].
3. *A First Course in Numerical Analysis.* McGraw-Hill, New York, 1965.

RALSTON, A., and WILF, H. S. (eds.)

1. *Mathematical Methods for Digital Computers.* Wiley, New York, Vol. I, 1960, Vol. II, 1967.

RAY, V. A., and MILLER, G. F.

1. Numerical evaluation of the downwash integral for a lifting rectangular platform, Rep. Maths 90, Nat. Phys. Lab., 1970.

REBOLIA, L.

1. Formule di quadratura di tipo gaussiano con valori delle derivate dell'integrando, *Calcolo* **3** (1966) 351–369.

REDDY, C. T.

1. Improved three point integration schemes for triangular finite elements, *IJNME* **12** (1978) 1890–1898.

REDDY, C. T., and SHIPPY, D. J.

1. Alternative integration formulae for triangular finite elements, *IJNME* **17** (1981) 133–139.

RICE, J. R.

1. On the conditioning of polynomial and rational forms, *Numer. Math.* **7** (1965) 426–435.
2. (Ed.), *Mathematical Software.* Academic Press, New York, 1971.
3. An educational adaptive quadrature algorithm, *SIGNUM Newsletter* **8**, No. 2 (1973) 27–41.
4. Adaptive quadrature: Convergence of parallel and sequential algorithms, *Bull. Amer. Math. Soc.* **80** (1974) 1250–1254.
5. Parallel algorithms for adaptive quadrature–convergence, in *Information Processing 74* (Rosenfeld, J. L., ed.), pp. 600–604. North Holland, London, 1974.
6. A metalgorithm for adaptive quadrature, *JACM* **22** (1975) 61–82.

7. Parallel algorithms for adaptive quadrature II. Metalgorithm correctness, *Acta Informatica* **5** (1975) 273–285.

8. Parallel algorithms for adaptive quadrature III. Program correctness, *TOMS* **2** (1976) 1–30.

9. *Numerical Methods, Software, and Analysis: IMSL Reference Edition*, McGraw-Hill, New York, 1983.

RICE, S. O.

1. Efficient evaluation of integrals of analytic functions by the trapezoidal rule, *Bell System Tech. J.* **52** (1973) 707–722.

2. Numerical evaluation of integrals with infinite limits and oscillating integrands, *Bell System Tech. J.* **54** (1975) 155–164.

RICHTER, N.

1. Properties of minimal integration rules, *SIAM J. Numer. Anal.* **7** (1970) 67–79.

2. Minimal interpolation and approximation in Hilbert spaces, *SIAM J. Numer. Anal.* **8** (1971) 583–597.

3. Properties of minimal integration rules, II, *SIAM J. Numer. Anal.* **8** (1971) 497–508.

RICHTMYER, R. D.

1. On the evaluation of definite integrals and a quasi-Monte Carlo method based on properties of algebraic numbers, Rep. LA-1342. Los Alamos Sci. Lab., Los Alamos, New Mexico, 1952.

2. A non-random sampling method, based on congruences for Monte Carlo problems, Rep. NYO-8674. Inst. of Math. Sci., New York Univ., New York, 1958.

RICHTMYER, R. D., DEVANEY, M., and METROPOLIS, N.

1. Continued fraction expansions of algebraic numbers, *Numer. Math.* **4** (1962) 68–84.

RIESS, R. D.

1. A note on error bounds for Gauss–Chebyshev quadrature, *SIAM J. Numer. Anal.* **8** (1971) 509–511.

RIESS, R. D., and JOHNSON, L. W.

1. Estimating Gauss–Chebyshev quadrature errors, *SIAM J. Numer. Anal.* **6** (1969) 557–559.

2. Error estimates for Clenshaw–Curtis quadrature, *Numer. Math.* **18** (1972) 345–353.

ROBINSON, I. G.

1. On numerical integration, *Proc. Austral. Comput. Conf., 4th, Adelaide, South Australia, 1969*, pp. 435–442.

2. Adaptive Gaussian integration, *Austral. Comput. J.* **3** (1971) 126–129.

3. Methods of Numerical Integration, Ph.D. Thesis, Tech. Rep. 1. Dept. of Infor. Sci., Univ. of Melbourne, 1973.

4. Adaptive multidimensional integration, Res. Rep. 32. School of Math. Sciences, Univ. of Melbourne, 1974.

5. On the exit criteria for the Romberg integration scheme—a pragmatic approach, *Austral. Comp. J.* **7** (1975) 113–115.

6. An algorithm for automatic integration using the adaptive Gaussian technique, *Austral. Comp. J.* **8** (1976) 106–115.

7. A comparison of numerical integration programs, *J. Comput. Appl. Math.* **5** (1979) 207–223.

8. A comparison of two automatic quadrature routines (QAGS and QSUBL), Rep. TW 43. App. Maths. and Prog. Div., Kath. Univ. Leuven, 1979.

ROBINSON, I., and DE DONCKER, E.
1. Algorithm 45. Automatic computation of improper integrals over a bounded or unbounded planar region, *Computing* **25** (1981) 253–284.

ROBINSON, S. M., and STROUD, A. H.
1. The approximate solution of an integral equation using high-order Gaussian quadrature formulas, *Math. Comp.* **15** (1961) 286–288.

RODIN, E. Y.
1. A new method of integration over polynomial finite element boundaries, *IJNME* **10** (1976) 1115–1124.

ROGHI, G.
1. Sul resto delle formule di quadratura di tipo gaussiano, *Matematiche* (*Catania*) **22** (1967) 143–159.

ROGOSINSKI, W. W.
1. On non-negative polynomials, *Ann. Univ. Sci. Budapest. Eötvös Sect. Math.* **3–4** (1961) 253–280.
2. Non-negative linear functionals, moment problems, and extremum problems in polynomial spaces, in *Studies in Mathematical Analysis and Related Topics* (Szegö, G., ed.), pp. 316–324. Stanford Univ. Press, Stanford, California, 1962.

ROMBERG, W.
1. Vereinfachte numerische Integration, *Norske Vid. Selsk. Forh.* (*Trondheim*) **28** (1955) 30–36.

ROOS, P., and ARNOLD, L.
1. Numerische Experimente zur mehrdimensionalen Quadratur, *Österreich. Akad. Wiss. Math.-Natur. Kl. S.-B. II* **172** (1963) 271–286.

ROOSE, D., and DE DONCKER, E.
1. Algorithm 022. Automatic integration over a sphere, *J. Comput. Appl. Math.* **7** (1981) 203–224.

ROOTHART, C. J., and FIOLET, H.
1. Quadrature procedures, Report MR 137/72. Mathematisch Centrum, Amsterdam, 1972.

ROSENBERG, L.
1. Bernstein polynomials and Monte Carlo integration, *SIAM J. Numer. Anal.* **4** (1967) 566–574.

ROSENFELD, A.
1. *Picture Processing by Computer*. Academic Press, New York, 1969.

ROSSER, J. B.
1. Transformations to speed the convergence of series, *J. Res. Nat. Bur. Standards* **46** (1951) 56–64.
2. A Runge–Kutta for all seasons, *SIAM Rev.* **9** (1967) 417–452.

ROTH, K. F.
1. On irregularities of distribution, *Mathematika* **1** (1954) 73–79.
2. *Rational Approximations to Irrational Numbers*. University College, London, 1962.

ROTHMANN, H. A.
1. Gaussian quadrature with weight function x^n on the interval $(-1, 1)$, *Math. Comp.* **15** (1961) 163–168.

ROWLAND, J. H., and VAROL, Y. L.
1. Exit criteria for Simpson's compound rule, *Math. Comp.* **26** (1972) 699–704.

RUBBERT, F. K.
1. Zur Praxis der numerischen Quadratur, *ZAMM* **29** (1949) 186–188.

RUDD, W. G., SALSBURG, Z. W., and MASINTER, L. M.
1. Methods of evaluating n-dimensional integrals with polytope bounds, *J. Comp. Phys.* **5** (1970) 125–138.

RUNGE, C.
1. Über die Zerlegung empirisch gegebener periodischer Funktionen in Sinuswellen. *Z. Math. Phys.* **48** (1903) 443–456.

RUNGE, C., and WILLERS, F. A.
1. Numerische und graphische Quadratur und Integration gewöhnlicher und partieller Differentialgleichungen, *Encyklopädie der Mathematischen Wissenschaften*, Vol. 2, Sect. 3, pp. 47–176. Teubner, Leipzig, 1909–1921.

RUSSELL, D. L.
1. Numerical integration by midpoint procedure with preferential interval placement, SHARE File No. 0704-1017 AND107.

RUTISHAUSER, H.
1. On a modification of the QD-algorithm with Graeffe-type convergence, *ZAMP* **13** (1962) 493–496.
2. Ausdehnung des Rombergschen Prinzips, *Numer. Math.* **5** (1963) 48–54.
3. *Description of ALGOL 60*. Springer-Verlag, Berlin and New York, 1967.

RYVKIN, V. B.
1. On the existence of quadrature formulas of Gauss type which are exact for functions of a Tchebyscheff–Haar system, *Dokl. Akad. Nauk BSSR* **10** (1966) 824–826.

SACK, R. A.
1. Newton–Cotes type quadrature formulas with terminal corrections, *Comput. J.* **5** (1962–1963) 230–237.

SACK, R. A., and DONOVAN, A. F.
1. An algorithm for Gaussian quadrature given modified moments, *Numer. Math.* **18** (1972) 465–478.

SAENGER, A.
1. On numerical quadrature of Fourier transforms, *J. Math. Anal. Appl.* **8** (1964) 1–3.

SAG, T. W., and SZEKERES, G.
1. Numerical evaluation of high-dimensional integrals, *Math. Comp.* **18** (1964) 245–253.

SALIHOV, G. N.
1. Cubature formulas for a hypersphere that are invariant with respect to the group of the regular 6000-face, *Soviet Math. Dokl.* **16** (1975) 1046–1050.
2. An error estimate for cubature formulas in the space $L_2^{(n)}(S)$, *Soviet Math. Dokl.* **16** (1975) 1098–1101.

SALTYKOV, A. I.
1. Tables for the computation of multiple integrals using the method of optimal coefficients, *U.S.S.R. Computational Math. and Math. Phys.* **3** (1963) 235–242.

SALZER, H. E.
1. A simple method for summing certain slowly convergent series, *J. Math. and Phys.* **33** (1954) 356–359.
2. Osculatory quadrature formulas, *J. Math. and Phys.* **34** (1955) 103–112.
3. Equally weighted quadrature formulas over semi-infinite and infinite intervals, *J. Math. and Phys.* **34** (1955) 54–63.
4. Orthogonal polynomials arising in the numerical evaluation of inverse Laplace transforms, *MTAC* **9** (1955) 164–177.
5. Formulas for the partial summation of series, *MTAC* **10** (1956) 149–156.
6. Tables for the numerical calculation of inverse Laplace transforms, *J. Math. and Phys.* **37** (1958) 89–109.
7. Additional formulas and tables for orthogonal polynomials originating from inversion integrals, *J. Math. and Phys.* **40** (1961) 72–86.
8. An interpolation formula for "nearly-odd" functions, with an application to the summation of even functions, *SIAM J. Appl. Math.* **9** (1961) 282–287.
9. Chebyshev interpolation and quadrature formulas of very high degree, *CACM* **12** (1969), 271; Errata, *CACM* **12** (1969) 385.
10. Some new divided difference algorithms for two variables, in Langer [1, pp. 61–98].

SALZER, H. E., and KIMBRO, G. M.
1. Improved formulas for complete and partial summation of certain series, *Math. Comp.* **15** (1961) 23–39.

SALZER, H. E., and LEVINE, N.
1. Table of a Weierstrass continuous non-differentiable function, *Math. Comp.* **15** (1961) 120–130.

SALZER, H. E., SHOULTZ, D. C., and THOMPSON, E. D.
1. *Tables of Osculatory Integration Coefficients.* Convair, San Diego, 1960.

SARD, A.
1. Best approximate integration formulas; best approximation formulas, *Amer. J. Math.* **71** (1949) 80–91.
2. *Linear approximation*, Math. Surveys No. 9. Amer. Math. Soc., Providence, Rhode Island, 1963.

SARMA, V. L. N.
1. Eberlein measure and mechanical quadrature formulae. I: Basic theory, *Math. Comp.* **22** (1968) 607–616.

SARMA, V. L. N., and STROUD, A. H.
1. Eberlein measure and mechanical quadrature formulae. II: Numerical results, *Math. Comp.* **23** (1969) 781–784.

SASAKI, T.
1. Multidimensional Monte Carlo integration based on factorized approximation functions, *SIAM J. Numer. Anal.* **15** (1978) 938–952.

SCHEMPP, W., and ZELLER, K. (Eds.)
1. *Multivariate Approximation Theory*, ISNM 51. Birkhäuser Verlag, Basel, 1979.

SCHERER, R.

1. A note on Radau and Lobatto formulae for O.D.E.'s, *BIT* **17** (1977) 235–243.

SCHIFF, B., LIFSON, H., PEKERIS, C. L., and RABINOWITZ, P.

1. $2^{1,3}p$, $3^{1,3}p$ and $4^{1,3}p$ states of He and the 2^1p state of Li^+, *Phys. Rev.* **140** (1965) A1104–1121.

SCHMEISSER, G.

1. Optimale Quadraturformeln mit semidefiniten Kernen, *Numer. Math.* **20** (1972) 32–53.

SCHMID, H. J.

1. On cubature formulae with a minimal number of knots, *Numer. Math.* **31** (1978) 281–297.
2. On Gaussian cubature formulae of degree $2k - 1$, in Hämmerlin [5, pp. 252–263].
3. Construction of cubature formulae using real ideals, in Schempp and Zeller [1, pp. 359–377].
4. Interpolatorische Kubaturformeln und reelle Ideale, *Math. Z.* **170** (1980) 267–282.
5. Interpolatory cubature formulae and real ideals, in De Vore and Scherer [1, pp. 245–254].
6. On the numerical solution of non-linear equations characterizing minimal cubature formulae, *Computing* **24** (1980) 251–257.

SCHMIDT, R.

1. Die allgemeine Newtonsche Quadraturformel und Quadraturformeln für Stieltjesintegrale, *J. Reine Angew. Math.* **173** (1935) 52–59.

SCHMITTROTH, L. A.

1. Numerical inversion of Laplace transforms, *CACM* **3** (1960) 171–173.

SCHNEIDER, C.

1. Vereinfachte Rekursionen zur Richardson-Extrapolation in Spezialfällen, *Numer. Math.* **24** (1975) 177–184.
2. Produktintegration mit nicht-äquidistanten Stützstellen, *Numer. Math.* **35** (1980) 35–43.

SCHOENBERG, I. J.

1. Monosplines and quadrature formulae, in *Theory and Applications of Spline Functions* (Greville, T. N. E., ed.), pp. 157–207. Academic Press, New York, 1969.
2. On monosplines of least square deviation and best quadrature formulae, *SIAM J. Numer. Anal.* **2** (1965) 144–170; **3** (1966) 321–328.
3. (Editor), *Approximations with Special Emphasis on Spline Functions.* Academic Press, New York, 1969.
4. A second look at approximate quadrature formulae and spline interpolation, *Advances in Math.* **4** (1970) 277–300.
5. On polynomial spline functions on the circle. II. Monosplines and quadrature formulae, Tech. Summary Rep. 1002. MRC, Madison, Wisconsin, 1971.
6. Remarks concerning a numerical inversion of the Laplace transform due to Bellman, Kalaba and Lockett, *J. Math. Anal. Appl.* **43** (1973) 823–828.

SCHOENBERG, I. J., and SHARMA, A.

1. The interpolatory background of the Euler–Maclaurin quadrature formula, *Bull. Amer. Math. Soc.* **77** (1971) 1034–1038.

SCHOENBERG, I. J., and SILLIMAN, S. D.

1. On semicardinal quadrature formulae, *Math. Comp.* **28** (1974) 483–497.

SCHÖNHAGE, A.

1. Mehrdimensionale Romberg-Integration, *Numer. Math.* **14** (1970) 299–304.
2. Zur Quadratur holomorpher periodischer Funktionen, *J. Approx. Theory* **13** (1975) 341–347.

SCHRADER, H.-J.
1. Zur numerischen Berechnung optimaler Quadraturformeln, NAM-Bericht 17. Univ. Göttingen, 1977.

SCHRAGE, L.
1. A more portable Fortran random number generator, *TOMS* **5** (1979) 132–138.

SCHUMAKER, L. L.
1. Some algorithms for the computation of interpolating and approximating spline functions, in *Theory and Applications of Spline Functions* (Greville, T. N. E., ed.), pp. 87–102. Academic Press, New York, 1969.
2. Approximation by splines, in *Theory and Applications of Spline Functions* (Greville, T. N. E., ed.), pp. 65–85. Academic Press, New York, 1969.
3. *Spline Functions: Basic Theory*. Wiley, New York, 1981.

SCHWARTZ, C.
1. Variational principles for integrals, *J. Comp. Phys.* **3** (1968–1969) 512–522.
2. Numerical integration of analytic functions, *J. Comp. Phys.* **4** (1969) 19–29.

SCHWARTZ, L.
1. *Théorie des distributions*, New edition. Hermann, Paris, 1966.

SCHWEIKERT, D. G.
1. Numerical parametric integration of continuous functions with a singular derivative, *Proc. ACM Nat. Conf., 23rd, 1968*, pp. 475–480.
2. A comparison of error improvement estimates for adaptive trapezoid integration, *CACM* **13** (1970) 163–166.

SECREST, D. H.
1. Numerical integration of arbitrarily spaced data and estimation of errors, *SIAM J. Numer. Anal.* **2** (1965) 52–68.
2. Best approximate integration formulas and best error bounds, *Math. Comp.* **19** (1965) 79–83.

SEGETH, K.
1. On quadrature formulae involving values of derivatives, *ZAMM* **48** (1968) T104–105.

SEIDMAN, T. I., and KORSAN, R. J.
1. Endpoint formulas for interpolatory cubic splines, *Math. Comp.* **26** (1972) 897–900.

SERBIN, H.
1. Numerical quadrature of some improper integrals, *Quart. Appl. Math.* **12** (1954–1955) 188–194.

SHAMPINE, L. F., and ALLEN, JR., R. C.
1. *Numerical Computing: An Introduction*. Saunders, Philadelphia, 1973.

SHAMPINE, L. F., and GORDON, M. K.
1. *Computer Solution of Ordinary Differential Equations: The Initial Value Problem*. Freeman, San Francisco, 1975.

SHANKS, D.
1. Non-linear transformations of divergent and slowly convergent sequences, *J. Math. and Phys.* **34** (1955) 1–42.

SHANKS, J. A.
1. Romberg tables for singular integrands, *Comput. J.* **15** (1972) 360–361.

SHAO, T. S., CHEN, T. C., and FRANK, R. M.
1. Tables of zeros and Gaussian weights of certain associated Laguerre polynomials and the related generalized Hermite polynomials, Tech. Rep. 00.1100. IBM Data Systems Div., Poughkeepsie, New York, 1964.
2. Tables of zeros and Gaussian weights of certain associated Laguerre polynomials and the related generalized Hermite polynomials, *Math. Comp.* **18** (1964) 598–616.

SHARIPKHODZHAEVA, F.
1. Cubature formulas of the 9-th and 10-degrees of accuracy for a sphere of n-dimensional space, *U.S.S.R. Comp. Math. Math. Phys.* **19** (1979) 248–251.

SHARYGIN, I. F.
1. The use of number theoretic methods of integration in the case of non-periodic functions, *Soviet Math. Dokl.* **1** (1960) 506–509.
2. A lower estimate for the error of quadrature formula for certain classes of functions, *U.S.S.R. Comp. Math. Phys.* **3**(2) (1963) 489–497.

SHAVITT, I.
1. The Gaussian function in calculations of statistical mechanics and quantum mechanics, in *Methods in Computational Phys.* **2** (1963) 1–45.

SHIRTLIFFE, C. J., and STEPHENSON, D. G.
1. A computer oriented adaption of Salzer's methods for inverting Laplace transforms. *J. Math. and Phys.* **40** (1961) 135–141.

SHIZGAL, B.
1. A Gaussian quadrature procedure for use in the solution of the Boltzman equation and related problems, *J. Comp. Phys.* **41** (1981) 309–328.

SHOHAT, J. A.
1. Sur la convergence des quadratures mécaniques dans un intervalle infini, *C. R. Acad. Sci. Paris* **186** (1928) 344–346.
2. On a certain formula of mechanical quadrature with non-equidistant ordinates, *Trans. Amer. Math. Soc.* **31** (1929) 448–463.
3. On mechanical quadratures, in particular, with positive coefficients, *Trans. Amer. Math. Soc.* **42** (1937) 461–496.

SHOHAT, J. A., and WINSTON, C.
1. On mechanical quadratures, *Rend. Circ. Mat. Palermo* **58** (1934) 153–160.

SHREIDER, Y. A. (ed.)
1. *The Monte Carlo Method.* Pergamon, Oxford, 1966.

SIDI, A.
1. Some properties of a generalization of the Richardson extrapolation process, *JIMA* **24** (1970) 327–346.
2. Numerical quadrature and nonlinear sequence transformations; Unified rules for efficient computation of integrals with algebraic and logarithmic endpoint singularities, *Math. Comp.* **35** (1980) 851–874.
3. Extrapolation methods for oscillatory infinite integrals, *JIMA* **26** (1980) 1–20.
4. Numerical quadrature rules for some infinite range integrals, *Math. Comp.* **38** (1982) 127–142.

5. The numerical evaluation of very oscillatory infinite integrals by extrapolation, *Math. Comp.* **38** (1982) 517–529.

6. An algorithm for a special case of a generalization of the Richardson extrapolation process, *Numer. Math.* **38** (1982) 299–307.

7. Euler–Maclaurin expansions for integrals over triangles and squares of functions having algebraic/logarithmic singularities along an edge, *J. Approx. Theory* **39** (1983) 39–53.

SIKORSKI, K.

1. Optimal quadrature algorithms in H_p spaces, *Numer. Math.* **39** (1982) 405–410.

SIKORSKI, K., and STENGER, F.

1. Optimal quadratures in H_p spaces. TOMS **10** (1984).

SILLIMAN, S. D.

1. The numerical evaluation by splines of the Fourier transform and the Laplace transform, Tech. Summary Rep. 1183. MRC, Madison, Wisconsin, 1972.

2. The numerical evaluation by splines of Fourier transforms, *J. Approx. Theory* **12** (1974) 32–51.

3. On complete semicardinal quadrature formulae, *Comput. Math. Appl.* **5** (1979) 41–49.

SILVERMAN, H. F.

1. Identification of linear systems using fast Fourier transform techniques, Ph.D. Thesis, Brown Univ., Providence, Rhode Island, 1970.

2. An introduction to programming the Winograd Fourier transform algorithm (WFTA), *IEEE Trans. Acoust. Speech Signal Process.* **ASSP-25** (1977) 152–165.

SILVERMAN, H. F., and PEARSON, A. E.

1. Impulse response identification from truncated input data using FFT techniques, *Proc. Internat. IEEE Conf. Systems, Networks, and Comput., 1971.*

SILVESTER, P.

1. Symmetric quadrature formulae for simplexes, *Math. Comp.* **24** (1970) 95–100.

SIMON, A. H.

1. Oscillatory integrals and contour integration. RCA Lab., Princeton, New Jersey.

SIMPSON, R. B., and YAZICI, A.

1. An organization of the extrapolation method of multi-dimensional quadrature for vector processing, Res. Rep. CS-78-37. Dept. of Computer Science, Univ. of Waterloo, Waterloo, Canada, 1978.

SINGLETON, R. S.

1. On computing the fast Fourier transform, *CACM* **10** (1967) 647–654.

SKOBLYA, N.

1. *Tables for the Numerical Inversion of the Laplace Transform* $f(x) = (2\pi i)^{-1} \int_{c-i\infty}^{c+i\infty} e^{xp}F(p)\,dp$, Izdat. "Nauka i Tehnika," Minsk, 1964.

SLAGLE, J. R.

1. A heuristic program that solves symbolic integration problems in freshman calculus. *JACM* **10** (1963) 507–520.

SLOAN, I. H.

1. The numerical evaluation of principal-value integrals, *J. Comp. Phys.* **3** (1968–1969) 332–333.

2. On the numerical evaluation of singular integrals, *BIT* **18** (1978) 91–102.
3. On choosing the points in product integration, *J. Math. Phys.* **21** (1980) 1032–1039.

SLOAN, I. H., and SMITH, W. E.
1. Product-integration with the Clenshaw–Curtis and related points. Convergence properties, *Numer. Math.* **30** (1978) 415–428.
2. Product-integration with the Clenshaw–Curtis points: Implementation and error estimates, *Numer. Math.* **34** (1980) 387–401.
3. Properties of interpolatory product integration rules, *SIAM J. Numer. Anal.* **19** (1982) 427–442.

SMIRNOV, V. I.
1. *A Course of Higher Mathematics*. Addison-Wesley, Reading, Massachusetts, 1964.

SMITH, F. J.
1. Quadrature methods based on the Euler–MacLaurin formula and on the Clenshaw–Curtis method of integration, *Numer. Math.* **7** (1965) 406–411.

SMITH, H. V.
1. A method for the integration of oscillatory functions, *BIT* **17** (1977) 338–343.
2. The evaluation of the error term in some numerical quadrature formulae, *IJNME* **14** (1979) 468–472. See also *IJNME* **15** (1980) 157–160.

SMITH, W. E., and LYNESS, J. N.
1. Applications of Hilbert transform theory to numerical quadrature, *Math. Comp.* **23** (1969) 231–252.

SMITH, W. E., and SLOAN, I. H.
1. Product-integration rules based on the zeros of Jacobi polynomials, *SIAM J. Numer. Anal.* **17** (1980) 1–13.

SNYDER, M. A.
1. *Chebyshev Methods in Numerical Approximation*. Prentice-Hall, Englewood Cliffs, New Jersey, 1966.

SOBOL, I. M.
1. On the distribution of points in a cube and the approximate evaluation of integrals, *U.S.S.R. Computational Math. and Math. Phys.* **7**, No. 4 (1967) 86–112.
2. Accurate estimate of the error of multidimensional quadrature formulas for functions of class S_p, *Soviet Math. Dokl.* **1** (1960) 726–729.
3. On the calculation of multi-dimensional integrals, *Soviet Math. Dokl.* **2** (1961) 1022–1025.
4. Calculation of improper integrals using uniformly distributed sequences, *Soviet Math. Dokl.* **14** (1973) 734–738.
5. Uniformly distributed sequences with an additional uniform property, *U.S.S.R. Comput. Math. Math. Phys.* **16**(5) (1976) 236–242.
6. On the systematic search in a hypercube, *SIAM J. Numer. Anal.* **16** (1979) 790–793.

SOBOLEV, S. L.
1. Cubature formulas on the sphere invariant under finite groups of rotations, *Soviet Math. Dokl.* **3** (1962) 1307–1310.
2. The formulas of mechanical cubature on the surface of a sphere, *Sibirsk. Mat. Ž.* **3** (1962) 769–796.
3. The number of nodes in cubature formulas on the sphere, *Soviet Math. Dokl.* **3** (1962) 1391–1394.
4. *Introduction to the Theory of Cubature Formulae* (Russian). Izdat. 'Nauka', Moscow, 1974.

5. Les coefficients optimaux des formules d'intégration approximative, *Ann. Scuola Norm. Sup. Pisa Cl. Sci.* (4) **5** (1978), 455–469; **6** (1979) 729.

SOLODOV, V. M.

1. Application of the method of optimal coefficients to numerical integration, *U.S.S.R. Computational Math. and Math. Phys.* **9**, No. 1 (1969) 14–34.

2. On the error involved in a numerical integration, *Soviet Math. Dokl.* **4** (1963) 85–88.

SOTTAS, G.

1. On the positivity of quadrature formulas with Jacobi abscissas, *Computing* **29** (1982) 83–88.

SPINELLI, R. A.

1. Numerical inversion of a Laplace transform, *SIAM J. Numer. Anal.* **3** (1966) 636–649.

SQUIRE, W.

1. Application of generalized Gauss–Laguerre quadrature to boundary-layer problems, *J. Aerosp. Sci.* **26** (1959) 540–541.

2. Some applications of quadrature by differentiation, *SIAM J. Appl. Math.* **9** (1961) 94–108.

3. Numerical evaluation of integrals using Moran transformation, Aerosp. Eng. TR-14. Univ. of West Virginia, Morgantown, 1969.

4. *Integration for Engineers and Scientists.* Amer. Elsevier, New York, 1970.

5. Comment on numerical calculation of Fourier-transform integrals, *Electron. Lett.* **9** (1973) 291.

6. Partition-extrapolation methods for numerical quadratures, *Internat. J. Comput. Math.* **5** (1975) 81–91.

7. An efficient iterative method for numerical evaluation of integrals over a semi-infinite range, *IJNME* **10** (1976) 478–484.

8. A quadrature method for finite intervals, *IJNME* **10** (1976) 708–712.

9. Numerical evaluation of a class of singular double integrals by symmetric pairing, *IJNME* **10** (1976) 703–708.

10. In defense of linear quadrature rules, *Comput. Math. Appl.* **7** (1981) 147–149.

STANCU, D. D.

1. Sur quelques formules générales de quadrature du type Gauss–Christoffel, *Mathematica (Cluj)* **1**(24) (1959) 167–182.

2. The remainder of certain linear approximation formulas in two variables, *SIAM J. Numer. Anal.* **1** (1964) 137–163.

STANCU, D. D., and STROUD, A. H.

1. Quadrature formulas with simple Gaussian nodes and multiple fixed nodes, *Math. Comp.* **17** (1963) 384–394.

STEEN, N. M., BYRNE, G. P., and GELBARD, E. M.

1. Gaussian quadratures for the integrals $\int_0^\infty e^{-x^2}f(x)\,dx$ and $\int_0^B e^{-x^2}f(x)\,dx$, *Math. Comp.* **23** (1969) 661–671.

STEFFENSEN, J. F.

1. *Interpolation.* Chelsea, New York, 1950.

STEKLOV, V. A.

1. On the approximate calculation of definite integrals with the aid of formulas of mechanical quadratures, *Izv. Akad. Nauk SSSR* [6], **10** (1916) 169–186.

2. Sopra la teoria delle quadrature dette mechaniche, *Reale Accad. Naz. Lincei. Rend. Cl. Sci. Fis. Mat. Nat.* [5] **32** (1923) 320–326.

STENGER, F.
1. Numerical integration in n dimensions, Master's Thesis, Dept. of Math., Univ. of Alberta, Edmonton, 1963.
2. Error bounds for the evaluation of integrals by repeated Gauss-type formulae, *Numer. Math.* **9** (1966) 200–213.
3. Bounds on the error of Gauss-type quadratures, *Numer. Math.* **8** (1966) 150–160.
4. Kronecker product extensions of linear operators, *SIAM J. Numer. Anal.* **5** (1968) 422–435.
5. Integration formulae based on the trapezoidal formula, *JIMA* **12** (1973) 103–114. See also *JIMA* **19** (1977) 145–147 and **21** (1978) 359–361.
6. Euler–Maclaurin type quadrature formulas for the finite and semi-infinite interval, to be published.
7. Tabulation of certain fully symmetric numerical integration formulas of degree 7, 9 and 11, *Math. Comp.* **25** (1971) 935 and Microfiche Supplement.
8. Optimal convergence of minimum norm approximations in H_p, *Numer. Math.* **29** (1978) 345–362.
9. Numerical methods based on Whittaker cardinal or sinc functions, *SIAM Rev.* **23** (1981) 165–224.

STERN, M. D.
1. Optimal quadrature formulae, *Comput. J.* **9** (1966–1967) 396–403.

STETTER, F.
1. Ableitungsfreie Schranken für die numerische Integration, *Computing* **2** (1967) 257–262.
2. Error bounds for Gauss–Chebyshev quadrature, *Math. Comp.* **22** (1968) 657–659.
3. On a generalization of the midpoint rule, *Math. Comp.* **22** (1968) 661–663.

STETTER, H. J.
1. Numerical approximation of Fourier-transforms, *Numer. Math.* **8** (1966) 235–249.

STETTER, H. J., and UEBERHUBER, C. W.
1. Proposal for population studies in numerical quadrature, *J. Comput. Appl. Math.* **1** (1975) 213–215.

STEWART, C. E.
1. On the numerical evaluation of singular integrals of Cauchy type, *SIAM J. Appl. Math.* **8** (1960) 342–353.

STIEFEL, E. L.
1. Altes und Neues über numerische Quadratur, *ZAMM* **41** (1961) 408–413.
2. *An Introduction to Numerical Mathematics.* Academic Press, New York, 1963.

STIEFEL, E. L., and RUTISHAUSER, H.
1. Remarques concernant l'intégration numérique, *C.R. Acad. Sci. Paris* **252** (1961) 1899–1900.

STOER, J.
1. Über zwei Algorithmen zur Interpolation mit rationalen Funktionen, *Numer. Math.* **3** (1961) 285–304.

STOER, J., and BULIRSCH, J.
1. *Introduction to Numerical Analysis.* Springer-Verlag, New York, 1980.

STRANG, G., and FIX, G. J.

1. *An Analysis of the Finite Element Method.* Prentice-Hall, Englewood Cliffs, N.J., 1973.

STRAUSS, C.

1. Computerized curve fairing. Div. Appl. Math., Brown Univ., Providence, Rhode Island, 1966.

STRÖM, T.

1. Strict error bounds in Romberg quadrature, *BIT* **7** (1967) 314–321.

STROUD, A. H.

1. Remarks on the disposition of points in numerical integration formulas, *MTAC* **11** (1957) 257–261.
2. Quadrature methods for functions of more than one variable. *Ann. N.Y. Acad. Sci.* **86** (1960) 776–791.
3. Numerical integration formulas of degree two, *Math. Comp.* **14** (1960) 21–26.
4. A bibliography on approximate integration, *Math. Comp.* **15** (1961) 52–80.
5. Numerical integration formulas of degree 3 for product regions and cones, *Math. Comp.* **15** (1961) 143–150.
6. Approximate integration formulas of degree 3 for simplexes, *Math. Comp.* **18** (1964) 590–597.
7. Error estimates for Romberg quadrature, *SIAM J. Numer. Anal.* **2** (1965) 480–488.
8. Estimating quadrature errors for functions with low continuity, *SIAM J. Numer. Anal.* **3** (1966) 420–424.
9. Some approximate integration formulas of degree 3 for an n-dimensional simplex, *Numer. Math.* **9** (1966) 38–45.
10. Some fifth degree integration formulas for symmetric regions, I. *Math. Comp.* **20** (1966) 90–97; II. *Numer. Math.* **9** (1967) 460–468.
11. Some seventh degree formulas for symmetric regions, *SIAM J. Numer. Anal.* **4** (1967) 37–44.
12. Approximate multiple integration, in Ralston and Wilf [1, Vol. II, pp. 145–155].
13. Numerical integration, in *Digital Computer User's Handbook* (Klerer, M., and Korn, G. A., eds.). McGraw-Hill, New York, 1967.
14. Integration formulas and orthogonal polynomials, *SIAM J. Numer. Anal.* **4** (1967) 381–389.
15. Some seventh degree integration formulas for the surface of an n-sphere, *Numer. Math.* **11** (1968) 273–276.
16. A fifth degree integration formula for the n-simplex, *SIAM J. Numer. Anal.* **6** (1969) 90–98.
17. Integration formulas and orthogonal polynomials for two variables, *SIAM J. Numer. Anal.* **6** (1969) 222–229.
18. *Approximate Calculation of Multiple Integrals.* Prentice-Hall, Englewood Cliffs, New Jersey, 1971.
19. *Numerical Quadrature and Solution of Ordinary Differential Equations.* Springer-Verlag, New York, 1974.
20. Some fourth degree integration formulas for simplexes, *Math. Comp.* **30** (1976) 291–294.

STROUD, A. H., and CHEN, K.-W.

1. Peano error estimates for Gauss–Laguerre quadrature formulas, *SIAM J. Numer. Anal.* **9** (1972) 333–340.

STROUD, A. H., and GOIT, JR., E. H.

1. Some extensions of integration formulas, *SIAM J. Numer. Anal.* **5** (1968) 243–251.

STROUD, A. H., and KOHLI, J. P.
1. Computation of $J_n(x)$ by numerical integration, *CACM* **12** (1969) 236–238.

STROUD, A. H., and SECREST, D. H.
1. Approximate integration formulas for certain spherically symmetric regions, *Math. Comp.* **17** (1963) 105–135.
2. *Gaussian Quadrature Formulas*. Prentice-Hall, Englewood Cliffs, New Jersey, 1966.

STROUD, A. H., and STANCU, D. D.
1. Quadrature formulas with multiple Gaussian nodes, *SIAM J. Numer. Anal.* **2** (1965) 129–143.

STRUBLE, G. W.
1. Orthogonal polynomials: Variable-signed weight functions, *Numer. Math.* **5** (1963) 88–94.

STUMMEL, F.
1. Rounding error analysis of elementary numerical algorithms, *Computing, Suppl.* **2** (1980) 169–195.
2. Perturbation theory for evaluation algorithms of arithmetic expressions, *Math. Comp.* **37** (1981) 435–473.

SUGIHARA, M., and MUROTA, K.
1. A note on Haselgrove's method for numerical integration, *Math. Comp.* **39** (1982) 549–554.

SYNGE, J. L.
1. *The Hypercircle in Mathematical Physics*. Cambridge Univ. Press, London and New York, 1957.

SZEGÖ, G.
1. Über gewisse orthogonale Polynome, die zu einer oszillierenden Belegungsfunktion gehören, *Math. Ann.* **110** (1935) 501–513.
2. *Orthogonal Polynomials*. Amer. Math. Soc., New York, 1959.

TAKAHASI, H., and MORI, M.
1. Error estimation in the numerical integration of analytic functions, *Rep. Comput. Centre Univ. Tokyo* **3** (1970) 41–108.
2. Estimation of errors in the numerical quadrature of analytic functions, *Applicable Anal.* **1** (1971–1972) 201–229.
3. Quadrature formulas obtained by variable transformation, *Numer. Math.* **21** (1973) 206–219.
4. Double exponential formulas for numerical integration, Publ. RIMS, Kyoto Univ. **9** (1974) 721–741.

TALBOT, A.
1. The accurate numerical inversion of Laplace transforms, *JIMA* **23** (1979) 97–120.

TASHIRO, Y.
1. On methods for generating uniform random points on the surface of a sphere. *Ann. Inst. Statist. Math.*, **29** (1977) 295–300.

TAURIAN, O. E.
1. A method and a program for the numerical evaluation of the Hilbert transform of a real function, *Comp. Phys. Comm.* **20** (1980) 291–307.

TAUSSKY, O., and TODD, J.

1. Generation and testing of pseudo-random numbers, in *Symposium on Monte Carlo Methods* (Meyer, H. A., ed.), pp. 15–28. Wiley, New York, 1956.

TCHAKALOFF, L.

1. Formules générales de quadrature mécanique du type de Gauss, *Colloq. Math.* **5** (1957) 69–73.

TCHAKALOFF, V.

1. Formules de cubature mécaniques à coefficients non negatifs, *Bull. Sci. Math.* [2] **81** (1957) 123–134.

TEMME, N. M.

1. The numerical computation of special functions by use of quadrature rules for saddle point integrals. I. Trapezoidal integration rules, Rep. TW 164/79; II. Gamma functions, modified Bessel functions and parabolic cylinder functions, Rep. TW 183/78. Mathematisch Centrum, Amsterdam, 1978–1979.

TEMPERTON, C.

1. Self-sorting mixed-radix fast Fourier transforms, *J. Comp. Phys.* **52** (1983) 1–23.

THACHER, JR., H. C.

1. Optimum quadrature formulas in s dimensions, *MTAC* **11** (1957) 189–194.
2. Certification of algorithm 32, MULTINT, *CACM* **6** (1963) 69. See also *CACM* **11** (1968) 826.
3. An efficient composite formula for multidimensional quadrature, *CACM* **7** (1964) 23–25.
4. Remark on algorithm 60, Romberg integration, *CACM* **7** (1964) 420–421.
5. An iterative method for quadratures, *Comput. J.* **5** (1962–1963) 228–229.
6. Generalization of concepts related to linear dependence, *SIAM J. Appl. Math.* **6** (1958) 288–300.
7. Derivation of interpolation formulas in several independent variables, *Ann. N.Y. Acad. Sci.* **86** (1960) 760–775.

THEOCARIS, P. S., IOAKIMIDIS, N. I., and KAZANTZAKIS, J. G.

1. On the numerical evaluation of two-dimensional principal value integrals, *IJNME* **15** (1980) 629–634.

THOMAS, K. S.

1. Improved convergence for product integration, in Hämmerlin [6, pp. 262–267].

TIKSON, M.

1. Tabulation of an integral arising in the theory of cooperative phenomena, *J. Res. Nat. Bur. Standards* **50** (1953) 177–178.

TOCHER, K. D.

1. The application of automatic computers to sampling experiments, *J. Roy. Statist. Soc. Ser. B* **16** (1954) 39–61.

TODA, H., and ONO, H.

1. Notes on effective usage of double exponential formulas for numerical integration, in Mori [2, pp. 21–47].

TODD, J.

1. Classical numerical analysis, in *Survey of Numerical Analysis* (Todd, J., ed.), pp. 27–118. McGraw-Hill, New York, 1962.
2. The problem of error in digital computation, in *Error in Digital Computation* (Rall, L. B., ed.), Vol. 1., pp. 3–41. Wiley, New York, 1965.

3. Evaluation of the exponential integral for large complex arguments, *J. Res. Nat. Bur. Standards* **52** (1954) 313–317.

TOMPA, H.
1. Gaussian numerical integration of a function depending on a parameter, *Comput. J.* **10** (1967–1968) 204–205.

TÖRN, A.
1. Crude Monte Carlo quadrature in infinite variance case and the central limit theorem, *BIT* **6** (1966) 339–346.

TRAUB, J. F., and WOŹNIAKOWSKI, H.
1. *A General Theory of Optimal Algorithms.* Academic Press, New York, 1980.

TRAUBOTH, H. H.
1. Recursive formulas for the evaluation of the convolution integral, *JACM* **16** (1969) 63–72.

TRICOMI, F. G.
1. Equazioni integrali contenenti il valor principale di un integrale doppio, *Math. Z.* **27** (1928) 87–133.
2. *Vorlesungen über Orthogonalreihen.* Springer-Verlag, Berlin and New York, 1955.

TSUDA, T.
1. Numerical integration of functions of very many variables, *Numer. Math.* **20** (1973) 377–391.

TUAN, P. D.
1. On the remainder in quadrature rules, *Math. Comp.* **25** (1971) 819–826.

TUCK, E. O.
1. A simple "Filon-trapezoidal" rule, *Math. Comp.* **21** (1967) 239–241.

TURÁN, P.
1. On the theory of the mechanical quadrature, *Acta Sci. Math. (Szeged)* **12** (1950) 30–37.

TURING, A. M.
1. A method for the calculation of the zeta-function, *Proc. London Math. Soc.* [2] **48** (1945) 180–197.

TYLER, G. W.
1. Numerical integration of functions of several variables, *Canad. J. Math.* **5** (1953) 393–412.

USPENSKY, J. V.
1. Sur les valeurs asymptotiques des coefficients de Cotes, *Bull. Amer. Math. Soc.* **31** (1925) 145–156.
2. On the convergence of quadrature formulas related to an infinite interval, *Trans. Amer. Math. Soc.* **30** (1928) 542–559.
3. On the expansion of the remainder in the Newton–Cotes formula, *Trans. Amer. Math. Soc.* **37** (1935) 381–396.

VALENTIN, R. A.
1. Applications of functional analysis to optimal numerical approximations for analytic functions, Ph.D. Thesis, Div. of Appl. Math., Brown Univ., Providence, Rhode Island, 1965.
2. The use of the hypercircle inequality in deriving a class of numerical approximation rules for analytic functions, *Math. Comp.* **22** (1968) 110–117.

VAN DE VOOREN, A. I., and VAN LINDE, H. J.

1. Numerical calculation of integrals with strongly oscillating integrand, *Math. Comp.* **20** (1966) 232–245.

VAN DER CORPUT, J. G.

1. Verteilungsfunktionen, *Nederl. Akad. Wetensch. Proc.* **38** (1935) 813–821.

VAN DER SLUIS, A.

1. The remainder term in quadrature formulae, *Numer. Math.* **19** (1972) 49–55.
2. Asymptotic expansions for quadrature errors over a simplex, in Hämmerlin [6, pp. 222–240].

VAN DER SLUIS, A., and ZWEERUS, J. R.

1. An appraisal of some methods for computing Cauchy principal values of integrals, in Hämmerlin [5, pp. 264–287].

VAN DOOREN, P., and DE RIDDER, L.

1. Algorithm 6. An adaptive algorithm for numerical integration over an n-dimensional cube, *J. Comput. Appl. Math.* **2** (1976) 207–217.

VERGNES, J.

1. Determination d'un pas optimum d'integration pour la methode de Simpson, *Math. Comput. Simulation* **22** (1980) 177–188.

VERGNES, J., and DUMONTET, J.

1. Finding an optimal partition for a numerical integration using the trapezoidal rule, *Math. Comput. Simulation* **21** (1979) 231–241.

VILLARS, D. S.

1. Use of the IBM 701 computer for quantum mechanical calculations II. Overlap integral, NAVORD Rep. 5257, U.S. Naval Ordnance Test Station, China Lake, California, 1956.
2. "Simultaneous Multiple Integration, Floating Point," SHARE File No. 0704-0240 NO SIG.

VON MISES, R.

1. Zur mechanischen Quadratur, *ZAMM* **13** (1933) 53–56.
2. Uber allgemeine Quadraturfomeln, *J. Reine Angew. Math.* **174** (1936) 56–67.
3. Formules de cubature, *Rev. Math. Union Interbalk.* **1** (1936) 17–27.
4. Numerische Berechnung mehrdimensionaler Integrale, *ZAMM* **34** (1954) 201–210.

VON SYDOW, B.

1. Error estimates for Gaussian quadrature formulae, *Numer. Math.* **29** (1977) 59–64.

WACTLAR, H. D., and BARNETT, M. P.

1. Mechanization of tedious algebra—the e coefficients of theoretical chemistry, *CACM* **7** (1964) 704–710.

WAGNER, H.-J.

1. Vergleich von Tschebyscheff-Integrationsmethoden, *ZAMM* **53** (1973) 1–8.

WALLICK, G. C.

1. Remark on algorithm 351, Modified Romberg quadrature, *CACM* **13** (1970) 374–375.

WALSH, J. L., and SEWELL, W. E.

1. Note on degree of approximation to an integral by Riemann sums, *Amer. Math. Monthly* **44** (1937) 155–160.

WALTER, G., and SCHULTZ, D.
1. Some eigenfunction methods for computing a numerical Fourier transform, *JIMA* **18** (1976) 279–293.

WANG, A. H., and KLEIN, R. L.
1. A numerical method to obtain optimal quadrature formulas, *IJNME* **12** (1978) 487–504.

WANG, P. S.
1. Symbolic evaluation of definite integrals by residue theory in *MACSYMA*, *Information Processing 74* (Rosenfeld, J. L., ed.), pp. 823–827. North-Holland, Amsterdam, 1974.

WANNER, G.
1. Zur Berechnung von Integralen mit asymptotischen Reihen, *ZAMM* **47** (1967) 201.

WARNOCK, T. T.
1. Computational investigations of low-discrepancy point sets, in Zaremba [5, pp. 319–343].

WATERMAN, P. C., YOS, J. M., and ABODEELY, R. J.
1. Numerical integration of non-analytic functions, *J. Math. and Phys.* **43** (1964) 45–50.

WATSON, G. S.
1. *Statistics on Spheres.* Wiley, New York, 1983.

WEBER, H.
1. Numerical computation of the Fourier transform using Laguerre functions and the fast Fourier transform, *Numer. Math.* **36** (1981) 197–209.

WEEG, G. P.
1. Numerical integration of $\int_0^\infty e^{-x} J_0(\eta x/\xi) J_1(x/\xi) x^{-n}\, dx$, *Math. Comp.* **13** (1959) 312–313.

WEEKS, W. T.
1. Numerical inversion of Laplace transforms using Laguerre functions, *JACM* **13** (1966) 419–426.

WEINBERGER, H. F.
1. Some remarks on good, simple, and optimal quadrature formulas, in *Recent Advances in Numerical Analysis* (deBoor, C., and Golub, G. H., eds.), pp. 207–229. Academic Press, New York, 1978.

WELCH, P. D.
1. The use of the fast Fourier transform for the estimation of power spectra: A method based on time averaging over short, modified periodograms, *IEEE Trans. Audio Electroacoust.* **AU-15** (1967) 70–73.

WERNER, H., and WUYTACK, L.
1. Nonlinear quadrature rules in the presence of a singularity, *Comput. Math. Appl.* **4** (1978) 237–245.

WERNER, H., and ZWICK, D.
1. Algorithms for numerical integration with regular splines, Report No. 27. Rechenzentrum der Universität Münster, 1977.

WESSELING, P.
1. An asymptotic expansion for product integration applied to Cauchy principal value integrals, *Numer. Math.* **24** (1975) 435–452.

WHEELER, J. C.
1. Modified moments and Gaussian quadratures, *Rocky Mountain J. Math.* **4** (1974) 287–296.

WHELCHEL, J. E., and GUINN, D. F.

1. FFT organizations for high speed digital filtering, *IEEE Trans. Audio Electroacoust.* **AU-18** (1970) 159–168.

WHITNEY, E. L.

1. Estimates of weights in Gauss-type quadrature, *Math. Comp.* **19** (1965) 277–286.

WIDDER, D. V.

1. Some mean-value theorems connected with Cotes's method of mechanical quadrature, *Bull. Amer. Math. Soc.* **31** (1925) 56–62.
2. Mechanical inversion of the Laplace transform, Rep. RM-187. Rand Corp., Santa Monica, California, 1949.
3. *An Introduction to Transform Theory.* Academic Press, New York, 1971.

WILF, H. S.

1. Exactness conditions in numerical quadrature, *Numer. Math.* **6** (1964) 315–319.
2. Advances in numerical quadrature, in Ralston and Wilf [1, Vol. II, pp. 133–144].
3. *Mathematics for the Physical Sciences.* Wiley, New York, 1962.

WILHELMSEN, D. R.

1. Nonnegative cubature on convex sets, *SIAM J. Numer. Anal.* **11** (1974) 332–346.

WILKINSON, J. H.

1. *Rounding Errors in Algebraic Process.* Prentice-Hall, Englewood Cliffs, N.J., 1963.

WILSON, M. W.

1. A general algorithm for nonnegative quadrature formulas, *Math. Comp.* **23** (1969) 253–258.
2. Necessary and sufficient conditions for equidistant quadrature formula, *SIAM J. Numer. Anal.* **7** (1970) 134–141.
3. Discrete least squares and quadrature formulas, *Math. Comp.* **24** (1970) 271–282.

WINOGRAD, S.

1. On computing the discrete Fourier transform, *Math. Comp.* **32** (1978) 175–199.
2. On the multiplicative complexity of the discrete Fourier transform, *Adv. Math.* **32** (1979) 83–117.
3. *Arithmetic Complexity of Computations*, CBMS-NSF Regional Conference Series in Applied Mathematics **33**. *SIAM*, Philadelphia, 1980.

WINSTON, C.

1. On mechanical quadrature formulae involving the classical orthogonal polynomials, *Ann. of Math.* **35** (1934) 658–677.

WOLBERG, J. R.

1. *Application of Computers to Engineering Analysis.* McGraw-Hill, New York, 1971.

WONG, R.

1. Error bounds for asymptotic expansions of integrals, *SIAM Rev.* **22** (1980) 401–435.
2. Quadrature formulas for oscillatory integral transforms, *Numer. Math.* **39** (1982) 351–360.

WRIGHT, K.

1. Series methods for integration, *Comput. J.* **9** (1966–1967) 191–199.

WYNN, P.

1. On a cubically convergent process for determining the zeros of certain functions, *MTAC* **10** (1956) 97–100.
2. On a device for computing the $e_m(S_n)$ transformation, *Math. Comp.* **10** (1956) 91–96.

YAVNE, R.

1. An economical method for calculating the discrete Fourier transform, *Proc. AFIPS Fall Joint Comput. Conf. 1968*, **33**, Pt. 1, pp. 115–125.

YOUNG, A.

1. Approximate product-integration, *Proc. Roy. Soc. London Ser. A* **224** (1954) 552–561.

2. The application of approximate product-integration to the numerical solution of integral equations, *Proc. Roy. Soc. London Ser. A* **224** (1954) 561–573.

YOUNG, D.

1. An error bound for the numerical quadrature of analytic functions, *J. Math. and Phys.* **31** (1952) 42–44.

ZAKRZEWSKA, K., DUDEK, J., and NAZAREWICZ, N.

1. A numerical calculation of multidimensional integrals, *Comp. Phys. Comm.* **14** (1978) 299–309.

ZAMFIRESCU, I.

1. An extension of Gauss' method for the calculation of improper integrals, *Acad. R. P. Romîne Stud. Cerc. Mat.* **14** (1963) 615–631.

ZAREMBA, S. K.

1. Good lattice points in the sense of Hlawka and Monte-Carlo integration, *Monatsh. Math.* **72** (1968) 264–269.

2. The mathematical basis of Monte Carlo and quasi-Monte Carlo methods, *SIAM Rev.* **10** (1968) 303–314.

3. Good lattice points, discrepancy, and numerical integration, *Ann. Mat. Pura Appl.* [4] **73** (1966) 293–317.

4. A quasi-Monte Carlo method for computing double and other multiple integrals, *Aequationes Math.* **4** (1970) 11–22.

5. (Editor), *Applications of Number Theory to Numerical Analysis*. Academic Press, New York, 1972.

6. La méthode des "Bons Treillis" pour le calcul des intégrales multiples, in Zaremba [5, pp. 39–116].

7. Good lattice points modulo primes and composite numbers, in Osgood [1, pp. 327–356].

8. Good lattice points modulo composite numbers, *Monatsh. Math.* **78** (1974) 446–460.

9. On Cartesian products of good lattices, *Math. Comp.* **30** (1976) 546–552.

ZIENKIEWICZ, O. C.

1. *The Finite Element Method in Engineering Science*. McGraw-Hill, New York, 1971.

ZHILEIKIN, Ya. M.

1. On the error in the approximate evaluation of integrals of rapidly oscillating functions, *U.S.S.R. Computational Math. and Math. Phys.* **11**, No. 1 (1971) 344–348.

ZHILEIKEN, YA. M. and KUKARKIN, A. B.

1. Optimal evaluation of integrals of rapidly oscillating functions, *U.S.S.R. Comput. Math. Math. Phys.* **18** (2) (1978) 15–21.

ZINTERHOF, P.

1. Einige zahlentheoretische Methoden zur numerischen Quadratur und Interpolation, *S.-B. Österr. Akad. Wiss. Math.-Natur. Kl.* II, **177** (1969) 51–77.

ZOHAR, S.

1. A prescription of Winograd's DFT algorithm, *IEEE Trans.* **ASSP-27** (1979) 409–421.

2. Winograd's discrete Fourier transform algorithm, in Huang [1, pp. 89–160].

Index

F

G

H

I

J

K

L

M

N

O